RADIO WAVES IN THE
IONOSPHERE

RADIO WAVES IN THE IONOSPHERE

THE MATHEMATICAL THEORY OF THE REFLECTION OF RADIO WAVES FROM STRATIFIED IONISED LAYERS

BY

K. G. BUDDEN
M.A. Ph.D. F. Inst P. A.M.I.E.E.
Fellow of St John's College and
Lecturer in Physics in the
University of Cambridge

CAMBRIDGE
AT THE UNIVERSITY PRESS
1966

CAMBRIDGE UNIVERSITY PRESS
Cambridge, New York, Melbourne, Madrid, Cape Town, Singapore, São Paulo, Delhi

Cambridge University Press
The Edinburgh Building, Cambridge CB2 8RU, UK

Published in the United States of America by Cambridge University Press, New York

www.cambridge.org
Information on this title: www.cambridge.org/9780521114394

First published 1961
Reprinted 1966
This digitally printed version 2009

A catalogue record for this publication is available from the British Library

ISBN 978-0-521-04363-2 hardback
ISBN 978-0-521-11439-4 paperback

To the memory of
DOUGLAS R. HARTREE

CONTENTS

Preface *page* xxiii

CHAPTER 1. INTRODUCTION

1.1 The composition of the ionosphere 1
1.2 Plane waves and spherical waves. The curvature of the earth 1
1.3 Effect of collisions and of the earth's magnetic field 2
1.4 Relation to other kinds of wave-propagation 2
1.5 The variation of electron density with height. The Chapman layer 3
1.6 Approximations to the electron density profile 5
1.7 The variation of collision-frequency with height 6
1.8 The structure of the ionosphere 7
1.9 Horizontal variations and irregularities 10

CHAPTER 2. THE BASIC EQUATIONS

2.1 Units 11
2.2 Harmonic waves and complex quantities 12
2.3 Definitions of electric intensity **E** and magnetic intensity **H** 13
2.4 The current density **J** and electric polarisation **P** 14
2.5 The electric displacement **D** and magnetic induction **B** 15
2.6 Maxwell's equations 16
2.7 Cartesian coordinate system 16
2.8 Progressive plane waves 17
2.9 Plane waves in free space 19
2.10 The notation $\mathscr{H}$ and **H** 20
2.11 The energy stored in a radio wave in the ionosphere 20
2.12 The flow of energy. Poynting's theorem 21
2.13 The Poynting vector 22

CHAPTER 3. THE CONSTITUTIVE RELATIONS

3.1 Introduction *page* 24
3.2 Free, undamped electrons 24
3.3 Electron collisions. Damping of the motion 25
3.4 Effect of the earth's magnetic field on motion of electrons 26
3.5 The effect of the magnetic field of the wave on the motion of electrons 28
3.6 The susceptibility matrix 29
3.7 The Lorentz polarisation term 30
3.8 The effect of small irregularities in the ionosphere 31
3.9 The effect of heavy ions 31
3.10 The energy stored in a radio wave in the ionosphere (continued) 33
3.11 The principal axes 35

CHAPTER 4. PROPAGATION IN A HOMOGENEOUS ISOTROPIC MEDIUM

4.1 Definition of the refractive index 38
4.2 The Maxwell equation derived from Faraday's law 38
4.3 Isotropic medium without damping 39
4.4 Isotropic medium with collision damping 40
4.5 The physical interpretation of a complex refractive index 41
4.6 Evanescent waves 41
4.7 Inhomogeneous plane waves 42
4.8 Energy flow in inhomogeneous plane waves 44
4.9 The case when $X = 1$, $n = 0$ 45

CHAPTER 5. PROPAGATION IN A HOMOGENEOUS ANISOTROPIC MEDIUM. MAGNETOIONIC THEORY

5.1 Introduction 47
5.2 The wave-polarisation 47
5.3 The polarisation equation 48

5.4 Properties of the polarisation equation *page* 49
5.5 Alternative measure of the polarisation. Axis ratio and tilt-angle 51
5.6 The Appleton–Hartree formula for the refractive index 52
5.7 The longitudinal component of the electric field 53
5.8 The flow of energy for a wave in a magnetoionic medium 54
5.9 The effect of heavy ions on polarisation and refractive index 56
Examples 57

CHAPTER 6. PROPERTIES OF THE APPLETON–HARTREE FORMULA

6.1 General properties. Zeros and infinity of the refractive index 59
6.2 Collisions neglected 60
6.3 Frequency above the gyro-frequency 60
6.4 Longitudinal propagation when $Y < 1$ 60
6.5 Transverse propagation when $Y < 1$ 61
6.6 Intermediate inclination of the field when $Y < 1$ 62
6.7 Frequency below the gyro-frequency 64
6.8 Longitudinal propagation when $Y > 1$ 64
6.9 Transverse propagation when $Y > 1$ 65
6.10 Intermediate inclination of the field when $Y > 1$ 65
6.11 Effect of collisions included 66
6.12 The critical collision-frequency 67
6.13 Longitudinal propagation when collisions are included 69
6.14 Transverse propagation when collisions are included 70
6.15 Intermediate inclination of the field 72
6.16 The 'quasi-longitudinal' approximation 76
6.17 The 'quasi-transverse' approximation 77
6.18 The effect of heavy ions 78

CHAPTER 7. DEFINITION OF THE REFLECTION AND TRANSMISSION COEFFICIENTS

7.1 Introduction *page* 85

7.2 The reference-level for reflection coefficients 85

7.3 The reference-level for transmission coefficients 87

7.4 The four reflection coefficients and the four transmission coefficients 88

7.5 The sign convention 88

7.6 The reflection coefficient matrix 90

7.7 Alternative forms of the reflection coefficients 90

7.8 Spherical waves 91

Examples 94

CHAPTER 8. REFLECTION AT A SHARP BOUNDARY

8.1 Introduction 96

8.2 The boundary conditions 96

8.3 Snell's law 97

8.4 Derivation of the Fresnel formulae for isotropic media 98

8.5 General properties of the Fresnel formulae 100

8.6 The Fresnel formulae when the electric vector is in the plane of incidence 101

8.7 The Fresnel formulae when the electric vector is horizontal 105

8.8 Reflection when $X = 1$, $Z = 0$, $\mathfrak{n} = 0$ 106

8.9 Normal incidence 108

8.10 Homogeneous ionosphere with parallel boundaries 108

8.11 Normal incidence on a parallel-sided slab 112

8.12 Reflection at normal incidence when the earth's magnetic field is allowed for 114

8.13 Earth's magnetic field horizontal. Normal incidence 115

8.14 Earth's magnetic field vertical. Normal incidence 116

8.15 Reflection when the earth's magnetic field is included. Approximate formulae for oblique incidence 116

8.16 The validity of the approximations *page* 119
8.17 Reflection at oblique incidence. The Booker quartic 120
8.18 Some properties of the Booker quartic 123
8.19 Reflection at oblique incidence for north–south or south–north propagation 124
8.20 Reflection at oblique incidence in the general case 126
Examples 127

CHAPTER 9. SLOWLY VARYING MEDIUM. THE W.K.B. SOLUTIONS

9.1 Introduction 128
9.2 The differential equations 128
9.3 The phase memory concept 130
9.4 Loss-free medium. Constancy of energy-flow 131
9.5 Derivation of the W.K.B. solution 131
9.6 Condition for the validity of the W.K.B. solutions 133
9.7 Properties of the W.K.B. solutions 134
9.8 The reflection coefficient 136
9.9 Coupling between upgoing and downgoing waves 137
9.10 Extension to oblique incidence 138
9.11 The differential equations for oblique incidence 140
9.12 The W.K.B. solutions for horizontal polarisation at oblique incidence 140
9.13 The W.K.B. solutions at oblique incidence when the electric vector is parallel to the plane of incidence 142
9.14 The effect of including the earth's magnetic field 143
9.15 Ray theory and 'full wave' theory 144

CHAPTER 10. RAY THEORY FOR VERTICAL INCIDENCE WHEN THE EARTH'S MAGNETIC FIELD IS NEGLECTED

10.1 The use of pulses 146
10.2 The group velocity 147
10.3 The equivalent height of reflection $h'(f)$ 149

10.4 The 'true height' and the 'phase height' *page* 150
10.5 The equivalent height of reflection for a linear profile of electron density 150
10.6 The equivalent height of reflection for an exponential variation of electron density 151
10.7 Equivalent height for a parabolic profile of electron density 152
10.8 Equivalent height for the 'sech^2' profile of electron density 156
10.9 Two separate parabolic layers 157
10.10 The effect of a 'ledge' in the electron density profile 158
10.11 The calculation of electron density $N(z)$, from $h'(f)$ data 160
10.12 Solution when $N(z)$ is monotonic 161
10.13 Partial solution when $N(z)$ is not monotonic 163
10.14 The shape of a pulse of radio waves 166
10.15 The effect of electron collisions on group refractive index 170
10.16 The effect of collisions on equivalent height $h'(f)$ and phase height $h(f)$ 171
10.17 Relation between equivalent height, phase height and absorption 172
Examples 174

CHAPTER 11. RAY THEORY FOR OBLIQUE INCIDENCE WHEN THE EARTH'S MAGNETIC FIELD IS NEGLECTED

11.1 Introduction. The ray path 175
11.2 Wave-packets 177
11.3 The equation for the ray when the earth's magnetic field is neglected 178
11.4 The ray path for a linear gradient of electron density 179
11.5 The ray path for exponential variation of electron density 180
11.6 The ray path for a parabolic profile of electron density 182
11.7 The skip distance 183

11.8 The equivalent path P' at oblique incidence *page* 185
11.9 Breit and Tuve's theorem. Martyn's theorem for equivalent path 186
11.10 The equivalent path at oblique incidence for a linear gradient of electron density 188
11.11 The equivalent path at oblique incidence for a parabolic profile of electron density 188
11.12 The dependence of signal on frequency near the maximum usable frequency 190
11.13 The prediction of maximum usable frequencies. Appleton and Beynon's method 191
11.14 The curvature of the earth 192
11.15 The prediction of maximum usable frequencies. Newbern Smith's method 194
11.16 The absorption of radio waves. Martyn's theorem for absorption 195
11.17 The effect of electron collisions on equivalent path 197
Examples 197

CHAPTER 12. RAY THEORY FOR VERTICAL INCIDENCE WHEN THE EARTH'S MAGNETIC FIELD IS INCLUDED

12.1 Introduction 199
12.2 Magnetoionic 'splitting' 199
12.3 The group refractive index—collisions neglected 200
12.4 The effect of collisions on the group refractive index 204
12.5 The equivalent height of reflection $h'(f)$—collisions neglected 205
12.6 The $h'(f)$ curves when collisions are neglected 206
12.7 The penetration-frequencies for the ordinary and extraordinary waves 208
12.8 The equivalent height for a parabolic layer 209
12.9 Two separate parabolic layers 210

12.10 The effect of a 'ledge' in the electron density profile *page* 212
12.11 The effect of collisions on equivalent height $h'(f)$ 212
12.12 The polarisation of waves in a wave-packet 214
12.13 The calculation of electron density $N(z)$ from $h'(f)$ 215
12.14 Example of the use of the method 218
12.15 Other versions of the foregoing method 221
12.16 Failure of the method when $N(z)$ is not monotonic 222
12.17 The use of $h'_x(f)$ for the extraordinary ray 223
Example 224

CHAPTER 13. RAY THEORY FOR OBLIQUE INCIDENCE WHEN THE EARTH'S MAGNETIC FIELD IS INCLUDED

13.1 Introduction 225
13.2 The variable q 225
13.3 Derivation of the Booker quartic 226
13.4 The transition to a continuous medium 228
13.5 The path of a wave-packet 229
13.6 The reversibility of the path 230
13.7 The reflection of a wave-packet 230
13.8 A simple example of ray paths at oblique incidence 231
13.9 Further properties of the Booker quartic 233
13.10 The Booker quartic for east–west and west–east propagation 236
13.11 The Booker quartic for north–south and south–north propagation 238
13.12 The Booker quartic in the general case when collisions are neglected 244
13.13 Lateral deviation at vertical incidence 246
13.14 Lateral deviation for propagation from (magnetic) east to west or west to east 248
13.15 Lateral deviation in the general case 250

13.16 Calculation of attenuation, using the Booker quartic *page* 250
13.17 The 'refractive index' surface in a homogeneous medium 252
13.18 The direction of the ray 253
13.19 The ray velocity and the ray surface 255
13.20 Whistlers 256
13.21 Determination of ray direction by Poeverlein's construction 258
13.22 Propagation in the magnetic meridian. The 'Spitze' 260
13.23 The refractive index surfaces for the extraordinary ray when $Y < 1$ 262
13.24 The refractive index surfaces for the extraordinary ray when $Y > 1$ 266
13.25 The second refractive index surface for the ordinary ray when $Y > 1$ 268
Examples 270

CHAPTER 14. THE GENERAL PROBLEM OF RAY TRACING

14.1 Introduction 271
14.2 Equations of the refractive index surface and the ray surface 272
14.3 The Eikonal function 274
14.4 The canonical equations for a ray, and the generalisation of Snell's law 276
14.5 Other relations between the equations for the ray surface and the refractive index surface 278
14.6 Fermat's principle 279
14.7 Equivalent path and absorption 279
14.8 The problem of finding the maximum usable frequency 281
Examples 282

CHAPTER 15. THE AIRY INTEGRAL FUNCTION, AND THE STOKES PHENOMENON

15.1 Introduction 283
15.2 Linear gradient of electron density associated with an isolated zero of q 283

15.3 The differential equation for horizontal polarisation and oblique incidence *page* 285
15.4 The Stokes differential equation 286
15.5 Qualitative discussion of the solutions of the Stokes equation 287
15.6 Solutions of the Stokes equation expressed as contour integrals 288
15.7 Solutions of the Stokes equation expressed as Bessel functions 291
15.8 Tables of the Airy integral functions 291
15.9 The W.K.B. solutions of the Stokes equation 292
15.10 The Stokes phenomenon of the 'discontinuity of the constants' 292
15.11 Stokes lines and anti-Stokes lines 293
15.12 The Stokes diagram 294
15.13 Definition of the Stokes constant 295
15.14 Furry's derivation of the Stokes constants for the Stokes equation 296
15.15 Asymptotic approximations obtained from the contour integrals 297
15.16 Summary of some important properties of complex variables 297
15.17 Integration by the method of steepest descents 300
15.18 Application of the method of steepest descents to solutions of the Stokes equation 302
15.19 Integration by the method of stationary phase 307
15.20 Method of stationary phase applied to the Airy integral function 309
15.21 Asymptotic expansions 310
15.22 The range of validity of asymptotic approximations 310
15.23 The choice of a fundamental system of solutions of the Stokes equation 312
15.24 Connection formulae, or circuit relations 313
15.25 The intensity of light near a caustic 313

CHAPTER 16. LINEAR GRADIENT OF ELECTRON DENSITY

16.1 Introduction *page* 319
16.2 Purely linear profile. Electron collisions neglected 319
16.3 Application to a slowly varying profile 322
16.4 The effect of electron collisions. The height z as a complex variable 326
16.5 Constant collision-frequency. Purely linear profile of electron density 327
16.6 The slowly varying profile when collisions are included. Derivation of the phase integral formula 329
16.7 Discussion of the phase integral formula 331
16.8 Effect of curvature of the electron density profile 333
16.9 Reflection at a discontinuity of gradient 334
16.10 Linear gradient between two homogeneous regions 336
16.11 Symmetrical ionosphere with double linear profile 340
16.12 The differential equation for oblique incidence applicable when the electric vector is parallel to the plane of incidence 343
16.13 The behaviour of the fields near a zero of the refractive index for 'vertical' polarisation at oblique incidence 346
16.14 The generation of harmonics in the ionosphere 347
16.15 The phase integral formula for 'vertical' polarisation at oblique incidence 348
16.16 Asymptotic approximations for the solutions of the differential equation for 'vertical' polarisation 349
16.17 Application of the phase integral formula 350

CHAPTER 17. VARIOUS ELECTRON DENSITY PROFILES WHEN THE EARTH'S MAGNETIC FIELD IS NEGLECTED

17.1 Introduction 353
17.2 Exponential profile. Constant collision-frequency 354
17.3 The phase integral formula applied to the exponential layer 357

17.4 The parabolic layer *page* 358
17.5 Partial penetration and reflection 363
17.6 The equivalent height of reflection for a parabolic layer 365
17.7 Electron density with square law increase 366
17.8 The sinusoidal layer 368
17.9 Circuit relations. Introduction to Epstein's theory 369
17.10 The hypergeometric differential equation 370
17.11 The circuit relations for the hypergeometric function 372
17.12 Application to the wave-equation 375
17.13 The reflection and transmission coefficients of an Epstein layer 377
17.14 Epstein profiles 378
17.15 Ionosphere with gradual boundary 380
17.16 The 'sech²' profile 380
17.17 Fixed electron density and varying collision-frequency 383

CHAPTER 18. ANISOTROPIC MEDIA. COUPLED WAVE-EQUATIONS AND W.K.B. SOLUTIONS

18.1 Introduction 385
18.2 The differential equations 385
18.3 The four characteristic waves 387
18.4 Matrix form of the equations 389
18.5 The differential equations for vertical incidence 391
18.6 The W.K.B. solution for vertical incidence on a loss-free medium 392
18.7 Introduction to W.K.B. solutions in the general case 394
18.8 Introduction to coupled wave-equations 394
18.9 Försterling's coupled equations for vertical incidence 396
18.10 Coupled equations in the general case, in matrix form 398
18.11 Expressions for the elements of **S**, $\mathbf{S}^{-1}$, and $-\mathbf{S}^{-1}\mathbf{S}'$ 399

18.12 The W.K.B. solutions in the general case *page* 401
18.13 The first-order coupled equations for vertical incidence 402
18.14 The W.K.B. solutions for vertical incidence 405
18.15 The first-order equations in other special cases 406
18.16 Second-order coupled equations 408
18.17 Condition for the validity of the W.K.B. solutions 410
Example 411

CHAPTER 19. APPLICATIONS OF COUPLED WAVE-EQUATIONS

19.1 Introduction 412
19.2 Properties of the coupling parameter ψ 412
19.3 Behaviour of the coefficients near a coupling point 417
19.4 Properties of the coupled differential equations near a reflection point and near a coupling point 418
19.5 The use of successive approximations 421
19.6 The phase integral formula for coupling 423
19.7 The Z-trace 424
19.8 The method of 'variation of parameters' 426
19.9 The 'coupling echo' 428
19.10 The transition through critical coupling 429
19.11 Introduction to limiting polarisation 432
19.12 The free space below the ionosphere 433
19.13 The differential equation for the study of limiting polarisation 434
Examples 436

CHAPTER 20. THE PHASE INTEGRAL METHOD

20.1 Introduction 437
20.2 The Riemann surface for the refractive index 438
20.3 The linear electron density profile 440
20.4 The parabolic electron density profile 443

20.5 A further example of the method *page* 446
20.6 Coupling branch points and their Stokes lines and anti-Stokes lines 450
20.7 The phase integral method for coupling 452
20.8 Further discussion of the transition through critical coupling (continued from § 19.10) 455
Example 457

CHAPTER 21. FULL WAVE SOLUTIONS WHEN THE EARTH'S MAGNETIC FIELD IS INCLUDED

21.1 Introduction 458
21.2 The differential equations 458
21.3 Vertical incidence and vertical magnetic field 459
21.4 Exponential profile of electron density. Constant collision-frequency 460
21.5 Exponential profile (continued). Incident wave linearly polarised 462
21.6 Other electron density profiles for vertical field and vertical incidence 464
21.7 Vertical magnetic field and oblique incidence. Introduction to Heading and Whipple's method 464
21.8 Regions O, I and I (*a*) 465
21.9 Reflection and transmission coefficients of region I 467
21.10 Regions II and II (*a*) 468
21.11 The reflection coefficients of region II 469
21.12 The combined effect of regions I and II 471
21.13 The effect of an infinity in the refractive index 472
21.14 Isolated infinity of refractive index 474
21.15 Refractive index having infinity and zero 476
21.16 The apparent loss of energy near an infinity of refractive index 479
Example 481

CHAPTER 22. NUMERICAL METHODS FOR FINDING REFLECTION COEFFICIENTS

21.1 Introduction *page* 482
22.2 Methods of integrating differential equations 483
22.3 The size of the step 484
22.4 The choice of dependent variable 484
22.5 The three parts of the calculation of reflection coefficients 486
22.6 The starting solutions at a great height 487
22.7 Calculation of the components of the reflection coefficient 489
22.8 The wave-admittance in an isotropic ionosphere 491
22.9 The wave-admittance matrix **A** for an anisotropic ionosphere 493
22.10 The starting value of **A** 495
22.11 Relation between the admittance matrix **A** and the reflection coefficient matrix **R** 496
22.12 The differential equation for **A** 498
22.13 Symmetry properties of the differential equations 499
22.14 Equivalent height of reflection 499
Example 501

CHAPTER 23. RECIPROCITY

23.1 Introduction 502
23.2 Aerials 502
23.3 Goubau's reciprocity theorem 505
23.4 One magnetoionic component 506
23.5 Reciprocity with full wave solutions 508

Appendix. The Stokes constant for the differential equation (16.98) for 'vertical' polarisation 510

Bibliography 512

Index of definitions of the more important symbols 525

Subject and name index 530

PREFACE

This book is based on a course of lectures given annually in Cambridge since 1956, and repeated in 1957 at the U.S. National Bureau of Standards Laboratories, Boulder, Colorado, U.S.A. Its object is to set out the mathematical basis of the theory of the propagation of radio waves in a horizontally stratified ionosphere. It is hoped that the book will serve both as a text-book, for those comparatively new to the subject, and as a reference book for more experienced readers. Some of the more advanced topics are printed in smaller type and could be omitted on a first reading. Throughout the book the stress is on the understanding of the mathematical methods rather than on their immediate practical use, since the radio engineer who really understands the mathematics is much better equipped to solve practical problems than one who does not.

The reader is assumed to be familiar with calculus, the theory of complex variables, vectors including the operators div, curl, grad, and electromagnetic theory as far as Maxwell's equations. Matrices are used in a few places, but the reader unfamiliar with them need not be deterred from studying the rest of the book.

For standard mathematical techniques there are references throughout the book to well-known mathematical treatises. One of the most useful of these is *Methods of Mathematical Physics* by Sir Harold and Lady Jeffreys which is a mine of valuable information.

It is inevitable that some important topics are omitted. Throughout the book it is assumed that the ionosphere is horizontally stratified, but this is only approximately true, since horizontal variations and irregularities play an important part in radio wave-propagation. The extensive recent work in this field is not covered here. Nor is there any discussion of reflection or scattering of radio waves from cylindrical structures such as meteor trails. The statistical mechanics of an ionised medium, and phenomena such as wave-interaction, which depend on electron temperature, are also excluded. In the theory of the propagation of radio waves to great distances, the space between the earth and the ionosphere is often treated as a wave-guide and the propagation constants of the 'wave guide modes' are found. This is sometimes called the 'mode theory' of radio wave-propagation. It is a large topic, which is beyond the scope of this book and really needs one to itself.

I am indebted first to Sir Edward Appleton whose work is the basis of most of the theory of radio waves in the ionosphere. I have made very extensive use both of his published papers and of notes taken at his lectures and colloquia at Cambridge in the years 1935–39. At this time, too, Professor H. G. Booker and Mr E. Cunningham lectured on radio waves at Cambridge. I attended both courses, and still make frequent use of the lecture notes. On returning to Cambridge after the war I received a great deal of help and encouragement from the late Professor D. R. Hartree.

There are many diagrams in this book which are the result of computations made on EDSAC, the automatic digital computer in the University Mathematical Laboratory, Cambridge. I am greatly indebted to the Director, Dr M. V. Wilkes, and his staff for permission to use the EDSAC, and for able instruction in its use. To Dr Wilkes I am further indebted for his share in my initiation into the radio-wave field before the war.

The writing of this book was started at Boulder, Colorado, U.S.A., in 1957, while I was on sabbatical leave from Cambridge. I should like to thank the Director and staff of the Boulder Laboratories of the U.S. National Bureau of Standards for making it possible for me to work in Boulder and for their encouragement and valuable discussions of many aspects of the work.

Numerous other colleagues have given valuable advice and help. It is impossible to list them all, but I should mention particularly Dr D. W. Barron, Dr B. L. Briggs, Dr P. C. Clemmow, Dr J. O. Thomas, Mr G. Millington, Dr M. L. V. Pitteway, Dr H. Poeverlein, Dr D. Shinn, and Dr K. Weekes.

But above all I am indebted to Mr J. A. Ratcliffe. I decided to write this book as a result of his suggestion, and some of the problems discussed in it were propounded by him. He, too, read the typescript and made valuable suggestions for improving it. Mr Ratcliffe's recent book on *The Magneto-ionic Theory* gives an excellent insight into the physical principles underlying radio wave-propagation, and should be studied by all readers of the present book.

K. G. BUDDEN

CAVENDISH LABORATORY
CAMBRIDGE
January 1959

CHAPTER 1

INTRODUCTION

1.1 The composition of the ionosphere

The ionosphere consists of a number of ionised regions above the earth's surface, which play a most important part in the propagation of radio waves. Our knowledge of it is derived almost entirely from radio measurements, and it is therefore important to understand the processes by which the waves are reflected. The ionosphere is believed to influence radio waves mainly because of the presence of free electrons. The early experiments showed that the electrons must be arranged approximately in horizontally stratified layers, so that the number density is a function only of the height above the earth's surface. The ionosphere must be almost electrically neutral, for if there were any appreciable space charge, it would give rise to large electric forces which would prevent stable layers from forming. There must therefore be at least as many positive ions as electrons, per unit volume. Besides negative electrons there may also be heavy negative ions formed by the attachment of electrons to air molecules. Heavy ions of both signs might play a part in the propagation of radio waves, and this problem is discussed in §§ 3.9, 5.9 and 6.18. A heavy ion, whether positive or negative, has a mass approximately 60,000 times that of an electron, and it is shown that, for all frequencies above a few hundred cycles, ions must be about 60,000 times more numerous than electrons if they are to have a detectable effect. If this could happen at all, it would only be in the very lowest regions of the ionosphere, but there seems to be no evidence that heavy ions give any observable effect. It is therefore assumed, in most of this book, that only the free electrons can affect radio propagation.

1.2 Plane waves and spherical waves. The curvature of the earth

Radio waves travelling from a transmitter to a receiver near the earth's surface may take one of several possible paths. A wave may travel over the earth's surface, and it is then known as the ground wave. The earth is an imperfectly conducting, curved surface, and the theory of the propagation of the ground wave involves many problems of the greatest mathematical interest, but they are beyond the scope of this book.

Another wave may travel up to the ionosphere, be reflected there, and return to the receiver. It is with a single reflection of this kind that the present book is mainly concerned. The wave originates at a source of small dimensions so that the wave front is approximately spherical, but by the time it reaches the ionosphere the radius of curvature is so large that the wave can be treated as plane. This involves an approximation which is examined in §7.8, and it is shown that the error is negligible except in certain special cases rarely met with in practice. Similarly, the ionospheric layers are curved because of the earth's curvature, but in most problems this curvature can be neglected.

1.3 Effect of collisions and of the earth's magnetic field

The motion of an electron in the ionosphere is affected by the earth's magnetic field and by the collisions which the electron makes with other particles. It was shown by Lorentz that the collisions have the same effect as a retarding force proportional to the velocity. In most of this book the retarding force is included, but it may be neglected in some problems, which arise at high frequencies and can be treated by 'ray theory' methods (chs. 10 to 14). The retarding force or damping force is most important at low frequencies, and here the wavelength is so long that ray-theory methods are inapplicable, and a 'full-wave' treatment must be used. The second half of the book is devoted to this.

The effect of the earth's magnetic field is to make the ionosphere a doubly refracting medium. This is a great complication, and often leads to a differential equation of the fourth order. It is convenient to neglect the earth's magnetic field for many purposes, so that the differential equation to be solved is only of the second order. This is done for considerably more than half of the problems discussed. One reason for this is that only in this way can the differential equations be reduced to a form whose solution has been studied by mathematicians. But by neglecting the earth's magnetic field, principles can be established which can then be extended to more general cases where the field is included.

1.4 Relation to other kinds of wave-propagation

The theory of radio wave-propagation in the ionosphere is closely related to other branches of physics which deal with wave-propagation in media whose properties vary from place to place. For example, in wave-mechanics a study is made of the propagation of electron waves in a potential field. The variation of potential is analogous to the variation of the square of the refractive index for radio waves. But the potential is real in nearly all problems of wave-mechanics, so that there is nothing analogous to the damping forces which, for radio waves, lead to a complex value of the squared refractive index. Moreover, the differential equations in wave-mechanics are nearly always of the second order, which means that there is nothing analogous to double refraction. Some of the material of chs. 9, 10 and 15 is of great importance when applied to wave-mechanics.

In seismology some study has been made of the propagation of elastic waves in media whose properties vary gradually from place to place. For example, some parts of the ocean bed are horizontally stratified and, for sound waves, behave very like an inverted ionosphere. Many of the results of chs. 9, 10, 11 and 15 could be applied to this case.

In solids three kinds of elastic wave can be propagated, two transverse and one longitudinal, so that this medium might be considered to be triply refracting. But seismologists are interested mainly in propagation through homogeneous solids, and in the reflection and transmission which occurs at the sharp boundary between two media. (See, for example, Bullen, *Theory of Seismology*; Jeffreys, *The Earth* (ch. II); Musgrave, 1959.) There appears to have been very little study of propagation of elastic waves through solids whose properties vary continuously from place to place.

It is therefore probable that the theory of wave propagation in continuously variable media has advanced farthest in the field of radio waves in the ionosphere.

1.5 The variation of electron density with height. The Chapman layer

Before the reflecting properties of the ionosphere can be calculated, it is necessary to know how the number density of electrons, N, varies with height above the earth's surface. To study this problem, some assumption must be made about how the ionospheric layers are formed. A most important contribution to this problem was made by Chapman (1931 *a*, *b*, 1939), who derived a law for the variation of N with height, which is now known as the Chapman law. The full theory of the formation of ionospheric layers has been refined and extended by Chapman and others, and is beyond the scope of this book. Only the simplest version of the Chapman theory is given here.

Assume that the earth's atmosphere is constant in composition, and at a constant temperature. Then the air density $\mathfrak{d}$ at height z above the ground is

$$\mathfrak{d} = \mathfrak{d}_0 e^{-z/H}, \tag{1.1}$$

where $\mathfrak{d}_0$ is the density at the ground, $H = RT/Mg$, and R is the gas constant, M the mean molecular weight, T the absolute temperature, and g the gravitational acceleration. The curvature of the earth is neglected, and g is assumed constant. H is called the 'scale height' of the atmosphere, and is approximately 10 km at the ground. The sun's radiation enters the atmosphere at an angle χ from the zenith. Let the mass absorption coefficient of the air for the radiation be σ, and assume that the rate of production of electrons, q, is proportional to the rate of absorption of radiation per unit volume. Let the flux of energy in the incident radiation be I_0 outside the earth's atmosphere, and I at a height z. The energy flux I decreases as the radiation passes down through the atmosphere, and it is clear that

$$dI = I\sigma\mathfrak{d} \sec\chi \, dz. \tag{1.2}$$

This is combined with (1.1) and integrated, which gives

$$I = I_0 \exp\{-\sigma\mathfrak{d}_0 H \sec\chi \, e^{-z/H}\}. \tag{1.3}$$

Now it is convenient to take $z_0 = H\log(\sigma\mathfrak{d}_0 H)$ so that (1.3) becomes

$$I = I_0 \exp\left[-\sec\chi \exp\left\{-\frac{z-z_0}{H}\right\}\right]. \tag{1.4}$$

The rate of absorption of energy at height z is $\cos\chi(dI/dz)$, and since this is proportional to the rate of production of electrons q, we have from (1.4)

$$q = q_0 \exp\left[1 - \frac{z-z_0}{H} - \sec\chi \exp\left\{-\frac{z-z_0}{H}\right\}\right], \tag{1.5}$$

where q_0 is a constant, namely I_0/eH (here e is the exponential). q has the maximum value $q_0 \cos\chi$ when $(z-z_0)/H = \log(\cos\chi)$, so that q_0 is the maximum rate of electron production when $\chi = 0$.

Next it is necessary to consider how electrons are removed. It is now believed that a number of processes contribute to the removal, but the effect is the same as if the electrons simply recombined† with positive ions. Assume that the only ions present are electrons and positive ions, and that the number of each per unit volume is N. Then the rate of removal of electrons is αN^2, where α is a constant called the recombination coefficient. The variation of N with time t is then given by

$$\frac{dN}{dt} = q - \alpha N^2. \tag{1.6}$$

If α is large enough, the term dN/dt can be neglected. When this happens, the processes of formation and removal of electrons come into equilibrium in a negligibly small time, and we then have

$$N = (q/\alpha)^{\frac{1}{2}}. \tag{1.7}$$

† This may not be true for the F-layer where it is believed that electrons are removed according to the attachment law $dN/dt = q - \beta N$, and the constant β is called the attachment coefficient.

This is a very much oversimplified picture of the actual mechanism of formation and removal of electrons in the ionosphere, but it does at least give a general guide as to how the electron density might vary with height. Combination of (1.7) with (1.5) gives

$$N = N_0 \exp \frac{1}{2}\left[1 - \frac{z-z_0}{H} - \sec\chi \exp\left\{-\frac{z-z_0}{H}\right\}\right]. \tag{1.8}$$

This expression will be called the 'Chapman law'. It assumes that the recombination coefficient α is independent of height. In Fig. 1.1 the expression (1.8) for N is plotted against the height z for various values of χ. It is seen that N has a maximum value $N_m = N_0(\cos\chi)^{\frac{1}{2}}$ at the level $z_m = z_0 + H\log(\sec\chi)$. It falls off quite steeply below this, and less steeply above it.

An alternative form of (1.8) is obtained by taking

$$\zeta = \frac{z-z_m}{H} = \frac{z-z_0}{H} - \log(\sec\chi), \tag{1.9}$$

so that ζH is height measured from the level of maximum N. Then

$$N = N_m \exp \tfrac{1}{2}(1 - \zeta - e^{-\zeta}), \tag{1.10}$$

which shows that the 'shape' of a Chapman layer is independent of the sun's zenith angle χ. The curvature of the $N(z)$ profile at the maximum is $N_m/2H^2$. This is the same as the curvature at the apex of a parabolic profile (see § 10.7) whose 'half thickness' is $2H$.

1.6 Approximations to the electron density profile

If N is assumed to vary with height z according to the Chapman law (1.8), then this expression would appear in the differential equation which has to be solved to find the reflection coefficient of the ionosphere. But such a differential equation would be so complicated that it could only be solved by numerical methods. Such methods have been used extensively, and are discussed in ch. 22. It is also useful, however, to select small ranges of z, and use approximate and simpler expressions for the electron density. This permits the differential equations to be reduced to simple standard forms whose solutions have well-known properties. For example, it is often possible to choose a range of z so small that the variation of N may be assumed to be linear. This case is of the greatest importance, and is the subject of chs. 15 and 16. Near the maximum values of N in Fig. 1.1, the linear law is not satisfactory, but the profile can be treated as a parabola. In the lower part of a Chapman layer it is often useful to treat the variation as approximately exponential over a small range. Other laws of variation of N with z are discussed because of their mathematical interest. An especially important case is the homogeneous medium with a sharp lower boundary, which is discussed in ch. 8, and which, for very low frequencies, may be a fair approximation to the true ionosphere.

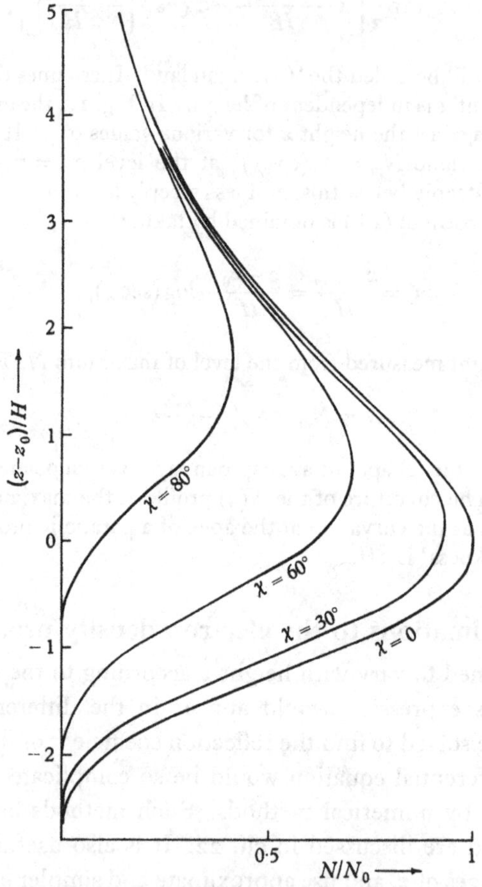

Fig. 1.1. Curves showing how the electron number density N varies with height z according to the simple Chapman theory for a flat earth, for various values of the sun's zenith angle χ.

1.7 The variation of collision-frequency with height

The average number of collisions ν which an electron makes per unit time with the air molecules depends upon the number density of the molecules, and therefore on the density and composition of the air. It also depends on the velocity of the electron, but for many purposes it is permissible to neglect this effect. Then, in an atmosphere which is constant in composition and temperature:

$$\nu = \nu_0 \exp(-z/H), \qquad (1.11)$$

where H is the scale height defined on p. 4, and ν_0 is constant. In practice H takes different values at different levels, and the law can only be expected to hold over ranges of z so small that H may be treated as constant. A useful summary of the factors which affect the value of ν have been made by Nicolet (1953). Fig. 1.2 shows how ν depends on the height z according to the best estimates at present available.

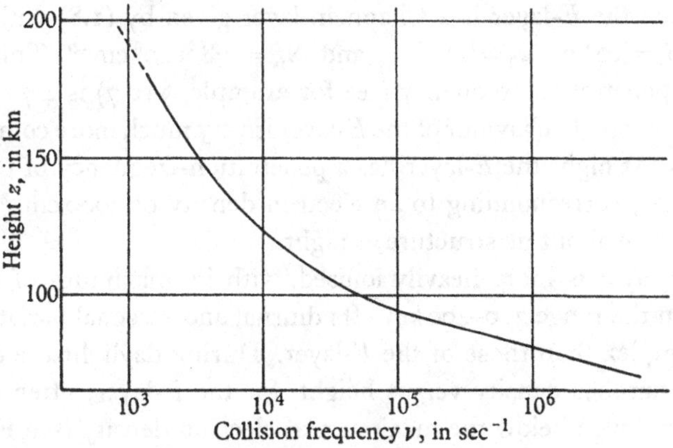

Fig. 1.2. The dependence of electron collision-frequency ν upon height z. This curve is based partly on the work of Crompton, Huxley and Sutton (1953). The author is greatly indebted to Dr K. Weekes for supplying the data from which it was plotted.

It is found that changes of the value of ν affect the propagation of radio waves far less than changes of the electron number density N. For many purposes it is therefore permissible to treat ν as constant over a small range of height z. This is especially true at high frequencies (greater than about 1 Mc/s), where the wavelength is small compared with the scale height H, which is about 10 km.

The dependence of ν on electron velocity gives rise to the phenomena of wave-interaction, which is beyond the scope of this book. Here ν is treated as a constant at each level.

1.8 The structure of the ionosphere

A great deal of information has been accumulated on the detailed structure of the ionosphere, but the topic is beyond the scope of this book, and only a brief outline is given here of the major features. For further details see Appleton (1935) and Rawer (1953).

The ionosphere consists of two main layers known as E and F. The E-layer has its maximum electron density at a height of about 110 km. Radio measurements have been used to study its height and penetration-frequency and their variation with time of day and season. The electron density just above the maximum of the E-layer is not easy to investigate, but it is probable that it is only slightly less than the maximum value for a range of height extending right up to the base of the F-layer. Thus the E- and F-layers are not really distinct. For the purposes of this book, however, it will often be sufficiently accurate to assume that during daylight hours the E-layer is a Chapman layer given by (1.8) with a scale height $H = 10$ km, $z_0 = 115$ km, and $N_0 = 2{\cdot}8 \times 10^5\,\text{cm}^{-3}$. This means that the penetration-frequency (see for example, § 10.7) is $4{\cdot}7 \times (\cos\chi)^{\frac{1}{4}}$ Mc/s. The actual behaviour of the E-layer is very much more complicated than this. At night the E-layer has a penetration-frequency of the order of 0·5 Mc/s, corresponding to an electron density of $3000\,\text{cm}^{-3}$. Much less is known about its structure at night.

The F-layer is more heavily ionised, with its maximum of electron density in the range 200–400 km. Its diurnal and seasonal variations are more complex than those of the E-layer. During daylight the curve of electron number density versus height, for the F-layer, often shows a subsidiary bulge below the maximum of electron density (see Fig. 1.3). This is known as the F_1-layer, and it may occasionally attain an actual maximum. The main maximum above it is known as the F_2-layer. There is now some evidence to show that the formation of the whole F-layer can be explained by a single ionising agency which would form a simple layer like a Chapman layer if the attachment coefficient β (see footnote to § 1.5) were constant for all heights. Bradbury (1937, 1938) has put forward the hypothesis that β decreases as the height z increases, so that the maximum of ionisation in the F_2-layer arises, not from a fast rate of production, but because of a slow rate of removal of electrons. In temperate latitudes the penetration frequency of the F-layer ranges from about 2 Mc/s at night, to 8 Mc/s in a summer day. These correspond to electron densities $5 \times 10^4\,\text{cm}^{-3}$ and $8 \times 10^5\,\text{cm}^{-3}$ respectively.

In chs. 10 and 12 some account is given of methods of finding the function $N(z)$ in the F-layer from radio observations. Work of this kind shows that, at the maximum, the curvature of the $N(z)$ curve is about the same as that at the apex of a parabolic layer of half-thickness 100 km, or that of a Chapman layer of scale height about 50 km. The

F-layer is therefore much thicker than the E-layer, as well as more heavily ionised.

There is some evidence that in the daytime there may be another layer lower down with its maximum of electron density near 80 km. This has been called the 'D-layer'. The evidence is not strong, and it is possible that even though $N(z)$ is enhanced below the E-layer, it is still a monotonically increasing function, as shown in Fig. 1.3. It is better to use the term 'D-region' to mean the part of the ionosphere below about 90 km.

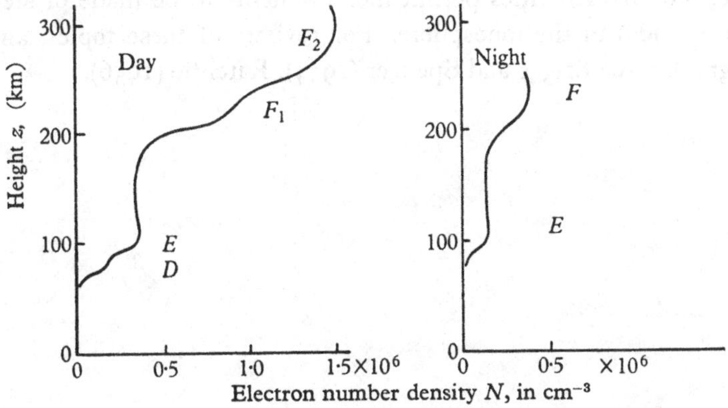

Fig. 1.3. The curves show very roughly how the electron number density N is believed to depend on height z. Actual electron density profiles vary over wide ranges and depend markedly on time of day, season, sun-spot number and whether or not the ionosphere is disturbed.

The experimental study of this region comes largely from radio observations at very low frequencies, and here the wavelength is so long that interpretation of the observations is much less direct than at high frequencies. One of the most important features of the full wave theory given in this book is that it may help to disentangle the numerous radio observations at very low frequencies. Here the procedure is to assume some profile for the electron density and work out its reflecting properties. If these do not agree with observations, some other profile must be tried until a satisfactory result is obtained. The process is difficult because of the complexity of the mathematics and the variability of the radio observations with time of day and season, and with ionospheric disturbances. Some more direct information about the structure of the D-region has been provided recently through the work of Gardner and Pawsey (1953) and Fejer and Vice (1959).

1.9 Horizontal variations and irregularities

Many of the results of radio observations can be explained by assuming that the ionosphere is horizontally stratified, that is, that the electron density and collision frequency are functions only of the height *z*. This assumption is adopted throughout this book.

Recent experimental work has been shown that there must also be horizontal variations of electron density in the ionosphere. These are often irregular, and are subject both to steady movements and random change. The irregularities permit measurements to be made of steady motion (winds) in the ionosphere. For reviews of these topics, and a bibliography, see Briggs and Spencer (1954), Ratcliffe (1956).

CHAPTER 2
THE BASIC EQUATIONS

2.1 Units

Electromagnetic theory is often based on the experimental results that the force F_q between two electric charges q_1, q_2, and the force F_m between two magnetic poles m_1, m_2 are given, respectively, by

$$F_q = \frac{q_1 q_2}{4\pi\epsilon_0 r^2}, \quad F_m = \frac{4\pi m_1 m_2}{\mu_0 r^2}, \tag{2.1}$$

where r is the separation in each case. Here ϵ_0 and μ_0 are known, respectively, as the electric and magnetic permittivities of free space. They are constants whose function is similar to that of the constant G in gravitation theory. Their numerical values depend on the particular system of units being used. The factors 4π appear in (2.1) for rationalised systems of units. Some authors argue that free magnetic poles do not exist, and base their definitions of magnetic quantities on the forces between currents. Others use a different version of the second equation (2.1), in which μ_0 is in the numerator, thereby implying a different meaning for the term 'magnetic pole'. These considerations are unimportant in the ionosphere, whose magnetic permeability is always taken as unity.

When ϵ_0 and μ_0 are left in a formula as symbols, the formula is valid in any self-consistent system of units. It is often implied that such formulae are restricted to m.k.s. units, but this is incorrect. Equations (2.1), and all other formulae in this book which do not use specific numerical values, are valid in any self-consistent rationalised units (including rationalised m.k.s. units). The symbol μ_0 must not be confused with μ which is used later for the real part of the refractive index.

A third dimensional constant κ_0 is sometimes introduced into electromagnetic theory as follows. Let $i\delta\mathbf{l}$ be a current element vector at a point P, and let Q be another point such that PQ is a vector $\mathbf{r}$. Then the contribution to the magnetic-intensity vector $\mathbf{H}$ at Q is given by

$$\delta\mathbf{H} = \kappa_0 i \frac{\delta\mathbf{l}\times\mathbf{r}}{r^3}, \tag{2.2}$$

where r is the magnitude of $\mathbf{r}$. Throughout this book we take $\kappa_0 = 1$, which is the correct value for most systems of units including rationalised m.k.s. units.

Some mathematical works use Gaussian units which are often defined by saying that electrical quantities, including electric intensity **E**, are in electrostatic units, and magnetic quantities, including magnetic intensity **H** are in electromagnetic units. But such a system of mixed units would not be self consistent, and it is bad practice to use symbols with different systems of units in the same formula. A better definition of Gaussian units is to say that $\epsilon_0 = \mu_0 = 1$, and it can then be shown that $\kappa_0 = c$, the velocity of electromagnetic waves in free space. Gaussian units have many advantages in mathematical work, but may not be very familiar to radio engineers. In this book, therefore, we use the more conventional system with $\kappa_0 = 1$. The advantages of the Gaussian system are achieved in a different way by using for magnetic intensity a measure $\mathscr{H}$ defined in §2.10. For a full discussion of system of units see Shire (1960).

2.2 Harmonic waves and complex quantities

In nearly all the problems discussed here, the field is assumed to arise from harmonic waves, which means that all field variables (components of **E**, **H**, **P**, **D**, and **J**) vary sinusoidally with time with the same angular frequency ω. The differential equations (2.8) to (2.11) are linear, and it will be shown in ch. 3 that the relation between **D** or **P** or **J** and **E** is also linear. Hence each field component may be assumed to include the time through a factor $e^{i\omega t}$ where $\omega = 2\pi f$ is called the 'angular wave-frequency' and f is called the 'wave-frequency'. Let F be some field component. Then with this convention $F = F_0 e^{i\omega t}$ where F_0 is a constant, in general complex. Let $F_0 = A e^{i\phi}$ where A and ϕ are real. Then $F = A e^{i(\omega t+\phi)}$, where $\omega t + \phi$ is called the 'argument' or 'phase' of F and A is called its modulus. A is also called the 'amplitude' of the wave, and F_0 is sometimes called the 'complex amplitude'.† At any instant the observed value of F must be real. It may be taken as equal to the real part of $F_0 e^{i\omega t}$, that is $A \cos(\omega t + \phi)$. For since all the equations are linear, they can be separated into real and imaginary parts which must be satisfied separately. This device of the complex time factor $e^{i\omega t}$ is widely used in electrical engineering and in many branches of physics. It cannot be used without modification in problems which involve non-linear terms. For example, problems of energy flow involve products of the field quantities, and complex quantities cannot be used immediately because the real part of the product of two complex numbers is not the same as the product

† In some books on the theory of complex variables the term 'amplitude' is used to mean the 'argument' of a complex number. It is never used with this meaning in the present book.

of their real parts. In this example, however, it is possible to extend the use of complex numbers so that products of two complex quantities can be handled. This is the theory of the complex Poynting vector, which is dealt with in § 2.13.

2.3 Definitions of electric intensity E and magnetic intensity H

The electric intensity **E** at a point in free space is defined as follows. A small charge δq is placed at the point and the force $\delta \mathbf{F}$ acting on it is measured. Then

$$\mathbf{E} = \lim_{\delta q \to 0} \frac{\delta \mathbf{F}}{\delta q}.$$

The magnetic intensity **H** is defined in a similar way using a small magnetic pole.

In the ionosphere there are on the average N free electrons per unit volume. The electric and magnetic intensities defined as above must vary markedly in the free space between them, that is over a distance of the order $N^{-\frac{1}{3}}$. But when **E** and **H** and the other field variables are used in Maxwell's equations, they are assumed to represent vector fields continuously distributed in space and approximately constant over distances which are large compared with $N^{-\frac{1}{3}}$ but small compared to a wavelength. The ionosphere is thus treated as a continuous medium, and the use of Maxwell's equations implies a 'smoothing-out' process over a distance large compared with $N^{-\frac{1}{3}}$. Within the ionosphere, therefore, definitions of **E** and **H** must be found which effect this smoothing out.

For a medium which is not free space, the electric intensity **E** is usually defined in terms of a long thin cavity. The component of **E** in a given direction is found as follows. A long thin cavity is imagined to be cut in the medium parallel to a given direction. Its length must be so small that the electric state of the medium does not change appreciably within it. For waves in a homogeneous ionosphere, this means that the length is small compared to a wavelength. An infinitesimal test charge δq is placed at the centre of the cavity, and the component δF of the force acting on it in the direction of the cavity is measured. The component of **E** in this direction is $\lim_{\delta q \to 0} \delta F/\delta q$. For a medium which contains discrete ions and electrons, this definition can still be used provided that the cross-section of the cavity is very large compared with $N^{-\frac{1}{3}}$. It is the electric intensity so defined which is to be used in Maxwell's equations. The cavity definition thus effects the 'smoothing out' process mentioned above. A similar argument applies to the magnetic intensity **H** which is defined in terms of the force on a small magnetic pole in the cavity.

2.4 The current density J and electric polarisation P

In the ionosphere there is a current density **J** which arises from the motion of charges. In most problems it will be sufficiently accurate to consider **J** as arising from the movement of electrons only, but in some cases the movement of heavy ions may also contribute (see §§3.9, 5.9, 6.18).

Let **r** be the average vector displacement of an electron from the position it would have occupied if there were no field. Then the average electron velocity is $\partial\mathbf{r}/\partial t$ and the current density is

$$\mathbf{J} = Ne\frac{\partial\mathbf{r}}{\partial t}, \tag{2.3}$$

where e is the charge on one electron, and is a negative number. The average number of electrons per unit volume, N, can only be defined for a volume large enough to contain many electrons, and (2.3) therefore implies a 'smoothing out' over a distance large compared with $N^{-\frac{1}{3}}$.

Now it is convenient to define the electric polarisation **P** thus:

$$\mathbf{P} = Ne\mathbf{r}, \tag{2.4}$$

so that

$$\mathbf{J} = \frac{\partial\mathbf{P}}{\partial t}. \tag{2.5}$$

The electric polarisation must not be confused with the wave-polarisation described in §5.2.

For dielectrics the electric polarisation is usually defined to be the electric dipole moment per unit volume, but this definition is not suitable for a medium containing free electrons. For if the electrons are randomly distributed and are then all given small equal displacements in the same direction, they remain randomly distributed and it is not obvious that the medium has become polarised.

The values of **r** and therefore of **J** and **P** depend on the electric field **E** and are found from the equation of motion of an electron. The relations are called the 'constitutive relations' of the ionosphere and are derived in ch. 3.

For harmonically varying fields in a loss-free dielectric, **P** is always in phase with the electric intensity **E**, so that $\partial\mathbf{P}/\partial t = \mathbf{J}_1$ is in quadrature with **E**. Then $\mathbf{J}_1$ arises from the displacement of bound charges and is part of the 'displacement current' (the other part is $\epsilon_0(\partial\mathbf{E}/\partial t)$). If the dielectric is also a conductor, there is in addition a conduction current $\mathbf{J}_2$ in phase with $\mathbf{E}$. In the ionosphere, however, $\mathbf{P}$ is not necessarily in phase with $\mathbf{E}$, but may be expressed as the sum of two components, one $\mathbf{P}_1$ in phase with $\mathbf{E}$ and the other $\mathbf{P}_2$ in quadrature with $\mathbf{E}$. The component $\mathbf{J}_2 = \partial\mathbf{P}_2/\partial t$ is then in phase with $\mathbf{E}$ (it cannot be in antiphase) and is like a conduction current, while $\mathbf{J}_1 = \partial\mathbf{P}_1/\partial t$ is part of the displacement current, as before. The use here of $\mathbf{P}$ and $\mathbf{J}$ is therefore more general than in the usual theory of dielectrics. Either can be used to describe the state of the electrons (and ions) in the ionosphere, and includes their contribution to both the conduction and displacement currents.

The definition (2.4) of $\mathbf{P}$ is somewhat ambiguous because the origin for $\mathbf{r}$ can never be specified, and it would be impossible to devise an experiment to measure $\mathbf{P}$ at a given point. But the velocity $\partial\mathbf{r}/\partial t$ has a definite meaning, and an experiment to measure

$$\mathbf{J} = Ne\frac{\partial\mathbf{r}}{\partial t} = \frac{\partial\mathbf{P}}{\partial t}$$

could easily be suggested. Strictly speaking, therefore, the whole theory should be formulated in terms of $\mathbf{J}$. But $\mathbf{P}$ appears in the last Maxwell equations (2.11) only through its time derivative, so that the ambiguity is removed. Hence we use $\mathbf{P}$ because it makes the mathematics more concise.

2.5 The electric displacement D and magnetic induction B

The electric displacement $\mathbf{D}$ is defined thus:

$$\mathbf{D} = \epsilon_0\mathbf{E} + \mathbf{P}, \tag{2.6}$$

where $\mathbf{E}$ is defined as in § 2.3. The time derivative $\partial\mathbf{D}/\partial t$ may be called the total current density and includes $\partial\mathbf{P}/\partial t$ which is made up of two parts, a conduction current, and a part of the displacement current, as shown in the last section.

An alternative definition of $\mathbf{D}$ equivalent to (2.6) uses the force normal to the plane of a flat plate-like cavity, on a test-charge in the cavity. If this definition is used for the ionosphere, it is important to remember that the cavity is imagined to be cut *before* the electrons are displaced; no electrons must cross the central plane of the cavity.

The magnetic permeability of the ionosphere is assumed to be unity so that the magnetic induction $\mathbf{B}$ is defined by

$$\mathbf{B} = \mu_0\mathbf{H}. \tag{2.7}$$

2.6 Maxwell's equations

The electromagnetic field in the ionosphere is governed by the four Maxwell equations:

$$\operatorname{div}\mathbf{D} = 0, \tag{2.8}$$

$$\operatorname{div}\mathbf{B} = 0, \tag{2.9}$$

$$\operatorname{curl}\mathbf{E} = -\frac{\partial \mathbf{B}}{\partial t} = -\mu_0\frac{\partial \mathbf{H}}{\partial t}, \tag{2.10}$$

$$\operatorname{curl}\mathbf{H} = \frac{\partial \mathbf{D}}{\partial t}. \tag{2.11}$$

The first, (2.8), is a differential form of Gauss's theorem and results from the inverse square law of force in electrostatics. It is assumed that there is no permanent space charge, but that the contribution to **P** of every electron and ion is included. The second equation, (2.9), is similarly a result of the inverse square law of force in magnetism.

The third equation, (2.10), is a differential form of Faraday's law of electromagnetic induction, which states that $\int \mathbf{E}.d\mathbf{l}$ for a closed circuit is minus the rate of change of the magnetic flux linking the circuit. The integral is proportional to the work done in taking a small charge round the circuit. This could be found by imagining a long thin cavity to be cut along the line of the circuit. The cross-section must be large compared with $N^{-\frac{1}{3}}$ but may still be so small that a negligible amount of material is removed, so that the disturbance of the fields is inappreciable. The test-charge is then moved in the cavity right round the circuit and the work measured. This argument shows that the electric intensity **E** used in (2.10) must be as defined by a long thin cavity, as on p. 13.

The fourth equation, (2.11), is a differential form of Ampère's circuital theorem using the *total* current density $\partial\mathbf{D}/\partial t$. This is made up of a part $\partial\mathbf{P}/\partial t$ arising from the movement of electrons (and possibly ions) and a part $\epsilon_0(\partial\mathbf{E}/\partial t)$ which is the contribution to the displacement current from the changing electric intensity.

2.7 Cartesian coordinate system

Let x, y, z be right-handed Cartesian coordinates, and let **i**, **j**, **k**, be unit vectors in the directions of the x-, y-, z-axes. Subscripts x, y, z will be used to denote the x, y, z components respectively of a vector. For example, the components of **F** are written F_x, F_y, F_z. There is a very useful expression for the operator curl, which may be written in the form of a determinant

$$\operatorname{curl}\mathbf{F} = \begin{vmatrix} \mathbf{i} & \mathbf{j} & \mathbf{k} \\ \partial/\partial x & \partial/\partial y & \partial/\partial z \\ F_x & F_y & F_z \end{vmatrix}. \tag{2.12}$$

When this is evaluated the operators in the second row must always precede the field components in the third row.

For harmonic waves all field variables contain the time t only through the factor $e^{i\omega t}$ (§ 2.2). Hence the operator $\partial/\partial t$ is equivalent to multiplication by $i\omega$. If this and (2.7), (2.12) are used, the last two Maxwell equations (2.10), (2.11) become:

$$\left.\begin{aligned} \frac{\partial E_z}{\partial y}-\frac{\partial E_y}{\partial z} &= -i\omega\mu_0 H_x, & \frac{\partial H_z}{\partial y}-\frac{\partial H_y}{\partial z} &= i\omega D_x, \\ \frac{\partial E_x}{\partial z}-\frac{\partial E_z}{\partial x} &= -i\omega\mu_0 H_y, & \frac{\partial H_x}{\partial z}-\frac{\partial H_z}{\partial x} &= i\omega D_y, \\ \frac{\partial E_y}{\partial x}-\frac{\partial E_x}{\partial y} &= -i\omega\mu_0 H_z, & \frac{\partial H_y}{\partial x}-\frac{\partial H_x}{\partial y} &= i\omega D_z. \end{aligned}\right\} \tag{2.13}$$

In most of chs. 2 to 6 we consider only a single plane wave in a homogeneous medium, and the z-axis is chosen to be the direction of the wave normal. For chs. 7 onwards there may be several plane waves with wave normals in different directions and the z-axis is then chosen to be vertically upwards.

2.8 Progressive plane waves

A plane wave is defined to be a disturbance in which there is no variation of any field component in any plane parallel to a fixed plane. The z-axis may be chosen to be normal to this fixed plane, and is called the 'wave normal'. The derivatives $\partial/\partial x$, $\partial/\partial y$ are then zero for all field components, so that (2.13) become

$$\frac{\partial E_y}{\partial z} = i\omega\mu_0 H_x, \qquad \frac{\partial H_x}{\partial z} = i\omega D_y, \tag{2.14}$$

$$\frac{\partial E_x}{\partial z} = -i\omega\mu_0 H_y, \qquad \frac{\partial H_y}{\partial z} = -i\omega D_x. \tag{2.15}$$

$$H_z = 0, \qquad D_z = 0. \tag{2.16}$$

The equations (2.16) show that, for plane waves, $\mathbf{D}$ and $\mathbf{H}$ are transverse to the wave normal. In an isotropic medium $\mathbf{D}$ is proportional to $\mathbf{E}$, so that **E** also is transverse. But the ionosphere is not isotropic because of the earth's magnetic field, and it is shown later, § 5.7, that **E** often has a longitudinal component, that is, a component in the direction of the wave normal.

For a homogeneous isotropic medium we may write

$$\mathbf{D} = \epsilon_0 \mathfrak{n}^2 \mathbf{E}, \tag{2.17}$$

where $\mathfrak{n}$ is the refractive index, derived later, chs. 4 and 5, and is a constant (in general complex) at a given frequency. This may be substituted in (2.14), (2.15). The equations (2.14) then involve E_y, H_x only, and (2.15) involve E_x, H_y only. These pairs are therefore independent, and the variables in one pair can vanish without affecting the other. If this happens, the wave is said to be linearly polarised. Elimination of H_y from (2.15) gives

$$\frac{\partial^2 E_x}{\partial z^2} + \omega^2 \mathfrak{n}^2 \epsilon_0 \mu_0 E_x = 0. \tag{2.18}$$

Two independent solutions of this are

$$E_x = E_x^{(1)} e^{-i\mathfrak{n}\omega z/c}, \tag{2.19}$$

$$E_x = E_x^{(2)} e^{+i\mathfrak{n}\omega z/c}, \tag{2.20}$$

where $E_x^{(1)}$, $E_x^{(2)}$ are constants, and $c = (\epsilon_0 \mu_0)^{-\frac{1}{2}}$. The expressions (2.19) and (2.20) represent waves travelling in the direction of positive and negative z, respectively. Any other solution of (2.18) can be expressed as the sum of multiples of the two independent solutions (2.19), (2.20).

Substitution of (2.19) into (2.15) gives

$$\mathfrak{n}\epsilon_0^{\frac{1}{2}} E_x = \mu_0^{\frac{1}{2}} H_y. \tag{2.21}$$

Thus the ratio of E_x to H_y is a constant so that E_x and H_y both depend on z only through the factor $e^{-i\omega\mathfrak{n}z/c}$. A plane wave of this kind, in which the dependence of all field quantities upon z is the same, is called a 'progressive plane wave'. The operator $\partial/\partial z$ must then be equivalent to multiplication by a constant (in this case $\partial/\partial z \equiv -i\omega\mathfrak{n}/c$). The ratio E_x/H_y is called the 'wave-impedance', so that a progressive wave is one for which the wave-impedance is a constant independent of z.

Similarly, substitution of (2.20) into (2.15) gives

$$\mathfrak{n}\epsilon_0^{\frac{1}{2}} E_x = -\mu_0^{\frac{1}{2}} H_y, \tag{2.22}$$

so that E_x and H_y both vary with z only through the factor $e^{i\mathfrak{n}\omega z/c}$, and $\partial/\partial z \equiv i\mathfrak{n}\omega/c$ for both field components. Thus (2.20) is also a progressive plane wave. Consider the expression

$$E_x = E_0 \cos(\mathfrak{n}\omega z/c), \tag{2.23}$$

which is a solution of (2.18). It is a plane wave, and can be expressed as the sum of terms like (2.19) and (2.20) with equal modulus. Substitution of (2.23) in (2.15) gives

$$\mu_0^{\frac{1}{2}} H_y = -i\epsilon_0^{\frac{1}{2}} \mathfrak{n} E_0 \sin(\mathfrak{n}\omega z/c). \tag{2.24}$$

It is at once clear that this does not have the properties of a progressive wave. In this case the two component progressive waves have equal modulus and the wave is called a 'standing wave'. If the moduli of the component progressive wave were unequal, the wave would be called a 'partial standing wave'.

It can be shown in a similar way that (2.14) lead to two linearly polarised progressive plane waves, with the electric vector parallel to the y-axis. It has thus been shown that for progressive plane waves travelling in, say, the positive z-direction, there are two independent solutions, each linearly polarised with their electric vectors at right angles. They may be combined with any moduli and relative phase to give a resultant wave which is, in general, elliptically polarised. Any polarisation ellipse can be produced by a suitable combination of the component plane waves and, once established, the ellipse does not change as long as the wave remains in the same homogeneous isotropic medium. In contrast to this, it will be shown later (§ 5.3) that in a homogeneous magnetoionic medium, progressive waves can only have one of two possible polarisations. A wave of any other polarisation must in general be made up of two component waves travelling with different velocities.

2.9 Plane waves in free space

In free space there are no electrons and $\mathbf{D} = \epsilon_0 \mathbf{E}$ so that $\mathfrak{n} = 1$. Then the progressive plane wave solution (2.19) gives

$$E_x = E_x^{(1)} e^{-i\omega z/c}, \tag{2.25}$$

where $c = (\epsilon_0 \mu_0)^{-\frac{1}{2}}$ is the velocity of the waves, and (2.21) gives

$$\epsilon_0^{\frac{1}{2}} E_x = \mu_0^{\frac{1}{2}} H_y. \tag{2.26}$$

Thus the ratio E_x/H_y is real which shows that E_x and H_y are in phase, and the vectors **E**, **H** and the wave normal, in that order, form a right-handed system. The wave-impedance is

$$E_x/H_y = (\mu_0/\epsilon_0)^{\frac{1}{2}} = Z_0, \tag{2.27}$$

which is called 'the characteristic impedance of free space'.

2.10 The notation $\mathscr{H}$ and H

It is now convenient to adopt for the magnetic field **H** a different measure which simplifies the equations and will be used throughout this book. Take

$$\mathscr{H} = Z_0\mathbf{H}. \tag{2.28}$$

Thus $\mathscr{H}$ measures the magnetic field in terms of the electric field that would be associated with it in a progressive plane wave in free space. It has the same effect as if **E** and **H** were measured in Gaussian units (see § 2.1). The vector $\mathscr{H}$ has the same physical dimensions as the electric intensity **E** (which is not true for **E** and **H** in Gaussian units).

Further let

$$k = \omega/c = 2\pi/\lambda, \tag{2.29}$$

where λ is the wavelength in free space. Then the last two Maxwell equations (2.10), (2.11) become

$$\operatorname{curl}\mathbf{E} = -ik\mathscr{H}, \quad \operatorname{curl}\mathscr{H} = \frac{ik}{\epsilon_0}\mathbf{D}, \tag{2.30}$$

or written in full, in Cartesian coordinates:

$$\left.\begin{aligned}
\frac{\partial E_z}{\partial y}-\frac{\partial E_y}{\partial z} &= -ik\mathscr{H}_x, & \frac{\partial \mathscr{H}_z}{\partial y}-\frac{\partial \mathscr{H}_y}{\partial z} &= \frac{ik}{\epsilon_0}D_x,\\
\frac{\partial E_x}{\partial z}-\frac{\partial E_z}{\partial x} &= -ik\mathscr{H}_y, & \frac{\partial \mathscr{H}_x}{\partial z}-\frac{\partial \mathscr{H}_z}{\partial x} &= \frac{ik}{\epsilon_0}D_y,\\
\frac{\partial E_y}{\partial x}-\frac{\partial E_x}{\partial y} &= -ik\mathscr{H}_z, & \frac{\partial \mathscr{H}_y}{\partial x}-\frac{\partial \mathscr{H}_x}{\partial y} &= \frac{ik}{\epsilon_0}D_z.
\end{aligned}\right\} \tag{2.31}$$

These equations are the starting-point for much of the later work in this book.

2.11 The energy stored in a radio wave in the ionosphere

It is shown in books on electromagnetic theory that when a small change $\delta\mathbf{D}$ is made in the electric displacement in a medium, the electric forces must supply energy per unit volume equal to

$$\delta W_E = \mathbf{E}.\delta\mathbf{D}. \tag{2.32}$$

A simple way of showing this for the ionosphere is to imagine that a homogeneous sample of the ionised medium is enclosed in a condenser with parallel plates of area A separated by a distance d. Let the charge per unit area on the plates be σ. Then $\sigma = D$ where D is the component of **D** perpendicular to the plates. It is assumed that **E** is normal to the plates and has magnitude E. Then the potential difference between the plates is Ed. If σ is now increased by a small amount $\delta\sigma$, the electric energy supplied is $Ed.A\,\delta\sigma$, that is energy $E\delta\sigma$ per unit volume. This is equal to $E.\delta D$ which is the vector product $\mathbf{E}.\delta\mathbf{D}$ in (2.32). A more general proof is given by Stratton (1941, ch. II). It can be shown that the proof still applies when $\mathbf{D}$ is defined by (2.6) in which $\mathbf{P}$ includes components of both displacement current and conduction current.

When a static electric field is applied to a dielectric in which $\mathbf{D}$ is proportional to $\mathbf{E}$, the expression (2.32) shows that the stored energy per unit volume of the dielectric is $\frac{1}{2}\mathbf{E}.\mathbf{D}$. This is not, however, in general true for varying fields such as those in a wave, and is incorrect for the ionosphere, as is shown by the examples given in § 3.10.

It can be shown that, in addition to the electric energy (2.32), some energy is supplied by the magnetic forces, given by

$$\delta W_M = \mathbf{H}.\delta\mathbf{B} \tag{2.33}$$

per unit volume. In the ionosphere $\mathbf{B} = \mu_0\mathbf{H}$, and if $\mathbf{H}$ is zero when $t = 0$, the total magnetic energy supplied in time t is

$$\mu_0\int_0^t \mathbf{H}\frac{\partial \mathbf{H}}{\partial t}.dt = \tfrac{1}{2}\mu_0\mathbf{H}^2,$$

which is the magnetic energy stored per unit volume.

2.12 The flow of energy. Poynting's theorem

The arguments of this section apply to fields which vary with time in any way whatever. They are not confined to harmonically varying fields, and hence the complex number convention of § 2.2 is not used here. Equations (2.32), (2.33) show that the rate at which energy is being supplied per unit volume of the ionosphere is

$$\frac{\partial W}{\partial t} = \frac{\partial}{\partial t}(W_E + W_M) = \mathbf{E}\frac{\partial \mathbf{D}}{\partial t} + \mathbf{H}\frac{\partial \mathbf{B}}{\partial t}. \tag{2.34}$$

This may be integrated over some volume V of the medium. The result is

$$\frac{\partial}{\partial t}\int_V W.dV = \int_V \left\{\mathbf{E}\frac{\partial \mathbf{D}}{\partial t} + \mathbf{H}\frac{\partial \mathbf{B}}{\partial t}\right\} dV, \tag{2.35}$$

which is the rate at which the electric and magnetic forces are supplying energy to the volume. Some of this energy may be undergoing conversion into heat within the volume, but this does not affect the present argument.

If, now, the Maxwell equations (2.10), (2.11) are used, the expression (2.35) becomes

$$\int_V \{\mathbf{E}\,\mathrm{curl}\,\mathbf{H}-\mathbf{H}\,\mathrm{curl}\,\mathbf{E}\}\,dV = -\int_V \mathrm{div}(\mathbf{E}\times\mathbf{H}).dV, \tag{2.36}$$

from a well-known theorem of vector analysis. The last integral can be expressed as a surface integral by using the 'divergence theorem'. The result is

$$\frac{\partial}{\partial t}\int_V W.dV = -\int_S (\mathbf{E}\times\mathbf{H})_\perp .dS, \tag{2.37}$$

where the integral is evaluated over the whole of the surface S enclosing the volume V, and the subscript $\perp$ indicates that the normal component of the vector $\mathbf{E}\times\mathbf{H}$ is to be taken at each point on the surface.

2.13 The Poynting vector

Equation (2.37) is Poynting's theorem. It suggests that the vector

$$\mathbf{\Pi} = \mathbf{E}\times\mathbf{H} \tag{2.38}$$

gives the flux of energy in the electromagnetic field, and $\mathbf{\Pi}$ is called the Poynting vector. This result cannot be regarded as proved, for in the second integral of (2.37) it would be permissible to add to the integrand any vector whose surface integral is zero. In this way other vectors can be defined which could equally well be said to give the flux of energy. It is often of interest, however, to evaluate the Poynting vector $\mathbf{\Pi}$ in particular problems. In this book it is assumed that $\mathbf{\Pi}$ gives the flux of energy, but it should be remembered that other interpretations are possible.

The Poynting vector (2.38) involves a product of the two field quantities $\mathbf{E}$ and $\mathbf{H}$. Hence the complex number convention cannot be used. Suppose, now, that the components of $\mathbf{E}$ and $\mathbf{H}$ are expressed as complex numbers for a wave varying harmonically in time. Then the real parts must be taken before the product is formed. Let a star * denote a complex conjugate. Then

$$\begin{aligned}\mathbf{\Pi} &= \tfrac{1}{4}(\mathbf{E}+\mathbf{E}^*)\times(\mathbf{H}+\mathbf{H}^*)\\ &= \tfrac{1}{4}\mathbf{E}\times\mathbf{H}+\tfrac{1}{4}\mathbf{E}^*\times\mathbf{H}^*+\tfrac{1}{4}(\mathbf{E}\times\mathbf{H}^*+\mathbf{E}^*\times\mathbf{H}).\end{aligned} \tag{2.39}$$

Here the first two terms contain factors $e^{2i\omega t}$, $e^{-2i\omega t}$ respectively, and their average values, over a long period of time, are zero. The last two terms are independent of time. A bar over a quantity will now be used to denote the average value of a quantity taken over many cycles of the oscillation. Then (2.39) gives:

$$\bar{\mathbf{\Pi}} = \tfrac{1}{4}(\mathbf{E} \times \mathbf{H}^* + \mathbf{E}^* \times \mathbf{H}) = \tfrac{1}{2}\mathscr{R}(\mathbf{E} \times \mathbf{H}^*), \tag{2.40}$$

where $\mathscr{R}$ denotes that the real part is to be taken. The product $\tfrac{1}{2}\mathbf{E} \times \mathbf{H}^*$ for a harmonic wave is called the complex Poynting vector. Its real part is assumed to be the time average of the flux of energy in the wave.

CHAPTER 3

THE CONSTITUTIVE RELATIONS

3.1 Introduction

Before (2.31) can be applied to the theory of wave-propagation in the ionosphere, it is necessary to express the electric displacement **D**, and therefore the electric polarisation **P**, in terms of the electric intensity **E**. The resulting expressions are called the constitutive relations of the ionosphere and are derived in this chapter. The subject of wave-propagation is resumed in ch. 4.

3.2 Free, undamped electrons

As in § 2.4 let **r** be the average vector displacement of an electron from the position it would occupy if there were no field. In magnetoionic theory only the time derivatives of **r** are used so that the origin for **r** is unimportant. The electrons have random velocities because of their thermal motions, but these are in all directions and the average over many electrons is zero. The thermal motions play a part in some phenomena, such as wave interaction and space-charge waves but they are neglected here. The displacement **r** is an additional displacement superimposed on the thermal motions and caused by the electric field **E**.

The force arising from this field is **E***e* for each electron. It is now assumed as a first approximation that all other forces on the electrons are negligible. In particular the force exerted by the magnetic field $\mathscr{H}$ of the wave is neglected. The justification for this is discussed in § 3.5. Then Newton's laws of motion give

$$\mathbf{E}e = \mathfrak{m}\frac{\partial^2 \mathbf{r}}{\partial t^2}, \tag{3.1}$$

where $\mathfrak{m}$ is the mass of an electron. Hence from (2.4)

$$\frac{Ne^2}{\mathfrak{m}}\mathbf{E} = \frac{\partial^2 \mathbf{P}}{\partial t^2}. \tag{3.2}$$

Now **E** varies with time through the factor $e^{i\omega t}$ and we are interested only in that component of **P** which also varies in this way. Hence $\partial/\partial t \equiv i\omega$ and

$$\mathbf{P} = -\frac{Ne^2}{\omega^2 \mathfrak{m}}\mathbf{E}. \tag{3.3}$$

This is the constitutive relation when electron collision damping is neglected. (The effect of damping is discussed in the next section.) Equation (3.3) may be written

$$\mathbf{P} = -\epsilon_0 X\mathbf{E}, \tag{3.4}$$

where

$$X = \frac{Ne^2}{\epsilon_0 \mathfrak{m}\omega^2}. \tag{3.5}$$

The quantity X is very important and appears throughout the theory in this book. In Appleton's original paper (1932), and in many important early papers on magnetoionic theory, the symbol x was used, but in recent years this has often been replaced by X, to avoid confusion with the coordinate x. Sometimes X is written

$$X = \frac{\omega_N^2}{\omega^2} = \frac{f_N^2}{f^2}, \quad \omega_N^2 = \frac{Ne^2}{\epsilon_0 \mathfrak{m}}, \quad \omega_N = 2\pi f_N. \tag{3.6}$$

Here f_N is called the 'plasma frequency' of the medium, and ω_N is the angular plasma frequency. Its square is proportional to the electron number density N. X also is proportional to N, and inversely proportional to the square of the wave-frequency. Note that its value is independent of the sign of the electronic charge.

The following are useful numerical values. The frequency f_N is in cycles per second, and the electron number density N is in cm.$^{-3}$:

$$f_N^2 = 8{\cdot}061 \times 10^7 N, \quad f_N = 8{\cdot}98 \times 10^3 N^{\frac{1}{2}}, \quad N = 1{\cdot}240 \times 10^{-8} f_N^2. \tag{3.7}$$

3.3 Electron collisions. Damping of the motion

Suppose that each electron makes, on the average, ν collisions per unit time with other particles. If τ is the time between two such successive collisions of one electron, then the average value of τ is $1/\nu$. It is shown in books on statistical mechanics that the probability that τ lies in the range τ to $\tau + d\tau$ is $\nu\, e^{-\nu\tau}\, d\tau$. Suppose that each electron is subjected to a steady force $\mathbf{F}$. In time τ an electron moves a distance $s = \frac{1}{2}(F/\mathfrak{m})\tau^2$ in the direction of $\mathbf{F}$. This movement is superimposed on the random thermal movements. The average value of s for many electrons is therefore

$$\bar{s} = \frac{1}{2}\frac{F}{\mathfrak{m}} \int_0^\infty \tau^2 \nu e^{-\nu\tau}\, d\tau. \tag{3.8}$$

This is easily evaluated and gives $\bar{s} = F/(\mathfrak{m}\nu^2)$, which is the average displacement in the direction of $\mathbf{F}$ between two collisions. Since there are on the average ν collisions per second, the average velocity is given by

$$\mathbf{v} = \mathbf{F}/\mathfrak{m}\nu. \tag{3.9}$$

Since this is a steady velocity, each electron behaves as if it were subjected to a retarding force $\mathfrak{m}\nu\mathbf{v}$, which is proportional to its velocity.

In the ionosphere $\mathbf{F}$ is a sinusoidally varying force. If ν is large compared with the wave-frequency, the above treatment might be expected to be approximately true, but when ν is comparable with, or much less than, the angular frequency ω, the analysis is more difficult. It can be shown, however, that in this case the relation (3.9) can still be expected to hold. A very illuminating discussion of electron collisions is given by Ratcliffe (1959). See also Huxley (1937*a*, *b*, 1938, 1940).

Equation (3.9) shows that an electron is subjected to a retarding or damping force, proportional to its velocity. This equation will be assumed to hold for all values of ν.

When the damping force is added to the equation of motion it becomes

$$\mathbf{E}e = \mathfrak{m}\frac{\partial^2\mathbf{r}}{\partial t^2} + \mathfrak{m}\nu\frac{\partial\mathbf{r}}{\partial t}, \tag{3.10}$$

whence

$$\mathbf{P} = \frac{1}{i\omega\nu - \omega^2}\frac{Ne^2}{\mathfrak{m}}\mathbf{E}. \tag{3.11}$$

This now replaces (3.3) and is the constitutive relation with damping included. It may be rewritten

$$\mathbf{P} = -\epsilon_0\frac{X}{1-iZ}\mathbf{E}, \tag{3.12}$$

where X is defined in (3.5) and

$$Z = \nu/\omega. \tag{3.13}$$

The symbol Z must not be confused with Z_0, the characteristic impedance of free space (see § 2.9). The quantity Z is a useful measure of the collision-frequency. In Appleton's original theory (1932) the symbol z was used, but this is now often replaced by Z, to avoid confusion with the coordinate z. In much of this book the notation $U = 1 - iZ$ is used. If $\nu = 0$, then $U = 1$.

3.4 Effect of the earth's magnetic field on motion of electrons

Let the magnetic induction of the earth's field be denoted by the vector $\mathfrak{B}$. A charge e, moving with velocity $\partial\mathbf{r}/\partial t$ through it, is subjected to a force $e(\partial\mathbf{r}/\partial t)\times\mathfrak{B}$. Here e is the true charge on the electron, actually a negative number. Then the equation of motion of an electron is

$$\mathbf{E}e + e\frac{\partial\mathbf{r}}{\partial t}\times\mathfrak{B} = \mathfrak{m}\frac{\partial^2\mathbf{r}}{\partial t^2} + \mathfrak{m}\nu\frac{\partial\mathbf{r}}{\partial t}. \tag{3.14}$$

The operator $\partial/\partial t$ is replaced by $i\omega$, and the equation is multiplied by $Ne/\mathfrak{m}\omega^2$. This gives (since $\mathbf{P} = Ne\mathbf{r}$)

$$\frac{Ne^2}{\mathfrak{m}\omega^2}\mathbf{E}+\frac{ie}{\mathfrak{m}\omega}\mathbf{P}\times\mathfrak{B} = -\mathbf{P}(1-iZ). \tag{3.15}$$

Let
$$\mathbf{Y} = \frac{e}{\mathfrak{m}\omega}\mathfrak{B}. \tag{3.16}$$

Because of the negative value of e, the vector $\mathbf{Y}$ is in the opposite direction to $\mathfrak{B}$. Now (3.15) is re-arranged to give

$$-\epsilon_0 X\mathbf{E} = \mathbf{P}(1-iZ)+i\mathbf{P}\times\mathbf{Y}. \tag{3.17}$$

Let $Y = |e\mathfrak{B}/\mathfrak{m}\omega|$, and let l, m, n be the direction cosines of the vector $\mathbf{Y}$ (opposite in direction to the earth's magnetic field because e is negative). Then (3.17) may be written out in Cartesian coordinates:

$$\left.\begin{aligned} -\epsilon_0 XE_x &= UP_x+inYP_y-imYP_z,\\ -\epsilon_0 XE_y &= -inYP_x+UP_y+ilYP_z,\\ -\epsilon_0 XE_z &= imYP_x-ilYP_y+UP_z,\end{aligned}\right\} \tag{3.18}$$

where
$$U = 1-iZ. \tag{3.19}$$

The right-hand side may be regarded as a matrix product and the equations may be written in matrix form, thus:

$$-\epsilon_0 X\begin{pmatrix}E_x\\E_y\\E_z\end{pmatrix} = \begin{pmatrix}U & inY & -imY\\ -inY & U & ilY\\ imY & -ilY & U\end{pmatrix}\begin{pmatrix}P_x\\P_y\\P_z\end{pmatrix}. \tag{3.20}$$

These are the constitutive relations with the effects of damping and the earth's magnetic field included. They will be used in ch. 4 to derive the important formulae of the magnetoionic theory.

In Appleton's original theory (1932) the symbol y was used for the quantity here denoted by Y. This quantity may be written

$$Y = \left|\frac{e\mathfrak{B}}{\mathfrak{m}\omega}\right| = \frac{\omega_H}{\omega}, \quad \omega_H = \left|\frac{e\mathfrak{B}}{\mathfrak{m}}\right| = 2\pi f_H. \tag{3.21}$$

An electron moving in a magnetic field traverses a helical path, and makes one turn of the helix in a time $1/f_H$. It is easy to show that this time is independent of the velocity of the electron provided this is low enough for relativistic effects to be neglected. f_H is called the 'gyro-frequency' for electrons, and ω_H is the 'angular gyro-frequency'.

3.5 The effect of the magnetic field of the wave on the motion of electrons

In the preceding sections the force exerted on an electron by the magnetic field $\mathscr{H}$ of the wave has been neglected. To justify this, it is interesting to estimate the order of magnitude of this force by considering the motion of one electron. Suppose a wave of angular frequency ω is travelling in the positive z-direction, and is linearly polarised with its electric vector in the x-direction. Let x denote the coordinate of one electron, and neglect damping forces. Let the electric field have amplitude E_0. The following argument involves products of harmonically varying quantities, so that complex numbers should be avoided. Hence we write

$$\mathfrak{m}\frac{\partial^2 x}{\partial t^2} = eE_0 \cos \omega t,$$

whence
$$\frac{\partial x}{\partial t} = \frac{eE_0}{\mathfrak{m}\omega}\sin \omega t.$$

Since damping is neglected, the refractive index $\mathfrak{n}$ is real (see §4.3). Hence the magnetic field $H_y = (\epsilon_0/\mu_0)^{\frac{1}{2}}\mathfrak{n}E_0 \cos \omega t$ (2.21). Then the instantaneous force exerted on the electron by this field is

$$\mu_0 e H_y \frac{\partial x}{\partial t} = \frac{\mathfrak{n}e^2E_0^2}{\mathfrak{m}\omega c}\sin \omega t \cos \omega t. \tag{3.22}$$

This is in the direction of the z-axis, that is, of the wave normal. Notice that it varies with twice the wave-frequency, and that its average value is zero.

If damping forces are included, the magnetic field H_y and the velocity $\partial x/\partial t$ are no longer in quadrature, and the force then has an average value different from zero. There is thus an average force in the direction of the wave normal, which gives rise to radiation pressure. It is usually neglected in the theory of radio waves in the ionosphere, but it is discussed in books on electromagnetic theory (see, for example, Shire, 1960).

The maximum value of the force (3.22) is $(\mathfrak{n}e^2E_0^2)/(2\mathfrak{m}\omega c) = F_M$. The ratio of this to the maximum electric force F_E is

$$F_M/F_E = \frac{\mathfrak{n}eE_0}{2\mathfrak{m}\omega c}. \tag{3.23}$$

The value of E_0 at 100 km from a transmitter radiating 10^6 W is of the order 0·08 V/m, and the magnetic field of the wave is about $2{\cdot}7\times10^{-6}$ oersted. The magnetic and electric fields encountered in the ionosphere from man-made radio transmitters will rarely reach these values, and the frequency is always greater than 10 kc/s. The refractive index n is of order unity. Hence the ratio (3.23) will not exceed about $3{\cdot}7\times10^{-4}$, and will usually be very much less than this, so that the effect of the magnetic field of the wave can be neglected.

Electromagnetic energy is radiated from a lightning flash, which may give electric fields in the ionosphere much greater than 0·1 V/m, and magnetic fields which are comparable with the earth's magnetic field. In this case the received signals are known as 'atmospherics', and they are studied with receivers which accept frequencies down to 100 c/s or lower. The magnetic field of the wave may play a part in the mechanism of reflection of these signals from the ionosphere, at points close to the flash.

3.6 The susceptibility matrix

In the matrix form (3.20) of the constitutive relations the components of **E** are expressed in terms of the components of **P**. In later chapters it will be necessary to have the components of **P** expressed in terms of those of **E**. Equation (3.20) is equivalent to three simultaneous equations for P_x, P_y, P_z. These can be solved, and the process is equivalent to inversion of the 3×3 matrix. The result is

$$\frac{1}{\epsilon_0}\begin{pmatrix}P_x\\P_y\\P_z\end{pmatrix}=-\frac{X}{U(U^2-Y^2)}$$

$$\times\begin{pmatrix}U^2-l^2Y^2 & -inYU-lmY^2 & imYU-lnY^2\\ inYU-lmY^2 & U^2-m^2Y^2 & -ilYU-mnY^2\\ -imYU-lnY^2 & ilYU-mnY^2 & U^2-n^2Y^2\end{pmatrix}\begin{pmatrix}E_x\\E_y\\E_z\end{pmatrix}. \qquad (3.24)$$

The expression multiplying (E_x, E_y, E_z) is called the susceptibility matrix of the ionosphere, and is denoted by **M**. It was given in this form by Banerjea (1947). Its components are denoted by M_{ij} $(i, j = x, y, z)$.

It can be shown that the elements of **M** have the following properties:

$$M_{xx}(M_{yz}+M_{zy}) = M_{xy}M_{zx}+M_{yx}M_{xz}, \qquad (3.25)$$

$$M_{xx}M_{yy}-M_{yz}M_{zy}=\frac{X^2}{U^2-Y^2}, \qquad (3.26)$$

and four other relations obtained by permuting the suffixes in (3.25) and (3.26).

$$M_{xx}M_{yy}M_{zz}+M_{xy}M_{yz}M_{zx}+M_{yx}M_{zy}M_{xz}-M_{xx}M_{yz}M_{zy}$$
$$-M_{yy}M_{xz}M_{zx}-M_{zz}M_{yx}M_{xy}=-\frac{X^3}{U(U^2-Y^2)}. \quad (3.27)$$

These relations are needed later.

If the medium is loss-free, $Z = 0$ and $U = 1$. Then $M_{ij} = M^*_{ji}$ (where a star denotes a complex conjugate) so that the matrix $\mathbf{M}$ is Hermitian. The average rate of conversion of electric energy into heat per unit volume is given by $\frac{1}{2}\mathscr{R}(\mathbf{E}^* . \partial\mathbf{P}/\partial t)$. It is easily verified that this is zero when $\mathbf{M}$ is Hermitian.

3.7 The Lorentz polarisation term

In the preceding sections it has been assumed that the only electric force acting on an electron is $\mathbf{E}e$ arising from the electric field $\mathbf{E}$ of the wave. In certain types of dielectric this is not true. Lorentz (*Theory of Electrons*, ch. IV) considered dielectrics in which all the electrons were assumed to be bound within atoms or molecules, and were displaced from their equilibrium positions by the applied electric field $\mathbf{E}$. In this case the electric polarisation $\mathbf{P}$ is defined as the electric dipole moment per unit volume, and the difficulty mentioned at the end of § 2.4 does not arise. Lorentz showed that, if the molecules of a dielectric are randomly arranged in space (as in glass, for example) or if they are arranged in a cubic lattice, then the average electric field acting on a molecule is $\mathbf{E}+\frac{1}{3}\mathbf{P}/\epsilon_0$. The second term of this expression is called the Lorentz polarisation term. Many authors have assumed that this expression applied also to the random arrangement of free electrons in the ionosphere. If this were so, the constitutive relations (3.20) would become

$$-X\epsilon_0\begin{pmatrix}E_x\\E_y\\E_z\end{pmatrix}=\begin{pmatrix}U+\frac{1}{3}X & inY & -imY\\ -inY & U+\frac{1}{3}X & ilY\\ imY & -ilY & U+\frac{1}{3}X\end{pmatrix}\begin{pmatrix}P_x\\P_y\\P_z\end{pmatrix}. \quad (3.28)$$

Some arguments for and against the inclusion of the Lorentz term have been given by Darwin (1934, 1943). The statistical problem involved is particularly difficult. There is now, however, some experimental evidence that the Lorentz polarisation term should not be included for the ionosphere. The strongest evidence is probably from the theory of whistling atmospherics or 'Whistlers', which is discussed in § 13.20. See also, Beynon (1947) and Newbern Smith (1941).

3.8 The effect of small irregularities in the ionosphere

The vector **P** used in the constitutive relation (2.54) was given by $\mathbf{P} = Ne\mathbf{r}$ (2.4). **P** is thus found by taking an average over a volume large enough to contain many electrons. When **E**, **P** and **D** are used later (chs. 4 and 5) in Maxwell's equations, they are assumed to represent vector fields continuously distributed in space, and approximately constant over distances which are small compared to a wavelength. There would be no meaning in speaking of the value of **P** at a specific point in the free space between the electrons. The use of Maxwell's equations implies a 'smoothing-out' process over a distance which must be large compared with $N^{-\frac{1}{3}}$.

There is now much evidence to show that the ionosphere is an irregular medium, that is, that the electron number density N varies from place to place. The effect of the irregularities on the propagation of waves through the medium may be treated by considering the energy scattered from them, or by studying their effect on the refraction and diffraction of the waves, but these topics are outside the scope of this book. There is, however, an alternative method for studying this problem in the special case when the irregularities are very small compared with one wavelength. The electric field **E** and the electric polarisation **P** depend on the electron number density N and vary from place to place because N varies, but it is possible to find their average values $\mathbf{E}_0$ and $\mathbf{P}_0$ by a 'smoothing-out' process similar to that already mentioned. Thus $\mathbf{E}_0$ and $\mathbf{P}_0$ are averages for a volume which is large compared with the size of the irregularities. It is these average values which must be used in Maxwell's equations, and it is therefore necessary to know how $\mathbf{P}_0$ depends on $\mathbf{E}_0$. A new set of constitutive relations is therefore required, giving $\mathbf{P}_0$ in terms of $\mathbf{E}_0$, and this is in general different from the commonly used relations (3.20). In this book, however, the effect of small irregularities is ignored. For a discussion of this subject see Budden (1959).

3.9 The effect of heavy ions

In deriving the constitutive relations (3.20) and (3.24) it was assumed that the current **J** and the electric polarisation **P** arose entirely from the motion of electrons in the ionosphere. The electric field of the wave also causes the heavy ions to move, and this gives an additional current, which is usually negligibly small, but may be important in special circumstances. The total electric polarisation **P** is then the sum of the contributions $\mathbf{P}_e$ and $\mathbf{P}_i$ from the electrons and heavy ions respectively. In this section subscripts e and i are used to indicate quantities which apply respectively to electrons and heavy ions.

The part $\mathbf{P}_e$ of the electric polarisation which is contributed by the electrons is given by (3.24) when X, Y, and U are replaced by X_e, Y_e, U_e, and the part $\mathbf{P}_i$ contributed by the heavy ions is given by the same expression with X, Y, U replaced by X_i, Y_i, U_i. Now (3.5) shows that

$$X_i = \frac{N_i e^2}{\epsilon_0 \mathfrak{m}_i \omega^2}, \tag{3.29}$$

where N_i is the number of heavy ions per unit volume, and $\mathfrak{m}_i$ is the mass of one ion. For a molecule of oxygen $\mathfrak{m}_i$ is about $59\,000\mathfrak{m}_e$ and for a molecule of hydrogen $\mathfrak{m}_i$ is about $3700\mathfrak{m}_e$. If N_i is of the same order as N_e it is clear that in both cases X_i is negligibly small compared with X_e, and the contribution $\mathbf{P}_i$ can be neglected entirely. Only if $N_i \gg N_e$ will the effect of heavy ions be important.

Since the ionosphere must be electrically neutral, it is only possible for N_i to be many thousands of times greater than N_e if there are roughly equal numbers of heavy positive and negative ions. The contributions which these make to $\mathbf{P}_i$ must be computed separately and added. The values of X_i and U_i are very closely the same for both. Now (from (3.16)) $\mathbf{Y}_i = e_i\mathbf{\mathfrak{B}}/(\mathfrak{m}_i\omega)$. Its direction cosines l, m, n, depend on the sign of the charge e_i and have opposite values for the positive and negative ions. Hence when the two contributions to $\mathbf{P}_i$ are added, terms containing odd powers of l, m, n, cancel, and the expression (3.24) for $\mathbf{P}_i$ becomes:

$$\frac{1}{\epsilon_0}\begin{pmatrix} P_x^{(i)} \\ P_y^{(i)} \\ P_z^{(i)} \end{pmatrix} = -\frac{X_i}{U_i(U_i^2-Y_i^2)}\begin{pmatrix} U_i^2 - l^2Y_i^2 & -lmY_i^2 & -lnY_i^2 \\ -lmY_i^2 & U_i^2-m^2Y_i^2 & -mnY_i^2 \\ -lnY_i^2 & -mnY_i^2 & U_i^2-n^2Y_i^2 \end{pmatrix}\begin{pmatrix} E_x \\ E_y \\ E_z \end{pmatrix}, \tag{3.30}$$

where X_i is given by (3.29) and N_i is the *total* number of heavy ions including both positive and negative ions. Notice that the part of $\mathbf{P}_i$ which depends on Y_i is a vector parallel to $\mathbf{Y}$ and the remaining part is a vector parallel to $\mathbf{E}$. Equation (3.21) shows that

$$Y_i = \frac{\omega_H^{(i)}}{\omega}, \quad \omega_H^{(i)} = \left|\frac{e\mathfrak{B}}{\mathfrak{m}_i}\right| = 2\pi f_H^{(i)}, \tag{3.31}$$

where $f_H^{(i)}$ is the gyro-frequency for heavy ions in the earth's magnetic field. For electrons $f_H^{(e)}$ is of the order of 1 Mc/s. Hence $f_H^{(i)}$ is of the order of 17 c/s for ionised oxygen molecules, or 300 c/s for ionised hydrogen molecules. Thus for all but the very lowest frequencies Y_i is very small compared with unity. If it is neglected, (3.30) may be written

$$\mathbf{P}_i = -\frac{\epsilon_0 X_i}{U_i}\mathbf{E}. \tag{3.32}$$

The neglect of Y_i is equivalent to assuming that the motion of the heavy ions is unaffected by the earth's magnetic field. This is justified for all frequencies used in radio communication. For some naturally occurring radio signals, however, such as 'whistlers' (§ 13.20) frequencies down to 100 c/s or less are observed, and it may not then be permissible to neglect Y_i.

The effect of heavy ions on the polarisation and refractive index of a wave is discussed in §§ 5.9 and 6.18.

3.10 The energy stored in a radio wave in the ionosphere (continued)

The expression (2.32) for the energy supplied per unit volume of the ionosphere by the electric forces can be used to find the energy stored in unit volume. It was mentioned on p. 21 that this is $\frac{1}{2}\mathbf{E}.\mathbf{D}$ for a dielectric in which $\mathbf{D}$ is proportional to $\mathbf{E}$. For varying fields, however, this is incorrect. For consider the simple example in which electron collisions and the earth's magnetic field are neglected. Equation (3.4) then shows that $\mathbf{E}$ and $\mathbf{P}$ are in antiphase. Consequently $\mathbf{E}$ and $\mathbf{D}$ are either in phase or in antiphase, so that they both pass through zero at the same instant. But (3.1) shows that the electron velocities are in quadrature with $\mathbf{E}$, and therefore with $\mathbf{D}$. The electrons have their greatest kinetic energy at the instant when both $\mathbf{E}$ and $\mathbf{D}$ are zero. Hence $\frac{1}{2}\mathbf{E}.\mathbf{D}$ cannot be the correct expression for the stored energy per unit volume.

It is of interest to derive the correct expression for the energy per unit volume in the ionosphere. This will be done first for the case when collisions and the earth's magnetic field are neglected. Suppose that an electric field $E_0 \sin \omega t$ is switched on at the instant $t = 0$. This is not a purely harmonic wave because of its sudden beginning. Moreover, the following argument uses products of field quantities, and therefore the complex number convention of §2.2 will not be used. At a time t the total energy that has been supplied per unit volume by the electric forces is

$$W_E = \int_0^t \mathbf{E}.\frac{\partial \mathbf{D}}{\partial t}\,dt. \tag{3.33}$$

Since there is no mechanism for absorbing energy, this must all be stored in the medium. Now (3.2) shows that

$$\frac{\partial \mathbf{P}}{\partial t} = \frac{Ne^2}{m}\int_0^t \mathbf{E}\,dt = \mathbf{E}_0\frac{Ne^2}{m\omega}(1-\cos\omega t) \tag{3.34}$$

and

$$\frac{\partial \mathbf{D}}{\partial t} = \epsilon_0\frac{\partial \mathbf{E}}{\partial t}+\frac{\partial \mathbf{P}}{\partial t} = \mathbf{E}_0\left\{\omega\epsilon_0\cos\omega t+\frac{Ne^2}{m\omega}(1-\cos\omega t)\right\}. \tag{3.35}$$

Then

$$\begin{aligned} W_E &= \mathbf{E}_0^2\int_0^t\left\{\epsilon_0\,\omega\cos\omega t\sin\omega t+\frac{Ne^2}{m\omega}(1-\cos\omega t)\sin\omega t\right\}dt \\ &= \tfrac{1}{2}\mathbf{E}_0^2\sin^2\omega t\left(\epsilon_0-\frac{Ne^2}{m\omega^2}\right)+\mathbf{E}_0^2\frac{Ne^2}{m\omega^2}(1-\cos\omega t) \\ &= \tfrac{1}{2}\mathbf{E}.\mathbf{D}+\mathbf{E}_0^2\frac{Ne^2}{m\omega^2}(1-\cos\omega t). \end{aligned} \tag{3.36}$$

In this rather artificial example the electrons are assumed to be at rest when the field is first switched on. They are accelerated during the first half-cycle and decelerated during the second, so that they are again at rest after one cycle, but they move always in the same direction. Hence there is a steady drift of the electrons superimposed on their oscillatory motion. After one whole cycle the drift velocity and the oscillatory motion exactly cancel, so that the total kinetic energy is zero. Both terms of (3.36) are then zero, which shows correctly that there is then no stored energy. After one half-cycle, the oscillatory motion and the drift-velocity are in the same direction. The electrons then have kinetic energy given by the second term of (3.36), which represents the whole of the stored energy at that instant, since the first term is zero.

The energy per unit volume when the electrons have no steady drift-velocity can be found in several ways, and two methods are given here; in both the earth's magnetic field is neglected. For the first method suppose that the electric field is switched on gradually so that

$$\mathbf{E} = \begin{cases} \mathbf{E}_0 \alpha t \sin \omega t & \text{for} \quad 0 \leqslant t \leqslant 1/\alpha, \\ \mathbf{E}_0 \sin \omega t & \text{for} \quad t \geqslant 1/\alpha, \end{cases} \tag{3.37}$$

where α is equal to $\omega/2\pi n$ where n is any integer. Then (3.34) shows that

$$\frac{\partial \mathbf{P}}{\partial t} = \begin{cases} \dfrac{N\alpha e^2 \mathbf{E}_0}{\mathfrak{m}\omega}\left\{\dfrac{\sin \omega t}{\omega} - t \cos \omega t\right\} & \text{for} \quad t \leqslant 1/\alpha, \\ -\dfrac{Ne^2 \mathbf{E}_0}{\mathfrak{m}\omega} \cos \omega t & \text{for} \quad t \geqslant 1/\alpha. \end{cases} \tag{3.38}$$

The energy supplied to unit volume by the electric forces is given by (3.33) and can now be evaluated. When $t > 1/\alpha$, it is given by

$$W_E = \tfrac{1}{2}\mathbf{E}_0^2\left\{\left(\epsilon_0 - \frac{Ne^2}{\mathfrak{m}\omega^2}\right)\sin^2 \omega t + \frac{Ne^2}{\mathfrak{m}\omega^2}\right\}$$

$$= \tfrac{1}{2}\mathbf{E}.\mathbf{D} + \tfrac{1}{2}\mathbf{E}_0^2 \frac{Ne^2}{\mathfrak{m}\omega^2}. \tag{3.39}$$

It can easily be shown that the second term is the kinetic energy of the electrons in unit volume at the instant when both **E** and **D** are zero.

For the second method, let the electric field $\mathbf{E}_0 \sin \omega t$ be switched on suddenly at $t = 0$, and suppose that there is some damping of the electron motion because of collisions. The equation of motion of an electron is (3.10) and the solution of this in the present case is

$$\frac{\partial \mathbf{P}}{\partial t} = Ne\frac{\partial \mathbf{v}}{\partial t} = \mathbf{E}_0 \frac{Ne^2}{\mathfrak{m}(\omega^2 + \nu^2)}\{\omega e^{-\nu t} + \nu \sin \omega t - \omega \cos \omega t\}. \tag{3.40}$$

The total energy supplied per unit volume by the electric forces in time t (from (3.33)) is then given by

$$W_E = \int_0^t \mathbf{E}.\left(\epsilon_0 \frac{\partial \mathbf{E}}{\partial t} + \frac{\partial \mathbf{P}}{\partial t}\right) dt,$$

which can be shown to lead to

$$W_E = \tfrac{1}{2}\epsilon_0 \mathbf{E}_0^2 \sin^2 \omega t + \mathbf{E}_0^2 \frac{Ne^2}{\mathfrak{m}(\omega^2+\nu^2)} \left\{ -\tfrac{1}{2}\sin^2\omega t + \tfrac{1}{2}\nu t - \tfrac{1}{4}(\nu/\omega)\sin 2\omega t + \frac{\omega^2}{\omega^2+\nu^2} - \frac{\omega}{\omega^2+\nu^2} e^{-\nu t}(\nu \sin\omega t + \omega\cos\omega t)\right\}. \quad (3{\cdot}41)$$

Now energy is being converted into heat at the rate $N\mathfrak{m}\nu\mathbf{v}^2$ per unit volume per unit time, and this cannot be regarded as part of the energy in the wave. The total energy lost as heat in time t is given by

$$N\mathfrak{m}\nu \int_0^t \mathbf{v}^2\, dt = W', \text{ say.}$$

This can be shown to be

$$W' = \mathbf{E}_0^2 \frac{Ne^2}{\mathfrak{m}(\omega^2+\nu^2)} \left\{ -\frac{\nu^2}{\omega^2+\nu^2}\sin^2\omega t + \tfrac{1}{2}\nu t + \tfrac{1}{4}(\nu/\omega)\frac{\omega^2-\nu^2}{\omega^2+\nu^2}\sin 2\omega t + \frac{1}{2}\frac{\omega^2}{\omega^2+\nu^2} - \frac{2\omega\nu}{\omega^2+\nu^2}e^{-\nu t}\sin\omega t - \frac{1}{2}\frac{\omega^2}{\omega^2+\nu^2}e^{-2\nu t}\right\}. \quad (3{\cdot}42)$$

The energy W_E'' stored per unit volume after time t is the difference between the energy supplied W_E, and the energy lost W'. Hence

$$W_E'' = W_E - W' = \tfrac{1}{2}\epsilon_0\mathbf{E}_0^2\sin^2\omega t + \mathbf{E}_0^2\frac{Ne^2}{2\mathfrak{m}(\omega^2+\nu^2)^2} \times\{(\nu^2-\omega^2)\sin^2\omega t + \omega^2 - \nu\omega\sin 2\omega t + 2\omega e^{-\nu t}(\nu\sin\omega t - \omega\cos\omega t) + \omega^2 e^{-2\nu t}\}. \quad (3{\cdot}43)$$

When a long time has elapsed after the field was switched on, $e^{-\nu t}$ is negligible. Then

$$W_E'' = \tfrac{1}{2}\epsilon_0\mathbf{E}_0^2\sin^2\omega t + \tfrac{1}{2}\mathbf{E}_0^2\frac{Ne^2}{\mathfrak{m}(\omega^2+\nu^2)} - \tfrac{1}{2}\mathbf{E}_0^2\frac{Ne^2}{\mathfrak{m}(\omega^2+\nu^2)^2}(\nu\cos\omega t + \omega\sin\omega t)^2. \quad (3{\cdot}44)$$

This is the energy stored per unit volume in the general case when collisions are included, and the steady state has been reached (earth's field neglected). If now ν is put equal to zero in (3.44), the result is identical with (3.39), which was obtained by the first method.

Expressions for the energy stored per unit volume could be derived in a similar way, when the effect of the earth's magnetic field is allowed for, but they are too complicated to be of much value.

3.11 The principal axes

In formulating the constitutive relations, (3.20), no special choice of co-ordinate axes was made. In later chapters it is often convenient to choose the z-axis as the direction of the wave-normal, but there is then still some freedom of choice for the x- and y-axes, and we choose them so that the earth's magnetic field is parallel to the x–z-plane. Then the direction cosine $m = 0$, and (3.20) becomes:

$$-\epsilon_0 X \begin{pmatrix} E_x \\ E_y \\ E_z \end{pmatrix} = \begin{pmatrix} U & inY & 0 \\ -inY & U & ilY \\ 0 & -ilY & U \end{pmatrix} \begin{pmatrix} P_x \\ P_y \\ P_z \end{pmatrix}. \tag{3.45}$$

If the 3×3 matrix in (3.20) or (3.45) were symmetric it would be possible to choose axes so that all non-diagonal elements are zero; this can be done for many crystalline dielectrics, and the resulting axes are called the principal axes of the system. The matrix is then said to be diagonalised and the required rotation of axes is effected by an orthogonal matrix transformation. For the ionosphere, however, the matrix is not symmetric and there are no principal axes of the kind used for crystalline dielectrics.

For the ionosphere the matrix in (3.20) or (3.45) is the sum of a constant, U, and a Hermitian matrix. Such matrices play an important part in quantum mechanics, and it is wel known that a Hermitian matrix can be diagonalised by a unitary transformation, which is conveniently thought of as effecting a complex rotation in Hilbert space (see, for example, Courant and Hilbert, *Methods of Mathematical Physics*, chs. I and II).

To find the required unitary transformation for the ionosphere, we first find a set of axes for which the matrix in (3.45) is diagonal. Let $\mathbf{P}^{(i)}$ be a vector parallel to one of these axes. Then the right side of (3.45) is a vector parallel to $\mathbf{P}^{(i)}$. Hence

$$\begin{pmatrix} U & inY & 0 \\ -inY & U & ilY \\ 0 & -ilY & U \end{pmatrix} \begin{pmatrix} P_x^{(i)} \\ P_y^{(i)} \\ P_z^{(i)} \end{pmatrix} = \lambda_i \begin{pmatrix} P_x^{(i)} \\ P_y^{(i)} \\ P_z^{(i)} \end{pmatrix}, \tag{3.46}$$

where the constant λ_i is an eigenvalue of the matrix. This is a set of three equations for $P_x^{(i)}$, $P_y^{(i)}$, $P_z^{(i)}$ which only have solutions if the determinant of the coefficients is zero, that is if

$$\begin{vmatrix} U-\lambda & inY & 0 \\ -inY & U-\lambda & ilY \\ 0 & -ilY & U-\lambda \end{vmatrix} = 0. \tag{3.47}$$

This gives for the eigenvalues:

$$\lambda_1 = U+Y, \quad \lambda_2 = U-Y, \quad \lambda_3 = U. \tag{3.48}$$

The equations (3.46) can now be solved to give the ratios $P_x^{(i)}:P_y^{(i)}:P_z^{(i)}$. Thus

$$\left.\begin{aligned} P_x^{(1)}:P_y^{(1)}:P_z^{(1)} &= in:1:-il, \\ P_x^{(2)}:P_y^{(2)}:P_z^{(2)} &= -in:1:il, \\ P_x^{(3)}:P_y^{(3)}:P_z^{(3)} &= l:0:n. \end{aligned}\right\} \tag{3.49}$$

These give the directions of the principal axes. Notice that they depend only on the *direction* of the earth's magnetic field, and not on its magnitude, nor on

X or Z. The directions of $\mathbf{P}^{(1)}$ and $\mathbf{P}^{(2)}$ are not real and are often described as complex directions in Hilbert space. $\mathbf{P}^{(3)}$ is a real direction parallel to $\mathbf{Y}$, that is to the earth's magnetic field. The vectors $\mathbf{P}^{(1)}$, $\mathbf{P}^{(2)}$, $\mathbf{P}^{(3)}$ are not orthogonal in the ordinary sense but are 'Hermitian orthogonal', that is $\mathbf{P}^{(i)}.\mathbf{P}^{(j)*} = 0$ for $i \neq j$ where a star * denotes a complex conjugate.

Now define† a unitary matrix $\mathbf{U}$ and its inverse $\mathbf{U}^{-1}$ thus:

$$\mathbf{U} = \begin{pmatrix} 2^{-\frac{1}{2}}in & -2^{-\frac{1}{2}}in & l \\ 2^{-\frac{1}{2}} & 2^{-\frac{1}{2}} & 0 \\ -2^{-\frac{1}{2}}il & 2^{-\frac{1}{2}}il & n \end{pmatrix}, \quad \mathbf{U}^{-1} = \begin{pmatrix} -2^{-\frac{1}{2}}in & 2^{-\frac{1}{2}} & 2^{-\frac{1}{2}}il \\ 2^{-\frac{1}{2}}in & 2^{-\frac{1}{2}} & -2^{-\frac{1}{2}}il \\ l & 0 & n \end{pmatrix}. \quad (3.50)$$

Equation (3.45) then gives

$$-\epsilon_0 X\mathbf{U}^{-1}\begin{pmatrix} E_x \\ E_y \\ E_z \end{pmatrix} = \mathbf{U}^{-1}\begin{pmatrix} U & inY & 0 \\ -inY & U & ilY \\ 0 & -ilY & U \end{pmatrix}\mathbf{U}\mathbf{U}^{-1}\begin{pmatrix} P_x \\ P_y \\ P_z \end{pmatrix}, \quad (3.51)$$

and on multiplying out the matrices it becomes

$$-\epsilon_0 X\begin{pmatrix} E_1 \\ E_2 \\ E_3 \end{pmatrix} = \begin{pmatrix} U+Y & 0 & 0 \\ 0 & U-Y & 0 \\ 0 & 0 & U \end{pmatrix}\begin{pmatrix} P_1 \\ P_2 \\ P_3 \end{pmatrix}, \quad (3.52)$$

where

$$\left.\begin{aligned} E_1 &= 2^{-\frac{1}{2}}\{-inE_x + E_y + ilE_z\}, \\ E_2 &= 2^{-\frac{1}{2}}\{inE_x + E_y - ilE_z\}, \\ E_3 &= lE_x + nE_z, \end{aligned}\right\} \quad (3.53)$$

with similar relations for the components of $\mathbf{P}$. The expressions (3.53) are called the 'principal axis components' of the electric intensity $\mathbf{E}$. Some authors have used them, or simple functions of them as the dependent variables in the differential equations used for finding reflection coefficients (Davids, 1953; Davids and Parkinson, 1955) (see § 22.4).

The theory of this section can be formulated without the simplification $m = 0$, but the algebra is more complicated. For a further discussion of principal axes see Lange-Hesse (1952), Westfold (1949), and the two references above.

† $\mathbf{U}$ has no direct connection with $U = 1 - iZ$.

CHAPTER 4

PROPAGATION IN A HOMOGENEOUS ISOTROPIC MEDIUM

4.1 Definition of the refractive index

In this and the next chapter solutions of Maxwell's equations (2.31) are considered which represent progressive plane waves in a medium which is not free space. As in ch. 2 the z-axis is chosen to be in the direction of the wave normal. Then if such solutions exist, the definition of a progressive wave (§ 2.8) requires that all field quantities shall contain z only through the factor $e^{-ik\mathfrak{n}z}$ where $\mathfrak{n}$ is some constant. In free space $\mathfrak{n} = 1$. In other media it may have other values, and it may possibly be complex. In a magnetoionic medium for a given direction of the wave-normal, two different values of $\mathfrak{n}$ are in general possible. If V is the velocity of the waves, then clearly $\mathfrak{n} = c/V$. $\mathfrak{n}$ is called the 'refractive index' of the medium. The interpretation of a complex refractive index is discussed in § 4.5.

4.2 The Maxwell equation derived from Faraday's law

The three left-hand equations in (2.31) were derived from the third Maxwell equation (2.10), which is based on Faraday's law of electromagnetic induction. These equations are now written for the progressive plane wave described in the last section, for which

$$\frac{\partial}{\partial x} = \frac{\partial}{\partial y} = 0, \quad \frac{\partial}{\partial z} = -ik\mathfrak{n}.$$

The result is

$$\mathfrak{n}E_y = -\mathscr{H}_x, \tag{4.1}$$

$$\mathfrak{n}E_x = \mathscr{H}_y, \tag{4.2}$$

$$\mathscr{H}_z = 0. \tag{4.3}$$

Equation (4.3) shows that the magnetic field is entirely in the plane of the wave-front (purely transverse), as for the plane wave in free space, but the electric field may in general have a longitudinal component. The relations (4.1) and (4.2) are true for a progressive wave travelling in the direction of positive z. For a progressive wave travelling in the opposite direction, the sign of one side of each relation must be reversed.

A more complicated disturbance, such as a partial standing wave, can be resolved into its component progressive waves by using (4.1) and (4.2) and the corresponding relations for a backwards travelling wave. An example of this will be found in § 9.9.

The results of the present section apply to any non-magnetic medium. To make further progress it is necessary to use the other Maxwell equations in (2.31), which involve the constitutive relation giving **P** and **D** in terms of **E**.

4.3 Isotropic medium without damping

When electron collisions and the earth's magnetic field are neglected, the constitutive relation for the ionosphere is given by (3.4). If this is inserted in the right-hand equations in (2.31) with

$$\frac{\partial}{\partial x} = \frac{\partial}{\partial y} = 0, \quad \frac{\partial}{\partial z} = -ik\mathfrak{n},$$

it gives

$$\mathfrak{n}\mathscr{H}_y = (1-X)E_x, \tag{4.4}$$

$$-\mathfrak{n}\mathscr{H}_x = (1-X)E_y, \tag{4.5}$$

$$0 = E_z. \tag{4.6}$$

From (4.4) and (4.2) the ratio $\mathscr{H}_y/E_x$ may be eliminated to give

$$\mathfrak{n}^2 = 1 - X. \tag{4.7}$$

The same result is obtained by eliminating $\mathscr{H}_x/E_y$ from (4.5) and (4.1). These two pairs of equations are independent, which shows that the two linearly polarised plane waves, with their electric vectors parallel to the x- and y-axes, respectively, are propagated independently, and (4.7) applies to both. They therefore travel with the same velocity $c/\mathfrak{n}$. Two such linearly polarised waves with any amplitudes and relative phase could be combined to give a wave which would in general be elliptically polarised.

The axis of z was chosen to be in the direction of the wave-normal. There is no restriction on the direction in which the wave may travel, and hence the refractive index $\mathfrak{n}$, given by (4.7), is independent of the direction of the wave-normal. A medium with this property is called 'isotropic'.

Equation (4.6) shows that the electric field is entirely in the plane of the wave-front (purely transverse) as it was for plane waves in free space. This result is true for any isotropic medium, but may not be true for an anisotropic medium.

Since X is proportional to the electron number density N, (4.7) shows that the square of the refractive index decreases linearly as N (or X) increases. It is zero when $X = 1$, that is when $\omega = \omega_N$ so that the wave-frequency is equal to the plasma-frequency. If $X > 1$, then $\mathfrak{n}^2$ is negative, and $\mathfrak{n}$ is purely imaginary. The physical interpretation of an imaginary refractive index is discussed in §4.6.

4.4 Isotropic medium with collision damping

If the effect of electron collisions is included, but the earth's magnetic field is neglected, the constitutive relation for the ionosphere is given by

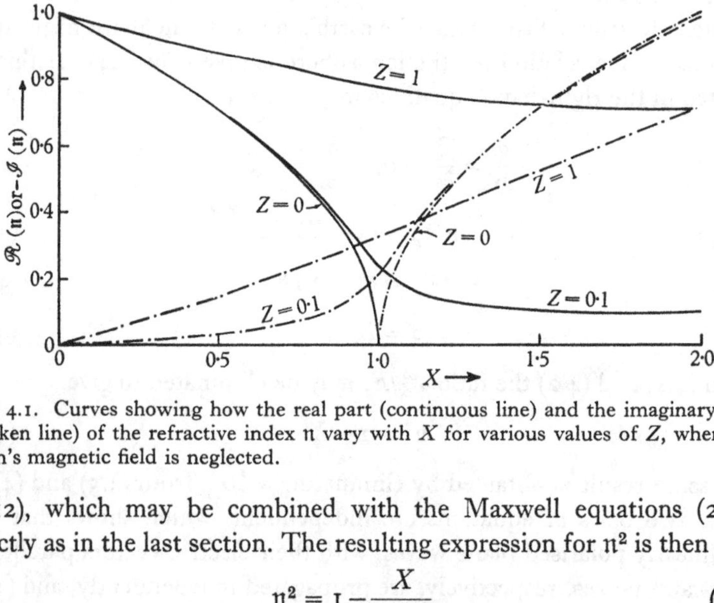

Fig. 4.1. Curves showing how the real part (continuous line) and the imaginary part (broken line) of the refractive index $\mathfrak{n}$ vary with X for various values of Z, when the earth's magnetic field is neglected.

(3.12), which may be combined with the Maxwell equations (2.31) exactly as in the last section. The resulting expression for $\mathfrak{n}^2$ is then

$$\mathfrak{n}^2 = 1 - \frac{X}{1 - iZ}. \tag{4.8}$$

All the results of the last section still apply except that $\mathfrak{n}^2$ and therefore $\mathfrak{n}$ are in general complex quantities. Fig. 4.1 shows how the real and imaginary parts of $\mathfrak{n}$ vary with X for several different values of Z. More extensive curves of this kind and a full discussion are given by Ratcliffe (1959). It is clear that if Z is non-zero and real, there is no real value of X that makes $\mathfrak{n}^2$ vanish. In later chapters, however, it will prove useful to assume that X and Z are analytic functions of the height z. They must be real and positive when z is real, but when z is complex, X and Z can take complex values. Then $\mathfrak{n}^2$ can vanish for one or more complex values of z. These zeros of $\mathfrak{n}^2$ in the complex z-plane play a very important part in the theory of reflection of radio waves.

4.5 The physical interpretation of a complex refractive index

In §2.8 a progressive plane wave was defined to be one in which all field quantities vary with the space coordinates only through a factor $e^{-ik\mathfrak{n}z}$, where z is measured in the direction of the wave-normal. In the mathematical application of this definition there was nothing to restrict $\mathfrak{n}$ to real values, and in fact it was shown in the last two sections that $\mathfrak{n}$ may be complex or imaginary. Let F be some field quantity in the wave, so that $F = F_0 e^{-ik\mathfrak{n}z}$, and let

$$\mathfrak{n} = \mu - i\chi, \tag{4.9}$$

where μ and χ are real. Then the expression for F becomes

$$F = F_0 e^{-ik\mu z} e^{-k\chi z}. \tag{4.10}$$

This represents a disturbance travelling with velocity c/μ and attenuated as it travels, at a rate depending on χ. The wave must be losing energy, which is absorbed by the medium. If χ were negative, the wave would grow as it travelled. Clearly this is physically impossible, since there is no source of energy in the medium, and the medium is said to be 'passive'. It is therefore to be expected that when μ is positive, χ must also be positive. Now in an isotropic medium with damping, $\mathfrak{n}^2$ is given by (4.8), and if the square root is chosen to have a positive real part, it is easy to show that the imaginary part must be negative, so that χ is positive. In fact all physically acceptable values of $\mathfrak{n}$, for all real heights in the ionosphere, must lie either in the second or fourth quadrants of the complex plane. This restriction does not apply to the values of $\mathfrak{n}$ at complex heights, mentioned at the end of the last section. (Values in the second quadrant have μ and χ both negative, and apply to waves travelling in the direction of negative z, and attenuated as they travel.)

4.6 Evanescent waves

If $\mathfrak{n}^2$ is negative and real, then $\mathfrak{n}$ is purely imaginary, so that $\mu = 0$, and (4.10) becomes

$$F = F_0 e^{-k\chi z}. \tag{4.11}$$

This appears to represent a wave travelling with infinite wave-velocity. Every field component varies harmonically in time, but there is no harmonic variation in space. The phase is the same for all values of z. A disturbance of this kind is called an 'evanescent' wave. It is still a 'progressive' wave according to the definition adopted here. Other examples of such waves are given in §9.7.

The expressions (4.1) to (4.3) are true whether $\mathfrak{n}$ is real or complex. If it is real, then the electric and magnetic fields of a linearly polarised wave are in phase (or antiphase). In an evanescent wave $\mathfrak{n}$ is purely imaginary, so that the electric and magnetic fields are in quadrature. This means that the time average value of the Poynting vector $\mathbf{\Pi} = \mathbf{E} \times \mathbf{H}$ is zero. Hence there is no net energy-flow for such a wave in an isotropic medium. For an evanescent wave in an anisotropic medium there can be some flow of energy perpedicular to the wave-normal (see § 5.8). If $\mathfrak{n}$ has the general complex value (4.9), the phase difference between the electric and magnetic fields is given by arc tan (χ/μ).

If $\mathfrak{n}^2$ is real and positive, a progressive plane wave is unattenuated, and there is no absorption of energy. If $\mathfrak{n}^2$ is real and negative, the only possible progressive wave is evanescent, and there is no energy flow and again no absorption of energy. A medium for which $\mathfrak{n}^2$ is real, whether positive or negative, is said to be 'loss-free'.

4.7 Inhomogeneous plane waves

For the progressive plane waves considered in previous sections, it was convenient to choose the axis of z to be in the direction of the wave-normal. This restriction is not necessary, and we now consider a progressive plane wave with its wave-normal in the x–z-plane in a direction at an angle θ to the z-axis. Then the distance of the point (x, y, z) from a wave-front through the origin is $x \sin\theta + z\cos\theta$, and every field component of the wave must vary in space according to the equation

$$F = F_0 \exp\{-ik\mathfrak{n}(x\sin\theta + z\cos\theta)\}. \tag{4.12}$$

This expression could be combined with Maxwell's equations and the constitutive relations, and an expression for $\mathfrak{n}$ derived exactly as in §§ 4.3 and 4.4. The algebra would be more complicated, but exactly the same expressions for $\mathfrak{n}^2$ (4.7) or (4.8) would result. This must obviously be so, since the physical laws are unaffected by the choice of axes. In (4.12) $\cos\theta$ and $\sin\theta$ must obey the relation

$$\cos^2\theta + \sin^2\theta = 1. \tag{4.13}$$

Now suppose that $\cos\theta$ and $\sin\theta$ have complex values. The analysis could still be carried out, and still the value of $\mathfrak{n}$ would be unaffected. A wave of this kind is called an 'inhomogeneous plane wave'. Such waves will be encountered later, and some of their properties are given here.

Let

$$\cos\theta = a + ib, \quad \sin\theta = f + ig, \tag{4.14}$$

where a, b, f and g are all real. Then (4.13) shows that

$$ab = -fg, \tag{4.15}$$

$$a^2 - b^2 + f^2 - g^2 = 1. \tag{4.16}$$

Suppose, first, that $\mathfrak{n}$ is real. Then the expression (4.12) may be written:

$$F = F_0 \exp\{-ik\mathfrak{n}(fx + az)\} \exp\{k\mathfrak{n}(gx + bz)\}. \tag{4.17}$$

Here the first exponential alone represents a plane wave with wave-fronts parallel to the plane $fx+az = 0$. These planes are not true wave-fronts in the mathematical sense, however (except when $b = g = 0$), and it is better to call them 'planes of constant phase'. The second exponential in (4.17) shows that the waves may change in amplitude as x or z varies, and that planes of constant amplitude are parallel to the plane $gx+bz = 0$. Now (4.15) shows that the planes of constant phase and planes of constant amplitude are at right angles. This result is still true if $\mathfrak{n}$ is purely imaginary. Hence it is always true when $\mathfrak{n}^2$ is real, whether positive or negative, that is for a loss-free medium. If $\mathfrak{n}$ is real, a wave with its normal in a real direction has constant amplitude everywhere, because the exponent in the second exponential in (4.17) is zero. Similarly, if $\mathfrak{n}$ is purely imaginary, a wave with its normal in a real direction has constant phase everywhere, and is called an 'evanescent' wave (§4.6): in this case the exponent in the second exponential of (4.17) is again zero, and the exponent in the first exponential is purely real.

Next suppose that $\mathfrak{n}$ is complex, so that $\mathfrak{n} = \mu - i\chi$ where μ and χ are real. Then the expression (3.13) may be written

$$F = F_0 \exp[-ik\{(f\mu+g\chi)x+(a\mu+b\chi)z\}]\exp[k\{(g\mu-f\chi)x+(b\mu-a\chi)z\}]. \tag{4.18}$$

The planes of constant phase and constant amplitude are no longer at right angles, and by using (4.15), (4.16) it can be shown that the angle between them is

$$\arctan\left\{\frac{\mu^2+\chi^2}{\mu\chi}(bf-ag)\right\}. \tag{4.19}$$

For the wave of (4.18) to be evanescent, it is necessary that

$$f\mu+g\chi = a\mu+b\chi = 0.$$

Now μ/χ cannot be negative, so that these conditions can only be satisfied if $\mu = 0$. The refractive index is then purely imaginary.

In the preceding examples, the wave-normal was inclined at complex angles to the x- and z-axes, but was at right angles to the y-axis. It might at first be thought that a more complicated type of inhomogeneous plane wave is possible, with its wave-normal at complex angles to all three coordinate axes. But suppose that such a wave-normal has direction cosines l_1+il_2, m_1+im_2, n_1+in_2, where $l_1, l_2, m_1, m_2, n_1, n_2$ are real. Choose a new Cartesian coordinate system with the axes in real directions. This can be formed by two rotations of the original axes, so that two conditions can be imposed upon it. Let L, M, N be the direction cosines of the new y-axis, where L, M, N, are real. Then the conditions we impose are

$$Ll_1+Mm_1+Nn_1 = 0, \quad Ll_2+Mm_2+Nn_2 = 0, \tag{4.20}$$

which ensure that the real direction of the new y-axis is at right angles to the complex direction of the wave-normal. Hence every inhomogeneous plane wave has the properties given in the first part of this section.

4.8 Energy flow in inhomogeneous plane waves

It has been shown that for progressive plane waves in isotropic media, the electric and magnetic vectors are at right angles to the wave-normal. This is still true for inhomogeneous plane waves, but since the direction of the wave normal is complex, the directions of the field vectors may also be complex. Consider the case of the last section in which the wave-normal is in the x–z-plane at an angle θ to the x-axis, and let the wave be linearly polarised with its electric vector in the (real) direction of the y-axis, so that the magnetic vector $\mathscr{H}$ is in the x–z-plane. The magnitude of $\mathscr{H}$, from (4.1) is equal to $\mathfrak{n}E_y$, so that its components are

$$\left.\begin{aligned}\mathscr{H}_x &= -\mathfrak{n}E_y \sin\theta,\\ \mathscr{H}_z &= \mathfrak{n}E_y\cos\theta.\end{aligned}\right\}\tag{4.21}$$

Now let $\cos\theta$ and $\sin\theta$ be complex numbers given by (4.14) and suppose that $\mathfrak{n}$ is real. Then (4.21) shows that $\mathscr{H}_x$ and $\mathscr{H}_y$ differ in phase, so that the magnetic vector traces out an ellipse in the x–z-plane. Let $E_y = E_0 e^{i\omega t}$ where E_0 is real. To find the direction of energy flow we compute the components Π_x, Π_y of the Poynting vector. These involve products of field components, and hence the complex number convention will not be used. Substitute for E_y in (4.21) and take real parts. Then

$$\left.\begin{aligned}E_y &= E_0\cos\omega t,\\ \mathscr{H}_x &= -\mathfrak{n}E_0(f\cos\omega t - g\sin\omega t),\\ \mathscr{H}_z &= \mathfrak{n}E_0(a\cos\omega t - b\sin\omega t).\end{aligned}\right\}\tag{4.22}$$

Whence

$$\left.\begin{aligned}Z_0\Pi_x &= E_y\mathscr{H}_z = \mathfrak{n}E_0^2(a\cos^2\omega t - b\cos\omega t\sin\omega t),\\ Z_0\Pi_z &= -E_y\mathscr{H}_x = \mathfrak{n}E_0^2(f\cos^2\omega t - g\cos\omega t\sin\omega t).\end{aligned}\right\}\tag{4.23}$$

The time average values of Π_x, Π_z are then given by

$$\left.\begin{aligned}Z_0\overline{\Pi}_x &= \tfrac{1}{2}\mathfrak{n}E_0^2 a,\\ Z_0\overline{\Pi}_z &= \tfrac{1}{2}\mathfrak{n}E_0^2 f,\end{aligned}\right\}\tag{4.24}$$

and the direction of the resultant average energy flow makes an angle arc tan (f/a) with the x-axis. It can be seen from (4.17) that this is at right angles to the planes of constant phase, and therefore parallel to the planes of constant amplitude. This result could have been predicted on other grounds. For, since $\mathfrak{n}$ is real, the medium cannot absorb energy from the wave. Moreover, planes of constant amplitude are also planes of constant average energy density. Hence the energy can only flow parallel to the planes of constant amplitude on the average, because any other direction of flow would result in a change of energy density.

This result shows further that the component of $\mathscr{H}$ perpendicular to the planes of constant phase is in quadrature with the electric field, and it can also be shown from (4.22) that the component of $\mathscr{H}$ parallel to the planes of constant phase is in phase with E_y. Hence the ellipse traced out by $\mathscr{H}$ has its major or minor axis in a plane of constant phase. The existence of a component of $\mathscr{H}$ perpendicular to the planes of constant phase does not violate the condition that $\mathscr{H}$ is perpendicular to the wave-normal; for the normal to the planes of constant phase is a *real* direction, and is not the true wave-normal. Notice that the ellipse is in a plane which contains the wave-normal. (It is very different from the polarisation ellipse of an elliptically polarised plane wave with its normal in a real direction, for then the magnetic vector traverses an ellipse which is perpendicular to the wave-normal.)

If the wave is polarised with its magnetic vector in the direction of the y-axis, the analysis is very similar, and again the average energy flux is parallel to the planes of constant amplitude. In this case the electric vector traverses an ellipse in the x–z-plane, with its major or minor axis parallel to the planes of constant phase. When $\mathfrak{n}$ is purely imaginary, the results are practically the same as those derived above. When $\mathfrak{n}$ is complex, the analysis is more complicated.

4.9 The case when $X = 1$, $\mathfrak{n} = 0$

The condition $X = 1$, $Z = 0$, makes $\mathfrak{n}^2 = 0$. This could never be attained in practice, because the collision frequency is never exactly zero, although it can be neglected for many purposes. Consideration of the case $\mathfrak{n} = 0$ is useful, however, since it helps to give an understanding of the physical meaning of some of the equations.

It has already been assumed that for a progressive plane wave with its wave-normal parallel to the z-axis,

$$\frac{\partial}{\partial x} = \frac{\partial}{\partial y} = 0, \quad \frac{\partial}{\partial z} \equiv -ik\mathfrak{n}.$$

If $\mathfrak{n} = 0$, the space derivatives of all field quantities vanish, so that at any one instant the electric field, for example, is constant in magnitude and direction throughout the medium. Equations (4.1) to (4.3) show that if the electric field is finite, the magnetic field vanishes everywhere, and (2.6) and (3.4) together lead to $\epsilon_0 \mathbf{E} = -\mathbf{P}$ or $\mathbf{D} = 0$. In this disturbance the whole medium is performing a natural oscillation of angular frequency $\omega = \omega_N$. It can be shown that if all the electrons in a homogeneous slab of ionised medium were displaced the same distance in a direction normal to the slab and then released, they would oscillate about their original positions with this angular frequency, and the electric field would be exactly as described above.

In such a disturbance it is possible for the electric field to have a complex direction. Suppose the electric field has amplitude E_0 and is in the x–z-plane at an angle θ to the z-axis. Then $E_z = E_0 \cos\theta$, $E_x = E_0 \sin\theta$. If $\cos\theta$ is real and greater than unity, then $\sin\theta$ is purely imaginary so that E_x and E_z are in quadrature. This system could equally be considered as formed by the superposition of two independent disturbances, one with the electric vector parallel to the x-axis, and the other with the electric vector parallel to the z-axis and in quadrature with the first.

Disturbances of this kind can hardly be called waves, and it is more correct to call them 'plasma oscillations'. Another kind of disturbance is possible when $\mathfrak{n} = 0$, however, which more nearly resembles an inhomogeneous plane wave.

Consider a plane progressive wave in a medium whose refractive index $\mathfrak{n}$ is very small but not quite zero. Let the wave-normal be in the x–z-plane at an angle θ to the x-axis, and let the electric field be in the x–z-plane and have amplitude $\mathfrak{n}E_0$. Then

$$E_x = \mathfrak{n}E_0 \cos\theta \exp\{-ik\mathfrak{n}(x\sin\theta + z\cos\theta)\}, \tag{4.25}$$

$$E_z = -\mathfrak{n}E_0 \sin\theta \exp\{-ik\mathfrak{n}(x\sin\theta + z\cos\theta)\}, \tag{4.26}$$

$$\mathscr{H}_y = \mathfrak{n}^2 E_0 \exp\{-ik\mathfrak{n}(x\sin\theta + z\cos\theta)\}. \tag{4.27}$$

Now let $\mathfrak{n} \to 0$ and $\sin\theta \to \infty$ in such a way that $\mathfrak{n}\sin\theta$ remains finite and equal to $\sin\theta_I$. Then

$$\mathfrak{n}\cos\theta \to \pm i\sin\theta_I. \tag{4.28}$$

The minus sign will be chosen. Then the expressions (4.25), (4.26), (4.27) become, in the limit:

$$\left.\begin{aligned} E_x &= -iE_0 \sin\theta_I \exp(-ikx\sin\theta_I)\exp(-kz\sin\theta_I),\\ E_z &= -E_0 \sin\theta_I \exp(-ikx\sin\theta_I)\exp(-kz\sin\theta_I),\\ \mathscr{H}_y &= 0. \end{aligned}\right\} \tag{4.29}$$

It can be verified that these formulae satisfy Maxwell's equations. They describe an inhomogeneous plane wave in which planes of constant phase are parallel to the y–z-plane, and planes of constant amplitude are parallel to the x–z-plane. Since the magnetic field vanishes, the Poynting vector is zero, and there is no flow of energy. A wave of this kind will be encountered in § 8.8.

As a second example, suppose that the electric field is parallel to the y-axis, and has amplitude E_0. Then instead of (4.25) to (4.27) we have:

$$E_y = E_0 \exp\{-ik\mathfrak{n}(x\sin\theta + z\cos\theta)\}, \tag{4.30}$$

$$\mathscr{H}_x = -\mathfrak{n}\cos\theta E_0 \exp\{-ik\mathfrak{n}(x\sin\theta + z\cos\theta)\}, \tag{4.31}$$

$$\mathscr{H}_z = \mathfrak{n}\sin\theta E_0 \exp\{-ik\mathfrak{n}(x\sin\theta + z\cos\theta)\}. \tag{4.32}$$

Again let $\mathfrak{n} \to 0$ and $\sin\theta \to \infty$, so that $\mathfrak{n}\sin\theta = \sin\theta_I$. Then

$$\mathfrak{n}\cos\theta = -i\sin\theta_I \tag{4.33}$$

as before, and the expressions become in the limit:

$$E_y = E_0 \exp(-ikx\sin\theta_I)\exp(-kz\sin\theta_I), \tag{4.34}$$

$$\mathscr{H}_x = iE_0 \sin\theta_I \exp(-ikx\sin\theta_I)\exp(-kz\sin\theta_I), \tag{4.35}$$

$$\mathscr{H}_z = E_0 \sin\theta_I \exp(-ikx\sin\theta_I)\exp(-kz\sin\theta_I). \tag{4.36}$$

They represent an inhomogeneous plane wave whose planes of constant phase and amplitude are as in the preceding example. This time the magnetic field does not vanish. Since $\mathbf{D} = 0$ it follows that $\operatorname{curl}\mathscr{H} = 0$, and this is satisfied by (4.35) and (4.36).

CHAPTER 5

PROPAGATION IN A HOMOGENEOUS ANISOTROPIC MEDIUM. MAGNETO-IONIC THEORY

5.1 Introduction

In the previous chapter the earth's magnetic field was neglected, that is, the medium was assumed to be isotropic. We now consider the propagation of plane waves in a homogeneous medium when the earth's magnetic field is included. This leads to the two important results: (*a*) progressive waves are possible only for certain values of the wave-polarisation; (*b*) the refractive index depends on the direction of the wave-normal and, for a given wave-normal direction, there are in general two refractive indices associated with two different wave-polarisations.

5.2 The wave-polarisation

It is convenient, as before, to choose the z-axis in the direction of the wave-normal. There is then still some freedom in the choice of the x- and z-axes, and they are chosen so that the earth's magnetic field lies in the x–z-plane, which is called the 'magnetic meridian plane'. This makes $m = 0$ ($-l$, $-m$, $-n$ are the direction cosines of the earth's magnetic field, see § 3.4).

The electric vector **E** in a progressive wave may be resolved into three components E_x, E_y, E_z. The wave-polarisation is now defined thus

$$\rho = \frac{E_y}{E_x}. \tag{5.1}$$

Equations (4.1) and (4.2) are valid in any (non-magnetic) medium and show that

$$\rho = -\frac{\mathscr{H}_x}{\mathscr{H}_y}. \tag{5.2}$$

The component E_z may not be zero, but it does not enter into the definition of ρ. The component $\mathscr{H}_z$ is always zero.

The wave-polarisation ρ is a complex number which shows how the transverse component of **E** varies with time. For example, if ρ is real, E_x and E_y have the same phase and the transverse component of **E** is always parallel to a fixed line; this is called linear polarization. If $\rho = i$, E_x and E_y are in quadrature and have the same amplitude, and the polarisation is circular. To an observer looking in the direction of the wave-normal (that is, in the direction of positive z) the electric vector would appear to rotate anticlockwise. This is called left-handed circular polarisation. If $\rho = -i$, the rotation is clockwise, and the polarisation is called right-handed circular polarisation. If ρ is complex, the polarisation is elliptical, and (5.1), (5.2) show that the magnetic vector $\mathscr{H}$ and the transverse component of **E**, traverse similar ellipses in the same sense. The ellipses for the electric and magnetic vectors have their major axes at right angles.

5.3 The polarisation equation

With the choice of axes adopted in the preceding section, $m = 0$ and the constitutive relations (3.20) become

$$-X\epsilon_0\begin{pmatrix} E_x \\ E_y \\ E_z \end{pmatrix} = \begin{pmatrix} U & inY & 0 \\ -inY & U & ilY \\ 0 & -ilY & U \end{pmatrix}\begin{pmatrix} P_x \\ P_y \\ P_z \end{pmatrix}. \tag{5.3}$$

We seek a solution of Maxwell's equations which represents a progressive plane wave, so that all field components vary in space only through the term $e^{-ik\mathrm{n}z}$ and $\partial/\partial x = \partial/\partial y \equiv 0$, $\partial/\partial z \equiv -ik\mathrm{n}$. Then the right-hand equations in (2.13) give:

$$ik\mathrm{n}\mathscr{H}_y = \frac{ik}{\epsilon_0}D_x, \tag{5.4}$$

$$-ik\mathrm{n}\mathscr{H}_x = \frac{ik}{\epsilon_0}D_y, \tag{5.5}$$

$$D_z = 0. \tag{5.6}$$

Hence from (5.1), (5.2), (5.4) and (5.5), and the definition (2.6) of **D**:

$$\frac{D_y}{D_x} = -\frac{\mathscr{H}_x}{\mathscr{H}_y} = \frac{E_y}{E_x} = \frac{P_y}{P_x} = \rho. \tag{5.7}$$

Equation (5.6) shows that **D** is in the plane of the wave-front, so that

$$\epsilon_0 E_z + P_z = 0. \tag{5.8}$$

It will be shown later that in general E_z and P_z do not vanish separately. Substitution in the third equation of (5.3) leads to

$$(U - X)P_z = ilYP_y. \tag{5.9}$$

The first two equations of (5.3), combined with (5.9), give:

$$\left.\begin{aligned} -\epsilon_0 X E_x &= U P_x + in Y P_y, \\ -\epsilon_0 X E_y &= -in Y P_x + \left\{U - \frac{l^2 Y^2}{U-X}\right\} P_y. \end{aligned}\right\} \tag{5.10}$$

The second of these equations is divided by the first, and (5.7) is used. The result is

$$\rho = \frac{-inY + \rho\{U - l^2Y^2/(U-X)\}}{U + inY\rho},$$

whence
$$\rho^2 - \frac{il^2 Y}{n(U-X)}\rho + 1 = 0. \tag{5.11}$$

This is a quadratic equation for ρ and shows that in a homogeneous anisotropic medium, a progressive plane wave must have a polarisation given by one of the two solutions of (5.11).

Let $Y_L = nY$, $Y_T = lY$. These are called the longitudinal and transverse components of the vector **Y**. Then (5.11) may be written (since $U = 1 - iZ$)

$$\rho^2 - \frac{iY_T^2}{Y_L(1-X-iZ)}\rho + 1 = 0, \tag{5.12}$$

and its solutions are

$$\rho = \frac{iY_T^2}{2Y_L(1-X-iZ)} \pm i\left\{\frac{Y_T^4}{4Y_L^2(1-X-iZ)^2} + 1\right\}^{\frac{1}{2}}. \tag{5.13}$$

Equation (5.12) is the polarisation equation of magnetoionic theory and plays an extremely important part in the theory of later chapters.

5.4 Properties of the polarisation equation

From (5.13) it is clear that when collisions are neglected, so that $Z = 0$, both values of ρ are purely imaginary. The two polarisation ellipses then have their major axes parallel to either the x- or the y-axis.

The condition that the two solutions of the quadratic shall be equal is

$$Y_T^4 + 4Y_L^2(1-X-iZ)^2 = 0. \tag{5.14}$$

When X and Z are real, this is only possible if

$$X = 1 \quad \text{and} \quad Z = Z_c, \tag{5.15}$$

where
$$Z_c = \left|\frac{Y_T^2}{2Y_L}\right|. \tag{5.16}$$

Equation (5.16) may also be written

$$\nu = \omega_c, \quad \text{where} \quad \omega_c = \left|\frac{\omega_H l^2}{2n}\right|. \tag{5.17}$$

Thus ω_c is a critical value of the collision-frequency, and is independent of the wave-frequency. It depends only on the strength of the earth's magnetic field and its inclination Θ to the wave-normal. Now $l = -\sin\Theta$, $n = -\cos\Theta$. Hence

$$\omega_c = \left|\frac{\omega_H \sin^2\Theta}{2\cos\Theta}\right|. \tag{5.18}$$

Suppose that Y_L is negative. Then, in a wave for which the conditions (5.15) hold, the value of ρ is $+1$, so that for both polarisations the electric vector is in the same plane, at 45° to the x-axis. In these conditions it is possible, however, for another wave to exist whose electric vector has a component at right angles to this plane. Its properties are described in §6.12.

Let the two values (5.13) of ρ be ρ_o and ρ_x. The subscripts O and X mean 'ordinary' and 'extraordinary' respectively. The meanings of these terms do not matter in the present section but are given later, in §6.6. Now (5.12) shows that

$$\rho_o \rho_x = 1. \tag{5.19}$$

Choose new axes, x', y', z', formed from the x-, y-, z-axes by rotation through an angle ψ about the z-axis, and let E'_x, E'_y, ρ'_o, ρ'_x, be the new values of E_x, E_y, ρ_o, ρ_x. Then, for the polarisation ρ_o, $E_y = \rho_o E_x$ and

$$\left.\begin{aligned} E'_x &= E_x(\cos\psi + \rho_o \sin\psi), \\ E'_y &= E_x(-\sin\psi + \rho_o\cos\psi), \end{aligned}\right\} \tag{5.20}$$

$$\rho'_o = \frac{\rho_o - \tan\psi}{1 + \rho_o \tan\psi}. \tag{5.21}$$

Similarly

$$\rho'_x = \frac{\rho_x - \tan\psi}{1 + \rho_x \tan\psi}. \tag{5.22}$$

Now (5.19) shows that if $\psi = \frac{1}{4}\pi$, then

$$\rho'_x = -\rho'_o. \tag{5.23}$$

Hence the two polarisation ellipses are mirror images of each other in a plane at 45° to the original axes, that is to the magnetic meridian plane. This is illustrated in Fig. 5.1.

The polarisation ρ is defined by (5.1), in terms of a right-handed system of axes in which the z-axis is the direction of the wave-normal. For a wave travelling in the opposite direction, the z-axis and either the x- or the y-axis would have to be reversed. The middle term of (5.11) then changes sign, so that ρ changes sign. The two polarisation ellipses for the two waves travelling in opposite directions are, however, exactly the same, and ρ changes sign, only because of the change of axes. This convention is used in this and the next chapter. In some problems discussed in later chapters, waves in both directions are present, but it is convenient to define ρ in terms of the *same* set of axes for both waves. Then, for ordinary waves travelling in either direction, ρ has the same value ρ_o, and similarly for extraordinary waves.

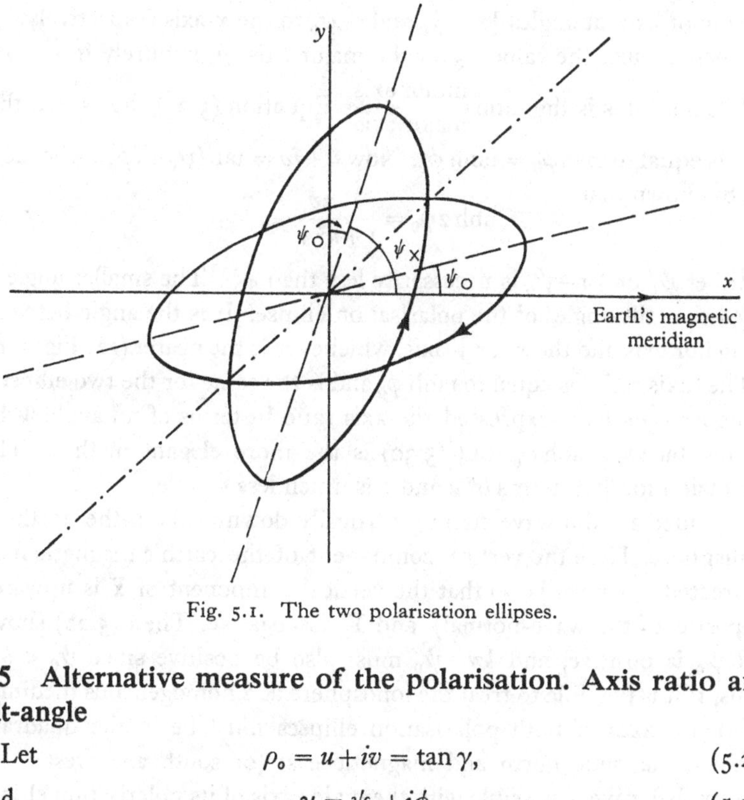

Fig. 5.1. The two polarisation ellipses.

5.5 Alternative measure of the polarisation. Axis ratio and tilt-angle

Let $$\rho_o = u + iv = \tan\gamma, \tag{5.24}$$

and $$\gamma = \psi_0 + i\phi_0, \tag{5.25}$$

where u, v, ψ_0, ϕ_0 are real. Thus γ is a complex angle, whose real part is chosen to be in the range $-\frac{1}{2}\pi$ to $\frac{1}{2}\pi$. Then (5.19) shows that

$$\rho_x = \tan(\tfrac{1}{2}\pi - \gamma). \tag{5.26}$$

Substitution of (5.24) in the quadratic (5.12) gives

$$\sin 2\gamma = \frac{-2Y_L Z - 2iY_L(1-X)}{Y_T^2}, \tag{5.27}$$

whence $$\sin 2\psi_0 \cosh 2\phi_0 = \frac{-2Y_L Z}{Y_T^2}. \tag{5.28}$$

The values ρ_o', ρ_x' referred to new axes, as in the preceding section, are from (5.21),

$$\rho_o' = \tan(\gamma - \psi),\ \rho_x' = \cot(\gamma + \psi) \tag{5.29}$$

which show that ρ_o' is purely imaginary when $\psi = \psi_0$, or when $\psi = \psi_0 + \frac{1}{2}\pi$. These are the inclinations to the x-axis of the major and minor axes respectively of the polarisation ellipse for the ordinary wave. Similarly, the polarisation ellipse of the extraordinary wave has its major and minor axes at angles $\frac{1}{2}\pi - \psi_0$ and $-\psi_0$ to the x-axis respectively.

When ψ takes the value ψ_0 for the major axis, ρ_o' is purely imaginary and its modulus is the ratio $\dfrac{\text{minor axis}}{\text{major axis}}$. Equation (5.24) shows that this ratio is equal to $\tan i\phi_0 = \tanh\phi_0$. Now $u + iv = \tan(\psi_0 + i\phi_0)$, whence it can be shown that

$$\tanh 2\phi_0 = \frac{2v}{1 + u^2 + v^2}. \tag{5.30}$$

Either ψ_0 or $\frac{1}{2}\pi - \psi_0$ is necessarily less than 45°. The smaller angle is called the 'tilt-angle' of the polarisation ellipse. It is the angle between the major axis and the x- or y-axis, whichever is the nearer (see Fig. 5.1).

The 'axis ratio' is equal to $\tanh\phi_0$ and is the same for the two ellipses. Some authors have expressed the axis ratio in terms of an angle Φ by writing $\tan\Phi = \tanh\phi_0$, but (5.30) is the more elegant method. The expression for Φ in terms of u and v is much less simple.

Consider a radio wave coming vertically downwards in the northern hemisphere. Here the vertical component of the earth's magnetic field is directed downwards, so that the vertical component of $\mathbf{Y}$ is upwards (opposite to the wave-normal), and Y_L is negative. Then (5.28) shows that ψ_0 is positive, and $\frac{1}{2}\pi - \psi_0$ must also be positive since $\psi_0 < \frac{1}{2}\pi$. Thus, if it is possible to treat the ionosphere as a homogeneous medium, the major axes of both polarisation ellipses must be in the quadrant between magnetic north and magnetic east (or south and west). No progressive wave is possible with the major axis of its polarisation ellipse in the north-west and south-east quadrants. A more detailed discussion of the polarisation of a downcoming radio wave is given in §§ 19.11 to 19.13.

5.6 The Appleton–Hartree formula for the refractive index

The Maxwell equations (5.4) and (4.2) give respectively $\epsilon_0 \mathfrak{n}\mathscr{H}_y = D_x$ and $\mathfrak{n}E_x = \mathscr{H}_y$. If $\mathscr{H}_y$ is eliminated

$$\epsilon_0 \mathfrak{n}^2 E_x = D_x. \tag{5.31}$$

Similarly, (5.5) and (4.1) give

$$\epsilon_0 \mathfrak{n}^2 E_y = D_y. \tag{5.32}$$

These equations are the same as would be obtained for an isotropic dielectric, in which case $\mathfrak{n}^2$, the square of the refractive index, is equal to the dielectric constant. It is doubtful if the result is of any physical significance here, however; it would not be true if the magnetic permeability were different from unity, and it is not true for the z-direction, because $D_z = 0$, but in general $E_z \neq 0$.

Now $D_x/\epsilon_0 = E_x + P_x/\epsilon_0$ so that (5.31) gives

$$P_x = \epsilon_0(\mathfrak{n}^2 - 1) E_x. \tag{5.33}$$

This may be combined with the first equation (5.10) to eliminate E_x. Since $P_y/P_x = \rho$, from (5.7), the result is

$$\frac{X}{\mathfrak{n}^2 - 1} = -U - i\mathfrak{n}Y\rho. \tag{5.34}$$

Hence
$$\mathfrak{n}^2 = 1 - \frac{X}{U + i\mathfrak{n}Y\rho}. \tag{5.35}$$

Since ρ has the two possible values (5.13), $\mathfrak{n}^2$ also has two values. The full expression for $\mathfrak{n}^2$ is obtained by substituting the values of ρ from (5.13), and is given later at (6.1). This is the Appleton–Hartree formula for the refractive index, and its properties are discussed in ch. 6.

The expressions (5.13) for ρ and (5.35) for $\mathfrak{n}^2$ depend on the parameters X, Y, Z of the ionosphere, and the two values of $\mathfrak{n}$ can therefore be regarded simply as properties of the ionosphere at a given point. In a medium which varies from point to point it may not be possible for progressive waves to exist, but still (5.35) can be used at each point to *define* the two values of $\mathfrak{n}$. The steps of the above derivation can then be carried out backwards and they lead to (5.31) and (5.32). The process is valid in a *variable* medium since the constitutive relations used in (5.33) and (5.34) depend only on the properties of the ionosphere at the point considered, and do not depend on how these properties vary from point to point.

We thus have the important result that equations (5.31) and (5.32) are true in a variable medium, for a given direction, where $\mathfrak{n}$ is one of the two refractive indices for a progressive wave in a fictitious homogeneous medium with the same parameters X, Y, Z as those of the variable medium at the point considered, and the wave-normal of the progressive wave is in the given direction. This result is used later (§ 18.9).

5.7 The longitudinal component of the electric field

It has already been mentioned in § 5.3 that both E_z and P_z may differ from zero, although $D_z = 0$. We now derive an expression for the ratio E_z/E_x.

Equation (5.9) was derived from the constitutive relation. By dividing it by P_x and using (5.7), (5.8) and (5.33), we obtain (since $lY = Y_T$)

$$\frac{U-X}{\mathfrak{n}^2-1}\frac{E_z}{E_x} = i\rho Y_T. \tag{5.36}$$

Hence

$$\frac{E_z}{E_x} = -\frac{i\rho Y_T(\mathfrak{n}^2-1)}{U-X} \tag{5.37}$$

and

$$\frac{E_z}{E_y} = -\frac{iY_T(\mathfrak{n}^2-1)}{U-X}. \tag{5.38}$$

If collisions are neglected, so that $Z = 0$, $U = 1$, then $\mathfrak{n}^2$ is purely real and ρ is purely imaginary. Then E_z/E_x is real, hence E_z and E_x are then in phase or in antiphase.

Equation (5.37) shows that the longitudinal component of the electric field vanishes only if either $Y_T = 0$, or $\mathfrak{n}^2 = 1$, or $\rho = 0$. The first of these conditions holds when the wave-normal is parallel to the earth's magnetic field (longitudinal propagation). The second condition applies only when $X = 0$, that is for free space. It can be shown that one value of $\mathfrak{n}^2$ is unity when $X = 1$, $Z = 0$, but in this case ρ is infinite, and (5.37) is not zero. The third condition, $\rho = 0$, requires that $Y_L = 0$, so that the earth's magnetic field is perpendicular to the wave-normal, and parallel to the electric field $\mathbf{E}$ of the wave. Hence the only movement imparted to the electrons is parallel to the earth's magnetic field, and the electron motion is unaffected by the field.

5.8 The flow of energy for a wave in a magnetoionic medium

It is interesting to evaluate the average value of the complex Poynting vector, given by (2.40), for a wave in a magnetoionic medium. This vector may be interpreted as the flux of energy in the wave, although it was indicated in § 2.13 that other interpretations are possible.

Since $\mathbf{H} = (1/Z_0)\mathscr{H}$, and the component $\mathscr{H}_z$ is always zero, the components of the vector $\bar{\mathbf{\Pi}}$ are as follows:

$$\left.\begin{aligned}\bar{\Pi}_x &= -\frac{1}{4Z_0}\{E_z\mathscr{H}_y^* + E_z^*\mathscr{H}_y\},\\ \bar{\Pi}_y &= \frac{1}{4Z_0}\{E_z\mathscr{H}_x^* + E_z^*\mathscr{H}_x\},\\ \bar{\Pi}_z &= \frac{1}{4Z_0}\{E_x\mathscr{H}_y^* + E_x^*\mathscr{H}_y - E_y\mathscr{H}_x^* - E_y^*\mathscr{H}_x\}.\end{aligned}\right\} \tag{5.39}$$

In an isotropic medium E_z is also zero, so that $\overline{\Pi}_x$, $\overline{\Pi}_y$ are zero and the energy flow is in the direction of the wave normal, but this is not in general true for an anisotropic medium. It is convenient to express the field vectors E_y, E_z, $\mathscr{H}_x$, $\mathscr{H}_y$, in terms of E_x. From (4.1), (4.2) and (5.7), we have:

$$E_y = \rho E_x, \quad \mathscr{H}_y = \mathfrak{n} E_x, \quad \mathscr{H}_x = -\mathfrak{n}\rho E_x, \tag{5.40}$$

and E_z is given by (5.37). Hence

$$\overline{\Pi}_x = -\frac{iY_T}{4Z_0}\left\{\frac{\mathfrak{n}\rho^*(\mathfrak{n}^{*2}-1)}{U^*-X} - \frac{\mathfrak{n}^*\rho(\mathfrak{n}^2-1)}{U-X}\right\}|E_x|^2, \tag{5.41}$$

$$\overline{\Pi}_y = \frac{iY_T\rho\rho^*}{4Z_0}\left\{\frac{(\mathfrak{n}^2-1)\,\mathfrak{n}^*}{U-X} - \frac{(\mathfrak{n}^{*2}-1)\,\mathfrak{n}}{U^*-X}\right\}|E_x|^2, \tag{5.42}$$

$$\overline{\Pi}_z = \frac{1}{4Z_0}(\mathfrak{n}+\mathfrak{n}^*)(1+\rho\rho^*)\,|E_x|^2. \tag{5.43}$$

These equations give the magnitude and direction of $\overline{\mathbf{\Pi}}$ in the general case, and the values of $\mathfrak{n}$ and ρ may be inserted for either of the two characteristic waves. The expressions are too complicated to be of much interest, but they are useful in the special case when electron collisions are neglected, so that $\mathfrak{n}^2$ is real, ρ is purely imaginary, and $U = U^* = 1$. Then $\rho^* = -\rho$. There are two cases of interest. The first is when $\mathfrak{n}$ is purely real, so that $\mathfrak{n}^* = \mathfrak{n}$. Then the equations give

$$\left.\begin{aligned} \overline{\Pi}_x &= \frac{i\rho\mathfrak{n}(\mathfrak{n}^2-1)\,Y_T}{2Z_0(1-X)}\,|E_x|^2, \\ \overline{\Pi}_y &= 0, \\ \overline{\Pi}_z &= \frac{\mathfrak{n}}{2Z_0}(1-\rho^2)\,|E_x|^2, \end{aligned}\right\} \tag{5.44}$$

which show that the vector $\overline{\mathbf{\Pi}}$ is in the plane containing the earth's magnetic field and the wave-normal, and makes an angle

$$\arctan\left\{\frac{Y_T\, i\rho(\mathfrak{n}^2-1)}{(1-X)(1-\rho^2)}\right\} \tag{5.45}$$

with the wave-normal. This expression is real, since ρ is purely imaginary. It is derived by a different method in § 13.13, and its significance is discussed there.

The second case of interest is when $\mathfrak{n}$ is purely imaginary, so that the wave is evanescent, and $\mathfrak{n}^* = -\mathfrak{n}$. Then (5.41) to (5.43) give

$$\left.\begin{aligned} \overline{\Pi}_x &= \overline{\Pi}_z = 0, \\ \overline{\Pi}_y &= \frac{i\mathfrak{n}\rho^2(\mathfrak{n}^2-1)\,Y_T}{2Z_0(1-X)}\,|E_x|^2, \end{aligned}\right\} \tag{5.46}$$

which shows that the vector $\overline{\mathbf{\Pi}}$ is perpendicular both to the wave-normal and to the earth's magnetic field.

5.9 The effect of heavy ions on polarisation and refractive index

The effect of heavy ions on the constitutive relations of the ionosphere was discussed in § 3.9, where subscripts or superscripts e and i were used to indicate quantities which apply to electrons and heavy ions respectively. The same notation will be used in this section. The possible wave-polarisations, when allowance is made for heavy ions, can be found by an extension of the method of § 5.3.

The contribution $\mathbf{P}^{(i)}$ made by the heavy ions to the electric polarisation $\mathbf{P}$ is assumed to be given by (3.32), which is applicable to all but the very lowest frequencies. It is thus assumed that Y_i (3.31) is negligibly small. Similarly, the contribution $\mathbf{P}^{(e)}$ made by the electrons is given by (5.3) which must now be written

$$-X_e\epsilon_0\begin{pmatrix}E_x\\E_y\\E_z\end{pmatrix}=\begin{pmatrix}U_e & inY & 0\\ -inY & U_e & ilY\\ 0 & -ilY & U_e\end{pmatrix}\begin{pmatrix}P_x^{(e)}\\P_y^{(e)}\\P_z^{(e)}\end{pmatrix}, \tag{5.47}$$

where Y refers to electrons. From (3.32) we have

$$\frac{P_y^{(i)}}{P_x^{(i)}}=\frac{E_y}{E_x}=\rho, \tag{5.48}$$

and hence, with (5.7), $$\rho=\frac{D_y}{D_x}=\frac{P_y^{(i)}+P_y^{(e)}}{P_x^{(i)}+P_x^{(e)}}=\frac{P_y^{(e)}}{P_x^{(e)}}. \tag{5.49}$$

Equation (5.6) gives $$\epsilon_0E_z+P_z^{(e)}+P_z^{(i)}=0, \tag{5.50}$$

and when the value of $P_z^{(i)}$ is inserted from (3.32) we obtain

$$P_z^{(e)}=-\epsilon_0E_z\left(1-\frac{X_i}{U_i}\right). \tag{5.51}$$

Now E_z may be eliminated from (5.51) and the last of the three equations in (5.47) which gives

$$P_z^{(e)}=-\frac{ilY(U_i-X_i)}{U_iX_e-U_e(U_i-X_i)}P_y^{(e)}. \tag{5.52}$$

The first two equations in (5.47) may now be written down, and the value (5.52) for $P_z^{(e)}$ substituted. This gives

$$-\epsilon_0E_x=\frac{U_e}{X_e}P_x^{(e)}+\frac{inY}{X_e}P_y^{(e)}, \tag{5.53}$$

$$-\epsilon_0E_y=-\frac{inY}{X_e}P_x^{(e)}+\frac{1}{X_e}\left\{U_e+\frac{l^2Y^2(U_i-X_i)}{U_iX_e-U_e(U_i-X_i)}\right\}P_y^{(e)}. \tag{5.54}$$

These equations should be compared with (5.10), obtained when the effect of heavy ions was neglected. The second equation is divided by the first and (5.49) is used. The result is

$$\rho = \frac{-inY+\rho\left\{U_e+\dfrac{l^2Y^2(U_i-X_i)}{U_iX_e-U_e(U_i-X_i)}\right\}}{U_e+inY\rho}, \tag{5.55}$$

whence

$$\rho^2-\frac{il^2Y}{n}\frac{U_i-X_i}{U_e(U_i-X_i)-U_iX_e}\rho+1=0, \tag{5.56}$$

which is a quadratic equation analogous to (5.11). It reduces to (5.11) when $X_i = 0$.

The two values of ρ given by (5.55) have many of the properties described in § 5.4, for the case when the effect of heavy ions was neglected. In particular, (5.19), $\rho_o\rho_x = 1$, still holds. If the collision damping for both ions and electrons is neglected, then ρ is purely imaginary. It can also be shown by an argument similar to that of § 5.5 that the major axes of both polarisation ellipses must still be in the quadrant between magnetic north and magnetic east (or south and west).

The refractive index $\mathfrak{n}$ may be found by the method of § 5.6. Equation (5.33) becomes

$$P_x = P_x^{(e)}+P_x^{(i)} = \epsilon_0(\mathfrak{n}^2-1)E_x \tag{5.57}$$

and (3.32) may be used to eliminate $P_x^{(i)}$. This gives

$$P_x^{(e)} = \epsilon_0\left(\mathfrak{n}^2-1+\frac{X_i}{U_i}\right)E_x \tag{5.58}$$

which, with (5.49) and (5.53), leads to

$$\mathfrak{n}^2 = 1-\frac{X_i}{U_i}-\frac{X_e}{U_e+inY\rho}. \tag{5.59}$$

This corresponds to the Appleton–Hartree formula. The two values of ρ from (5.56) lead to two values of $\mathfrak{n}^2$ as before. Some properties of this formula are given in § 6.18.

At extremely low frequencies (in the audible range) it may not be permissible to neglect Y_i, and it would then be necessary to use the more complicated formula (3.30) for the relation between $\mathbf{P}_i$ and $\mathbf{E}$. This would lead to a much more complicated theory.

Examples

1. A homogeneous medium contains N free electrons per unit volume with negligible collision damping, and there is a constant superimposed magnetic field. A linearly polarised plane electromagnetic wave of angular frequency ω travels with its wave-normal and its electric vector both perpendicular to the magnetic field. Show that the refractive index μ for this wave is given by

$$\mu^2 = 1-\frac{X(1-X)}{1-X-Y^2},$$

where $Y = \omega_H/\omega$, $X = Ne^2/\epsilon_0 \mathfrak{m}\omega^2$ (rationalised units), ω_H is the angular gyro-frequency for electrons, e, $\mathfrak{m}$ are the charge and mass of the electron, and ϵ_0 is the electric permittivity of free space.

Show that the electric field has a longitudinal component in quadrature with the transverse component and that the ratio of the longitudinal to the transverse amplitudes is $XY/(1-X-Y^2)$.

Use the Poynting vector to find the instantaneous direction of the energy flow. Hence find the direction of average energy flow when

$$(a)\ X < 1-Y \text{ or } X > 1+Y, \quad (b)\ 1-Y < X < 1+Y.$$

(Math. Tripos, 1958, Part III.)

2. Discuss the propagation of electromagnetic waves in an isotropic material medium, treating the electric field at each point as the resultant of the electric fields of the incident wave and of the wavelets scattered by elements of volume of the medium.

What extensions of the theory would you expect to have to make to deal with (*a*) anisotropic substances, (*b*) magnetic rotation of the plane of polarisation?

(Natural Sciences Tripos, 1952, Part II, Physics.
See Darwin (1924), Hartree (1929, 1931*b*).)

CHAPTER 6

PROPERTIES OF THE APPLETON–HARTREE FORMULA

6.1 General properties. Zeros and infinity of the refractive index

The refractive index of a medium containing free electrons, with a superimposed steady magnetic field, is given by (5.35). If the value of ρ is substituted from (5.13), the result is

$$\mathfrak{n}^2 = 1 - \frac{X}{1 - iZ - \dfrac{\frac{1}{2}Y_T^2}{1 - X - iZ} \pm \left\{\dfrac{\frac{1}{4}Y_T^4}{(1 - X - iZ)^2} + Y_L^2\right\}^{\frac{1}{2}}}. \tag{6.1}$$

This is the Appleton–Hartree formula which dominates much of the theory of radio wave propagation, and some familiarity with its properties is essential to an understanding of the subject. A very thorough discussion is given by Ratcliffe (*The Magneto-ionic Theory*, C.U.P. 1959), whose book is strongly recommended. See also Booker (1934). The treatment given here is necessarily much briefer.

It can be shown directly from (6.1) that one value of $\mathfrak{n}^2$ is zero when

$$X + iZ = 1, \tag{6.2}$$

or

$$X + iZ = 1 + Y, \tag{6.3}$$

or

$$X + iZ = 1 - Y. \tag{6.4}$$

If Z is not zero, these conditions cannot be satisfied for real values of X, but they are important when $X(z)$ is treated as a function of the height z regarded as a complex variable.

One value of $\mathfrak{n}^2$ is infinite when

$$X = (1 - iZ)\frac{(1 - iZ)^2 - Y^2}{(1 - iZ)^2 - Y_L^2}. \tag{6.5}$$

The effect of this infinity on wave-propagation is discussed in §§ 21.13 to 21.16.

The properties of the Appleton–Hartree formula will now be illustrated by discussing some special cases.

6.2 Collisions neglected

At high frequencies (greater than about 1 Mc/s), it is permissible for many purposes to neglect collisions. In this case $Z = 0$, and the formula (6.1) becomes

$$\mathfrak{n}^2 = 1 - \frac{X}{1 - \frac{Y_T^2}{2(1-X)} \pm \left\{\frac{Y_T^4}{4(1-X)^2} + Y_L^2\right\}^{\frac{1}{2}}}. \tag{6.6}$$

Clearly $\mathfrak{n}^2$ is always real and it is useful to plot curves showing how it depends upon X (proportional to the electron number density). Now the three zeros of $\mathfrak{n}^2$ (6.2) to (6.4) occur where

$$X = 1, \quad X = 1 \pm Y, \tag{6.7}$$

and the infinity (6.5) occurs where

$$X = \frac{1 - Y^2}{1 - Y_L^2}. \tag{6.8}$$

It can also be shown that when $X = 1$, one of the values of $\mathfrak{n}^2$ is unity.

6.3 Frequency above the gyro-frequency

In the ionosphere above the British Isles the gyro-frequency is about 1·2 Mc/s. Most radio investigation of the F-region uses frequencies greater than this. In this case $Y < 1$.

6.4 Longitudinal propagation when $Y < 1$

If the wave-normal is anti-parallel to the earth's magnetic field, $Y_T = 0$, $n = +1$, and the formula becomes

$$\mathfrak{n}^2 = 1 - \frac{X}{1 \pm Y}. \tag{6.9}$$

This case was discussed by Lorentz (*Theory of Electrons*, ch. IV). It can easily be shown (from (5.35)) that for the plus sign $\rho = -i$, and for the minus sign $\rho = +i$. The formula gives two straight lines, which are shown in Fig. 6.1. For the line through $X = 1 + Y$ the wave is circularly polarised with a right-handed sense, and for the line through $X = 1 - Y$, it is circularly polarised with a left-handed sense. The formula (6.9) does not display the zero at $X = 1$, nor the infinity which (6.8) shows is also at $X = 1$. This is because (6.9) was obtained from (6.6) by setting $Y_T^2 = 0$, but this process is not valid when $X = 1$. The true state of affairs is made clearer in § 6.6 below.

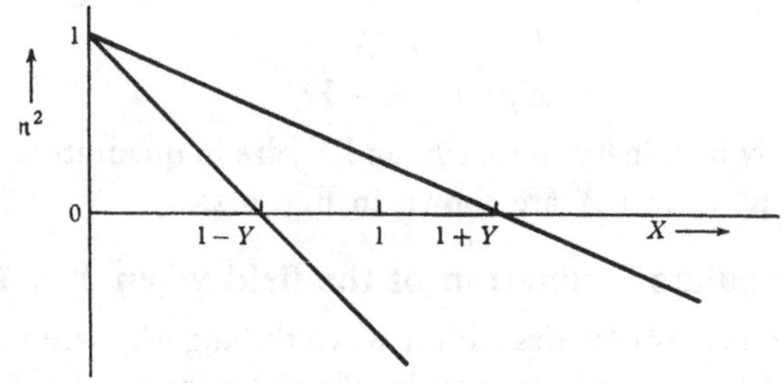

Fig. 6.1. Variation of $\mathfrak{n}^2$ with X for purely longitudinal propagation, when $Y < 1$.

6.5 Transverse propagation when $Y < 1$

If the wave-normal is perpendicular to the earth's magnetic field, $Y_L = 0$, and the formula (6.6) becomes

$$\mathfrak{n}^2 = 1 - X, \tag{6.10}$$

or

$$\mathfrak{n}^2 = 1 - \frac{X(1-X)}{1-X-Y^2}. \tag{6.11}$$

Equation (6.10) is the same as (4.7) which was obtained when the earth's magnetic field was neglected. For this case $\rho = 0$, so that the wave is linearly polarised with its electric field parallel to the x-axis (it was shown

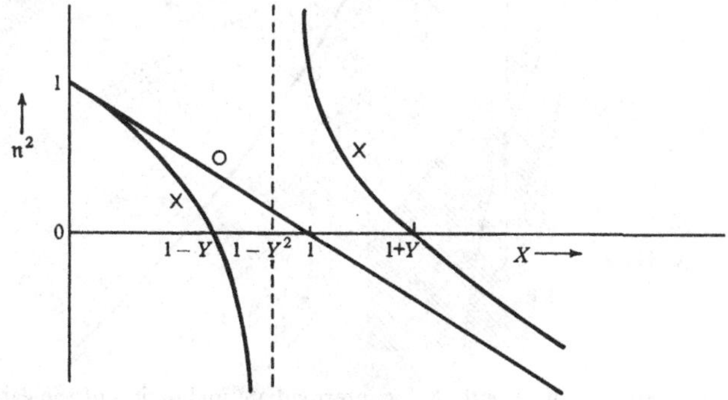

Fig. 6.2. Variation of $\mathfrak{n}^2$ with X for purely transverse propagation when $Y < 1$.

at the end of § 5.7 that $E_z = 0$ in this case). Hence the only movement imparted to the electrons is parallel to the earth's magnetic field, and the wave behaves as it would if the field were absent.

Equation (6.11) has zeros where $X = 1 \pm Y$, and an infinity where $X = 1 - Y^2$. This is between $X = 1 - Y$ and $X = 1$, since $Y < 1$.

For this case $\rho = \infty$ so that $E_x = 0$. The formula (5.38) gives

$$\frac{E_z}{E_y} = \frac{iYX}{1 - X - Y^2}, \tag{6.12}$$

which is purely imaginary so that E_z and E_y are in quadrature.

Curves of $\mathfrak{n}^2$ against X are shown in Fig. 6.2.

6.6 Intermediate inclination of the field when $Y < 1$

Fig. 6.3 shows how $\mathfrak{n}^2$ varies with X when the angle between the earth's magnetic field and the wave-normal is different from 0 or $\frac{1}{2}\pi$. The dotted lines show the limiting positions for longitudinal and transverse propagation discussed in §§ 6.4 and 6.5 respectively. One value of $\mathfrak{n}^2$ is now

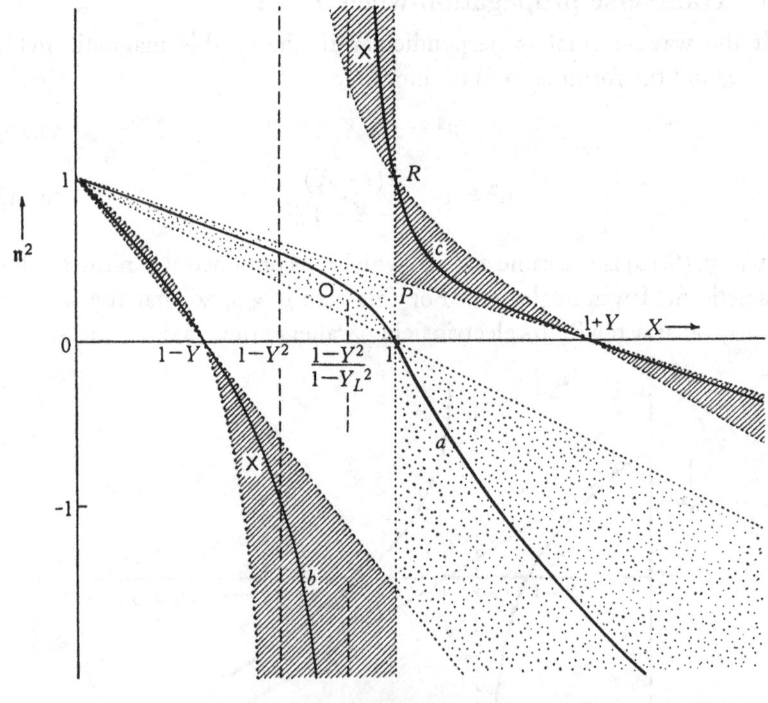

Fig. 6.3. Variation of $\mathfrak{n}^2$ with X for intermediate inclination of the earth's magnetic field, when $Y < 1$. Electron collisions are neglected.

infinite where X is given by (6.8). This is between $1 - Y^2$ and 1. The vertical broken line at $X = 1 - Y^2$ and the vertical dotted line at $X = 1$ mark the positions between which the infinity occurs. The thick lines are typical curves, and it can be shown that they must always lie in the shaded regions bounded by the curves for the longitudinal and transverse cases, and by the line $X = 1$.

If the relative directions of the wave-normal and the earth's magnetic field were changed continuously, so that they became perpendicular, the curve marked O in Fig. 6.3 would deform continuously into the straight line through $X = 1$. This line gives $\mathfrak{n}^2$ for the wave in which the electrons are unaffected by the earth's magnetic field for transverse propagation. The wave is therefore called the 'ordinary wave'. The other wave, which is always affected by the earth's magnetic field for all directions of the wave-normal, is called the 'extraordinary wave', and the curve is marked X.

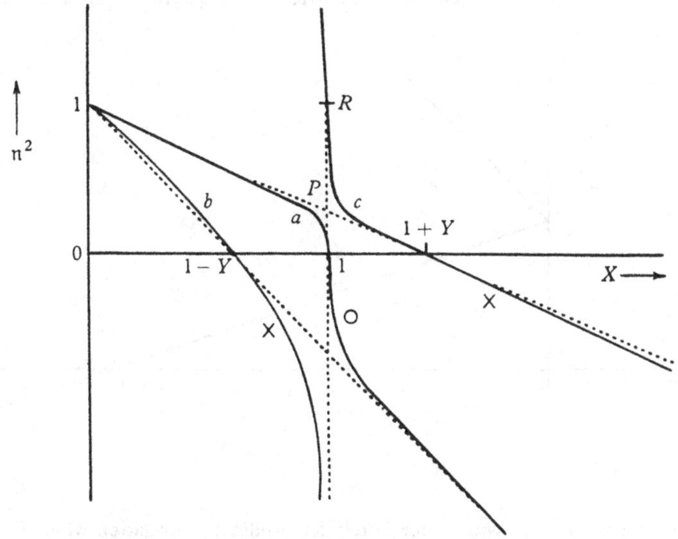

Fig. 6.4. Variation of $\mathfrak{n}^2$ with X when the propagation is almost longitudinal and $Y < 1$.

When the angle between the earth's magnetic field and the wave-normal is very small, the curves are as shown in Fig. 6.4. The infinity is close to $X = 1$, and both curves have sharp bends near this value of X. The curve for the ordinary wave may be followed from left to right. For $X < 1$, the corresponding value of the polarisation ρ is close to $-i$, but when the almost vertical part of the curve near $X = 1$ is traversed, ρ changes rapidly and goes over to a value close to $+i$, which is retained for $X > 1$. It will be shown later (ch. 19) that this rapid change of polarisation is associated with strong coupling between the ordinary and extraordinary waves. If an ordinary wave entered a medium of increasing electron density (increasing X) under the conditions of Fig. 6.4, it would cause some of the extraordinary wave to be generated near $X = 1$. In the limit when the propagation is entirely longitudinal, the extraordinary wave takes over completely at $X = 1$, and for $X > 1$, the only wave present would be the extraordinary wave. This explains how the curves of Fig. 6.4 go over, in the limit, into the curves of Fig. 6.1.

6.7 Frequency below the gyro-frequency

When the frequency is less than the gyro-frequency (about 1·2 Mc/s), the form of the curves of $\mathfrak{n}^2$ against X is somewhat different. At these low frequencies the effect of collisions becomes important, but it is nevertheless of interest to consider first the consequences of neglecting

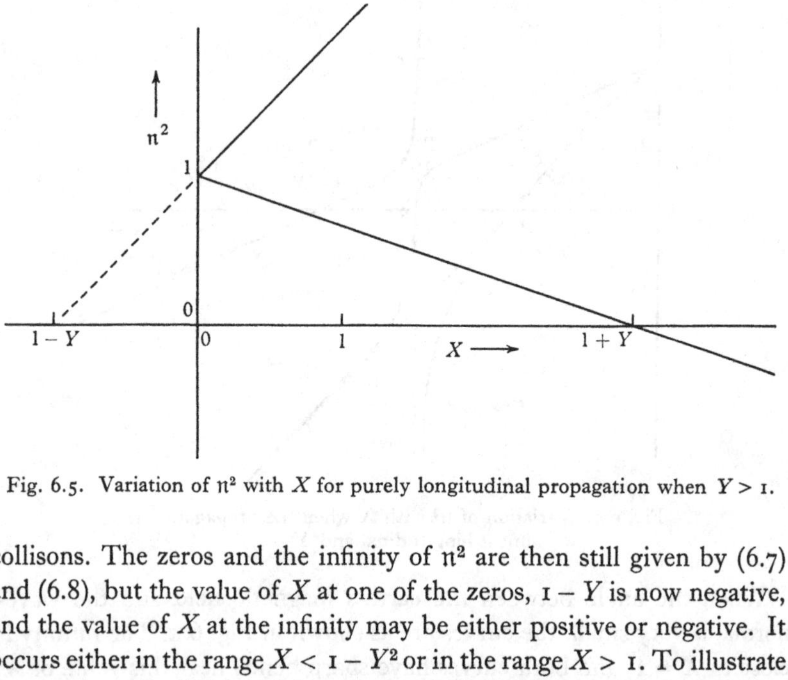

Fig. 6.5. Variation of $\mathfrak{n}^2$ with X for purely longitudinal propagation when $Y > 1$.

collisons. The zeros and the infinity of $\mathfrak{n}^2$ are then still given by (6.7) and (6.8), but the value of X at one of the zeros, $1 - Y$ is now negative, and the value of X at the infinity may be either positive or negative. It occurs either in the range $X < 1 - Y^2$ or in the range $X > 1$. To illustrate these properties, it is useful to draw the curves for negative as well as positive values of X, although negative values cannot occur at real heights in the ionosphere.

6.8 Longitudinal propagation when $Y > 1$

When the wave-normal is parallel to the earth's magnetic field, $Y_T = 0$, and $\mathfrak{n}^2$ is given by (6.9). This represents two straight lines which are shown in Fig. 6.5. Notice that one value of $\mathfrak{n}^2$ never becomes zero for any real positive value of X.

6.9 Transverse propagation when $Y>1$

When the wave-normal is perpendicular to the earth's magnetic field, $Y_L = 0$, and $\mathfrak{n}^2$ is given by (6.10) and (6.11). The equation (6.10) refers to the wave in which the electron motions are unaffected by the earth's magnetic field, and in which $\rho = 0$. This is the ordinary wave. Equation (6.11) refers to the extraordinary wave, for which $\rho = \infty$. The corresponding curve in Fig. 6.6 has zeros at $X = 1 \pm Y$, and an infinity at $X = 1 - Y^2$.

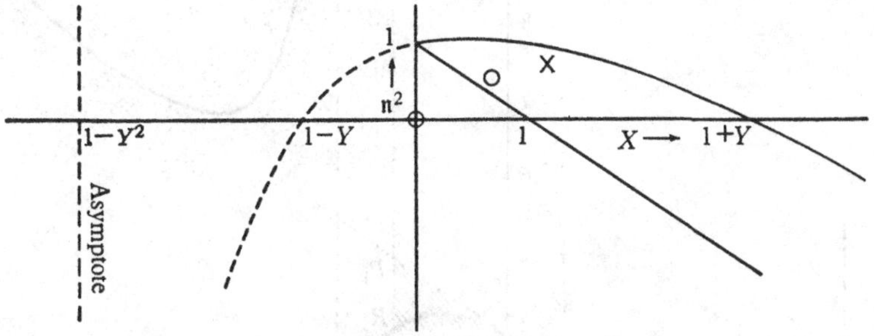

Fig. 6.6. Variation of $\mathfrak{n}^2$ with X for purely transverse propagation when $Y > 1$.

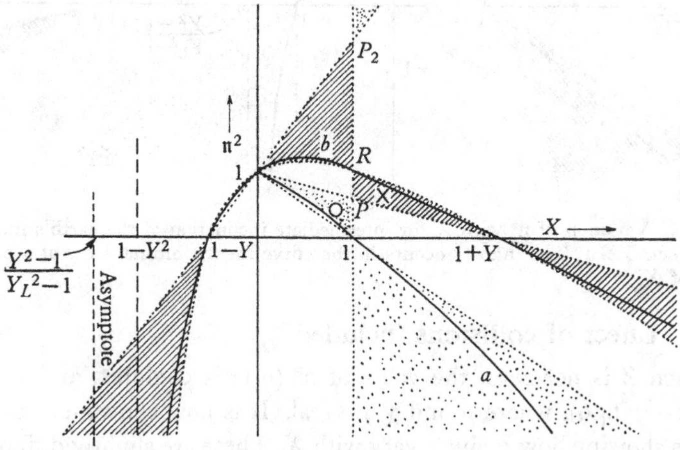

Fig. 6.7. Variation of $\mathfrak{n}^2$ with X for intermediate inclination of earth's magnetic field, when $Y > 1$. The infinity occurs in the curve for the extraordinary ray at a negative value of X.

6.10 Intermediate inclination of the field when $Y > 1$

Figs. 6.7 and 6.8 show how $\mathfrak{n}^2$ varies with X when the angle between the earth's magnetic field and the wave-normal is different from 0 or $\frac{1}{2}\pi$. The dotted lines show the limiting positions for longitudinal and transverse propagation. One value of $\mathfrak{n}^2$ is infinite where X is given by (6.8). The vertical broken line at $X = 1 - Y^2$, and the vertical dotted line at $X = 1$ mark the positions outside which the infinity occurs. The thick lines are typical curves. Figs. 6.7 and 6.8 are for the cases when the infinity occurs for negative and positive values of X, respectively. It can be shown that the curves must always lie in the shaded regions.

The transition to the longitudinal case occurs in a similar way to that described in § 6.6 above.

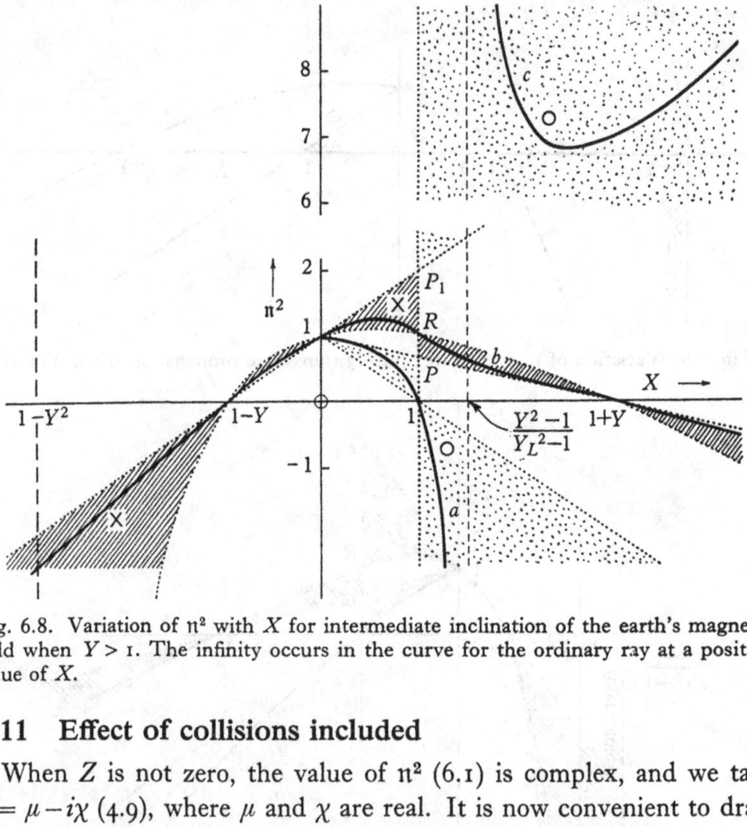

Fig. 6.8. Variation of $\mathfrak{n}^2$ with X for intermediate inclination of the earth's magnetic field when $Y > 1$. The infinity occurs in the curve for the ordinary ray at a positive value of X.

6.11 Effect of collisions included

When Z is not zero, the value of $\mathfrak{n}^2$ (6.1) is complex, and we take $\mathfrak{n} = \mu - i\chi$ (4.9), where μ and χ are real. It is now convenient to draw curves showing how μ and χ vary with X. These are slightly different in form from curves of $\mathfrak{n}^2$ against X, used previously.

It was mentioned in § 4.5 that for a passive medium μ and χ must always be both positive (or both negative). This is still true when the earth's magnetic field is included. It is obvious on physical grounds that the medium must always absorb energy from the wave, but it can also be proved formally as follows. $\mathfrak{n}^2$ is given by (5.35) which may be written $\mathfrak{n}^2 = 1 - X/D$ where $D = 1 - iZ + in Y(u + iv)$. Here u and v are the real and imaginary parts of the polarisation ρ, and n is the cosine of the angle between the vector $\mathbf{Y}$ and the wave-normal. Now it follows from (5.24), (5.25) and (5.28) that u and n have opposite signs, so that the imaginary part of D is always negative. Hence the imaginary part of $\mathfrak{n}^2$ is always negative and cannot be zero unless $Z = 0$.

6.12 The critical collision-frequency

It was shown in § 5.4 that the two possible values of the polarisation ρ are both equal to ± 1 when $X = 1, Z = Z_c$, where $Z_c = |Y_T^2/2Y_L|$ by (5.15) and (5.16). This is simply the condition that the square root, in the Appleton–Hartree formulae (6.1), shall be zero. (If $Z = 0$, this can only happen if $Y = 0$ also.) Equation (5.35) shows that when this condition holds, the refractive indices are equal. It was shown in § 6.11 that $\mathscr{R}(\rho)$ and Y_L must have opposite signs. Hence the refractive index is given by

$$\mathfrak{n}^2 = 1 - \frac{X}{1 - i(Z + |Y_L|)}. \tag{6.13}$$

From (5.37) the longitudinal component of the electric field is given by

$$\frac{E_z}{E_x} = \frac{iY_T X}{(1 - X - iZ)\{1 - i(Z + |Y_L|)\}}. \tag{6.14}$$

For downgoing waves in the northern hemisphere Y_L is negative. For the critical condition, therefore, $\rho = +1$. This means that for both waves the electric vector is parallel to the plane $x = y$.

It is shown later (§ 18.3) that the differential equations which govern the propagation of waves in a given direction (here taken to be parallel to the z-axis) in a homogeneous medium, are equivalent to a single differential equation of the fourth order. This should have four independent solutions, and it has been shown that in general this is so, for there are two solutions which represent progressive waves travelling in the direction of positive z, and two more for waves travelling in the opposite direction. For the critical condition described above, however, it might appear that there are only two independent solutions representing waves travelling in opposite directions, both with $\rho = +1$ and with $\mathfrak{n}$ given by (6.13). But in this case two more solutions can be found, in which the electric vector has a component at right angles to the plane $x = y$. This may be shown as follows.

Suppose that the medium is very nearly, but not quite, at the critical condition, so that $\rho = 1 \pm \epsilon$ where ϵ is so small that its square and higher powers may be neglected, and let the associated values of the refractive index be $\mathfrak{n}_+$ and $\mathfrak{n}_-$. Then (5.35) gives

$$\mathfrak{n}_+^2 - \mathfrak{n}_-^2 = -\frac{2inYX\epsilon}{(U-inY)^2}, \tag{6.15}$$

$$\mathfrak{n}_+ - \mathfrak{n}_- = -\frac{inYX\epsilon}{\mathfrak{n}(U-inY)^2}, \tag{6.16}$$

where $\mathfrak{n}$ (without subscript) is written for the average value of $\mathfrak{n}_+$ and $\mathfrak{n}_-$. Choose new axes x', y', z formed from the original axes (of § 5.2) by rotation through 45° about the z-axis, and let E'_x, E'_y be the components of $\mathbf{E}$ parallel to the x'- and y'-axes, and let ρ' denote the new value of the polarisation ρ. Then (5.21) and (5.22) show that $\rho' = \pm\frac{1}{2}\epsilon$. Let the two waves with these polarisations have equal amplitudes, so that their components E'_x annul each other when $z = 0$. Then

$$E'_x = A\{\exp(-ik\mathfrak{n}_+z) - \exp(-ik\mathfrak{n}_-z)\},$$

where A is a constant. The other field components may be found from (5.7) and (4.2). To the first order in ϵ they are given by:

$$\left.\begin{aligned} E'_x &= -ikz(\mathfrak{n}_+ - \mathfrak{n}_-)A\,e^{-ik\mathfrak{n}z},\\ E'_y &= A\epsilon\,e^{-ik\mathfrak{n}z},\\ \mathscr{H}'_x &= -\mathfrak{n}A\epsilon\,e^{-ik\mathfrak{n}z},\\ \mathscr{H}'_y &= (\mathfrak{n}_+ - \mathfrak{n}_-)A\,e^{-ik\mathfrak{n}z} - ik\mathfrak{n}z(\mathfrak{n}_+ - \mathfrak{n}_-)A\,e^{-ik\mathfrak{n}z}.\end{aligned}\right\} \tag{6.17}$$

Now let A tend to ∞, and ϵ tend to zero in such a way that $A\epsilon$ remains constant and equal to B, say. Then by using (6.16) the field components may be written:

$$\left.\begin{aligned} E'_x &= -\frac{knYX}{\mathfrak{n}(U-inY)^2}Bz\,e^{-ik\mathfrak{n}z},\\ E'_y &= B\,e^{-ik\mathfrak{n}z},\\ \mathscr{H}'_x &= -\mathfrak{n}B\,e^{-ik\mathfrak{n}z},\\ \mathscr{H}'_y &= -\frac{nYX}{(U-inY)^2}B\left\{\frac{i}{\mathfrak{n}} + kz\right\}e^{-ik\mathfrak{n}z}.\end{aligned}\right\} \tag{6.18}$$

The wave-field described by (6.18) has been derived by superimposing two progressive waves, travelling in the direction of positive z. The result is not a progressive wave since E'_x and $\mathscr{H}'_y$ have components with a factor z, but it is a solution of Maxwell's equations in this critical case. A fourth solution could be found which represented a similar wave travelling in the direction of negative z.

The wave (6.18) is mainly of mathematical interest. The following is an example of a problem where it would be needed. Consider a homogeneous medium with a plane sharp boundary, and suppose that the critical condition holds for a direction normal to the boundary. A plane wave is incident normally on the boundary, and it is required to find the amplitude and polarisation of the reflected wave. The boundary conditions cannot in general be satisfied unless there are two waves in the medium, namely a linearly polarised progressive wave with refractive index given by (6.13), and wave of the type (6.18). The detailed solution of this problem is complicated, however, and of no special practical interest.

6.13 Longitudinal propagation when collisions are included

When $Y_T = 0$ the formulae (5.13) and (5.35) give

$$\rho = \mp i, \quad \mathfrak{n}^2 = 1 - \frac{X}{1 - iZ \pm Y}. \tag{6.19}$$

The equations (5.37) and (5.38) show that in this case the electric field of the wave has no longitudinal component. In the northern hemisphere

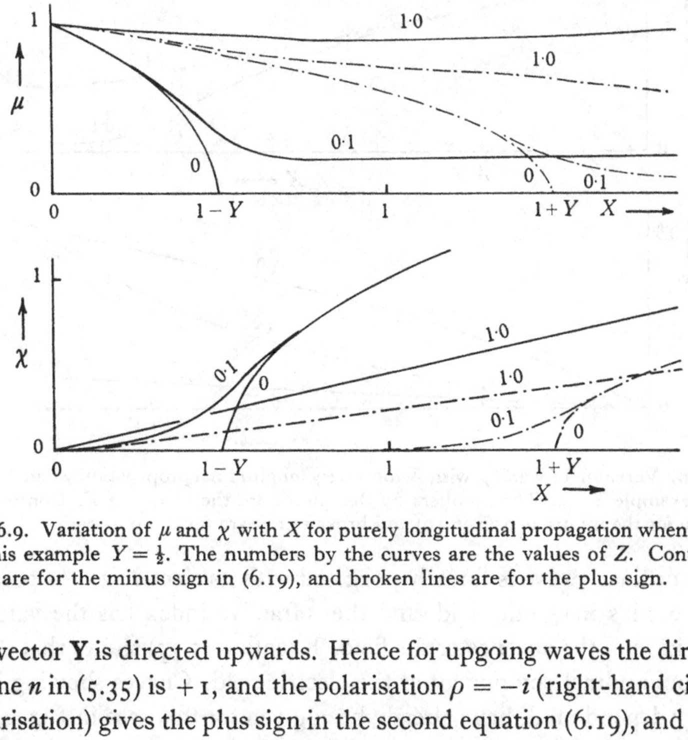

Fig. 6.9. Variation of μ and χ with X for purely longitudinal propagation when $Y < 1$. In this example $Y = \frac{1}{2}$. The numbers by the curves are the values of Z. Continuous lines are for the minus sign in (6.19), and broken lines are for the plus sign.

the vector **Y** is directed upwards. Hence for upgoing waves the direction cosine n in (5.35) is $+1$, and the polarisation $\rho = -i$ (right-hand circular polarisation) gives the plus sign in the second equation (6.19), and corresponds to the ordinary wave when $X < 1$. (When $X > 1$ there is an ambiguity of nomenclature which is explained in § 6.15.) Some curves for this case, showing how μ and χ vary with X for fixed values of Z, are given in Fig. 6.9 for a frequency above the gyro-frequency ($Y < 1$), and in Fig. 6.10 for a frequency below the gyro-frequency ($Y > 1$).

6.14 Transverse propagation when collisions are included

When $Y_L = 0$ the formulae (5.13) and (5.35) give for the ordinary wave

$$\rho = 0, \quad n^2 = 1 - \frac{X}{1 - iZ}, \tag{6.20}$$

and (5.13), (6.1) give for the extraordinary wave

$$\rho = \infty, \quad n^2 = 1 - \frac{X(1 - X - iZ)}{(1 - iZ)(1 - X - iZ) - Y^2}. \tag{6.21}$$

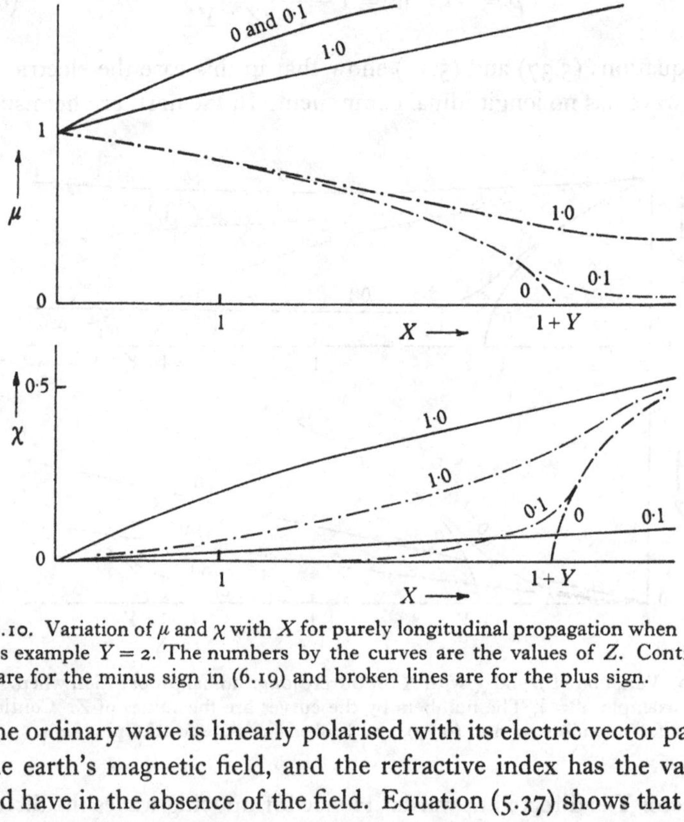

Fig. 6.10. Variation of μ and χ with X for purely longitudinal propagation when $Y > 1$. In this example $Y = 2$. The numbers by the curves are the values of Z. Continuous lines are for the minus sign in (6.19) and broken lines are for the plus sign.

The ordinary wave is linearly polarised with its electric vector parallel to the earth's magnetic field, and the refractive index has the value it would have in the absence of the field. Equation (5.37) shows that there is no longitudinal component of the electric field. Curves showing how μ and χ depend on X have already been given for this case in Fig. 4.1.

The extraordinary wave is linearly polarised with its electric vector E_y perpendicular to the earth's field, and (5.38) shows that there is a longitudinal component E_z of the electric field given by

$$\frac{E_z}{E_y} = \frac{iYX}{(1 - iZ)(1 - X - iZ) - Y^2}. \tag{6.22}$$

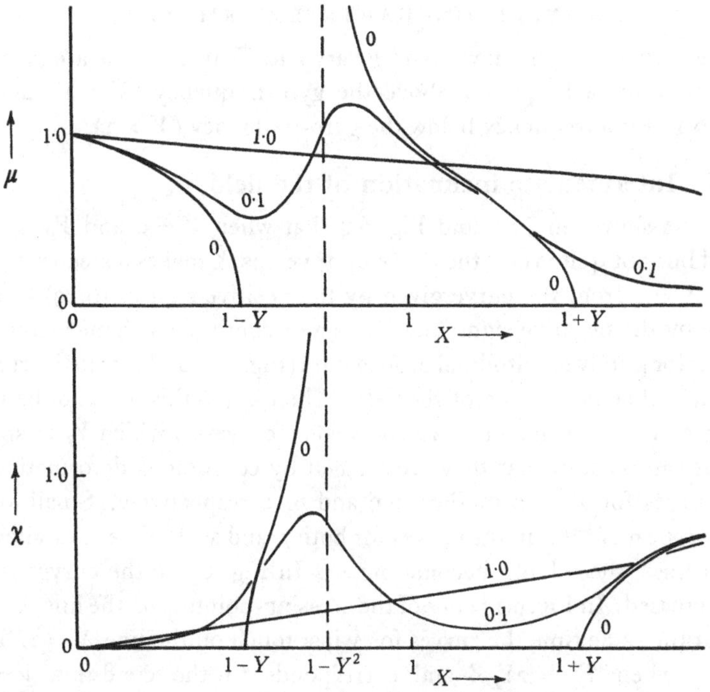

Fig. 6.11. Variation of μ and χ with X for the extraordinary wave and purely transverse propagation when $Y < 1$. In this example $Y = \frac{1}{2}$. The numbers by the curves are the values of Z.

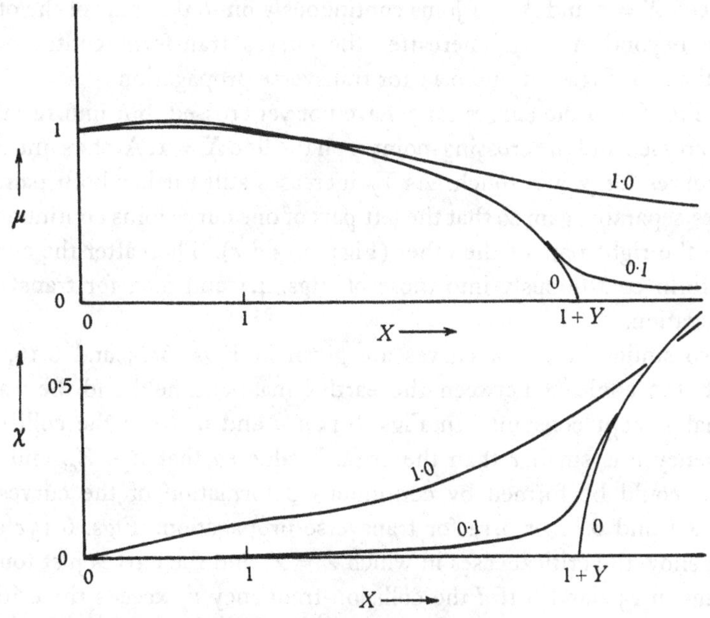

Fig. 6.12. Variation of μ and χ with X for the extraordinary wave and purely transverse propagation when $Y > 1$. In this example $Y = 2$. The numbers by the curves are the values of Z.

Some curves showing how μ and χ vary with X in this case are given in Fig. 6.11 for a frequency above the gyro-frequency ($Y < 1$), and in Fig. 6.12 for a frequency below the gyro-frequency ($Y > 1$).

6.15 Intermediate inclination of the field

It was shown in §6.6 and Fig. 6.4 that when $Z = 0$ and Y_T is very small but not quite zero, the curve of $\mathfrak{n}^2$ versus X makes a steep traverse near $X = 1$ from the curve given by the positive sign in (6.19) to that given by the negative sign. There was a discontinuous change from the curves for purely longitudinal propagation (Figs. 6.1 and 6.5) to the curves for any other inclination of the field. When $Z \neq 0$ this does not happen at once. Figs. 6.13*a* and 6.14*a* show some curves in which Y_T is small, and it can be seen that they are formed by continuous deformation of the curves for $Y_T = 0$ in Figs. 6.9 and 6.10 respectively. Small kinks appear, near $X = 1$, in the curves for both μ and χ. If Y_T is now allowed to increase, these kinks become bigger. In Fig. 6.13*b* the curves for μ have crossed. In Fig. 6.13*c* one of the crossing-points is on the line $X = 1$, and at the same time the curves for χ just touch on the line $X = 1$. This occurs when $Y_T^2 = 2Y_LZ$ and corresponds to the condition $\nu = \omega_c$ mentioned in §6.12. If Y_T is now allowed to increase still further, both pairs of curves separate again in such a way that the part of one curve between $X = 0$ and $X = 1$ joins continuously on to that part of the other curve beyond $X = 1$. Thereafter the curves transform continuously into those of Figs. 4.1 and 6.11 for transverse propagation.

In Fig. 6.14*b* the curves for χ have not yet crossed, but in 6.14*c* they have crossed and the crossing-point is on the line $X = 1$. At the same time the curves for μ just touch. As Y_T increases still further both pairs of curves separate again so that the left part of one curve joins continuously on to the right part of the other (Figs. 6.14*d*, *e*). Thereafter the curves transform continuously into those of Figs. 4.1 and 6.12 for transverse propagation.

Two similar series of curves are given in Figs. 6.15 and 6.16, for which the angle Θ between the earth's magnetic field and the wave-normal is kept constant. In Figs. 6.15*a*, *b* and 6.16*a*, *b* the collision-frequency ν is smaller than the critical value so that $Z < Z_c$, and the curves could be formed by continuous deformation of the curves in Figs. 4.1 and 6.11 or 6.12 for transverse propagation. Figs. 6.15*c* and 6.16*c* show the critical cases in which $Z = Z_c$ and the curves just touch. In Figs. 6.15*d* and 6.16*d* the collision-frequency ν exceeds the critical value ω_c so that $Z > Z_c$ and the curves could be formed by continuous

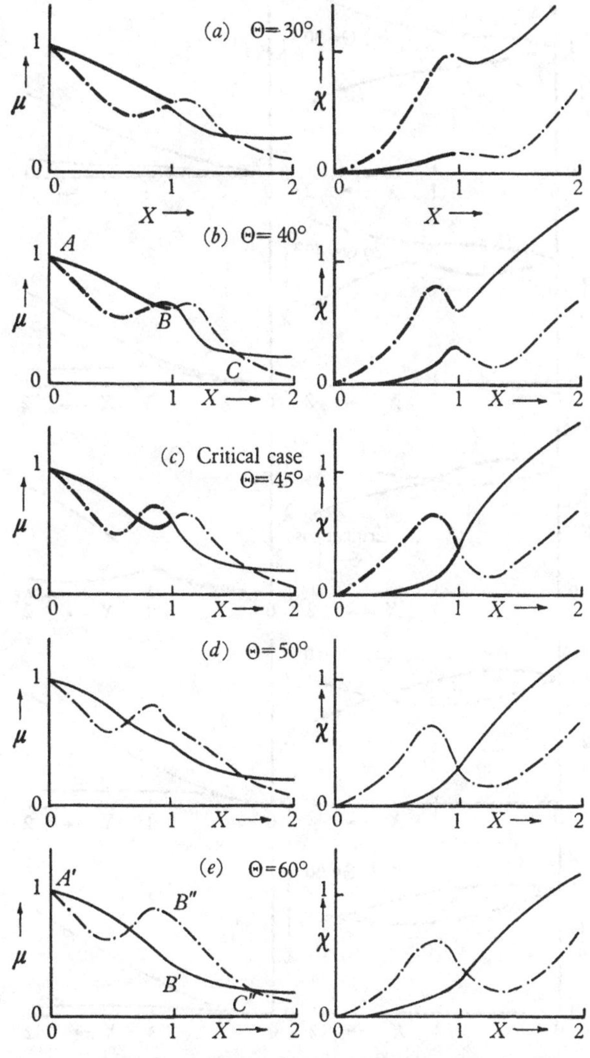

Fig. 6.13. Variation of μ and χ with X for different inclinations, Θ, of the wave-normal to the earth's magnetic field when Z is a constant. In this example $Y = \frac{1}{2}$ and $Z = 0{\cdot}18$.

deformation of the curves in Figs. 6.9 and 6.10 for longitudinal propagation.

The phenomenon just described leads to the ambiguity of nomenclature already mentioned in §6.13. Consider the curve ABC in Fig. 6.13*b*, for which $Y_T^2 < 2Y_L Z$. As Y_T is increased, the part AB of the curve deforms continuously into the part $A'B'$ of one curve in Fig. 6.13*e* for the nearly transverse case. This curve is for the ordinary wave, and so

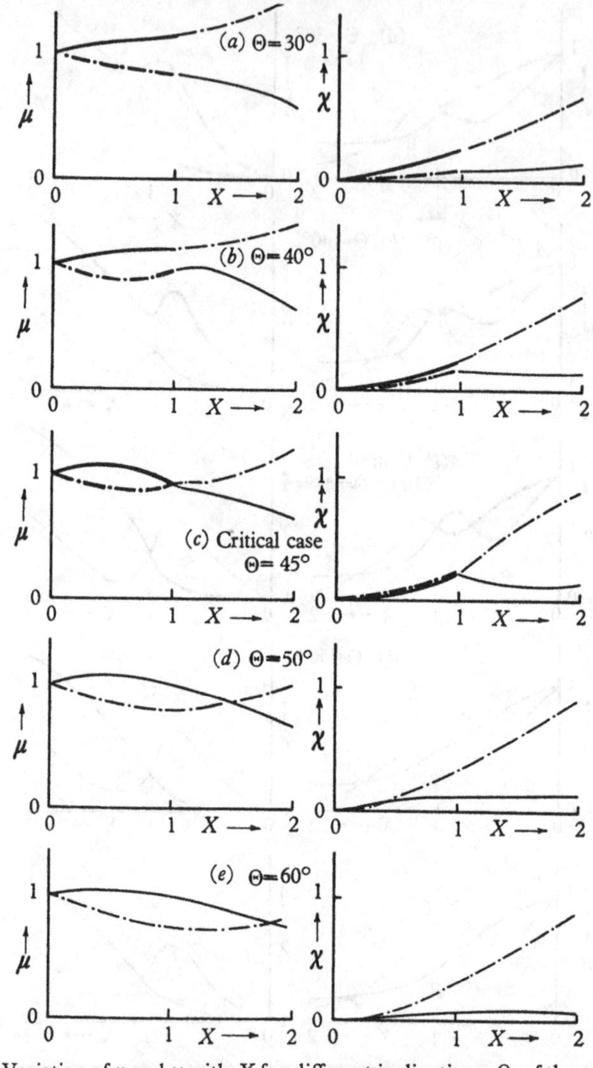

Fig. 6.14. Variation of μ and χ with X for different inclinations, Θ, of the wave-normal to the earth's magnetic field when Z and Y are constant. In this example $Z = 0{\cdot}707$, $Y = 2$.

the original segment AB in Fig. 6.13*b* should be assigned to the ordinary wave, in accordance with the definition given in § 6.6. Now consider the part BC of the curve in Fig. 6.13*b*. As Y_T increases, this part deforms continuously into the part $B''C''$ of one curve in Fig. 6.13*e* (nearly transverse case), but this is now the curve for the extraordinary wave. The definition of § 6.6 would therefore require that in Fig. 6.13*b* (for

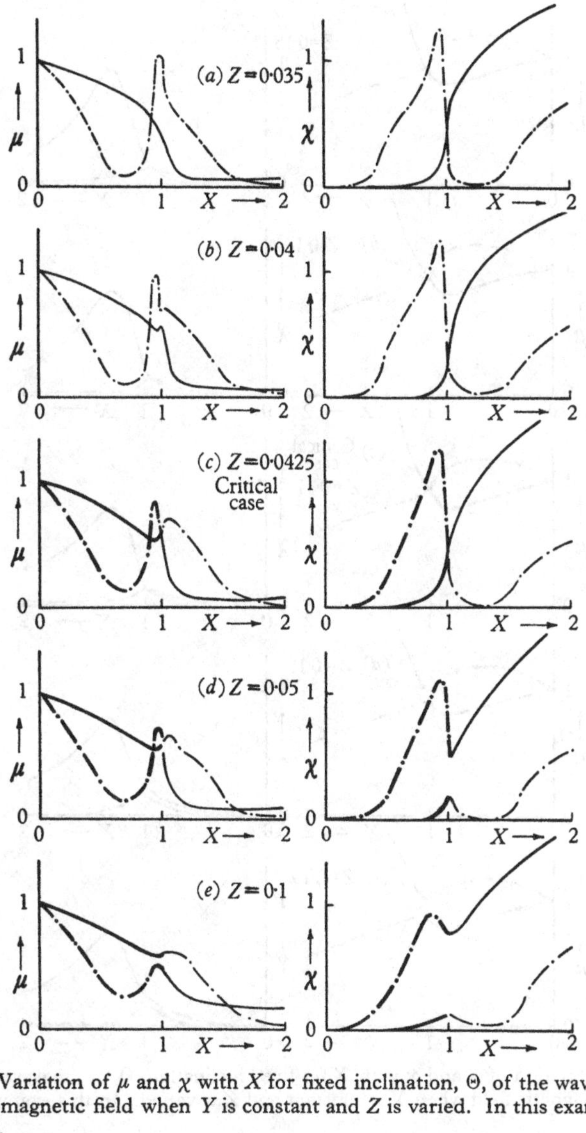

Fig. 6.15. Variation of μ and χ with X for fixed inclination, Θ, of the wave-normal to the earth's magnetic field when Y is constant and Z is varied. In this example $Y = \frac{1}{2}$, $\Theta = 23^\circ 16'$.

$Y_T^2 < 2Y_L Z$) the segment AB is assigned to the ordinary wave, and the segment BC to the extraordinary wave, even though these two segments form a continuous curve. Some authors, however, prefer to use the same terms for the whole curve. This may be either 'ordinary' or 'extraordinary' depending on whether the usual definition is made to apply for $X < 1$ or $X > 1$.

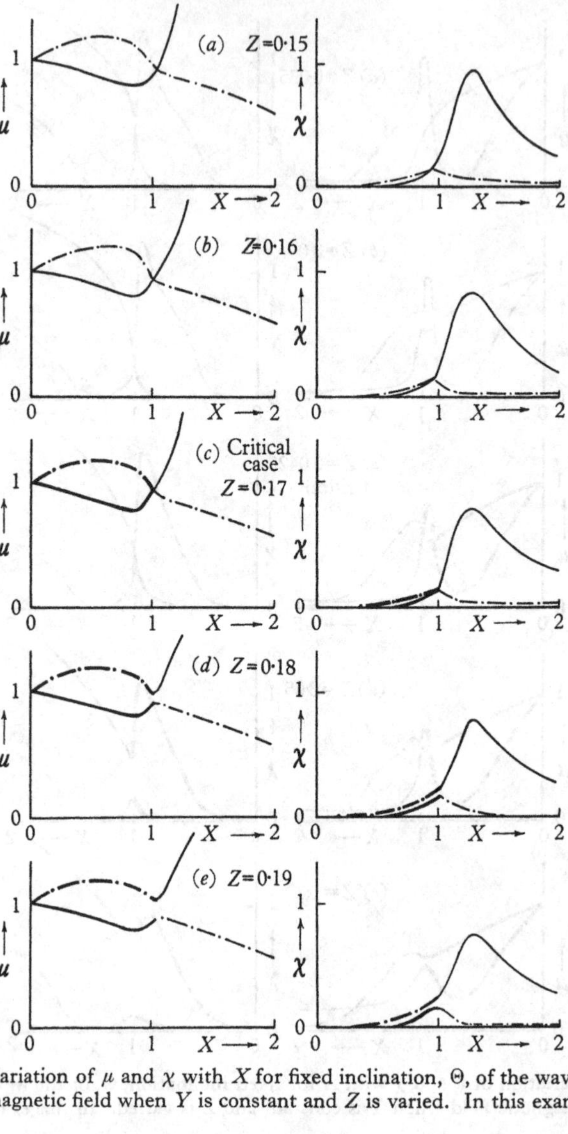

Fig. 6.16. Variation of μ and χ with X for fixed inclination, Θ, of the wave-normal to the earth's magnetic field when Y is constant and Z is varied. In this example $Y = 2$, $\Theta = 23° 16'$.

Curves similar to those of Figs. 6.13 to 6.16 have been given by Booker (1934), Goubau (1935 *a*, *b*) and Ratcliffe (1959).

6.16 The 'quasi-longitudinal' approximation

The full Appleton–Hartree formula (6.1) is so complicated that it is laborious to use it for calculating the refractive index n. There are two cases, however, when it can be simplified by making approximations. The first is when

$$|Y_T^2/2Y_L| \ll |1-X-iZ|. \tag{6.23}$$

From (5.13) and (5.35) this leads to

$$\rho = \mp i, \quad \mathfrak{n}^2 = 1 - \frac{X}{1-iZ \pm Y_L}. \tag{6.24}$$

($\rho = +i$ gives the minus sign in the second equation, and $\rho = -i$ gives the plus sign. Y_L is a positive number for upgoing waves in the northern hemisphere.) This is the same as for the purely longitudinal case (6.19), except that Y is replaced by Y_L. Hence the condition (6.23) is called the 'quasi-longitudinal' approximation. It has been fully discussed by Booker (1935) and Ratcliffe (1959). From (5.37) the longitudinal component of the electric field is given by

$$\frac{E_z}{E_x} = \frac{\pm Y_T X}{(1-iZ \pm Y_L)(1-iZ-X)}. \tag{6.25}$$

An example of the use of the quasi-longitudinal approximation is given in §8.15.

6.17 The 'quasi-transverse' approximation

The second case when an approximate form of the Appleton–Hartree formula can be used is when

$$|Y_T^2/2Y_L| \gg |1-X-iZ|. \tag{6.26}$$

For this case (5.13) and (5.35) give for the ordinary wave:

$$\rho = \frac{-iY_L(1-X-iZ)}{Y_T^2}, \quad \mathfrak{n}^2 = 1 - \frac{X}{1-iZ+(1-X-iZ)\,Y_L^2/Y_T^2}, \tag{6.27}$$

and (5.37) gives for the longitudinal component of the electric field

$$\frac{E_z}{E_x} = \frac{XY_LY_T}{Y_T^2(1-iZ)+Y_L^2(1-X-iZ)}. \tag{6.28}$$

Equation (6.26) shows that the values of ρ and $\mathfrak{n}^2$ given by (6.27) are very close to those in (6.20) for the purely transverse case. Hence the condition (6.26) is called the 'quasi-transverse' approximation. In particular the value of ρ is very small, so that the wave is almost linearly polarised with the electric field in the magnetic meridian.

For the extraordinary wave, (5.13) and (5.35) give

$$\rho = \frac{iY_T^2}{Y_L(1-X-iZ)}, \quad n^2 = 1 - \frac{X(1-X-iZ)}{(1-iZ)(1-X-iZ)-Y_T^2}. \tag{6.29}$$

These also are close to the values (6.21) for the purely transverse case. Equation (5.36) gives for the longitudinal component of the electric field

$$\frac{E_z}{E_y} = \frac{iY_T X}{(1-iZ)(1-X-iZ)-Y_T^2}. \tag{6.30}$$

If we put $X = 1$ in (6.26), the condition for the quasi-transverse approximation to be valid is $Z \ll Z_c$ or $\nu \ll \omega_c$ where Z_c and ω_c are as defined in § 5.4. For waves vertically incident on the ionosphere in England, the condition $\nu = \omega_c$ is believed to hold at about 95 km above the earth's surface. For greater heights than this, $\nu < \omega_c$, and in the E- and F-regions the quasi-transverse approximation would be expected to apply when $X = 1$. This occurs at the level of reflection for the ordinary wave, for frequencies above about 1 Mc/s.

6.18 The effect of heavy ions

The effect of heavy ions was discussed in § 3.9, where subscripts e and i were used to indicate quantities which refer to electrons and heavy ions respectively. This notation will be used again here. In § 5.9 the results were used to find the effect of heavy ions on the polarisation ρ and refractive index n, and it was shown ((5.56) and (5.59)) that:

$$\rho^2 - \frac{iY_T^2}{Y_L}\frac{U_i - X_i}{U_e(U_i-X_i)-U_iX_e}\rho + 1 = 0, \tag{6.31}$$

$$n^2 = 1 - \frac{X_i}{U_i} - \frac{X_e}{U_e + iY_L\rho}. \tag{6.32}$$

Equation (6.31) replaces the quadratic equation (5.12) for the polarisation, and (6.32) replaces the Appleton–Hartree formula (6.1). Some properties of the formulae (6.31) and (6.32) will now be given.

The refractive index (6.32) is zero when

$$-iY_L\rho = U_e - \frac{X_e}{1-X_i/U_i}. \tag{6.33}$$

If the right side of (6.33) is zero, the denominator of the middle term of (6.31) is zero, and therefore one value of ρ is zero. This satisfies (6.33) and hence one value of n^2 is zero when

$$\frac{X_i}{U_i} + \frac{X_e}{U_e} = 1. \tag{6.34}$$

This leads to (6.2) when X_i is zero. The other zeros of $\mathfrak{n}^2$ can be found by substituting the value of ρ from (6.33) into (6.31). This gives

$$U_e - \frac{X_e}{1 - X_i/U_i} = \pm Y, \tag{6.35}$$

which leads to (6.3) and (6.4) when X_i is zero. The three zeros given by (6.34) and (6.35) are independent of the angle between the wave-normal and the earth's magnetic field.

The refractive index is infinite when the denominator in (6.32) is zero, that is when

$$\rho = -\frac{U_e}{iY_L}. \tag{6.36}$$

When this is substituted in (6.31) it leads to

$$\frac{X_e/U_e}{1 - X_i/U_i} = \frac{U_e^2 - Y^2}{U_e^2 - Y_L^2}, \tag{6.37}$$

which should be compared with (6.5) to which it reduces when $X_i = 0$.

For purely longitudinal propagation $Y_T = 0$, and the two values of ρ are $\pm i$. Then the two values of $\mathfrak{n}$ are given by

$$\mathfrak{n}^2 = 1 - \frac{X_i}{U_i} - \frac{X_e}{U_e \mp Y}. \tag{6.38}$$

According to (6.34) and (6.37) there should be both a zero and an infinity where $X_i/U_i + X_e/U_e = 1$, but these are not displayed by the formula (6.38). This is because purely longitudinal propagation shows a degeneracy. This phenomenon was discussed in § 6.6 for the case when there were no heavy ions.

For purely transverse propagation $Y_L = 0$, and the two values of ρ are 0 and ∞. When $\rho = 0$, the value of $\mathfrak{n}$ is given by

$$\mathfrak{n}^2 = 1 - \frac{X_i}{U_i} - \frac{X_e}{U_e}, \tag{6.39}$$

which is the value it would have if the earth's magnetic field were neglected. This is because the electric field is parallel to the earth's magnetic field, and the electrons and ions therefore move only parallel to the earth's field and are unaffected by it. The expression (6.39) has one zero given by (6.34), and no infinity.

When $\rho = \infty$, the last term of (6.32) is not zero because the product $Y_L\rho$ is indeterminate. We therefore multiply (6.31) by $(iY_L)^2$ which gives a quadratic equation for $iY_L\rho$. When Y_L tends to zero, the solutions are $(iY_L\rho) = 0$, and $-Y^2(U_i - X_i)/\{U_e(U_i - X_i) - U_i X_e\}$. Substitution of the last of these in (6.32) gives

$$\mathfrak{n}^2 = 1 - \frac{X_i}{U_i} - \frac{X_e\{U_e(U_i - X_i) - U_i X_e\}}{U_e\{U_e(U_i - X_i) - U_i X_e\} - Y^2(U_i - X_i)}. \tag{6.40}$$

This reduces to (6.21) when $X_i = 0$. It has zeros given by (6.35), and an infinity given by (6.37) with $Y_L = 0$.

It has already been mentioned that when $X_i = X_e$ the heavy ions must be about 60 000 times more numerous than electrons. Hence heavy ions are not likely to play any important part in the propagation of radio waves except possibly in the very lowest parts of the ionosphere, and here the collision-frequencies of both ions and electrons with other particles are probably high.

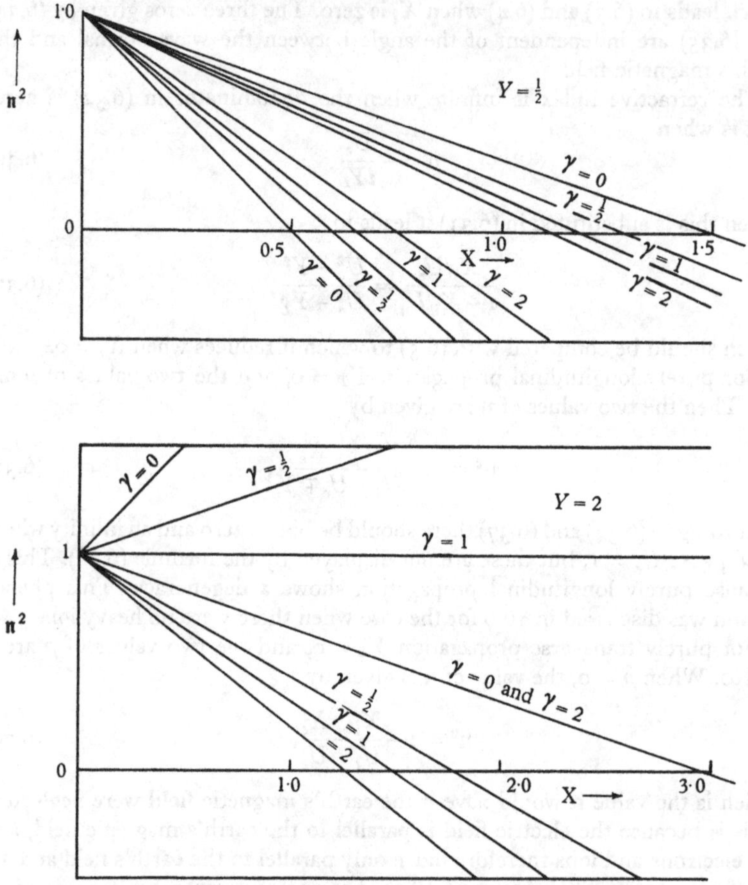

Fig. 6.17. Variation of $\mathfrak{n}^2$ with X for purely longitudinal propagation when heavy ions are present and collisions are neglected. $\gamma = X_i/X_e$.

This means that Z_i and Z_e cannot be neglected at levels where the formulae (6.31) and (6.32) are used. In spite of this it is of interest to discuss the properties of the formulae when Z_i and Z_e are both zero, for in this way curves can be plotted showing how $\mathfrak{n}^2$ depends on X_e and X_i and they can be compared with the corresponding curves of Figs. 6.1 to 6.8 for electrons alone.

There are now two independent variables X_e and X_i, and it is convenient to assume that their ratio is constant, so that

$$X_i = \gamma X_e. \tag{6.41}$$

We also take
$$X = X_e + X_i. \tag{6.42}$$
The figures show how n^2 depends on X for different values of γ.

For purely longitudinal propagation the formula (6.38) is used, and it reduces to
$$n^2 = 1 - X\left\{\frac{\gamma}{1+\gamma} + \frac{1}{(1+\gamma)(1 \mp Y)}\right\}, \tag{6.43}$$
which gives two straight lines intersecting the axis $n^2 = 0$ where
$$X = \frac{(1+\gamma)(1 \pm Y)}{1+\gamma(1 \pm Y)}. \tag{6.44}$$
Fig. 6.17 shows these lines for $Y = 2$, $Y = \frac{1}{2}$, and for $\gamma = 0, \frac{1}{2}, 1, 2$.

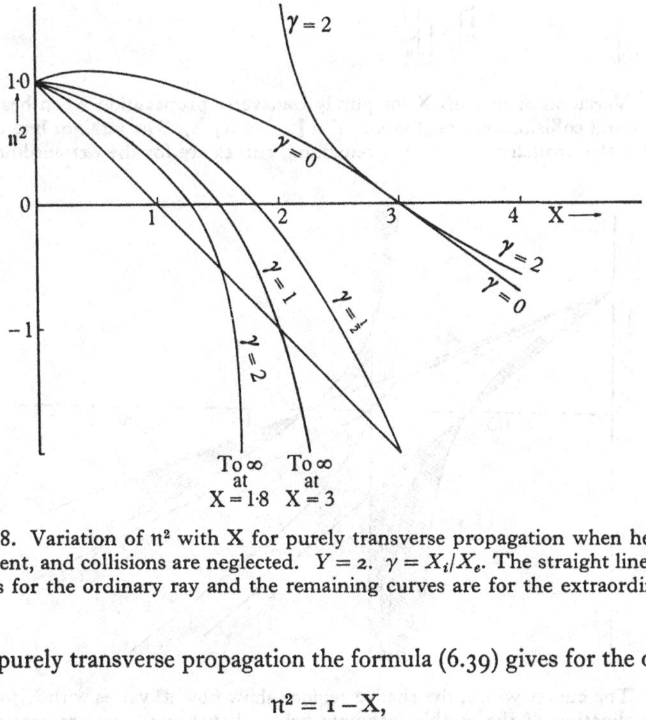

Fig. 6.18. Variation of n^2 with X for purely transverse propagation when heavy ions are present, and collisions are neglected. $Y = 2$. $\gamma = X_i/X_e$. The straight line through $X = 1$ is for the ordinary ray and the remaining curves are for the extraordinary ray.

For purely transverse propagation the formula (6.39) gives for the ordinary wave
$$n^2 = 1 - X, \tag{6.45}$$
which is the same for all values of γ and represents a straight line with a zero where $X = 1$. For the extraordinary wave the formula (6.40) reduces to
$$n^2 = 1 - \frac{\gamma X}{1+\gamma} - \frac{X(1-X)}{(1+\gamma)(1-X) - Y^2(1+\gamma-\gamma X)}, \tag{6.46}$$
which has zeros where
$$X = \frac{1 \pm Y}{1 \pm Y\gamma/(1+\gamma)}, \tag{6.47}$$

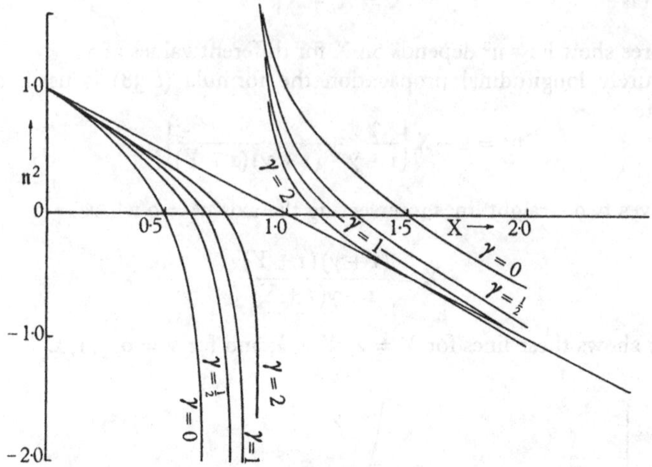

Fig. 6.19. Variation of $\mathfrak{n}^2$ with X for purely transverse propagation when heavy ions are present and collisions are neglected. $Y = \frac{1}{2}$. $\gamma = X_i/X_e$. The straight line through $X = 1$ is for the ordinary ray and the remaining curves are for the extraordinary ray.

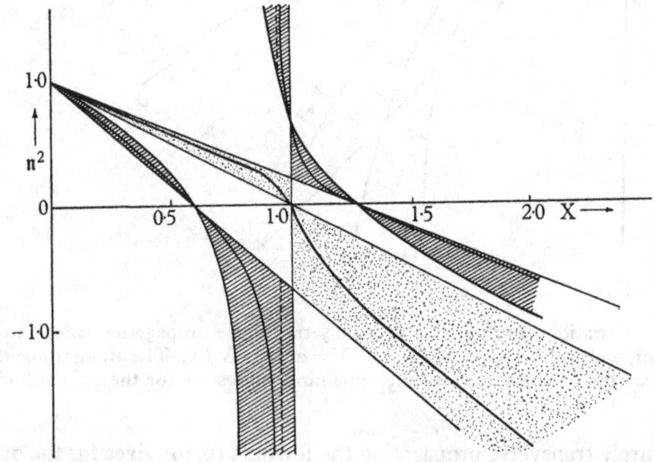

Fig. 6.20. The curves within the shaded regions show how $\mathfrak{n}^2$ varies with X for intermediate inclination of the earth's magnetic field, when heavy ions are present, and collisions are neglected. $Y = \frac{1}{2}$. $\gamma = X_i/X_e = \frac{1}{2}$. The earth's magnetic field is at 30° to the wave normal.

and an infinity where
$$X = \frac{(1-Y^2)(1+\gamma)}{1+\gamma(1-Y^2)}. \tag{6.48}$$

Curves showing how $\mathfrak{n}^2$ depends on X for this case are given with $\gamma = 0, \frac{1}{2}, 1, 2$, in Figs. 6.18 and 6.19, for $Y = 2$, $Y = \frac{1}{2}$ respectively. It should be noticed that when $\gamma \neq 0$, the two values of n^2 given by (6.45) and (6.46) are equal when

$$X = \frac{1+\gamma}{\gamma}. \tag{6.49}$$

This does not occur for any finite value of X when only electrons are effective.

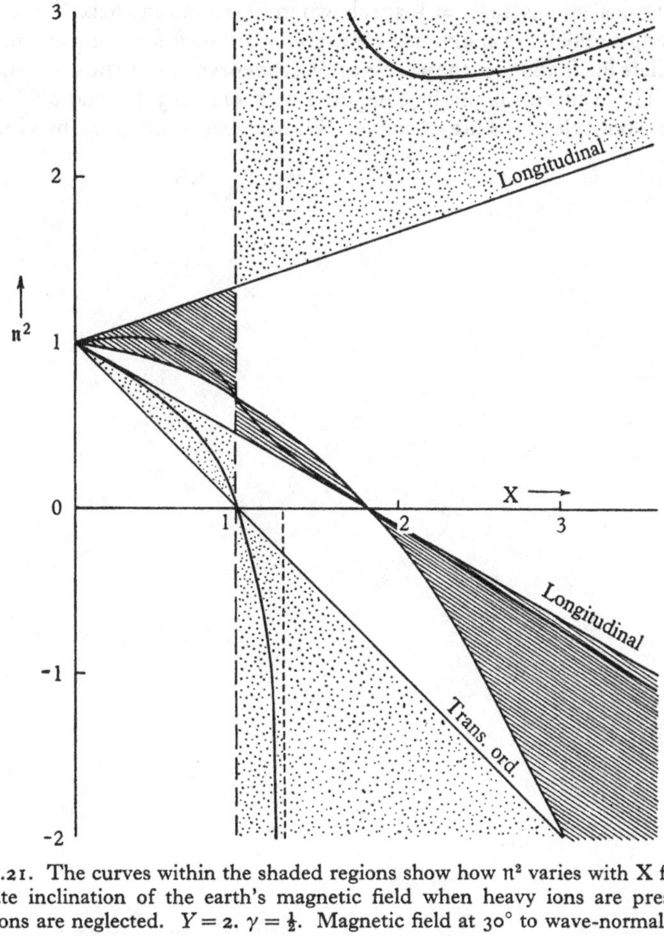

Fig. 6.21. The curves within the shaded regions show how n^2 varies with X for intermediate inclination of the earth's magnetic field when heavy ions are present and collisions are neglected. $Y = 2$. $\gamma = \frac{1}{2}$. Magnetic field at 30° to wave-normal.

For the more general case when the angle between the wave-normal and the earth's magnetic field is not zero or 90°, the formulae (6.31) and (6.32) are used with $Z_i = Z_e = 0$. They become:

$$\rho^2 - \frac{iY_T^2}{Y_L} \frac{1 - \frac{\gamma}{1+\gamma}X}{1-X} \rho + 1 = 0, \tag{6.50}$$

$$n^2 = 1 - \frac{\gamma}{1+\gamma}X - \frac{X}{(1+\gamma)(1+iY_L\rho)}. \tag{6.51}$$

Now $\mathfrak{n}^2$ has three zeros, given by (6.47) and by $X = 1$. When $X = 1$, the other value of $\mathfrak{n}^2$ is $1/(1+\gamma)$. There is an infinity where

$$X = \frac{(1+\gamma)(1-Y^2)}{1-Y_L^2+\gamma(1-Y^2)}. \tag{6.52}$$

Some typical curves, for $\gamma = \frac{1}{2}$, are shown in Figs. 6.20 and 6.21. The curves for the longitudinal and transverse cases are also shown for comparison. For intermediate inclinations of the wave-normal, the curves always lie in the shaded regions. These curves should be compared with Figs. 6.3, 6.7 and 6.8 for the case when only electrons are effective. Similar curves are given by Goubau (1935*a*).

CHAPTER 7

DEFINITION OF THE REFLECTION AND TRANSMISSION COEFFICIENTS

7.1 Introduction

In the preceding chapters a Cartesian coordinate system was used in which the z-axis was the direction of the wave-normal of a plane wave. The following chapters deal with the reflection of radio waves from the ionosphere, and it is convenient to use a different Cartesian coordinate system in which the z-axis is vertically upwards so that the x- and y-axes are horizontal. The two systems are different except for waves travelling vertically upwards.

The ionosphere is assumed to be horizontally stratified, which means that the electron number density $N(z)$ and collision frequency $\nu(z)$ are functions only of the vertical coordinate z. The origin is often taken to be at the ground so that z is height above the earth's surface. Sometimes, however, it is convenient to take the origin of z at another level.

It is assumed that the incident wave is a plane wave travelling upwards, in general obliquely, and the direction of the x-axis is chosen so that the wave-normal is in the x–z-plane and is pointing in the positive directions of both x and z, at an angle θ_I to the z-axis. Then the x–z-plane is called the 'plane of incidence'. If F is any field component of the incident wave then in the free space below the ionosphere

$$F = F_1 e^{-ik(Cz+Sx)}, \tag{7.1}$$

where $S = \sin\theta_I$, $C = \cos\theta_I$, and F_1 is a constant which is in general complex. This wave gives rise to a reflected wave below the ionosphere and possibly to a transmitted wave above it. The main problem in this book is to find the reflection and transmission coefficients, and the purpose of this chapter is to set out their definitions.

7.2 The reference-level for reflection coefficients

Consider, first, a wave vertically incident on the ionosphere, so that $\theta_I = 0$, and let some field component be given by

$$F = F_1 e^{-ikz}, \tag{7.2}$$

where z is height above the ground, and F_1 is a complex constant. This gives rise to a reflected wave travelling downwards for which

$$F = F_2 e^{ikz}, \tag{7.3}$$

where F_2 is another complex constant. If the ratio of these two field components is measured at the ground the result is

$$R_0 = F_2/F_1, \tag{7.4}$$

and the complex number R_0 is called the reflection coefficient 'at the ground' or 'with the ground as reference-level', or 'referred to the level $z = 0$'. (The particular field component F is specified later, § 7.4.)

If the ratio were measured at some level $z = z_1$, the result would be

$$R_1 = F_2 e^{ikz_1}/F_1 e^{-ikz_1} = R_0 e^{2ikz_1}, \tag{7.5}$$

where R_1 is the reflection coefficient referred to the level $z = z_1$. Thus the effect of changing the reference-level is to alter the argument but not the modulus of the reflection coefficient. Similarly, the reflection coefficient referred to some other level z_2 is

$$R_2 = R_0 e^{2ikz_2} = R_1 e^{2ik(z_2 - z_1)}. \tag{7.6}$$

This gives the rule for changing the reference-level.

The downgoing wave (7.3) may arise by reflection in an ionosphere in which N varies gradually with height and at a level, $z = z_3$ say, where N is appreciably different from zero, the upgoing and downgoing waves are no longer given by (7.2) and (7.3). Yet it is still possible to use the level $z = z_3$ as the reference-level, and to write

$$R_3 = F_2 e^{ikz_3}/F_1 e^{-ikz_3} = R_0 e^{2ikz_3}. \tag{7.7}$$

Here R_3 is the reflection coefficient referred to the level $z = z_3$ and is calculated *as though this level were in free space*, even though the ratio in (7.7) could never be observed.

For many purposes the ground is the best reference-level since this is where reflection coefficients are measured. But in some cases other levels are more convenient. For example, in the next chapter where the ionosphere is assumed to be a sharply bounded homogeneous medium, the reference-level is taken at the boundary.

Suppose, next, that the incident wave is oblique so that F is given by (7.1). Then the reflected wave is

$$F = F_2 e^{ik(Cz - Sx)}. \tag{7.8}$$

Both incident and reflected waves depend on x through the same factor e^{-ikSx} (see § 9.10). The reflection coefficient is defined to be the ratio of these fields *measured at the same point*, whose coordinates are (x, z_1), say, so that

$$R_1 = \frac{F_2}{F_1} e^{2ikCz_1}. \tag{7.9}$$

Thus R_1 is independent of the horizontal coordinate x, but it is still necessary to specify the reference-level for the vertical coordinate. If this is changed to a new level z_2 then the new reflection coefficient R_2 is

$$R_2 = R_1 e^{2ikC(z_2 - z_1)}. \tag{7.10}$$

If C is real, the effect of changing the reference-level is to alter the argument but not the modulus of the reflection coefficient. Problems sometimes arise in which C is complex, for example, when the space between the earth and the ionosphere is treated as a wave-guide (Budden, 1951 *a*, 1957). The incident and reflected waves are then inhomogeneous plane waves (§ 4.7) and changing the reference-level can change both the modulus and argument of the reflection coefficient.

In observations at oblique incidence the incident and reflected waves are usually measured at different points on the earth's surface. Suppose that these are separated by a horizontal distance D. Then the ratio of the fields in the reflected and incident waves at the ground is

$$R_D = F_2 e^{-ikSD}/F_1 = R_0 e^{-ikSD}. \tag{7.11}$$

7.3 The reference-level for transmission coefficients

Suppose that there is free space above the ionosphere and that the wave (7.1) incident from below gives rise to a wave above the ionosphere whose field component F is

$$F = F_3 e^{-ik(Cz+Sx)}. \tag{7.12}$$

These two waves could never be observed at the same point because the ionosphere intervenes between them. The field (7.12), however, can be extrapolated backwards to some level $z = z_1$ below the top of the ionosphere, as though the wave were travelling in free space. Similarly, the field (7.1) can be extrapolated upwards to the same level $z = z_1$. The transmission coefficient T is defined to be the ratio of the fields (7.12) to (7.1) extrapolated in this way to the same point. Thus

$$T = F_3/F_1. \tag{7.13}$$

This ratio is independent of x and z, so that in this example it is unnecessary to specify a reference-level for the transmission coefficient.

In some problems, however, there is, above the ionosphere, a homogeneous medium which is not free space. Then the transmitted field is no longer given by (7.12) but by

$$F = F_3 e^{-ik(qz+Sx)}, \tag{7.14}$$

where q is a constant different from C. This field is now extrapolated down to the level $z = z_1$, as though the wave were travelling in the same homogeneous medium throughout, and the transmission coefficient is

$$T = \frac{F_3 e^{-ikqz_1}}{F_1 e^{-ikCz_1}} = \frac{F_3}{F_1} e^{ik(C-q)z_1}. \tag{7.15}$$

Since this depends on z_1, the reference level must now be specified. Often q is complex even when C is real, so that a change of reference-level changes both the modulus and argument of T.

7.4 The four reflection coefficients and the four transmission coefficients

The most important reflection and transmission coefficients are those deduced when the incident wave is linearly polarised with its electric vector either parallel to the plane of incidence or perpendicular to it. The reflected wave, in general, is elliptically polarised, but may be resolved into linearly polarised components whose electric vectors also are parallel and perpendicular to the plane of incidence. It is convenient to introduce four coefficients denoted by ${}_{\parallel}R_{\parallel}$, ${}_{\parallel}R_{\perp}$, ${}_{\perp}R_{\parallel}$, ${}_{\perp}R_{\perp}$ to indicate the complex ratio of a specified electric field in the wave after reflection to a specified electric field in the wave before reflection. The first subscript denotes whether the electric field specified in the incident wave is parallel ($\parallel$) or perpendicular ($\perp$) to the plane of incidence, and the second subscript refers in the same way to the electric field in the reflected wave. In an exactly similar way the four transmission coefficients ${}_{\parallel}T_{\parallel}$, ${}_{\parallel}T_{\perp}$, ${}_{\perp}T_{\parallel}$, ${}_{\perp}T_{\perp}$ are introduced to indicate the ratios of the electric field components in the transmitted wave to those in the incident wave.

7.5 The sign convention

To complete the definition of the reflection coefficients it is necessary to adopt some convention with regard to sign. Let the incident wave be linearly polarised with its electric vector parallel to the plane of incidence (the x–z-plane), so that the magnetic vector is parallel to the y-axis. Then the electric vector is inclined at an angle θ_I to the x-axis. At the instant when the magnetic vector has its maximum positive value, the electric vector points in a direction between the positive x-axis and the negative z-axis, that is, away from the ionosphere. Now consider that component of the reflected wave whose electric field is in the x–z-plane. When the magnetic field of this component has its maximum positive value, the electric vector points in a direction between the negative x-axis and the negative z-axis, that is, again, away from the ionosphere. Hence the following sign convention is adopted: the electric field components parallel to the plane of incidence in the incident and reflected waves are taken to be positive when they point obliquely downwards away from the ionosphere. That of the transmitted wave is taken to be positive when it points obliquely downwards towards the ground. This sign convention can be used for normal incidence by considering this as a limiting case of slightly oblique incidence. The electric field components perpendicular to the plane of incidence, in all three waves, are taken to be positive when they point in the direction of positive y.

This sign convention means, for example, that if ${}_{\|}R_{\|} = 1$ at normal incidence, then the electric vectors in the incident and reflected waves have opposite directions, and the magnetic vectors have the same direction. On the other hand, if ${}_{\perp}R_{\perp} = 1$ at normal incidence, the electric vectors in the incident and reflected waves have the same direction, but the magnetic vectors have opposite directions.

It is sometimes convenient to use alternative forms of the definitions of the reflection and transmission coefficients. The superscripts (I), (R), (T) will be used to denote the incident, reflected, and transmitted waves respectively. The equations (4.1) and (4.2) give the relations between the electric and magnetic fields in a progressive wave, and by using them it can be shown that the above definitions are equivalent to the following, where $\mathfrak{n} = 1$ below the ionosphere, and $\mathfrak{n} = \mathfrak{n}_2$ above it:

Let $E_y^{(I)} = \mathscr{H}_x^{(I)} = 0$. Then

$$\left.\begin{aligned} {}_{\|}R_{\|} &= \frac{\mathscr{H}_y^{(R)}}{\mathscr{H}_y^{(I)}}, & {}_{\|}R_{\perp} &= \frac{E_y^{(R)}}{\mathscr{H}_y^{(I)}}, \\ {}_{\|}T_{\|} &= \frac{\mathscr{H}_y^{(T)}}{\mathfrak{n}_2 \mathscr{H}_y^{(I)}}, & {}_{\|}T_{\perp} &= \frac{E_y^{(T)}}{\mathscr{H}_y^{(I)}}. \end{aligned}\right\} \tag{7.16}$$

Let $E_x^{(I)} = \mathscr{H}_y^{(I)} = 0$. Then

$$\left.\begin{aligned} {}_{\perp}R_{\|} &= \frac{\mathscr{H}_y^{(R)}}{E_y^{(I)}}, & {}_{\perp}R_{\perp} &= \frac{E_y^{(R)}}{E_y^{(I)}}, \\ {}_{\perp}T_{\|} &= \frac{\mathscr{H}_y^{(T)}}{\mathfrak{n}_2 E_y^{(I)}}, & {}_{\perp}T_{\perp} &= \frac{E_y^{(T)}}{E_y^{(I)}}. \end{aligned}\right\} \tag{7.17}$$

7.6 The reflection coefficient matrix

The four coefficients ${}_{\|}R_{\|}$, ${}_{\|}R_{\perp}$, ${}_{\perp}R_{\|}$, ${}_{\perp}R_{\perp}$ are sufficient to determine completely the reflecting properties of the ionosphere for an incident wave of any polarisation, because the incident wave can be resolved into two linearly polarised components with electric vectors parallel and perpendicular to the plane of incidence, and the reflected waves produced by these components can be found separately and then combined. This process can be expressed very simply as a matrix product as follows:

Let the components of the electric field in the incident wave be denoted by $E_{\|}^{(I)}$, $E_{\perp}^{(I)}$, so that $\mathscr{H}_y^{(I)} = E_{\|}^{(I)}$ and $E_y^{(I)} = E_{\perp}^{(I)}$. These may be arranged as a column matrix of two elements $\mathbf{e}^{(I)}$ and the corresponding fields in the reflected wave may be denoted by a column matrix $\mathbf{e}^{(R)}$ so that

$$\mathbf{e}^{(I)} = \begin{pmatrix} E_{\|}^{(I)} \\ E_{\perp}^{(I)} \end{pmatrix}, \quad \mathbf{e}^{(R)} = \begin{pmatrix} E_{\|}^{(R)} \\ E_{\perp}^{(R)} \end{pmatrix}. \tag{7.18}$$

Similarly, the reflection coefficients may be written as a 2×2 matrix thus:

$$\mathbf{R} = \begin{pmatrix} {}_{\|}R_{\|} & {}_{\perp}R_{\|} \\ {}_{\|}R_{\perp} & {}_{\perp}R_{\perp} \end{pmatrix}, \tag{7.19}$$

which is called the 'reflection coefficient matrix'.

Then it follows at once from the definitions of the R's that

$$\mathbf{e}^{(R)} = \mathbf{R}\mathbf{e}^{(I)}. \tag{7.20}$$

If a wave is produced by two successive reflection processes whose reflection coefficient matrices are $\mathbf{R}_1$, $\mathbf{R}_2$, then it follows from (7.20) that the fields in the second reflected wave are given by $\mathbf{R}_2\mathbf{R}_1\mathbf{e}^{(I)}$. The resultant reflection coefficient matrix is the matrix product of the matrices for the separate reflections. Its value depends on the order in which the reflections occur. This result can clearly be extended to any number of reflections.

7.7 Alternative forms of the reflection coefficients

Instead of resolving the incident and reflected waves into linearly polarised components, they might be resolved into components in a different way. For example, they could be resolved into right-handed and left-handed circularly polarised components. Four new reflection coefficients ${}_rR_r$, ${}_rR_l$, ${}_lR_r$, ${}_lR_l$ are then defined as before. The first subscript denotes whether the electric field in the incident wave is right-handed (r) or left-handed (l) circularly polarised, and the second subscript refers in the same way to the reflected wave. The four new coefficients form a matrix

$$\mathbf{R}_0 = \begin{pmatrix} {}_rR_r & {}_lR_r \\ {}_rR_l & {}_lR_l \end{pmatrix}, \tag{7.21}$$

which can be derived by a transformation of the original matrix **R**, as follows:

Let the electric fields in the right-handed and left-handed circularly polarised components of the incident wave be $E_r^{(I)}$, $E_l^{(I)}$, respectively, and let the corresponding components in the reflected wave be $E_r^{(R)}$, $E_l^{(R)}$. These may be written as the two column matrices

$$\mathbf{e}_0^{(I)} = \begin{pmatrix} E_r^{(I)} \\ E_l^{(I)} \end{pmatrix}, \quad \mathbf{e}_0^{(R)} = \begin{pmatrix} E_r^{(R)} \\ E_l^{(R)} \end{pmatrix}, \tag{7.22}$$

and from the definition of $\mathbf{R}_0$ $\quad \mathbf{e}_0^{(R)} = \mathbf{R}_0 \mathbf{e}_0^{(I)}$. (7.23)

Now let the incident wave be resolved into linearly polarised components. Then

$$E_\parallel^{(I)} = E_r^{(I)} + E_l^{(I)}, \quad E_\perp^{(I)} = i(E_r^{(I)} - E_l^{(I)}). \tag{7.24}$$

This can be written in matrix form

$$\mathbf{e}^{(I)} = 2^{\frac{1}{2}}\mathbf{U}\,\mathbf{e}_0^{(I)}, \tag{7.25}$$

where $\quad \mathbf{U} = 2^{-\frac{1}{2}}\begin{pmatrix} 1 & 1 \\ i & -i \end{pmatrix}.$ (7.26)

Similarly for the reflected wave

$$\mathbf{e}^{(R)} = 2^{\frac{1}{2}}\mathbf{U}\,\mathbf{e}_0^{(R)}. \tag{7.27}$$

Now substitute from (7.20), (7.23) and (7.25). Then

$$\mathbf{RU}\,\mathbf{e}_0^{(I)} = \mathbf{UR}_0\mathbf{e}_0^{(I)}. \tag{7.28}$$

This must be true for all possible pairs $\mathbf{e}_0^{(I)}$. Hence

$$\mathbf{R}_0 = \mathbf{U}^{-1}\mathbf{RU}. \tag{7.29}$$

The matrices **R** and $\mathbf{R}_0$ have other interesting properties, some of which are given in the exercises at the end of this chapter.

7.8 Spherical waves

For the definitions of the reflection coefficients given in the preceding sections it was assumed that the incident wave was a plane wave. In practice the incident wave originates at a source of small dimensions, namely the transmitting aerial, so that the wave-front is approximately spherical, but by the time it reaches the ionosphere the radius of curvature is so large that the wave is very nearly a plane wave. In treating it as exactly plane an approximation is used, whose validity must now be examined.

For the simplest transmitting aerials there is some component F of the radiated field for which

$$F = \frac{e^{-ikr}}{r}, \tag{7.30}$$

where r is radial distance from the aerial. For example, if the aerial is a Hertzian dipole, F is the component of the Hertz vector parallel to the dipole axis (see, for example, Stratton, 1941, ch. VIII). For more complicated aerials the expression (7.30) must be multiplied by a function of the polar coordinate angles, but it can be shown that the field can then be expressed as the sum of a number of terms, of which (7.30) is the first and is called the dipole term, and the next higher terms are the quadrupole terms, etc. (see, for example, Jeffreys and Jeffreys, 1956, §24.22). Each of these terms can be treated separately by the following method.

The expression (7.30) can be written as an integral, thus:

$$F = \frac{k}{2\pi i}\int_{-\frac{1}{2}\pi-i\infty}^{\frac{1}{2}\pi+i\infty}\int_{-\frac{1}{2}\pi-i\infty}^{\frac{1}{2}\pi+i\infty} \exp\left[-kr\{\sin\theta\sin u\cos(\phi-v) + \cos\theta\cos u\}\right]\sin u\,du\,dv, \tag{7.31}$$

where r, θ, ϕ are the polar coordinates of the point of observation, with the transmitting aerial as origin, and the coordinate $z = r\cos\theta$ is measured vertically upwards. The integrand represents a plane wave with its wave-normal in the direction $\theta = u$, $\phi = v$, and (7.31) therefore expresses the field as the sum of a doubly infinite set of plane waves. The integrals are contour integrals in the complex u and complex v planes, and since the limits are complex, the integrand includes inhomogeneous plane waves (§4.7) that is, waves whose normals are at complex angles to the real axes. Equation (7.31) is said to resolve the field into an 'angular spectrum of plane waves'. This kind of resolution is important in the theory of the propagation of radio waves over a finitely conducting earth (see, for example, Weyl, 1919; Booker and Clemmow, 1950; Stratton, 1941, p. 578).

Now each component plane wave in the integrand of (7.31) is reflected from the ionosphere, and gives another plane wave travelling obliquely downwards. The reflection coefficient R is in general a function of the direction u, v of the incident wave-normal. Suppose that it is referred to the level $r\cos\theta = z_1$ (§7.2). Then the reflected field is

$$F_R = \frac{k}{2\pi i}\iint R(u,v)\exp\left[-ikr\{\sin\theta\sin u\cos(\phi-v) - \cos\theta\cos u\} - 2ikz_1\cos u\right]\sin u\,du\,dv, \tag{7.32}$$

where the limits of integration are as in (7.31). It often happens that $R(u,v)$ is independent of the azimuth angle v, and is a function only of the angle of incidence u. Then the v integration in (7.32) may be effected approximately by the method of stationary phase (§15.19). The approximation is good provided $kr \gg 1$ which is always the case in practice. The phase is stationary where $v = \phi$, and the result is

$$F_R \sim (k/2\pi)^{\frac{1}{2}} e^{-i(\frac{1}{4}\pi)} r^{-\frac{1}{2}} \int R(u)\exp\{ikr\cos(\theta+u) - 2ikz_1\cos u\}\sin^{\frac{1}{2}} u \sin^{\frac{1}{2}}\theta\,du. \tag{7.33}$$

This field is to be observed at the ground, that is where $\theta = \frac{1}{2}\pi$. With this simplification (7.33) gives

$$F_R \sim (k/2\pi)^{\frac{1}{2}} e^{-i(\frac{1}{4}\pi)} r^{-\frac{1}{2}} \int R(u) \exp\{-ik(r \sin u + 2z_1 \cos u)\} \sin^{\frac{1}{2}} u \, du. \tag{7.34}$$

The further substitutions

$$\mathrm{r}_1 = (r^2 + 4z_1^2)^{\frac{1}{2}}, \quad \theta_1 = \arctan(r/2z_1) \tag{7.35}$$

give $$F_R \sim (k/2\pi)^{\frac{1}{2}} e^{-i(\frac{1}{4}\pi)} r^{-\frac{1}{2}} \int R(u) \exp\{-ik\mathrm{r}_1 \cos(u - \theta_1)\} \sin^{\frac{1}{2}} u \, du. \tag{7.36}$$

Now suppose first that $R(u)$ varies only very slowly with u. This could happen, for example, if there were a reflecting metal sheet at the reference level $z = z_1$. Then the behaviour of the integrand in (7.36) depends predominantly on the exponential, and the factor $R(u) \sin^{\frac{1}{2}} u$ may be treated as a constant. This integral can then be evaluated approximately by the method of stationary phase. The phase is stationary where $u = \theta_1$ and the value of the integral is

$$F_R \sim \frac{1}{\mathrm{r}_1} R(\theta_1) e^{-ik\mathrm{r}_1}. \tag{7.37}$$

Now r_1 (7.35) is simply the distance from the receiver to the image of the transmitter in the plane $z = z_1$. Hence the received field (7.37) is the field (spherical wave) of a transmitter at this image point, multiplied by the reflection coefficient $R(\theta_1)$ for a *plane* wave, incident at the angle θ_1 made by a line drawn from the receiver to the image point.

Next suppose that $R(u)$ depends on u through a factor

$$\exp\{-2ik(z_2 - z_1) \cos u\}$$

that is $$R(u) = R_1(u) \exp\{-2ik(z_2 - z_1) \cos u\}, \tag{7.38}$$

where $R_1(u)$ varies very slowly with u. This could happen, for example, if the reflecting metal sheet mentioned above were at the level $z = z_2$. If (7.38) is inserted in (7.36) and the integration carried out as before, the result is

$$F_R \sim \frac{1}{\mathrm{r}_2} R_1(\theta_2) e^{-ik\mathrm{r}_2}, \tag{7.39}$$

where r_2 is the distance from the receiver to the image of the transmitter in the plane $z = z_2$ and θ_2 is the corresponding angle of incidence. If $R(u)$ had been referred to the level z_2, the exponential factor would not appear in (7.38) and the result (7.39) would have been reached more simply.

These examples show that if $R(u)$ is a slowly varying function of u it is possible to find a reference-level for which the received field is given by (7.37). In such cases the use of a reflection coefficient $R(\theta_1)$ for a plane wave is justified, and the component plane wave with its normal at θ_1 to the vertical is called the 'predominant' plane wave. The required reference-level $z = z_1$ must satisfy the condition

$$\frac{\partial}{\partial u}\{(\arg R(u)\} = 0 \quad \text{when} \quad u = \theta_1. \tag{7.40}$$

The level z_1 is then called the 'equivalent height' of reflection. Alternative definitions are given in §§ 10.3 and 11.9, and it can be shown that they are effectively the same (Appleton, 1928, 1930).

It has been assumed that $|R(u)|$ is a slowly varying function of the angle u. This condition may fail if $R(u)$ has a zero or a branch point for some real value of u. These cases have been treated by various authors. Thus if $R(u_0)$ is zero, the angle u_0 is the Brewster angle and is an important feature in the theory of propagation of radio waves over a finitely conducting earth (see references earlier in this section). $R(u)$ may have a branch point at the critical angle and this case has been discussed by Ott (1942). In both cases a wave can be generated which is 'guided' by the reflecting surface. It is possible that the ionosphere can sustain such guided waves but the topic is beyond the scope of this book.

Sometimes the reflection coefficient $R(u, v)$ in (7.32) is not independent of the azimuth angle v. Then it is still possible to evaluate the integrals by the method of stationary phase, but the final formula (7.37) uses an image transmitter which is not at the geometrical image point. Examples where this occurs are given in ch. 13. They arise from 'lateral deviation' of a ray, and are more easily handled by the ray-tracing methods of §§ 13.14, 13.15 and ch. 14.

Examples

1. It is observed that a wave reflected from the ionosphere has the same polarisation no matter what the polarisation of the incident wave. Show that the reflection coefficient matrix **R** is singular, i.e. $\det \mathbf{R} = 0$, and find the polarisation of the reflected wave in terms of the elements of **R**. Show that for one polarisation of the incident wave there is no reflection.

2. The surface of the sea may be regarded as a perfect conductor, so that the horizontal components of the total electric field close to the surface are zero. Show that for all angles of incidence the reflection coefficient matrix at the surface as reference level, is

$$\mathbf{R} = \begin{pmatrix} 1 & 0 \\ 0 & -1 \end{pmatrix}.$$

What is the value of $\mathbf{R}_0$ (§ 7.7)?

3. The polarisation of the incident wave is to be adjusted so that the reflected wave has the same polarisation. Show that in general this can be done in two different ways. What condition does **R** satisfy when there is only one way?

Answer: $({}_{\parallel}R_{\parallel} - {}_{\perp}R_{\perp})^2 = -4\,{}_{\parallel}R_{\perp} \cdot {}_{\perp}R_{\parallel}$.

4. State Maxwell's electromagnetic equations in vector form for free space. Express them in terms of the components $E_x, E_y, E_z, H_x, H_y, H_z$ of the electric and magnetic fields respectively in a right-handed system of Cartesian coordinates x, y, z, for a plane electromagnetic wave travelling with its normal parallel to the z-axis.

A plane wave in free space is moving from left to right along the z-axis and is incident on the plane surface $z = a$, where a is positive, and is partially reflected there. In the plane $z = 0$ the components E_x, E_y, H_x, H_y of the total field are measured and found to have the amplitudes and phases given by the following table:

	Amplitude	Phase (in degrees)
E_x	13	22·6
E_y	5	0
$Z_0 H_x$	5	0
$Z_0 H_y$	13	22·6

Here Z_0 is the characteristic impedance of free space. Show that the incident wave is plane-polarised, and find its amplitude and the plane of its electric field. Find also the amplitude and polarisation of the reflected wave, and hence deduce the reflection coefficient.

(Natural Sciences Tripos, 1955. Part II, Physics: Theoretical Option.)

CHAPTER 8

REFLECTION AT A SHARP BOUNDARY

8.1 Introduction

For some purposes it is useful to assume that the ionosphere is a sharply bounded homogeneous medium. This idea has been particularly valuable in the theory of the propagation of very low frequency waves (see, for example, Barber and Crombie, 1959; Bremmer, 1949; Budden, 1951*b*; Rydbeck, 1944; Yokoyama and Namba, 1932). Moreover, many of the properties of an ionosphere in which the electron density varies continuously with height can be derived by considering it to be divided, by a series of parallel planes, into slabs each of which is homogeneous, and then proceeding to the limit when the slabs are infinitely thin. This device is used in §§ 9.10 and 13.2. It is therefore important to establish the laws of reflection and transmission at a sharp boundary.

8.2 The boundary-conditions

Consider a horizontal plane boundary between two homogeneous media, and let the x- and y-axes be in the boundary plane, so that z is measured vertically upwards. The value of a field quantity F at $z = 0$ in the top medium will be denoted by $F(+0)$, and the corresponding value in the bottom medium by $F(-0)$. These are the values immediately adjacent to the boundary, on opposite sides of it. Then the following boundary-conditions must hold:

$$E_x(+0) = E_x(-0), \qquad E_y(+0) = E_y(-0); \tag{8.1}$$

$$\mathscr{H}_x(+0) = \mathscr{H}_x(-0), \quad \mathscr{H}_y(+0) = \mathscr{H}_y(-0). \tag{8.2}$$

The equations (8.1) are derived by applying Faraday's law of electromagnetic induction to a thin rectangular circuit suitably oriented with its two long sides on opposite sides of the boundary, and then allowing the short dimension of the rectangle to tend to zero. The equations (8.2) are derived in a similar way by applying Ampère's circuital theorem to a rectangular circuit embracing the boundary. There are two other boundary-conditions, one relating $D_z(+0)$ and $D_z(-0)$, and the other relating $B_z(+0)$ and $B_z(-0)$. These are equivalent to (8.1) and (8.2) and can be derived from them, but they will not be needed in this book.

8.3 Snell's law

Let a plane wave be incident obliquely on the boundary from below, with its wave-normal in the x–z-plane at an angle θ_I to the z-axis in a clockwise direction. Then any field component F in this wave is given by

$$F = F^{(I)} \exp\{-ik\mathfrak{n}_1(x \sin\theta_I + z \cos\theta_I)\}, \tag{8.3}$$

where $F^{(I)}$ is a constant, and $\mathfrak{n}_1$ is the refractive index of the lower medium. In general there will be a reflected wave in the lower medium, and a

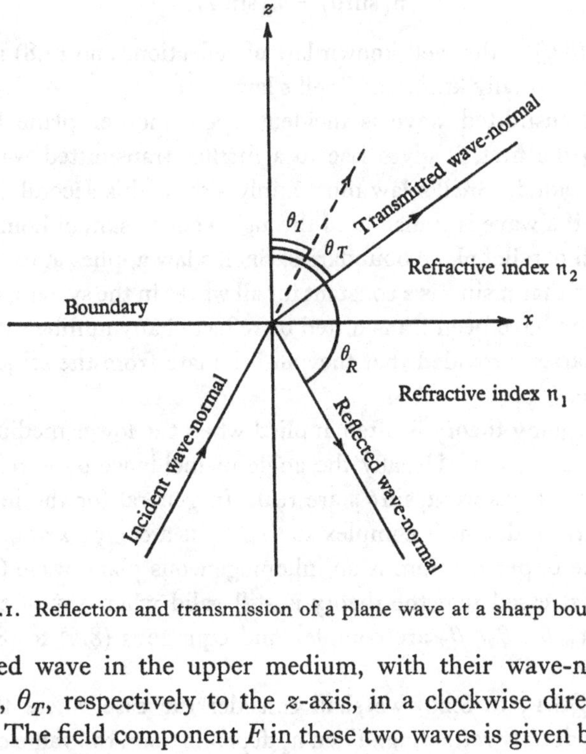

Fig. 8.1. Reflection and transmission of a plane wave at a sharp boundary.

transmitted wave in the upper medium, with their wave-normals at angles θ_R, θ_T, respectively to the z-axis, in a clockwise direction (see Fig. 8.1). The field component F in these two waves is given by:

reflected wave: $$F = F^{(R)} \exp\{-ik\mathfrak{n}_1(x \sin\theta_R + z \cos\theta_R)\}, \tag{8.4}$$

transmitted wave: $$F = F^{(T)} \exp\{-ik\mathfrak{n}_2(x \sin\theta_T + z \cos\theta_T)\}, \tag{8.5}$$

where $\mathfrak{n}_2$ is the refractive index of the upper medium (terms in y are omitted for the reason given below).

Now the field components must satisfy boundary conditions such as (8.1) and (8.2), and these apply at $z = 0$ for all values of x and y. Hence the dependence of any field component upon x and y must be the same for all three waves. The field of the incident wave does not depend on y, so that the fields of the reflected and transmitted waves cannot do so either. Further, the dependence upon x is only the same for all three waves if

$$\mathfrak{n}_1 \sin\theta_I = \mathfrak{n}_1 \sin\theta_R = \mathfrak{n}_2 \sin\theta_T. \tag{8.6}$$

Clearly, θ_R cannot be equal to θ_I. Hence

$$\theta_R = \pi - \theta_I. \tag{8.7}$$

Similarly, θ_T must tend to θ_I when $\mathfrak{n}_2 \to \mathfrak{n}_1$, hence

$$\mathfrak{n}_1 \sin\theta_I = \mathfrak{n}_2 \sin\theta_T. \tag{8.8}$$

Equation (8.7) is the well-known law of reflection, and (8.8) is the law of refraction usually known as Snell's law.

If the transmitted wave is incident upon another plane boundary parallel to the first, it gives rise to a further transmitted wave in the medium beyond. Snell's law must apply also at this second boundary. Similarly, if a wave is transmitted through a succession of homogeneous media with parallel plane boundaries, Snell's law applies at each. Hence (8.6) shows that $\mathfrak{n} \sin\theta$ is a constant for all waves in the system, including waves which have been transmitted or reflected any number of times at the boundaries, provided that they all originate from the same incident plane wave.

The foregoing theory is often applied when the lower medium is free space so that $\mathfrak{n}_1 = 1$. Usually the angle of incidence θ_I is real so that $\mathfrak{n}_1 \sin\theta_I$ and therefore $\mathfrak{n}_2 \sin\theta_T$ are real. In general for the ionosphere the refractive index $\mathfrak{n}_2$ is complex so that θ_T is a complex angle and the wave in the upper medium is an inhomogeneous plane wave (§ 4.7). It should be stressed that the theory is still valid when any of the quantities $\mathfrak{n}_1$, $\mathfrak{n}_2$, θ_I, θ_T, θ_R are complex and equations (8.6) to (8.8) must still apply.

Later (§§ 8.15 to 8.20) we shall consider the case where the upper medium is doubly refracting. Then $\mathfrak{n}_2$ depends upon the value of θ_T, and it is in general possible to find two pairs of values of $\mathfrak{n}_2$ and θ_T which satisfy Snell's law. There are then two transmitted waves in the upper medium, and they may both be inhomogeneous plane waves.

8.4 Derivation of the Fresnel formulae for isotropic media

The reflection and transmission coefficients will now be deduced in some important cases. In this section the relations (4.1) and (4.2) are used repeatedly. They give the relation between the electric and magnetic fields for a plane wave in an isotropic medium. It should be remembered that they apply only for a medium whose magnetic permeability is unity.

Let the incident wave be linearly polarised with its electric vector in the x–z-plane, so that the magnetic vector is parallel to the y-axis. Then for the incident wave, $E_y = \mathscr{H}_x = 0$, and it can be shown† that this must also be true for the reflected and transmitted waves. The superscripts (I), (R), (T) will now be used to denote the incident, reflected, and transmitted waves respectively. In each of these waves $\mathbf{E}$ is perpendicular to the wave-normal, and (4.2) shows that its amplitude is equal to $\mathscr{H}_y/\mathfrak{n}$. Hence:

lower medium:

incident wave $$\mathfrak{n}_1 E_x^{(I)} = \mathscr{H}_y^{(I)} \cos\theta_I, \tag{8.9}$$

reflected wave $$\mathfrak{n}_1 E_x^{(R)} = \mathscr{H}_y^{(R)} \cos\theta_R = -\mathscr{H}_y^{(R)} \cos\theta_I; \tag{8.10}$$

upper medium:

transmitted wave $$\mathfrak{n}_2 E_x^{(T)} = \mathscr{H}_y^{(T)} \cos\theta_T. \tag{8.11}$$

The boundary condition (8.1) for E_x gives

$$\frac{1}{\mathfrak{n}_1}(\mathscr{H}_y^{(I)} - \mathscr{H}_y^{(R)}) \cos\theta_I = \frac{1}{\mathfrak{n}_2} \mathscr{H}_y^{(T)} \cos\theta_T, \tag{8.12}$$

and the boundary condition (8.2) for $\mathscr{H}_y$ gives

$$\mathscr{H}_y^{(I)} + \mathscr{H}_y^{(R)} = \mathscr{H}_y^{(T)}. \tag{8.13}$$

Elimination of $\mathscr{H}_y^{(T)}$ from these equations gives

$${}_{\parallel}R_{\parallel} \equiv \frac{\mathscr{H}_y^{(R)}}{\mathscr{H}_y^{(I)}} = \frac{\mathfrak{n}_2 \cos\theta_I - \mathfrak{n}_1 \cos\theta_T}{\mathfrak{n}_2 \cos\theta_I + \mathfrak{n}_1 \cos\theta_T}, \tag{8.14}$$

and elimination of $\mathscr{H}_y^{(R)}$ gives

$${}_{\parallel}T_{\parallel} \equiv \frac{\mathfrak{n}_1 \mathscr{H}_y^{(T)}}{\mathfrak{n}_2 \mathscr{H}_y^{(I)}} = \frac{2\mathfrak{n}_1 \cos\theta_I}{\mathfrak{n}_2 \cos\theta_I + \mathfrak{n}_1 \cos\theta_T}. \tag{8.15}$$

It is convenient to express ${}_{\parallel}R_{\parallel}$ in a different form by using Snell's law (8.8). The result is

$${}_{\parallel}R_{\parallel} = \frac{\mathfrak{n}_2^2 \cos\theta_I - \mathfrak{n}_1(\mathfrak{n}_2^2 - \mathfrak{n}_1^2 \sin^2\theta_I)^{\frac{1}{2}}}{\mathfrak{n}_2^2 \cos\theta_I + \mathfrak{n}_1(\mathfrak{n}_2^2 - \mathfrak{n}_1^2 \sin^2\theta_I)^{\frac{1}{2}}}. \tag{8.16}$$

Now consider the case where the incident wave is linearly polarised with its electric vector in the direction of the y-axis, that is horizontal.

† The reader should verify that the boundary conditions cannot be satisfied if $E_y \neq 0$ for the reflected and transmitted waves only.

Then $E_x = \mathscr{H}_y = 0$ for the incident wave, and it can be shown that this must also be true for the reflected and transmitted waves. The equation (4.1) shows that in this case:

lower medium:

incident wave $\mathscr{H}_x^{(I)} = -\mathfrak{n}_1 E_y^I \cos\theta_I$, (8.17)

reflected wave $\mathscr{H}_x^{(R)} = -\mathfrak{n}_1 E_y^{(R)} \cos\theta_R = \mathfrak{n}_1 E_y^{(R)} \cos\theta_I$; (8.18)

upper medium:

transmitted wave $\mathscr{H}_y^{(T)} = -\mathfrak{n}_2 E_y^{(T)} \cos\theta_T$. (8.19)

The boundary condition (8.1) for E_y gives

$$E_y^{(I)} + E_y^{(R)} = E_y^{(T)}, \tag{8.20}$$

and the boundary condition (8.2) for $\mathscr{H}_x$, with (8.17), (8.18) and (8.19) gives

$$\mathfrak{n}_1(E_y^{(I)} - E_y^{(R)})\cos\theta_I = \mathfrak{n}_2 E_y^{(T)} \cos\theta_T. \tag{8.21}$$

Elimination of $E_y^{(T)}$ leads to

$$_\perp R_\perp \equiv \frac{E_y^{(R)}}{E_y^{(I)}} = \frac{\mathfrak{n}_1\cos\theta_I - \mathfrak{n}_2\cos\theta_T}{\mathfrak{n}_1\cos\theta_I + \mathfrak{n}_2\cos\theta_T}, \tag{8.22}$$

and elimination of $E_y^{(R)}$ gives

$$_\perp T_\perp \equiv \frac{E_y^{(T)}}{E_y^{(I)}} = \frac{2\mathfrak{n}_1\cos\theta_I}{\mathfrak{n}_1\cos\theta_I + \mathfrak{n}_2\cos\theta_T}. \tag{8.23}$$

$_\perp R_\perp$ and $_\perp T_\perp$ are the reflection and transmission coefficients in this case. Snell's law may be used to eliminate θ_T from (8.22), and the resulting expression is

$$_\perp R_\perp = \frac{\mathfrak{n}_1\cos\theta_I - (\mathfrak{n}_2^2 - \mathfrak{n}_1^2\sin^2\theta_I)^{\frac{1}{2}}}{\mathfrak{n}_1\cos\theta_I + (\mathfrak{n}_2^2 - \mathfrak{n}_1^2\sin^2\theta_I)^{\frac{1}{2}}}. \tag{8.24}$$

The formulae (8.14), (8.15), (8.22) and (8.23) were given by Fresnel. They are often used in text-books of Optics for the cases when $\mathfrak{n}_1$ and $\mathfrak{n}_2$ are real, but they are not restricted to real variables, and will be applied here to cases when the refractive indices and the angles are in general complex. When both media are isotropic, the other two reflection coefficients $_\parallel R_\perp$ and $_\perp R_\parallel$ are zero, and so also are $_\parallel T_\perp$ and $_\perp T_\parallel$.

8.5 General properties of the Fresnel formulae

In this section the lower medium is assumed to be free space so that $\mathfrak{n}_1 = 1$, and the refractive index of the upper medium will be denoted simply by $\mathfrak{n}$, the subscript 2 being omitted.

Consider first the formula (8.16) which gives the reflection coefficient ${}_{\parallel}R_{\parallel}$ when the electric vector is in the plane of incidence.

${}_{\parallel}R_{\parallel}$ is zero when $\mathfrak{n}^2 \cos\theta_I = (\mathfrak{n}^2 - \sin^2\theta_I)^{\frac{1}{2}}$, that is, when

$$\tan\theta_I = \mathfrak{n}. \tag{8.25}$$

When $\mathfrak{n}$ is real, this value of θ_I is a real angle and is called the 'Brewster angle'. When $\mathfrak{n}$ is complex, this value of θ_I is also complex, and is called the 'complex Brewster angle', and in this case the incident and transmitted waves are both inhomogeneous plane waves (see § 4.7). An interesting application of the complex Brewster angle was given by Zenneck (1907), who showed that an inhomogeneous plane wave, incident on a flat imperfectly conducting earth at the complex Brewster angle, behaves as though it is guided by the earth's surface. (A discussion of the problem is beyond the scope of this book, but see, for example, Pedersen, 1927.)

Now consider the formula (8.24) which gives the reflection coefficient ${}_{\perp}R_{\perp}$ when the electric vector is horizontal. The condition for ${}_{\perp}R_{\perp}$ to be zero is $\cos^2\theta_I = \mathfrak{n}^2 - \sin^2\theta_I$. This cannot be satisfied for any values of θ_I, real or complex, if $\mathfrak{n} \neq 1$. Hence in this case there is no Brewster angle.

When $\cos\theta_T$ is zero, $\sin\theta_T$ is unity and Snell's law gives

$$\sin\theta_I = \mathfrak{n}. \tag{8.26}$$

Then for both polarisations $R = 1$. This value of θ_I is called the 'critical angle', by analogy with the corresponding case in Optics. If $\mathfrak{n}$ is complex, then the critical angle is also complex.

8.6 The Fresnel formulae when the electric vector is in the plane of incidence

We now consider the case where the upper medium is the ionosphere, containing free electrons of constant number density, and the effect of the earth's magnetic field is neglected. If collisions are also neglected, the refractive index $\mathfrak{n}$ is given by $\mathfrak{n}^2 = 1 - X$ (equation (4.7)), which is always real and less than unity. If $\mathfrak{n}^2$ is positive, then $\mathfrak{n}$ is in the range $0 < \mathfrak{n} < 1$, and both the Brewster angle θ_B and the critical angle θ_c are real, and $0 < \theta_B < \theta_c < \frac{1}{2}\pi$. We now consider how ${}_{\parallel}R_{\parallel}$ varies as θ_I varies from 0 to $\frac{1}{2}\pi$. Typical curves are given in Fig. 8.2. The formula (8.16) shows that when $\theta_I < \theta_B$, then ${}_{\parallel}R_{\parallel}$ is real and negative, and when $\theta_B < \theta_I < \theta_c$ then ${}_{\parallel}R_{\parallel}$ is real and positive. Now Snell's law gives

$$\sin\theta_T = \sin\theta_I / \sin\theta_c.$$

When $\theta_I > \theta_c$, this shows that $\sin\theta_T > 1$, so that $\cos\theta_T$ is purely imaginary. Then the numerator and denominator in (8.16) are complex conjugates, and $|_{\|}R_{\|}| = 1$. When $\theta_I = \theta_c$, then $\arg {}_{\|}R_{\|} = 0$, and when $\theta_I = \frac{1}{2}\pi$, $\arg {}_{\|}R_{\|} = \pi$. Hence in the range $\theta_c < \theta_I < \frac{1}{2}\pi$ the value of $\arg {}_{\|}R_{\|}$ changes continuously from 0 to π. This is shown in the curves of Fig. 8.2.

When $X > 1$, so that $\mathfrak{n}^2 < 0$, $\mathfrak{n}$ must be purely imaginary, so that both the Brewster angle and the critical angle are purely imaginary. When

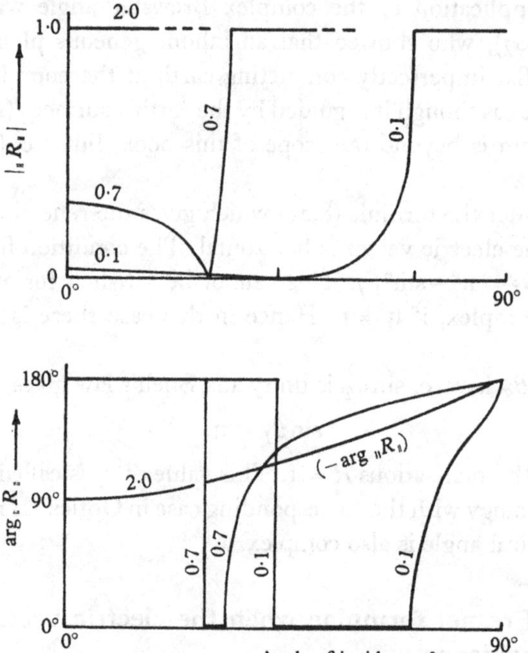

Fig. 8.2. Variation of modulus and argument of reflection coefficient ${}_{\|}R_{\|}$ with angle of incidence θ_I for sharply bounded homogeneous ionosphere. Collisions and the earth's magnetic field are neglected. The numbers by the curves are the values of X.

θ_I is real and in the range $0 < \theta_I < \frac{1}{2}\pi$, $\sin\theta_T$ is purely imaginary, and $\cos\theta_T$ is real. Then (8.16) shows that $|_{\|}R_{\|}| = 1$ for the whole range, and $\arg {}_{\|}R_{\|}$ changes continuously from $2 \arctan |\mathfrak{n}|$ when $\theta_I = 0$, to π when $\theta_I = \frac{1}{2}\pi$. A curve for this case is included in Fig. 8.2. The case $X = 1$ is considered separately in § 8.7.

When the effect of electron collisions is allowed for, $\mathfrak{n}$ is always complex, and the Brewster angle and critical angle are complex. Again it is of interest to draw curves showing how ${}_{\|}R_{\|}$ varies as θ_I goes from 0 to $\frac{1}{2}\pi$, and some examples are given in Figs. 8.3 and 8.4. When $Z \ll 1$, the effect is simply to round off the discontinuities in the curves for $Z = 0$. In Fig. 8.3 the curve which showed a zero of ${}_{\parallel}R_{\parallel}$ at the Brewster angle now shows a minimum at some real value of θ_I. This is sometimes called the 'quasi-Brewster angle', since the true Brewster angle is complex.

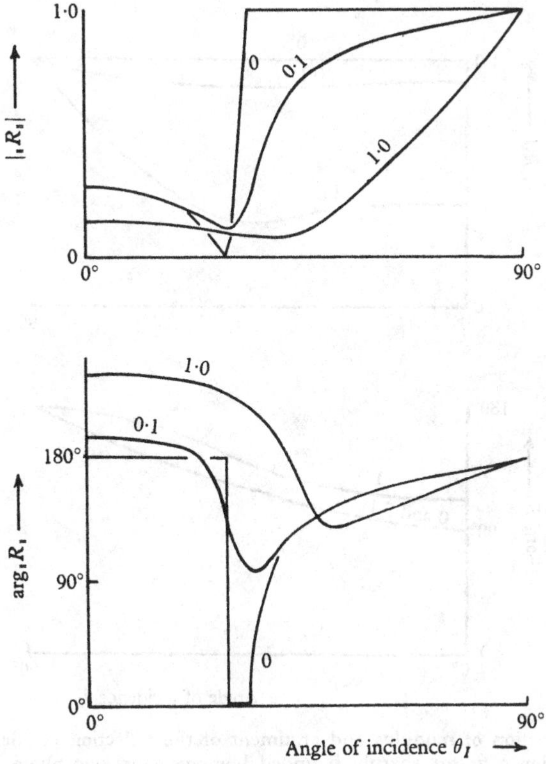

Fig. 8.3. Variation of modulus and argument of reflection coefficient ${}_{\parallel}R_{\parallel}$ with angle of incidence θ_I for sharply bounded homogeneous ionosphere. The earth's magnetic field is neglected. $X = 0{\cdot}7$. The numbers by the curves are the values of Z.

At very low frequencies $Z \gg 1$, and it is permissible to neglect 1 in comparison with iZ in the denominator of the formula (4.8) for $\mathfrak{n}^2$, which then becomes

$$\mathfrak{n}^2 \approx 1 - i\frac{X}{Z} = 1 - i\frac{\omega_r}{\omega}, \tag{8.27}$$

where

$$\omega_r = \frac{Ne^2}{\epsilon_0 m\nu} = \frac{\omega_N^2}{\nu}. \tag{8.28}$$

This assumption is equivalent to treating the ionosphere as an ordinary electric conductor with conductivity $\epsilon_0 \omega_r$ and unit dielectric constant. For then $\mathbf{J} = \epsilon_0 \omega_r \mathbf{E}$, so that

$$\mathbf{D} = \epsilon_0 \mathbf{E}\left(1 - i\frac{\omega_r}{\omega}\right),$$

which leads at once to (8.27).

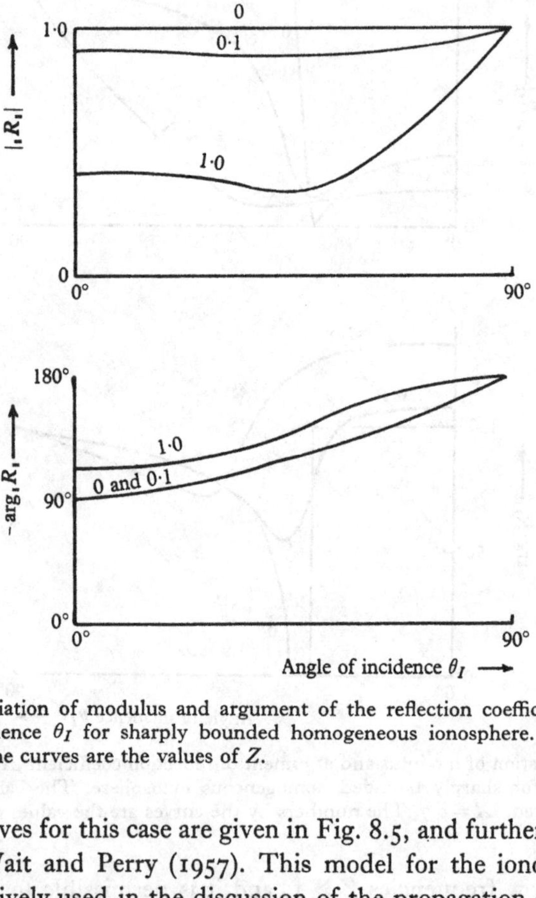

Fig. 8.4. Variation of modulus and argument of the reflection coefficient $_\parallel R_\parallel$ with angle of incidence θ_I for sharply bounded homogeneous ionosphere. $X = 2$. The numbers by the curves are the values of Z.

Some curves for this case are given in Fig. 8.5, and further curves are given by Wait and Perry (1957). This model for the ionosphere has been extensively used in the discussion of the propagation of very low frequency waves to great distances, treating the space between the earth and the ionosphere as a wave-guide (see, for example, Budden, 1951*b*, 1952*b*, 1953, 1957; and Wait, 1957), but the topic is beyond the scope of this book.

8.7 The Fresnel formulae when the electric vector is horizontal

Again the ionosphere is assumed to be homogeneous, and the earth's magnetic field is neglected. When collisions are neglected, and $X < 1$, $\mathfrak{n}$ is real and positive and less than one. Since the electric vector is horizontal, the reflection coefficient ${}_{\perp}R_{\perp}$ is given by (8.24) with $\mathfrak{n}_1 = 1$, $\mathfrak{n}_2 = \mathfrak{n}$,

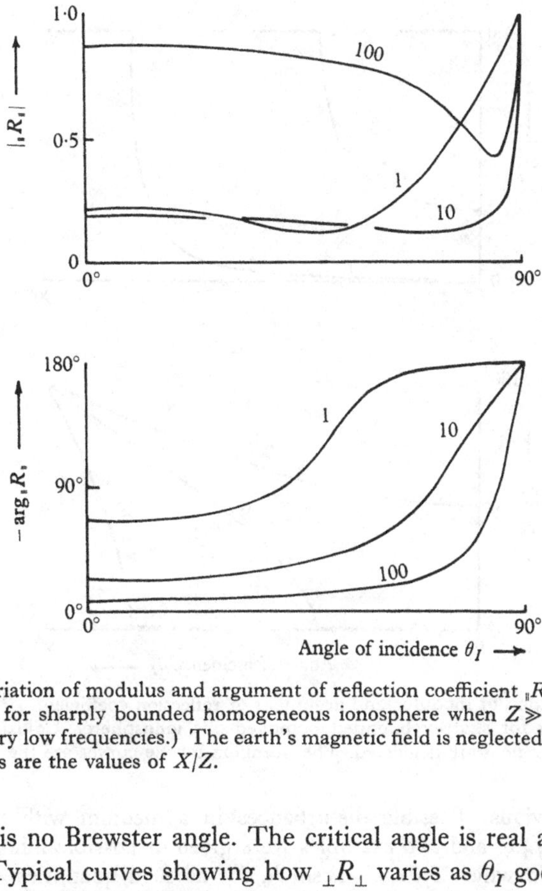

Fig. 8.5. Variation of modulus and argument of reflection coefficient ${}_{\parallel}R_{\parallel}$ with angle of incidence θ_I for sharply bounded homogeneous ionosphere when $Z \gg 1$. (Mainly of interest at very low frequencies.) The earth's magnetic field is neglected. The numbers by the curves are the values of X/Z.

and there is no Brewster angle. The critical angle is real and equal to $\arcsin \mathfrak{n}$. Typical curves showing how ${}_{\perp}R_{\perp}$ varies as θ_I goes from 0 to $\frac{1}{2}\pi$ are given in Fig. 8.6. When $\theta_I < \theta_c$, ${}_{\perp}R_{\perp}$ is real and positive. When $\theta_I > \theta_c$, $\cos\theta_T$ is purely imaginary and $|{}_{\perp}R_{\perp}| = 1$. In the range $\theta_c < \theta_I < \frac{1}{2}\pi$ the value of $\arg {}_{\perp}R_{\perp}$ changes continuously from 0 to π. This applies also to $\arg {}_{\parallel}R_{\parallel}$, but the curves for $\arg {}_{\parallel}R_{\parallel}$ and $\arg {}_{\perp}R_{\perp}$ are different, as is seen by comparing Figs. 8.2 and 8.6.

When $X > 1$, the curves are similar in general form to those for ${}_{\parallel}R_{\parallel}$.

A typical curve is included in Fig. 8.6. When the effect of collisions is allowed for, the discontinuities are rounded off as they were for $_{\parallel}R_{\parallel}$. Fig. 8.7 shows some examples.

8.8 Reflection when $X = 1$, $Z = 0$, $\mathfrak{n} = 0$

If we attempt to apply Snell's law when $\mathfrak{n} = 0$, it appears to require that $\sin\theta_T$ is infinite, and the nature of the wave in the top medium is not im-

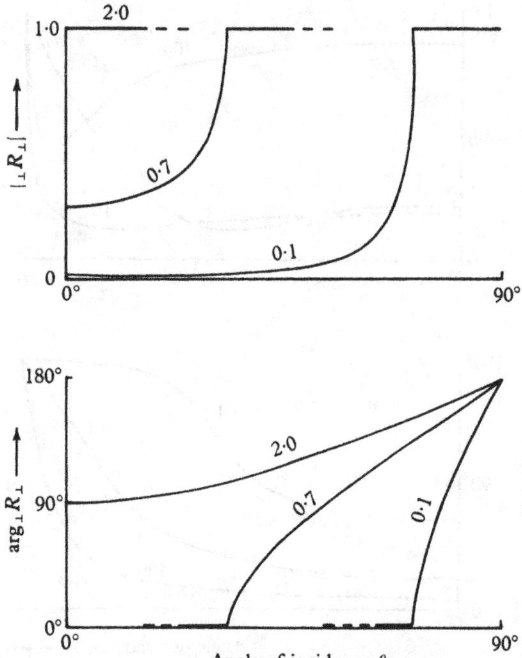

Fig. 8.6. Variation of modulus and argument of reflection coefficient $_{\perp}R_{\perp}$ with angle of incidence θ_I for sharply bounded homogeneous ionosphere. Collisions and the earth's magnetic field are neglected. The numbers by the curves are the values of X.

mediately obvious. Possible disturbances in a medium with $\mathfrak{n} = 0$ were described in §4.9, and two examples were given of possible inhomogeneous plane waves in which $\partial/\partial x \equiv -ik\sin\theta_I$, for all field quantities, which is a necessary condition when the incident wave has its normal at an angle θ_I to the z-axis.

In the first example the electric vector was in the x–z-plane, and the wave in the top medium had $\mathscr{H}_y = 0$ (equations (4.29)). Hence the boundary condition (8.2) shows that the magnetic fields $\mathscr{H}_y$ for the incident and reflected waves must exactly balance, so that $_{\parallel}R_{\parallel} = -1$ (from (7.16)) for all angles of incidence. Let the magnetic field adjacent to the boundary for the incident wave be $\mathscr{H}_y^{(I)}$.

Then the horizontal component of the total electric field just below the boundary is equal to $2\mathscr{H}_y^{(I)}\cos\theta_I$. This must be equal to $-iE_0\sin\theta_I$ (from (8.1)), where E_0 is the electric field in the top medium. Hence $E_0 = 2i\mathscr{H}_y^{(I)}\cot\theta_I$, and

$$_{\parallel}T_{\parallel} = 2i\cot\theta_I. \tag{8.29}$$

In the second example of §4.9, the electric vector was horizontal, and the

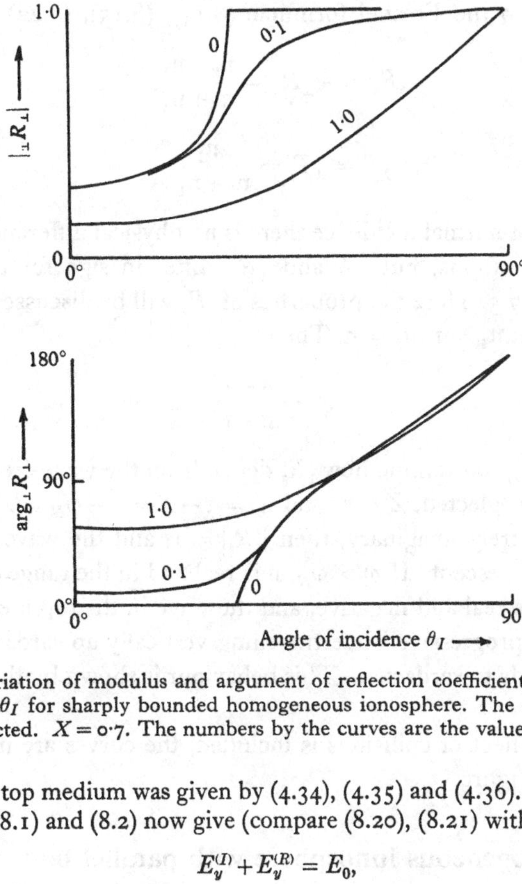

Fig. 8.7. Variation of modulus and argument of reflection coefficient $_{\perp}R_{\perp}$ with angle of incidence θ_I for sharply bounded homogeneous ionosphere. The earth's magnetic field is neglected. $X = 0{\cdot}7$. The numbers by the curves are the values of Z.

wave in the top medium was given by (4.34), (4.35) and (4.36). The boundary conditions (8.1) and (8.2) now give (compare (8.20), (8.21) with $\mathfrak{n}_1 = 1$):

$$E_y^{(I)} + E_y^{(R)} = E_0, \tag{8.30}$$

$$(E_y^{(I)} - E_y^{(R)})\cos\theta_I = -iE_0\sin\theta_I. \tag{8.31}$$

Elimination of E_0 gives

$$_{\perp}R_{\perp} \equiv \frac{E_y^{(R)}}{E_y^{(I)}} = \frac{\cos\theta_I + i\sin\theta_I}{\cos\theta_I - i\sin\theta_I} = e^{2i\theta_I}, \tag{8.32}$$

and elimination of $E_y^{(R)}$ gives

$$_{\perp}T_{\perp} \equiv \frac{E_0}{E_y^{(I)}} = \frac{2\cos\theta_I}{\cos\theta_I - i\sin\theta_I} = 1 + e^{2i\theta_I}. \tag{8.33}$$

These results could have been predicted from the formula (8.22), by using (4.33).

The expression (8.32) shows that when θ_I is real, $|{}_\perp R_\perp| = 1$, and there is a phase change on reflection equal to twice the angle of incidence.

8.9 Normal incidence

When $\theta_I = 0$ the Fresnel formulae (8.14), (8.15), (8.22) and (8.23) reduce to:

$$ {}_\parallel R_\parallel = -{}_\perp R_\perp = \frac{\mathfrak{n}_2 - \mathfrak{n}_1}{\mathfrak{n}_2 + \mathfrak{n}_1}, \tag{8.34} $$

$$ {}_\parallel T_\parallel = {}_\perp T_\perp = \frac{2\mathfrak{n}_1}{\mathfrak{n}_2 + \mathfrak{n}_1}. \tag{8.35} $$

For this case of normal incidence there is no physical difference between the two polarisations, but ${}_\parallel R_\parallel$ and ${}_\perp R_\perp$ differ in sign for the reasons explained in §7.5. Here the properties of ${}_\parallel R_\parallel$ will be discussed.

As before let $\mathfrak{n}_1 = 1$, $\mathfrak{n}_2 = \mathfrak{n}$. Then

$$ {}_\parallel R_\parallel = \frac{\mathfrak{n} - 1}{\mathfrak{n} + 1}. \tag{8.36} $$

It is interesting to examine how ${}_\parallel R_\parallel$ depends on the wave-frequency. If collisions are neglected, $Z = 0$, and $\mathfrak{n}^2 = 1 - X = 1 - \omega_N^2/\omega^2$. If $\omega < \omega_N$ so that $\mathfrak{n}$ is purely imaginary, then $|{}_\parallel R_\parallel| = 1$, and the wave in the top medium is evanescent. If $\omega > \omega_N$, $\mathfrak{n}$ is real and in the range $0 < \mathfrak{n} < 1$, and then ${}_\parallel R_\parallel$ is real and negative, and the wave in the top medium is an unattenuated progressive wave travelling vertically upwards. As ω gets larger, ${}_\parallel R_\parallel$ tends towards zero. This behaviour is shown by the curves of Fig. 8.8.

When the effect of collisions is included, the curves are modified as shown in the figure.

8.10 Homogeneous ionosphere with parallel boundaries

As a rough approximation an ionospheric layer can be pictured as a homogeneous medium with two horizontal boundaries a distance d apart. Expressions will now be deduced for the reflection and transmission coefficients of this model, when the electric vector is in the plane of incidence.

A wave-incident from below on the lower boundary is partially transmitted and partially reflected. The transmitted wave is partially reflected at the top boundary, and the reflected wave returns to the lower boundary where it is again partially reflected, and so on. There is thus an infinite series of internal reflections in the layer, and the resultant reflected wave below the ionosphere could be found by adding the first reflected wave and the series of waves transmitted through the lower boundary. This method is sometimes used in Optics, for example in the theory of the Fabry–Perot interferometer, but here a different method will be used.

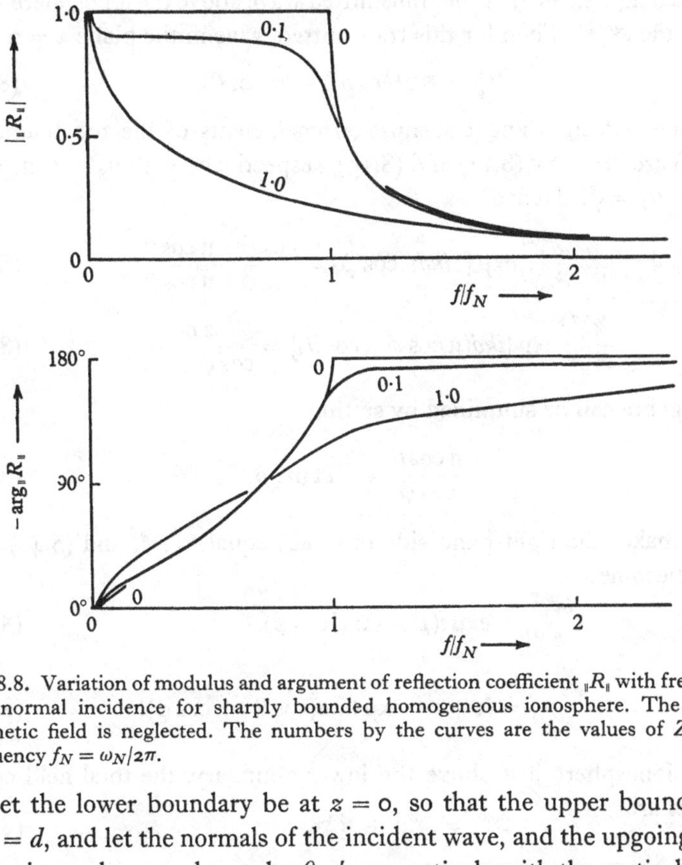

Fig. 8.8. Variation of modulus and argument of reflection coefficient $_{\|}R_{\|}$ with frequency f at normal incidence for sharply bounded homogeneous ionosphere. The earth's magnetic field is neglected. The numbers by the curves are the values of Z at the frequency $f_N = \omega_N/2\pi$.

Let the lower boundary be at $z = 0$, so that the upper boundary is at $z = d$, and let the normals of the incident wave, and the upgoing wave in the ionosphere, make angles θ, ϕ, respectively with the vertical.

The upgoing waves in the ionosphere all have their wave-normals at the same inclination, ϕ, to the vertical and so they combine to give a single resultant plane wave. Let its magnetic field have amplitude $\mathscr{H}_y^{(1)}$ at the point $x = 0$, $z = 0$. Then for the upgoing wave in the plane $x = 0$:

$$\mathscr{H}_y = \mathscr{H}_y^{(1)} \exp(-ik\mathfrak{n}z\cos\phi), \tag{8.37}$$

$$E_x = \frac{1}{\mathfrak{n}} \mathscr{H}_y^{(1)} \cos\phi \exp(-ik\mathfrak{n}z\cos\phi). \tag{8.38}$$

Similarly, let the magnetic field of the downgoing wave in the ionosphere have amplitude $\mathscr{H}_y^{(2)}$. Then for the downgoing wave in the plane $x = 0$:

$$\mathscr{H}_y = \mathscr{H}_y^{(2)} \exp(ik\mathfrak{n}z\cos\phi),$$

$$E_x = -\frac{1}{\mathfrak{n}}\mathscr{H}_y^{(2)} \exp(ik\mathfrak{n}z\cos\phi).$$

Let the magnetic field of the transmitted wave above the ionosphere have amplitude $\mathscr{H}_y^{(T)}$. Then for this transmitted wave in the plane $x = 0$

$$\mathscr{H}_y = \mathscr{H}_y^{(T)} \exp(-ikz\cos\theta). \tag{8.39}$$

Now the reflection and transmission coefficients of the top boundary ($z = d$) are given by (8.14) and (8.15), respectively, with $\mathfrak{n}_2 = 1$, $\mathfrak{n}_1 = \mathfrak{n}$, $\theta_T = \theta$, $\theta_I = \phi$. Hence

$$\frac{\mathscr{H}_y^{(2)}}{\mathscr{H}_y^{(1)}}\exp\{2ikd\mathfrak{n}\cos\phi\} = \frac{\cos\phi - \mathfrak{n}\cos\theta}{\cos\phi + \mathfrak{n}\cos\theta} \tag{8.40}$$

and

$$\frac{\mathscr{H}_y^{(T)}}{\mathscr{H}_y^{(1)}}\exp\{ikd(\mathfrak{n}\cos\phi - \cos\theta)\} = \frac{2\cos\phi}{\cos\phi + \mathfrak{n}\cos\theta}. \tag{8.41}$$

The algebra can be simplified by setting

$$\frac{\mathfrak{n}\cos\theta}{\cos\phi} = -i\tan\tfrac{1}{2}\beta, \tag{8.42}$$

which makes the right-hand side of (8.40) equal to $e^{i\beta}$, and (8.40) and (8.41) become:

$$\frac{\mathscr{H}_y^{(2)}}{\mathscr{H}_y^{(1)}} = \exp i(\beta - 2k\mathfrak{n}d\cos\phi), \tag{8.43}$$

$$\frac{\mathscr{H}_y^{(T)}}{\mathscr{H}_y^{(1)}} = (1 + e^{i\beta})\exp\{ikd(\cos\theta - \mathfrak{n}\cos\phi)\}. \tag{8.44}$$

In the ionosphere just above the lower boundary, the total field components are:

$$\mathscr{H}_y = \mathscr{H}_y^{(1)} + \mathscr{H}_y^{(2)}, \tag{8.45}$$

$$E_x = \frac{1}{\mathfrak{n}}\cos\phi\,(\mathscr{H}_y^{(1)} - \mathscr{H}_y^{(2)}). \tag{8.46}$$

Let the magnetic fields of the incident and resultant reflected waves below the ionosphere have amplitudes $\mathscr{H}_y^{(I)}$, $\mathscr{H}_y^{(R)}$ respectively. Then the total field components just below the boundary are:

$$\mathscr{H}_y = \mathscr{H}_y^{(I)} + \mathscr{H}_y^{(R)}, \tag{8.47}$$

$$E_x = \cos\theta\,(\mathscr{H}_y^{(I)} - \mathscr{H}_y^{(R)}). \tag{8.48}$$

The boundary conditions (8.1) and (8.2) now require that:

$$\mathscr{H}_y^{(I)} + \mathscr{H}_y^{(R)} = \mathscr{H}_y^{(1)} + \mathscr{H}_y^{(2)}, \tag{8.49}$$

$$\cos\theta\,(\mathscr{H}_y^{(I)} - \mathscr{H}_y^{(R)}) = \frac{1}{\mathfrak{n}}\cos\phi\,(\mathscr{H}_y^{(1)} - \mathscr{H}_y^{(2)}). \tag{8.50}$$

The terms $\mathscr{H}_y^{(1)}$, $\mathscr{H}_y^{(2)}$ are eliminated from (8.49), (8.50) and (8.43), and the result is

$$_{\parallel}R_{\parallel} \equiv \frac{\mathscr{H}_y^{(R)}}{\mathscr{H}_y^{(I)}} = \frac{\sin(k\mathfrak{n}d\cos\phi)}{\sin(\beta - k\mathfrak{n}d\cos\phi)}. \tag{8.51}$$

Similarly, $\mathscr{H}_y^{(R)}$, $\mathscr{H}_y^{(1)}$, $\mathscr{H}_y^{(2)}$ may be eliminated from (8.49), (8.50), (8.43) and (8.44) to give

$$T \equiv \frac{\mathscr{H}_y^{(T)}}{\mathscr{H}_y^{(I)}} = \frac{\sin\beta\exp(ikd\cos\theta)}{\sin(\beta - k\mathfrak{n}d\sin\phi)}. \tag{8.52}$$

The symbol T without subscripts is here used for the transmission coefficient of the whole ionosphere, to avoid confusion with the transmission coefficients at a single boundary.

The above theory could be applied to the reflection of light from a parallel-sided glass plate. In this case the refractive index $\mathfrak{n}$ of the glass is real, and the formula (8.51) shows that the reflection coefficient is zero when $k\mathfrak{n}d\cos\phi = n\pi$, where n is an integer. This is equivalent to the condition $2\mathfrak{n}d\cos\phi = n\lambda$, which means that the optical path difference between two successive reflected waves (mentioned at the beginning of this section) is an integral multiple of one wavelength, so that these waves have the same phase and reinforce each other. It is shown in text-books on Optics (see, for example, Houstoun, *Treatise on Light* (1938), ch. IX) that in this case the first reflected wave from the lower boundary exactly annuls the resultant of the remaining reflected waves.

It can easily be shown that if the ionosphere is very slightly absorbing, so that $\mathfrak{n}$ has a small imaginary part, and if d becomes indefinitely large, the formula (8.51) approaches the value given by (8.14) for the reflection coefficient of a sharply bounded semi-infinite medium. As a further check of the results (8.51) and (8.52), it may be verified that when $\mathfrak{n} = 1$, so that the 'ionosphere' is just free space, $_{\parallel}R_{\parallel} = 0$, $T = 1$. Similarly, when $d = 0$, the same result is obtained. It can also be shown that if the ionosphere is loss free, so that $\mathfrak{n}^2$ is real, then $|_{\parallel}R_{\parallel}|^2 + |T|^2 = 1$, which shows that the sum of the energies in the reflected and transmitted waves is equal to the energy in the incident wave.

8.11 Normal incidence on a parallel-sided slab

For waves which are normally incident on a homogeneous ionosphere between parallel boundaries, we have $\theta = \phi = 0$, and (8.42) becomes

$$-i \tan \tfrac{1}{2}\beta = \mathfrak{n}. \tag{8.53}$$

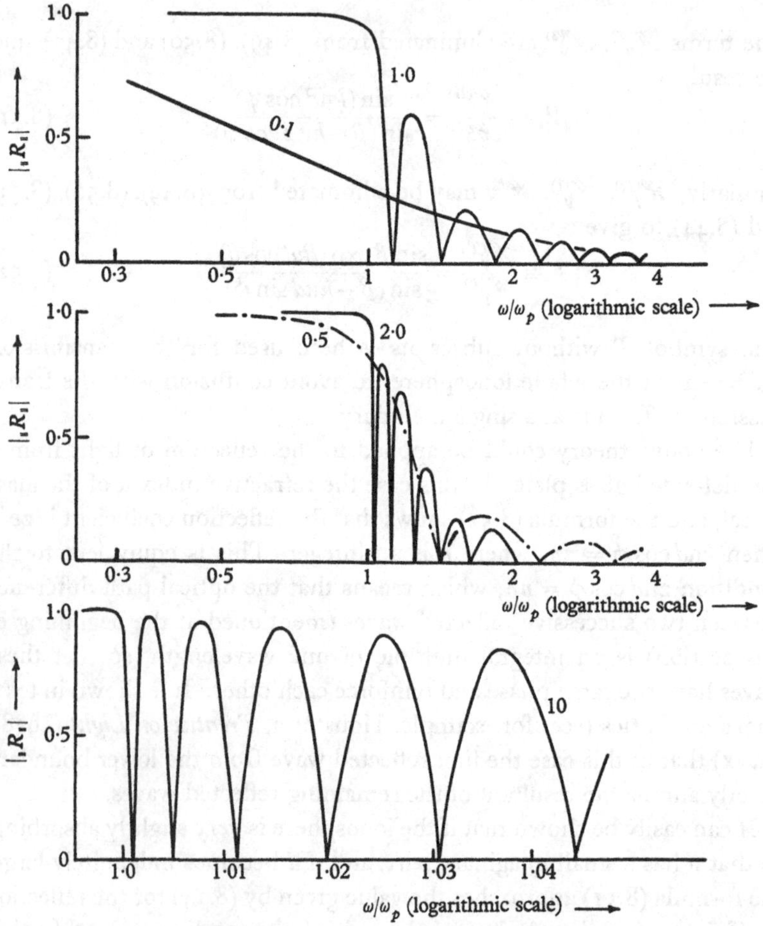

Fig. 8.9. Reflection coefficient for a parallel-sided slab of thickness d, at normal incidence, as a function of angular frequency ω. The angular penetration-frequency is ω_p, and λ_p is the corresponding free space wavelength. The numbers by the curves are values of d/λ_p. Electron collisions and the earth's magnetic field are neglected.

It can be shown that formulae (8.51) and (8.52) give for the reflection and transmission coefficients:

$$_{\parallel}R_{\parallel} = \frac{(1-\mathfrak{n}^2)\tan k\mathfrak{n}d}{2i\mathfrak{n} - (1+\mathfrak{n}^2)\tan k\mathfrak{n}d}, \tag{8.54}$$

$$T = \frac{2in \sec(knd)\, e^{ikd}}{2in - (1+n^2)\tan knd}. \tag{8.55}$$

It is interesting to study how these coefficients depend upon the wave-frequency. An actual ionospheric layer has a penetration frequency $f_p = \omega_p/2\pi$ which is equal to the plasma frequency at the maximum of the layer. Waves of frequency less than f_p are strongly reflected, so that $|{}_{\parallel}R_{\parallel}| \doteqdot 1$, $|T| \doteqdot 0$, and waves of frequency above f_p penetrate the layer almost completely at normal incidence, so that $|{}_{\parallel}R_{\parallel}| \doteqdot 0$, $|T| \doteqdot 1$.

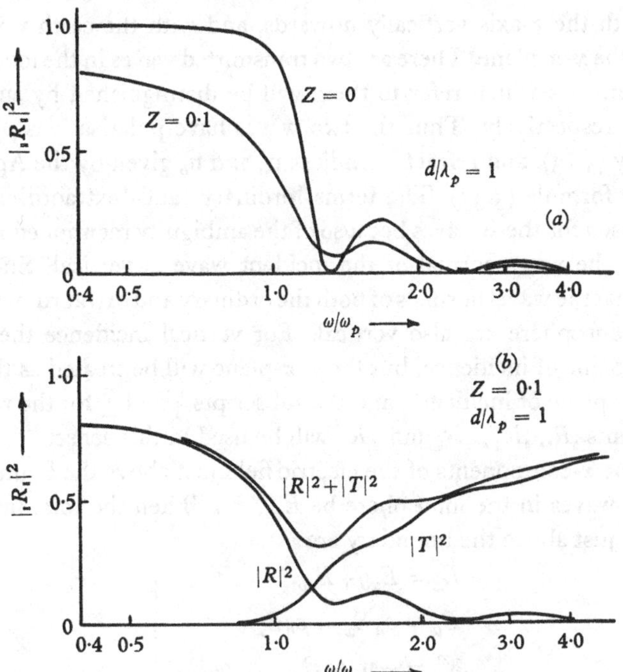

Fig. 8.10. Effect of electron collisions on reflection and transmission coefficients for a parallel-sided slab, at normal incidence, as a function of angular frequency ω.

There is a small range of frequencies near f_p, in which the transition occurs, and here there may be partial penetration and reflection. The size of the range depends on the layer thickness; it is large for thin layers and small for thick layers. Curves of $|{}_{\parallel}R_{\parallel}|$ as a function of frequency calculated from (8.54) are given in Fig. 8.9 for various values of the thickness d. They should be compared with Fig. 16.8 (§ 16.11), Fig. 17.3 (§ 17.5) and Fig. 17.5 (§ 17.8) which give similar curves for other models of the ionosphere. In preparing Fig. 8.9 electron collisions were neglected. The effect of collisions is illustrated in Fig. 8.10 which gives curves for a fixed value of d, and of the collision frequency ν.

8.12 Reflection at normal incidence when the earth's magnetic field is allowed for

When the effect of the earth's magnetic field on the ionosphere is allowed for, the method of finding the reflection coefficients is very much more complicated. It will be illustrated first by considering a linearly polarised wave to be normally incident from below. Cartesian axes are used with the z-axis vertically upwards, and with the earth's magnetic field in the x–z-plane. There are two transmitted waves in the ionosphere, and quantities which refer to them will be distinguished by subscripts a and b respectively. Thus the two waves have polarisations ρ_a and ρ_b given by (5.13), and refractive indices $\mathfrak{n}_a$ and $\mathfrak{n}_b$ given by the Appleton–Hartree formula (5.35). The terms 'ordinary' and 'extraordinary' will not be used for these waves because of the ambiguity mentioned in § 6.15.

Since the wave-normal of the incident wave is vertical, Snell's law shows that the wave-normals of both the ordinary and extraordinary waves in the ionosphere are also vertical. For vertical incidence there is no unique plane of incidence, but the x–z-plane will be treated as though it were the plane of incidence, and the subscripts $\parallel$ and $\perp$ for the reflection coefficients ${}_\parallel R_\parallel$, ${}_\parallel R_\perp$, ${}_\perp R_\parallel$ and ${}_\perp R_\perp$ will be used in this sense.

Let the x-components of the electric field just above the boundary for the two waves in the ionosphere be E_{xa}, E_{xb}. Then the total field components just above the boundary are:

$$\left.\begin{aligned} E_x &= E_{xa}+E_{xb},\\ E_y &= \rho_a E_{xa}+\rho_b E_{xb},\\ \mathscr{H}_x &= -\mathfrak{n}_a\rho_a E_{xa}-\mathfrak{n}_b\rho_b E_{xb},\\ \mathscr{H}_y &= \mathfrak{n}_a E_{xa}+\mathfrak{n}_b E_{xb}. \end{aligned}\right\} \tag{8.56}$$

There is also in general a component E_z but its value is not needed.

Now suppose that the incident wave is linearly polarised with its electric vector in the plane of incidence, and of magnitude $E_x^{(I)}$. Then the total field components just below the boundary are:

$$\left.\begin{aligned} E_x &= (1-{}_\parallel R_\parallel)E_x^{(I)},\\ E_y &= {}_\parallel R_\perp E_x^{(I)},\\ \mathscr{H}_x &= {}_\parallel R_\perp E_x^{(I)},\\ \mathscr{H}_y &= (1+{}_\parallel R_\parallel)E_x^{(I)}. \end{aligned}\right\} \tag{8.57}$$

The boundary conditions (8.1) and (8.2) require that these four quantities are equated to those in (8.56) respectively. This gives four homogeneous equations in $E_x^{(I)}$, E_{xa}, E_{xb}, which may therefore be eliminated to give two equations for ${}_{\parallel}R_{\parallel}$ and ${}_{\parallel}R_{\perp}$. The results are:

$$ {}_{\parallel}R_{\parallel} = \frac{\rho_b}{\rho_b - \rho_a}\frac{\mathfrak{n}_a - 1}{\mathfrak{n}_a + 1} - \frac{\rho_a}{\rho_b - \rho_a}\frac{\mathfrak{n}_b - 1}{\mathfrak{n}_b + 1}, \tag{8.58} $$

$$ {}_{\parallel}R_{\perp} = \frac{2}{\rho_b - \rho_a}\left[\frac{1}{\mathfrak{n}_a + 1} - \frac{1}{\mathfrak{n}_b + 1}\right] \tag{8.59} $$

(the relation $\rho_a\rho_b = 1$ has been used here: see § 5.4). To find the other two reflection coefficients, let the incident wave be linearly polarised with its electric vector parallel to the y-axis and of magnitude $E_y^{(I)}$. Then the total field components below the boundary are:

$$ \left.\begin{aligned} E_x &= -{}_{\perp}R_{\parallel}E_y^{(I)}, \\ E_y &= (1 + {}_{\perp}R_{\perp})E_y^{(I)}, \\ \mathscr{H}_x &= (-1 + {}_{\perp}R_{\perp})E_y^{(I)}, \\ \mathscr{H}_y &= {}_{\perp}R_{\parallel}E_y^{(I)}. \end{aligned}\right\} \tag{8.60} $$

These four quantities must be equated to those in (8.56) respectively, and this gives four homogeneous equations in $E_y^{(I)}$, E_{xa}, E_{xb}. These are eliminated, and the result is:

$$ {}_{\perp}R_{\parallel} = \frac{2}{\rho_b - \rho_a}\left[\frac{1}{\mathfrak{n}_a + 1} - \frac{1}{\mathfrak{n}_b + 1}\right] = {}_{\parallel}R_{\perp}, \tag{8.61} $$

$$ {}_{\perp}R_{\perp} = \frac{\rho_a}{\rho_b - \rho_a}\frac{\mathfrak{n}_a - 1}{\mathfrak{n}_a + 1} - \frac{\rho_b}{\rho_b - \rho_a}\frac{\mathfrak{n}_b - 1}{\mathfrak{n}_b + 1}. \tag{8.62} $$

The formulae (8.58), (8.59), (8.61) and (8.62) will now be examined in some special cases.

8.13 Earth's magnetic field horizontal. Normal incidence

When the earth's magnetic field is horizontal, the terms 'ordinary' and 'extraordinary' may be used without ambiguity. The polarisation for the ordinary wave, $\rho_a = \rho_o$, is zero, and the refractive index $\mathfrak{n}_a = \mathfrak{n}_o$ is the value that it would have if the earth's field were absent, given by (6.20). Similarly, $\rho_b = \rho_x = \infty$ and $\mathfrak{n}_b = \mathfrak{n}_x$ is given by (6.21). These results lead to

$$ {}_{\parallel}R_{\perp} = {}_{\perp}R_{\parallel} = 0, \quad {}_{\parallel}R_{\parallel} = \frac{\mathfrak{n}_o - 1}{\mathfrak{n}_o + 1}, \quad {}_{\perp}R_{\perp} = \frac{1 - \mathfrak{n}_x}{1 + \mathfrak{n}_x}. $$

The last two are just the Fresnel formulae (8.14) and (8.22) for normal incidence.

8.14 Earth's magnetic field vertical. Normal incidence

Suppose that the earth's field is vertical and directed downwards, as in the northern hemisphere. Then the vector **Y** is directed upwards, and its direction cosine $n = +1$ for upgoing waves. For one upgoing wave $\rho_a = -i$, which corresponds to right-handed circular polarisation. Similarly, $\rho_b = +i$. Hence

$$\mathfrak{n}_a^2 = 1 - \frac{X}{U+Y}, \quad \mathfrak{n}_b^2 = 1 - \frac{X}{U-Y}. \tag{8.63}$$

The formulae for the reflection coefficients become:

$$_\parallel R_\parallel = -{}_\perp R_\perp = \frac{1}{2}\left\{\frac{\mathfrak{n}_a - 1}{\mathfrak{n}_a + 1} + \frac{\mathfrak{n}_b - 1}{\mathfrak{n}_b + 1}\right\}, \tag{8.64}$$

$$_\parallel R_\perp = {}_\perp R_\parallel = i\left\{\frac{1}{\mathfrak{n}_b + 1} - \frac{1}{\mathfrak{n}_a + 1}\right\}. \tag{8.65}$$

These formulae may be simplified by using the alternative representation described in §7.7, in which the incident and reflected waves are resolved into circularly polarised components. The four reflection coefficients $_rR_r$, $_rR_l$, $_lR_r$, $_lR_l$ were defined on p. 90. Their values may be obtained from (8.64) and (8.65) by using the transformation (7.29), and are as follows:

$$_rR_r = {}_lR_l = 0, \quad _rR_l = \frac{\mathfrak{n}_a - 1}{\mathfrak{n}_a + 1}, \quad _lR_r = \frac{\mathfrak{n}_b - 1}{\mathfrak{n}_b + 1}. \tag{8.66}$$

These results show that the two waves in the ionosphere are associated with independent reflections. For example, an incident wave which is circularly polarised with a right-handed sense would give a wave in the ionosphere with the same polarisation, and the reflected wave would be circularly polarised with a left-handed sense. This apparent reversal of the sense occurs because the directions of the wave-normals are opposite for the incident and reflected waves. The absolute direction of rotation is the same for both waves, namely clockwise when seen by an observer looking upwards.

8.15 Reflection when the earth's magnetic field is included. Approximate formulae for oblique incidence

Let a plane wave be incident upon the ionosphere from below, with its wave-normal at an angle θ_I to the vertical. The ionosphere is assumed to be a sharply bounded homogeneous medium, and the effect of the earth's magnetic field is included, so that there are two transmitted waves in the ionosphere. For one of these, let the wave-normal make an angle θ_a with the vertical, and let the polarisation and refractive index be ρ_a, $\mathfrak{n}_a$ respectively. Similarly, let θ_b, ρ_b, $\mathfrak{n}_b$ be the corresponding quantities for the other transmitted wave. The terms 'ordinary' and 'extraordinary' will not be used for these waves because of the ambiguity mentioned in § 6.15. Snell's law must apply for both waves, and hence

$$\sin\theta_I = \mathfrak{n}_a \sin\theta_a = \mathfrak{n}_b \sin\theta_b. \tag{8.67}$$

The unknown angles θ_a and θ_b could be found from (8.67), if $\mathfrak{n}_a$ and $\mathfrak{n}_b$ were known. But the values of $\mathfrak{n}_a$ and $\mathfrak{n}_b$ depend upon Y_L, Y_T which in turn depend on θ_a and θ_b. Hence (8.67) cannot immediately be used to give θ_a and θ_b. This difficulty can be overcome by using the Booker quartic equation, as described later, § 8.17.

There is an interesting case, however, when $\mathfrak{n}_a$ and $\mathfrak{n}_b$ are approximately independent of θ_a and θ_b, and then the simpler method described below can be used. This case occurs at very low frequencies (less than 100 kc/s), so that $Y \gg 1$.

Suppose that in the ionosphere it is permissible to make the quasi-longitudinal approximation (6.23). Then (6.24) show that:

$$\left.\begin{aligned} \rho_a = -i, \quad \mathfrak{n}_a^2 &= 1 - \frac{X}{1 - iZ + Y_L}, \\ \rho_b = i, \quad \mathfrak{n}_b^2 &= 1 - \frac{X}{1 - iZ - Y_L}. \end{aligned}\right\} \tag{8.68}$$

Here $\mathfrak{n}_a$ and $\mathfrak{n}_b$ still depend upon θ_a and θ_b respectively because of the term Y_L, which would have different values in the two cases. It will be assumed, however, that Y_L is a constant, and the validity of this assumption will be discussed later. Since $Y \gg 1$, the formulae for $\mathfrak{n}_a$, $\mathfrak{n}_b$ may be written

$$\left.\begin{matrix} \mathfrak{n}_a^2 \\ \mathfrak{n}_b^2 \end{matrix}\right\} = 1 - \frac{iX}{Z \pm iY_L} = 1 - i\left(\frac{\omega_r}{\omega}\right)\exp(\mp i\tau). \tag{8.69}$$

If Y_L is purely real, then

$$\frac{\omega_r}{\omega} = X(Z^2 + Y_L^2)^{-\frac{1}{2}} = \frac{\omega_N^2}{\omega(\nu^2 + \omega_L^2)^{\frac{1}{2}}}, \tag{8.70}$$

$$\tan\tau = \frac{Y_L}{Z} = \frac{\omega_L}{\nu}. \tag{8.71}$$

In general $\mathfrak{n}_a$ and $\mathfrak{n}_b$ are complex, so that θ_a and θ_b are complex, and therefore Y_L is complex. The correct expressions for ω_r and τ are therefore more complicated. Each wave has a component of the electric field in the direction of its wave-normal, but it will be shown later that this is negligibly small in the cases discussed.

Let the electric field of the incident wave just below the boundary have components $E_\parallel^{(I)}$ in the x–z-plane, directed obliquely away from the boundary, and $E_y^{(I)}$ parallel to the y-axis. Similarly, let the electric field of the reflected wave have components $E_\parallel^{(R)}$, $E_y^{(R)}$, and let the electric fields of the two transmitted waves just above the boundary have components $E_\parallel^{(a)}$, $E_y^{(a)}$, and $E_\parallel^{(b)}$, $E_y^{(b)}$. Then the total field components just below the boundary are:

$$\left.\begin{aligned} E_x &= (E_\parallel^{(I)} - E_\parallel^{(R)})\cos\theta_I, \\ E_y &= E_y^{(I)} + E_y^{(R)}, \\ \mathscr{H}_x &= (E_y^{(R)} - E_y^{(I)})\cos\theta_I, \\ \mathscr{H}_y &= E_\parallel^{(I)} + E_\parallel^{(R)}, \end{aligned}\right\} \tag{8.72}$$

and those just above the boundary are

$$\left.\begin{aligned} E_x &= E_\parallel^{(a)}\cos\theta_a + E_\parallel^{(b)}\cos\theta_b, \\ E_y &= -iE_\parallel^{(a)} + iE_\parallel^{(b)}, \\ \mathscr{H}_x &= i\mathfrak{n}_a E_\parallel^{(a)}\cos\theta_a - i\mathfrak{n}_b E_\parallel^{(b)}\cos\theta_b, \\ \mathscr{H}_y &= \mathfrak{n}_a E_\parallel^{(a)} + \mathfrak{n}_b E_\parallel^{(b)}. \end{aligned}\right\} \tag{8.73}$$

The boundary conditions (8.1) and (8.2) require that the four quantities (8.73) are equated to those in (8.72) respectively. Thence the reflection coefficients can be evaluated. For example, to find ${}_\parallel R_\perp$, put $E_y^{(I)} = 0$, and eliminate $E_\parallel^{(R)}$, $E_\parallel^{(a)}$, $E_\parallel^{(b)}$, from the equations. This gives an equation for $E_y^{(R)}/E_\parallel^{(I)} = {}_\parallel R_\perp$. The results are as follows:

$${}_\parallel R_\parallel = \{(\mathfrak{n}_a + \mathfrak{n}_b)(\cos^2\theta_I - \cos\theta_a\cos\theta_b) + (\mathfrak{n}_a\mathfrak{n}_b - 1)(\cos\theta_a + \cos\theta_b)\cos\theta_I\}/D, \tag{8.74}$$

$${}_\parallel R_\perp = \{2i\cos\theta_I(\mathfrak{n}_a\cos\theta_a - \mathfrak{n}_b\cos\theta_b)\}/D, \tag{8.75}$$

$${}_\perp R_\parallel = \{2i\cos\theta_I(\mathfrak{n}_a\cos\theta_b - \mathfrak{n}_b\cos\theta_a)\}/D, \tag{8.76}$$

$${}_\perp R_\perp = \{(\mathfrak{n}_a + \mathfrak{n}_b)(\cos^2\theta_I - \cos\theta_a\cos\theta_b) - (\mathfrak{n}_a\mathfrak{n}_b - 1)(\cos\theta_a + \cos\theta_b)\cos\theta_I\}/D, \tag{8.77}$$

where

$$D = (\mathfrak{n}_a + \mathfrak{n}_b)(\cos^2\theta_I + \cos\theta_a\cos\theta_b) + (\mathfrak{n}_a\mathfrak{n}_b + 1)(\cos\theta_a + \cos\theta_b)\cos\theta_I. \tag{8.78}$$

The equations can also be used to find the amplitudes $E_\parallel^{(a)}$ and $E_\parallel^{(b)}$ of the transmitted waves in the ionosphere. Four transmission coefficients may be defined as follows:

$$\left.\begin{aligned} {}_\parallel T_a &= E_\parallel^{(a)}/E_\parallel^{(I)} \\ {}_\parallel T_b &= E_\parallel^{(b)}/E_\parallel^{(I)} \end{aligned}\right\} \text{ when } E_y^{(I)} = 0, \qquad \left.\begin{aligned} {}_\perp T_a &= E_\parallel^{(a)}/E_y^{(I)} \\ {}_\perp T_b &= E_\parallel^{(b)}/E_y^{(I)} \end{aligned}\right\} \text{ when } E_\parallel^{(I)} = 0. \tag{8.79}$$

Their values are:

$${}_\parallel T_a = \{2\cos\theta_I(\cos\theta_I + \mathfrak{n}_b\cos\theta_b)\}/D, \tag{8.80}$$

$${}_\parallel T_b = \{2\cos\theta_I(\cos\theta_I + \mathfrak{n}_a\cos\theta_a)\}/D, \tag{8.81}$$

$${}_\perp T_a = \{2i\cos\theta_I(\mathfrak{n}_b\cos\theta_I + \cos\theta_b)\}/D, \tag{8.82}$$

$${}_\perp T_b = \{-2i\cos\theta_I(\mathfrak{n}_a\cos\theta_I + \cos\theta_a)\}/D. \tag{8.83}$$

As a partial check of these formulae, it may be verified that they reduce to the Fresnel formulae (8.14), (8.15), (8.22) and (8.23), for a homogeneous medium, in which case $\mathfrak{n}_a = \mathfrak{n}_b$ and $\theta_a = \theta_b$.

If τ and ω_r/ω are assumed to be known, and independent of θ_a and θ_b, then the above formulae may be used to calculate the reflection and transmission coefficients. The refractive indices $\mathfrak{n}_a$ and $\mathfrak{n}_b$ are given by (8.69) whence θ_a and θ_b can be calculated using Snell's law (8.67). Their values are then inserted in the formulae. It is shown in the next section that in some cases of practical interest, the values of the reflection coefficients are independent of the horizontal direction of the transmission path.

The value $\tau = 0$ gives the reflection coefficients when the earth's field is neglected. The value $\tau = \frac{1}{2}\pi$ occurs when the collision frequency ν is zero.

The formulae (8.74) and (8.75) have been used by Budden (1951*b*) and by Wait and Perry (1957) to plot curves showing how the moduli and arguments of the reflection coefficients vary with the angle of incidence θ_I, for various values of ω_r/ω and τ. The values which gave best agreement with experimental results at 16 kc/s (Bracewell *et al.* 1951) were found to be $\omega_r/\omega = 2$, $\tau = 60°$. These correspond to the values

$$X = 167, \quad Z = 42, \quad N = 530\,\text{cm}^{-3}, \quad \nu = 4{\cdot}2 \times 10^6\,\text{sec}^{-1},$$

which are not unreasonable for the lowest part of the ionosphere (65–75 km) where very low-frequency radio waves are known to be reflected in the daytime.

8.16 The validity of the approximations

The condition (6.23) for the validity of the quasi-longitudinal approximation is equivalent to

$$|Y_T^4/\{4Y_L^2(1-X-iZ)^2\}| \ll 1. \tag{8.84}$$

In the northern hemisphere this would be least accurate for propagation from north to south in the magnetic meridian. Budden (1951*b*) gave a table of values of the quantity in (8.84) for this case, using various values of ω_r/ω and τ that were of practical interest. He showed that it is greatest for large angles of incidence, for large values of τ and for small values of ω_r/ω. The quasi-longitudinal approximation was shown to be valid for most of the range covered by the curves in his paper.

It was assumed in § 8.15 that Y_L is a constant independent of θ_a and θ_b, and the same for both. The validity of this assumption must now be examined. Let Θ be the angle between the earth's magnetic field and the vertical, and let ϕ be the angle between the magnetic meridian plane and the plane of incidence (x–z-plane) measured east from magnetic north. Then Y_L is given by

$$Y_L = (\sin\theta \sin\Theta \cos\phi + \cos\theta \cos\Theta)\, Y, \tag{8.85}$$

where θ may be either θ_a or θ_b. It was mentioned in § 8.15 that when Snell's law (8.67) is used to derive θ_a and θ_b, these angles are complex because the refractive indices $\mathfrak{n}_a$ and $\mathfrak{n}_b$ are complex. As a result Y_L is complex, and the denominator of (8.69) must be written

$$Z \mp \mathscr{I}(Y_L) \pm i\mathscr{R}(Y_L). \tag{8.86}$$

Both $\mathscr{R}(Y_L)$ and $\mathscr{I}(Y_L)$ vary with angle of incidence, and the result is a change in the values of ω_r/ω and τ. The modification is most serious at very oblique incidence, but even then the associated changes are small. For example, suppose that $\tau = 60°$ and $\omega_r/\omega = 4$ when the angle of incidence θ_I is 30°. Then for west to east transmission in England ($\phi = 90°$, $\Theta = 30°$ approx.), and for a frequency of 16 kc/s, the correct values of τ and ω_r/ω for other angles of incidence are as shown in the following table:

Angle of incidence ...	0°	30°	90°	0°	30°	90°
Wave		a			b	
τ	61° 4′	60°	56° 23′	58° 8′	60°	64° 40′
ω_r/ω	3·95	4	4·18	4·05	4	3·89

These variations are not enough to affect the general form of the results. It is sufficiently accurate, for most purposes, to regard fixed values of ω_r/ω and τ as referring to fixed values of N and ν for all angles of incidence. This is equivalent to treating Y_L as a constant. An important consequence of this is that the reflection coefficients are independent of the horizontal direction ϕ of the transmission path.

The theory of § 8.15 was worked out on the assumption that the waves in the ionosphere had only transverse electric fields. It has been shown, however, that even when the quasi-longitudinal approximation is valid, the electric fields also have components in the direction of the wave-normal, given by (6.25). These components will be denoted by $E_L^{(a)} E_L^{(b)}$ for the two waves. In general they have components parallel to the boundary of the ionosphere, which would give the additional terms $E_L^{(a)} \sin\theta_a + E_L^{(b)} \sin\theta_b$, in the first expression (8.73) for the fields just above the boundary. The ratio of the first term $E_L^{(a)} \sin\theta_a$ to the term $E_\parallel^{(a)} \cos\theta_a$ in (8.73) is $E_L^{(a)}/E_\parallel^{(a)} \tan\theta_a$, and there is a similar expression for terms labelled 'b'. The ratios are therefore small for nearly vertical incidence, for then θ_a and θ_b are small, and the longitudinal electric fields are nearly perpendicular to the boundary. A table of values of $|E_L/E_\parallel \tan\theta|$ for some other cases of interest is given by Budden (1951 *b*), which shows that errors which arise by neglecting $E_L^{(a)}$, $E_L^{(b)}$ would become serious for about the same values of the parameters as those for which the quasi-longitudinal approximation becomes inaccurate. For very large values of ω_r/ω, the errors are small, which provides encouragement for using the theory in discussing the behaviour of the lowest frequencies (10 kc/s and lower).

8.17 Reflection at oblique incidence. The Booker quartic

A plane wave is assumed to be incident upon the ionosphere from below, and it gives rise to a reflected wave and to two transmitted waves in the ionosphere, as described at the beginning of § 8.15, and the same notation will be used as in that section. We now consider the problem of finding the refractive indices $\mathfrak{n}_a$ and $\mathfrak{n}_b$ for the two waves in the ionosphere, and the angles θ_a and θ_b between their wave-normals and the vertical, when no approximations are made. The ionosphere is again assumed to be a sharply bounded homogeneous medium, and the effect of the earth's magnetic field is included. Cartesian axes are used with the z-axis vertically upwards, and the wave-normal of the incident wave is in the x–z-plane. The direction cosines of the vector **Y** (opposite in direction to the earth's magnetic field) are l, m, n.

Snell's law must hold, so that

$$\sin\theta_I = \mathfrak{n}\sin\theta, \tag{8.87}$$

where $\mathfrak{n}$ and θ may refer to either transmitted wave. The following argument applies to both waves, so the subscripts a and b will be omitted temporarily. Both $\mathfrak{n}$ and θ are unknown, but the product $\mathfrak{n}\sin\theta$ is known. Now let

$$q = \mathfrak{n}\cos\theta. \tag{8.88}$$

The quantity q was first introduced into magneto-ionic theory by Booker (1936, 1939), and it plays a most important part in much of the theory in this book. To understand its significance, it is useful to treat the refractive index $\mathfrak{n}$ as though it is a vector inclined at an angle θ to the vertical, and of magnitude $\mathfrak{n}$ (see Fig. 8.11). Then q is the vertical component of $\mathfrak{n}$, and $\sin\theta_I$ is its horizontal component. Clearly if q can be found, then $\mathfrak{n}$ and θ can be derived by the relations

$$\mathfrak{n}^2 = q^2 + \sin^2\theta_I, \quad \tan\theta = \sin\theta_I/q. \tag{8.89}$$

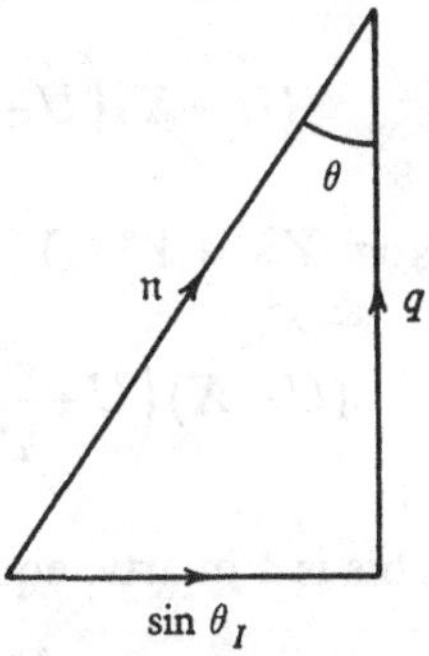

Fig. 8.11. Relation between the refractive index $\mathfrak{n}$, the variable q, and the angle θ between the wave-normal and the vertical.

It will now be shown that q is one root of a quartic equation known as the 'Booker quartic'. An alternative derivation of the Booker quartic is given later in § 13.3.

It is convenient to introduce the notation

$$\sin\theta_I = S, \quad \cos\theta_I = C. \tag{8.90}$$

Within the ionosphere the direction cosines of the wave-normal are $\sin\theta$, 0, $\cos\theta$, and (8.89) shows that these are equal to

$$S(q^2+S^2)^{-\frac{1}{2}}, \quad 0, \quad q(q^2+S^2)^{-\frac{1}{2}}. \tag{8.91}$$

q and θ are in general complex, so that these direction cosines are complex.

The cosine of the angle between the wave-normal and the vector **Y** is equal to $(lS+qn)(q^2+S^2)^{-\frac{1}{2}}$, and hence the component of $\mathbf{Y}$ in the direction of the wave-normal is

$$Y_L = Y(lS+qn)(q^2+S^2)^{-\frac{1}{2}}. \tag{8.92}$$

Equation (8.89) shows that

$$\mathfrak{n}^2-1 = q^2-C^2. \tag{8.93}$$

The notation $1-iZ = U$ will be used here. Then, with a slight rearrangement, the Appleton–Hartree formula (6.1) may be written

$$U-\tfrac{1}{2}Y_T^2(U-X)^{-1}+X(q^2-C^2)^{-1} = \{\tfrac{1}{4}Y_T^4(U-X)^{-2}+Y_L^2\}^{\frac{1}{2}}. \tag{8.94}$$

Both sides of this equation are squared, $\frac{1}{4}Y_T^4(U-X)^{-2}$ is subtracted from both sides, and the equation is multiplied by $U-X$. The result is

$$(U-X)\left(U+\frac{X}{q^2-C^2}\right)^2-Y_T^2\left(U+\frac{X}{q^2-C^2}\right) = (U-X)\,Y_L^2. \tag{8.95}$$

Now $Y_T^2 = Y^2-Y_L^2$, and when (8.92) is used, (8.95) gives

$$(U-X)\left(U+\frac{X}{q^2-C^2}\right)^2-Y^2\left(U+\frac{X}{q^2-C^2}\right)+XY^2\frac{(lS+qn)^2}{q^2-C^2} = 0. \tag{8.96}$$

This is a quartic equation in q, which may be written

$$F(q) \equiv \alpha q^4+\beta q^3+\gamma q^2+\delta q+\epsilon = 0, \tag{8.97}$$

where

$$\left.\begin{aligned}
\alpha &= U(U^2-Y^2)+X(n^2Y^2-U^2),\\
\beta &= 2lnSXY^2,\\
\gamma &= -2U(U-X)(C^2U-X)+2Y^2(C^2U-X)+XY^2(1-C^2n^2+S^2l^2),\\
\delta &= -2C^2lnSXY^2,\\
\epsilon &= (U-X)(C^2U-X)^2-C^2Y^2(C^2U-X)-l^2S^2C^2XY^2.
\end{aligned}\right\} \tag{8.98}$$

The quartic gives four values of q, and it will be shown later that two of them belong to upgoing waves, and two to downgoing waves. To find the reflection coefficient of a sharply bounded ionosphere, the two values q_a and q_b which belong to upgoing waves must be selected. From them the values of $\mathfrak{n}_a$, $\mathfrak{n}_b$, and θ_a, θ_b, may be found by using (8.89). The method of finding the reflection coefficients in this case is given in §§ 8.19 and 8.20.

8.18 Some properties of the Booker quartic

In three special cases the quartic (8.97) reduces to a quadratic equation for q^2. The first is for vertical incidence, so that $S = 0$, $C = 1$, and $q^2 = \mathfrak{n}^2$. The solutions of the equation are then simply the Appleton–Hartree formula (6.1) for $\mathfrak{n}^2$.

The second case is when $l = 0$, which means that the earth's magnetic field is in the y–z-plane. The plane of propagation is the x–z-plane, so that the waves travel from magnetic east to west, or west to east. The solutions of the quadratic are then

$$q^2 = C^2 - \frac{X}{U - \dfrac{Y^2(1-C^2n^2)}{2(U-X)} \pm \left\{\dfrac{Y^4(1-C^2n^2)^2}{4(U-X)^2} + \dfrac{Y^2n^2(C^2U-X)}{U-X}\right\}^{\frac{1}{2}}}, \tag{8.99}$$

which may be regarded as an extension of the Appleton–Hartree formula (6.1).

The third case is when $n = 0$, which means that the earth's magnetic field is horizontal. This therefore applies to propagation at the magnetic equator. The solutions of the quadratic are then

$$q^2 = C^2 - \frac{X}{U - \dfrac{Y^2(1-l^2S^2)}{2(U-X)} \pm \left\{\dfrac{Y^4(1-l^2S^2)^2}{4(U-X)^2} + \dfrac{Y^2l^2S^2U}{U-X}\right\}^{\frac{1}{2}}}. \tag{8.100}$$

If in addition $l = 0$, so that propagation is east–west or west–east at the magnetic equator, the solutions (8.100) become

$$q^2 = C^2 - \frac{X}{U}, \quad q^2 = C^2 - \frac{X}{U - Y^2/(U-X)}. \tag{8.101}$$

The first is independent of Y, which means that the waves are unaffected by the earth's magnetic field. It can be shown that in this case the electric vector is horizontal, so that the electrons are moved only in directions parallel to the earth's field and are unaffected by it.

The above three cases are the only ones for which the coefficients β and δ (see (8.98)) are zero, so that the quartic is a quadratic for q^2, and the roots of the quartic occur in pairs with equal values but opposite signs.

The condition that one root of the quartic is infinite is that the coefficient $\alpha = 0$. This requires that

$$X = \frac{U(U^2 - Y^2)}{U^2 - n^2Y^2}, \tag{8.102}$$

which is independent of S and C.

In the general case when Z is not zero, no root of the Booker quartic can be real. For if q were real, then (8.89) shows that $\mathfrak{n}^2$ and θ are both real. But it was shown in § 6.11 that when the wave-normal is in a real direction, $\mathfrak{n}^2$ cannot be purely real unless Z is zero.

When Z is not zero, two of the roots of the Booker quartic must have negative imaginary parts, and the other two must have positive imaginary parts. This can be proved as follows. Consider some set of fixed values of X, U, Y, l, m, n, for which α is not zero. Starting at $S = 0$, let S be slowly increased (so that C decreases). No root of the quartic can be real for any value of S, so that the imaginary parts of the roots cannot change sign. Now when $S = 0$, the values of q are the four values of $\mathfrak{n}$ given by the Appleton–Hartree formula, and $\mathfrak{n}^2$ always has a negative imaginary part (see § 6.11). Hence two values of $\mathfrak{n}$ have negative imaginary parts (for the upgoing waves), and two have positive imaginary parts (for the downgoing waves). This must also apply for other values of S, and therefore always applies to the four roots of the Booker quartic.

For oblique incidence, it is convenient to define 'upgoing' waves as those for which $\mathscr{I}(q)$ is negative, and 'downgoing' waves as those for which $\mathscr{I}(q)$ is positive. It will be shown later that for an upgoing wave the real part of q may be either positive or negative. Conservation of energy requires that when $\mathscr{I}(q)$ is negative, the direction of energy flow must be upwards, but a negative value of $\mathscr{R}(q)$ means that the wave-fronts are travelling downwards. This apparent contradiction is explained by the fact that the directions of the wave-normal and of the 'ray' are in general different (see chs. 13 and 14).

Since q is in general complex, both $\mathfrak{n}$ and θ are complex. A complex value of θ means that the wave in the ionosphere is an inhomogeneous plane wave of the type described in §§ 4.7 and 4.8. The algebra of ch. 5 can, however, be applied to such a wave, and in particular the state of polarisation can be found just as if the angle θ were real. This property is used in the two following sections.

When Z is zero, the coefficients in the quartic (8.97) are all real, and the roots are therefore either real or in conjugate complex pairs. To decide whether a real value of q refers to an upgoing or downgoing wave, it is convenient to give Z a very small non-zero value, and to examine the sign of $\mathscr{I}(q)$. If it is negative, the wave is an upgoing wave.

8.19 Reflection at oblique incidence for north–south or south–north propagation

We now return to the problem of finding the reflection coefficient of a homogeneous sharply bounded ionosphere when no approximations are made. The case of propagation from magnetic north to south or south to north is considered first, since the algebra is slightly simpler, and serves to illustrate the method to be followed in the more general case. It is not possible to give formulae like (8.74) to (8.77) for the reflection coefficients, but the following method could be used in numerical calculations.

A plane wave is incident on the ionosphere from below with its wave-normal at an angle θ_I to the vertical. It gives rise to a reflected wave and to two transmitted waves in the ionosphere. The notation of § 8.15 will be used, and the total field components just below the boundary may be derived as in that section, and are given by (8.72). Subscripts a and b will be used to distinguish the two transmitted waves in the ionosphere. The Booker quartic is solved (with $S = \sin\theta_I$), and the two roots q_a, q_b, with negative imaginary parts are selected, since these refer to the two upgoing waves. Consider one of these two waves, denoted by the subscript a, and let $E_{\parallel}^{(a)}$, $E_L^{(a)}$, denote the components of its electric field in the x–z-plane, perpendicular and parallel respectively to the wave-normal. Thus $E_L^{(a)}$ is the longitudinal component of the electric field described in § 5.7. Similarly, let $E_y^{(a)}$ denote the component of the electric field parallel to the y-axis. From (8.89) $\mathfrak{n}_a$ and θ_a can be found. The state of polarisation of the wave is given by

$$E_y^{(a)}/E_{\parallel}^{(a)} = \rho_a = \frac{U+X/(\mathfrak{n}_a^2-1)}{-iY_L^{(a)}}. \tag{8.103}$$

The last equation is derived from (5.35), and $Y_L^{(a)}$ is the component of $\mathbf{Y}$ in the direction of the wave-normal. It is given by

$$Y_L^{(a)} = Y(l\sin\theta_a + n\cos\theta_a). \tag{8.104}$$

The longitudinal component $E_L^{(a)}$ is given by (5.38) which becomes

$$E_L^{(a)}/E_y^{(a)} = -iY_T^{(a)}(\mathfrak{n}_a^2-1)/(U-X), \tag{8.105}$$

where

$$Y_T^{(a)} = Y(l\cos\theta_a - n\sin\theta_a). \tag{8.106}$$

The components of the electric field parallel to the x- and z-axes are:

$$\left.\begin{aligned} E_x^{(a)} &= E_{\parallel}^{(a)}\cos\theta_a + E_L^{(a)}\sin\theta_a, \\ E_z^{(a)} &= -E_{\parallel}^{(a)}\sin\theta_a + E_L^{(a)}\cos\theta_a. \end{aligned}\right\} \tag{8.107}$$

The component $\mathscr{H}_x^{(a)}$ of the magnetic field can be found from the Maxwell equation $\operatorname{curl}\mathbf{E} = -ik\boldsymbol{\mathscr{H}}$. All field components in the wave contain the factor $\exp\{-ik(Sx+q_a z)\}$, so that the differential operators in the curl are equivalent to

$$\frac{\partial}{\partial x} \equiv -ikS, \quad \frac{\partial}{\partial y} \equiv 0, \quad \frac{\partial}{\partial z} \equiv -ikq_a, \tag{8.108}$$

and the equation gives

$$\mathscr{H}_x^{(a)} = -q_a E_y^{(a)}. \tag{8.109}$$

Equation (4.2) shows that

$$\mathscr{H}_y^{(a)} = \mathfrak{n}_a E_{\parallel}^{(a)}. \tag{8.110}$$

Hence the four components $E_x^{(a)}$, $E_y^{(a)}$, $\mathscr{H}_x^{(a)}$, and $\mathscr{H}_y^{(a)}$ can now all be expressed in terms of the single quantity $E_y^{(a)}$, thus:

$$\left.\begin{aligned} E_x^{(a)} &= E_y^{(a)}\left\{\frac{\cos\theta_a}{\rho_a} - i\sin\theta_a\, Y_T^{(a)}(\mathrm{n}_a^2-1)/(U-X)\right\},\\ \mathscr{H}_x^{(a)} &= -q_a E_y^{(a)}, \quad \mathscr{H}_y^{(a)} = \frac{\mathrm{n}_a}{\rho_a}E_y^{(a)}. \end{aligned}\right\} \tag{8.111}$$

An exactly similar set of relations is obtained for the other wave, with the subscript or superscript b, and the total field components just above the boundary are found by adding the two sets. They must be equal to the set (8.72) which gives the corresponding components just below the boundary. The four reflection coefficients can now be found by the method of § 8.15.]

8.20 Reflection at oblique incidence in the general case

When the earth's magnetic field is not in the x–z-plane, the problem of finding the field components just above the boundary is a little more complicated. As before, we consider one of the two transmitted waves in the ionosphere, denoted by the subscript a.

Choose new axes, x', y', z', so that the z'-axis is the wave-normal, and the x'–z'-plane contains the earth's magnetic field. Then the direction cosines of these new axes, referred to the orginal x, y, z, axes are:

$$\left.\begin{aligned} &x'\!: \quad (l\cos\theta_a - n\sin\theta_a)\cos\theta_a/G_a, \quad m/G_a, \quad -(l\cos\theta_a - n\sin\theta_a)\sin\theta_a/G_a;\\ &y'\!: \quad -m\cos\theta_a/G_a, \quad (l\cos\theta_a - n\sin\theta_a)/G_a, \quad m\sin\theta_a/G_a;\\ &z'\!: \quad \sin\theta_a, \quad 0, \quad \cos\theta_a; \end{aligned}\right\} \tag{8.112}$$

where
$$G_a = \{1-(l\sin\theta_a + n\cos\theta_a)^2\}^{\frac{1}{2}}. \tag{8.113}$$

The longitudinal and transverse components of $\mathbf{Y}$ are given by

$$\left.\begin{aligned} Y_L^{(a)} &= Y(l\sin\theta_a + n\cos\theta_a),\\ Y_T^{(a)} &= YG_a. \end{aligned}\right\} \tag{8.114}$$

Let the field components of the wave in these new coordinates be $E_{x'}^{(a)}$, $E_{y'}^{(a)}$, $E_{z'}^{(a)}$, $\mathscr{H}_{x'}^{(a)}$, $\mathscr{H}_{y'}^{(a)}$, $\mathscr{H}_{z'}^{(a)}$. Then

$$\left.\begin{aligned} &E_{x'}^{(a)} = E_{y'}^{(a)}/\rho_a, \quad \mathscr{H}_{y'}^{(a)} = E_{y'}^{(a)}\mathrm{n}_a/\rho_a, \quad \mathscr{H}_{x'}^{(a)} = -\mathrm{n}_a E_{y'}^{(a)},\\ &E_{z'}^{(a)} = -E_{y'}^{(a)} Y_T^{(a)}(\mathrm{n}_a^2-1)/(U-X), \quad \mathscr{H}_{z'}^{(a)} = 0; \end{aligned}\right\} \tag{8.115}$$

where ρ_a is given by (8.103). Thus the components are all expressed in terms of the single variable $E_{y'}^{(a)}$. The components parallel to the original x- and y-axes may now be found by using the direction cosines (8.112). In this way the components $E_x^{(a)}$, $E_y^{(a)}$, $\mathscr{H}_x^{(a)}$, $\mathscr{H}_y^{(a)}$ just above the boundary may all be expressed as multiples of the single quantity $E_{y'}^{(a)}$. The components for the other transmitted wave, with subscript or superscript b, may be found in a similar way, and the total field components just above the boundary are found by adding the two sets. They must be equal to the set (8.72) which are the corresponding components just below the boundary. The four reflection coefficients can now be found by the method of § 8.15.

Calculations by a method similar to the above have been made by Yabroff (1957), who gives curves showing how the reflection coefficients vary with angle of incidence for various directions of the earth's magnetic field.

The reflection coefficients in this case can also be found by a completely different method, involving the numerical solution of a differential equation. Details are given in §§ 22.10 and 22.11. See also, Barron and Budden (1959).

Examples

1. Show how Maxwell's equations can be used to deduce the main features of a plane linearly polarised electromagnetic wave in a homogeneous medium. Explain carefully the meaning of the terms 'refractive index' and 'characteristic impedance' for such a medium. What significance is attached to complex values of these quantities?

A plane electromagnetic wave in free space is incident normally on the plane boundary of a homogeneous loss-free medium of magnetic permeability $\mu = 2$, and dielectric constant $\epsilon = \frac{1}{2}$. Find the reflection coefficient.

(Natural Sciences Tripos, 1955. Part II, Physics.)

2. Derive the relations between the components of the electric and magnetic fields on the two sides of the plane boundary between two homogeneous media. What form do these boundary conditions take if one of the media is a perfect conductor?

A piece of glass is partially silvered. Find the reflection coefficient for electromagnetic waves incident normally on the surface, if the thickness of the silver is negligible compared with one wavelength. Assume that the resistivity of silver is the same at high frequencies as for steady currents. Find how thick the silver must be to make the (intensity) reflection coefficient equal to one half, and comment on your result. (Refractive index of glass = 1·5. Resistivity of silver = $1{\cdot}5 \times 10^{-6}$ ohm cm.)

(Natural Sciences Tripos, 1953. Part II, Physics.)

3. Find an expression for the reflection coefficient of a sharply bounded homogeneous ionosphere for propagation from (magnetic) east to west or west to east at the magnetic equator, when the electric vector is in the plane of incidence. Show that the expression is not reciprocal, i.e. it is changed when the sign of the angle of incidence is reversed.

(For the solution see Barber and Crombie, 1959.)

CHAPTER 9

SLOWLY VARYING MEDIUM. THE W.K.B. SOLUTIONS

9.1 Introduction

In this chapter we shall discuss the propagation of radio waves in a slowly varying horizontally stratified ionosphere. The effect of the earth's magnetic field will be neglected except for a brief reference in § 9.14. The corresponding theory, taking into account the earth's magnetic field, is given in §§ 18.12 and 18.14. The meaning of the term 'slowly varying' will be made clear during the discussion. The z-axis is taken vertically upwards, and the ionosphere is horizontally stratified, so that the electron number density N, the collision frequency ν, and hence the refractive index $\mathfrak{n}$, are functions only of z. The case of vertical incidence is considered first. Oblique incidence is considered in §§ 9.11 to 9.13.

9.2 The differential equations

Let a plane radio wave be vertically incident from below, on the horizontally stratified ionosphere. Since the wave normal is vertical, all field components are functions only of z, and

$$\frac{\partial}{\partial x} \equiv \frac{\partial}{\partial y} \equiv 0 \tag{9.1}$$

for the incident wave below the ionosphere. The ionosphere does not vary in the x- and y-directions, and so it could not impress any such variations on the wave. This suggests that we seek solutions for which (9.1) applies to all field components at all heights.

The total field of a radio wave must satisfy the Maxwell equations (2.31) with $\mathbf{D} = \epsilon_0 \mathfrak{n}^2 \mathbf{E}$, where $\mathfrak{n}^2 = 1 - X/(1 - iZ)$ (equation (4.8)). If these are written in full, using (9.1), they give:

$$-\frac{\partial E_y}{\partial z} = -ik\mathscr{H}_x, \tag{9.2}$$

$$\frac{\partial E_x}{\partial z} = -ik\mathscr{H}_y, \tag{9.3}$$

$$\mathscr{H}_z = 0, \tag{9.4}$$

$$\frac{\partial \mathscr{H}_y}{\partial z} = -ik\mathfrak{n}^2 E_x, \tag{9.5}$$

$$\frac{\partial \mathscr{H}_x}{\partial z} = ik\mathfrak{n}^2 E_y, \tag{9.6}$$

$$E_z = 0. \tag{9.7}$$

Equations (9.4) and (9.7) were true for plane waves in a homogeneous medium (§§ 4.2 to 4.4). They evidently still apply for the more complicated wave-system discussed here. This is because here the wave-normal is parallel to the z-axis. For oblique incidence they are not both true.

Equations (9.2) and (9.6) involve only E_y and $\mathscr{H}_x$. They are independent of (9.3) and (9.5) which involve only E_x, $\mathscr{H}_y$. Hence the two pairs can be discussed independently. It is only necessary to discuss the first pair, because the solution for the second pair is similar. The wave is therefore assumed to be linearly polarised with its electric vector parallel to the y-axis. The second pair, (9.3) and (9.5), would give a solution for linear polarisation at right angles to this, and the two solutions could be combined to give, in general, an elliptically polarised wave. Because of the identical form of the two pairs of equations, the state of polarisation would then be the same at all heights.

It is at once clear that there is no solution of (9.2) and (9.6) which represents a 'progressive' wave. For according to § 2.8, in a progressive wave all field quantities vary with z only through a factor $e^{\phi(z)}$ which is the same for all. Now (9.2) would require that $d\phi/dz$ is a constant, while (9.6) would require that it is proportional to $\mathfrak{n}^2$. This is only possible if $\mathfrak{n}^2$ is constant. In spite of this, it is possible to derive *approximate* solutions of the equations which have many of the properties of progressive waves. These are the W.K.B. solutions. At every place where $\mathfrak{n}$ is varying, the reflection-process is going on, and it is this which prevents the occurrence of a true progressive wave. Except in certain places, however, the process is very weak, so that the W.K.B. solutions are very good approximations. This argument is illustrated more mathematically in § 9.9.

The W.K.B. solutions play a most important part in the theory of reflection of waves. In the theory of radio waves they were used first by Gans (1915). For other kinds of wave (sound, water waves), they were used earlier. Their history is given by Jeffreys and Jeffreys (1956).

Let $\mathscr{H}_x$ be eliminated from (9.2) and (9.6). This gives

$$\frac{d^2E_y}{dz^2} + k^2\mathfrak{n}^2 E_y = 0, \tag{9.8}$$

which is a very important differential equation satisfied by the total electric field E_y, and most of chs. 16 and 17 are devoted to its solutions for various forms of the function $\{\mathfrak{n}(z)\}^2$. Since E_y is a function of z only, we now use the total derivative sign d/dz instead of the partial derivative sign $\partial/\partial z$.

9.3 The phase memory concept

Suppose that the variation of $\mathfrak{n}$ with height z is so slow that a range of z can be chosen in which $\mathfrak{n}$ is nearly constant. Consider an upgoing progressive wave in this region, which is linearly polarised with its electric vector parallel to the y-axis. Then its field components would be given by

$$E_y = A\,e^{-ik\mathfrak{n}z}, \tag{9.9}$$

$$\mathscr{H}_x = -\mathfrak{n}A\,e^{-ik\mathfrak{n}z}, \tag{9.10}$$

where A is constant, to a first approximation. Now (9.9) ought to satisfy the differential equation (9.8), but by direct substitution it can be seen that the fit is not very good, unless $d\mathfrak{n}/dz$ and $d^2\mathfrak{n}/dz^2$ are small.

If $\mathfrak{n}$ were real, the part $k\mathfrak{n}z$ of the exponents in (9.9) and (9.10) would be a real angle called the 'phase' of the wave. Since $\mathfrak{n}$ is in general complex, the angle $k\mathfrak{n}z$ is complex but will still be called the (complex) phase. Thus the real part of the complex phase is the same as the phase angle as ordinarily understood, and the imaginary part gives an additional term in the exponent. This term is real because of the factor $-i$, and therefore affects only the amplitude of the wave.

When the wave represented by (9.9) and (9.10) passes through an infinitesimal thickness δz of the ionosphere, the change in its (complex) phase is $k\mathfrak{n}\,\delta z$, which therefore depends on $\mathfrak{n}$. When it passes through a finite thickness z, the change of phase is $k\int_0^z \mathfrak{n}\,dz$. This suggests that a better solution than (9.9) and (9.10) might be

$$E_y = A\exp\left(-ik\int_0^z \mathfrak{n}\,dz\right), \tag{9.11}$$

$$\mathscr{H}_x = -\mathfrak{n}A\exp\left(-ik\int_0^z \mathfrak{n}\,dz\right). \tag{9.12}$$

These satisfy (9.2) exactly, but (9.6) is approximately satisfied only when $d\mathfrak{n}/dz$ is small. However, the agreement is better than for (9.9) and (9.10). The integral in (9.11) and (9.12) constitutes what is sometimes called the 'phase memory' concept. It expresses the idea that a change of phase is cumulative for a wave passing through a slowly varying medium. The solution (9.11) and (9.12) is adequate for the discussion of many problems in the theory of radio wave-propagation. A still more accurate solution, the 'W.K.B. solution', is derived in the following sections.

9.4 Loss-free medium. Constancy of energy-flow

To obtain a solution which is more accurate than (9.11) and (9.12), it is convenient to treat A as a function of z. One method of deriving an approximation to this function can be applied if the medium is loss-free. Then in a progressive wave the flow of energy must be the same for all z. The energy flux is proportional to the time average value of the Poynting vector. If $\mathfrak{n}$ is positive, the exponents in (9.11) and (9.12) are purely imaginary, and the complex Poynting vector $\frac{1}{2}(E_y\mathscr{H}_x^* + E_y^*\mathscr{H}_x)$ is proportional to $\mathfrak{n}A^2$, which must be constant. Hence A is proportional to $\mathfrak{n}^{-\frac{1}{2}}$, so that the solution becomes

$$E_y = A_0\mathfrak{n}^{-\frac{1}{2}}\exp\left\{-ik\int_0^z \mathfrak{n}\,dz\right\}, \tag{9.13}$$

$$\mathscr{H}_y = -A_0\mathfrak{n}^{\frac{1}{2}}\exp\left\{-ik\int_0^z \mathfrak{n}\,dz\right\}, \tag{9.14}$$

where A_0 is a constant. The argument applies also to a downcoming wave, so that another solution is

$$E_y = A_0\mathfrak{n}^{-\frac{1}{2}}\exp\left\{ik\int_0^z \mathfrak{n}\,dz\right\}, \tag{9.15}$$

$$\mathscr{H}_x = A_0\mathfrak{n}^{\frac{1}{2}}\exp\left\{ik\int_0^z \mathfrak{n}\,dz\right\}. \tag{9.16}$$

These two pairs are the two W.K.B. solutions. They are approximate because the reflection process is occurring at all levels, and so the energy-flow is not quite constant.

The above derivation applies only when $\mathfrak{n}$ is real and positive, but it is useful because it illustrates the physical significance of the W.K.B. solutions in this case. Another derivation is given in the next section, which applies for any value of $\mathfrak{n}$ including complex and purely imaginary values.

9.5 Derivation of the W.K.B. solution

In a homogeneous medium there are two solutions of (9.2) to (9.7) which represent progressive waves. Both the fields E_y and $\mathscr{H}_x$ in these waves depend on z only through a factor $e^{i\phi(z)}$ where $\phi(z) = \mp ik\mathfrak{n}z$. The function $\phi(z)$ may be called the generalised phase of the wave. It is in general complex because $\mathfrak{n}$ is complex. In a homogeneous medium $d\phi/dz = \mp ik\mathfrak{n}$, and $d^2\phi/dz^2 = 0$. This suggests that we try to find the function $\phi(z)$ for a slowly varying medium, and we may expect that $d^2\phi/dz^2$ is very small, so that its square and higher powers may be neglected. This idea is the basis of the following method. The function $\phi(z)$ is an example of the Eikonal function which is used later (§ 14.3).

Let

$$E_y = A\,e^{i\phi(z)}, \tag{9.17}$$

where A is a constant. Then

$$\frac{d^2E_y}{dz^2} = A\left\{i\frac{d^2\phi}{dz^2} - \left(\frac{d\phi}{dz}\right)^2\right\}e^{i\phi}. \tag{9.18}$$

When this is substituted in the differential equation (9.8), it shows that $\phi(z)$ must satisfy the equation:

$$\left(\frac{d\phi}{dz}\right)^2 = k^2\mathfrak{n}^2 + i\frac{d^2\phi}{dz^2}. \tag{9.19}$$

This is a non-linear differential equation which would be very difficult to solve exactly, but an approximate solution may be obtained as follows. Since $d^2\phi/dz^2$ is small, it may be neglected to a first approximation. Then

$$\frac{d\phi}{dz} \approx \mp k\mathfrak{n} \tag{9.20}$$

and

$$\frac{d^2\phi}{dz^2} \approx \mp k\frac{d\mathfrak{n}}{dz}. \tag{9.21}$$

The approximate value (9.21) is now substituted in (9.19) which gives, as a second approximation,

$$\frac{d\phi}{dz} \approx \left\{k^2\mathfrak{n}^2 \mp ik\frac{d\mathfrak{n}}{dz}\right\}^{\frac{1}{2}}. \tag{9.22}$$

The square root may be expanded using the binomial theorem, and since the second term is small, only two terms of the expansion need be retained. The result is

$$\frac{d\phi}{dz} \approx \mp k\mathfrak{n}\left\{1 \mp \frac{i}{2k\mathfrak{n}^2}\frac{d\mathfrak{n}}{dz}\right\}. \tag{9.23}$$

Now the second minus sign in (9.23) was obtained from the minus sign in the first approximation (9.20), and for this case the first sign in (9.23) must be minus. Similarly, if the second sign is plus, the first sign must be plus also. Hence (9.23) becomes

$$\frac{d\phi}{dz} \approx \mp k\mathfrak{n} + \frac{i}{2\mathfrak{n}}\frac{d\mathfrak{n}}{dz}, \tag{9.24}$$

which may be integrated at once, to give:

$$\phi \approx \mp k\int^{z} \mathfrak{n}\, dz + i\log(\mathfrak{n}^{\frac{1}{2}}). \tag{9.25}$$

The constant of integration may be made zero by a suitable choice of the lower limit of the integral, which is left unspecified here, since its value does not affect the way in which $\phi(z)$ depends on z. Substitution of (9.25) into (9.17) gives

$$E_y \approx A\mathfrak{n}^{-\frac{1}{2}} \exp\left\{\mp ik\int^{z} \mathfrak{n}\, dz\right\}. \tag{9.26}$$

These two solutions are the W.K.B. solutions. If (9.26) is substituted in the Maxwell equation (9.2), it gives

$$\mathscr{H}_x \approx \mp A\mathfrak{n}^{\frac{1}{2}} \exp\left\{\mp ik\int^{z} \mathfrak{n}\, dz\right\} - \tfrac{1}{2}\mathfrak{n}^{-\frac{3}{2}}\frac{d\mathfrak{n}}{dz} \exp\left\{\mp ik\int^{z} \mathfrak{n}\, dz\right\}. \tag{9.27}$$

Here the second term is small compared with the first, and can usually be neglected.

It should be stressed that the foregoing derivation of the W.K.B. solutions is valid in the most general case when the refractive index is complex.

9.6 Condition for the validity of the W.K.B. solutions

To test whether the W.K.B. solutions are good approximations to solutions of the Maxwell equations, the expression (9.26) may be substituted in the left side of the differential equation (9.8). The result is

$$A\left\{\frac{3}{4}\left(\frac{1}{\mathfrak{n}}\frac{d\mathfrak{n}}{dz}\right)^2 - \frac{1}{2}\frac{1}{\mathfrak{n}}\frac{d^2\mathfrak{n}}{dz^2}\right\} \exp\left\{\mp ik\int^{z} \mathfrak{n}\, dz\right\}. \tag{9.28}$$

This must be very small compared with either term in (9.8). Hence the condition is

$$\frac{1}{k^2}\left|\frac{3}{4}\left(\frac{1}{\mathfrak{n}^2}\frac{d\mathfrak{n}}{dz}\right)^2 - \frac{1}{2}\frac{1}{\mathfrak{n}^3}\frac{d^2\mathfrak{n}}{dz^2}\right| \ll 1. \tag{9.29}$$

The relation (9.29) is a quantitative definition of the term 'slowly varying'. It requires that the derivatives $d\mathfrak{n}/dz$ and $d^2\mathfrak{n}/dz^2$ shall be sufficiently small, and that $\mathfrak{n}$ is not too small. On account of the factor $1/k^2$, the condition is most easily satisfied at high frequencies, but no matter how large the frequency nor how small the derivatives $d\mathfrak{n}/dz$ and $d^2\mathfrak{n}/dz^2$, the condition is certain to fail at a level where $\mathfrak{n}$ passes through a zero. Then one W.K.B. solution can generate some of the other, and this constitutes the process of 'reflection'. The approximations made in § 9.5 no longer apply, and to study the reflection process it is necessary to use more accurate solutions of the differential equations.

The condition (9.29) fails if the derivatives $d\mathfrak{n}/dz$ and $d^2\mathfrak{n}/dz^2$ are large, even when $\mathfrak{n}$ is not small. An extreme example of this is the reflection at the sharp boundary between two homogeneous media (see ch. 8), where the derivatives are infinite. It is shown in § 20.5 that reflection at a steep gradient, and reflection near a zero of $\mathfrak{n}$ are really different aspects of the same phenomenon.

The application of (9.29) may be illustrated by considering a special case. Suppose that $\mathfrak{n}^2$ varies linearly with z, and that the effect of collisions is negligible. Let $z = z_0$ where $\mathfrak{n}^2$ is zero, so that $\mathfrak{n}^2 = -\mathfrak{a}(z - z_0)$. Then the condition (9.29) in this special case becomes

$$\tfrac{5}{16}\mathfrak{a}^2 k^{-2} |\mathfrak{n}|^{-6} \ll 1 \quad \text{or} \quad \tfrac{5}{16} k^{-2} \mathfrak{a}^{-1} |z - z_0|^{-3} \ll 1. \tag{9.30}$$

It may be shown that in the lower part of a Chapman layer, the maximum value of $\mathfrak{a}$ is $1{\cdot}4/H$, where H is the scale height (see § 1.5). Consider a frequency of 1 Mc/s, so that $k = 20\pi/3\,\text{km}^{-1}$, and take $H = 10\,\text{km}$. Then (9.30) gives:

$$|z - z_0| \gg 0{\cdot}2\,\text{km}. \tag{9.31}$$

Now the level where $\mathfrak{n} = 0$ may be regarded as the level where reflection occurs. Hence (9.31) shows that the W.K.B. solutions are good approximations at all levels except within about one free-space wavelength of the level of reflection. This result is true for all frequencies above about 1 Mc/s, and for nearly all vertical gradients that are likely to occur in the ionosphere.

9.7 Properties of the W.K.B. solutions

The two W.K.B. solutions (9.26) were derived by starting with the idea of a progressive wave in a small region of an inhomogeneous medium. There is no exact solution of the differential equation (9.8) which represents a purely progressive wave, except when $\mathfrak{n}^2$ is a constant. In an inhomogeneous medium, the W.K.B. solutions are the analogues of the upgoing and downgoing progressive waves. The solution in (9.26) with the plus sign is the downgoing wave, and that with the minus sign is the upgoing wave. Equation (9.27) shows that (if the small second term is neglected):

$$\left.\begin{aligned} \mathscr{H}_x &= -\mathfrak{n}E_y \quad \text{for the upgoing wave,} \\ \mathscr{H}_x &= \mathfrak{n}E_y \quad \text{for the downgoing wave.} \end{aligned}\right\} \tag{9.32}$$

These relations are the same as in a homogeneous medium, for all values of $\mathfrak{n}$. In particular, if $\mathfrak{n}$ is real and positive, $-\mathscr{H}_x$ and E_y are in phase for the upgoing wave, and $\mathscr{H}_x$ and E_y are in phase for the downgoing wave.

If collisions are neglected, and the electron density increases continuously as the height increases, then $\mathfrak{n}$ decreases continuously. Equation (9.26) shows that the electric field in either wave increases, because of the term $\mathfrak{n}^{-\frac{1}{2}}$, and (9.27) shows that the magnetic field decreases because of the term $\mathfrak{n}^{\frac{1}{2}}$. As the level where $\mathfrak{n} = 0$ is approached, the electric field would become indefinitely large according to (9.26), but before that happens a level is reached where the W.K.B. solutions can no longer be used, because the condition (9.29) is violated. When this level is passed, however, (9.29) is again valid. Hence two W.K.B. solutions are possible above the level of reflection. Here $\mathfrak{n}^2$ is negative and $\mathfrak{n}$ is purely imaginary, and one W.K.B. solution is given by:

$$\left.\begin{aligned} E_y &= A\mathfrak{n}^{-\frac{1}{2}} \exp\left\{-k\int^z |\mathfrak{n}|\, dz\right\}, \\ \mathscr{H}_x &= -\mathfrak{n}E_y. \end{aligned}\right\} \tag{9.33}$$

The electric and magnetic fields are now in quadrature so that the time-average value of the vertical component of the Poynting vector is zero. This indicates that on the average there is no vertical flow of energy. There is some energy stored in the field, which simply pulsates up and down with twice the wave-frequency. Both fields decrease as the height increases, and for both of them the phase is independent of z. A wave of the type (9.33) is called 'evanescent'. An example of a plane evanescent wave in a homogeneous medium was given in §4.6. The wave (9.33) is more complicated, because the medium is not homogeneous.

The other W.K.B. solution above the level of reflection is given by

$$\left.\begin{aligned} E_y &= A\mathfrak{n}^{-\frac{1}{2}} \exp\left\{k\int^z |\mathfrak{n}|\, dz\right\}, \\ \mathscr{H}_x &= \mathfrak{n}E_y. \end{aligned}\right\} \tag{9.34}$$

It has similar properties to the wave in (9.33), but both fields increase indefinitely as the height increases. This wave, therefore, could not be excited by waves incident upon the ionosphere from below, but it could occur if waves came into the ionosphere from above.

When the effect of collisions is allowed for, $\mathfrak{n}^2$ is complex and $\mathfrak{n} = \mu - i\chi$ where μ and χ are always positive (see §4.5). Then $\mathfrak{n}$ never becomes exactly zero, but in many cases of interest there is still a region where $|\mathfrak{n}|$ is so small that the condition (9.29) is violated. Then the

W.K.B. solutions fail and reflection occurs. If the collision frequency ν is so large that (9.29) is valid at all levels, there is no appreciable reflection, and an upgoing wave continues to travel upwards until it is absorbed. Its fields are given with good accuracy by the upgoing W.K.B. solutions at all levels.

9.8 The reflection coefficient

Suppose that an upgoing radio wave of unit amplitude is generated at the ground, where $z = 0$. Then the upgoing W.K.B. solution

$$E_y = \mathfrak{n}^{-\frac{1}{2}} \exp\left\{-ik\int_0^z \mathfrak{n}\, dz\right\} \tag{9.35}$$

gives the complex amplitude at any other level. It is required to find the amplitude R of the resulting reflected wave when it reaches the ground. Its field at any other level is given by the downgoing W.K.B. solution, and is therefore equal to

$$E_y = R\mathfrak{n}^{-\frac{1}{2}} \exp\left\{ik\int_0^z \mathfrak{n}\, dz\right\}. \tag{9.36}$$

If reflection occurs where $\mathfrak{n} = 0$, and if the fields in the two waves tend to equality as $\mathfrak{n} \to 0$, then clearly

$$R = \exp\left\{-2ik\int_0^{z_0} \mathfrak{n}\, dz\right\}, \tag{9.37}$$

where z_0 is the value of z which makes $\mathfrak{n} = 0$. In fact this treatment is incorrect, and a more exact version given in § 16.6 shows that the right-hand side of (9.37) must be multiplied by a factor i. For many purposes this is unimportant, however, and many useful results can be obtained by using the form (9.37) for the reflection coefficient R. The modulus of R is given by

$$|R| = \exp\left\{-2k\int_0^{z_0} \chi\, dz\right\}. \tag{9.38}$$

This quantity is often observed in radio measurements.

The integral in (9.37) expresses the total change of the complex phase of the wave during its passage from the ground to the reflection level and back. It is therefore called the 'phase integral'. If the effect of collisions is included, the condition $\mathfrak{n} = 0$ does not hold for any real value of z. Often, however, $\mathfrak{n}$ is a known analytic function of z, and is zero for a complex value z_0 of z. Then the phase integral is a contour integral in the complex z-plane. This idea is the basis of the 'phase integral' method.

9.9 Coupling between upgoing and downgoing waves

A wave represented by one W.K.B. solution may give rise as it travels to another wave represented by the other W.K.B. solution. The magnitude of this process is measured by the left-hand side of (9.29) and is negligible except near the level of reflection. This idea is illustrated by the following alternative derivation of the W.K.B. solutions.

In a homogeneous medium a progressive wave travelling in the direction of positive z, has field components E_y and $\mathscr{H}_x$ related by $\mathscr{H}_x = -\mathfrak{n}E_y$ (from equation (4.1)), and a progressive wave travelling in the direction of negative z has $\mathscr{H}_x = \mathfrak{n}E_y$. Both waves are assumed to be linearly polarised. In an inhomogeneous medium these relations may be used to define the analogues of the progressive waves. Hence let

$$\left.\begin{aligned} E_y &= E_y^{(1)} + E_y^{(2)},\\ \text{and}\qquad \mathscr{H}_x &= \mathscr{H}_x^{(1)} + \mathscr{H}_x^{(2)}, \end{aligned}\right\} \tag{9.39}$$

where

$$\left.\begin{aligned} \mathscr{H}_x^{(1)} &= -\mathfrak{n}E_y^{(1)},\\ \mathscr{H}_x^{(2)} &= \mathfrak{n}E_y^{(2)}. \end{aligned}\right\} \tag{9.40}$$

Now the Maxwell equations (9.2) and (9.6) give:

$$\left.\begin{aligned} \frac{dE_y^{(1)}}{dz} + \frac{dE_y^{(2)}}{dz} &= -ik\mathfrak{n}E_y^{(1)} + ik\mathfrak{n}E_y^{(2)},\\ -\mathfrak{n}\frac{dE_y^{(1)}}{dz} + \mathfrak{n}\frac{dE_y^{(2)}}{dz} &= ik\mathfrak{n}^2E_y^{(1)} + ik\mathfrak{n}^2E_y^{(2)} + E_y^{(1)}\frac{d\mathfrak{n}}{dz} - E_y^{(2)}\frac{d\mathfrak{n}}{dz}. \end{aligned}\right\} \tag{9.41}$$

Multiply the second of these by $1/\mathfrak{n}$ and subtract it from, or add to it, the first. This gives the two equations:

$$\left.\begin{aligned} \frac{dE_y^{(1)}}{dz} + ik\mathfrak{n}E_y^{(1)} + \frac{1}{2\mathfrak{n}}\frac{d\mathfrak{n}}{dz}E_y^{(1)} &= \frac{1}{2\mathfrak{n}}\frac{d\mathfrak{n}}{dz}E_y^{(2)},\\ \frac{dE_y^{(2)}}{dz} - ik\mathfrak{n}E_y^{(2)} + \frac{1}{2\mathfrak{n}}\frac{d\mathfrak{n}}{dz}E_y^{(2)} &= \frac{1}{2\mathfrak{n}}\frac{d\mathfrak{n}}{dz}E_y^{(1)}, \end{aligned}\right\} \tag{9.42}$$

which may be solved by successive approximations. The right-hand sides both contain the factor $(1/2\mathfrak{n})(d\mathfrak{n}/dz)$ which is small in a slowly varying medium, except near $\mathfrak{n} = 0$. As a first approximation, the right-hand sides are neglected† and (9.42) then gives two independent differential equations for $E_y^{(1)}$ and $E_y^{(2)}$ whose solutions are:

$$E_y^{(1)} = \mathfrak{n}^{-\frac{1}{2}}\exp\left\{-ik\int^z \mathfrak{n}\,dz\right\},\quad E_y^{(2)} = \mathfrak{n}^{-\frac{1}{2}}\exp\left\{ik\int^z \mathfrak{n}\,dz\right\}. \tag{9.43}$$

† This is in effect the same assumption as was made in §9.5 where $d^2\phi/dz^2$ was neglected.

These are simply the W.K.B. solutions. They may be substituted in the right-hand sides of (9.42) to give a second approximation. In the first equation (9.42), let

$$E_y^{(1)} = A(z)\mathfrak{n}^{-\frac{1}{2}}\exp\left\{-ik\int^z \mathfrak{n}\,dz\right\}.$$

Then the equation gives

$$\frac{dA(z)}{dz} = \frac{1}{\mathfrak{n}}\frac{d\mathfrak{n}}{dz}\exp\left\{2ik\int^z \mathfrak{n}\,dz\right\}. \tag{9.44}$$

If $\mathfrak{n}$ is real, the right-hand side of (9.44) is a rapidly oscillating function of z. Hence when $(1/\mathfrak{n})(d\mathfrak{n}/dz)$ is small, $A(z)$ differs from a constant only by a small oscillating function of z. A second approximation for $E_y^{(2)}$ may be found in a similar way.

The case when $\mathfrak{n}$ is nearly purely imaginary usually occurs in the ionosphere above the level of reflection, where the second W.K.B. solution (9.43) has zero amplitude. Hence to a first approximation $E_y^{(2)} = 0$, and the above process for obtaining the second approximation leaves the other W.K.B. solution unchanged.

The equations (9.42) could be used to obtain third- and higher-order approximations, but the algebra becomes very complicated.

9.10 Extension to oblique incidence

The preceding sections have discussed the W.K.B. solutions for a slowly varying ionosphere for the case of vertical incidence only, but the methods may easily be extended to apply to oblique incidence.

The electron density and collision frequency in the ionosphere vary continuously with the height z, but we may imagine that the ionosphere is replaced by a number of thin discrete strata, in each of which the medium is homogeneous. By making these strata thin enough and numerous enough, we may approximate as closely as we please to the actual ionosphere. A plane wave is incident upon the ionosphere from below, with its normal in the x–z-plane at an angle θ_I to the vertical. At the lower boundary of the first stratum, it is partially reflected and partially transmitted. The transmitted wave is partially reflected and transmitted at the second boundary between the strata, and so on. In any one stratum there are in general two waves which are the resultants of all the partially reflected and transmitted waves entering the stratum. In the nth stratum let $\mathfrak{n}_n$ be the refractive index, and let θ_n, θ_n', be the angles between the wave-normals and the vertical. Then for the boundary between the $(n-1)$th and nth strata Snell's law, (8.6) gives

$$\mathfrak{n}_{n-1}\sin\theta_{n-1} = \mathfrak{n}_n \sin\theta_n = \mathfrak{n}_n \sin\theta'_n. \tag{9.45}$$

Hence $\theta'_n = \pi - \theta_n$, and $\mathfrak{n}_n \sin\theta_n$ is the same for all strata. Below the ionosphere $\mathfrak{n} = 1$, and hence

$$\mathfrak{n}_n \sin\theta_n = \sin\theta_I. \tag{9.46}$$

For a wave in the nth stratum, any field quantity depends on z and x only through factors

$$A\exp\{-ik\mathfrak{n}_n(z\cos\theta_n + x\sin\theta_n)\} + B\exp\{ik\mathfrak{n}_n(z\cos\theta_n - x\sin\theta_n)\}, \tag{9.47}$$

where A and B are the amplitudes of the resultant upgoing and downgoing waves. If the operator $\partial/\partial x$ operates upon this expression, it is equivalent to multiplication by $-ik\mathfrak{n}_n\sin\theta_n = -ik\sin\theta_I$ (from (9.46)), which is independent of x and z. Similarly, the operator $\partial/\partial y$ is equivalent to multiplication by zero. It is convenient to use the notation $S = \sin\theta_I$, $C = \cos\theta_I$. Hence we may write symbolically:

$$\frac{\partial}{\partial x} \equiv -ikS, \quad \frac{\partial}{\partial y} \equiv 0. \tag{9.48}$$

Now this result is true no matter how thin are the strata, and it may therefore be expected to hold in the limit when the strata are infinitesimally thin, so that the electron density and collision-frequency vary continuously with height.

The normals to the waves in the nth stratum are horizontal when $\theta_n = \theta'_n = \frac{1}{2}\pi$, so that $\sin\theta_n = 1$, and Snell's law (9.46) gives $\mathfrak{n}_n = \sin\theta_I$. If a wave is followed on its upward path until its normal becomes horizontal, we should expect that thereafter it would travel downwards, so that reflection would occur where $\mathfrak{n} = S$. This is shown to be so in § 9.12, and more rigorously in ch. 16.

The meaning of the result (9.48) may be summarised as follows. The horizontal variations of all field quantities in the incident wave are given by (9.48). The ionosphere is horizontally stratified, so that its properties are independent of x and y, and Snell's law shows that it is incapable of impressing any new horizontal variations on the waves within it. Hence (9.48) must apply at all levels. For vertical incidence the relations (9.1) were used, but for oblique incidence they are now to be replaced by (9.48).

9.11 The differential equations for oblique incidence

To find the differential equations to be satisfied by the wave-fields at oblique incidence, Maxwell's equations (2.31) must be written in full, using (9.48). They give:

$$-\frac{\partial E_y}{\partial z} = -ik\mathscr{H}_x, \tag{9.49}$$

$$\frac{\partial E_x}{\partial z} + ikSE_z = -ik\mathscr{H}_y, \tag{9.50}$$

$$-ikSE_y = -ik\mathscr{H}_z, \tag{9.51}$$

$$-\frac{\partial \mathscr{H}_y}{\partial z} = ik\mathrm{n}^2 E_x, \tag{9.52}$$

$$\frac{\partial \mathscr{H}_x}{\partial z} + ikS\mathscr{H}_z = ik\mathrm{n}^2 E_y, \tag{9.53}$$

$$-ikS\mathscr{H}_y = ik\mathrm{n}^2 E_z. \tag{9.54}$$

These equations may be separated into two independent sets. Equations (9.49), (9.51), (9.53) contain E_y, $\mathscr{H}_x$, $\mathscr{H}_z$, and (9.50), (9.52), (9.54) contain E_x, E_z, $\mathscr{H}_y$. One of these sets of field variables may be set equal to zero without affecting the other, so that the corresponding waves are propagated independently.

For the first set the electric field is everywhere parallel to the y-axis, that is, horizontal, and the waves are therefore said to be horizontally polarised. For the second set the electric field is everywhere parallel to the x–z-plane, that is, to the plane of incidence, and here the waves are sometimes said to be 'vertically polarised', but it should be stressed that the electric vector is not in general vertical.

9.12 The W.K.B. solutions for horizontal polarisation at oblique incidence

The field component $\mathscr{H}_z$ may be eliminated from (9.51) and (9.53) to give

$$\frac{\partial \mathscr{H}_x}{\partial z} = ik(\mathrm{n}^2 - S^2)\, E_y. \tag{9.55}$$

Now it is convenient to put $\mathrm{n}^2 - S^2 = q^2$, as was done in § 8.17 (equation (8.89)). There the effect of the earth's magnetic field was allowed for, and q was one solution of a quartic equation. Here the earth's magnetic field is neglected, so that $\mathfrak{n}^2$ is given by (4.8) and is independent of θ_I. Hence the expression for q is given simply by

$$q^2 = \mathfrak{n}^2 - S^2 = 1 - \frac{X}{1-iZ} - S^2 = C^2 - \frac{X}{1-iZ}. \tag{9.56}$$

The physical significance of q was explained in §8.17 (see Fig. 8.11).

The equations (9.48) show that all field quantities contain a factor e^{-ikSx}. This is assumed to be omitted, in the same way as the time factor $e^{i\omega t}$ is omitted. Then the terms which remain are functions of z only, and so the partial differentiation sign $\partial/\partial z$ may be replaced by the total differentiation sign d/dz, and equations (9.49) and (9.55) give

$$\frac{dE_y}{dz} = ik\mathscr{H}_x, \quad \frac{d\mathscr{H}_x}{dz} = ikq^2E_y. \tag{9.57}$$

Elimination of $\mathscr{H}_x$ gives $$\frac{d^2E_y}{dz^2} + k^2q^2E_y = 0. \tag{9.58}$$

This is the same as (9.8), except that $\mathfrak{n}^2$ is replaced by q^2. The whole of the arguments of §9.5 can now be used to derive the W.K.B. solutions, which are as follows:

$$E_y = Aq^{-\frac{1}{2}} \exp\left\{\mp ik\int^z q\,dz\right\}, \tag{9.59}$$

$$\mathscr{H}_x = \mp Aq^{\frac{1}{2}} \exp\left\{\mp ik\int^z q\,dz\right\}, \tag{9.60}$$

where A is a constant. In (9.60) a small second term has been omitted. It is analogous to the second term of (9.27). Both the expressions (9.59) and (9.60) should contain a factor $\exp\{i(\omega t - kSx)\}$, but this is often omitted because we are here mainly interested in the dependence on z.

The goodness of the approximation of (9.59) to true solutions of Maxwell's equations may be tested by substituting in (9.58) as was done in §9.6 for the corresponding equations for vertical incidence. The condition that (9.59) shall be a good approximation is

$$\frac{1}{k^2}\left|\frac{3}{4}\left(\frac{1}{q^2}\frac{dq}{dz}\right)^2 - \frac{1}{2}\frac{1}{q^3}\frac{d^2q}{dz^2}\right| \ll 1. \tag{9.61}$$

This will fail near the level where $q = 0$, no matter how small are the derivatives dq/dz and d^2q/dz^2. Hence $q = 0$, or $\mathfrak{n}^2 = S^2$ gives the level of reflection in this case.

It may be inferred that the reflection coefficient of the ionosphere is given by

$$R = \exp\left\{-2ik\int_0^{z_0} q\,dz\right\}, \tag{9.62}$$

where z_0 is the value of z which makes $q = 0$. This would follow from an argument similar to that given in § 9.8 for vertical incidence. A more exact treatment (§ 16.6) shows that the right-hand side of (9.62) must be multiplied by a factor i, but this is unimportant for many purposes. When the effect of collisions is allowed for, q is complex at all levels, and is never zero for any real value of z. Often, however, q is a known analytic function of z, and is zero for a complex value z_0 of z. The phase integral in (9.62) is then a contour integral in the complex z-plane (see ch. 16).

The W.K.B. solutions (9.59) and (9.60) can be derived by an alternative method exactly analogous to that given in § 9.9.

9.13 The W.K.B. solutions at oblique incidence when the electric vector is parallel to the plane of incidence

When the electric vector is parallel to the x–z-plane, the appropriate equations are (9.50), (9.52) and (9.54), and these are not so simple as for horizontal polarisation. The field component E_z may be eliminated from (9.50) and (9.54), and the resulting equations are:

$$\left.\begin{aligned} \frac{dE_x}{dz} &= -ik\frac{q^2}{\mathfrak{n}^2}\mathscr{H}_y, \\ \frac{d\mathscr{H}_y}{dz} &= -ik\mathfrak{n}^2 E_x, \end{aligned}\right\} \tag{9.63}$$

which corresponds to (9.57) for horizontal polarisation. It is not now possible to obtain a second-order differential equation of the simple form (9.58). Elimination of E_x from (9.63) gives

$$\frac{d^2\mathscr{H}_y}{dz^2} - \frac{1}{\mathfrak{n}^2}\frac{d(\mathfrak{n}^2)}{dz}\frac{d\mathscr{H}_y}{dz} + k^2q^2\mathscr{H}_y = 0. \tag{9.64}$$

The W.K.B. solutions may be found by the method of § 9.5. Let

$$\mathscr{H}_y = e^{i\phi(z)}, \tag{9.65}$$

where $\phi(z)$ is the generalised complex phase. Then the differential equation which ϕ must satisfy is

$$i\frac{d^2\phi}{dz^2} - \left(\frac{d\phi}{dz}\right)^2 - \frac{i}{\mathfrak{n}^2}\frac{d(\mathfrak{n}^2)}{dz}\frac{d\phi}{dz} + k^2q^2 = 0. \tag{9.66}$$

Since the medium is slowly varying, the first and third terms are small, and hence to a first approximation:

$$\frac{d\phi}{dz} = \mp kq, \quad \frac{d^2\phi}{dz^2} = \mp k\frac{dq}{dz}. \tag{9.67}$$

These values are inserted in the first and third terms of (9.65) which gives for the second approximation

$$\frac{d\phi}{dz} = \mp \left\{k^2q^2 \mp ik\left(\frac{dq}{dz} - \frac{2q}{\mathfrak{n}}\frac{d\mathfrak{n}}{dz}\right)\right\}^{\frac{1}{2}}, \tag{9.68}$$

where either the upper or lower sign must be used throughout, for the same reasons as in §9.5. The last two terms on the right of (9.68) are small, and their squares and products will be neglected. Then expanding the right-hand side by the binomial theorem gives

$$\frac{d\phi}{dz} = \mp kq + \frac{1}{2}\frac{i}{q}\frac{dq}{dz} - \frac{i}{\mathfrak{n}}\frac{d\mathfrak{n}}{dz}, \tag{9.69}$$

which may be integrated at once, and the result inserted in (9.65). This gives

$$\mathscr{H}_y = \mathfrak{n}q^{-\frac{1}{2}}\exp\left\{\mp ik\int^z q\,dz\right\}. \tag{9.70}$$

A corresponding expression for E_x may now be derived from the second equation (9.63). If derivatives of q and $\mathfrak{n}$ are neglected:

$$E_x = \pm\mathfrak{n}^{-1}q^{\frac{1}{2}}\exp\left\{\mp ik\int^z q\,dz\right\}. \tag{9.71}$$

The expressions (9.70) and (9.71) are the W.K.B. solutions when the electric field is parallel to the plane of incidence. They should contain a factor $\exp\{i(\omega t - kSx)\}$, but this is often omitted because we are here mainly interested in the dependence on z.

The goodness of the approximation of (9.70) to true solutions of Maxwell's equations may be tested by substituting in (9.64). The condition that (9.70) shall be a good approximation is

$$\frac{1}{k^2}\left|\frac{3}{4}\left(\frac{1}{q^2}\frac{dq}{dz}\right)^2 - \frac{1}{2}\frac{1}{q^3}\frac{d^2q}{dz^2} + \frac{1}{q^2}\left\{\frac{1}{\mathfrak{n}}\frac{d^2\mathfrak{n}}{dz^2} - 2\left(\frac{1}{\mathfrak{n}}\frac{d\mathfrak{n}}{dz}\right)^2\right\}\right| \ll 1. \tag{9.72}$$

This will fail near the level where $q = 0$, no matter how small are the derivatives dq/dz, d^2q/dz^2, $d\mathfrak{n}/dz$ and $d^2\mathfrak{n}/dz^2$. Hence $q = 0$ gives the level of reflection. The condition will also fail near the level when $\mathfrak{n} = 0$, which is not a level of reflection. If the electron density increases monotonically with height in the ionosphere, then the level where $\mathfrak{n} = 0$ is above the reflection level where $q = 0$. If the two levels are well separated, the reflection process is unaffected by the failure of (9.72) at the level $\mathfrak{n} = 0$, and the reflection coefficient is given by (9.62) just as for horizontal polarisation. If, however, the level where $\mathfrak{n} = 0$ is close to the level where $q = 0$, the reflection coefficient may be affected, and a more detailed study of the differential equation is needed to find its true value. This is given in §§ 16.12 to 16.17.

9.14 The effect of including the earth's magnetic field

The differential equations have not yet been formulated for waves in the ionosphere when the effect of the earth's magnetic field is included.

They are given in ch. 18, and it is shown there that approximate W.K.B. solutions exist in this case also. In a W.K.B. solution for oblique incidence, each field component is expressed in the form:

$$F = F_0(z)\exp\left\{-ik\int^z q\,dz\right\}, \tag{9.73}$$

where q is one of the four roots of the Booker quartic equation (8.97), and $F_0(z)$ is some function of z analogous to $q^{-\frac{1}{2}}$ in (9.59) or $\mathfrak{n}^{-1}q^{\frac{1}{2}}$ in (9.71), but more complicated (see § 18.12). There are thus four W.K.B. solutions in general. It will be shown later that two of them represent upgoing waves, and the other two represent downgoing waves. For vertical incidence q becomes equal to $+\mathfrak{n}$ and $-\mathfrak{n}$, where $\mathfrak{n}$ has one of the values given by the Appleton–Hartree formula (6.1). The W.K.B. solutions are found to be good approximations at most levels in a slowly varying medium, but they fail near levels where two roots of the Booker quartic become equal. These are levels of reflection if the two roots refer to waves of which one travels upwards and the other downwards. It is also possible that the two equal roots both refer to upgoing waves (or both to downgoing waves), and at the level where this occurs there is said to be strong 'coupling' between the two waves.

The level where $q = 0$ is not necessarily a level of reflection or coupling. It may be a level of reflection, but this only happens if the quartic has a double root there. If the quartic has only one root at this level, then the W.K.B. solution may be a good approximation there. This subject is treated fully in § 18.12.

9.15 Ray theory and 'full wave' theory

In the preceding sections it has been shown that the W.K.B. solutions may be used at nearly all levels in a slowly varying medium. They fail near certain levels where reflection is occurring, but reflection coefficients may be deduced on the assumption that the wave represented by the upgoing W.K.B. solution is completely converted into that represented by the downgoing W.K.B. solution at the reflection point. If the effect of collisions is included, the variables $\mathfrak{n}$ and q are complex, and the reflection point must then be assumed to be at some complex value of the height.

Every W.K.B. solution is of the form (9.73). It is the product of the exponential which expresses the 'phase memory' idea, and the term $F_0(z)$ which varies much more slowly than the exponential. The exponent is equal to $-i$ multiplied by the generalised or complex phase of the wave, which is calculated as though the wave travelled at each level according to the laws of geometrical optics. A W.K.B. solution is, therefore, often said to be a mathematical expression of 'ray' theory. A more precise meaning for the term 'ray' is given in ch. 11.

When the W.K.B. solutions fail, it is necessary to make a further study of the differential equations, and to find a solution which cannot in general be interpreted in terms of geometrical optics. Such a solution is sometimes called a 'full wave' solution. For high frequencies, above about 1 Mc/s, most of the important aspects of radio wave-propagation in the ionosphere can be handled by 'ray theory'. At levels of reflection or coupling, the W.K.B. solutions fail, but an investigation of the reflection or coupling process by 'full wave' theory shows that in general the 'ray theory' can still be considered to hold, with only trivial modification. An exception to this is the phenomenon of partial penetration and reflection for frequencies near the penetration frequency of an ionised layer (§ 17.5). Here the ray theory is inadequate, and a 'full wave' treatment is necessary.

For lower frequencies (below about 1 Mc/s), and especially for very low frequencies (below about 100 kc/s) the ionosphere can change appreciably within a distance of one wavelength, and cannot always be regarded as a slowly varying medium. Then the W.K.B. solutions fail to be good approximations, and a full wave-solution is required for nearly every problem. Chs. 10 to 14 deal with problems that can be treated by ray-theory methods. The later chapters are almost entirely concerned with 'full wave' solutions.

CHAPTER 10

RAY THEORY FOR VERTICAL INCIDENCE WHEN THE EARTH'S MAGNETIC FIELD IS NEGLECTED

10.1 The use of pulses

In the commonest and most powerful method of using radio waves to investigate the ionosphere, pulses of radio energy are generated by a transmitter on the ground. Each pulse travels upwards and is reflected from the ionosphere. It returns to a receiver on the ground, and the time of travel is measured by using a cathode-ray oscillograph with a linear time base. The pulses are transmitted at regular intervals, often about $\frac{1}{50}$ sec, and the time base makes a traverse starting just before each transmitted pulse. This is the technique used in radar. In this and the next chapter the propagation of pulses of radio waves is discussed for the special case when the earth's magnetic field is neglected. This chapter deals with vertical incidence, and it is assumed that a pulse is composed of plane waves in which all the wave-normals are vertical. Such a pulse would be infinite in horizontal extent. An actual transmitter can radiate a pulse of waves within a narrow cone. Then the wave-fronts are spherical, and the pulse or 'wave packet' is limited in extent in all directions. The propagation of wave-packets at both oblique and vertical incidence is discussed in ch. 11. Chs. 12 to 14 deal with these same problems when the effect of the earth's magnetic field is included.

The methods of 'ray theory' (see § 9.15) are used throughout this and the following chapters as far as ch. 14. This means that the W.K.B. solutions are assumed to be good approximations at all levels. The wave represented by the upgoing W.K.B. solution is assumed to be converted into that represented by the downgoing W.K.B. solution at the level of reflection. Some examples where this assumption fails are given in later chapters.

The present chapter is not concerned with the polarisation of the waves, and the field of a wave will be denoted simply by the symbol E, which may be thought of as one component of the electric field of the wave, although the arguments could be applied equally well to a component of the magnetic field. The analysis in this chapter will usually be expressed in terms of the wave-frequency f, rather than the angular-frequency ω.

In most of the present chapter the effect of electron collisions is neglected, and the electron density is assumed to be less than $\epsilon_0 \mathfrak{m}\omega^2/e^2$ so that the frequency f is always greater than the plasma-frequency f_N (see § 3.2). Thus the refractive index $\mathfrak{n}$ is real and positive and is the same as its real part μ. To emphasise this the symbol μ will be used for the refractive index. Discussion of problems involving a complex $\mathfrak{n}$ is resumed in § 10.15.

10.2 The group velocity

A pulsed radio signal of finite duration can be represented by a Fourier integral thus:

$$E(t) = \int_{-\infty}^{\infty} F(f)\, e^{2\pi i f t}\, df. \tag{10.1}$$

The form of the function $F(f)$ depends upon the shape of the pulse, which is usually a signal of constant frequency f_1, and roughly constant amplitude lasting for a time T, which is often long compared with $1/f_1$. (T may be 200 μs or less, and f_1 may be anything from 50 kc/s to greater than 20 Mc/s.) $F(f)$ then has maxima at or near f_1 and $-f_1$, and f_1 is therefore called the predominant frequency. If $T \gg 1/f_1$, the maxima are narrow so that only frequencies very close to f_1 play a part in the propagation of the pulse. It can be shown that because $E(t)$ is real, $|F(f)|$ is symmetric, and $\arg F(f)$ is antisymmetric about $f = 0$. In practice, $\arg F(f)$ does not vary appreciably in a small range near f_1. In this and the following sections the exact form of $F(f)$ is unimportant. It is considered in more detail in § 10.14.

Each component-frequency gives a wave which travels up into the ionosphere independently, and its field is given by the W.K.B. expression (9.13). Here the factor $\mathfrak{n}^{-\frac{1}{2}} = \mu^{-\frac{1}{2}}$ varies slowly with z. Below the ionosphere it is unity for all frequencies, for both the incident and reflected waves. In the ionosphere it may be thought of as causing a temporary modification of the function $F(f)$. It does not affect the results of this chapter, and will therefore be omitted. Only the exponential or 'phase memory' term in (9.13) need be retained. Equation (10.1) gives the signal at the ground, where $z = 0$. At height z the signal becomes

$$E(t, z) = \int_{-\infty}^{\infty} F(f) \exp\left\{2\pi i f\left(t - \frac{1}{c}\int_0^z \mu\, dz\right)\right\} df, \tag{10.2}$$

which could be evaluated for any values of t and z, but only has appreciable magnitude for those values which give the position z of the pulse at a given time t. Clearly the value of the integral is large when frequencies near the predominant frequency f_1 all have the same phase. This requires that

$$\frac{\partial}{\partial f}\left\{ft - \frac{1}{c}\int_0^z \mu f\, dz\right\} = 0 \quad \text{when} \quad f = f_1. \tag{10.3}$$

Hence the position of the pulse at time t is given by

$$t - \frac{1}{c}\int_0^z \left\{\frac{\partial}{\partial f}(\mu f)\right\}_{f=f_1} dz = 0. \tag{10.4}$$

Let U_z be the upward velocity of the pulse. Then (10.4) shows that

$$\frac{c}{U_z} = \left\{\frac{\partial}{\partial f}(\mu f)\right\}_{f=f_1}. \tag{10.5}$$

Equation (10.5) gives the upward vertical component of the velocity of the pulse. If the velocity has any horizontal component, this could not be detected in the present example, since the pulse was assumed to be infinite in horizontal extent. To study the horizontal component of the velocity, it is convenient to consider a wave-packet which has only limited horizontal extent. This is done later in §§ 11.2 and 11.3, where it is shown that, when the earth's magnetic field is neglected, so that the refractive index μ is independent of the direction of the wave-normal, a wave-packet which is vertically incident upon the ionosphere continues to travel vertically, and U_z is its true velocity. When the earth's magnetic field is included, however, the value of μ depends upon the direction of the wave-normal, so that the wave-packet does not in general travel vertically, and U_z is the vertical component of its velocity.

The true velocity U of a wave-packet is called the 'group velocity'. Thus, for an isotropic medium, (10.5) gives the value $U_z = U$ of the group velocity. For an anisotropic medium it gives simply the vertical component U_z of the group velocity U. The ratio c/U_z is sometimes called the 'group refractive' index, and denoted by μ', by analogy with the 'wave refractive' index $\mu = c/V$, where V is the wave-velocity. It is important to remember that μ' is not in general equal to c/U.

Equation (10.5) gives

$$\mu' = \mu + f\frac{\partial \mu}{\partial f} = \frac{\partial}{\partial f}(\mu f). \tag{10.6}$$

When the earth's magnetic field and electron collisions are neglected, μ is the same as n in (4.8), and with (3.6) this leads to

$$\mu = \left(1 - \frac{f_N^2}{f^2}\right)^{\frac{1}{2}}, \tag{10.7}$$

so that

$$\mu' = \left(1 - \frac{f_N^2}{f^2}\right)^{-\frac{1}{2}} = \frac{1}{\mu}. \tag{10.8}$$

In this special case U_z and U are the same, and (10.8) shows that

$$UV = c^2. \tag{10.9}$$

It is interesting to note that several other types of wave-motion obey a relation similar to (10.9). For example, it is true for waves in a wave-guide. For electron waves in a potential field, the product of the wave-velocity, and the particle (or group) velocity is a constant.

10.3 The equivalent height of reflection $h'(f)$

Let τ be the time taken for the pulse to travel from the ground to the reflection level and back to the ground. The product $\frac{1}{2}c\tau$ is called the equivalent height of reflection and denoted by $h'(f)$, since it is a function of the frequency f. It is the height to which the pulse would have to go if it travelled always with the velocity c.

It is assumed here that each component harmonic wave in the pulse is reflected at the level $z = z_0$, where $\mu = 0$ as indicated in §9.6. Since μ depends upon the frequency f, $z_0(f)$ is also a function of frequency. The field of the reflected wave reaching the ground is given by

$$E(t) = \int_{-\infty}^{\infty} F(f) \exp\left\{2\pi i f\left(t - \frac{2}{c}\int_0^{z_0(f)} \mu\, dz\right)\right\} df. \tag{10.10}$$

The value of τ is that value of t which makes the integrand have stationary phase for frequencies near the predominant frequency f_1. Hence

$$\left[\frac{\partial}{\partial f}\left\{f\left(\tau - \frac{2}{c}\int_0^{z_0} \mu\, dz\right)\right\}\right]_{f=f_1} = 0, \tag{10.11}$$

so that

$$c\tau = 2\int_0^{z_0} \left\{\frac{\partial}{\partial f}(f\mu)\right\}_{f=f_1} dz + 2f_1\mu(z_0)\left\{\frac{\partial z_0}{\partial f}\right\}_{f=f_1}. \tag{10.12}$$

Now $\mu(z_0)$ is zero, and $\partial z_0/\partial f$ is finite except when f is near to the penetration frequency of a layer, and this case needs special treatment by full wave-theory (see §17.6). Apart from this, the last term of (10.12) is zero, and the remaining term (with (10.6)), gives

$$h'(f) = \int_0^{z_0(f)} \mu'\, dz. \tag{10.13}$$

This indicates that the pulse travels with the velocity U_z right up to the level of reflection. The frequency f is the predominant frequency f_1 in the pulse, but the subscript 1 may now be omitted. The dependence of z_0 upon frequency does not affect this result. This is because, near the level of reflection, μ is close to zero and the wavelength in the medium is very large. Hence small variations of reflection level do not affect the phases of the component harmonic waves.

Equation (10.8) shows that μ' is infinite for an isotropic medium at the level of reflection, and this is true also for an anisotropic medium as is shown in § 12.5, but the integral (10.13) nevertheless converges. Some examples of the form of the function $h'(f)$ in special cases are given in §§ 10.5 to 10.10.

10.4 The 'true height' and the 'phase height'

The equivalent height of reflection $h'(f)$ was defined in the last section for a pulse of radio waves which was actually reflected at the level $z = z_0$, where $\mu = 0$ for the predominant frequency in the pulse. The value $z_0(f)$ is called the 'true height' of reflection.

If a transmitter on the ground emits continuous waves of frequency f, the phase of the waves returning to the ground after reflection is given by the phase integral

$$2k\int_0^{z_0}\mu\,dz = \frac{4\pi}{\lambda}\int_0^{z_0}\mu\,dz = 2kh(f) \qquad (10.14)$$

as explained in § 9.8. The integral is denoted by $h(f)$ which is called the 'phase height' of reflection. If the wave were travelling entirely in free space, it would have to reach the height $h(f)$ in order to make the reflected wave have the correct phase. Thus $2h(f)/c$ is the time that one wave-crest takes to travel up to the reflection level and back to the ground. Equation (10.6) shows that

$$h'(f) = h(f) + f\frac{\partial h(f)}{\partial f}. \qquad (10.15)$$

It can be seen from (10.8) that for an isotropic medium below the level of reflection $\mu \leqslant 1 \leqslant \mu'$. Hence in this case:

$$h(f) < z_0(f) < h'(f).$$

10.5 The equivalent height of reflection for a linear profile of electron density

Suppose that the electron number density N increases linearly with height z in the ionosphere. Let the base of the ionosphere be at height $z = h_0$ above the ground. Then, since f_N^2 is proportional to N,

$$f_N^2 = \alpha(z - h_0) \quad \text{when} \quad z > h_0, \qquad (10.16)$$

where α is a constant. The level of reflection z_0 is given by

$$z_0 = h_0 + f^2/\alpha. \tag{10.17}$$

Hence (10.13) and (10.8) together give

$$h'(f) = h_0 + \int_{h_0}^{z_0} \{1 - \alpha(z - h_0)/f^2\}^{-\frac{1}{2}}\, dz, \tag{10.18}$$

whence

$$h'(f) = h_0 + 2f^2/\alpha. \tag{10.19}$$

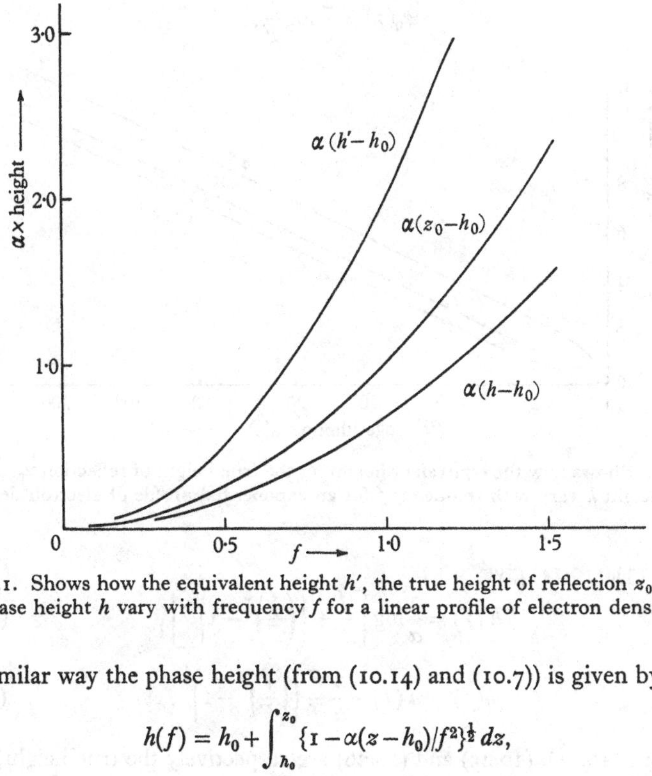

Fig. 10.1. Shows how the equivalent height h', the true height of reflection z_0 and the phase height h vary with frequency f for a linear profile of electron density.

In a similar way the phase height (from (10.14) and (10.7)) is given by

$$h(f) = h_0 + \int_{h_0}^{z_0} \{1 - \alpha(z - h_0)/f^2\}^{\frac{1}{2}}\, dz, \tag{10.20}$$

whence

$$h(f) = h_0 + \tfrac{2}{3} f^2/\alpha. \tag{10.21}$$

The quantities $h'(f)$ (10.19), $z_0(f)$ (10.17), and $h(f)$ (10.21) are shown plotted as functions of the frequency f for this case in Fig. 10.1.

10.6 The equivalent height of reflection for an exponential variation of electron density

Suppose that the electron number density N increases exponentially with height z. Then

$$f_N^2 = F^2 e^{\alpha z}, \tag{10.22}$$

where F and α are constants. In this case N is never zero; it must have a small non-zero value even at the ground. F is the value at the ground of the plasma-frequency f_N, and waves of frequency less than F could not be propagated as pulses. Hence the following results apply only for frequencies f greater than F. From (10.8)

$$\mu = \frac{1}{\mu'} = \left\{1 - e^{z\alpha}\frac{F^2}{f^2}\right\}^{\frac{1}{2}}. \tag{10.23}$$

The true height of reflection is given by

$$z_0(f) = \frac{2}{\alpha}\log\frac{f}{F}, \tag{10.24}$$

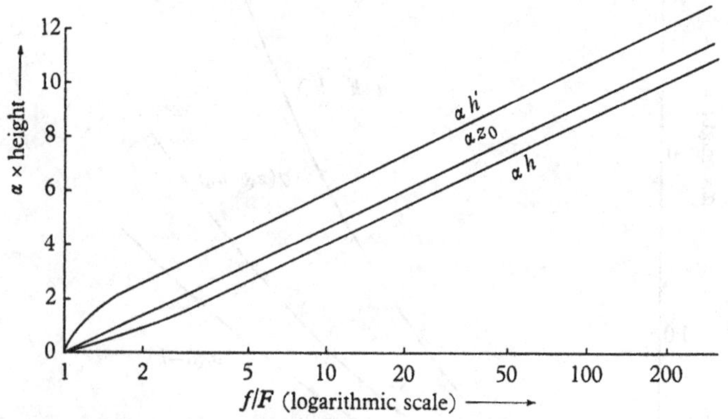

Fig. 10.2. Shows how the equivalent height h', the true height of reflection z_0, and the phase height h vary with frequency f for an exponential profile of electron density.

and (10.13), (10.14) give

$$h'(f) = \frac{2}{\alpha}\log\left[\frac{f}{F} + \left\{\left(\frac{f}{F}\right)^2 - 1\right\}^{\frac{1}{2}}\right], \tag{10.25}$$

$$h(f) = h'(f) - \frac{2}{\alpha}\frac{F}{f}\left\{\left(\frac{f}{F}\right)^2 - 1\right\}^{\frac{1}{2}}. \tag{10.26}$$

Equations (10.24), (10.25) and (10.26) are respectively the true height, equivalent height, and phase height. They are shown plotted as functions of frequency in Fig. 10.2.

10.7 Equivalent height for a parabolic profile of electron density

Suppose that the electron number density N is given by the parabolic law:

$$\left.\begin{aligned} N &= N_m\{1 - (z - z_m)^2/a^2\} \quad &\text{for}\quad |z - z_m| \leqslant a, \\ N &= 0 \quad &\text{for}\quad |z - z_m| \geqslant a. \end{aligned}\right\} \tag{10.27}$$

One curve in Fig. 10.4 shows how N varies with height z in this case. N has the maximum value N_m where $z = z_m$. The constant a is called the 'half thickness' of the ionosphere. This example is of particular importance in the study of layers which have a maximum of electron density. It is shown later that some important properties of such a layer depend on the curvature of the curve of N versus z at the maximum. By studying the properties of a parabolic layer with the same curvature at the maximum, it is possible to derive results for the actual layer.

The plasma-frequency which corresponds to the maximum electron density N_m will be denoted by f_p, and is called the 'penetration frequency' because waves of frequency less than f_p are reflected below z_m, whereas waves of greater frequency travel right through the layer without reflection. It is convenient to use a new height variable

$$\zeta = (z - z_m)/a. \tag{10.28}$$

Then (10.27) leads to

$$\left.\begin{aligned} f_N^2 &= f_p^2(1-\zeta^2) \quad \text{for} \quad |\zeta| \leqslant 1, \\ f_N^2 &= 0 \qquad\qquad\quad \text{for} \quad |\zeta| \geqslant 1, \end{aligned}\right\} \tag{10.29}$$

and the refractive index and group refractive index are given, from (10.8), by

$$\mu = \frac{1}{\mu'} = \begin{cases} \{1-(1-\zeta^2)f_p^2/f^2\}^{\frac{1}{2}} & \text{for} \quad |\zeta| \leqslant 1, \\ 1 & \text{for} \quad |\zeta| \geqslant 1. \end{cases} \tag{10.30}$$

Two cases must now be studied. The first is for $f < f_p$, so that reflection occurs within the parabolic layer. The value ζ_0 of ζ at the level of reflection is given by

$$\zeta_0(f) = -\{1 - f^2/f_p^2\}^{\frac{1}{2}}. \tag{10.31}$$

The equivalent height of reflection is

$$h'(f) = \int_0^{z_0} \mu'\,dz = z_m - a + \frac{af}{f_p}\int_{-1}^{\zeta_0}\left\{\zeta^2 - \left(1 - \frac{f^2}{f_p^2}\right)\right\}^{-\frac{1}{2}} d\zeta, \tag{10.32}$$

which leads to
$$h'(f) = z_m - a + \tfrac{1}{2}a\frac{f}{f_p}\log\frac{f_p+f}{f_p-f}. \tag{10.33}$$

Similarly, the phase height is

$$h(f) = \int_0^{z_0} \mu\,dz = z_m - a + a\frac{f_p}{f}\int_{-1}^{\zeta_0}\left\{\zeta^2 - \left(1 - \frac{f^2}{f_p^2}\right)\right\}^{\frac{1}{2}} d\zeta, \tag{10.34}$$

which leads to
$$h(f) = z_m - \tfrac{1}{2}a - \tfrac{1}{4}a\left(\frac{f_p}{f} - \frac{f}{f_p}\right)\log\frac{f_p+f}{f_p-f}. \tag{10.35}$$

The true height (10.31), equivalent height (10.33), and phase height (10.35), are shown as functions of frequency in the left half of Fig. 10.3.

The second case is for $f > f_p$ so that the waves travel right through the

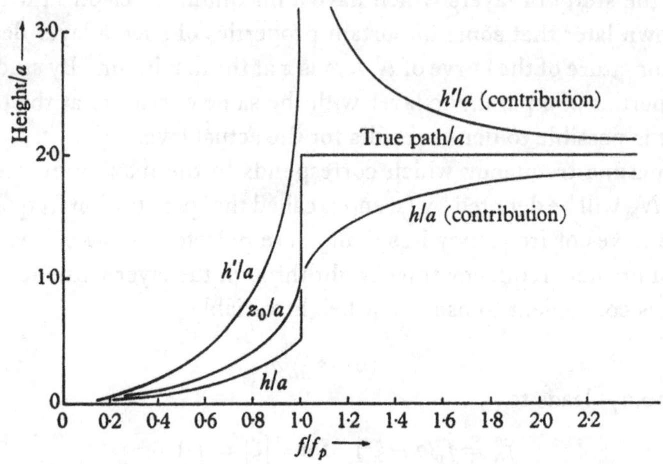

Fig. 10.3. Left half shows how the equivalent height h', the true height of reflection z_0 and the phase height h vary with frequency f for a parabolic profile of electron density. Right half shows the contributions to h', h and the true path for one passage through the layer.

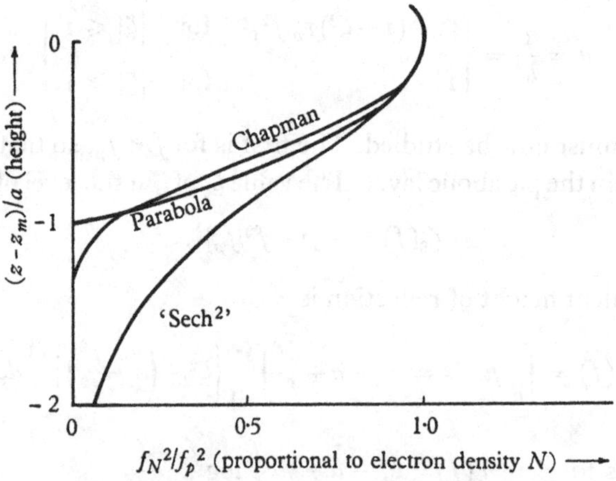

Fig. 10.4. Variation of electron density with height for parabolic and 'sech²' profiles, and Chapman layer (with $H = \frac{1}{2}a$).

parabolic layer without reflection. They may, however, be reflected from a higher layer, and the parabolic layer can contribute to the equivalent height of reflection. We shall now calculate the contribution made by the region between the ground and the top of the parabolic layer. The contribution to the true height is clearly $z_m + a$. The contribution to the equivalent height is

$$h'(f) = \int_0^{z_m+a} \mu' \, dz = z_m - a + 2a\frac{f}{f_p}\int_0^1 \left\{\zeta^2 + \left(\frac{f^2}{f_p^2} - 1\right)\right\}^{-\frac{1}{2}} d\zeta, \tag{10.36}$$

which leads to

$$h'(f) = z_m - a + \frac{f}{f_p}\log\frac{f+f_p}{f-f_p}. \tag{10.37}$$

The contribution to the phase height is

$$h(f) = \int_0^{z_m+a} \mu \, dz = z_m - a + 2a\frac{f_p}{f}\int_0^1 \left\{\zeta^2 + \left(\frac{f^2}{f_p^2} - 1\right)\right\}^{\frac{1}{2}} d\zeta, \tag{10.38}$$

which leads to

$$h(f) = z_m + \tfrac{1}{2}a\left(\frac{f}{f_p} - \frac{f_p}{f}\right)\log\frac{f+f_p}{f-f_p}. \tag{10.39}$$

The contributions to the true height, $z_m + a$, the equivalent height (10.37), and the phase height (10.39), are shown plotted as functions of frequency in the right half of Fig. 10.3.

The above results show that the equivalent height approaches infinity as the frequency f approaches the penetration frequency f_p. The group velocity of a pulse is smallest near the top of its trajectory. For a frequency just below penetration, the region of low group velocity is thicker than it is for lower frequencies, because the gradient of electron number density is small near the level of reflection. This explains why the equivalent height is large for frequencies close to penetration. The pulse spends a relatively long time near the level of maximum electron density, and for a small range of frequencies below f_p the shape of the $h'(f)$ curve is determined predominantly by the curvature of the $N(z)$ profile at the maximum. The shape of the $N(z)$ curve at lower levels is less important for these frequencies, because the pulse spends relatively little time there. This property is illustrated in Fig. 10.5 which shows the $h'(f)$ curves for three different electron density profiles, all with the same curvature at the maximum. These are the parabolic profile, the 'sech^2' profile discussed in the next section, and the Chapman layer given by (1.10) with $H = \frac{1}{2}a$.

It was shown in § 10.3 that the expression (10.13) for the equivalent height cannot be used for frequencies very close to the penetration-frequency, and in this case the problem must be treated by full-wave theory. This is done in § 17.6, where it is shown that the effective value of $h'(f)$ near the penetration frequency remains finite but attains a high maximum value. The amplitude of the reflected wave is then small, however, and the maximum is not easy to observe.

10.8 Equivalent height for the 'sech²' profile of electron density

The parabolic profile of the last subsection has discontinuities in the gradient of electron number density at the top and bottom of the layer ($z = z_m \pm a$). In this respect it differs from actual ionospheric layers, in which the transitions are believed to be gradual. A profile which does not have this drawback is given by

$$f_N^2 = f_p^2 \operatorname{sech}^2\left(\frac{z-z_m}{a}\right). \tag{10.40}$$

Fig. 10.5. Variation of equivalent height of reflection $h'(f)$ with frequency for 'sech²' profile of electron density. $z_m/a = 6$. The curves for a parabolic profile and a Chapman layer are also shown for comparison.

One curve in Fig. 10.4 shows how the electron density N varies with height z in this case. Like the parabola there is a maximum of electron density where $z = z_m$, and the penetration-frequency is f_p. The electron density is never zero; it must have small non-zero value even at the ground. In this respect it resembles the exponential profile of § 10.6 above. It is convenient to use the variable ζ defined by (10.28). Then the equivalent height, when $f < f_p$, is given by

$$h'(f) = \int_0^{z_0} \mu' \, dz = a \int_{-z_m/a}^{\zeta_0} \left\{1 - \left(\frac{f_p^2}{f^2}\right) \operatorname{sech}^2 \zeta\right\}^{-\frac{1}{2}} d\zeta, \tag{10.41}$$

where

$$\operatorname{sech} \zeta_0 = \frac{f}{f_p}.$$

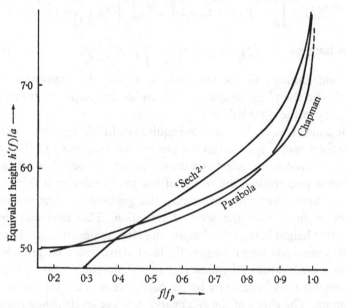

This leads to

$$h'(f) = a\log\left[\left(\frac{f_p^2}{f^2}-1\right)^{-\frac{1}{2}}\sinh\frac{z_m}{a}+\left\{\left(\frac{f_p^2}{f^2}-1\right)^{-1}\sinh^2\frac{z_m}{a}-1\right\}^{\frac{1}{2}}\right]. \tag{10.42}$$

The curvature of the profile (10.40) at the maximum is the same as that of the parabolic profile (10.27) for the same values of f_p and a. Fig. 10.5 shows how $h'(f)$ in (10.42) varies with frequency, and the corresponding curve for the parabola is also shown. At high frequencies $h'(f)$ is greater for the 'sech²' profile. Although the true height of reflection must be lower, the larger number of electrons at low levels causes greater group retardation, which more than compensates for the shorter true path. At low frequencies, however, $h'(f)$ is lower for the 'sech²' layer, because the true height falls rapidly as the frequency decreases.

10.9 Two separate parabolic layers

During the daytime there are two main ionised layers known as E and F. If these are assumed to have separate parabolic profiles, as shown in Fig. 10.6, then $h'(f)$ can be calculated by using (10.33) and (10.37). Let $f_p^{(E)}$, $z_m^{(E)}$, $a^{(E)}$ be the penetration-frequency, height of maximum and half-thickness respectively, for the E-layer, and let $f_p^{(F)}$, $z_m^{(F)}$, $a^{(F)}$ be the corresponding quantities for the F-layer. It is assumed that $f_p^{(F)} > f_p^{(E)}$ and that $z_m^{(F)} - z_m^{(E)} > a^{(F)} + a^{(E)}$ so that the layers do not overlap.† Then

$$h'(f) = \begin{cases} z_m^{(E)} - a^{(E)} + \frac{1}{2}a^{(E)}\dfrac{f}{f_p^{(E)}}\log\dfrac{f_p^{(E)}+f}{f_p^{(E)}-f} & \text{for } f < f_p^{(E)}, \\[2ex] z_m^{(F)} - 2a^{(E)} - a^{(F)} + a^{(E)}\dfrac{f}{f_p^{(E)}}\log\dfrac{f+f_p^{(E)}}{f-f_p^{(E)}} & \\[2ex] \qquad + \frac{1}{2}a^{(F)}\dfrac{f}{f_p^{(F)}}\log\dfrac{f_p^{(F)}+f}{f_p^{(F)}-f} & \text{for } f_p^{(E)} < f < f_p^{(F)}. \end{cases} \tag{10.43}$$

This function is shown plotted for typical values in Fig. 10.7. The full wave-correction (§ 17.6) near $f_p^{(E)}$ and $f_p^{(F)}$ is not used.

Although the earth's magnetic field has been neglected in deriving (10.43), the curve of Fig. 10.7 is very similar in general form to some $h'(f)$ records obtained in practice. A characteristic feature of these records is the infinity at the penetration frequency $f_p^{(E)}$ of the E-layer. When such an infinity appears, it can usually be attributed to penetration of one layer. For frequencies just above it, the waves are being reflected

† It is now believed that the E- and F-layers are not distinct, but that the region between them is ionised with an electron density nearly as great as in the E-layer itself. See § 1.8 and Fig. 1.3.

from a higher layer, but are retarded in their passage through the layer which has just been penetrated. For greater frequencies the retardation is less because the group velocity in the lower layer is not so small as for frequencies near the penetration-frequency.

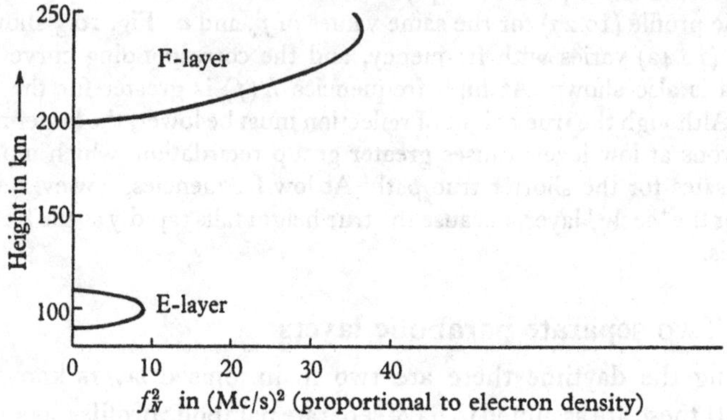

Fig. 10.6. Example of variation of electron density with height when the ionosphere consists of two separate parabolic layers.

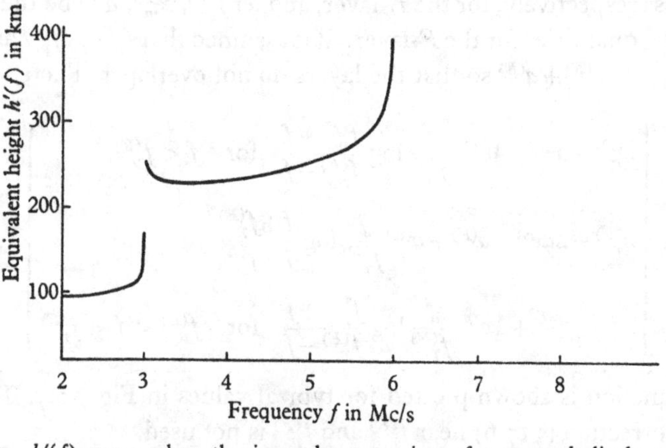

Fig. 10.7. $h'(f)$ curve when the ionosphere consists of two parabolic layers as in Fig. 10.6. The earth's magnetic field and electron collisions are neglected.

10.10 The effect of a 'ledge' in the electron density profile

A possible form for the electron density profile $N(z)$ is shown in Fig. 10.8. Here $N(z)$ increases monotonically as z increases, but there is a part of the curve where dN/dz is small for a considerable range. This feature is often called a 'ledge' in the profile. Suppose that in the middle of the ledge the plasma frequency f_N is denoted by F_N. The form of the

$h'(f)$ curve in this case is shown in Fig. 10.9. For frequencies near F_N the gradient dN/dz is small at the level of reflection, and the group retardation is large, so that $h'(f)$ is large. For greater frequencies, the reflection level is higher and the group retardation in the ledge is less. Consequently $h'(f)$ is less. For still greater frequencies, $h'(f)$ increases again, and becomes infinite at the penetration-frequency.

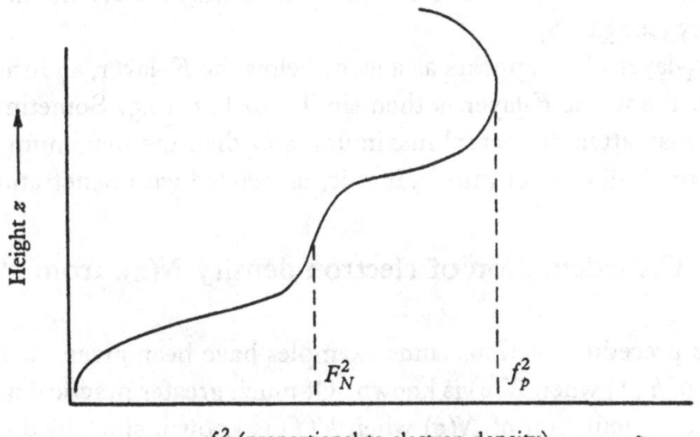

Fig. 10.8. Electron density profile for ionospheric layer with a ledge.

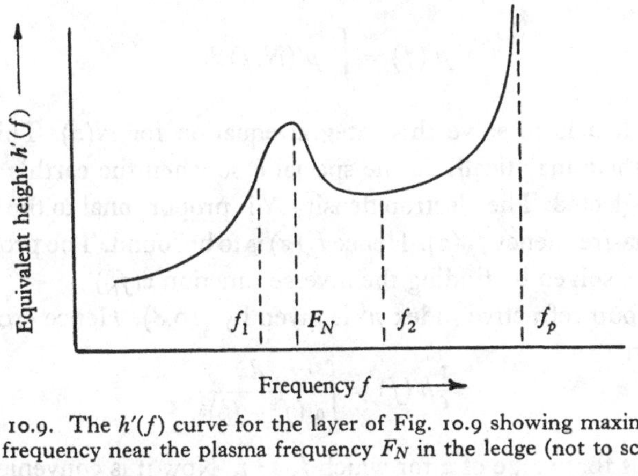

Fig. 10.9. The $h'(f)$ for the layer of Fig. 10.9 showing maximum at a frequency near the plasma frequency F_N in the ledge (not to scale).

The effect of a ledge is therefore to give a maximum in the $h'(f)$ curve. The size of the maximum increases if the gradient dN/dz in the ledge decreases, but $h'(f)$ is always finite as long as $N(z)$ increases monotonically. Hence, when $N(z)$ is a monotonically increasing function, the corresponding part of the $h'(f)$ curve is finite and continuous. If $N(z)$ has a maximum, the $h'(f)$ curve shows an infinity at the penetration-frequency, and is discontinuous there. When corrected by the full-wave theory (§ 17.6), the $h'(f)$ curve does not go to infinity but has a maximum near the penetration-frequency, although it is still discontinuous. The maximum is rarely observed on actual $h'(f)$ records, and for practical purposes the curve may be said to go to infinity. When the effect of the earth's magnetic field is allowed for, a discontinuity of a different kind can appear in the $h'(f)$ curve for the extraordinary wave at the gyro-frequency (see § 12.6).

The F_1-layer often appears as a ledge below the F_2-layer, and the $h'(f)$ curve for the whole F-layer is then similar to Fig. 10.9. Sometimes the F_1-layer may attain an actual maximum, and then the maximum in the $h'(f)$ record will go over into an infinity associated with penetration.

10.11 The calculation of electron density $N(z)$, from $h'(f)$ data

In the preceding sections some examples have been given of the calculation of $h'(f)$ when $N(z)$ is known. Of much greater practical importance is the calculation of $N(z)$ when $h'(f)$ is known, since in the commonest form of radio sounding of the ionosphere it is $h'(f)$ that is observed. The relation between $h'(f)$ and $N(z)$ may be written

$$h'(f) = \int_0^{z_0} \mu'(N,f)\,dz. \tag{10.44}$$

The problem is to solve this integral equation for $N(z)$. This can be accomplished analytically in the special case when the earth's magnetic field is neglected. The electron density N is proportional to the square of the plasma-frequency $f_N(z)$. Hence $f_N(z)$ is to be found. The problem will actually be solved by finding the inverse function $z(f_N)$.

The group refractive index μ' is given by (10.8). Hence (10.44) may be written

$$\frac{1}{f}h'(f) = \int_0^{z_0} \frac{dz}{(f^2-f_N^2)^{\frac{1}{2}}}, \tag{10.45}$$

where z_0 is that value of z for which $f_N = f$. Now it is convenient to use new variables

$$u = f_N^2, \quad v = f^2, \tag{10.46}$$

so that

$$v^{-\frac{1}{2}}h'(v^{\frac{1}{2}}) = \int_0^{z_0} \frac{dz}{\{v-u(z)\}^{\frac{1}{2}}}. \tag{10.47}$$

Here $u(z)$ is the unknown function which is to be found.

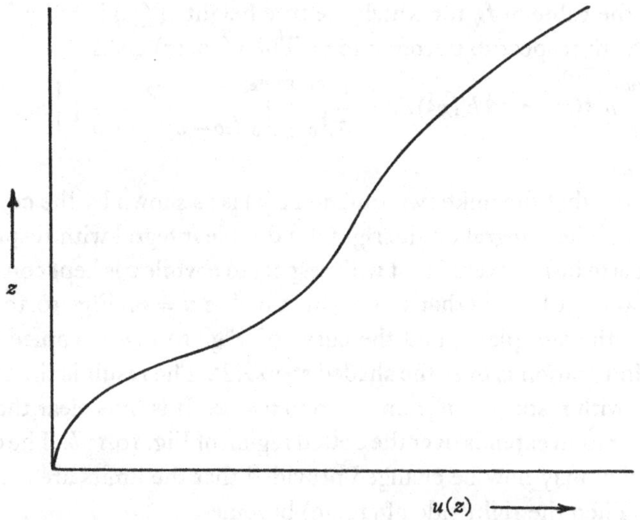

Fig. 10.10*a*. The unknown function $u(z)$, proportional to electron density.

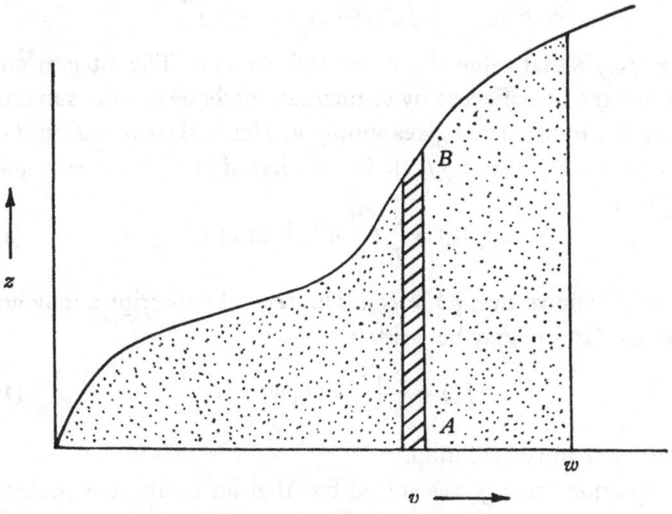

Fig. 10.10*b*. The v–z-plane.

10.12 Solution when $N(z)$ is monotonic

When $u(z)$ is a monotonically increasing function of z, (10.47) can be solved by a standard method (see, for example, Whittaker and Watson, 1935, § 11.8). An alternative method using Laplace transforms has been given by Manning (1947, 1949). Multiply both sides by $(1/\pi)(w-v)^{-\frac{1}{2}}$, where

$$w = f_{N0}^2, \tag{10.48}$$

and f_{N0} is the value of f_N for which the true height $z(f_{N0})$ is to be found. Integrate with respect to v from 0 to w. Then (10.47) gives

$$\frac{1}{\pi}\int_0^w v^{-\frac{1}{2}}(w-v)^{-\frac{1}{2}}h'(v^{\frac{1}{2}})\,dv = \frac{1}{\pi}\int_0^w \left[\int_0^{z_0} \frac{dz}{(w-v)^{\frac{1}{2}}(v-u)^{\frac{1}{2}}}\right]dv. \tag{10.49}$$

Now suppose that the unknown function $u(z)$ is as shown by the curve of Fig. 10.10*a*. The integral on the right is a double integral with respect to z and v. It is to be integrated first with respect to z while v is kept constant, and the top limit for z is that value which makes $u = v$. Fig. 10.10*b* is a diagram of the v–z-plane, and the curve of Fig. 10.10*a* is copied in it. The first integration is over the shaded strip AB. The result is then to be integrated with respect to v from $v = 0$ to $v = w$. It is thus clear that the whole integration extends over the dotted region of Fig. 10.10*b*. The order of integration may now be changed provided that the limits are suitably modified. Then the right side of (10.49) becomes

$$\frac{1}{\pi}\int_0^{z_0(f_{N0})}\left[\int_u^w \frac{dv}{(w-v)^{\frac{1}{2}}(v-u)^{\frac{1}{2}}}\right]dz, \tag{10.50}$$

where $z_0(f_{N0})$ is that value of z which makes $u = w$. The integration with respect to v may be effected by elementary methods (by the substitution $v = w\cos^2\theta + u\sin^2\theta$) and gives simply π. Hence the expressions (10.49) and (10.50) are equal to $z_0(f_{N0})$. On the left of (10.49) put $v = w\sin^2\alpha$, which gives

$$z_0(f_{N0}) = \frac{2}{\pi}\int_0^{\frac{1}{2}\pi} h'(w^{\frac{1}{2}}\sin\alpha)\,d\alpha. \tag{10.51}$$

Now (10.48) shows that $w^{\frac{1}{2}} = f_{N0}$. The second subscript 0 may now be omitted and (10.51) may be written

$$z(f_N) = \frac{2}{\pi}\int_0^{\frac{1}{2}\pi} h'(f_N\sin\alpha)\,d\alpha, \tag{10.52}$$

which is the required solution.

The equation (10.47) was solved by Abel for a different problem. A particle is projected horizontally with velocity V on to a frictionless slope, for which the height $H(x)$ is an unknown function of the horizontal distance x. The slope is assumed to be so small that the horizontal component of the velocity may be taken to be the same as the actual oblique velocity. The time T taken for the particle to come to rest and then return to its starting-point is given by

$$T(V) = 2\int_0^{x_0} \frac{dx}{\{V^2 - 2gH(x)\}^{\frac{1}{2}}}, \tag{10.53}$$

where g is the gravitational acceleration, and x_0 is the value of x for which $H(x) = \frac{1}{2}V^2/g$. The function $T(V)$ can be determined experimentally. Abel's problem was to find the function $H(x)$. Clearly (10.53) is of the same form as (10.47) and can be solved in the same way (see de Groot, 1930). Ratcliffe (1954) has used this analogy by constructing a mechanical model in which a steel ball-bearing is projected up an incline shaped like the electron density profile $N(z)$ in the ionosphere. This exhibits many of the features of the reflection of a pulse of radio waves projected up into the ionosphere.

10.13 Partial solution when $N(z)$ is not monotonic

Suppose now that $u(z)$ is not a monotonically increasing function of z, but has one maximum and one minimum, as shown in Fig. 10.11 *a*. At the maximum let $z = z_1$, $u = u_p$ where

$$u_p = f_p^2, \tag{10.54}$$

and f_p is the penetration frequency. Let $z = z_3$ be the height where u again has the value u_p (see Fig. 10.11 *a*). It was shown in § 10.7 that the $h'(f)$ curve must have an infinity where $f = f_p$, so that the curve is as in Fig. 10.11 *c*. It is now assumed that this curve is given, and we require to find the curve of Fig. 10.11 *a*. This problem cannot be solved completely. Only the part of the curve below z_1 can be found unambiguously.

The equation (10.47) still holds. To solve it, we apply the method of the last section. Multiply both sides by $(1/\pi)(w-v)^{-\frac{1}{2}}$ where w is given by (10.48). Integrate with respect to v from 0 to w, which gives (10.49), whose right side is a double integral with respect to z and v. Fig. 10.11 *b* is a diagram of the v–z-plane and the curve of Fig. 10.11 *a* is copied in it. Suppose first that w is less than u_p and is given by the ordinate CD in the figure. Then the double integral extends over the area $OACDB$. The order of integration may be changed as before and the final result is

$$z(f_N) = \frac{2}{\pi}\int_0^{\frac{1}{2}\pi} h'(f_N \sin\alpha)\, d\alpha \quad \text{for} \quad f_N < f_p, \tag{10.55}$$

which is the same as (10.52), and enables the part of the curve below $z = z_1$ in Fig. 10.11 *a* to be found. Next suppose that w is greater than u_p, and is given by the ordinate $C'D'$ in Fig. 10.11 *b*. In the double integral on the right of (10.49) the first integration with respect to z extends from $z = 0$ up to the curve. When v is just less than u_p, the upper limit is on the part DP of the curve. When v becomes just greater than u_p, the upper limit suddenly increases to the part QB' of the curve. It is thus clear that the double integral extends over the dotted area of Fig. 10.11 *b*, and the area PQR is not included. Now the order of integration may be changed, as in the last section, but if the right side of (10.49) were written in the form of (10.50), the integration would

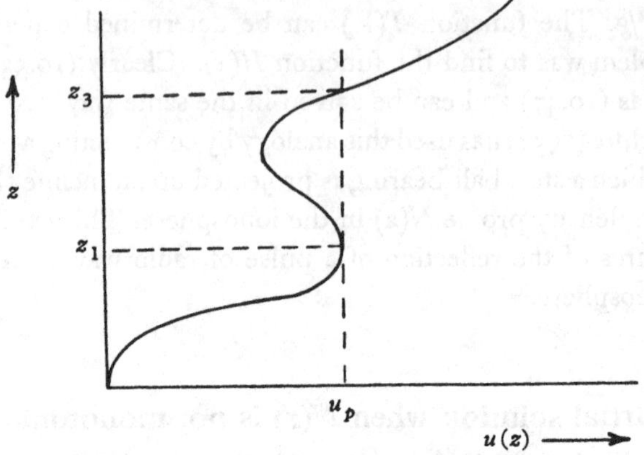

Fig. 10.11 *a*. The unknown function $u(z)$, proportional to electron density.

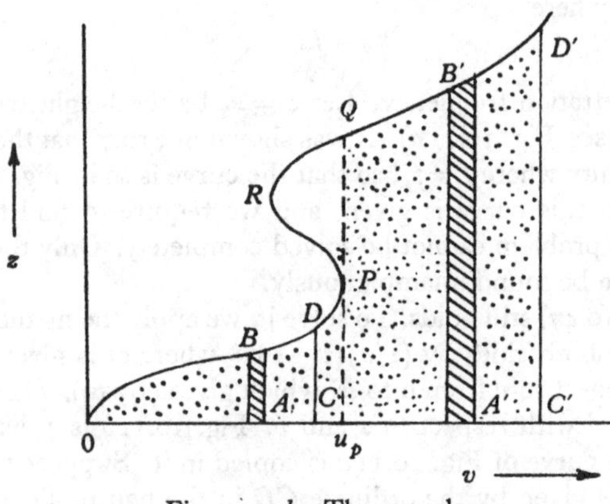

Fig. 10.11 *b*. The v–z-plane.

extend over the dotted area plus the area PQR. Equation (10.50) is therefore no longer correct. The following integral must be subtracted from it:

$$\frac{1}{\pi}\int_{z_1}^{z_3}\left[\int_u^{u_p}\frac{dv}{(w-v)^{\frac{1}{2}}(v-u)^{\frac{1}{2}}}\right]dz. \tag{10.56}$$

This is the integral over the area PQR. Now (10.50) can be integrated as before, and gives simply $z_0(f_N)$. The integration with respect to v in (10.56) can be effected in the same way as before, and gives for this term

$$\zeta(f_{N0}) = \frac{1}{\pi}\int_{z_1}^{z_3}\text{arc cos}\left(\frac{w-2u_p+u}{w-u}\right)dz. \tag{10.57}$$

Hence (10.49) becomes

$$z(f_N) = \frac{2}{\pi}\int_0^{\frac{1}{2}\pi}h'(f_N\sin\alpha)\,d\alpha + \zeta(f_N). \tag{10.58}$$

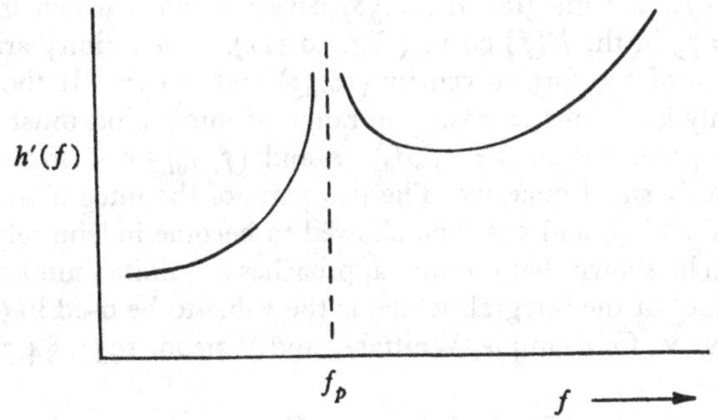

Fig. 10.11 *c*. The $h'(f)$ curve from which $u(z)$ is to be found.

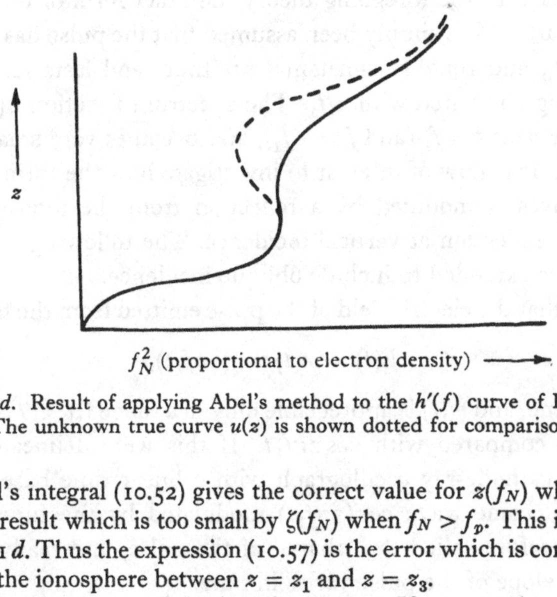

Fig. 10.11 *d*. Result of applying Abel's method to the $h'(f)$ curve of Fig. 10.11 *c*. The unknown true curve $u(z)$ is shown dotted for comparison.

Hence Abel's integral (10.52) gives the correct value for $z(f_N)$ when $f_N < f_p$, but gives a result which is too small by $\zeta(f_N)$ when $f_N > f_p$. This is illustrated in Fig. 10.11 *d*. Thus the expression (10.57) is the error which is contributed by the part of the ionosphere between $z = z_1$ and $z = z_3$.

In the range $z_1 < z < z_3$ it is clear that $u < u_p$. If $w \gg u_p$, the angle

$$\arccos\left(\frac{w-2u_p+u}{w-u}\right)$$

must be very small, and hence the error $\zeta(f_N)$ is small. This means that the error is not serious for frequencies which greatly exceed the penetration-frequency f_p. In particular the E-layer (for which f_p is of the order 2 to 4 Mc/s in the daytime) would not seriously affect results for the densest part of the F-layer which may reflect frequencies of the order 5 to 10 Mc/s. There is now good evidence (see, for example, Piggott, 1954; Seddon and Jackson, 1955) to show that the minimum of the electron density between the E- and F-layers is very shallow, and this would mean that the error ζ is always small.

When $f_N > f_p$, the integral in (10.58) covers a range which includes the infinity at $f = f_p$ in the $h'(f)$ curve (Fig. 10.11 *c*). This infinity arises from a logarithm term of the form given by (10.33) and (10.37). If the integral is evaluated analytically in this case, the range of integration must be divided into the two parts $0 \leqslant \sin\alpha \leqslant (f_p/f_0) - \epsilon$ and $(f_p/f_0) + \epsilon \leqslant \sin\alpha \leqslant 1$, where ϵ is an arbitrarily small quantity. The two parts of the integral are evaluated separately and added, and ϵ is then allowed to become indefinitely small. In this way it can be shown that the sum approaches a definite limit known as the 'principal value' of the integral, which is the value to be used in (10.58) (for further details see, for example, Whittaker and Watson, 1935, §4.5).

10.14 The shape of a pulse of radio waves

A pulsed radio signal or wave-packet was represented by the Fourier integral (10.1). In the foregoing theory the exact form of the pulse has played no part. It has simply been assumed that the pulse has a constant frequency f_1, and roughly constant amplitude, and lasts for a time T which is long compared with $1/f_1$. The spectrum function $F(f)$, (10.1), is then large near $f = f_1$ (and $f = -f_1$), and becomes very small at other frequencies. It is now of interest to investigate how the form of a pulse of radio waves is modified by a reflection from the ionosphere. We discuss only reflection at vertical incidence. The following analysis can very easily be extended to include oblique incidence.

Suppose that the electric field of the pulse emitted from the transmitter is

$$E(t) = m(t)\cos(2\pi f_1 t), \tag{10.59}$$

where t is time, and $m(t)$ is appreciable only in a range $|t| < T$, and varies very slowly compared with $\cos 2\pi f_1 t$. If this were delineated on the screen of a cathode-ray oscillograph with a linear time base, it would appear as a cosine wave $\cos(2\pi f_1 t)$ modulated by the function $m(t)$, which is therefore called the 'shape' of the pulse, and really describes how the envelope of the pulse varies in time.

Now let $m(t)$ be expressed as a Fourier integral

$$m(t) = \int_{-\infty}^{\infty} M(f)\, e^{2\pi i f t}\, df, \tag{10.60}$$

where
$$M(f) = \int_{-\infty}^{\infty} m(t)\, e^{-2\pi i f t}\, dt. \tag{10.61}$$

Since $m(t)$ is necessarily real, $|M(f)|$ is symmetric in f, and $\arg M(f)$ is antisymmetric. Then a well-known theorem in Fourier analysis shows that

$$E(t) = \mathscr{R}\int_{-\infty}^{\infty} M(f - f_1)\, e^{2\pi i f t}\, df, \tag{10.62}$$

which is the signal emitted from the transmitter and corresponds to (10.1). When the waves have travelled to a height z, the signal is given by (10.2). When they have returned to the ground the integral in (10.2) must be replaced by $2\int_0^{z_0} \mu\, dz$ where z_0 is the level of reflection. This integral is simply the phase height of reflection $h(f)$ given by (10.14). Hence the reflected signal at the ground is

$$E_r(t) = \mathscr{R}\int_{-\infty}^{\infty} M(f-f_1)\exp\left[2\pi i f\left\{t-\frac{2}{c}h(f)\right\}\right]df. \qquad (10.63)$$

The time taken for the pulse to return to the ground is $(2/c)\,h'(f_1)$, where $h'(f_1)$ is the equivalent height of reflection for the predominant frequency f_1. Hence we take

$$t = \frac{2}{c}h'(f_1)+\tau, \qquad (10.64)$$

where τ is a new measure of the time.

$M(f-f_1)$ is only appreciable when $f-f_1$ is small. This suggests that the term $fh(f)$ in the exponent of (10.63) be expanded in a Taylor series about $f = f_1$. Let

$$f-f_1 = \sigma. \qquad (10.65)$$

Now (10.15) shows that $h'(f) = (\partial/\partial f)\{fh(f)\}$. Hence

$$fh(f) = f_1 h(f_1)+\sigma h'(f_1)+\tfrac{1}{2}\sigma^2 h_1'+\tfrac{1}{6}\sigma^3 h_2'+\ldots, \qquad (10.66)$$

where h_1', h_2', are written for $(\partial/\partial f)\{h'(f)\}$, $(\partial^2/\partial f^2)\{h'(f)\}$ respectively when $f = f_1$. With these substitutions the integral (10.63) becomes

$$E_r(t) = \mathscr{R}\exp\left[2\pi i f_1\left(\tau-\frac{2}{c}\{h(f_1)-h'(f_1)\}\right)\right]$$

$$\times\int_{-\infty}^{\infty} M(\sigma)\exp\left[2\pi i\left(\sigma\tau-\frac{1}{c}\sigma^2 h_1'-\frac{1}{3c}\sigma^3 h_2'\ldots\right)\right]d\sigma. \qquad (10.67)$$

Here the first exponential is the high-frequency oscillation. The phase shift term $(4\pi/c)f_1\{h(f_1)-h'(f_1)\}$ is of no particular interest. The integral in (10.67) is the function which modulates the high frequency, and therefore gives the pulse shape. Clearly if h_1', h_2', and higher derivatives are negligible, the integral is the same as the original pulse (10.60) which is then undistorted.

Two special cases are now of interest. The first is when the frequency f_1 is chosen so that the $h'(f)$ curve has a positive slope and only small curvature (for example, at a frequency f_1 in Fig. 10.9). Then h_1' has a positive value and h_2' is small. If the original pulse is wide enough $M(\sigma)$

falls quickly to zero, as σ increases from zero, and we therefore neglect the term involving σ^3 in (10.67). Then the pulse shape is given by

$$\int_{-\infty}^{\infty} M(\sigma) \exp\left(-2\pi i h_1' \frac{\sigma^2}{c}\right) \exp(2\pi i \sigma \tau)\, d\sigma. \tag{10.68}$$

This may be transformed by the convolution theorem of Fourier analysis. The Fourier transform of $\exp(-2\pi i h_1' \sigma^2/c)$ is

$$\int_{-\infty}^{\infty} \exp\left(-2\pi i h_1' \frac{\sigma^2}{c} + 2\pi i \sigma t\right) d\sigma = e^{-i(\frac{1}{4}\pi)} \left(\frac{2h_1'}{c}\right)^{-\frac{1}{2}} \exp\left(\frac{\pi i c t^2}{2h_1'}\right), \tag{10.69}$$

and hence (10.68) becomes (apart from a constant factor)

$$\int_{-\infty}^{\infty} m(\tau - t) \exp\left\{\frac{\pi i c t^2}{2h_1'}\right\} dt. \tag{10.70}$$

Integrals of this type are used in optics in finding the Fresnel diffraction pattern of an aperture. Suppose that a parallel beam of monochromatic light of wavelength λ is incident on a slit which allows light to pass with amplitude $m(x)$ where x is measured at right angles to the length of the slit. A screen is placed at a distance y from the slit, and z is distance measured parallel to x in the plane of the screen. Then the amplitude of light arriving at the screen is proportional to

$$\int_{-\infty}^{\infty} m(z - \xi) \exp\left\{\frac{\pi i \xi^2}{\frac{1}{2}\lambda y}\right\} d\xi, \tag{10.71}$$

and this is called the Fresnel diffraction pattern of the slit. Hence we may say that when a pulse of radio waves is reflected from the ionosphere, in conditions where the $h'(f)$ curve has a finite slope and small curvature, the pulse shape is modified to that of its own Fresnel diffraction pattern. Fig. 10.12 shows this modification for a pulse which is initially rectangular. The distorted pulses show oscillations whose amplitude is constant, but their scale in time depends on the value of h_1'. Further examples are given by Rydbeck (1942).

The second special case is when the $h'(f)$ curve has zero slope, but appreciable curvature, so that $h'(f)$ is a maximum or a minimum (for example, at a frequency F_N or f_2 in Fig. 10.9). Then the pulse shape is given by

$$\int_{-\infty}^{\infty} M(\sigma) \exp\left(-\frac{2}{3c}\pi i h_2' \sigma^3\right) \exp(2\pi i \sigma \tau)\, d\sigma. \tag{10.72}$$

The Fourier transform of the exponential is

$$\int_{-\infty}^{\infty} \exp\left\{\pi i\left(2\sigma t - \frac{2}{3c} h_2' \sigma^3\right)\right\} d\sigma = \left(\frac{4\pi^2 c}{h_2'}\right)^{\frac{1}{3}} \mathrm{Ai}\left\{-2\left(\frac{c\pi^2}{2h_2'}\right)^{\frac{1}{3}} t\right\} \tag{10.73}$$

(see (15.16): Ai denotes the Airy integral function). The convolution theorem may now be used as before and gives for (10.72) (apart from a constant factor)

$$\int_{-\infty}^{\infty} m(\tau - t)\,\mathrm{Ai}\left\{-2\left(\frac{c\pi^2}{2h_2'}\right)^{\frac{1}{3}} t\right\} dt. \tag{10.74}$$

This expression has no simple interpretation like the Fresnel diffraction pattern used in the preceding example. Suppose that the original pulse is

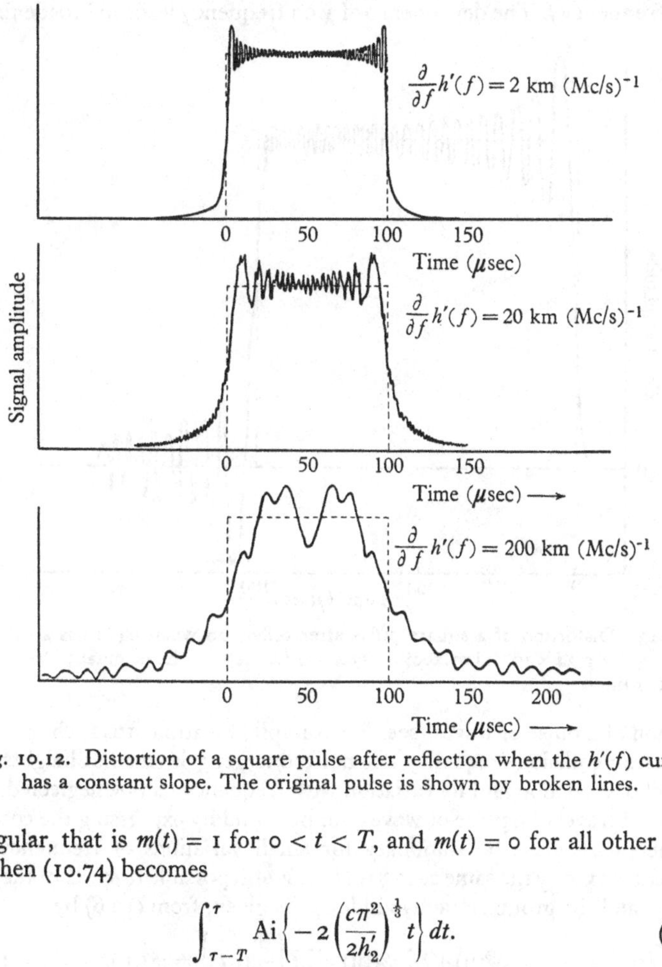

Fig. 10.12. Distortion of a square pulse after reflection when the $h'(f)$ curve has a constant slope. The original pulse is shown by broken lines.

rectangular, that is $m(t) = 1$ for $0 < t < T$, and $m(t) = 0$ for all other values of t. Then (10.74) becomes

$$\int_{\tau - T}^{\tau} \mathrm{Ai}\left\{-2\left(\frac{c\pi^2}{2h_2'}\right)^{\frac{1}{3}} t\right\} dt. \tag{10.75}$$

Fig. 10.13 shows an example of this function. (Another example is given by Rydbeck, 1942.) If $h_2'(f)$ is positive, as at $f = f_2$ in Fig. 10.9, then the curve is as shown. If $h_2'(f)$ is negative as at $f = F_N$ in Fig. 10.9, then the curve of Fig. 10.13 is reversed in time and the 'tail' becomes a precursor. This occurs because the 'side-band' frequencies arrive before the predominant frequency.

10.15 The effect of electron collisions on group refractive index

So far in this chapter the effect of electron collisions has been neglected. The effect of including collisions is to make the refractive index $\mathfrak{n}$ (4.8) complex, and its imaginary part $-\chi$ is negative when the real part μ is positive. The phase of the wave is then determined by μ and the presence of a non-zero χ means that the wave is attenuated as it travels. In general, both μ and χ depend on the frequency f. The dependence of χ on frequency leads to broadening and

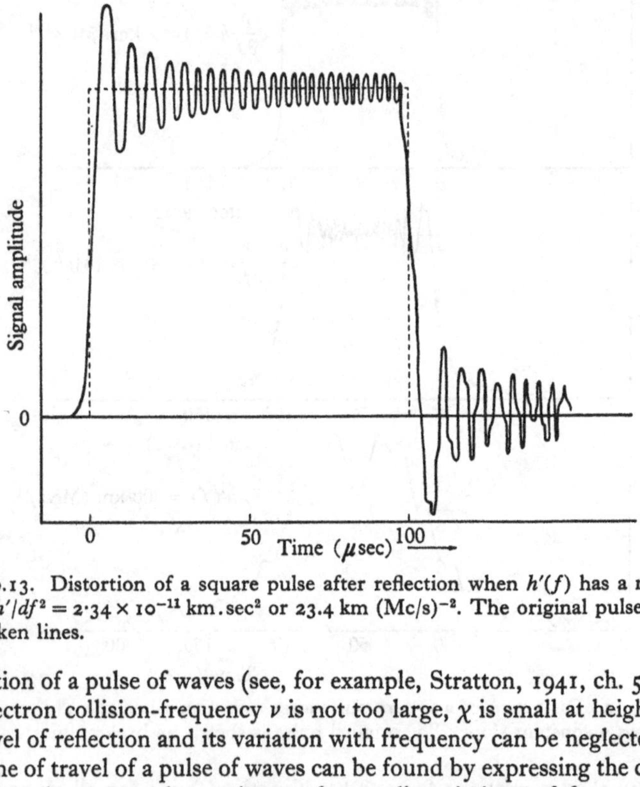

Fig. 10.13. Distortion of a square pulse after reflection when $h'(f)$ has a minimum, and $d^2h'/df^2 = 2{\cdot}34 \times 10^{-11}$ km.sec^2 or 23.4 km (Mc/s)$^{-2}$. The original pulse is shown by broken lines.

distortion of a pulse of waves (see, for example, Stratton, 1941, ch. 5). But if the electron collision-frequency ν is not too large, χ is small at heights below the level of reflection and its variation with frequency can be neglected. Then the time of travel of a pulse of waves can be found by expressing the condition that the phase must be stationary for small variations of frequency. The argument is exactly the same as in § 10.2. It is still possible to speak of the group velocity, and the group refractive index μ' is given, from (10.6) by

$$\mu' = \mathscr{R}(\mathfrak{n}) + f\frac{\partial}{\partial f}\mathscr{R}(\mathfrak{n}) = \mathscr{R}\left\{\frac{\partial}{\partial f}(\mathfrak{n}f)\right\} = \mathscr{R}(\mathfrak{n}'). \tag{10.76}$$

Thus $\mathfrak{n}'$ is a complex number whose real part μ' is the group refractive index. From (4.8) it is easily shown that

$$\mathfrak{n}' = \frac{1}{\mathfrak{n}}\left\{1 + \frac{iZX}{2(1-iZ)^2}\right\} = \frac{1}{\mathfrak{n}}\frac{1-\frac{1}{2}iZ}{1-iZ} - \mathfrak{n}\frac{iZ}{2(1-iZ)}. \tag{10.77}$$

Some curves showing how $\mu' = \mathscr{R}(\mathfrak{n}')$ depends on X are given in Fig. 10.14 for various values of Z.

10.16 The effect of collisions on equivalent height $h'(f)$ and phase height $h(f)$

The time τ for a pulse to travel from the ground to the reflection level $z = z_0$ and back was found in § 10.3 by expressing the condition that the phase of the reflected wave shall be stationary for small variations of frequency. This led to equations (10.11) to (10.13). It might seem that these can be used also when

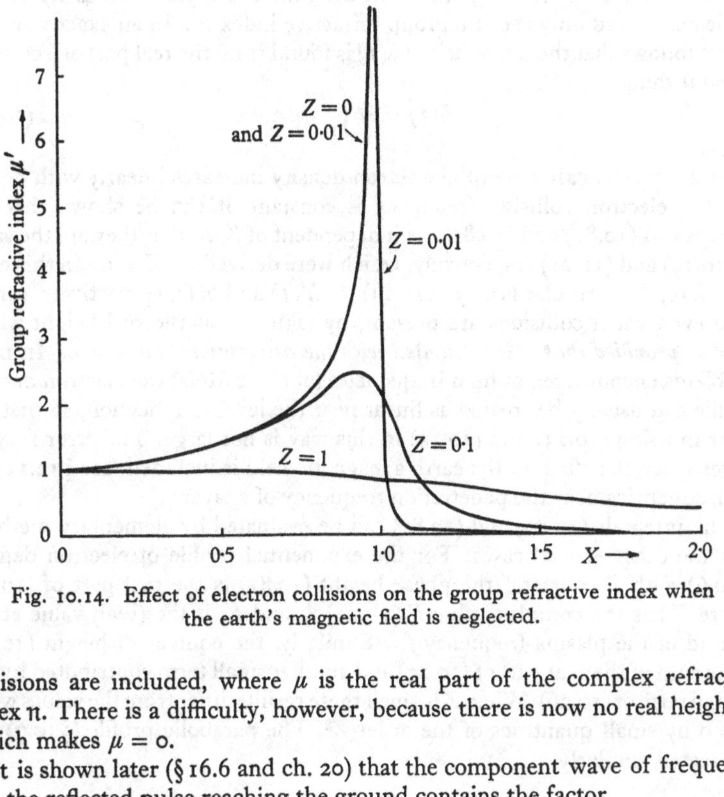

Fig. 10.14. Effect of electron collisions on the group refractive index when the earth's magnetic field is neglected.

collisions are included, where μ is the real part of the complex refractive index $\mathfrak{n}$. There is a difficulty, however, because there is now no real height z_0 which makes $\mu = 0$.

It is shown later (§ 16.6 and ch. 20) that the component wave of frequency f in the reflected pulse reaching the ground contains the factor

$$\exp\left\{2\pi i f\left(t - \frac{2}{c}\int_0^{z_0} \mathfrak{n}\,dz\right)\right\}, \qquad (10.78)$$

where the integral is a contour integral in the complex z-plane, and z_0 is a point in the complex z-plane which makes $\mathfrak{n} = 0$. Equation (10.78) is an example of the phase integral formula. The phase of the reflected wave is given by the real part of the integral in (10.78), and the time of travel τ of a pulse can be found from the condition that this phase shall be stationary for small variations of frequency. This requires that

$$\mathscr{R}\left[\frac{\partial}{\partial f}\left\{f\left(\tau-\frac{2}{c}\int_0^{z_0}\mathfrak{n}\,dz\right)\right\}\right]=0, \qquad (10.79)$$

which with (10.77) gives

$$h'(f)=\tfrac{1}{2}c\tau=\mathscr{R}\int_0^{z_0}\mathfrak{n}'\,dz, \qquad (10.80)$$

where again the integral is a contour integral in the complex z-plane. Thus to find the equivalent height we calculate the complex number $\mathfrak{n}'$ and evaluate the integral (10.80) to the point z_0 in the complex z-plane. It is no longer sufficient to find only the real group refractive index μ'. In an exactly similar way it follows that the phase height $h(f)$ is found from the real part of a contour integral thus:

$$h(f)=\mathscr{R}\int_0^{z_0}\mathfrak{n}\,dz. \qquad (10.81)$$

In the special case when the electron density increases linearly with height and the electron collision frequency is constant, it can be shown that the expressions (10.80) and (10.81) are independent of Z, so that they are the same as (10.19) and (10.21) respectively, which were derived for $Z=0$. In this case, therefore, the formulae (10.13), (10.14) for $h'(f)$ and $h(f)$, respectively, can be used even when collisions are present, by taking z_0 as the real height where $X=1$, *provided that* μ *is calculated with the assumption that* $Z=0$. In most problems encountered at high frequencies (above 2 Mc/s) the electron density profile can usually be treated as linear near the level of reflection, so that the error in using (10.13) and (10.14) in this way is not large. The error may be larger when the effect of the earth's magnetic field is included (see § 12.11), or for frequencies near the penetration frequency of a layer.

The integrals (10.80) and (10.81) can be evaluated by elementary methods in some other special cases. For the exponential profile of electron density (§ 10.6) with Z constant, the phase height (10.81) is the real part of (10.26) where F has the complex value $F_0(1-iZ)^{-\frac{1}{2}}$, and F_0 is the (real) value at the ground of the plasma-frequency f_N. Similarly, the equivalent height (10.80) is the sum of the real part of (10.25) and another small term contributed by the second term of (10.26). When Z is small these results differ from the values when $Z=0$ by small quantities of the order Z^2. The parabolic profile (§ 10.7) can be treated similarly.

10.17 Relation between equivalent height, phase height and absorption

The refractive index $\mathfrak{n}$ is equal to $\mu-i\chi$ where μ and χ are real, and (4.8) shows that for real heights, where X and Z are real,

$$\mu^2-\chi^2=1-\frac{X}{1+Z^2},\quad 2\mu\chi=\frac{ZX}{1+Z^2}, \qquad (10.82)$$

whence

$$\chi = \frac{Z}{2\mu}(1-\mu^2+\chi^2). \tag{10.83}$$

It can be shown (see §4.4 and Fig. 4.1) that if Z is small (less than about 0·1) χ is small compared to $1-\mu^2$ for levels where $X \leqslant 1$. Hence (10.83) gives approximately

$$\chi \approx \tfrac{1}{2}Z\left(\frac{1}{\mu}-\mu\right). \tag{10.84}$$

Further, $1/\mu$ is the value that the group refractive index μ' would have if there were no collisions, and Fig. 10.14 shows that it is still close to μ' when collisions are allowed for except near levels where $X = 1$. If we set $1/\mu = \mu'$ in (10.84) it becomes

$$\chi = \tfrac{1}{2}Z(\mu'-\mu). \tag{10.85}$$

Now assume that Z is constant and integrate (10.85) with respect to z over some range z_1 to z_2 of the path of the wave-packet. Then $\int_{z_1}^{z_2}\mu'\,dz$ is the contribution to the equivalent path P', and $\int_{z_1}^{z_2}\mu\,dz$ is the contribution to the phase path P, and we have, on multiplication by k,

$$k\int_{z_1}^{z_2}\chi\,dz = \tfrac{1}{2}kZ(P'-P), \tag{10.86}$$

provided the path does not go near the level where $X = 1$. Some authors assume without proof that (10.86) is true for the whole path including the level $X = 1$. This is correct for a linear profile of electron density (see below) but is not correct for other profiles, although the error is small in practical cases. If (10.86) is applied to the whole path it gives

$$2k\int_0^{z_0'}\chi\,dz = kZ(h'-h), \tag{10.87}$$

where z_0' is the (real) height where $X = 1$. Now the left side gives the total attenuation of the wave assuming that reflection occurs where $z = z_0'$. It is therefore approximately equal to $-\log|R|$, where R is the reflection coefficient. Hence

$$-\log|R| = \frac{\nu}{c}(h'-h). \tag{10.88}$$

This formula has been used by some workers (Appleton, 1935; Farmer and Ratcliffe, 1935) for estimating the electron collision-frequency ν.

To test the validity of (10.88) the expressions used above for h' and h, should be replaced by the more accurate contour integrals (10.80) and (10.81) respectively, and $-\log|R|$ should be set equal to $2k\mathscr{I}\int_0^{z_0} \mathrm{n}\,dz$, where again the integral is a contour integral and z_0 is the complex height which makes $\mathrm{n} = 0$. It can then be shown that the formula (10.88) is correct if the electron density profile is linear, but for other profiles there are small errors arising from terms in Z^2 and higher powers of Z.

Examples

1. What electron density profiles would give the following $h'(f)$ curves, if electron collisions and the earth's magnetic field were neglected?—

$$(a)\begin{cases} h'(f) = h_0 \quad \text{when} \quad f \leqslant f_1, \\ h'(f) = h_0 + \dfrac{2f}{\alpha}(f^2 - f_1^2)^{\frac{1}{2}} \quad \text{when} \quad f \geqslant f_1; \end{cases}$$

$$(b)\begin{cases} h'(f) = h_0 + \dfrac{2f^2}{\alpha} \quad \text{when} \quad f \leqslant f_1, \\ h'(f) = h_0 + \dfrac{2f^2}{\alpha} + 2f(f^2 - f_1^2)^{\frac{1}{2}}\left(\dfrac{1}{\beta} - \dfrac{1}{\alpha}\right) \quad \text{when} \quad f \geqslant f_1. \end{cases}$$

2. The electron density in the ionosphere is given by $F_0 e^{\alpha z}$ where z is height above the ground, α is a constant and F_0 is the plasma-frequency at the ground. The earth's magnetic field is to be neglected. Calculate the equivalent height of reflection $h'(f)$ for a frequency f when there are no collisions, and when the collision-frequency has a constant value ν. Show that, when collisions are allowed for, the equivalent height is less by approximately $\frac{1}{2}Z^2/\alpha + O(Z^4)$, where $Z = \nu/2\pi f$. (Assume that $F_0/f \ll 1$.)

CHAPTER 11

RAY THEORY FOR OBLIQUE INCIDENCE WHEN THE EARTH'S MAGNETIC FIELD IS NEGLECTED

11.1 Introduction. The ray path

We shall now consider some applications of 'ray theory' for waves which are obliquely incident on the ionosphere. Although in most of the chapter the earth's magnetic field is neglected, the introductory material on 'wave-packets' in this section and §§ 11.2 and 11.3 applies quite generally and is used in later chapters where the effect of the earth's field is included.

A radio transmitter may be thought of as a point source emitting waves whose wave-fronts are spherical. Such waves can be expressed as the sum of an infinite number of component plane waves. (See § 7.8 and the references given there.) In a Cartesian coordinate system with the origin at the transmitter, each field quantity of a component plane wave in free space is proportional to

$$\exp\{ik(ct - S_1x - S_2y - Cz)\},$$

where S_1, S_2, C are the direction cosines of its wave-normal. For the whole wave the field is then

$$\int_{-\infty}^{\infty}\int_{-\infty}^{\infty} A(S_1, S_2)\exp\{ik(ct - S_1x - S_2y - Cz)\}\,dS_1 dS_2, \qquad (11.1)$$

where the function $A(S_1, S_2)$ depends on the kind of transmitting aerial and on the field component used. For example, for a spherical wave of constant amplitude, the expression (11.1) is proportional to e^{-ikr}/r where r is distance from the origin, and A is then proportional to $(1 - S_1^2 - S_2^2)^{-\frac{1}{2}}$. In many cases of importance A is a very slowly varying function of S_1 and S_2 and may be treated as constant when compared with the exponential in (11.1), which is in general a much more rapidly varying function.

When a component plane wave enters the ionosphere, the exponential is replaced by

$$\exp\left\{ik\left(ct - S_1x - S_2y - \int_0^z q\,dz\right)\right\},$$

where q is given by

$$q^2 = \mathrm{n}^2 - S_1^2 - S_2^2. \qquad (11.2)$$

This follows from § 9.12, which applied to the special case where $S_2 = 0$, but can be easily extended. There is also another factor (such as $q^{-\frac{1}{2}}$ in (9.58) or $\mathfrak{n}^{-1}q^{\frac{1}{2}}$ in (9.71)) which is a slowly varying function of S_1, S_2 and z. Like the function A in (11.1) its effect on the wave is much less important that the exponential term, and as it does not affect the results of this and the following sections, it will be omitted.

We shall now suppose that (11.1) is some component E of the field. Within the ionosphere this becomes

$$E(t, x, y, z)$$
$$= \int_{-\infty}^{\infty}\int_{-\infty}^{\infty} A(S_1, S_2) \exp\left\{ik\left(ct - S_1 x - S_2 y - \int_0^z q\,dz\right)\right\} dS_1 dS_2. \tag{11.3}$$

This also gives the field after reflection from the ionosphere provided that the range of the integral $\int q\,dz$ extends from the ground to the level of reflection (in general different for each component wave) and then back to the receiving point with the sign of q reversed. The double integral (11.3) expresses the addition of many harmonically varying quantities. The contributions to the integral will only be appreciable near values of S_1, S_2, for which the phase is stationary. The phase is given by

$$\phi = k\left(ct - S_1 x - S_2 y - \int_0^z q\,dz\right), \tag{11.4}$$

and for this to be stationary we must have

$$-\frac{\partial\phi}{\partial S_1} \equiv x + \int_0^z \frac{\partial q}{\partial S_1}\,dz = 0, \quad -\frac{\partial\phi}{\partial S_2} \equiv y + \int_0^z \frac{\partial q}{\partial S_2}\,dz = 0. \tag{11.5}$$

These might be solved for S_1, S_2 at a given position (x, y, z) of the receiver, and would give $S_1 = S_1^0$, $S_2 = S_2^0$, say. Since A is slowly varying it may be replaced by the constant $A(S_1^0, S_2^0)$. Thus the value of A is important only for a small range of values of S_1 and S_2 near S_1^0, S_2^0 respectively. If A fell to zero outside this range, the signal at the receiver would not be appreciably affected.

It is interesting to trace the path traversed by the energy arriving at the receiver. Since this comes only from component waves with S_1 and S_2 near S_1^0 and S_2^0 respectively, we must find a series of points for which the signal is a maximum when $S_1 = S_1^0$, $S_2 = S_2^0$. The locus of these points is called the 'ray path', or simply the 'ray'. The signal is a maximum when the phase (11.4) is stationary with respect to variations of S_1

and S_2. Hence (11.5) are the equations of the ray, provided that in $\partial q/\partial S_1$, $\partial q/\partial S_2$ we put $S_1 = S_1^0$, $S_2 = S_2^0$ after differentiation.

In general it is not easy to solve (11.5) for S_1^0 and S_2^0 when x, y, z are given. The more usual procedure is to assume a pair of values S_1^0 and S_2^0, and trace the ray corresponding to them. Successive pairs could be tried until a pair is found which gives a ray passing through a given receiving point.

11.2 Wave-packets

On a given ray the signal is the resultant of waves whose original direction cosines S_1, S_2 are within narrow ranges near S_1^0, S_2^0 respectively. Suppose, now, that the transmitter radiates only these waves. Then the function $A(S_1, S_2)$ in (11.1) is no longer nearly constant but has a narrow maximum near $S_1 = S_1^0$, $S_2 = S_2^0$. Thus A is no longer a slowly varying function, and the stationary phase condition (11.5) must be replaced by

$$x + \int_0^z \frac{\partial q}{\partial S_1}\, dz + \frac{\partial}{\partial S_1}(\arg A) = 0, \quad y + \int_0^z \frac{\partial q}{\partial S_2}\, dz + \frac{\partial}{\partial S_2}(\arg A) = 0.$$

If in these we put $S_1 = S_1^0$, $S_2 = S_2^0$, they give the equation of the ray. But the ray must pass through the transmitter, which is at the origin. Hence

$$\frac{\partial}{\partial S_1}(\arg A) = \frac{\partial}{\partial S_2}(\arg A) = 0,$$

when $S_1 = S_1^0$, $S_2 = S_2^0$. Therefore, although A is not now slowly varying, this adds nothing to the stationary phase condition. The resulting signal leaving the transmitting aerial is a pencil of radiation with its maximum in the direction (S_1^0, S_2^0).

Suppose, further, that the signal is a pulse like that described in § 10.1. Then it must contain a range of frequencies, with maximum amplitude near the centre of the range. The variable $k = 2\pi f/c$ is proportional to frequency f. Hence the pulsed signal may be expressed by assuming that A is a function also of k, and the expression (11.3) becomes

$$E(t, x, y, z) = \int_{-\infty}^{\infty}\int_{-\infty}^{\infty}\int_{-\infty}^{\infty} A(k, S_1, S_2) \exp\left\{ik\left(ct - S_1 x - S_2 y - \int_0^z q\, dz\right)\right\} dk\, dS_1\, dS_2, \tag{11.6}$$

where A has a narrow maximum near $k = k_0$, $S_1 = S_1^0$, $S_2 = S_2^0$, and these will be called the 'predominant' values of k, S_1, S_2. This represents a signal confined to a narrow pencil, and of short duration. It is therefore a small packet of waves, and at a given time t it has maximum amplitude at some point (x, y, z). As t increases the locus of this point is the path of the wave-packet. The wave-packet is assumed to leave the transmitter when $t = 0$, and this requires that $(\partial/\partial k)(\arg A) = 0$ when $k = k_0$.

The signal (11.6) is a maximum when the phase is stationary with respect to variations of k, S_1, S_2, at the predominant values k_0, S_0^1, S_0^2. This gives the two equations (11.5) with $S_1 = S_1^0$, $S_2 = S_2^0$, and in addition

$$ct - S_1 x - S_2 y - \int_0^z \frac{\partial(kq)}{\partial k} dz = 0 \quad (k = k_0). \qquad (11.7)$$

This equation is used later to find the time of travel of a wave-packet.

Although most actual transmitters do not radiate wave-packets, we have shown that the problem of finding the path of the energy reaching a given receiver is the same as that of finding the path of a wave-packet. In future, therefore, we shall speak of a ray and of the path of a wave-packet as the same thing.

11.3 The equation for the ray when the earth's magnetic field is neglected

In this chapter and in ch. 13 we shall discuss rays which leave the transmitter in the x–z-plane, so that the predominant value S_2^0 is zero. The equations (11.5) for the ray then give

$$x = -\int_0^z \left(\frac{\partial q}{\partial S_1}\right) dz \quad (S_1 = S_1^0), \qquad (11.8)$$

$$y = -\int_0^z \left(\frac{\partial q}{\partial S_2}\right) dz \quad (S_2 = 0). \qquad (11.9)$$

When the earth's magnetic field is neglected, q is given by (11.2) which shows that $\partial q/\partial S_2 = 0$ when $S_2 = 0$. Hence (11.9) gives $y = 0$, which shows that the wave-packet remains in the x–z-plane throughout its path. This result is usually expressed by saying that there is no lateral deviation of the wave-packet when the earth's magnetic field is neglected. When it is allowed for, the expression for q is more complicated, and there can then in general be some lateral deviation. In the present chapter, however, we may assume that the wave-packet extends indefinitely in the y-direction, and the term $S_2 y$ in (11.6) may be omitted. The electric field of the wave-packet therefore becomes

$$E(t, x, z) = \int_{-\infty}^{\infty}\int_{-\infty}^{\infty} A(k, S) \exp\left\{ik\left(ct - Sx - \int_0^z q\,dz\right)\right\}, \qquad (11.10)$$

and the equation of its path in the x–z-plane is

$$x = -\int_0^z \left(\frac{\partial q}{\partial S}\right) dz, \tag{11.11}$$

where S_1 and S_1^0 are both now written simply as S, and q is given by (9.56).

In most of the present chapter the effect of electron collisions is neglected and the frequency f is assumed to be greater than the plasma-frequency f_N (§ 3.2). Thus the refractive index $\mathfrak{n}$ is real and positive and is the same as its real part μ. To emphasise this the symbol μ will be used for the refractive index. (Discussion of problems involving a complex $\mathfrak{n}$ is resumed in § 11.16.) Equations (8.87), (8.88) show that q is also real, and $q = \mu \cos\theta$, $q^2 = \mu^2 - S^2$.

The direction ξ which the ray makes with the vertical is given by $\tan\xi = dx/dz$ and (11.11) and (9.56) show that

$$\tan\xi = -\frac{\partial q}{\partial S} = \frac{S}{q}, \tag{11.12}$$

so that

$$\mu \sin\xi = S. \tag{11.13}$$

Now Snell's law shows that $S = \mu \sin\theta$, where θ is the inclination of the wave-normal to the vertical. Hence $\theta = \xi$ and the ray and the wave-normal have the same direction. This results from the fact that the refractive index μ is independent of the direction of the wave-normal. This is not in general true when the effect of the earth's magnetic field is included, and then the ray and the wave-normal may have different directions.

In this chapter the earth's magnetic field is neglected, and (11.11) and (9.56) then give for the equation of the ray path

$$x = S\int_0^z \frac{dz}{q}. \tag{11.14}$$

11.4 The ray path for a linear gradient of electron density

Let the base of the ionosphere be at height $z = h_0$ above the ground, and suppose that the electron number density N increases linearly with height above this. Then f_N^2, which is proportional to N, is given by (10.16), so that

$$\left.\begin{aligned} q^2 &= C^2 - \alpha(z-h_0)/f^2 \quad \text{when} \quad z \geqslant h_0, \\ q &= C \quad \text{when} \quad z \leqslant h_0, \end{aligned}\right\} \tag{11.15}$$

where $C = \cos\theta_I$, and θ_I is the angle of incidence of the wave-packet on the ionosphere. In this case (11.14) becomes

$$x = h_0 \tan\theta_I + \frac{f^2 \sin 2\theta_I}{\alpha} - \frac{2f^2 \sin\theta_I}{\alpha}\left\{\cos^2\theta_I - \frac{\alpha(z-h_0)}{f^2}\right\}^{\frac{1}{2}}, \tag{11.16}$$

which shows that within the ionosphere the ray path is a parabola. It is now of interest to find the total horizontal range D traversed by the wave-packet, when it returns to the ground. The top of its trajectory is where $dx/dz = \infty$, that is where $q = 0$, and at this point

$$x = h_0 \tan \theta_I + \frac{f^2 \sin 2\theta_I}{\alpha}.$$

The horizontal range is twice this, and hence

$$D = 2h_0 \tan \theta_I + \frac{2f^2 \sin 2\theta_I}{\alpha}. \tag{11.17}$$

The first term is contributed by the two straight sections of the path below the ionosphere, and the second term is the part within the ionosphere. It has been assumed that the earth is flat.

Fig. 11.1 shows how D/h_0 depends on θ_I for some typical cases. It can easily be shown that if $f^2/\alpha h_0 > 4$, the curve has a maximum and a minimum so that for some values of D (11.17) is satisfied by three different values of θ_I. This means that wave-packets can travel from the transmitter to the receiver by three different paths, as illustrated in Fig. 11.2. This is only possible if $D > 6\sqrt{3}\, h_0$. The proof is left as an exercise for the reader.

11.5 The ray path for exponential variation of electron density

Suppose that the electron number density N increases exponentially with height z, so that the plasma frequency f_N is given by (10.22). Then

$$q^2 = C^2 - \frac{F^2}{f^2} e^{\alpha z}, \tag{11.18}$$

where F is the value of f_N at the ground, and waves of frequency less than F cannot be propagated as wave-packets. In this case N is never zero; it must have a small non-zero value even at the ground. Hence θ_I is not the inclination of the wave-normal at the ground, but is the inclination it would have if the wave ever reached a region of free space.

The equation of a ray path is

$$x = S\int_0^z \left\{C^2 - \frac{F^2}{f^2} e^{\alpha z}\right\}^{-\frac{1}{2}} dz = \frac{2S}{\alpha C} \log\left\{\frac{\tan\frac{1}{2}\psi_2}{\tan\frac{1}{2}\psi_1}\right\}, \tag{11.19}$$

where

$$\sin\psi_2 = \frac{F}{fC} e^{\frac{1}{2}\alpha z}, \quad \sin\psi_1 = \frac{F}{fC}. \tag{11.20}$$

At the top of the trajectory $q = 0$ and $\psi_2 = \frac{1}{2}\pi$. Hence the horizontal range is given by

$$D = \frac{4S}{\alpha C} \log\left\{\frac{fC}{F} + \left(\frac{f^2C^2}{F^2} - 1\right)^{\frac{1}{2}}\right\}. \tag{11.21}$$

Here again it is assumed that the earth is flat. The value of C cannot be less than F/f. When $C = F/f$ the level $z = 0$ is already the top of the trajectory, so

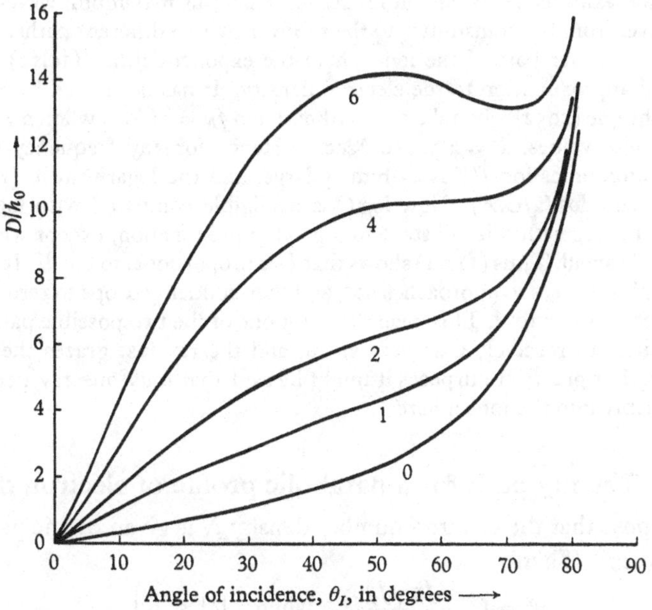

Fig. 11.1. Shows how the horizontal range D varies with angle of incidence θ_I for a linear profile of electron density. The earth is assumed to be flat, and the earth's magnetic field and electron collisions are neglected. The numbers by the curves are the values of $f^2/\alpha h_0$ where h_0 is the height of the base of the ionosphere.

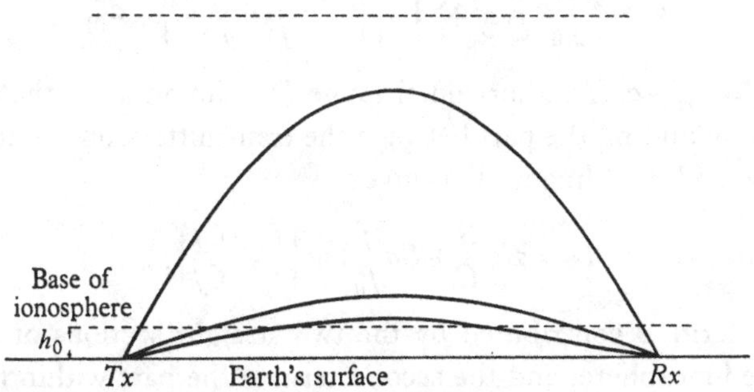

Fig. 11.2. Diagram, approximately to scale, showing the three possible ray paths for a linear profile of electron density. The earth is assumed to be flat, and the earth's magnetic field and electron collisions are neglected. $D/h_0 = 18$. $f^2/\alpha h_0 = 10$. The upper broken line is the level where $\mathfrak{n} = 0$ that would be reached by a ray at vertical incidence.

that no ray can travel upwards, and $D = 0$. For each pair of values of α and f/F, D has a maximum value, and at ranges less than this maximum, wave-packets can travel from the transmitter to the receiver by two different paths.

For the lower part of the ionosphere the exponential law (10.22) is often a useful approximation to the electron density. It has been used by Heading and Whipple (1952) who take $\alpha = 0{\cdot}6\,\text{km}^{-1}$ and $f_N = 16\,\text{kc/s}$ when $z = 60\,\text{km}$. With these values, $F = 2{\cdot}5 \times 10^{-4}\,\text{sec}^{-1}$. Hence for any frequency used in radio communication f/F is extremely large, and the logarithm in (11.21) is very closely $\log(2fC/F)$. Now $\log C$ is negligible compared with $\log(2f/F)$, so that the logarithm is constant to a good approximation, except when C is extremely small. Thus (11.21) shows that D is proportional to $\tan\theta_I$. It reaches very high values as θ_I approaches 90°, and then suddenly drops to zero when C becomes of order 10^{-4}. This means that for one of the two possible paths from transmitter to receiver, C is very small, and the ray just grazes the earth's surface. For practical purposes it might be said that only one ray penetrates appreciably into the ionosphere.

11.6 The ray path for a parabolic profile of electron density

Suppose that the electron number density N is given by the parabolic law (10.27). Then

$$\left.\begin{aligned} q^2 &= C^2 - \frac{f_p^2}{f^2} + \frac{f_p^2}{f^2}\zeta^2 && \text{when } |\zeta| \leqslant 1, \\ q &= C && \text{when } |\zeta| \geqslant 1, \end{aligned}\right\} \tag{11.22}$$

where f_p is the penetration frequency of the layer, and ζ is given by (10.28). The equation of a ray path is

$$x = S\int_0^{h_0} \frac{dz}{C} + aS\int_{-1}^{\zeta} \left\{C^2 - \frac{f_p^2}{f^2} + \frac{f_p^2}{f^2}\zeta^2\right\}^{-\frac{1}{2}} d\zeta, \tag{11.23}$$

where $h_0 = z_m - a$. The horizontal range D is found as in the previous examples by finding the part between the transmitter and the top of the trajectory, and doubling it. This gives

$$D = 2h_0\frac{S}{C} + Sa\frac{f}{f_p}\log\left\{\frac{f_p + Cf}{f_p - Cf}\right\}. \tag{11.24}$$

The first term is contributed by the two straight sections of the path below the ionosphere, and the second term is the part within the ionosphere. It has been assumed that the earth is flat.

Fig. 11.3 shows how D/h_0 depends on θ_I for some typical cases. When $f \leqslant f_p$ any wave-packet is reflected whatever the value of θ_I, and D varies between 0 and ∞. When $f > f_p$, however, wave-packets which are near vertical incidence can penetrate the ionosphere, and do not return to the ground. For reflection it is necessary that $C < f_p/f$ which gives a minimum value of θ_I, and (11.24) shows that D tends to ∞ as C tends to f_p/f and as C tends to zero. This behaviour is clearly shown by the two curves for which $f > f_p$ in Fig. 11.3. Between these values there is a minimum value of D, and for any greater value of D wave-packets can travel from the transmitter to the receiver by two different paths. The path for which θ_I is the smaller is known as the Pedersen ray. It is shown in § 11.11 that this ray has the greater time of travel.

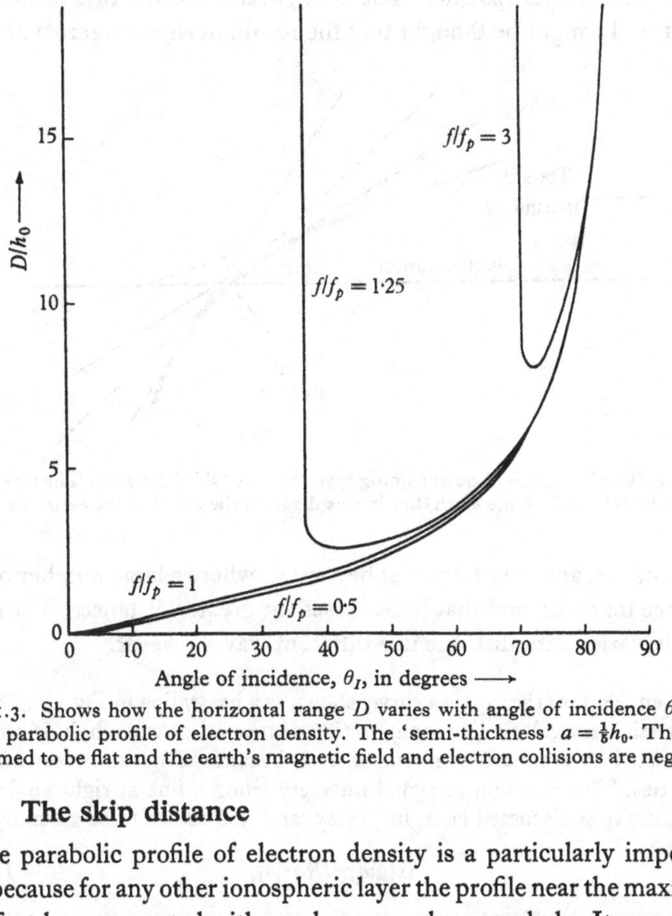

Fig. 11.3. Shows how the horizontal range D varies with angle of incidence θ_I for a typical parabolic profile of electron density. The 'semi-thickness' $a = \frac{1}{8}h_0$. The earth is assumed to be flat and the earth's magnetic field and electron collisions are neglected.

11.7 The skip distance

The parabolic profile of electron density is a particularly important one, because for any other ionospheric layer the profile near the maximum can often be represented with good accuracy by a parabola. It was shown in the last section that for a frequency which exceeds the penetration-frequency, there is a minimum distance D from the transmitter within which no ray travelling via the ionosphere can return to the ground. This behaviour is often observed in practice. It is found that the signals from some transmitters are weak because they travel only over the ground and are heavily attenuated, whereas signals from more distant transmitters on the same or nearly the same frequency are strongly received because they are reflected from the ionosphere. The minimum distance at which a ray can return to the ground from the ionosphere is called the 'skip' distance.

A ray which returns to the ground at the skip distance is travelling at a certain angle to the vertical. Rays at neighbouring angles all reach the ground at a greater distance. The configuration of the rays is shown in Fig. 11.4. It might be thought that the resultant signal is zero within the

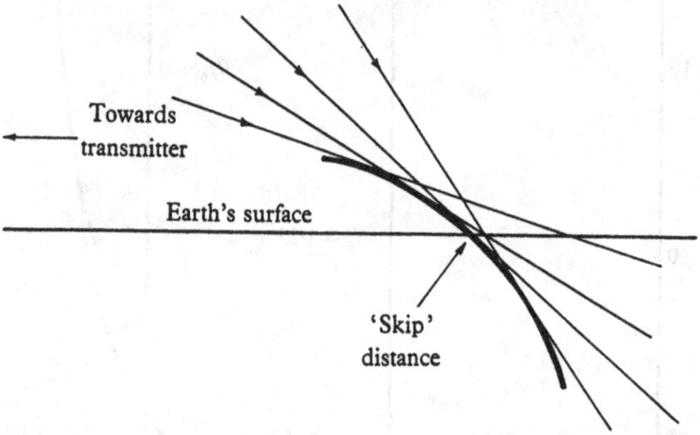

Fig. 11.4. Configuration of downcoming rays near the 'skip' distance. The rays would actually be reflected by the earth, but in this diagram the effect of the earth is ignored.

skip distance, and very large just beyond it, where a large number of rays are close together, and that it decreases for greater distances. The signal actually varies with distance in a different way, however.

The envelope of the rays is a curve, shown as a heavy line in Fig. 11.4, known as a caustic curve. A similar configuration of rays arises when light is reflected or refracted by a spherical mirror or lens, because of its imperfect focusing properties. The variation of signal intensity along a line at right angles to a caustic curve is discussed later, in § 15.25, and it is shown to be given by

$$\mathrm{Ai}\{2u(\pi^2/R\lambda^2)^{\frac{1}{3}}\}, \tag{11.25}$$

where Ai denotes the Airy integral function, u is distance measured perpendicular to the caustic, R is the radius of curvature of the caustic, and λ is the wavelength. The Airy integral function is shown in Fig. 15.5. It decreases rapidly when u is positive, that is within the skip distance. When u is negative, that is on the illuminated side of the caustic, it shows oscillations which arise because the two rays through any point may interfere or reinforce, depending on the value of u.

The skip distance is the minimum value of the horizontal range D, and occurs for some particular value of θ_I, say θ_0. Then $dD/d\theta_I = 0$ when $\theta_I = \theta_0$, and it can easily be shown that $R = (d^2D/d\theta_I^2)\cos\theta_0$. If x is distance measured along the ground from the skip distance, towards the transmitter, then $u = x\cos\theta_0$. Hence (11.25) becomes

$$\mathrm{Ai}\left\{2x\left(\pi^2\cos^2\theta_0\Big/\lambda^2\frac{d^2D}{d\theta_I^2}\right)^{\frac{1}{3}}\right\}. \tag{11.26}$$

Two of the curves in Fig. 11.3 show how D varies with θ_I for a parabolic profile of electron density, when the frequency exceeds the penetration frequency. When these curves were computed, the values of $(d^2/d\theta_I^2)(D/a)$ at the minima were computed at the same time. In the curve for $f = 3f_p$, the skip distance is about 1600 km, and $(d^2/d\theta_I^2)(D/a)$ is then about 5000, and $\theta_I = \theta_0 = 72°$. The curves were plotted for the case $h/a = 5$. We take $h = 200$ km, $a = 40$ km as values roughly representative of the F-layer. Let $f_p = 5$ Mc/s, so that $f = 15$ Mc/s and $\lambda = 0{\cdot}02$ km. Then (11.26) becomes approximately $\mathrm{Ai}\{0{\cdot}46x\}$ where x is in km. Now for the first two zeros of $\mathrm{Ai}(\zeta)$, the difference of the values of ζ is $1{\cdot}75$, which corresponds to a range of 3·8 km for x. Further, $\mathrm{Ai}(\zeta)$ is one-tenth of $\mathrm{Ai}(0)$ when $\zeta = 1{\cdot}99$ so that $x = 4{\cdot}3$ km. These figures give some idea of the scale of the field pattern on the ground near the skip distance.

For the other curve of Fig. 11.3, $f = \frac{5}{4}f_p$ and $(d^2/d\theta_I^2)(D/a) \doteqdot 180$, $\theta_0 = 40°$. The skip distance is now about 500 km. With the same values of a and f_p, the expression (11.26) becomes $\mathrm{Ai}\{1{\cdot}4x\}$ which shows that in this case the scale of the pattern is about three times smaller than in the first example.

11.8 The equivalent path P' at oblique incidence

In § 11.2 the field of a wave-packet was expressed as the integral (11.6), and the position of the wave-packet in space was found from the condition that the phase ϕ of the integrand shall be stationary for variations of the directions S_1, S_2 of the component waves. This led to the equations of the ray path (11.8) and (11.9). The position of the wave-packet at a given time t can be found in a similar way from the condition that the phase ϕ shall be stationary for variations of frequency f, which is proportional to k. This gives (11.7) which may be written

$$ct = S_1x + S_2y + \int_0^z \frac{\partial(kq)}{\partial k}\,dz, \tag{11.27}$$

where t is the time taken for the wave-packet to travel from the origin to the point (x, y, z) and x, y are given in terms of z by (11.8) and (11.9). In time t the distance that the wave-packet could travel in free space is ct which is called the equivalent path P'. In (11.27) S_2, S_1 and k must have the values which make $A(k, S_1, S_2)$ a maximum, namely S_0, 0 and k_0. The subscripts 0 will be omitted, so that (11.27) becomes

$$P' = Sx + \int_0^z \frac{\partial(fq)}{\partial f} dz. \tag{11.28}$$

The equation is true, in general, whether or not the wave-packet is deviated out of the plane $y = 0$.

11.9 Breit and Tuve's theorem. Martyn's theorem for equivalent path

When the earth's magnetic field and electron collisions are neglected q is given by (9.56) with $Z = 0$, which may be written

$$q^2 = C^2 - f_N^2/f^2, \tag{11.29}$$

where f_N is the 'plasma frequency', and is proportional to $\sqrt{N}$. Then $\partial(fq)/\partial f = C^2/q$ and (11.28) may be written

$$P' = Sx + \frac{C^2}{S}\int_0^z \frac{S}{q} dz. \tag{11.30}$$

Now (11.14) shows that the integral here is simply x. Hence

$$P' = x/S, \tag{11.31}$$

which means that when a wave-packet starts out at an angle θ_I to the vertical, and has travelled along some path TA (Fig. 11.5), its equivalent path is TB obtained by producing its initial path along a straight line to a point B vertically above A. In particular if the wave-packet returns to the ground at a distance $x = D$, the equivalent path is the sum of the two oblique sides TC, CR of the triangle formed by producing the initial and final directions of the wave-packet until they intersect at C. Hence

$$P' = D/S. \tag{11.32}$$

This is Breit and Tuve's theorem (1926). The fictitious path formed by the two straight lines TC, CR is called the triangulated path. The theorem is true only when the earth's magnetic field is neglected, and when the electron collision-frequency is small.

An alternative proof of Breit and Tuve's theorem is based on the result given at the end of § 11.3 that, when the earth's magnetic field is neglected, the directions of the ray and the wave-normal are the same. Hence, if θ is the inclination of the ray or wave-normal to the vertical, Snell's law gives $\mu \sin\theta = \sin\theta_I = S$. Let δs be an element of the ray path, and δx

its horizontal projection. Then $\delta x = \delta s \sin\theta = S(\delta s/\mu) = S\mu'\delta s$ where μ' is the 'group refractive index' defined in § 10.2. Now the wave-packet travels with the group velocity c/μ', and hence $\mu'\delta s = c\delta t$ where δt is the time taken to travel the distance δs. Hence $\delta x = Sc\delta t$, and when this is integrated, we obtain $x = S.P'$, which is the same as (11.31).

Equations (11.14) and (11.31) may be combined to give

$$P' = \int_0^z \frac{dz}{q}, \tag{11.33}$$

and when the value (11.29) for q is inserted, this becomes

$$P'(f) = \frac{1}{C}\int_0^z \frac{dz}{\{1 - f_N^2/(fC)^2\}^{\frac{1}{2}}}, \tag{11.34}$$

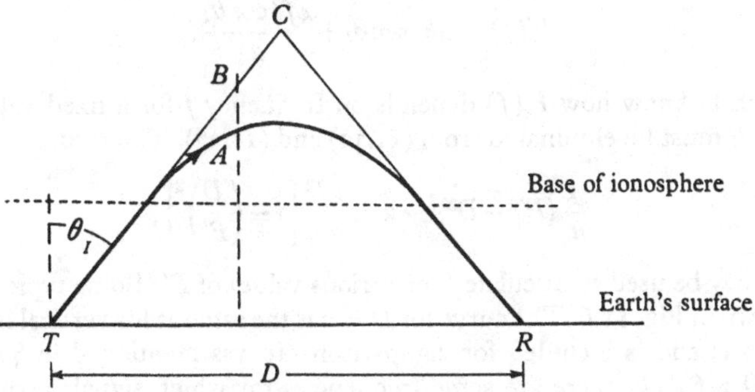

Fig. 11.5. The path of a ray in the ionosphere, and the associated 'triangulated path'.

where P' is measured at the particular frequency f. Now the integrand is the group refractive index $\mu'(fC)$ at the frequency fC. A wave of frequency fC vertically incident on the ionosphere would be reflected at the level z_0 where $f_N = fC$, and this is also the level of reflection of the frequency f at oblique incidence. Suppose that the wave-packet returns to the ground. Then the upper limit of the integral must be set equal to z_0, and a factor 2 introduced to allow for the upward and downward paths. This gives

$$P'(f) = \frac{2}{C}\int_0^{z_0} \mu'(fC)\,dz. \tag{11.35}$$

Here the integral is the equivalent height of reflection $h'(fC)$ for a frequency fC at vertical incidence. Hence

$$CP'(f) = 2h'(fC). \tag{11.36}$$

In words: 'the equivalent path for a frequency fC at vertical incidence is the vertical projection of the triangulated path for a frequency f at oblique incidence'. This is Martyn's theorem for the equivalent path (Martyn, 1935). It must not be confused with Martyn's theorem for absorption (§ 11.16).

11.10 The equivalent path at oblique incidence for a linear gradient of electron density

Let the base of the ionosphere be at a height $z = h_0$ above the ground, and suppose that the electron number density N increases linearly with height above this. Then f_N^2, which is proportional to N, is given by (10.16). This case was discussed in § 11.4, where it was shown that a wave-packet leaving the ground at an angle θ_I to the vertical returns to the ground at a distance D given by (11.17). The equivalent path P' may now be found by Breit and Tuve's theorem (11.32), whence

$$P'(f) = 2h_0 \sec\theta_I + \frac{4f^2\cos\theta_I}{\alpha}. \tag{11.37}$$

We wish to know how $P'(f)$ depends on frequency f for a fixed value of D. Hence θ_I must be eliminated from (11.32) and (11.37). The result is

$$\frac{1}{h_0}\{P'^2 - D^2\}^{\frac{1}{2}} - 2 = \frac{4f^2}{\alpha h_0}\left\{1 - \left(\frac{D}{P'}\right)^2\right\}, \tag{11.38}$$

which may be used to calculate f for various values of P'. Some typical curves are shown in Fig. 11.6. The curve for $D = 0$ is the same as for vertical incidence (Fig. 10.1) and is included for comparison. It was mentioned in § 11.4 that when $D > 6\sqrt{3}\,h_0$ there are some frequencies for which signals arrive at the receiver by three different paths. This is illustrated by the curve in Fig. 11.6 for $D = 20h_0$, which shows that for some frequencies there are three different values of $P'(f)$. This does not happen when $D/h_0 = 10$, which is slightly less than $6\sqrt{3}$.

11.11 The equivalent path at oblique incidence for a parabolic profile of electron density

Suppose that the electron number density N is given by the parabolic law (10.27). This case was discussed in § 11.6, where it was shown that a wave-packet leaving the ground at an angle θ_I to the vertical returns to the ground at a distance D given by (11.24). Breit and Tuve's theorem (11.32) holds in this case, and $P'(f)$ may be found by elimination of θ_I from (11.24) and (11.32). In view of the complexity of (11.24), this appears to be complicated. One method of procedure is as follows. Equation (11.24) may be written

$$\frac{D}{h_0 \tan\theta_I} = 2 + \frac{a}{h_0}\frac{fC}{f_p}\log\left\{\frac{1 + fC/f_p}{1 - fC/f_p}\right\}. \tag{11.39}$$

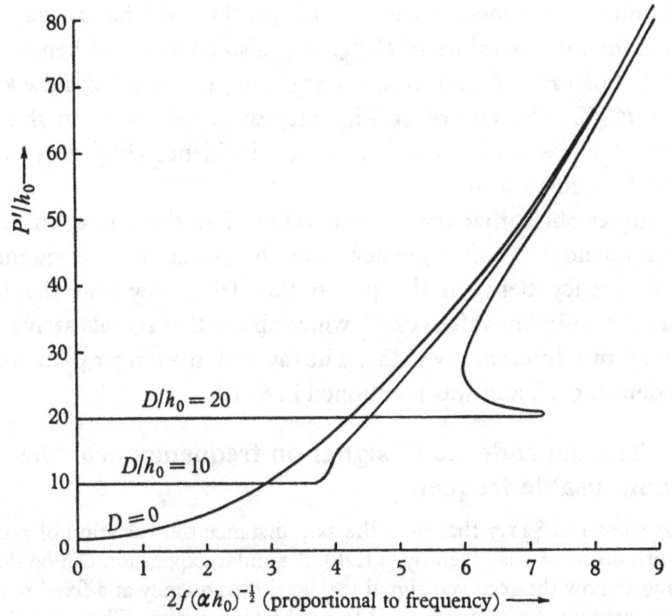

Fig. 11.6. Typical $P'(f)$ curves for oblique incidence and linear profile of electron density. The earth is assumed to be flat and the earth's magnetic field and electron collisions are neglected.

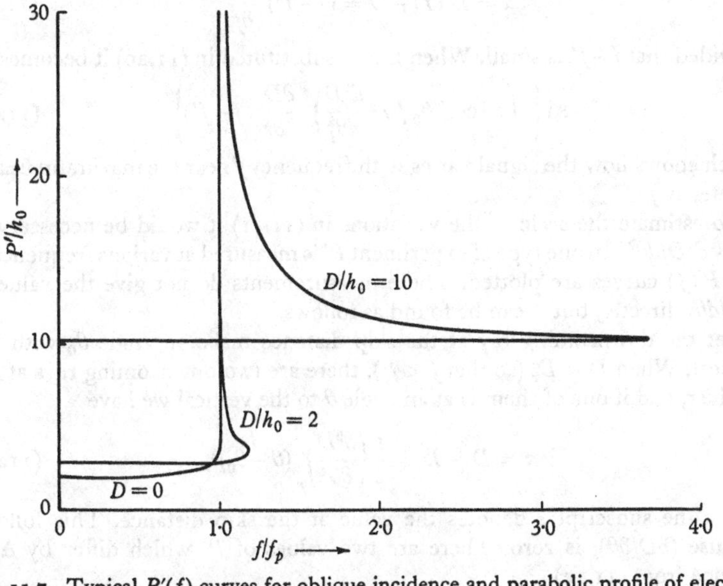

Fig. 11.7. Typical $P'(f)$ curves for oblique incidence and parabolic profile of electron density. The 'semi-thickness' $a = \frac{1}{8}h_0$. The earth is assumed to be flat and the earth's magnetic field and electron collisions are neglected.

Now a and h_0 are known constants. Hence the right-hand side can be tabulated for various values of fC/f_p. D is also known, and hence θ_I can be found. Then P' is found from (11.32), and f is found from the known value of fC/f_p. The curves of Fig. 11.7 were calculated in this way. That for $D = 0$ is the same as for vertical incidence (Fig. 10.3), and is included for comparison.

The curves show that for a given value of D there is a 'maximum usable frequency' (M.U.F.) greater than the penetration frequency f_p. For a frequency between the penetration frequency and the M.U.F. there are two different values of P', which shows that signals arrive at the receiver by two different ray paths. The ray with the longer path is called the 'Pedersen ray', and was mentioned in § 11.6.

11.12 The dependence of signal on frequency near the maximum usable frequency

It was shown in § 11.7 that near the skip distance the variation of received signal with distance x is given by (11.26). A similar expression can be derived which shows how the received signal varies with frequency at a fixed receiver.

Let the receiver be at distance D from the transmitter. The skip distance $D_s(f)$ depends on frequency f, and is equal to D when the frequency is equal to the M.U.F. f'. For any other frequency we may write

$$x = D_s(f) - D \doteqdot (f-f')\frac{\partial D_s}{\partial f}, \tag{11.40}$$

provided that $f-f'$ is small. When this is substituted in (11.26) it becomes

$$\mathrm{Ai}\left\{2\left(\pi^2\cos^2\theta_0 \Big/ \lambda^2\frac{d^2D}{d\theta_I^2}\right)^{\frac{1}{3}}\frac{\partial D_s}{\partial f}(f-f')\right\}, \tag{11.41}$$

which shows how the signal varies with frequency f near the maximum usable frequency f'.

To estimate the scale of the variations in (11.41) it would be necessary to know $d^2D/d\theta_I^2$. In one type of experiment P' is measured at various frequencies, and $P'(f)$ curves are plotted. These measurements do not give the value of $d^2D/d\theta_I^2$ directly, but it can be found as follows.

Let the downcoming ray at the skip distance make an angle θ_0 with the vertical. When $D < D_s$ (so that $f < f'$), there are two downcoming rays at the receiver, and if one of them is at an angle θ to the vertical we have

$$x = D_s - D \doteqdot \frac{1}{2}\left(\frac{\partial^2 D}{\partial\theta^2}\right)_s (\theta-\theta_0)^2, \tag{11.42}$$

where the subscript s denotes the value at the skip distance. This follows because $(\partial D/\partial\theta)_s$ is zero. There are two values of P' which differ by $\Delta P'$, whence from (11.32)

$$\Delta P' = -\frac{CD}{S^2}\Delta\theta. \tag{11.43}$$

Here $\Delta\theta$ is the difference in the values of θ for the two rays and is equal to $2(\theta-\theta_0)$. Now $(\theta-\theta_0)$ and x are eliminated from (11.40), (11.42) and (11.43), whence

$$(\Delta P')^2 = \frac{8C^2D^2}{S^4}(f-f')\frac{\partial D_s}{\partial f}\bigg/\frac{\partial^2 D}{\partial\theta^2}. \tag{11.44}$$

If $(\Delta P')^2$ is plotted against f and the slope of the curve is measured where it meets the line $(\Delta P')^2 = 0$, then $(\partial D_s/\partial f)/(\partial^2 D/\partial\theta^2)$ can be found. The method of finding $\partial D_s/\partial f$ is mentioned below.

The use of the formula (11.41) may be illustrated by the same examples as were used in § 11.7 which were for a parabolic profile of electron density (10.27) with $h = 200$ km, $a = 40$ km, $f_p = 5$ Mc/s. In the first example $D = 1600$ km, $\theta_0 = 72°$, $f = 3f_p$, $\lambda = 0{\cdot}02$ km. The value of $\partial D_s/\partial f$ can be found from the curves of Fig. 11.8 (see later, § 11.13). It is equal to $1{\cdot}2\times 10^{-4}$ km/s in this case, and the expression (11.41) becomes $\mathrm{Ai}\{55(f-f')\}$ where $f-f'$ is in Mc/s. Now $\mathrm{Ai}(\zeta)$ is one-tenth of $\mathrm{Ai}(0)$ when $\zeta = 1{\cdot}99$, so that $f-f' = 0{\cdot}036$ Mc/s. In the second example, $D = 500$ km, $\theta_0 = 40°$, $f = \frac{5}{4}f_p$, $\lambda = 0{\cdot}05$ km, and it can be shown that $\partial D_s/\partial f = 1{\cdot}9\times 10^{-4}$ km/s, so that (11.41) becomes $\mathrm{Ai}\{266(f-f')\}$, and the variations are about five times more rapid than in the first example.

In observations of $P'(f)$ at oblique incidence it is often found that appreciable signals are received at frequencies well beyond the point where the two branches of the $P'(f)$ curves come together, that is, at the extreme right of the curves in Fig. 11.7. From the above arguments it is clear that these cannot be explained in terms of wave-propagation near the caustic curve at the skip distance. They may possibly be explained by scattering of the waves by horizontal irregularities in the ionosphere, either in the reflecting layer or in some lower layer through which the waves pass.

11.13 The prediction of maximum usable frequencies. Appleton and Beynon's method

In planning the best use of a given radio communication link, it is important to know in advance what frequencies can be used. One method of predicting the maximum usable frequency for given transmitting and receiving stations was described by Appleton and Beynon (1940, 1947) and has been extensively used. In this section an outline is given of a method essentially equivalent to theirs.

By measurements of $h'(f)$ at vertical incidence it is possible to find how the electron number density N varies with height z. Methods are described in §§ 10.11 and 12.13 to 12.17. It has already been mentioned that a parabolic profile of electron density is often a good approximation to an ionospheric layer, and from the measurements of $h'(f)$ it is possible to find the penetration frequency f_p, the height h_0 and the semi-thickness a of the parabola which best represents the layer. It is not necessary to find the complete $N(z)$ profile in order to estimate a and h. Quicker methods have been described by Booker and Seaton (1940), and by Ratcliffe (1951). Extensive studies have been made of how f_p, a and h_0 vary with time of day, season and position on the earth, and it is now possible to predict their values for a given place at some future date and time with fair accuracy.

If the transmitter and receiver are less than about 2500 km apart, a radio wave travelling between them can make a single reflection at the ionosphere. For more widely separated stations, a ray making a single reflection cannot reach the receiver because of the curvature of the earth, and there may have to be two or more reflections at the ionosphere. The first step in finding the M.U.F. is to predict the values of f_p, a, and h_0 at the points in the ionosphere where reflection occurs. For a single reflection this will be a point midway between the transmitter and receiver. We need only discuss a single reflection. For paths with multiple reflections, each section of the path can be treated separately.

For given values of f_p, a, h_0 and the horizontal range D, it is now necessary to find the maximum frequency for which a ray can travel from transmitter to receiver. The range D is given by (11.24). When the frequency f is greater than f_p, D has a minimum for some value of the angle of incidence θ_I, as is shown by two of the curves in Fig. 11.3. To find when D is a minimum, we set $\partial D/\partial\theta_I = 0$. This gives

$$2h_0 + C^2 a\frac{fC}{f_p}\log\left\{\frac{1+fC/f_p}{1-fC/f_p}\right\} = \frac{2aS^2f^2C^2/f_p^2}{1-f^2C^2/f_p^2}, \tag{11.45}$$

where $S = \sin\theta_I$, $C = \cos\theta_I$. If this equation could be solved for θ_I and the value inserted in (11.24), it would give the minimum value of D as a function of frequency. A method of solution is as follows. Let

$$\frac{fC}{f_p} = \Omega, \quad A = \Omega\log\frac{1+\Omega}{1-\Omega}, \quad B = \frac{2\Omega^2}{1-\Omega^2}.$$

Then (11.45) becomes $\quad 2h_0 + aC^2A = aS^2B$

or

$$\tan^2\theta_I = \frac{A+2h_0/a}{B-2h_0/A}. \tag{11.46}$$

Now A and B may be tabulated for various values of Ω, and θ_I may be found from (11.46). Hence C is found, and from it $f/f_p = \Omega/C$. Equation (11.24) is written

$$\frac{D}{a} = \left(\frac{2h_0}{a}+A\right)\tan\theta_I,$$

whence the minimum value D_s of D for the frequency f is at once found. D_s/a may now be plotted against f/f_p. Typical curves are shown in Fig. 11.8 for two values of h_0/a. A more extensive set of curves is given by Appleton and Beynon (1947). At the M.U.F. the actual range must be just the skip distance D_s. Now a and h_0 can be predicted, and the curve for the appropriate value of h_0/a is selected. For the ordinate D_s/a the abscissa f/f_p is read. Since the penetration frequency f_p can also be predicted, the M.U.F. f can be found.

11.14 The curvature of the earth

In the analysis at the end of the last section, no allowance was made for the curvature of the earth. Fig. 11.9 represents a cross-section of the earth and the ionosphere, and $TABR$ is a ray travelling from transmitter to receiver. Only

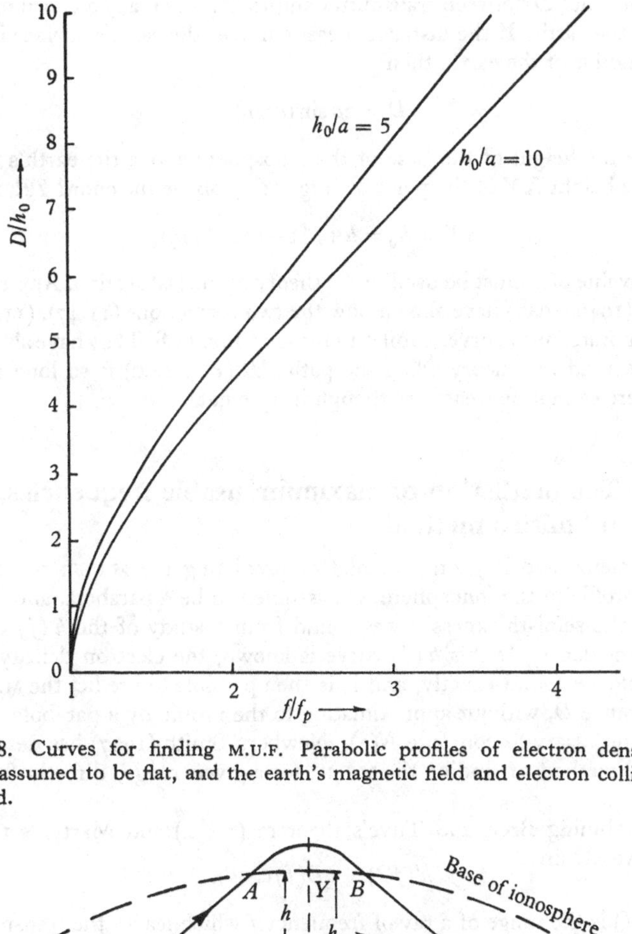

Fig. 11.8. Curves for finding M.U.F. Parabolic profiles of electron density. The earth is assumed to be flat, and the earth's magnetic field and electron collisions are neglected.

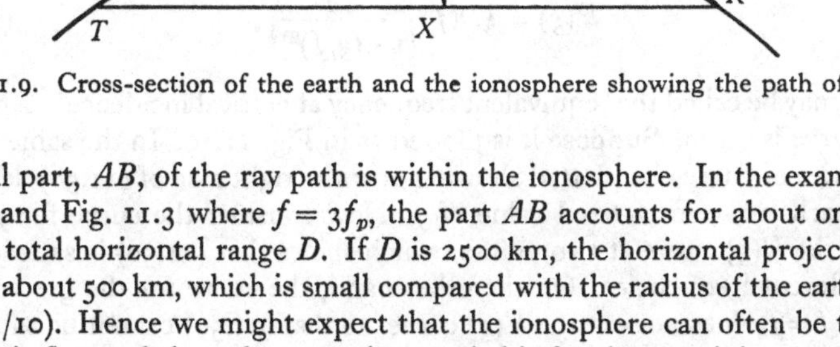

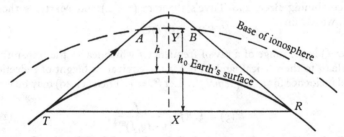

Fig. 11.9. Cross-section of the earth and the ionosphere showing the path of a ray.

a small part, AB, of the ray path is within the ionosphere. In the example of § 9.12 and Fig. 11.3 where $f = 3f_p$, the part AB accounts for about one-fifth of the total horizontal range D. If D is 2500 km, the horizontal projection of AB is about 500 km, which is small compared with the radius of the earth (less than 1/10). Hence we might expect that the ionosphere can often be treated as if it is flat, and the only correction needed is for the remaining part of the path.

The distance D between transmitter and receiver is really a chord of a great circle of the earth. If the distance measured over the earth's surface is d, and ρ is the radius of the earth, then

$$D = 2\rho \sin(d/2\rho). \tag{11.47}$$

Let h be the height of the base of the ionosphere above the earth's surface. Then the height XY of the point Y (Fig. 11.9), above the chord TR, is

$$XY = h_0 = h + \rho\{1 - \cos(d/2\rho)\}, \tag{11.48}$$

and this value of h must be used in the theory of the last section. Appleton and Beynon (1940, 1947) have shown how the two corrections (11.47), (11.48) can be incorporated into curves similar to those of Fig. 11.8. They have also shown how to extend the theory when the path AB (Fig. 11.9) is so long that the ionosphere cannot be treated as though it were flat.

11.15 The prediction of maximum usable frequencies. Newbern Smith's method

In Appleton and Beynon's method of predicting the M.U.F., the electron density profile in the ionosphere was assumed to be a parabola, and the best value of the semi-thickness a was found from a study of the $h'(f)$ curve at vertical incidence. If this $h'(f)$ curve is known, the electron density profile $N(z)$ could be found exactly, and it is then possible to predict the M.U.F. for a given range D, without approximating to the profile by a parabola. In fact it is not necessary to compute $N(z)$. Newbern Smith (1937) has described a graphical method of finding the M.U.F., for a given range, directly from the $h'(f)$ curve.

By combining Breit and Tuve's theorem (11.32) and Martyn's theorem (11.36) we obtain

$$h'(fC) = \tfrac{1}{2}D(f)\cot\theta_I. \tag{11.49}$$

Here $D(f)$ is the range of a ray of frequency f which leaves the transmitter at an angle θ_I to the vertical, and $h'(fC)$ is the equivalent height of reflection at vertical incidence for a frequency fC. Let $fC = g$. Then (11.49) may be written

$$h'(g) = \tfrac{1}{2}D(f)\frac{g/f}{\{1-(g/f)^2\}^{\frac{1}{2}}}. \tag{11.50}$$

Here g may be called the 'equivalent frequency at vertical incidence'. Now the $h'(g)$ curve is given. Suppose it is plotted as in Fig. 11.10. In the same figure let a series of curves be plotted showing how the right side of (11.50) depends on g, each curve for a fixed value of f. Now consider the curve for $f = f_1$. It cuts the $h'(g)$ curve at two points, which shows that (11.50) is satisfied for two different frequencies. On the other hand, the curve for the greater frequency $f = f_3$ does not cut the $h'(g)$ curve, so that (11.50) cannot be satisfied. Between them is the curve for an intermediate frequency $f = f_2$, which just touches the $h'(g)$ curve, and therefore f_2 must be the M.U.F.

This method can be applied more simply as follows. The $h'(g)$ curve is plotted with $\log g$ as abscissa, and the right side of (11.50) is plotted on a sheet of transparent material with $\log(g/f)$ as abscissa, so that it has a vertical asymptote where $\log(g/f) = 0$. The curves are superimposed, so that the horizontal axes coincide, and the transparent sheet is slid horizontally until the two curves just touch. The point where the line $\log(g/f) = 0$ cuts the axis of the $h'(g)$ diagram is then noted. At this point $g = f$, so that it gives the M.U.F., which is simply read off from the horizontal axis. In this way one curve on the transparent sheet can be used for all values of f, but it is necessary to have a different curve for each value of the horizontal range D.

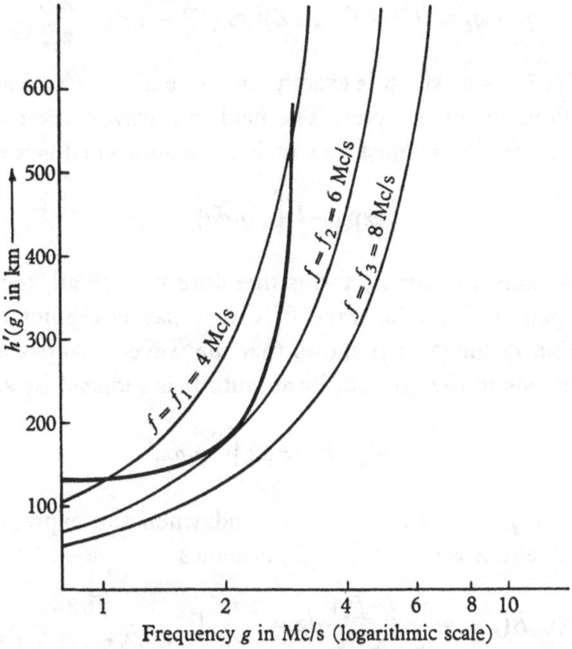

Fig. 11.10. Determination of M.U.F. The heavy line is an example of an observed $h'(f)$ curve, showing penetration at 3 Mc/s. The thin lines are the right side of (11.50) with $D = 1000$ km. In this example the M.U.F. is 6 Mc/s.

In the above description of Newbern Smith's method it has been assumed that the earth is flat. Refinements to allow for the earth's curvature have been described by Newbern Smith (1938). For a full description of the method see National Bureau of Standards Circular 462 (1948), ch. 6.

11.16 The absorption of radio waves. Martyn's theorem for absorption

In this chapter electron collisions have so far been neglected completely. If the electron collision-frequency ν is small enough, most of the foregoing theory can be used without modification, and in addition some new results can be established. When electron collisions are allowed for, the quantity q is given by (9.56) which may be written

$$q^2 = C^2 - \frac{X}{1+Z^2} - \frac{iXZ}{1+Z^2}, \tag{11.51}$$

so that q is a complex quantity and may be written $q_r - iq_i$ where q_r and q_i are real and positive. If $Z\,(=\nu/\omega)$ is so small that its squares and higher powers may be neglected, (11.51) gives

$$q_r - iq_i \approx (C^2 - X - iXZ)^{\frac{1}{2}} \approx (C^2-X)^{\frac{1}{2}} - i\frac{XZ}{2q_r}. \tag{11.52}$$

Thus q_r is $(C^2-X)^{\frac{1}{2}}$ which is exactly the same as the value of q used in earlier sections of this chapter. The field in a wave-packet is given by (11.6) or (11.10). These must now include an additional factor

$$\exp\left\{-k\int_0^z q_i\,dz\right\}. \tag{11.53}$$

Since q_i contains a factor Z, and is therefore very small, the equations giving the path and time of travel of a wave-packet are not appreciably affected. The factor (11.53) shows that the wave-packet is attenuated. When it returns to the ground, its amplitude is reduced by a factor $|R|$ where

$$-\log_e|R| = 2k\int_0^{z_0} q_i\,dz. \tag{11.54}$$

The value of q_i is given by (11.52), and when the expressions (3.6), (3.13), for X and Z are used, (11.54) becomes

$$-\log_e|R(f)| = \frac{2\pi}{c}\int_0^{z_0}\frac{f_N^2\nu}{f^2 q_r}\,dz = \frac{2\pi C}{c}\int_0^{z_0}\frac{f_N^2\,\nu\,dz}{C^2f^2\{1-f_N^2/f^2C^2\}^{\frac{1}{2}}}, \tag{11.55}$$

which gives the effect of absorption for a wave-packet of frequency f obliquely incident on the ionosphere at an angle $\arccos C$. If a wave-packet of frequency $g = fC$ were vertically incident on the ionosphere, the effect of absorption would be given by

$$-\log_e|R_0(g)| = \frac{2\pi}{c}\int_0^{z_0}\frac{f_N^2\,\nu\,dz}{g^2\{1-f_N^2/g^2\}^{\frac{1}{2}}}, \tag{11.56}$$

where R_0 is used to denote a reflection coefficient at vertical incidence. The levels of reflection z_0 are the same in these two cases. Hence

$$\log_e|R(f)| = C\log_e|R_0(fC)|, \tag{11.57}$$

which shows that the logarithm of the reflection coefficient for a frequency f at oblique incidence is C times the logarithm of the reflection coefficient for a frequency fC at vertical incidence. This is Martyn's theorem for absorption (Martyn, 1935; Appleton and Beynon, 1955). It applies only when the effect of the earth's magnetic field is neglected.

11.17 The effect of electron collisions on equivalent path

It was shown in § 10.17 that if the electron collision-frequency ν is small and constant, the contributions P' and P to the equivalent path and phase path respectively are related by (10.86). This must still apply for any section of the curved path traversed by a ray at oblique incidence, since the wave-normal is in the direction of the ray when the earth's magnetic field is neglected. Hence for the whole path

$$-\log|R| = \frac{1}{2}\frac{\nu}{c}(P'-P), \qquad (11.58)$$

where R is the reflection coefficient.

When the earth's magnetic field is allowed for, there is no simple formula corresponding to this. The attenuation then has to be computed numerically. Methods of doing this are given in §§ 13.6 and 14.7.

Examples

1. Suppose that the refractive index μ in the ionosphere is given by

$$\mu^2 = 1-\alpha(z-h) \quad \text{for} \quad z \geqslant h,$$

$$\mu^2 = 1 \quad \text{for} \quad z \leqslant h,$$

where α is a real constant, and the earth's magnetic field is neglected. A point transmitter sends up a wave-packet at an angle θ to the vertical. Find the distance at which it returns to the earth's surface.

The transmitter emits a pulse of radio waves uniformly in all directions. Show that the pulse first returns to the earth at a distance $4h(2/h\alpha-1)^{\frac{1}{2}}$ from the transmitter, provided that $\alpha < 2/h$.

(Math. Tripos, 1957. Part III.)

2. The electron number density N in the ionosphere increases monotonically with height z above the earth's surface, which is assumed to be flat. The earth's magnetic field and electron collisions are to be neglected. A radio 'wave-packet' with 'predominant' frequency f leaves the ground at an angle θ to the vertical and returns at a distance D away, taking a time $P'(f)/c$, where $P'(f)$ is the equivalent path for frequency f, and c is the velocity of electro-

magnetic waves in free space. Show how to find D and $P'(f)$ when the function $N(z)$ is known, and prove Martyn's theorem that

$$P'(f)\cos\theta = 2h'(f\cos\theta),$$

where $h'(f)$ is the equivalent height of reflection for frequency f at vertical incidence (equal to $\frac{1}{2}P'(f)$ at vertical incidence).

The base of the ionosphere is at height h_0, and above this N is proportional to $(z-h_0)^2$. Prove that

$$P'(f)\sin\theta = D = 2h_0\tan\theta + Kf\sin\theta,$$

where K is a constant.

(Math. Tripos, 1958. Part III.)

CHAPTER 12

RAY THEORY FOR VERTICAL INCIDENCE WHEN THE EARTH'S MAGNETIC FIELD IS INCLUDED

12.1 Introduction

The use of pulses of radio waves was described in § 10.1, and chs. 10 and 11 were devoted to a discussion of the behaviour of these pulses or 'wave-packets' when the earth's magnetic field is neglected. In this and the next two chapters the theory is extended to allow for the earth's magnetic field, and the mathematics is more complicated and less complete. The methods of 'ray theory' (see § 9.15) are used, which means that the W.K.B. solutions are assumed to be good approximations at all levels. These solutions, for the case when the earth's magnetic field is included, are derived later, in § 18.12, where it is shown that each W.K.B. solution is of the form (9.73). For the purposes of chs. 12 to 14 only the exponential or 'phase memory' term is important. The other factor $F_0(z)$ in (9.73), varies only slowly with z, and may be omitted for the reasons given in § 10.2.

In much of chs. 12 to 14 the effect of electron collisions is neglected, and only real positive values of the refractive indices are discussed. When this applies it will be emphasised by using the symbols μ_o, μ_x for the refractive indices of the ordinary and extraordinary waves respectively. When electron collisions are allowed for, the refractive indices are complex, and the symbols $\mathfrak{n}_o$, $\mathfrak{n}_x$ will be used. Then $\mathfrak{n}_o = \mu_o - i\chi_o$, $\mathfrak{n}_x = \mu_x - i\chi_x$ where μ_o, χ_o, μ_x and χ_x are all real and positive.

12.2 Magnetoionic 'splitting'

A pulse of radio waves which is vertically incident on the ionosphere travels up to the level where $\mu = 0$, and is reflected there. When the earth's magnetic field is allowed for, there are two waves which can travel vertically upwards, and these have different polarisations and different refractive indices, and are reflected at different levels. When the incident pulse enters the ionosphere, therefore, in general it divides into two pulses which are propagated independently, and their relative amplitudes depend on the state of polarisation of the incident wave. It is possible to choose this so that only one magnetoionic component wave is generated. For this purpose, at moderate or high magnetic latitudes, the wave must be nearly circularly polarised. In practice it is more usual for the incident wave to be linearly polarised, and it then gives the two magnetoionic components with roughly equal amplitudes. After reflection the two pulses return to the ground, often at different times and with different polarisations. (The theory of the 'limiting polarisation' of a downcoming pulse is discussed in §§ 19.11 to 19.13.) When the pulse-sounding technique described in § 10.1 is used, the two reflected pulses may give separate reflections or 'echos' on the cathode-ray oscillograph, and the echo is said to be 'split'.

12.3 The group refractive index—collisions neglected

Most of the arguments of § 10.2 still apply when the earth's magnetic field is allowed for. When a pulsed radio signal starts from the ground, its electric field is given by (10.1), and at a height z the signal is (10.2). The upward velocity of the pulse, U_z, is given by (10.5). When the earth's magnetic field is neglected, U_z is the same as the group velocity U, but this is not in general true when the earth's magnetic field is allowed for. It is customary, however, to write

$$\mu' = c/U_z, \tag{12.1}$$

where μ' is called the 'group refractive index', and is given by (10.6). The wave refractive index μ is given by the Appleton–Hartree formula (6.6) (with $\mathfrak{n} = \mu$, since $\mathfrak{n}$ is real), which may be used with (10.6) to find μ'. At any one frequency there are two values of μ' corresponding to the two characteristic waves, ordinary and extraordinary.

It is not now possible to give a simple formula like (10.8) for μ'. It must be calculated numerically, which can be done by the following method (Shinn and Whale, 1951).

The Appleton–Hartree formula (6.6) may be written

$$\mu^2 = 1 - \frac{X(1-X)}{D}, \tag{12.2}$$

where

$$D = (1-X) - \tfrac{1}{2}Y^2\sin^2\Theta \pm \{\tfrac{1}{4}Y^4\sin^4\Theta + Y^2(1-X)^2\cos^2\Theta\}^{\frac{1}{2}} \tag{12.3}$$

and Θ is the angle between the earth's magnetic field and the vertical. Equation (12.3) gives

$$f^2D = (f^2-f_N^2) - \tfrac{1}{2}f_H^2\sin^2\Theta \pm \{\tfrac{1}{4}f_H^4\sin^4\Theta + f_H^2(f^2-2f_N^2+f_N^4/f^2)\cos^2\Theta\}^{\frac{1}{2}}, \tag{12.4}$$

whence

$$\frac{\partial}{\partial f}(f^2D) = 2f \pm \frac{f_H^2(f-f_N^4/f^3)\cos^2\Theta}{\{\frac{1}{4}f_H^4\sin^4\Theta + f_H^2f^2(1-X)^2\cos^2\Theta\}^{\frac{1}{2}}}$$

$$= 2f \pm f(1-X^2)\,Y_L^2\{Y_L^2(1-X)^2 + \tfrac{1}{4}Y_T^4\}^{-\frac{1}{2}}. \qquad (12.5)$$

A re-arrangement of (12.2) leads to

$$(f^2\mu^2 - f^2)f^2D = f_N^2(f_N^2 - f^2), \qquad (12.6)$$

which is differentiated with respect to f, and (10.6) is used. The result is

$$(2f\mu\mu' - 2f)f^2D + (f^2\mu^2 - f^2)\frac{\partial}{\partial f}(f^2D) = -2f_N^2 f. \qquad (12.7)$$

This is combined with (12.5) to give

$$D(\mu\mu' - 1) = 1 - X - \mu^2 \pm \tfrac{1}{2}Y_L^2(1-X^2)(1-\mu^2)\{Y_L^2(1-X)^2 + \tfrac{1}{4}Y_T^4\}^{-\frac{1}{2}}, \qquad (12.8)$$

from which μ' can be calculated. When the plus sign is used in (12.3) it must be used throughout the above formulae, which then apply to the ordinary wave. Similarly, the minus sign refers to the extraordinary wave. Shinn and Whale used a slightly modified form of (12.8), and their calculations were made on the EDSAC (the digital computer in the University Mathematical Laboratory, Cambridge). They gave curves showing how μ' depends on X for various values of Y, and some of their results, together with some further calculations on EDSAC, have been used to plot the curves of Figs. 12.1 and 12.2. Tables of μ' for the ordinary ray have been published by Shinn (1955) and Becker (1956).

For the ordinary wave μ_o is zero, and μ_o' is infinite when $X = 1$. It is important to know how μ_o' approaches infinity when X approaches unity, and for this purpose it is convenient to put $\epsilon = 1 - X$ where ϵ is small. Then it can be shown that

$$\tfrac{1}{2}D = \epsilon + \epsilon^2\cot^2\Theta + O(\epsilon^4), \qquad (12.9)$$

$$\mu_o^2 = \operatorname{cosec}^2\Theta\{\epsilon - \epsilon^2\cot^2\Theta + O(\epsilon^3)\}, \qquad (12.10)$$

$$\mu_o = \operatorname{cosec}\Theta\,\epsilon^{\frac{1}{2}}\{1 - \tfrac{1}{2}\epsilon\cot^2\Theta + O(\epsilon^2)\}, \qquad (12.11)$$

$$\mu_o' = \operatorname{cosec}\Theta\,\epsilon^{-\frac{1}{2}}\{1 - \tfrac{3}{2}\epsilon\cot^2\Theta + O(\epsilon^2)\}. \qquad (12.12)$$

The series (12.12) is useful for computing μ_o' when ϵ is small, and further terms are given by Shinn and Whale.

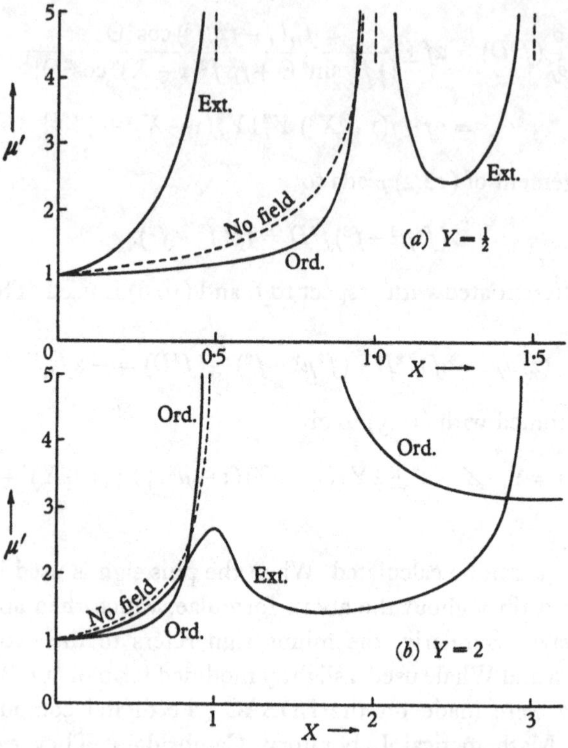

Fig. 12.1. The group refractive index μ' as a function of X (proportional to electron density) when collisions are neglected. Earth's magnetic field inclined to the vertical at an angle $\Theta = 23° 16'$.

For the extraordinary wave μ_x is zero, and μ'_x is infinite when $X = 1 \pm Y$. To find how μ'_x approaches infinity when X approaches $1 - Y$, we set $\epsilon_1 = 1 - Y - X$ where ϵ_1 is small. Then the formulae (12.2), (12.3) and (12.8) lead to

$$\tfrac{1}{2}D = Y - Y^2 + \epsilon_1\left(1 - \frac{2Y}{1+\sec^2\Theta}\right) - \epsilon_1^2\frac{\tan^4\Theta}{(1+\sec^2\Theta)^3} + O(\epsilon_1^3), \quad (12.13)$$

$$\mu_x^2 = \frac{1}{(1-Y)(1+\cos^2\Theta)}\left[2\epsilon_1 - \frac{\epsilon_1^2\tan^2\Theta}{(1-Y)(1+\sec^2\Theta)^2}\right.$$

$$\left.\times\left\{\frac{3+\sec^2\Theta}{Y} + \tan^2\Theta\right\} + O(\epsilon_1^3)\right], \quad (12.14)$$

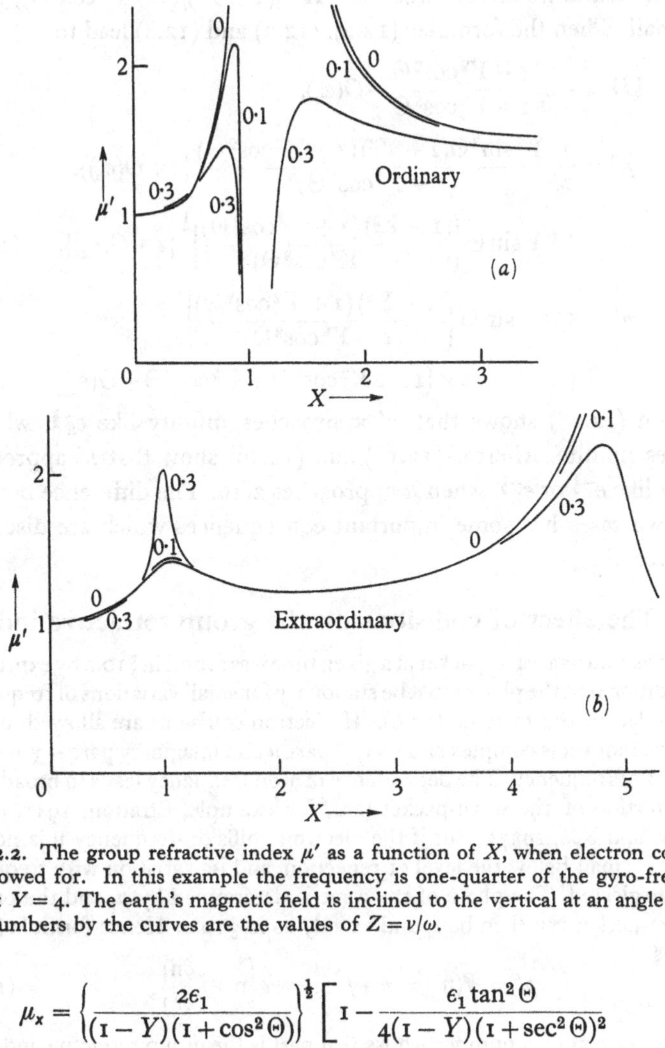

Fig. 12.2. The group refractive index μ' as a function of X, when electron collisions are allowed for. In this example the frequency is one-quarter of the gyro-frequency so that $Y = 4$. The earth's magnetic field is inclined to the vertical at an angle 23° 16′. The numbers by the curves are the values of $Z = \nu/\omega$.

$$\mu_x = \left\{\frac{2\epsilon_1}{(1-Y)(1+\cos^2\Theta)}\right\}^{\frac{1}{2}}\left[1 - \frac{\epsilon_1 \tan^2\Theta}{4(1-Y)(1+\sec^2\Theta)^2}\right.$$
$$\left.\times\left\{\frac{3+\sec^2\Theta}{Y} + \tan^2\Theta\right\} + O(\epsilon_1^2)\right], \quad (12.15)$$

$$\mu_x' = \{2\epsilon_1(1-Y)(1+\cos^2\Theta)\}^{-\frac{1}{2}}\,[2 - Y + O(\epsilon_1)]. \quad (12.16)$$

If Y is replaced by $-Y$ in the last four formulae, they give the values when X approaches $1 + Y$.

One value of the refractive index μ is infinite when X is given by (6.8). This infinity occurs for the extraordinary wave if $Y < 1$, and for the ordinary wave if $Y > 1$. It is important to know how μ' depends on X

in this case, and we therefore set $\epsilon_2 = X-(1-Y^2)/(1-Y^2\cos^2\Theta)$, where ϵ_2 is small. Then the formulae (12.2), (12.3) and (12.8) lead to

$$\tfrac{1}{2}D = -\epsilon_2\frac{1-Y^2\cos^2\Theta}{1+Y^2\cos^2\Theta}+O(\epsilon_2^2),$$

$$\mu^2 = \frac{1}{\epsilon_2}\frac{Y^2\sin^2\Theta(1-Y^2)(1+Y^2\cos^2\Theta)}{(1-Y^2\cos^2\Theta)^3}\{1+O(\epsilon_2)\},$$

$$\mu = \epsilon_2^{-\frac{1}{2}}\,Y\sin\Theta\left\{\frac{(1-Y^2)(1+Y^2\cos^2\Theta)}{(1-Y^2\cos^2\Theta)^3}\right\}^{\frac{1}{2}}\{1+O(\epsilon_2)\}, \qquad (12.17)$$

$$\mu' = \epsilon_2^{-\frac{3}{2}}\,Y\sin\Theta\left\{\frac{(1-Y^2)(1+Y^2\cos^2\Theta)}{(1-Y^2\cos^2\Theta)^7}\right\}^{\frac{1}{2}}$$
$$\times\{1-2Y^2\cos^2\Theta+Y^4\cos^2\Theta+O(\epsilon_2)\}. \qquad (12.18)$$

Equation (12.18) shows that μ' approaches infinity like $\epsilon_2^{-\frac{3}{2}}$, when μ becomes infinite, whereas (12.12) and (12.16) show that μ' approaches infinity like $\epsilon^{-\frac{1}{2}}$ or $\epsilon_1^{-\frac{1}{2}}$, when μ approaches zero. The difference between these two cases has some important consequences which are discussed in § 21.16.

12.4 The effect of collisions on the group refractive index

The position of a wave-packet at a given time was found in § 10.2 by expressing the condition that the phase must be stationary for small variations of frequency, and this led to the formula (10.6). If electron collisions are allowed for, the refractive index $\mathfrak{n}$ is complex and its real part μ and imaginary part $-\chi$ are both functions of frequency. The dependence of χ on frequency leads to broadening and distortion of the wave-packet (see, for example, Stratton, 1941, ch. V; Gibbons and Rao, 1957). But if the electron collision-frequency ν is not too large, χ is small below the level of reflection, and its variation with frequency can be neglected. The phase of the wave is determined by μ, and the path of the wave-packet can then be found exactly as in § 10.2. Hence formula (10.6) becomes

$$\mu' = \mathscr{R}(\mathfrak{n}') = \mu+f\frac{\partial\mu}{\partial f} = \mathscr{R}\left\{\mathfrak{n}+f\frac{\partial\mathfrak{n}}{\partial f}\right\}, \qquad (12.19)$$

where $\mathfrak{n}'$ is a complex number whose real part is the group refractive index μ'. The algebra is similar to that in § 12.3. Equations (12.2), (12.3) and (12.8) are replaced by the following:

$$\left.\begin{aligned}\mathfrak{n}^2 &= 1-\frac{X(1-X-iZ)}{D},\\ D &= (1-iZ)(1-X-iZ)-\tfrac{1}{2}Y_T^2\pm\{Y_L^2(1-X-iZ)^2+\tfrac{1}{4}Y_T^4\}^{\frac{1}{2}},\\ \mathfrak{n}\mathfrak{n}'-1 &= \frac{1}{D}\big(-X+\tfrac{1}{2}iXZ+(1-\mathfrak{n}^2)[1-iZ-\tfrac{1}{2}iXZ\\ &\qquad+\tfrac{1}{2}(1-X-iZ)(1+X)\,Y_L^2\{Y_L^2(1-X-iZ)^2+\tfrac{1}{4}Y_T^4\}^{-\frac{1}{2}}]\big).\end{aligned}\right\} \qquad (12.20)$$

Some authors (for example Gibbons and Rao, 1957) have given approximate methods of calculating μ'. Such methods were extremely valuable when the calculations had to be done with desk machines, but with modern high-speed digital computers it is now possible to compute a formula like (12.20) without approximation in a time which is negligible compared with the time required to print the answer.

Some of the curves in Fig. 12.2 show how the group refractive index is altered when the effect of electron collisions is included.

12.5 The equivalent height of reflection $h'(f)$—collisions neglected

For pulses of radio waves vertically incident upon the ionosphere, each pulse travels with a velocity whose vertical component is $U_z = c/\mu'$. Let τ be the time taken for the pulse to travel from the ground to the reflection level and back to the ground. Then $\frac{1}{2}c\tau$ is called the equivalent height of reflection and is denoted by $h'(f)$. It is the height to which the pulse would have to go if it travelled always with the velocity c. The reflection of a pulse of radio waves was discussed in §§ 10.3 and 10.4, and the theory given still applies when the effect of the earth's magnetic field is included. Hence $h'(f)$ is given by (10.13). The true height of reflection $z_0(f)$ is where the refractive index μ is zero, and the phase height $h(f)$ is given by (10.14). These expressions apply separately for the ordinary and extraordinary rays. When necessary, subscripts O and X will be used to distinguish the two values of μ, of μ', and of $h'(f)$.

The integrand in (10.13) is the group refractive index μ', which is a function of the frequency f and the electron density $N(z)$. Instead of using N it is more convenient to use the plasma-frequency f_N given by (3.6), so that (10.13) may be written

$$h'(f) = \int_0^{z_0} \mu'(f, f_N)\, dz. \tag{12.21}$$

In evaluating (12.21) it is sometimes convenient to use f_N as the independent variable. The upper limit of the integral is then different for the ordinary and extraordinary rays. For the ordinary ray (12.21) gives

$$h'(f) = \int_0^{f} \mu_o'(f, f_N) \frac{dz}{df_N} df_N. \tag{12.22}$$

This transformation can only be made when $z(f_N)$ is a monotonic function of f_N in the range of integration. A similar transformation for the extraordinary ray is used later (see § 12.17).

The expressions (12.12) and (12.16) show that μ' approaches infinity when z approaches the level of reflection, for both the ordinary and extraordinary waves. Nevertheless, the integrals (12.21) or (12.22) converge to a finite limit. For example, consider the ordinary wave, and let

$$f_N/f = \sin\phi. \qquad (12.23)$$

Then (12.22) becomes

$$h'(f) = \int_0^{\frac{1}{2}\pi} \mu_o'(f, f\sin\phi)\frac{dz}{df_N} f\cos\phi\, d\phi. \qquad (12.24)$$

Equation (12.12) shows that near the level of reflection μ_o' is proportional to $\epsilon^{-\frac{1}{2}}$ where $\epsilon = 1 - X = \cos^2\phi$. Hence $\mu_o' \cos\phi$ approaches unity and the integrand in (12.24) remains finite, provided that (dz/df_N) is finite. The transformation (12.23) is useful when $h'(f)$ is computed numerically. It has been used by Shinn and Whale (1951). A similar transformation can be found for the extraordinary wave.

Near an infinity of the refractive index, the integrand μ' behaves in a different way, shown by (12.18). This topic is discussed later, in § 21.16.

12.6 The $h'(f)$ curves when collisions are neglected

In ch. 10 where the earth's magnetic field was neglected, it was possible to give analytic expressions for $h'(f)$ in some special cases. This is no longer possible when the earth's magnetic field is included, because of the complexity of the formula (12.8) for μ', and the integral (12.21) must be computed numerically. Some typical results are shown in Figs. 12.3 to 12.7 where, for the full curves, the electron collision-frequency ν was neglected. Further curves are given by Goubau (1934), Kelso (1954) and Millington (1943).

Fig. 12.3 shows the form of the $h'(f)$ curves when the electron density increases linearly with height, as given by (10.16). For the ordinary wave the curve is continuous, but for the extraordinary wave there is a discontinuity at the gyro-frequency, because when $f < f_H$ the extraordinary wave is reflected where $X = 1 + Y$, whereas when $f > f_H$ it is reflected where $X = 1 - Y$. This also explains why $h_x'(f)$ is greater than $h_o'(f)$ when $f < f_H$ and $h_o'(f)$ is the greater when $f > f_H$. (The subscripts O, X are here used to distinguish the values of $h'(f)$ for the ordinary and extraordinary ray respectively.)

For frequencies just less than the gyro-frequency, $h_x'(f)$ is large and tends to infinity as f approaches f_H, but this infinity in $h_x'(f)$ is quite different from that which occurs just before the wave penetrates a layer. For frequencies just greater than the gyro-frequency, $h_x'(f)$ tends to a finite value as f approaches f_H. It has been shown (Shinn, unpublished report: see example at end of this chapter) that this limiting value is inversely proportional to the gradient of electron density at the base of the ionosphere.

Fig. 12.4 shows the form of the $h'(f)$ curve when the electron density increases exponentially with height, as given by (10.22). Now there is no

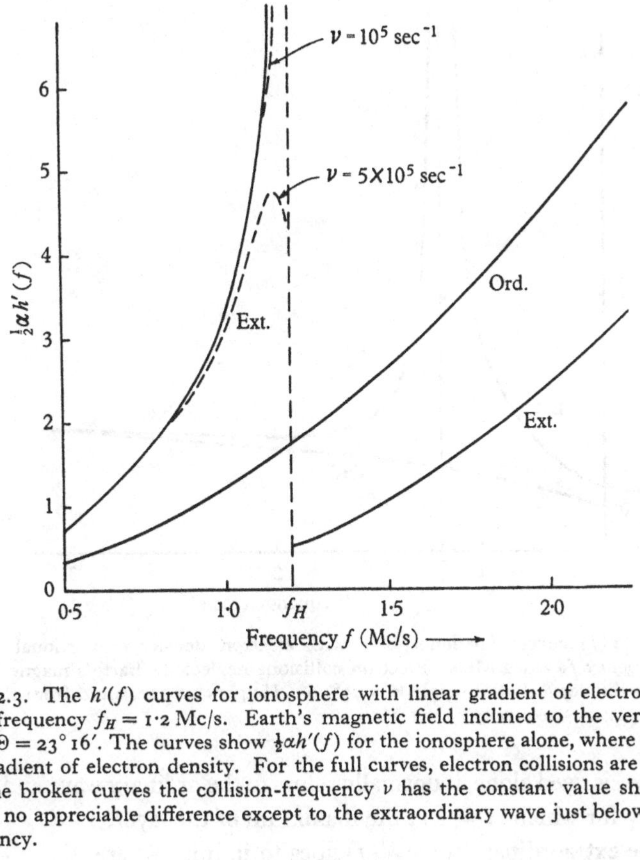

Fig. 12.3. The $h'(f)$ curves for ionosphere with linear gradient of electron density. Gyro-frequency $f_H = 1{\cdot}2$ Mc/s. Earth's magnetic field inclined to the vertical at an angle $\Theta = 23^\circ 16'$. The curves show $\frac{1}{2}\alpha h'(f)$ for the ionosphere alone, where α specifies the gradient of electron density. For the full curves, electron collisions are neglected. For the broken curves the collision-frequency ν has the constant value shown. This makes no appreciable difference except to the extraordinary wave just below the gyro-frequency.

true base of the ionosphere, but the gradient of electron density gets indefinitely smaller as the height decreases. When $f > f_H$, $h'_x(f)$ does not tend to a finite value as f approaches f_H, but passes through a minimum value and then tends to infinity, in agreement with Shinn's result.

12.7 The penetration-frequencies for the ordinary and extraordinary waves

In an ionospheric layer with a maximum of electron density N, the value of N at the maximum determines the penetration-frequency f_p. This is different for the ordinary and extraordinary waves, and its values will be denoted by $f_p^{(o)}$, $f_p^{(x)}$, respectively. When the term 'penetration-frequency' is used alone, it normally refers to the ordinary wave and is the frequency for which $X = 1$ at the maximum of the layer.

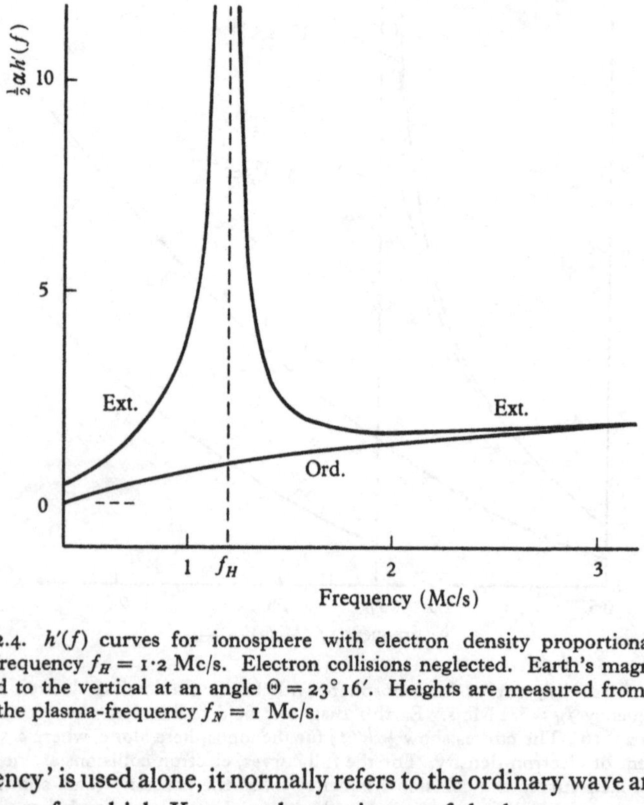

Fig. 12.4. $h'(f)$ curves for ionosphere with electron density proportional to $e^{\alpha z}$. Gyro-frequency $f_H = 1{\cdot}2$ Mc/s. Electron collisions neglected. Earth's magnetic field inclined to the vertical at an angle $\Theta = 23^\circ 16'$. Heights are measured from the level where the plasma-frequency $f_N = 1$ Mc/s.

For the extraordinary wave $h'_x(f)$ goes to infinity where $X = 1 - Y$ at the maximum of the layer. This requires that $(f_p^{(o)}/f_p^{(x)})^2 = 1 - f_H/f_p^{(x)}$, that is

$$f_p^{(o)2} = f_p^{(x)2} - f_H f_p^{(x)}, \tag{12.25}$$

which is a quadratic equation for $f_p^{(x)}$ whose solution is

$$f_p^{(x)} = \tfrac{1}{2}\{(f_H^2 + 4f_p^{(o)2})^{\frac{1}{2}} + f_H\}. \tag{12.26}$$

The other solution is negative and therefore of no practical interest. If

$f_p^{(o)}$ and $f_p^{(x)}$ are both large compared with f_H, so that $f_p^{(x)} - f_p^{(o)}$ is small compared with either $f_p^{(x)}$ or $f_p^{(o)}$, then (12.25) becomes approximately

$$f_p^{(x)} - f_p^{(o)} \approx \tfrac{1}{2} f_H. \tag{12.27}$$

This result was used by Appleton and Builder (1933) to measure f_H in the ionosphere, and to demonstrate that it is electrons which are responsible for the reflection of radio waves.

For the extraordinary wave $h_x'(f)$ also goes to infinity where $X = 1 + Y$ at the maximum of the layer. This requires that

$$f_p^{(o)2} = f_p^{(x)2} + f_H f_p^{(x)}, \tag{12.28}$$

which is a quadratic equation for $f_p^{(x)}$ whose solution is

$$f_p^{(x)} = \tfrac{1}{2}\{(f_H^2 + 4f_p^{(o)2})^{\frac{1}{2}} - f_H\}. \tag{12.29}$$

It might be expected that this infinity would be observed only when $f_p^{(x)}$, as given by (12.29), is less than f_H, for otherwise the extraordinary wave would be reflected where $X = 1 - Y$, and could not reach the level where $X = 1 + Y$. For frequencies greater than f_H it is possible, however, for reflection to occur where $X = 1 + Y$, and the resulting branch of the $h'(f)$ curve is called the 'Z-trace'. This phenomenon is described later, in § 19.7, but will be ignored here. If, then, the solution (12.29) is less than f_H, it is necessary that

$$f_p^{(o)} < f_H \sqrt{2}. \tag{12.30}$$

If (12.30) is satisfied, the $h_x'(f)$ curve will show two infinities at the penetration-frequencies given by (12.26) and (12.29). If, however, $f_p^{(o)} > f_H\sqrt{2}$ the $h_x'(f)$ curve will still show two infinities, one of which is at the penetration-frequency (12.26), but the other is at the gyro-frequency f_H, and is not associated with penetration. In Fig. 12.5 the curves show how the two penetration-frequencies for the extraordinary wave, (12.26) and (12.29), depend on the penetration-frequency for the ordinary wave.

12.8 The equivalent height for a parabolic layer

Fig. 12.6 shows the form of the $h'(f)$ curves for two different parabolic layers. In Fig. 12.6*a* the penetration-frequency $f_p^{(o)}$ is greater than $\sqrt{2} f_H$. The curve for $h_o'(f)$ is similar to that of Fig. 12.3 but goes to infinity at the penetration-frequency $f_p^{(o)}$. The curve for $h_x'(f)$ goes to infinity at the penetration-frequency $f_p^{(x)}$ given by (12.26). It also goes to infinity just below the gyro-frequency, but this is not associated with penetration (for experimental results showing this, see Appleton, Farmer and Ratcliffe, 1938). In Fig. 12.6*b* the penetration-frequency $f_p^{(o)}$ is less than $\sqrt{2} f_H$. The curve for the extraordinary wave now shows two infinities at the two penetration-frequencies, (12.26) and (12.29). The right-hand branch should always occur at frequencies just above the gyro-frequency, but as far as the author is aware it has never been observed for a layer whose penetration-frequency $f_p^{(o)}$ is less than f_H.

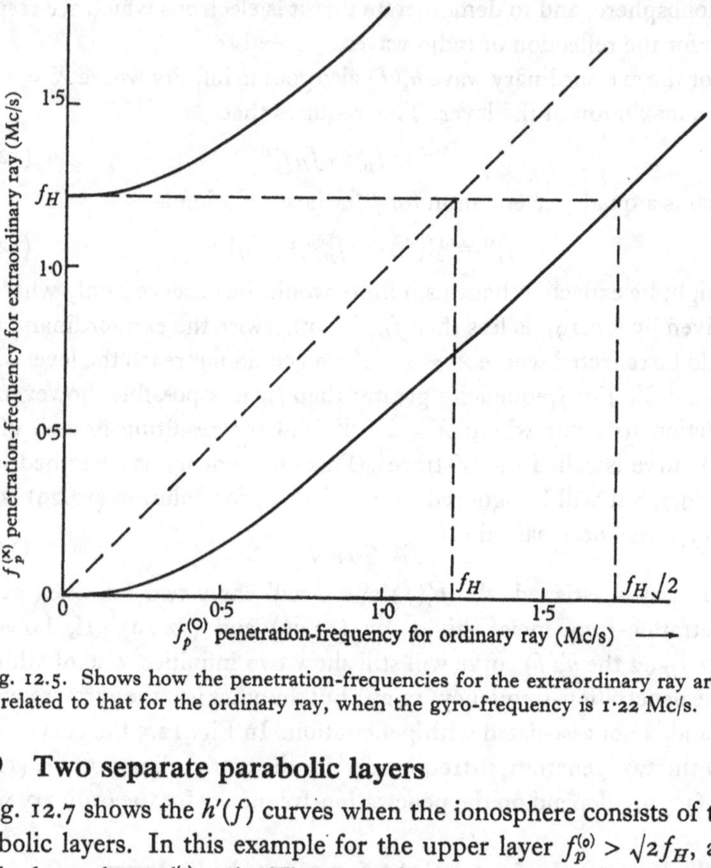

Fig. 12.5. Shows how the penetration-frequencies for the extraordinary ray are related to that for the ordinary ray, when the gyro-frequency is 1·22 Mc/s.

12.9 Two separate parabolic layers

Fig. 12.7 shows the $h'(f)$ curves when the ionosphere consists of two parabolic layers. In this example for the upper layer $f_p^{(o)} > \sqrt{2} f_H$, and for the lower layer $f_p^{(o)} < f_H$. The curve for the ordinary wave is similar to that obtained when the earth's magnetic field was neglected (§ 10.9 and Fig. 10.7). The curve for the extraordinary wave is now in four parts. The left branch (*a*) is for reflection from the lower layer and goes to infinity at the penetration-frequency given by (12.29). The next branch (*b*) is for reflection from the upper layer. It shows high values at the left end because the waves have just penetrated the lower layer but are still retarded as they pass through it. At the right end it goes to infinity as the frequency approaches the gyro-frequency, but this is not associated with penetration of any layer.

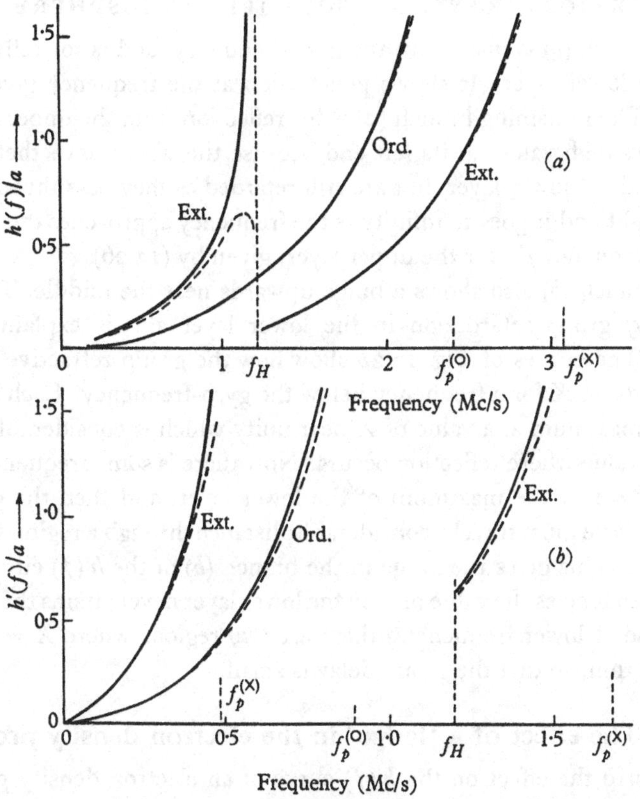

Fig. 12.6. Example of $h'(f)$ curve for parabolic layer. The contribution from the free space below the ionosphere is not included. $f_H = 1{\cdot}2$ Mc/s. In (a) $f_p^{(o)} = 2{\cdot}4$ Mc/s so that $f_p^{(o)} > \sqrt{2}f_H$. In (b) $f_p^{(o)} = 0{\cdot}9$ Mc/s so that $f_p^{(o)} < \sqrt{2}f_H$. For the full curves collisions are neglected and for the broken curves the collision-frequency $\nu = 10^6$ sec^{-1}.

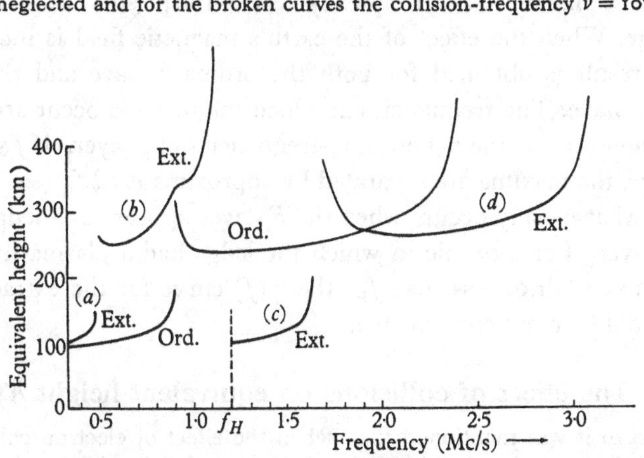

Fig. 12.7. An example of $h'(f)$ curves when the ionosphere consists of two parabolic layers. For the lower layer, $f_p^{(o)} = 0{\cdot}9$ Mc/s, semi-thickness $a = 20$ km, height of maximum = 110 km. For the upper layer $f_p^{(o)} = 2{\cdot}4$ Mc/s, semi-thickness $a = 50$ km, height of maximum = 260 km.

The branch (*c*) is just above the gyro-frequency, and is for reflection from the lower layer. It shows penetration at the frequency given by (12.26). The remaining branch (*d*) is for reflection from the upper layer, and shows high values at its left end because the waves have then just penetrated the lower layer, but are still retarded as they pass through it. At the right end it goes to infinity as the frequency approaches the penetration-frequency $f_p^{(x)}$ for the upper layer, given by (12.26).

The branch (*b*) also shows a bulge upwards near the middle. This is caused by group retardation in the lower layer and is explained as follows. The curves of Fig. 12.2*b* show how the group refractive index μ'_x depends on X for a frequency below the gyro-frequency. Each curve shows a maximum at a value of X near unity which is considerably less than the value where reflection occurs. Now there is some frequency for which $X = 1$ at the maximum of the lower layer, and then the extraordinary wave must travel a considerable distance through a region where μ'_x is large. This gives the bulge in the branch (*b*) of the $h'(f)$ curve. At higher frequencies the value of X in the lower layer never attains the value unity, and at lower frequencies there are two regions where $X = 1$ but these are thin, so that the group delay is small.

12.10 The effect of a 'ledge' in the electron density profile

In § 10.10 the effect on the $h'(f)$ curve of an electron density profile like that of Fig. 10.8 was considered. If there is a region, called a 'ledge' in the $N(z)$ profile, where dN/dz is small, then it was shown that the $h'(f)$ curve has a maximum at a frequency near to the plasma frequency f_N in the ledge. When the effect of the earth's magnetic field is included, a similar result is obtained for both the ordinary wave and the extraordinary wave. The frequencies at which the maxima occur are related in the same way as the penetration-frequencies of a layer. If $f_N \gg f_H$ in the ledge, the maxima are separated by approximately $\frac{1}{2}f_H$ (see (12.27)). This is what usually occurs when the F_1-layer appears as a ledge below the F_2-layer. For a profile in which the ledge had a plasma-frequency comparable with or less than f_H, the $h'(f)$ curve for the extraordinary wave could be more complicated.

12.11 The effect of collisions on equivalent height $h'(f)$

In § 10.16 it was mentioned that, when the effect of electron collisions is included, the equivalent height $h'(f)$ and the phase height $h(f)$ are given by the formulae (10.80) and (10.81) respectively, where the integrals are contour integrals in the complex z-plane, and the upper limit z_0 is the complex value of

z which makes $\mathfrak{n} = 0$. These results were based on the phase integral formula (§§ 16.6, 16.7 and ch. 20) which still applies when the effect of the earth's magnetic field is included (see end of § 19.2). Hence the formulae (10.80) and (10.81) may be used for both the ordinary and extraordinary waves. The integrals, however, are very complicated and must be evaluated by numerical methods.

In the special case when the collision-frequency is constant, the integral (10.80) can be converted to a more convenient form as follows. For the ordinary wave the upper limit z_0 is where $X = 1 - iZ$, and the lower limit is where $X = 0$. Define a new variable, x, thus:

$$X = \frac{f_N^2}{f^2} = (1 - x^2)(1 - iZ), \tag{12.31}$$

so that

$$dz = -(1 - iZ)X^{-\frac{1}{2}} f \frac{dz}{df_N} x\, dx. \tag{12.32}$$

Then (10.80) becomes

$$h_o'(f) = \int_0^1 \mathscr{R}[(1 - iZ)\mu_o' X^{-\frac{1}{2}}] f \frac{dz}{df_N} x\, dx. \tag{12.33}$$

Here the variable x is real over the whole range of integration so that the integral may be evaluated by one of the standard integration methods. (In the examples mentioned below a 12-point Gauss formula was used, and the calculations were done on EDSAC 2, the automatic digital computer in the University Mathematical Laboratory, Cambridge.)

Similarly, for the extraordinary wave the upper limit z_0 is where

$$X = 1 \pm Y - iZ$$

(plus sign for $f < f_H$, minus sign for $f > f_H$) and we take

$$X = (1 - x^2)(1 \pm Y - iZ), \tag{12.34}$$

so that

$$h_x'(f) = \int_0^1 \mathscr{R}[(1 \pm Y - iZ)\mu_x' X^{-\frac{1}{2}}] f \frac{dz}{df_N} x\, dx. \tag{12.35}$$

Notice that in (12.33) and (12.35) X is in general complex.

As an example of the use of these formulae suppose that the electron density increases linearly with height above the base of the ionosphere. Then $f_N^2 = \alpha z$ where α is a constant, and $f(dz/df_N) = (2f^2/\alpha)X^{\frac{1}{2}}$ so that (12.33) and (12.35) give

$$h_o'(f) = h_1 + \frac{2f^2}{\alpha}\int_0^1 \mathscr{R}[(1 - iZ)\mu_o']\, x\, dx \quad \text{with} \quad X = (1 - iZ)(1 - x^2), \tag{12.36}$$

$$h_x'(f) = h_1 + \frac{2f^2}{\alpha}\int_0^1 \mathscr{R}[(1 \pm Y - iZ)\mu_x']\, x\, dx \quad \text{with} \quad X = (1 \pm Y - iZ)(1 - x^2), \tag{12.37}$$

where h_1 is the height of the base of the ionosphere. For some results computed with these formulae see Fig. 12.3.

As a second example suppose that the electron density has a parabolic profile so that $f_N^2 = f_p^2[(2z/a)-(z^2/a^2)]$ for $0 \leqslant z \leqslant 2a$, where f_p is the penetration-frequency for the ordinary wave and a is the 'semi-thickness'. Then

$$f\frac{dz}{df_N} = \frac{af^2X^{\frac{1}{2}}}{f_p^2(1-Xf^2/f_p^2)^{\frac{1}{2}}}, \tag{12.38}$$

and (12.33) and (12.35) give

$$h_o'(f) = h_1 + \frac{af^2}{f_p^2}\int_0^1 \mathscr{R}[(1-iZ)(1-Xf^2/f_p^2)^{-\frac{1}{2}}\mu_o']\,x\,dx$$
$$\text{with}\quad X = (1-iZ)(1-x^2), \tag{12.39}$$

$$h_x'(f) = h_1 + \frac{af^2}{f_p^2}\int_0^1 \mathscr{R}[(1\pm Y-iZ)(1-Xf^2/f_p^2)^{-\frac{1}{2}}\mu_x']\,x\,dx$$
$$\text{with}\quad X = (1\pm Y-iZ)(1-x^2), \tag{12.40}$$

where h_1 is the height of the base of the ionosphere. Some results computed with these formulae are shown in Fig. 12.6.

12.12 The polarisation of waves in a wave-packet

It is interesting to consider how the polarisation of a radio wave changes as it travels to the reflection level and back. In the lowest part of the ionosphere where the electron density is small, the changes of polarisation are complicated and must be studied by 'full wave' methods. This problem is discussed later, in ch. 19. In the present section it is assumed that the medium is slowly varying, so that the equations of the magnetoionic theory for a homogeneous medium may be used. The polarisation is then given by the quadratic equation (5.12), and the longitudinal component of the electric field is given by (5.37) or (5.38). We continue to neglect electron collisions, so that $Z = 0$, and the two values (5.13) of the polarisation ρ become

$$\rho = \frac{iY_T^2}{2Y_L(1-X)} \pm i\left\{\frac{Y_T^4}{4Y_L^2(1-X)^2}+1\right\}^{\frac{1}{2}}. \tag{12.41}$$

These are purely imaginary, so that each polarisation ellipse has its major or minor axis in the magnetic meridian. In England the value of $Y_T^2/2Y_L$ is about $\frac{1}{7}Y$. If $Y < 1$, and X is very small, the axis ratio is between 0·86 and 1. In the lowest part of the ionosphere, therefore, both waves are very nearly circularly polarised. This result is not much affected if electron collisions are allowed for.

When $X = 1$, the two values (12.41) of ρ are 0 and ∞. The value $\rho = 0$ gives $\mu^2 = 0$, and therefore applies to the ordinary wave. At the level where it is reflected, therefore, the ordinary wave is linearly polarised with its electric field in the magnetic meridian. If $1-X=\epsilon$ where ϵ is small, the expression (12.41) gives for the ordinary wave

$$\rho = -i\epsilon Y_L/Y_T^2 + O(\epsilon^2).$$

This may be inserted in (5.37) (with $U=1$), whence

$$\frac{E_z}{E_x} = \frac{Y_L}{Y_T} + O(\epsilon), \tag{12.42}$$

which shows that at the level of reflection the electric field of the ordinary wave is parallel to the earth's magnetic field.

When $X = 1+Y$, the two values (12.41) of ρ are $-iY/Y_L$ and iY_L/Y. The first of these gives $\mu^2 = 0$, and therefore applies to the extraordinary wave. When inserted in (5.37), with $U=1$, and $\mu^2=0$, it gives

$$\frac{E_z}{E_x} = -\frac{Y_T}{Y_L}. \tag{12.43}$$

At the level of reflection of the extraordinary wave, therefore, the electric vector describes an ellipse in a plane at right angles to the earth's magnetic field, with its major axis horizontal. This applies also when $X = 1-Y$, for the conclusions are unaffected if Y is replaced by $-Y$.

12.13 The calculation of electron density $N(z)$ from $h'(f)$

Some examples have been given of $h'(f)$ curves when $N(z)$ is known. Of much greater practical importance is the calculation of $N(z)$ when $h'(f)$ is known. This problem has already been discussed (§§ 10.11 to 10.13) for the case when the earth's magnetic field is neglected, and it was shown that the integral equation (10.44) which related $h'(f)$ to $N(z)$ can then be solved analytically provided that $N(z)$ is a monotonically increasing function. When the earth's magnetic field is allowed for, however, the integral equation is much more complicated, and can only be solved by numerical methods. A possible method is as follows.

The equation is used in the form (12.22), and will be solved by finding the inverse function $z(f_N)$. It is assumed that $h_0'(f)$ is measured for the ordinary wave, and electron collisions are neglected. The method of solution consists in treating the integral in (12.22) as a sum of discrete terms. This can be done in several ways, of which the following are examples.

Suppose that $h_0'(f)$ is given as a tabulated function at equal intervals Δf of the frequency, and let

$$h_0'(n\Delta f) = h_n', \tag{12.44}$$

where n is an integer. The range of integration in (12.22) is divided up into discrete intervals Δf, so that for the mth interval

$$(m-1)\,\Delta f < f_N < m\Delta f,$$

where $m \leqslant n$. It is assumed that in each interval dz/df_N is constant and given by

$$\frac{dz}{df_N} \approx \frac{z_m - z_{m-1}}{\Delta f}, \tag{12.45}$$

where z_m is written for $z(m\Delta f)$. Now let

$$\left.\begin{aligned} M_{nm} &= \left(\frac{1}{\Delta f}\right)\int_{(m-1)\Delta f}^{m\Delta f} \mu_o'(n\Delta f, f_N)\, df_N \quad (m \leqslant n), \\ M_{nm} &= 0 \qquad\qquad\qquad\qquad\qquad\qquad\quad (m > n). \end{aligned}\right\} \tag{12.46}$$

Then equation (12.22) becomes, for all n,

$$h_n' = \sum_{m=1}^{n} M_{nm}(z_m - z_{m-1}), \tag{12.47}$$

which may be written

$$\begin{bmatrix} h_1' \\ h_2' \\ \vdots \\ \vdots \\ h_n' \end{bmatrix} = \begin{bmatrix} M_{11} & 0 & 0 & 0 & . \\ M_{21} & M_{22} & 0 & 0 & . \\ M_{31} & M_{32} & M_{33} & 0 & . \\ \vdots & \vdots & \vdots & \vdots & \vdots \\ M_{n1} & M_{n2} & M_{n3} & . & . \end{bmatrix} \begin{bmatrix} 1 & 0 & 0 & 0 & . \\ -1 & 1 & 0 & 0 & . \\ 0 & -1 & 1 & 0 & . \\ . & . & . & . & . \\ 0 & . & . & . & . \end{bmatrix} \begin{bmatrix} z_1 \\ z_2 \\ z_3 \\ \vdots \\ z_n \end{bmatrix}, \tag{12.48}$$

or more concisely in matrix notation

$$\mathbf{h}' = \mathbf{MDz}. \tag{12.49}$$

The sum of the elements in the rth row of the matrix **MD** is M_{r1} which is given by (12.46) with $m = 1$, $n = r$. Now if r is large compared to unity the plasma-frequency f_N is small compared with $r\Delta f$ so that μ_o' is close to unity in the whole range of integration in (12.46). Thus M_{r1} is unity when r is large enough. In practice, the values of r used usually exceed 5 and the important terms have much greater values of r (see § 12.14). Hence the sum of the elements in any row of the matrix **MD** may be taken as unity.

The matrix **M** is a lower triangular matrix, that is, all terms above the leading diagonal are zero. The matrix **MD** is also lower triangular and is non-singular. It could therefore be inverted to give

$$\mathbf{z} = (\mathbf{MD})^{-1}\mathbf{h}', \tag{12.50}$$

which is the required solution. The elements of $\mathbf{z}$ are the successive values z_m of the height where the plasma-frequency takes the values $m\Delta f$. If the matrix $(\mathbf{MD})^{-1}$ is known, the problem can be solved by evaluating the matrix product (12.50). It is unnecessary, however, to invert the matrix **MD**. The equation (12.48) can be solved by a method proposed by Mr D. T. Caminer (of Leo Computers Ltd., private communication). Let **MD** = **A**, and let the elements of **A** be A_{nm}. Then the equations (12.48) give

$$\left.\begin{aligned} h'_1 &= A_{11}z_1, \\ h'_2 &= A_{21}z_1 + A_{22}z_2, \\ h'_3 &= A_{31}z_1 + A_{32}z_2 + A_{33}z_3, \\ &\cdots\cdots\cdots\cdots\cdots\cdots \\ h'_n &= A_{n1}z_1 + A_{n2}z_2 + \ldots + A_{nn}z_n. \end{aligned}\right\} \qquad (12.51)$$

These may be solved in succession thus

$$\left.\begin{aligned} z_1 &= \frac{1}{A_{11}}h'_1, \\ z_2 &= \frac{1}{A_{22}}h'_2 - \frac{A_{21}}{A_{22}}z_1, \\ z_3 &= \frac{1}{A_{33}}h'_3 - \frac{A_{32}}{A_{33}}z_2 - \frac{A_{31}}{A_{33}}z_1, \\ &\cdots\cdots\cdots\cdots\cdots\cdots \\ z_n &= \frac{1}{A_{nn}}h'_n - \frac{A_{n,n-1}}{A_{nn}}z_{n-1} - \ldots - \frac{A_{n,1}}{A_{nn}}z_1. \end{aligned}\right\} \qquad (12.52)$$

To use this result it is first necessary to compute the coefficients $1/A_{11}$, $1/A_{22}$, $-A_{21}/A_{22}$, $1/A_{33}$, etc. These are related to the elements $M_{n,m}$ as follows:

$$\left.\begin{aligned} &\frac{1}{A_{11}} = \frac{1}{M_{11}}, \\ &\frac{1}{A_{22}} = \frac{1}{M_{22}}, \quad -\frac{A_{21}}{A_{22}} = 1 - \frac{M_{21}}{M_{22}}, \\ &\frac{1}{A_{33}} = \frac{1}{M_{33}}, \quad -\frac{A_{32}}{A_{33}} = 1 - \frac{M_{32}}{M_{33}}, \quad -\frac{A_{31}}{A_{33}} = \frac{M_{32} - M_{31}}{M_{33}}, \\ &\text{etc.} \end{aligned}\right\} \qquad (12.53)$$

These terms are all positive since $M_{n,m} > M_{n,m-1}$. The elements $M_{n,m}$ are evaluated from the integral (12.46), which may be transformed by using (12.23), and then gives

$$M_{n,m} = n\int_{\arcsin\{(m-1)\Delta f\}}^{\arcsin(m\Delta f)} \mu'_0(n\Delta f, n\Delta f\sin\phi)\cos\phi\, d\phi. \qquad (12.54)$$

The group refractive index μ'_0 depends on the strength and dip angle of the earth's magnetic field. The integrals (12.54) and the quantities (12.53) must therefore be evaluated separately for each observing station. They have been evaluated for Slough, England, by Thomas, Haslegrove and Robbins (1958), who have also prepared tables of the quantities (12.53) for other ionospheric observing stations.

For any row after about the fifth, the sum of the coefficients in (12.53) is unity, since M_{r1} is unity. The smaller numbered rows are not used in practice for reasons explained in the next section.

12.14 Example of the use of the method

The quantities (12.53) are given in Table 12.1 for the frequency range 2–6 Mc/s, and for a frequency interval $\Delta f = 0{\cdot}1$ Mc/s. The values of the earth's magnetic field and dip angle are for a height of 300 km in England.

Table 12.1. *Calculation of $z(f_N)$ from $h'(f)$ for the ordinary ray: earth's magnetic field and dip for a height of 300 km in England*

The author is greatly indebted to Dr Thomas, Dr Haslegrove and Miss Robbins for supplying the data from which this table was prepared. (See Thomas, Haslegrove and Robbins, 1958.)

Plasma frequency f_N in Mc/s	$\frac{1}{A_{nn}}$	$-\frac{A_{n,n-1}}{A_{nn}}$, $-\frac{A_{n,n-2}}{A_{nn}}$, etc., to $-\frac{A_{n,20}}{A_{nn}}$	Residuals
2·0	0·08658		0·91342
2·1	0·08407	0·81500	0·10093
2·2	0·08176	0·81509, 0·05303	0·05012
2·3	0·07963	0·81514, 0·05371, 0·01715	0·03438
2·4	0·07765	0·81516, 0·05434, 0·01738, 0·00882	0·02666
2·5	0·07580	0·81514, 0·05493, 0·01759, 0·00893, 0·00566	0·02194
2·6	0·07407	0·81510, 0·05548, 0·01779, 0·00904, 0·00574, 0·00410	0·01867
2·7	0·07245	0·81504, 0·05600, 0·01798, 0·00914, 0·00580, 0·00415, 0·00319	0·01624
2·8	0·07093	0·81497, 0·05648, 0·01816, 0·00923, 0·00586, 0·00420, 0·00323, 0·00259	0·01434
2·9	0·06950	0·81489, 0·05644, 0·01833, 0·00932, 0·00592, 0·00424, 0·00327, 0·00262, 0·00217	0·01280
3·0	0·06815	0·81479, 0·05737, 0·01849, 0·00940, 0·00597, 0·00428, 0·00330, 0·00265, 0·00220, 0·00186	0·01153
3·1	0·06688	0·81469, 0·05777, 0·01864, 0·00948, 0·00602, 0·00432, 0·00333, 0·00268, 0·00223, 0·00189, 0·00162	0·01045
3·2	0·06567	0·81458, 0·05816, 0·01878, 0·00955, 0·00607, 0·00436, 0·00336, 0·00271, 0·00225, 0·00191, 0·00164, 0·00143	0·00953
3·3	0·06452	0·81447, 0·05852, 0·01891, 0·00962, 0·00612, 0·00439, 0·00339, 0·00274, 0·00227, 0·00193, 0·00166, 0·00145, 0·00127	0·00873
3·4	0·06343	0·81435, 0·05887, 0·01904, 0·00969, 0·00616, 0·00442, 0·00341, 0·00276, 0·00230, 0·00195, 0·00168, 0·00147, 0·00129, 0·00114	0·00804
3·5	0·06239	0·81423, 0·05920, 0·01916, 0·00975, 0·00620, 0·00445, 0·00344, 0·00278, 0·00232, 0·00197, 0·00170, 0·00149, 0·00131, 0·00116, 0·00103	0·00743
3·6	0·06140	0·81411, 0·05951, 0·01927, 0·00981, 0·00624, 0·00448, 0·00346, 0·00280, 0·00234, 0·00199, 0·00172, 0·00150, 0·00132, 0·00118, 0·00105, 0·00094	0·00689
3·7	0·06045	0·81399, 0·05980, 0·01938, 0·00987, 0·00627, 0·00451, 0·00349, 0·00282, 0·00235, 0·00200, 0·00173, 0·00152, 0·00134, 0·00119, 0·00106, 0·00096, 0·00086	0·00641
3·8	0·05955	0·81387, 0·06009, 0·01949, 0·00992, 0·00631, 0·00454, 0·00351, 0·00284, 0·00237, 0·00202, 0·00175, 0·00153, 0·00135, 0·00120, 0·00108, 0·00097, 0·00087, 0·00079	0·00597
3·9	0·05868	0·81375, 0·06036, 0·01959, 0·00997, 0·00634, 0·00456, 0·00353, 0·00286, 0·00238, 0·00203, 0·00176, 0·00155, 0·00137, 0·00122, 0·00109, 0·00098, 0·00089, 0·00080, 0·00073	0·00559
4·0	0·05785	0·81363, 0·06061, 0·01968, 0·01002, 0·00637, 0·00458, 0·00354, 0·00287, 0·00240, 0·00205, 0·00177, 0·00156, 0·00138, 0·00123, 0·00110, 0·00099, 0·00090, 0·00081, 0·00074, 0·00067	0·00523
4·1	0·05705	0·81351, 0·06086, 0·01977, 0·01007, 0·00640, 0·00460, 0·00356, 0·00289, 0·00241, 0·00206, 0·00179, 0·00157, 0·00139, 0·00124, 0·00111, 0·00101, 0·00091, 0·00083, 0·00075, 0·00069, 0·00063	0·00491
4·2	0·05629	0·81339, 0·06110, 0·01986, 0·01011, 0·00643, 0·00463, 0·00358, 0·00290, 0·00243, 0·00207, 0·00180, 0·00158, 0·00140, 0·00125, 0·00113, 0·00102, 0·00092, 0·00084, 0·00076, 0·00070, 0·00064, 0·00058	0·00462

Plasma frequency f_N in Mc/s	$\frac{1}{A_{nn}}$	$-\frac{A_{n,n-1}}{A_{nn}}$, $-\frac{A_{n,n-2}}{A_{nn}}$, etc., to $-\frac{A_{n,20}}{A_{nn}}$	Residuals
4·3	0·05555	0·81327, 0·01632, 0·01994, 0·01015, 0·00646, 0·00464, 0·00360, 0·00291, 0·00244, 0·00208, 0·00181, 0·00159, 0·00141, 0·00126, 0·00114, 0·00103, 0·00093, 0·00085, 0·00077, 0·00071, 0·00064, 0·00059, 0·00054	0·00435
4·4	0·05484	0·81315, 0·06154, 0·02002, 0·01020, 0·00648, 0·00466, 0·00361, 0·00293, 0·00245, 0·00120, 0·00182, 0·00160, 0·00142, 0·00127, 0·00114, 0·00104, 0·00094, 0·00086, 0·00078, 0·00072, 0·00066, 0·00060, 0·00055, 0·00051	0·00410
4·5	0·05416	0·81304, 0·06175, 0·02010, 0·01023, 0·00651, 0·00468, 0·00363, 0·00294, 0·00246, 0·00211, 0·00183, 0·00161, 0·00143, 0·00128, 0·00115, 0·00104, 0·00095, 0·00087, 0·00079, 0·00073, 0·00067, 0·00061, 0·00056, 0·00052, 0·00048	0·00389
4·6	0·05351	0·81292, 0·06195, 0·02017, 0·01027, 0·00653, 0·00470, 0·00364, 0·00295, 0·00247, 0·00211, 0·00184, 0·00162, 0·00144, 0·00129, 0·00116, 0·00105, 0·00096, 0·00087, 0·00080, 0·00073, 0·00067, 0·00062, 0·00057, 0·00053, 0·00049, 0·00045	0·00367
4·7	0·05287	0·81281, 0·06214, 0·02024, 0·01031, 0·00655, 0·00472, 0·00365, 0·00296, 0·00248, 0·00213, 0·00185, 0·00163, 0·00145, 0·00130, 0·00117, 0·00106, 0·00097, 0·00088, 0·00081, 0·00074, 0·00068, 0·00063, 0·00058, 0·00054, 0·00049, 0·00046, 0·00042	0·00349
4·8	0·05226	0·81270, 0·06232, 0·02031, 0·01034, 0·00658, 0·00473, 0·00367, 0·00298, 0·00249, 0·00213, 0·00186, 0·00164, 0·00146, 0·00131, 0·00118, 0·00107, 0·00097, 0·00089, 0·00081, 0·00075, 0·00069, 0·00064, 0·00059, 0·00054, 0·00050, 0·00047, 0·00043, 0·00040	0·00331
4·9	0·05166	0·81259, 0·06250, 0·02038, 0·01037, 0·00660, 0·00475, 0·00378, 0·00299, 0·00250, 0·00214, 0·00187, 0·00165, 0·00147, 0·00131, 0·00119, 0·00108, 0·00098, 0·00090, 0·00082, 0·00076, 0·00070, 0·00064, 0·00060, 0·00055, 0·00051, 0·00047, 0·00044, 0·00041, 0·00038	0·00315
5·0	0·05109	0·81249, 0·06267, 0·02044, 0·01041, 0·00662, 0·00476, 0·00369, 0·00300, 0·00251, 0·00215, 0·00188, 0·00165, 0·00147, 0·00132, 0·00119, 0·00108, 0·00099, 0·00090, 0·00083, 0·00076, 0·00070, 0·00065, 0·00060, 0·00056, 0·00052, 0·00048, 0·00045, 0·00041, 0·00038, 0·00036	0·00299
5·1	0·05054	0·81238, 0·06284, 0·02050, 0·01044, 0·00664, 0·00477, 0·00370, 0·00301, 0·00252, 0·00216, 0·00188, 0·00166, 0·00148, 0·00133, 0·00120, 0·00109, 0·00099, 0·00091, 0·00084, 0·00077, 0·00071, 0·00066, 0·00061, 0·00057, 0·00053, 0·00049, 0·00045, 0·00042, 0·00039, 0·00036, 0·00034	0·00285
5·2	0·05000	0·81227, 0·06300, 0·02056, 0·01047, 0·00666, 0·00479, 0·00371, 0·00301, 0·00253, 0·00217, 0·00189, 0·00167, 0·00149, 0·00133, 0·00121, 0·00110, 0·00100, 0·00092, 0·00084, 0·00078, 0·00072, 0·00066, 0·00061, 0·00057, 0·00053, 0·00049, 0·00046, 0·00043, 0·00040, 0·00037, 0·00035, 0·00032	0·00272
5·3	0·04948	0·81217, 0·06315, 0·02062, 0·01050, 0·00667, 0·00480, 0·00372, 0·00302, 0·00254, 0·00217, 0·00190, 0·00167, 0·00149, 0·00134, 0·00121, 0·00110, 0·00101, 0·00092, 0·00085, 0·00078, 0·00072, 0·00067, 0·00062, 0·00058, 0·00054, 0·00050, 0·00047, 0·00043, 0·00040, 0·00037, 0·00035, 0·00033, 0·00030	0·00260
5·4	0·04897	0·81208, 0·06330, 0·02067, 0·01052, 0·00669, 0·00481, 0·00373, 0·00303, 0·00254, 0·00218, 0·00190, 0·00168, 0·00150, 0·00135, 0·00122, 0·00111, 0·00101, 0·00093, 0·00085, 0·00078, 0·00073, 0·00068, 0·00063, 0·00058, 0·00054, 0·00050, 0·00047, 0·00044, 0·00041, 0·00038, 0·00036, 0·00033, 0·00031, 0·00029	0·00248
5·5	0·04848	0·81198, 0·06344, 0·02072, 0·01055, 0·00671, 0·00483, 0·00374, 0·00304, 0·00255, 0·00219, 0·00191, 0·00169, 0·00150, 0·00135, 0·00122, 0·00111, 0·00102, 0·00093, 0·00086, 0·00079, 0·00073, 0·00068, 0·00063, 0·00059, 0·00055, 0·00051, 0·00048, 0·00045, 0·00042, 0·00039, 0·00036, 0·00034, 0·00032, 0·00030, 0·00028	0·00237
5·6	0·04810	0·81188, 0·06358, 0·02078, 0·01058, 0·00672, 0·00484, 0·00375, 0·00305, 0·00256, 0·00220, 0·00192, 0·00169, 0·00151, 0·00136, 0·00123, 0·00112, 0·00102, 0·00094, 0·00086, 0·00080, 0·00074, 0·00069, 0·00064, 0·00059, 0·00055, 0·00052, 0·00048, 0·00045, 0·00042, 0·00040, 0·00037, 0·00035, 0·00032, 0·00030, 0·00028, 0·00026	0·00227
5·7	0·04754	0·81179, 0·06372, 0·02082, 0·01060, 0·00674, 0·00485, 0·00376, 0·00306, 0·00256, 0·00220, 0·00192, 0·00170, 0·00151, 0·00136, 0·00123, 0·00112, 0·00103, 0·00094, 0·00087, 0·00080, 0·00074, 0·00069, 0·00064, 0·00060, 0·00056, 0·00052, 0·00049, 0·00046, 0·00043, 0·00040, 0·00037, 0·00035, 0·00033, 0·00031, 0·00029, 0·00027, 0·00025	0·00217
5·8	0·04709	0·81169, 0·06385, 0·02087, 0·01063, 0·00675, 0·00486, 0·00377, 0·00306, 0·00257, 0·00221, 0·00193, 0·00170, 0·00152, 0·00137, 0·00124, 0·00113, 0·00103, 0·00095, 0·00087, 0·00081, 0·00075, 0·00070, 0·00065, 0·00060, 0·00056, 0·00053, 0·00049, 0·00046, 0·00043, 0·00040, 0·00038, 0·00036, 0·00033, 0·00031, 0·00029, 0·00028, 0·00026, 0·00024	0·00209
5·9	0·04666	0·81160, 0·06397, 0·02092, 0·01065, 0·00677, 0·00487, 0·00378, 0·00307, 0·00258, 0·00221, 0·00193, 0·00171, 0·00152, 0·00137, 0·00124, 0·00113, 0·00104, 0·00095, 0·00088, 0·00081, 0·00075, 0·00070, 0·00065, 0·00061, 0·00057, 0·00053, 0·00050, 0·00047, 0·00044, 0·00041, 0·00039, 0·00036, 0·00034, 0·00032, 0·00030, 0·00028, 0·00026, 0·00025, 0·00023	0·00200
6·0	0·04623	0·81151, 0·06409, 0·02097, 0·01067, 0·00678, 0·00488, 0·00378, 0·00308, 0·00258, 0·00222, 0·00194, 0·00171, 0·00153, 0·00138, 0·00125, 0·00114, 0·00104, 0·00096, 0·00088, 0·00082, 0·00076, 0·00070, 0·00066, 0·00061, 0·00057, 0·00054, 0·00050, 0·00047, 0·00044, 0·00041, 0·00039, 0·00037, 0·00034, 0·00032, 0·00030, 0·00028, 0·00027, 0·00025, 0·00024, 0·00022	0·00192

The meaning of the entries in the last column of the table is explained below.

It is not easy to measure h' at very low frequencies, although apparatus has been constructed which records $h'(f)$ curves down to 50 kc/s (Blair, Brown and Watts, 1953). At most ionospheric observing stations, however, the data extend down only to about 1 or 2 Mc/s. Now for the F-layer in the daytime the penetration frequency $f_p^{(o)}$ is of the order of 4–8 Mc/s in temperate latitudes, and at low frequencies, up to about 2 Mc/s, the $h'(f)$ curve for the ordinary wave is almost horizontal, and $h'_o(f)$ is there approximately equal to the height of the base of the layer. This is because the smaller frequencies require comparatively few electrons to reflect the ordinary wave, and we should expect the value of $h'(f)$ to stay the same for all lower frequencies. Hence $h'_o(f)$ for all f in the range 0–2 Mc/s is roughly constant and equal to $z(f_N)$ where f_N may take any value in the range 0–2 Mc/s. Now consider (12.52). In the expression for z_{20} all the terms after the first on the right side contain the same value of z (z_1 to z_{10}) as a factor. There is therefore no need to tabulate the terms $-A_{20,19}/A_{20,20}$, $-A_{20,18}/A_{20,20}$, ..., $-A_{20,1}/A_{20,20}$ separately. All that is needed is their sum. This applies to the equations for all values of z after z_{20}, and the first 19 equations are not used. Hence in the table the last column, labelled 'residuals', is the sum of the 19 terms $-A_{n,19}/A_{n,n}$, $-A_{n,18}/A_{n,n}$, ..., $-A_{n,1}/A_{n,n}$. To compute this sum it is not necessary to find the terms separately. When the terms $1/A_{nn}$, $-A_{n,n-1}/A_{n,n}$, ..., $-A_{n,20}/A_{n,n}$ have been found, the residual can be computed by using the result (§ 12.13) that the sum of all coefficients in each of the equations (12.52) must be unity. When using the table to compute the products (12.52), the residual is multiplied by the value of $h'_o(f)$ for $f = 1{\cdot}9$ Mc/s, and added to the other terms.

This procedure may introduce errors because $h'_o(f)$ and $z(f_N)$ are not exactly equal when f and f_N are equal to 2 Mc/s. The table shows, however, that the residuals become smaller as f_N gets larger, so that the error is small for large values of f_N. The error arises because of our uncertainty of the value of $N(z)$ in the lowest parts of the ionosphere, but at high frequencies the group refractive index μ' is practically unity when $N(z)$ is small, and hence the error is small. In practice it is found that the error is negligible when the $h'_o(f)$ curve has zero slope, within the error of measurement, at the lowest frequency observed (2 Mc/s in the above example). For a layer with a penetration frequency $f_p^{(o)}$ only slightly greater than 2 Mc/s, this criterion would not hold, and it would be necessary to prepare a table extending down to smaller frequencies, and to obtain $h'(f)$ data for smaller values of f.

The table is used as follows. From the observed $h'(f)$ data the values of h'_0 are tabulated at intervals of 0·1 Mc/s from 1·9 Mc/s up to the penetration-frequency. The values of $h'(f)$ in the notation of the last section are h'_{19}, h'_{20}, h'_{21}, etc., for frequencies of 1·9, 2·0, 2·1, etc., Mc/s. Similarly, z_1, z_2, z_3, ..., z_{20}, etc., are the values of the height z for plasma-frequencies f_N equal to 0·1, 0·2, 0·3, ..., 2·0, etc., Mc/s. As explained above we take

$$z_{19} = z_{18} = \ldots = z_1 = h'_{19}.$$

Then the 20th equation in (12.52) is used with the first row in the table to give

$$z_{20} = 0{\cdot}08658h'_{20} + 0{\cdot}91342h'_{19}$$

and successive equations, with corresponding rows of the table, then give

$$z_{21} = 0{\cdot}08407h'_{21} + 0{\cdot}81500z_{20} + 0{\cdot}10093h'_{19},$$
$$z_{22} = 0{\cdot}08176h'_{22} + 0{\cdot}81509z_{21} + 0{\cdot}05303z_{20} + 0{\cdot}05012h'_{19}$$

and so on. The last of this series of equations gives a value of z which is the height where the plasma-frequency f_N is just below the penetration-frequency for the ordinary ray.

12.15 Other versions of the foregoing method

In the preceding sections it has been assumed that the values of $h'_0(f)$ were tabulated at equal intervals Δf of the frequency. Instead of using the frequency itself, any monotonic function of frequency can be used. For example, King (1957) has used the logarithm of the frequency. This has the advantage that for $h'(f)$ curves which extend up to high frequencies (10 or 15 Mc/s) the number of values of h' which must be recorded is smaller than for equal frequency intervals. The measurement of the records is, however, somewhat more laborious.

It is convenient to imagine that the ionosphere extends right down to the ground so that $N(z)$ is monotonic and extremely small at low levels, but never quite zero. Let f_1 be the value of the plasma-frequency at the ground. Then f_1 is very small and certainly much smaller than the smallest frequency at which $h'(f)$ is measured. The variable f_1 does not enter the final calculations and it is introduced merely to help in understanding the method. Let

$$\log_e(f/f_1) = l, \quad \log_e(f_N/f_1) = l_N. \tag{12.55}$$

Instead of using f_N as the independent variable in (12.22), we now use l_N so that (12.22) becomes

$$h'(f) = \int_0^l \mu'(f, f_N) \frac{dz}{dl_N} \, dl_N. \tag{12.56}$$

Suppose that $h'(f)$ is given as a tabulated function at equal intervals Δl of l and let

$$h'\{f, \exp(n\Delta l)\} = h'_n, \tag{12.57}$$

where n is an integer. The range of integration in (12.56) is divided into discrete intervals, so that for the mth interval $(m-1)\Delta l < l_N < m\Delta l$ where $m \leqslant n$. It is assumed that in each interval dz/dl_N is constant and given by

$$\frac{dz}{dl_N} = \frac{z_m - z_{m-1}}{\Delta l}, \tag{12.58}$$

where z_m is written for $z(f, \exp m\Delta l)$. Now let

$$N_{nm} = \frac{1}{\Delta l}\int_{(m-1)\Delta l}^{m\Delta l} \mu'(f_1 e^{n\Delta l}, f_N)\, dl_N. \tag{12.59}$$

Then equation (12.56) becomes for all n

$$h'_n = \sum_{m=0}^{n} N_{nm}(z_m - z_{m-1}). \tag{12.60}$$

The argument may now be continued in the same way in § 12.13, and a set of equations obtained corresponding exactly to (12.52). The quantities $-A_{nm}/A_{mm}$, etc., would have to be tabulated in a form similar to the table in § 12.14, but the columns and rows would be labelled with values of f and f_N respectively at equal intervals of $\log f_N$ and $\log f$. The last column of residuals would be found exactly as described in § 12.14. The frequency f_1 and the absolute values of the integers n, m, do not enter into the calculations of A_{nm} and the residuals. They were introduced merely to help explain the method.

12.16 Failure of the method when $N(z)$ is not monotonic

In § 10.13, where the earth's magnetic field was neglected, it was shown that if $N(z)$ and therefore $f_N(z)$ is not a monotonically increasing function, the $f_N(z)$ profile cannot be found completely from the function $h'(f)$. A similar limitation occurs when the effect of earth's magnetic field is allowed for. For example, if the ionosphere consists of two layers (see, for example, Fig. 12.7), the $f_N(z)$ profile can be found for values of f_N up to the penetration-frequency f_p of the lower layer. At this frequency, however, there is an infinity in the $h'(f)$ curve, and for greater frequencies the process of inverting the integral equation (12.26) may give an error in the value of $f_N(z)$. It was shown in § 10.13, however, that the error is not serious for frequencies which greatly exceed the penetration-frequency f_p, and similar arguments would lead to the same conclusion when the earth's magnetic field is included.

12.17 The use of $h'_x(f)$ for the extraordinary ray

The preceding sections have described how the $N(z)$ profile can be calculated when the function $h'_0(f)$ for the ordinary wave is known. A similar process can be used with $h'_x(f)$ for the extraordinary wave, but there are some important differences. Suppose that the function $h'_x(f)$ is known only for frequencies f above the gyro-frequency f_H. Then instead of using the plasma-frequency $f_N(z)$ as a measure of the electron density it is more convenient to use the frequency $f_x(z)$ given by

$$f_N^2 = f_x^2 - f_x f_H, \quad \text{or} \quad f_x = \tfrac{1}{2}\{f_H + (f_H^2 + 4f_N^2)^{\frac{1}{2}}\}. \tag{12.61}$$

At any level, f_x is the frequency for which the extraordinary wave would be reflected at that level. The group refractive index μ'_x is a function of f_x and the wave-frequency f, and is denoted by $\mu'_x(f, f_x)$. The integral (12.21) may now be transformed by using f_x as the independent variable, instead of z. It becomes

$$h'(f) = \int_{f_H}^{f} \mu'_x(f, f_x) \frac{dz}{df_x} df_x. \tag{12.62}$$

It is important to notice that the lower limit here is f_H and not zero, because (12.61) shows that f_N is zero when $f_x = f_H$.

The argument of § 12.13 may now be applied with (12.62) used instead of (12.22) and f_x instead of f_N. The matrix elements M_{nm} in (12.46) are zero when $m\Delta f < f_H$, and the corresponding values of $A_{n,m}/A_{nn}$ in (12.52) are also zero. Hence the table of these quantities begins at the gyro-frequency f_H. If the available values of $h'_x(f)$ extend down to f_H, the complete table may be used, and it is not necessary to compute the residuals. The complete function $N(z)$ could be found, if it were monotonic, from observations of $h'_x(f)$ for the extraordinary wave only, above the gyro-frequency. In practice, however, observations of $h'_x(f)$ are less satisfactory than for $h'_0(f)$, partly because the extraordinary wave is more heavily absorbed in the ionosphere, and in most work of this kind the ordinary wave is used and the extraordinary wave is often ignored.

For frequencies below the gyro-frequency it is convenient to use, as a measure of the electron density, the frequency $f'_x(z)$ given by

$$f_N^2 = f_x'^2 + f'_x f_H, \quad \text{or} \quad f'_x = \tfrac{1}{2}\{(f_H^2 + 4f_N^2)^{\frac{1}{2}} - f_H\}. \tag{12.63}$$

At any level f'_x is now the frequency for which the extraordinary wave would be reflected at that level. The group refractive index μ'_x is a function of f'_x and the wave-frequency f and is denoted by $\mu'_x\{f, f'_x\}$, where curly brackets are used to indicate that the functional dependence of μ' on f and f'_x is not the same as on f and f_x in (12.62). The integral (12.21) is transformed as before and becomes

$$h'(f) = \int_0^{f} \mu'_x\{f, f'_x\} \frac{dz}{df'_x} df'_x. \tag{12.64}$$

Here the lower limit is zero. The argument of § 12.13 can now be used exactly as for the ordinary wave. The quantities $-A_{nm}/A_{nn}$ could be tabulated for freqencies up to the gyro-frequency f_H, and could only be used with values of $h'_x(f)$ for which $f < f_H$.

For frequencies less than the gyro-frequency the effect of electron collisions is too large to be ignored, but it would be possible to modify (12.64) to allow for collisions, as indicated in § 12.11. In this way curves of $h'_x(f)$ for $f<f_H$ should give some interesting new information about the lower parts of the ionosphere.

Recently Titheridge (1959) has made a very thorough study of methods of calculating the electron density $N(z)$ from the equivalent height $h'(f)$, and has proposed methods which are both faster and more accurate than those described in §§ 12.13 to 12.17. His methods can be used for both the ordinary and the extraordinary rays, and his work includes a study of the effect of electrons in the lowest parts of the ionosphere and in the 'valley' between two ionospheric layers.

Example

Show that if the earth's magnetic field is vertical and electron-collisions are neglected, then for waves travelling vertically, the refractive index μ and the group refractive index μ' are related thus:

$$\mu\mu' = 1 - \frac{XY}{2(1+Y)^2} \quad \text{or} \quad 1 + \frac{XY}{2(1-Y)^2}.$$

In the second case express the equivalent height of reflection $h'(f)$ as an integral, for a frequency slightly greater than the gyro-frequency f_H, and show that

$$\lim_{\epsilon \to 0} h'\{f_H(1+\epsilon)\} = \frac{2}{3} \bigg/ \left(\frac{dX}{dz}\right)_0,$$

where $(dX/dz)_0$ denotes the value of dX/dz at the base of the ionosphere. (A useful transformation is $X = \epsilon(1-t^2)$. See Millington (1938*a*). The author is indebted to Dr D. H. Shinn of Marconi's Wireless Telegraph Co., Ltd., for bringing this and some similiar problems to his attention.)

CHAPTER 13

RAY THEORY FOR OBLIQUE INCIDENCE WHEN THE EARTH'S MAGNETIC FIELD IS INCLUDED

13.1 Introduction

In this chapter we shall consider the propagation of a wave-packet in a horizontally stratified slowly varying ionosphere when the effect of the earth's magnetic field is included. The propagation of wave-packets was discussed in §§ 11.1 to 11.3, and it is assumed that the reader is familiar with these sections.

A wave-packet consists of many component plane waves whose normals extend over a range of angles. Before discussing wave-packets, therefore, it is necessary to consider the behaviour of a single component plane wave incident obliquely on the ionosphere. This is done in §§ 13.2 to 13.4 which discuss the Booker quartic equation.

In most of this chapter, electron collisions are neglected. The effect of a small collision-frequency is discussed briefly in § 13.16.

13.2 The variable q

Let a plane wave be incident on the ionosphere from below, and let its wave-normal have direction cosines S_1, S_2, C, so that it makes an angle θ_I with the vertical, where $\cos\theta_I = C$, $\sin^2\theta_I = S_1^2+S_2^2$, and each field component contains a factor

$$\exp\{-ik(S_1x+S_2y+Cz)\}. \tag{13.1}$$

Imagine that the ionosphere is replaced by a number of thin discrete strata, in each of which the medium is homogeneous. The wave is partially transmitted and reflected at the successive boundaries, so that in any one stratum there are upgoing and downgoing waves which are the resultants of all the partially reflected and transmitted waves entering the stratum. Because the medium is doubly refracting, there are actually two upgoing waves and two downgoing waves. At each boundary there are boundary conditions, so that the fields must depend on x and y in the same way on the two sides of the boundary. Hence the waves depend on x and y only through the factor $\exp\{-ik(S_1x+S_2y)\}$ at all levels. In the nth stratum, let the wave-normal of one of the resultant waves make an angle θ_n with the vertical and let the associated refractive index be $\mathfrak{n}_n$. Then each field component contains a factor

$$\exp\{-ik(S_1 x + S_2 y + \mathfrak{n}_n \cos\theta_n z)\}, \tag{13.2}$$

and Snell's law gives

$$\mathfrak{n}_n \sin\theta_n = \sin\theta_I = (S_1^2 + S_2^2)^{\frac{1}{2}}. \tag{13.3}$$

Now the value of $\mathfrak{n}_n$ depends on the direction of the wave-normal, that is on θ_n, S_1 and S_2. The unknown angle θ_n could be found from (13.3) if $\mathfrak{n}_n$ were known, but this depends on Y_L and Y_T which in turn depend on θ_n. Hence (13.3) cannot immediately be used to give θ_n. This difficulty can be overcome by using the quantity q defined in § 8.17. The theory given there was for reflection at a single boundary and for the special case $S_2 = 0$. As in (8.88) we set

$$q = \mathfrak{n}_n \cos\theta_n, \tag{13.4}$$

whence (with (13.3))

$$\mathfrak{n}_n^2 = q^2 + S_1^2 + S_2^2. \tag{13.5}$$

The expression (13.2) now becomes

$$\exp\{-ik(S_1 x + S_2 y + qz)\}, \tag{13.6}$$

which is a factor appearing in all field quantities. It was shown in § 8.17 that q is one root of a quartic equation, the Booker quartic. This equation is derived in the next section for the more general case when $S_2 \neq 0$.

The direction cosines of the wave-normal are

$$S_1(S_1^2+S_2^2+q^2)^{-\frac{1}{2}}, \quad S_2(S_1^2+S_2^2+q^2)^{-\frac{1}{2}}, \quad q(S_1^2+S_2^2+q^2)^{-\frac{1}{2}}, \tag{13.7}$$

so that when q is known, both the refractive index and the direction cosines may be found.

13.3 Derivation of the Booker quartic

Consider a plane wave in the nth stratum, for which all field quantities depend on x, y, z only through the factor (13.6). Then we may write symbolically

$$\frac{\partial}{\partial x} \equiv -ikS_1, \quad \frac{\partial}{\partial y} \equiv -ikS_2, \quad \frac{\partial}{\partial z} \equiv -ikq. \tag{13.8}$$

The fields must satisfy the Maxwell equations (2.31) in which the constitutive relations (3.24) are to be used. It is convenient to write these equations in matrix form, using (13.8). Maxwell's equations then give:

$$\left.\begin{aligned}
\begin{pmatrix} 0 & -q & S_2 \\ q & 0 & -S_1 \\ -S_2 & S_1 & 0 \end{pmatrix}\begin{pmatrix} E_x \\ E_y \\ E_z \end{pmatrix} &= \begin{pmatrix} \mathscr{H}_x \\ \mathscr{H}_y \\ \mathscr{H}_z \end{pmatrix}, \\
\begin{pmatrix} 0 & -q & S_2 \\ q & 0 & -S_1 \\ -S_2 & S_1 & 0 \end{pmatrix}\begin{pmatrix} \mathscr{H}_x \\ \mathscr{H}_y \\ \mathscr{H}_z \end{pmatrix} &= -\begin{pmatrix} E_x \\ E_y \\ E_z \end{pmatrix} - \frac{1}{\epsilon_0}\begin{pmatrix} P_x \\ P_y \\ P_z \end{pmatrix},
\end{aligned}\right\} \tag{13.9}$$

which may be combined to give

$$\begin{pmatrix} 1-q^2-S_2^2 & S_1S_2 & S_1q \\ S_1S_2 & 1-q^2-S_1^2 & S_2q \\ S_1q & S_2q & 1-S_1^2-S_2^2 \end{pmatrix}\begin{pmatrix} E_x \\ E_y \\ E_z \end{pmatrix}+\frac{1}{\epsilon_0}\begin{pmatrix} P_x \\ P_y \\ P_z \end{pmatrix}=0. \quad (13.10)$$

Now **P** is given in terms of **E** by (3.24), which uses the susceptibility matrix **M**, whose elements are M_{ij} $(i, j = x, y, z)$. Hence (13.10) becomes:

$$\begin{pmatrix} 1-q^2-S_2^2+M_{xx} & S_1S_2+M_{xy} & S_1q+M_{xz} \\ S_1S_2+M_{yx} & 1-q^2-S_1^2+M_{yy} & S_2q+M_{yz} \\ S_1q+M_{zx} & S_2q+M_{zy} & C^2+M_{zz} \end{pmatrix}\begin{pmatrix} E_x \\ E_y \\ E_z \end{pmatrix}=0, \quad (13.11)$$

which is equivalent to three homogeneous equations in E_x, E_y, E_z. These must be consistent so that the determinant of the 3×3 matrix in (13.11) must be zero. This gives the quartic equation for q. When written out in full using the relations (3.25), (3.26) and (3.27) it becomes:

$$\begin{aligned} &q^4(1+M_{zz})+q^3\{S_1(M_{xz}+M_{zx})+S_2(M_{yz}+M_{zy})\} \\ &+q^2\Big\{S_1S_2(M_{xy}+M_{yx})-2C^2-\frac{2X^2}{U^2-Y^2} \\ &\quad -M_{xx}(1-S_1^2)-M_{yy}(1-S_2^2)-M_{zz}(C^2+1)\Big\} \\ &-qC^2\{S_1(M_{xz}+M_{zx})+S_2(M_{yz}+M_{zy})\} \\ &+C^4+C^2\{M_{xx}(1-S_1^2)+M_{yy}(1-S_2^2)+M_{zz}\}-S_1S_2C^2(M_{xy}+M_{yx}) \\ &+\frac{(2C^2+1)X^2}{U^2-Y^2}-\frac{X^3}{U(U^2-Y^2)}=0. \end{aligned} \quad (13.12)$$

This is multiplied by $U(U^2-Y^2)$ and rearranged using the expressions (3.24) for the elements of **M**, and gives finally:

$$F(q) \equiv \alpha q^4+\beta q^3+\gamma q^2+\delta q+\epsilon=0, \quad (13.13)$$

where

$$\left.\begin{aligned} \alpha &= U(U^2-Y^2)+X(n^2Y^2-U^2), \\ \beta &= 2nXY^2(S_1l+S_2m), \\ \gamma &= -2U(U-X)(C^2U-X)+2Y^2(C^2U-X) \\ &\qquad +XY^2\{1-C^2n^2+(S_1l+S_2m)^2\}, \\ \delta &= -2C^2nXY^2(S_1l+S_2m), \\ \epsilon &= (U-X)(C^2U-X)^2-C^2Y^2(C^2U-X) \\ &\qquad -C^2XY^2(S_1l+S_2m)^2. \end{aligned}\right\} \quad (13.14)$$

This should be compared with (8.98) to which it reduces when $S_2 = 0$.

The quartic could have been derived equally well by an extension of the method of § 8.17. In the derivation just given the algebra is rather more complicated, but the method has the slight advantage that it does not use the Appleton–Hartree formula, being based directly on Maxwell's equations and the constitutive relations.

13.4 The transition to a continuous medium

In the last section the Booker quartic equation (13.13) was derived by making the assumption that the ionosphere consists of thin discrete strata in each of which the medium is homogeneous, but no lower limit was set to the thickness of the strata, which may therefore be taken to be as thin and as numerous as desired. This suggests that the roots of the quartic equation can be used to describe the propagation of four different waves in a continuous ionosphere. For one such wave travelling in one direction, say upwards, the change of phase in a small distance δz would then be $-kq\delta z$, for constant x and y, where q is the appropriate root of the quartic. Similarly, the phase change in a larger distance z would be $-k\int_0^z q\,dz$. This is simply an extension of the 'phase memory' concept described in § 9.3 and suggests that any field component in the wave is proportional to

$$\exp\left\{ik\left(ct - S_1 x - S_2 y - \int_0^z q\,dz\right)\right\}. \tag{13.15}$$

This argument assumes that if the ionosphere consists of numerous discrete strata, the waves are completely transmitted through each boundary and the reflected waves are neglected. This is not quite true, and the expression (13.15) should contain another factor $F_0(z)$ (see § 9.14 and (9.73)). The resulting expression is then the W.K.B. solution when the earth's magnetic field is included. The factor $F_0(z)$ could be found by taking account of the reflections at the boundaries of the discrete strata before proceeding to the limit of infinitesimally thin strata (this has been done by Bremmer, 1949*b*). It is derived by a different method in § 18.12. In a slowly varying ionosphere the function $F_0(z)$ itself varies only slowly with z, and its effect on the wave is much less important than the exponential term (13.15). It does not affect the results of this chapter, and will be omitted here.

In most of the rest of this chapter the effect of electron collisions is neglected so that $U = 1$, and the coefficients α, β, γ, δ, ϵ of the Booker quartic (13.14) become:

$$\left.\begin{aligned}\alpha &= 1-X-Y^2+Xn^2Y^2,\\ \beta &= 2nXY^2(S_1l+S_2m),\\ \gamma &= -2(1-X)(C^2-X)+2Y^2(C^2-X)\\ &\qquad\qquad +XY^2\{1-C^2n^2+(S_1l+S_2m)^2\},\\ \delta &= -2C^2nXY^2(S_1l+S_2m),\\ \epsilon &= (1-X)(C^2-X)^2-C^2Y^2(C^2-X)-C^2XY^2(S_1l+S_2m)^2.\end{aligned}\right\} \quad (13.16)$$

13.5 The path of a wave-packet

The expression (13.15) represents a single wave, which extends indefinitely in the horizontal (x and y) directions and has a single frequency. A radio signal which is limited in lateral extent and in time is known as a wave-packet, and it was shown in § 11.2 that such a signal is formed by adding together an infinite number of waves like (13.15) with various values of S_1, S_2 and k. The field of a wave-packet is then given by (11.6). The only difference is that q is no longer given by a simple expression like (9.56) but is one root of the quartic equation (13.13).

It is assumed that the wave-packet leaves the transmitter in the x–z-plane at an angle arc sin S_0 to the z-axis. This means that the amplitude function $A(k, S_1, S_2)$ in (11.5) has a maximum when $S_1 = S_0$ and $S_2 = 0$, which are called the predominant values of S_1, S_2. It was then shown in § 11.3 that the path of a wave-packet is given by (11.8) and (11.9). When the earth's magnetic field is neglected, it was shown that (11.9) leads to $y = 0$, but this is no longer true in general. A wave-packet can be deviated out of the x–z-plane, and this is called 'lateral deviation'.

Equations (11.8) and (11.9) give, for an element of the path of a wave-packet,

$$\frac{dx}{dz} = -\left(\frac{\partial q}{\partial S_1}\right)_{S_1=S_0}, \quad \frac{dy}{dz} = -\left(\frac{\partial q}{\partial S_2}\right)_{S_2=0}. \quad (13.17)$$

A wave-packet contains many component plane waves like (13.15), and the quartic equation (13.13) must hold for each. Thus it must be true at any level for all values of S_1 and S_2, which may be expressed by writing

$$\frac{dF(q)}{dS_1} = \frac{dF(q)}{dS_2} = 0. \quad (13.18)$$

Now

$$\frac{dF(q)}{dS_1} = \frac{\partial F(q)}{\partial q}\frac{\partial q}{\partial S_1} + \frac{\partial \alpha}{\partial S_1}q^4 + \frac{\partial \beta}{\partial S_1}q^3 + \frac{\partial \gamma}{\partial S_1}q^2 + \frac{\partial \delta}{\partial S_1}q + \frac{\partial \epsilon}{\partial S_1} = 0, \quad (13.19)$$

and there is a similar equation with S_1 replaced by S_2. Equation (13.16) shows that $\partial\alpha/\partial S_1 = \partial\alpha/\partial S_2 = 0$. Hence, for the path of a wave-packet, from (13.17),

$$\left.\begin{aligned}\frac{dx}{dz} &= \left(\frac{\partial\beta}{\partial S_1}q^3+\frac{\partial\gamma}{\partial S_1}q^2+\frac{\partial\delta}{\partial S_1}q+\frac{\partial\epsilon}{\partial S_1}\right)\Big/\frac{\partial F(q)}{\partial q},\\ \frac{dy}{dz} &= \left(\frac{\partial\beta}{\partial S_2}q^3+\frac{\partial\gamma}{\partial S_2}q^2+\frac{\partial\delta}{\partial S_2}q+\frac{\partial\epsilon}{\partial S_2}\right)\Big/\frac{\partial F(q)}{\partial q},\end{aligned}\right\}\quad \begin{aligned}&\text{when } S_1 = S_0\\ &\text{and } S_2 = 0.\end{aligned} \tag{13.20}$$

Curves have been given by Booker (1949) showing how these quantities depend on X in some special cases.

13.6 The reversibility of the path

A wave-packet leaves the transmitter, is deviated by the ionosphere and returns to the ground. Suppose now that every component plane wave in the downcoming wave-packet is reversed in direction. This means that both S_1 and S_2 are reversed in sign, and (13.16) shows that the only effect on the quartic is to reverse the signs of the coefficients of q and q^3. Thus q also changes sign, so that $\partial q/\partial S_1$ and $\partial q/\partial S_2$ remain unchanged and (13.17) shows that the direction of the path is the same as before. The wave-packet therefore simply retraces its original path back to the transmitter. The state of polarisation of the wave in a wave-packet can be found from (13.11), which gives the ratios of the three components $E_x:E_y:E_z$. These depend on the elements of the 3×3 matrix in (13.11), which remain unaltered when q, S_1 and S_2 all change sign. Hence the polarisation, referred to *fixed* axes x, y, z is the same for the original and the reversed wave-packet. This has an important bearing on the study of reciprocity (§ 23.4).

13.7 The reflection of a wave-packet

If the ray path is horizontal, both dx/dz and dy/dz must be infinite. Now (13.14) shows that the numerators of (13.20) cannot be infinite unless q is infinite. Hence the denominators in (13.20) must be zero and the condition that dx/dz and dy/dz are both infinite is

$$\frac{\partial F(q)}{\partial q} = 0, \tag{13.21}$$

provided that neither numerator in (13.20) is zero at the same level. Equation (13.21) is therefore the condition that must be satisfied at a level of reflection. It is also the condition that two roots of the quartic shall be equal.

It is possible for the quartic to have two equal roots which differ from zero. Where this happens the path of the ray is horizontal, but the direction of the wave-normal is not necessarily horizontal. The wave-packet does not in general travel in the direction of the wave-normal. Similarly, at a level where one value of q is zero, the associated wave-packet is not necessarily travelling horizontally, although the wave-normal is horizontal. Thus the level where $q = 0$ is not a level of reflection, except in the special case where there are two roots which are both zero at the same level.

The first expression (13.20) is $-dx/dz$ for the ray path. It can change sign at a level where the numerator is zero and the path of the wave-packet is then parallel to the y–z-plane, and the x-component of its motion is reversed. Some examples of this can be seen in Fig. 13.30. The sign of the vertical component of its motion is unaltered, however, and can only be reversed where $\partial F(q)/\partial q$ is zero.

The coefficients (13.16) of the quartic are all real when electron collisions are neglected, so that the roots of the quartic are either (*a*) all real, or (*b*) two are real and two form a conjugate complex pair, or (*c*) the four roots are in two conjugate complex pairs. In case (*a*) two roots correspond to upgoing waves and two to downgoing waves. For if X is zero, the four roots are given by $q = \pm C$ and are equal in pairs. The waves are in free space and two are upgoing and two downgoing. Now let X be increased continuously until a zero of $\partial F/\partial q$ is first reached. In this range $\partial F/\partial q$ cannot change sign, so that the solutions continue to represent two upgoing and two downgoing waves. When $\partial F/\partial q$ is zero, two of the roots are equal, and the corresponding ray paths are horizontal. For a slightly smaller X the two values of $\partial F/\partial q$ for these roots must in general have opposite signs so that one path is upgoing and the other downgoing. Hence the other two roots must correspond one to an upgoing and one to a downgoing ray. If X is further increased the first two roots are a conjugate complex pair, and we pass to case (*b*). The remaining two roots represent one an upgoing and the other a downgoing ray. This continues until another zero of $\partial F/\partial q$ is reached, and we pass either to case (*c*), for which the four roots are complex, or back to case (*a*), when the roots must again represent two upgoing and two downgoing rays.

13.8 A simple example of ray paths at oblique incidence

As an example of ray paths at oblique incidence, suppose that the frequency is above the gyro-frequency, and consider a ray which leaves the earth obliquely in the magnetic meridian. It is shown later (§ 13.15) that there is no lateral deviation in this case, so the ray remains in the magnetic meridian throughout its path. The x–z-plane is the plane of incidence and also the magnetic meridian plane. We may take $S_2 = 0$, and consider how a wave-packet travels in the x–z-plane. Fig. 13.1*a* shows how the four roots of the quartic equation vary with X in this case. When the ray enters the ionosphere it splits into ordinary and extraordinary rays, exactly as for vertical incidence. The direction of the ordinary ray is given at each level by (13.17) (first equation), and the ray can thus be plotted as shown in Fig. 13.1 *b*. The wave-normal is not in general parallel to the ray, and its direction is indicated by arrows in the figure. When the level where $q = 0$ is reached (A in Figs. 13.1 *a* and *b*),

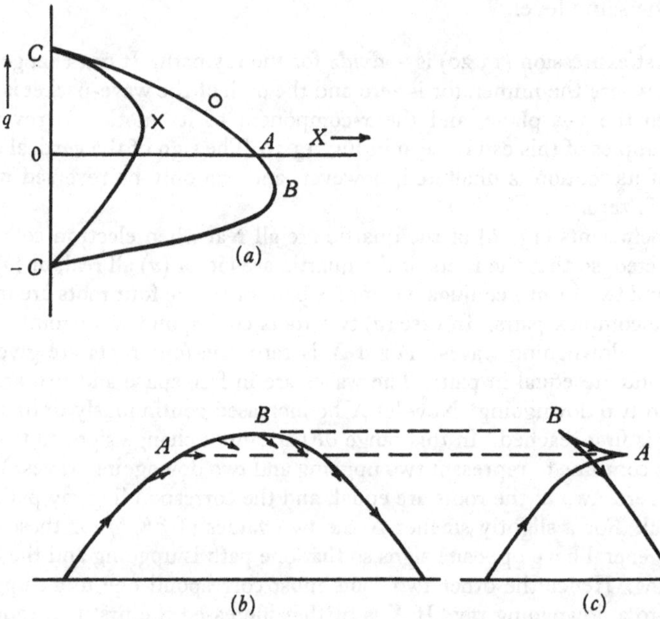

Fig. 13.1. Finding ray paths at oblique incidence. In (*a*) the curves show how the four roots q of the Booker quartic vary with X. (*b*) shows the path of the ordinary ray and the arrows show the direction of the wave-normal at each level. (*c*) is a curve which shows the direction of the wave-normal at each level.

the wave-normal is horizontal but the ray is still directed obliquely upwards. At B two roots of the quartic become equal, and the ray is horizontal, but q is not zero and the wave-normal is directed obliquely downwards, because q is negative. Thereafter the ray travels downwards and leaves the ionosphere. The behaviour of the extraordinary ray is similar, but the level of reflection is lower than for the ordinary ray, and the wave-normal becomes horizontal after reflection has occurred.

It is sometimes instructive to plot, for a given ray, a curve which shows the wave-normal direction at each level. This is done, for the above example, in Fig. 13.1*c*. At the point A this curve becomes horizontal. At higher levels it must point downwards, and this can only be done by continuing the curve backwards. It therefore has a cusp with a horizontal tangent at the level of A. At the reflection level (B in the figure), the wave-normal is directed downwards, and there is another cusp with an oblique tangent. Curves of this kind have been given by Booker (1939). It is important to remember that they do not represent the path of a wave-packet, but simply indicate the *direction* of the wave-normal.

13.9 Further properties of the Booker quartic

Some properties of the Booker quartic (13.13) were given in §8.18. It was shown, in particular, that there are three important special cases where the quartic reduces to a quadratic equation for q^2. These are: (*a*) vertical incidence, when q is the same as $\mathfrak{n}$, and the quartic gives simply the Appleton–Hartree formula; (*b*) propagation from magnetic east to west, or west to east; (*c*) propagation at the magnetic equator.

The Booker quartic performs the same role for oblique incidence as does the Appleton–Hartree formula for vertical incidence, and for a full understanding it would be necessary to make a complete study similar to that of ch. 6. The properties of the quartic have not been investigated so fully, however, and this account of them is necessarily much briefer. For further details see Booker (1939), Millington (1951, 1954) and Chatterjee (1952, 1953).

The effect of collisions is neglected and the coefficients α, β, γ, δ, ϵ are therefore given by (13.16). These are all real, so that the roots of the quartic are either (*a*) all real, or (*b*) two are real and two form a conjugate complex pair, or (*c*) the four roots are in two conjugate complex pairs.

One root of the quartic is infinite when $\alpha = 0$, which occurs when X is given by (8.102), and when $U = 1$ this becomes

$$X = \frac{1-Y^2}{1-n^2Y^2}, \tag{13.22}$$

which is independent of S_1, S_2 and C.

It is only necessary to include both S_1 and S_2 when considering lateral deviation, and if we set $S_2 = 0$, the resulting quartic equation can be used to find the projection of the ray path on the plane of incidence, that is, on the x–z-plane. With $S_2 = 0$ and $S_1 = S$, the quartic becomes:

$$F(q) \equiv \alpha q^4 + \beta q^3 + \gamma q^2 + \delta q + \epsilon = 0, \tag{13.23}$$

where

$$\left.\begin{aligned}\alpha &= 1-X-Y^2+Xn^2Y^2,\\ \beta &= 2SlnXY^2,\\ \gamma &= -2(1-X)(C^2-X)+2Y^2(C^2-X)+XY^2(1-C^2n^2+S^2l^2),\\ \delta &= -2SC^2lnXY^2,\\ \epsilon &= (1-X)(C^2-X)^2-C^2Y^2(C^2-X)-C^2XY^2l^2S^2.\end{aligned}\right\} \quad (13.24)$$

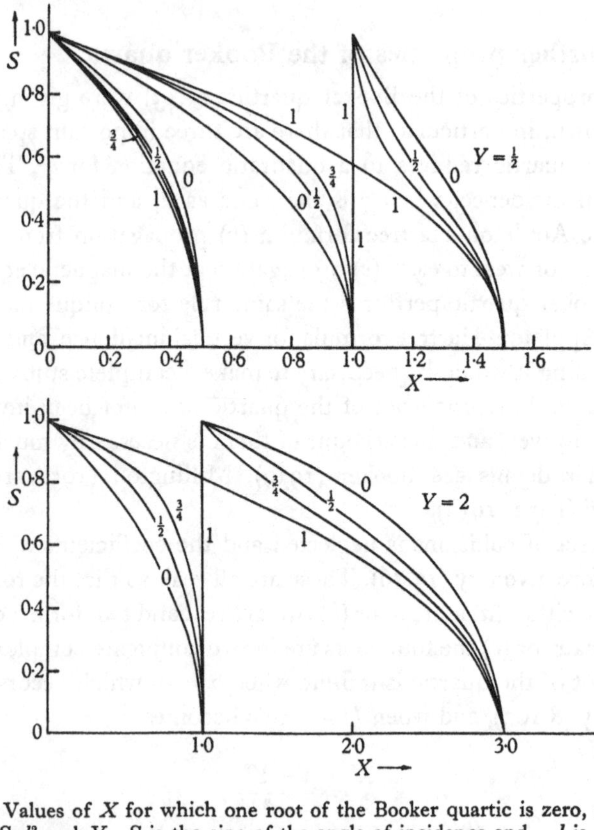

Fig. 13.2. Values of X for which one root of the Booker quartic is zero, for various values of S, l^2 and Y. S is the sine of the angle of incidence and $-l$ is x-direction cosine of the earth's magnetic field. The numbers by the curves are the values of l^2.

The zeros of q are given by $\epsilon = 0$ which is a cubic equation for X, whose three solutions give three different zeros of q. These solutions are plotted as functions of S in Fig. 13.2 for various values of Y and l. When $S = 0$ they become $X = 1$, $X = 1 \pm Y$ which are the three zeros of $\mathfrak{n}$ as given by the Appleton–Hartree formula. When $S = 1$ the three solutions are given by $X^2 = 0$, $X = 1$. It should be remembered that the levels where q is zero are not necessarily levels of reflection.

The quartic has a double root at levels where simultaneously $F(q) = 0$ and $\partial F(q)/\partial q = 0$. If q is eliminated from these equations the result is

$$\begin{vmatrix} 4\alpha\epsilon - 2\beta\delta - \gamma^2 & 6\beta\epsilon - \gamma\delta & 6\alpha\delta - \beta\gamma \\ 6\beta\epsilon - \gamma\delta & 8\gamma\epsilon - 3\delta^2 & 16\alpha\epsilon - \beta\delta \\ 6\alpha\delta - \beta\gamma & 16\alpha\epsilon - \beta\delta & 8\alpha\gamma - 3\beta^2 \end{vmatrix} = 0. \qquad (13.25)$$

This equation has been studied recently by Pitteway (1959).

When $X = 1$, the coefficients (13.24) all contain Y^2 as a factor. This may be cancelled from the quartic which then becomes:

$$(1-n^2)\,q^4 - 2Slnq^3 + (S^2 - C^2 + C^2n^2 - S^2l^2)\,q^2$$
$$+ 2SC^2lnq - C^2S^2(1-l^2) = 0. \qquad (13.26)$$

Hence at the level $X = 1$, the solutions of the quartic depend on the direction of the earth's magnetic field, but not on its magnitude. The expression (13.26) can be resolved into factors, and it is easy to show that the four solutions are

$$q = \pm C, \quad q = S\frac{ln \pm im}{1-n^2}. \qquad (13.27)$$

The last two are real only when $m = 0$, which is for propagation in the magnetic meridian, and then the quartic has a double root equal to Sn/l (since $1 - n^2 = l^2$). Three roots of the quartic can become equal at the level $X = 1$ if $\pm C = Sn/l$.

It is useful to display the solutions of the quartic as curves showing how q depends on X. Since the quartic is valid for all values of X, we must have $dF/dX = 0$ which gives

$$\frac{\partial F}{\partial q}\frac{\partial q}{\partial X} + \frac{\partial \alpha}{\partial X}q^4 + \frac{\partial \beta}{\partial X}q^3 + \frac{\partial \gamma}{\partial X}q^2 + \frac{\partial \delta}{\partial X}q + \frac{\partial \epsilon}{\partial X} = 0, \qquad (13.28)$$

whence
$$\frac{\partial q}{\partial X} = -\left(\frac{\partial \alpha}{\partial X}q^4 + \frac{\partial \beta}{\partial X}q^3 + \frac{\partial \gamma}{\partial X}q^2 + \frac{\partial \delta}{\partial X}q + \frac{\partial \epsilon}{\partial X}\right)\Bigg/\frac{\partial F}{\partial q}. \qquad (13.29)$$

Now the denominator of this is zero where the quartic has a double root, and then $\partial q/\partial X$ is infinite, unless the numerator is also zero. In general the numerator is not zero so that the curves have a vertical tangent where there is a double root. An exception is when $X = 1$, $m = 0$. The double root is then $q = Sn/l$ and the numerator of (13.29) is equal to

$$\{Y^2(S^2 - l^2)^2 - S^4\}/l^4, \qquad (13.30)$$

which is zero when
$$S = \pm l\left(\frac{Y}{Y\pm 1}\right)^{\frac{1}{2}}. \tag{13.31}$$

This important special case is related to the phenomenon of the 'Spitze' and is discussed in § 13.22.

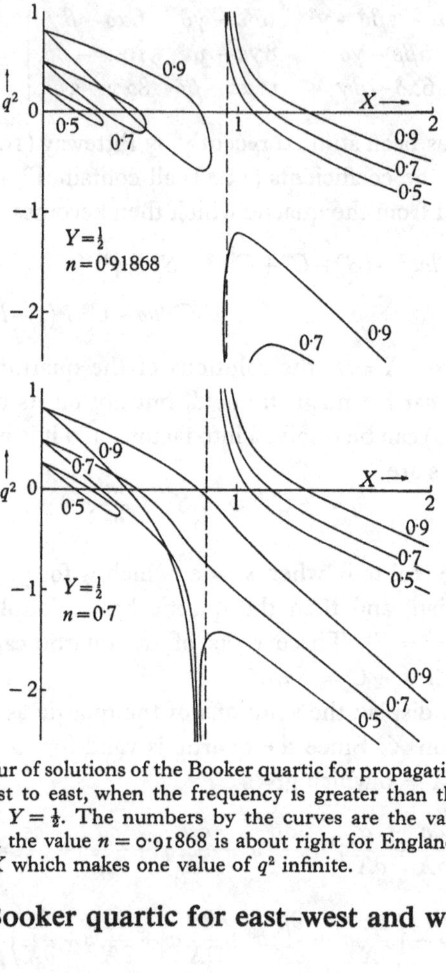

Fig. 13.3. Behaviour of solutions of the Booker quartic for propagation from (magnetic) east to west or west to east, when the frequency is greater than the gyro-frequency. In these examples $Y = \frac{1}{2}$. The numbers by the curves are the values of $C = \cos\theta_I$. In the top diagram the value $n = 0{\cdot}91868$ is about right for England. The broken line is at the value of X which makes one value of q^2 infinite.

13.10 The Booker quartic for east–west and west–east propagation

For propagation from (magnetic) east to west or west to east, the direction cosine $l = 0$, so that $\beta = \delta = 0$, and the quartic is a quadratic equation for q^2. Its solutions are given by (8.99), and when $U = 1$, this becomes

$$q^2 = C^2 - \frac{X}{1 - \dfrac{Y^2(1-C^2n^2)}{2(1-X)} \pm \left\{\dfrac{Y^4(1-C^2n^2)^2}{4(1-X)^2} + \dfrac{Y^2n^2(C^2-X)}{1-X}\right\}^{\frac{1}{2}}}. \tag{13.32}$$

This has zeros where

$$X = C^2, \quad X = \tfrac{1}{2}\{1 + C^2 \pm (S^4 + 4C^2Y^2)^{\frac{1}{2}}\}, \tag{13.33}$$

and each of these is a double root and a level of reflection.

It is now useful to draw curves showing how the two values of q^2 vary with X. Some typical curves are given in Figs. 13.3 and 13.4, where only

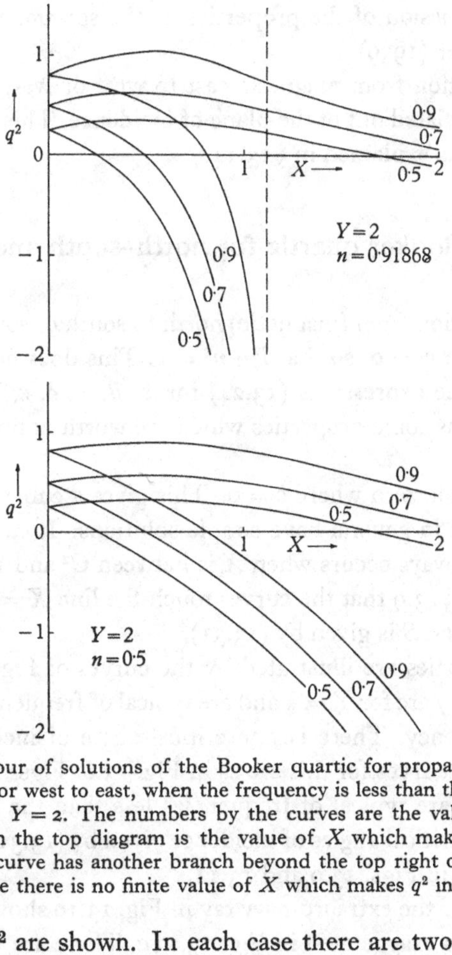

Fig. 13.4. Behaviour of solutions of the Booker quartic for propagation from (magnetic) east to west or west to east, when the frequency is less than the gyro-frequency. In these examples $Y = 2$. The numbers by the curves are the values of $C = \cos \theta_I$. The broken line in the top diagram is the value of X which makes one value of q^2 infinite, and each curve has another branch beyond the top right of the diagram. In the second example there is no finite value of X which makes q^2 infinite.

real values of q^2 are shown. In each case there are two curves, for the ordinary and extraordinary rays, starting from the value $q^2 = C^2$ where $X = 0$. The diagrams also show branches of the curves farther to the right, but these are associated with rays which, according to the 'ray theory', could not be formed from rays incident on the ionosphere from below at real angles of incidence, for they are not connected to the branches which extend to $X = 0$.

For each real and positive value of q^2 there are two values of q which are equal but with opposite signs. If the real values of q were plotted against X the curves would be symmetrical about the line $q = 0$, and each curve would have a vertical tangent where $q = 0$. Thus the levels where this happens are levels of reflection, and in this special case the wave-normal and the ray are both horizontal where reflection occurs.

A fuller discussion of the properties of the solution (13.32) has been given by Booker (1939).

For propagation from magnetic east to west or west to east, the ray is in general deviated out of the plane of incidence. This is called 'lateral deviation' and is explained in § 13.14.

13.11 The Booker quartic for north–south and south–north propagation

For propagation from (magnetic) north to south or south to north, the direction cosine $m = 0$, so that $l^2 + n^2 = 1$. This does not give any simplification of the expressions (13.24) for α, β, γ, δ, ϵ, but the quartic, nevertheless, has some properties which are worth noting in this special case.

One solution is zero where $\epsilon = 0$. This gives a cubic equation for X, which does not in general have simple solutions. It can be shown that one zero of q always occurs when X is between C^2 and 1. It has already been shown in § 13.9 that the curves touch the line $X = 1$, except in the critical case when S is given by (13.31).

These properties are illustrated by the curves of Figs. 13.5 to 13.12. Figs. 13.5 to 13.7 are for $Y = \frac{1}{2}$ and are typical of frequencies greater than the gyro-frequency. There is one critical angle of incidence given by (13.31) and the curves for this are as in Fig. 13.6. Figs. 13.8 to 13.12 are for $Y = 2$ and are typical of frequencies less than the gyro-frequency. There are two critical angles of incidence given by (13.31) and the curves for these are as in Figs. 13.9 and 13.11.

The curve for the extraordinary ray in Fig. 13.10 shows an interesting feature which is exaggerated in Fig. 13.13*a*. There are vertical tangents at the three points A, B, C, so that the ray path is horizontal at the three corresponding levels. Fig. 13.13*b* shows the ray path for this case.

It is possible for the Booker quartic to have three equal roots. An example is when $Y = 2$, the earth's field is at 30° to the vertical and the angle of incidence $\theta_I = 30°$. The curves are not shown but are very similar to those of Fig. 13.10. The curve for the extraordinary ray then

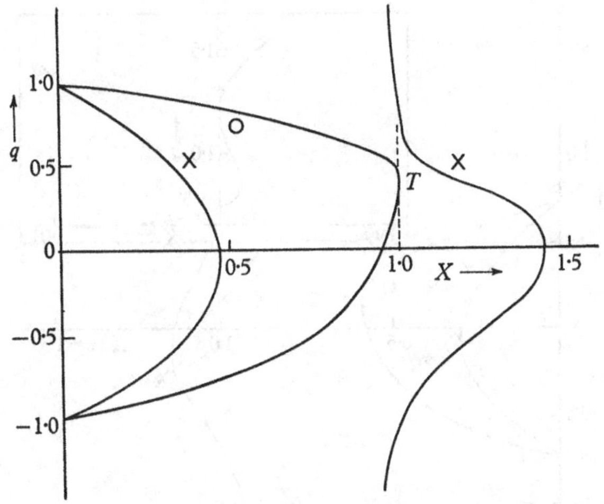

Fig. 13.5. Solutions of the Booker quartic for north–south propagation when $Y = \frac{1}{2}$. Earth's field at 30° to the vertical. Angle of incidence $\theta_I = 15°$. The line $X = 1$ touches the curve at the point T.

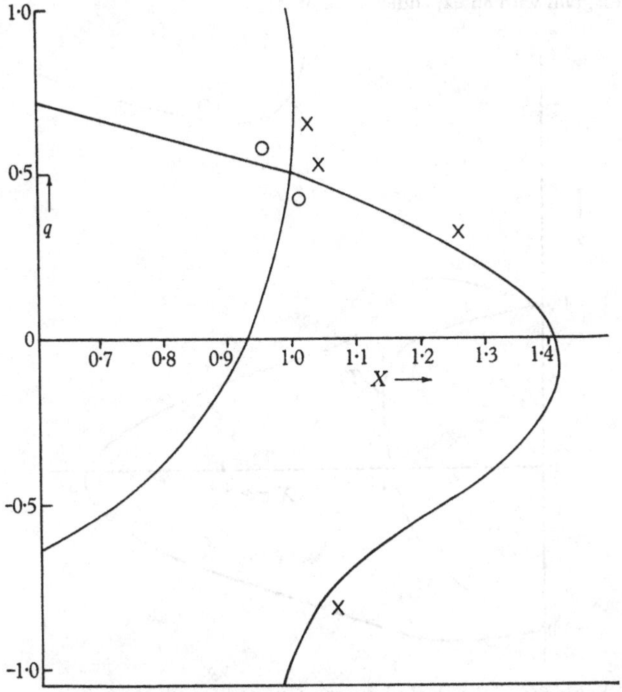

Fig. 13.6. Solutions of the Booker quartic for north–south propagation when $Y = \frac{1}{2}$. Earth's field at 30° to the vertical. The angle of incidence has the critical value 16·6° (13.31) so that the curves for the ordinary and extraordinary rays meet on the line $X = 1$.

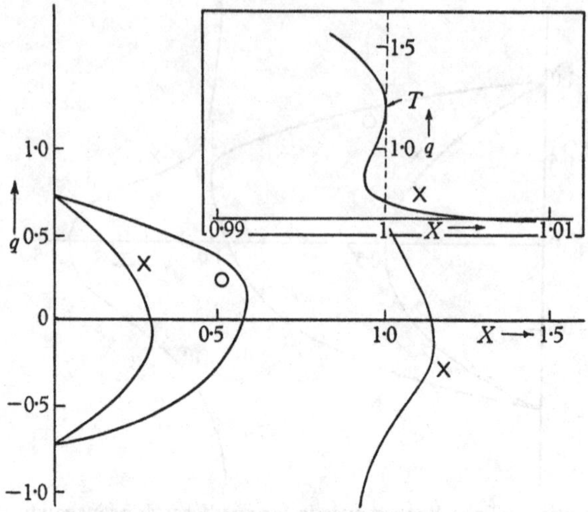

Fig. 13.7. Solutions of the Booker quartic for north–south propagation when $Y = \frac{1}{2}$. Earth's field at 30° to the vertical. Angle of incidence $\theta_I = 45°$. The right branch of the curve for the extraordinary wave touches the line $X = 1$ at the point T as shown in the inset diagram with an expanded scale of X.

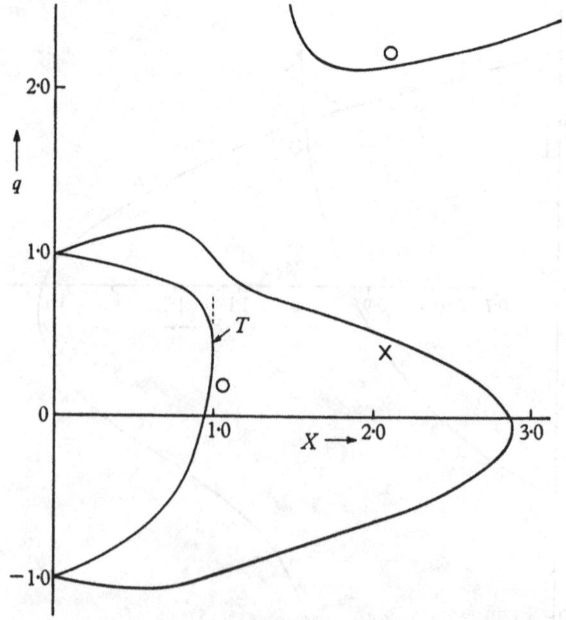

Fig. 13.8. Solutions of the Booker quartic for north–south propagation when $Y = 2$. Earth's field at 30° to the vertical. Angle of incidence $\theta_I = 15°$. Another branch of the curve for the ordinary wave lies beyond the bottom right corner and is not shown. The line $X = 1$ touches the curve at the point T.

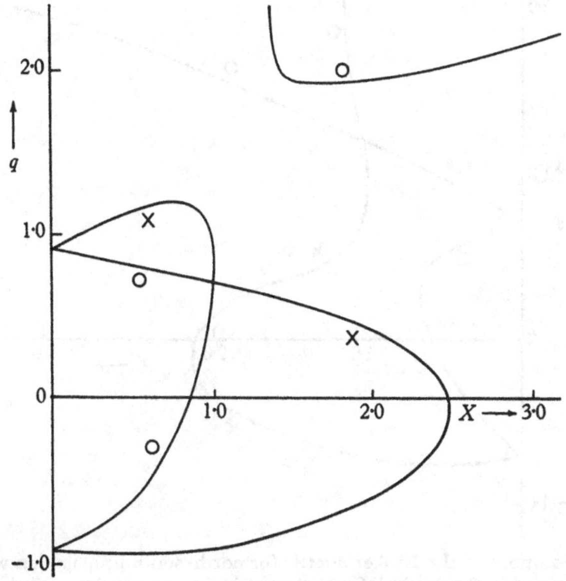

Fig. 13.9. Solutions of the Booker quartic for north–south propagation when $Y = 2$. Earth's field at 30° to the vertical. The angle of incidence has the critical value 24·2° so that the curves for the ordinary and extraordinary rays meet on the line $X = 1$. Another branch of the curve for the ordinary ray lies beyond the bottom right corner and is not shown.

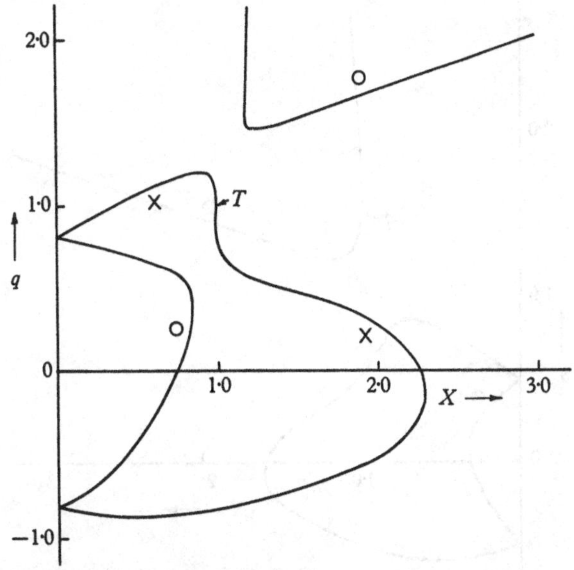

Fig. 13.10. Solutions of the Booker quartic for north–south propagation when $Y = 2$. Earth's field at 30° to the vertical. Angle of incidence $\theta_I = 35°$. Another branch of the curve for the ordinary ray lies beyond the bottom right corner and is not shown. The line $X = 1$ touches the curve at the point T.

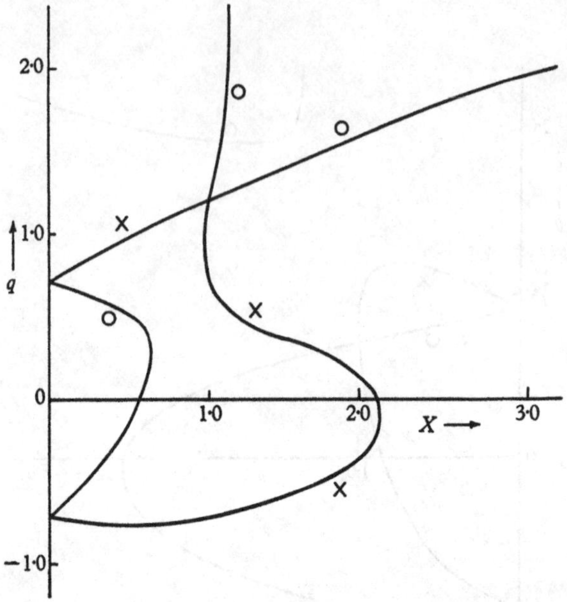

Fig. 13.11. Solutions of the Booker quartic for north–south propagation when $Y = 2$. Earth's field at 30° to the vertical. The angle of incidence has the critical value 45° so that the curves for the ordinary and extraordinary rays meet on the line $X = 1$. Another branch of the curve for the ordinary ray lies beyond the bottom right corner and is not shown.

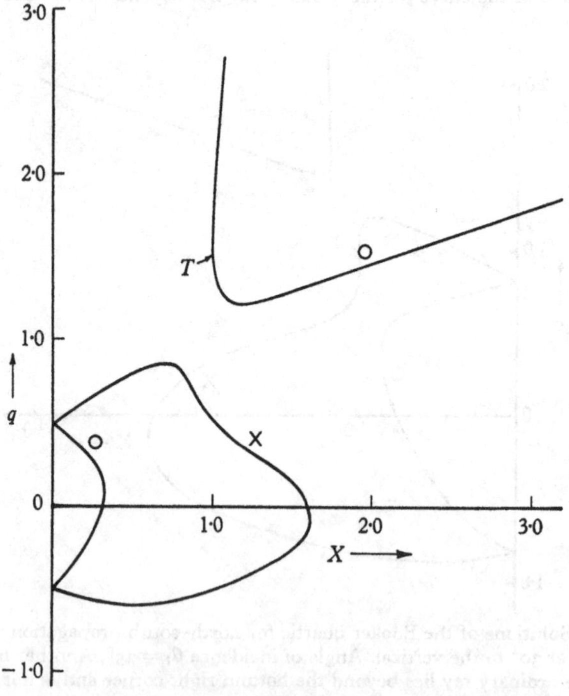

Fig. 13.12. Solutions of the Booker quartic for north–south propagation when $Y = 2$. Earth's field at 30° to the vertical. Angle of incidence $\theta = 60°$. Another branch of the curve for the ordinary ray lies beyond the bottom right corner and is not shown. The line $X = 1$ touches the curve at the point T.

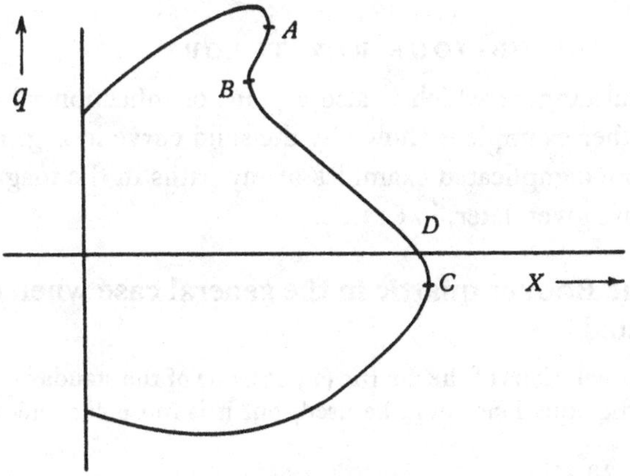

Fig. 13.13a. Solutions of the Booker quartic.

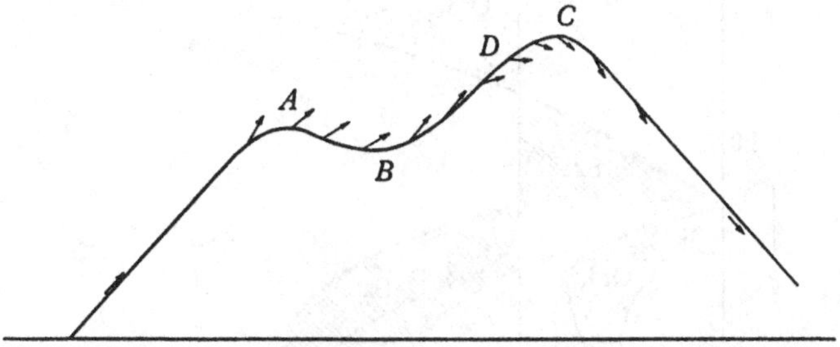

Fig. 13.13b. The corresponding ray path. The arrows indicate the direction of the wave-normal.

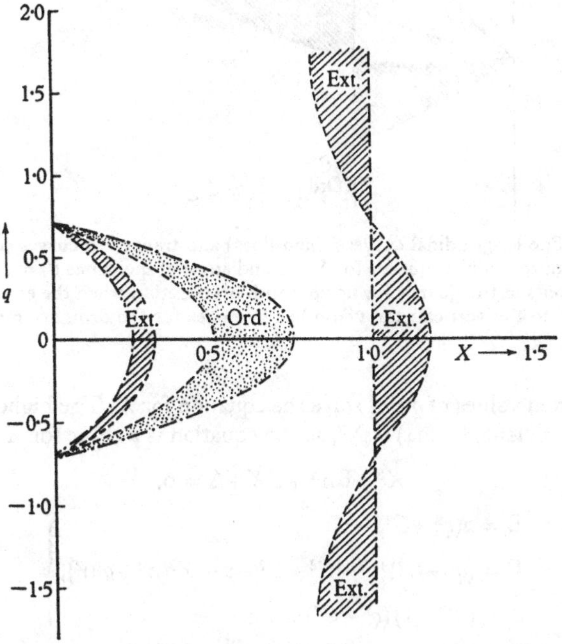

Fig. 13.14. The longitudinal curves (chain lines) and transverse curves (broken lines) associated with the Booker quartic for $Y = \frac{1}{2}$ and angle of incidence $\theta_I = 45°$.

has a vertical tangent which is also a point of inflection where $X = 1$, $q = C$. Another example is shown by the solid curve in Fig. 13.15.

Some more complicated examples of ray paths in the magnetic meridian plane are given later, in § 13.24.

13.12 The Booker quartic in the general case when collisions are neglected

To find the solutions of the quartic (13.23), one of the standard methods for solving quartic equations could be used, but it is more convenient to assign

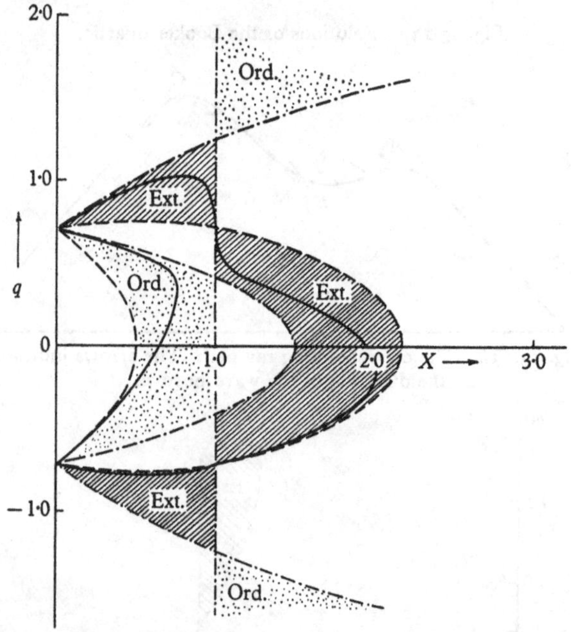

Fig. 13.15. The longitudinal curves (chain lines) and transverse curves (broken lines) associated with the Booker quartic for $Y = 2$ and angle of incidence $\theta_I = 45°$. The solid curve gives roots of the quartic for north–south propagation when the earth's magnetic field is at 45° to the vertical: the two other branches for the ordinary ray are outside the diagram.

a series of fixed values of q and solve the equation for X. The highest power of X in the coefficients (13.24) is X^3, so the equation is a cubic for X which may be written:

$$X^3 + \mathrm{B}X^2 + \Gamma X + \Delta = 0, \tag{13.34}$$

where

$$\left.\begin{aligned} \mathrm{B} &= 2(q^2 - C^2) - 1, \\ \Gamma &= (q^2 - C^2)\{q^2 - C^2 + Y^2 - 2 - Y^2(Sl + qn)^2\}, \\ \Delta &= (Y^2 - 1)(q^2 - C^2)^2. \end{aligned}\right\} \tag{13.35}$$

The cubic (13.34) can also be used to illustrate the general behaviour of solutions of the Booker quartic, as shown by Booker (1949). Consider the expression

$$(X-1)\{X-(Y-1)(q^2-C^2)\}\{X+(Y+1)(q^2-C^2)\}. \qquad (13.36)$$

It is equal to

$$X^3+\mathrm{B}X^2+\Gamma X+\Delta-XY^2(q^2-C^2)\{q^2+S^2-(Sl+qn)^2\}. \qquad (13.37)$$

Now equations (13.5) and (13.7) show that when $S_2 = 0$, the direction cosines of the wave-normal are S/μ, 0, q/μ, where μ is written for n, since collisions are neglected and we consider only real values of the refractive index. Let ϑ be the angle between the wave-normal and the earth's magnetic field. Then

$$\cos\vartheta = \frac{lS+qn}{\mu}, \qquad (13.38)$$

and (13.37) becomes

$$X^3+\mathrm{B}X^2+\Gamma X+\Delta-XY^2(\mu^2-1)\mu^2\sin^2\vartheta. \qquad (13.39)$$

Now the first four terms are the cubic (13.34) and are together equal to zero for any point on a ray. The last term is zero when the wave-normal is parallel to the earth's magnetic field. Hence (13.39), and therefore (13.36), is zero at those points (if any) of the ray path where the propagation is purely longitudinal. If (13.36) is set equal to zero, and the solutions q of the resulting equation are plotted against X, the curves may be called 'longitudinal curves'. They do not, in general, give solutions of the Booker quartic.

Similarly, consider the expression

$$(X+q^2-C^2)\{X^2+(q^2-C^2-1)X+(Y^2-1)(q^2-C^2)\}, \qquad (13.40)$$

which is equal to

$$X^3+\mathrm{B}X^2+\Gamma X+\Delta+XY^2(\mu^2-1)\mu^2\cos^2\vartheta. \qquad (13.41)$$

Again the first four terms are the cubic (13.34) and are together equal to zero for any point on a ray. The last term is zero when the wave-normal is perpendicular to the earth's magnetic field, and hence (13.40) and (13.41) are zero at those points (if any) of the ray path where the propagation is purely transverse. If (13.40) is set equal to zero and the solutions q of the resulting equation are plotted against X, the curves may be called 'transverse curves'. Now for any value of q, and therefore of μ, the last terms of (13.39) and (13.41) have opposite signs. Hence, between the longitudinal and transverse curves there must be a curve of q versus X, which represents a solution of the cubic (13.34), that is of the Booker quartic.

The longitudinal and transverse curves for some typical cases are shown in Figs. 13.14 and 13.15. They depend upon Y and upon the angle of incidence but are independent of the angle Θ between the earth's magnetic field and the vertical. The curve representing the solution of the Booker quartic must always lie in the shaded regions between the two sets of curves. The longitudinal and transverse curves therefore play the same role for the Booker quartic as did the curves of μ^2 versus X in Figs. 6.3, 6.7 and 6.8, for the longitudinal and transverse cases of the Appleton–Hartree formula (see §§6.6 and 6.10).

The curve of q versus X need not touch either the longitudinal or transverse curves but it may do so in some special cases. For example, for propagation from (magnetic) east to west or west to east, the wave-normal is horizontal, and therefore perpendicular to the earth's field, when q is zero, and the curve for q must touch the transverse curve at that point. Again, for propagation from north to south or south to north, suppose that the earth's field is at 45° to the vertical, and that the angle of incidence is also 45°. A curve for this case is shown in Fig. 13.15. The wave-normal of the incident wave is longitudinal, and that of the reflected wave is transverse, and the curve for q starts out along the longitudinal curve and moves over so that it ends in contact with the transverse curve.

13.13 Lateral deviation at vertical incidence

The Booker quartic may be used to find the path of a wave-packet which is vertically incident on the ionosphere from below. It is convenient to choose the x-axis to be in the magnetic meridian, pointing north, so that the direction cosine $m = 0$. The coefficient α in (13.16) is independent of S_1 and S_2, and the coefficients $\beta, \gamma, \delta, \epsilon$ depend on S_2 only through the term C^2, so that the partial derivatives $\partial\alpha/\partial S_2$, etc., with respect to S_2 all vanish for the 'predominant' value $S_2 = 0$. Hence $\partial q/\partial S_2$ and dy/dz (13.17) are zero, which shows that the wave-packet remains in the x–z-plane, that is in the magnetic meridian.

By differentiating (13.16) it can be shown that

$$\frac{\partial\alpha}{\partial S_1} = 0, \quad \frac{\partial\beta}{\partial S_1} = 2XlnY^2,$$

$$\frac{\partial\gamma}{\partial S_1} = \frac{\partial\epsilon}{\partial S_1} = 0, \quad \frac{\partial\delta}{\partial S_1} = -2XlnY^2 \quad \text{when} \quad S_1 = 0. \qquad (13.42)$$

Further, at vertical incidence q is the same as the refractive index μ, for the quartic then reduces to the Appleton–Hartree formula. Hence (13.17) and (13.20) give

$$\frac{dx}{dz} = -\left(\frac{\partial q}{\partial S_1}\right)_{S_1=0} = \frac{XlnY^2(\mu^2-1)}{2\alpha\mu^2+\gamma}, \qquad (13.43)$$

where a factor 2μ has been cancelled from numerator and denominator. This is the tangent of the angle of inclination of the ray path to the vertical. By using the appropriate value of μ it may be applied to either the ordinary or the extraordinary ray. Now μ must satisfy the equation

$\alpha\mu^4 + \gamma\mu^2 + \epsilon = 0$, which shows that the denominator in (13.43) is $\pm(\gamma^2 - 4\alpha\epsilon)^{\frac{1}{2}}$, and

$$\frac{dx}{dz} = \pm \frac{ln Y^2(1-\mu^2)}{\{4Y_L^2(1-X)^2 + Y_T^4\}^{\frac{1}{2}}}, \tag{13.44}$$

where the minus sign applies to the ordinary ray and the plus sign to the extraordinary ray. This has the same value for an upgoing or downgoing wave-packet so that the upward and downward paths are the same. In the northern hemisphere n is positive and l is negative so that dx/dz is positive for the ordinary ray. In the southern hemisphere both l and n are negative and dx/dz is negative for the ordinary ray. Hence the ordinary ray is always deviated towards the nearest magnetic pole and the extraordinary ray towards the equator. For frequencies above the gyro-frequency, $\mu_o^2 > \mu_x^2$ when $X < 1 - Y$. Hence $(1-\mu_o^2) < (1-\mu_x^2)$ and the inclination of the ray path to the vertical is smaller for the ordinary ray than for the extraordinary ray.

The ordinary ray is reflected where $X = 1$, and there

$$\frac{dx}{dz} = -\frac{n}{l} = -\cot\Theta,$$

where Θ is the angle between the earth's magnetic field and the vertical. Hence at reflection the ordinary ray path is perpendicular to the earth's magnetic field. It should be remembered, however, that the wave-normal is always vertical. It was explained in § 13.8 that the ray path and the wave-normal are not necessarily parallel. (See § 12.12 for remarks on the polarisation of the ordinary wave at its reflection level.)

Fig. 13.16 shows the form of the two ray paths for vertical incidence. An important result of this lateral deviation is that the reflection points for the ordinary and extraordinary rays are at different latitudes. The gyro-frequency is sometimes assessed by measuring the difference of the penetration frequencies of an ionospheric layer for the two rays. If the electron density at the maximum of the layer varies appreciably with latitude the measurement will give a wrong value for the gyro-frequency (discussed by Millington, 1951).

Equation (13.44) may also be derived by finding the direction of average energy flow as given by the complex Poynting vector (§ 2.13). This method was used in § 5.8 for a homogeneous medium. The argument would be expected still to apply in a medium which varies slowly enough with height. It was shown that the vector $\bar{\mathbf{\Pi}}$, giving the average energy flow, is in the plane containing the earth's magnetic field and the wave-normal, and is inclined to the wave-normal at an angle given by (5.45).

In the present example the wave-normal is vertical, and $\mathfrak{n}$ is real and therefore equal to its real part μ. Hence $\overline{\Pi}$ is in the magnetic meridian plane and is inclined to the vertical at an angle whose tangent is

$$\frac{\Pi_x}{\Pi_z} = \frac{Y_T(1-\mu^2)}{1-X}\,\frac{i\rho}{\rho^2-1}. \tag{13.45}$$

Now the wave-polarisation ρ is given by (12.41) whence it is easily shown that (13.45) is the same as (13.44).

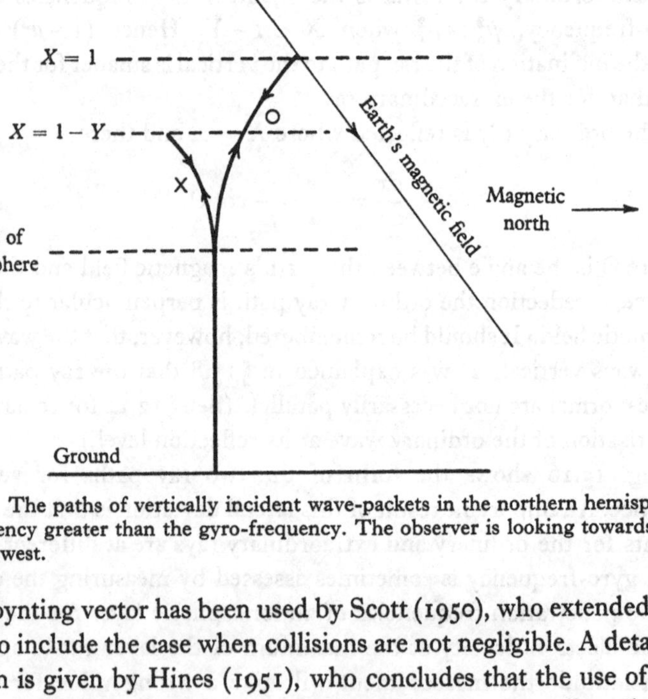

Fig. 13.16. The paths of vertically incident wave-packets in the northern hemisphere for a frequency greater than the gyro-frequency. The observer is looking towards the (magnetic) west.

The Poynting vector has been used by Scott (1950), who extended the analysis to include the case when collisions are not negligible. A detailed discussion is given by Hines (1951), who concludes that the use of the Poynting vector can lead to errors when collisions are allowed for. (See also Furutsu, 1952.)

13.14 Lateral deviation for propagation from (magnetic) east to west or west to east

In § 13.10 it was shown that when propagation is from (magnetic) east to west or west to east, the coefficients β and δ in the Booker quartic are zero and the quartic is a quadratic equation for q^2. In finding the path of a wave-packet, however, it is important to notice that β and δ must not be set equal to zero until after the differential coefficients in (13.20) have been found. This is because a wave-packet may include some component plane waves whose normals are not in the x–z-plane.

In (13.16) put $l = 0$. Then

$$\frac{\partial \alpha}{\partial S_2} = 0, \quad \frac{\partial \beta}{\partial S_2} = 2nmXY^2, \quad \frac{\partial \gamma}{\partial S_2} = 0, \quad \frac{\partial \delta}{\partial S_2} = -2C^2 nmXY^2, \quad \frac{\partial \epsilon}{\partial S_2} = 0, \tag{13.46}$$

when $S_2 = 0$, and (13.17), (13.20) give

$$\frac{dy}{dz} = -\left(\frac{\partial q}{\partial S_2}\right)_{S_2=0} = \frac{nmXY^2(q^2 - C^2)}{2\alpha q^2 + \gamma}, \tag{13.47}$$

where a factor $2q$ has been cancelled from numerator and denominator. This shows that the wave-packet is deviated out of the x–z-plane. If its path is

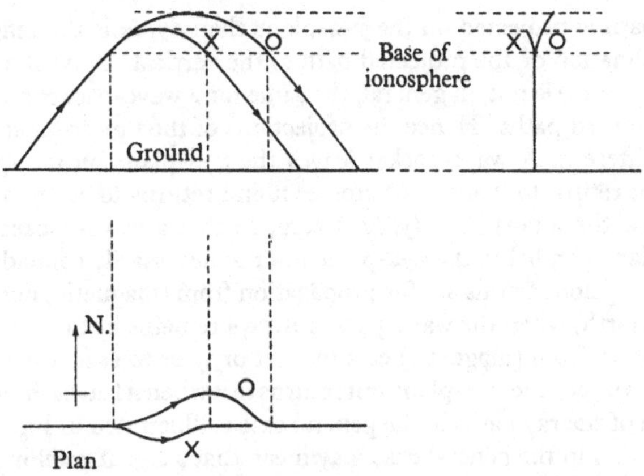

Fig. 13.17. The paths of wave-packets in the northern hemisphere when the plane of incidence is (magnetic) west to east.

projected on to the y–z-plane then dy/dz is the tangent of the angle of inclination of the projected path to the vertical. By using the appropriate value of q, (13.47) may be applied to either the ordinary or the extraordinary ray. It is the same for an upgoing or downgoing wave-packet so that the upward and downward projected paths are the same. Thus a wave-packet may leave the x–z-plane on its upward journey but returns to it again on its downward journey. This is illustrated in Fig. 13.17 which shows the projection of the ray paths on to the horizontal plane and on to two vertical planes.

Now q must satisfy the equation $\alpha q^4 + \gamma q^2 + \epsilon = 0$ which shows that the denominator in (13.47) is $\pm(\gamma^2 - 4\alpha\epsilon)^{\frac{1}{2}}$, and

$$\frac{dy}{dx} = \pm \frac{nmY^2(C^2 - q^2)}{\{Y^4(1 - C^2n^2)^2 + 4Y^2n^2(1 - X)(C^2 - X)\}^{\frac{1}{2}}}. \tag{13.48}$$

This is similar to (13.44) to which it reduces when $C = 1$ (with y for x and m for l). (See Millington, 1951.)

13.15 Lateral deviation in the general case

In the general case the derivatives with respect to S_2 of the coefficients (13.16) of the Booker quartic are

$$\left.\begin{aligned}\frac{\partial\alpha}{\partial S_2} = 0, \quad \frac{\partial\beta}{\partial S_2} = 2mnXY^2, \quad \frac{\partial\gamma}{\partial S_2} = 2lmSXY^2,\\ \frac{\partial\delta}{\partial S_2} = -2C^2nmXY^2, \quad \frac{\partial\epsilon}{\partial S_2} = -2C^2lmSXY^2,\end{aligned}\right\} \tag{13.49}$$

when $S_2 = 0$, where S is written for $S_1 = S_0$. The formulae (13.17) and (13.20) now give

$$\frac{dy}{dz} = -\left(\frac{\partial q}{\partial S_2}\right)_{S_2=0} = \frac{2mXY^2(q^2-C^2)(nq+lS)}{4\alpha q^3+3\beta\delta q^2+2\gamma q+\delta}. \tag{13.50}$$

If the ray path is projected on the y–z-plane then dy/dz is the tangent of the angle of inclination of the projected path to the vertical. Now at a given level in the ionosphere q is not, in general, the same for a wave-packet on its upward and its downward paths. Hence the projections of the upward and downward paths are different. A wave-packet leaves the x–z-plane on its upward path and does not return to it again, or crosses it and returns to earth in a different plane. Below the ionosphere dy/dz is zero, so that a wave-packet returns to earth in a plane parallel to the x–z-plane but not necessarily coincident with it. The only exceptions to this are for propagation from (magnetic) north to south or south to north, when the wave-packet always remains in the x–z-plane, and for propagation from (magnetic) east to west or west to east, when the wave-packet moves out of the x–z-plane but returns to it when it leaves the ionosphere.

The form of the ray paths in the general case is illustrated in Fig. 13.18. The lateral deviation in the general case may mean that a signal reaching a receiver from a sender which is not in the direction of one of the four (magnetic) cardinal points, arrives in a vertical plane different from that which contains the receiver and sender. The wave-normals of this signal are in the plane of arrival, so that if a direction-finding aerial is used, the apparent bearing of the sender will differ from the true bearing. This theory has been discussed by Booker (1949) and Millington (1954), who showed that the bearing error would not be serious except in unusual conditions.

13.16 Calculation of attenuation, using the Booker quartic

In the preceding sections the effect of electron collisions has been neglected, and the path of a wave-packet has been studied in those parts of the ionosphere where the associated root q of the quartic is purely real. We now consider the effect of a small collision-frequency ν. The quantity q depends on ν only through $U = 1 - iZ$. Let q_0 be its value when $Z = 0$. Then

$$q = q_0 - iZ\frac{\partial q}{\partial U} - \tfrac{1}{2}Z^2\frac{\partial^2 q}{\partial U^2} + \dots, \tag{13.51}$$

where the derivatives have their values for $Z = 0$. It can easily be shown that $\partial q/\partial U$ is then real when q is real. If Z is small, only two terms in (13.51) need be used. Then the real part of q is the same as if there were no collisions, and determines the path of the wave-packet, which is the same as before. There is now an imaginary part $-iZ(\partial q/\partial U)$ of q, which gives attenuation of the wave. Equation (13.15) shows that the amplitude is reduced by a factor

$$\exp\left\{-k\int_0^z Z\frac{\partial q}{\partial U}dz\right\},$$

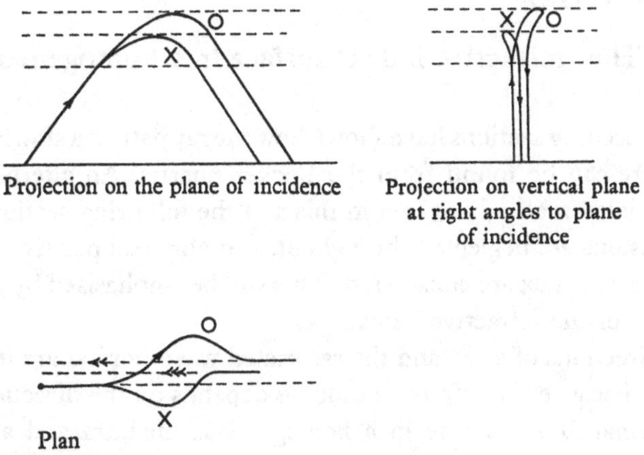

Fig. 13.18. The paths of wave-packets in the northern hemisphere when the plane of incidence is not in the direction of one of the four (magnetic) cardinal points. The double and triple arrows show the apparent direction of the transmitter when using the ordinary and extraordinary rays respectively.

where the range of integration extends over the path of the wave-packet. For a ray reflected from the ionosphere the range would extend from the ground to the level of reflection and back to the ground, and the integrand would in general be different for the upward and downward paths.

The quantity $\partial q/\partial U$ can be found from the quartic (13.13) using the form (13.14) for the coefficients. Since the quartic is satisfied for all values of U, we have $dF(q)/dU = 0$, whence

$$\frac{\partial q}{\partial U} = -\frac{q^4\dfrac{\partial \alpha}{\partial U}+q^2\dfrac{\partial \gamma}{\partial U}+\dfrac{\partial \epsilon}{\partial U}}{4q^3\alpha+3q^2\beta+2q\gamma+\delta} \tag{13.52}$$

(β and δ do not contain U and therefore do not appear in the numerator). Booker (1949) has given curves showing how $\partial q/\partial U$ depends on X in some special cases. For another method of finding the attenuation see § 14.7.

In § 13.6 it was shown that when every component wave in a wave-packet is reversed in direction, the wave-packet simply retraces its original path back to the transmitter. At each point of the path q, S_1 and S_2 are reversed in sign. Now (13.52) shows that $\partial q/\partial U$ is also simply reversed in sign, but the path of integration used in determining the attenuation is also reversed so that the total attenuation is the same for the original and the reversed paths. This result has an important bearing on the study of reciprocity (§ 23.4).

13.17 The 'refractive index' surface in a homogeneous medium

The preceding sections have shown how the ray path in a slowly varying ionosphere can be found from the Booker quartic. An alternative but closely related method is given in this and the following sections. Electron collisions are neglected throughout, and only real positive values of the refractive index are considered. This will be emphasised by using the symbol μ for the refractive index.

The directions of a ray and the associated wave-normal are in general different, because the refractive index μ depends on the direction of the wave-normal. This is true in a homogeneous medium and a relation between the two directions can be found. The results can then be used to trace a ray path in an inhomogeneous but slowly varying medium.

Let ϑ be the angle between a wave-normal and the superimposed magnetic field in a homogeneous ionised medium, and let $\mu(\vartheta)$ be the refractive index for this wave. We use a coordinate system in which the ζ-axis is the direction of the earth's magnetic field, and r, ϑ, φ are spherical polar coordinates. Consider a surface in which the radius r to each point is equal to $\mu(\vartheta)$. For any medium there are two such surfaces, one for each of the two characteristic waves. Since $\mu(\vartheta)$ depends only on ϑ, and $\mu(\pi-\vartheta) = \mu(\vartheta)$, each surface is a surface of revolution about the ζ-axis, and the plane $\vartheta = \frac{1}{2}\pi$ is a plane of symmetry. These surfaces are called 'refractive index surfaces'. They are used in the theory of crystal optics, but then the surfaces are often more complicated because the radius may depend on φ as well as ϑ. For an isotropic medium μ is independent of ϑ, and the refractive index surface is simply a sphere. A convenient method of plotting the refractive index surfaces for the ionosphere has been described by Clemmow and Mullaly (1955). (Some authors, e.g. Storey, 1953, use a surface in which the radius is the wave-velocity, proportional to $1/\mu$; this is called a 'wave-surface', but it will not be used here.) Some examples of refractive index surfaces are given later in Figs. 13.20 to 13.31.

13.18 The direction of the ray

Now suppose that a wave-packet starts out from the origin of co-ordinates at time $t = 0$. It travels in the homogeneous medium and splits into two components, the ordinary and extraordinary waves. We consider only one of these, say the ordinary wave. The field is given by

$$\iiint A(k, \vartheta, \varphi) \exp ik[ct - \mu(\vartheta)\{(\xi\cos\varphi + \eta\sin\varphi)\sin\vartheta + \zeta\cos\vartheta\}]\,dk\,d\vartheta\,d\varphi \tag{13.53}$$

(see § 11.2) where the amplitude function $A(k, \vartheta, \varphi)$ has a maximum for the predominant values k_0, ϑ_0, φ_0. The angles ϑ_0, φ_0 give the direction in which the transmitting aerial at the origin would radiate most strongly if it were in free space. The position (ξ, η, ζ) of the wave-packet is found from the condition that the exponent in (13.53) must be stationary with respect to variations of ϑ, φ and k, for the values ϑ_0, φ_0, k_0. Differentiation of the exponent with respect to φ and ϑ gives, respectively,

$$\eta\cos\varphi_0 = \xi\sin\varphi_0 \tag{13.54}$$

and

$$\{\zeta\cos\vartheta_0 + (\xi\cos\varphi_0 + \eta\sin\varphi_0)\sin\vartheta_0\}\left(\frac{d\mu}{d\vartheta}\right)_0 + \mu(\vartheta_0)\{(\xi\cos\varphi_0 + \eta\sin\varphi_0)\cos\vartheta_0 - \zeta\sin\vartheta_0\} = 0. \tag{13.55}$$

Equation (13.54) shows that the wave-packet remains in the plane $\varphi = \varphi_0$, so that the ray, the wave-normal, and the earth's magnetic field are coplanar. We may therefore choose the origin of φ so that $\varphi_0 = 0$. Let the direction of travel of the wave-packet make an angle α with the wave-normal, that is an angle $\vartheta_0 - \alpha$ with the ζ-axis, and let r_0 be its distance from the origin. Then $\zeta = r_0\cos(\vartheta_0 - \alpha)$, $\xi = r_0\sin(\vartheta_0 - \alpha)$, $\eta = 0$ and (13.55) becomes

$$\{1 + \tan(\vartheta_0 - \alpha)\tan\vartheta_0\}\left(\frac{\partial\mu}{\partial\vartheta}\right)_0 + \mu\{\tan(\vartheta_0 - \alpha) - \tan\vartheta_0\} = 0, \tag{13.56}$$

whence

$$\tan\alpha = \left(\frac{1}{\mu}\frac{\partial\mu}{\partial\vartheta}\right)_{\vartheta = \vartheta_0}. \tag{13.57}$$

Now $(1/\mu)(\partial\mu/\partial\vartheta)$ is the tangent of the angle between the radius and the normal to the refractive index surface. This can be seen from Fig. 13.19, which represents a cross-section of the surface by the plane $\varphi = 0$. Angle NCX = angle ACB, and $\tan ACB = AB/BC = \delta\mu/\mu\delta\vartheta$. Hence the direction $\vartheta_0 - \alpha$ of the ray is parallel to the normal to the refractive index surface.

Differentiation of the exponent in (13.53) with respect to k gives (when $\varphi_0 = 0$, $\zeta = r_0 \cos(\vartheta_0 - \alpha)$, $\xi = r_0 \sin(\vartheta_0 - \alpha)$ as above):

$$ct = r_0 \cos\alpha \frac{\partial}{\partial f}(f\mu), \tag{13.58}$$

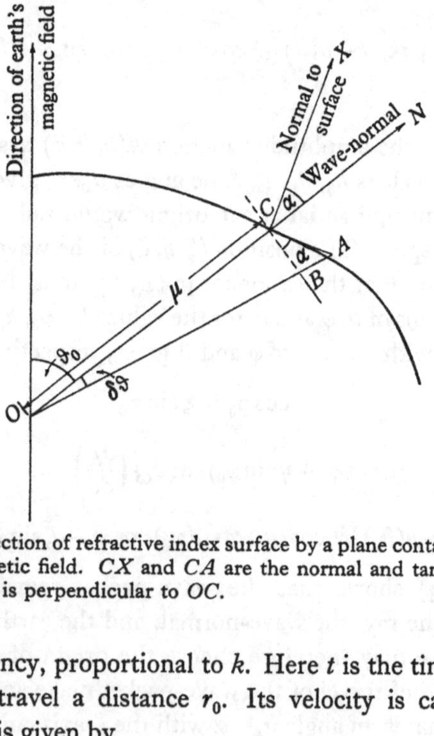

Fig. 13.19. Cross-section of refractive index surface by a plane containing the direction of the earth's magnetic field. CX and CA are the normal and tangent, respectively, at the point C. CB is perpendicular to OC.

where f is frequency, proportional to k. Here t is the time taken for the wave-packet to travel a distance r_0. Its velocity is called the group velocity, U, and is given by

$$U = \frac{c}{\mu'} \sec\alpha, \tag{13.59}$$

where $\mu' = (\partial/\partial f)(f\mu)$ is the 'group refractive index' as defined in § 12.3.

It is sometimes instructive to draw another kind of surface for which the radius in a given direction is the velocity U of a wave-packet travelling in that direction. This will be called the 'group velocity surface'. An example is given in § 13.20.

13.19 The ray velocity and the ray surface

The path of a wave-packet is called the 'ray path' or simply the 'ray'. In the last section it was shown that the direction of the ray is inclined to the wave-normal at an angle α given by (13.57). The wave-fronts within the wave-packet move in the direction of the wave-normal with the wave-velocity $V = c/\mu$, which is the velocity of some particular feature of the wave, for example a wave-crest. We shall now find the velocity of such a wave-crest in the direction of the ray. Clearly its point of intersection with the ray direction moves sideways as the wave-front advances, and its velocity is therefore

$$V_R = V \sec\alpha = c/(\mu\cos\alpha). \tag{13.60}$$

This is called the 'ray velocity'.† It is not the velocity U of the wave-packet. Usually $V_R > U$ so that successive wave-crests appear at the back of the wave-packet, move through it and disappear again at the front. It is convenient also to define a quantity

$$\mathscr{M} = \mu\cos\alpha = c/V_R \tag{13.61}$$

called the 'ray refractive index'.

As in the last section we use a coordinate system in which the ζ-axis is the direction of the earth's magnetic field, and r, ϑ, φ are spherical polar coordinates. Let $\vartheta_R = \vartheta_0 - \alpha$ be the direction of a given ray and let $V_R(\vartheta_R)$ be the ray velocity. Consider a surface in which the radius r to each point is equal to $V_R(\vartheta_R)$. This is called the 'ray surface' for the medium; for any medium there are two such surfaces, one for each of the two characteristic waves. Each is a surface of revolution about the ζ-axis, and has the plane $\vartheta = \frac{1}{2}\pi$ as a plane of symmetry.

For any given direction the point on the ray surface is clearly the point reached by a wave-front after unit time. The complete ray surface is thus simply the wave-front after unit time for a point source at the origin. The normal to the ray surface at any point is therefore the wave-normal.

It is now clear that the ray surface and the refractive index surface have an important reciprocal property which may be stated thus: To every point R on the ray surface there corresponds a point P on the refractive index surface. The radius OR (the ray direction) is parallel to the normal to the refractive index surface at P. The radius OP (the wave-normal) is parallel to the normal to the ray surface at R.

† Some authors (e.g. Storey, 1953) use a different terminology from that used here.

The ray surfaces are often more complicated than the associated refractive index surfaces. Some examples are shown in Figs. 13.22*b*, 13.25*b* and 13.27 which are cross-sections of the surfaces in the magnetic meridian plane.

The results in this and the preceding section have been established for a homogeneous medium. They are extended to a 'slowly varying' medium in § 14.2, where the reciprocal properties of the surfaces are expressed in more mathematical form.

13.20 Whistlers

In this section it will be assumed that the theory of § 13.18 may be applied to a 'slowly varying' medium.

The use of the refractive index surface, the ray surface and the group velocity surface may be illustrated by considering propagation in a homogeneous medium at very low frequencies. It is assumed that the frequency f is much less than the plasma frequency f_N and the gyro-frequency f_H, so that $X \gg 1$, $Y \gg 1$. Collisions are neglected and the Appleton–Hartree formula then gives

$$\mu^2 \approx 1 - \frac{X}{\pm Y_L + \frac{1}{2} Y_T^2/X}. \tag{13.62}$$

Now X is proportional to $1/f^2$, whereas Y is proportional to $1/f$. Hence if the frequency is low enough we may take $X \gg Y$, and then

$$\mu^2 = \pm X/Y_L. \tag{13.63}$$

Only the plus sign gives a real value for μ, and for this case

$$\mu \approx f_N \sec^{\frac{1}{2}} \vartheta (f f_H)^{-\frac{1}{2}}, \tag{13.64}$$

where ϑ is the angle between the wave-normal and the earth's magnetic field. This formula applies to the branch *c* of the curves in Fig. 6.8.

For a wave-packet in which $\vartheta = \vartheta_0$ for the predominant wave-normal, the direction of the ray may be found as in § 13.18. Let the ray make an angle $\vartheta_0 - \alpha$ with the earth's magnetic field. Then (13.57) gives

$$\tan \alpha = \tfrac{1}{2} \tan \vartheta_0, \tag{13.65}$$

whence

$$\tan(\vartheta_0 - \alpha) = \frac{\frac{1}{2} \tan \vartheta_0}{1 + \frac{1}{2} \tan^2 \vartheta_0}. \tag{13.66}$$

It can easily be shown that the maximum value of $\tan(\vartheta_0 - \alpha)$ is $1/\sqrt{8}$ and then $\vartheta_0 = 54° 44'$ and $\vartheta_0 - \alpha = 19° 29'$. Further $\mu' = \frac{1}{2}\mu$ and

$$\tfrac{1}{2} U = V_R = \frac{c}{\mu \cos \alpha}, \tag{13.67}$$

so that the ray surface and the group velocity surface have the same shape but differ in scale by a factor 2. For the direction of the earth's magnetic field $\vartheta_0 = \alpha = 0$, and U has the value

$$U_0 \approx 2c(ff_H)^{\frac{1}{2}}/f_N. \tag{13.68}$$

It is therefore small for the smallest frequencies. The refractive index surface and the ray surface for this case are shown in Fig. 13.20. It is noteworthy that the ray direction must be within about 20° of the direc-

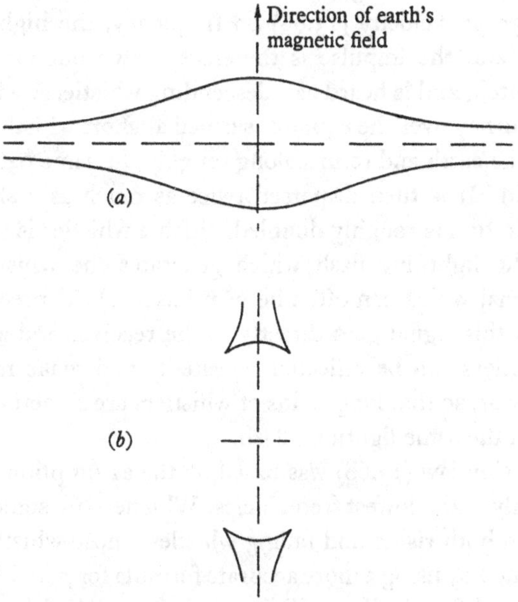

Fig. 13.20. Cross-section in the magnetic meridian plane of (a) the refractive index surface, and (b) the ray surface, for very low frequencies ('whistlers').

tion of the earth's magnetic field. Any signal of very low frequency in an ionised medium is therefore 'guided' by the magnetic field, and will tend to travel almost along the lines of force. This is the suggested mechanism for the production of the very low-frequency atmospherics known as 'whistlers' (see Burton and Boardman, 1933; Eckersley, 1935; Storey, 1953).

If an audio frequency amplifier, with a loud-speaker, is connected to a suitable aerial, such as a very long wire as high as possible, various atmospheric noises can be heard, and sometimes these include a whistle of falling pitch lasting for several seconds. A full description has been given by Storey (1953). These signals are known as 'whistlers'. It is now believed that a whistler originates in a lightning flash near the earth's surface. This is an impulse signal which contains a wide range of frequencies including those in the audible range. Some energy in this range can penetrate the ionosphere. It was suggested by Storey (1953) that there is enough ionisation beyond the F-layer for the signal to be guided by the earth's magnetic field, as described at the beginning of this section. It therefore travels roughly in the direction of the magnetic field, over the equator, and returns to earth at a point somewhere near the other end of the line of force along which it started out. Because of the dependence of the group velocity (13.68) on frequency, the high frequencies travel faster, and the impulse is therefore drawn out into a signal of descending pitch, and is heard as a descending whistle. A whistler which makes one journey over the equator is called a 'short whistler'. It can be reflected by the earth and return along roughly the same path to a region near its origin. It is then dispersed twice as much as a short whistler so that its duration is roughly doubled. Such a whistler is called a 'long whistler'. The lightning flash which generates the whistler gives an impulsive signal which can often be heard as a 'click' preceding a long whistler, and this signal goes directly to the receiver and so is not dispersed. Whistlers can be reflected repeatedly and make many transits over the equator, so that long trains of whistlers are sometimes heard, all originating in the same lightning flash.

The dispersion law (13.68) was based on the assumption that $X \ll Y$, and is true only at the lowest frequencies. Whistlers are sometimes heard which contain both rising and falling whistles ('nose whistlers'). These can be explained by using a more accurate formula for μ and U (Helliwell, Crary, Pope and Smith, 1956). Other whistles, both rising and falling, are also heard occasionally. These probably originate from disturbances other than lightning flashes.

A whistler does not follow a line of magnetic force exactly, but the path can be calculated (see, for example, Maeda and Kimura, 1956, and § 14.4). It is possible that the path and dispersion of a whistler may be affected by heavy ions. This has been considered by Hines (1957).

13.21 Determination of ray direction by Poeverlein's construction

A method of finding ray directions in the ionosphere, by using the Booker quartic, was described in §§ 13.5 to 13.8. An alternative method, using the refractive index surface, has been given by Poeverlein (1948, 1949, 1950), and will now be described. Electron collisions are neglected, and the refractive index n is the same as its real part μ. To emphasise this the symbol μ will be used for the refractive index.

A wave-packet is incident on a slowly varying ionosphere from below, with its wave-normal in the x–z-plane at an angle θ_I to the vertical. Imagine the refractive index surfaces for one of the two characteristic waves to be drawn with a common origin, for various values of the electron density, and oriented correctly with respect to the axes. Fig. 13.21

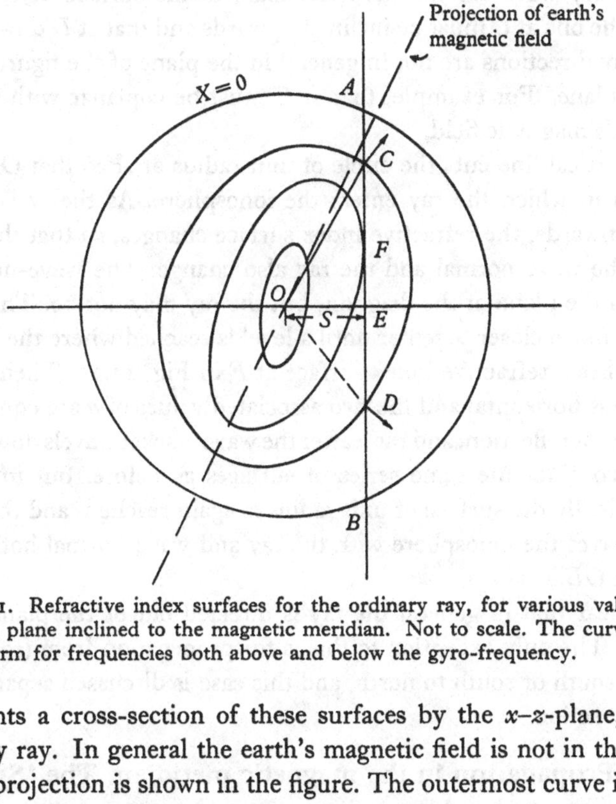

Fig. 13.21. Refractive index surfaces for the ordinary ray, for various values of X, and for a plane inclined to the magnetic meridian. Not to scale. The curves have a similar form for frequencies both above and below the gyro-frequency.

represents a cross-section of these surfaces by the x–z-plane, for the ordinary ray. In general the earth's magnetic field is not in this plane, but its projection is shown in the figure. The outermost curve is for the free space below the ionosphere and is a circle of unit radius. As the electron density increases the curves get smaller until they shrink to a point where $X = 1$.

Now draw a vertical line (AB in Fig. 13.21) in the x–z-plane at a distance $S = \sin\theta$ from the origin. For some level let it cut the refractive index surface at the two points C and D, and let OC make an angle θ with the z-axis. Then OC is the refractive index μ for a wave whose normal is in the direction OC. Clearly $OC \sin\theta = \mu \sin\theta = S$, which is simply Snell's law. Hence OC is one possible direction for the wave-normal of the ordinary wave. Another possible direction is OD, and two other directions are found from the intersection of the line AB with the refractive index surface for the extraordinary wave. It is also clear from the figure that $CE = \mu \cos\theta = q$. In fact the triangle OEC is the same as the triangle of Fig. 8.11.

Now it was shown in § 13.18 that the ray direction is perpendicular to the refractive index surface. Hence the two possible ray directions for the ordinary wave are the perpendiculars to the surface at C and D. Clearly the one at C must be inclined upwards and that at D downwards. These ray directions are not in general in the plane of the figure, that is the x–z-plane. For example, that at C must be coplanar with OC and the earth's magnetic field.

The vertical line cuts the circle of unit radius at A so that OA is the direction in which the ray enters the ionosphere. As the wave-packet travels upwards, the refractive index surface changes, so that the direction of the wave-normal and the ray also change. The wave-normal is always in the plane of the diagram, but the ray may not be. The points C and D move closer together until a level is reached where the line AB just touches a refractive index surface at F in Fig. 13.21. Then the ray direction is horizontal and the two associated values of q are equal. This is the level of reflection, and thereafter the wave-packet travels downwards and we consider the same series of surfaces as before, but in reverse order. Finally the surface of unit radius is again reached, and the wave-packet leaves the ionosphere with the ray and wave-normal both in the direction OB.

It is clear that in general the ray is directed out of the plane of the diagram. The only exception to this is for propagation from (magnetic) north to south or south to north, and this case is discussed separately in the next section.

13.22 Propagation in the magnetic meridian. The 'Spitze'

Fig. 13.21 shows a series of cross-sections of the refractive index surfaces by the plane of incidence, that is the x–z-plane. When this plane is also the magnetic meridian, these cross-sections have a slightly different form, of which an example is shown in Fig. 13.22 which gives the curves for the ordinary ray. As the electron density increases, the curves get smaller and smaller as before, but instead of shrinking to a point when $X = 1$, they shrink to the line PQ. This shows that the refractive index μ for the ordinary ray is zero when $X = 1$ for all directions except the direction of the earth's magnetic field. For this case of purely longitudinal propagation there is an ambiguity in the value of μ when $X = 1$, which was mentioned in §6.6.

Poeverlein's construction may now be used as described in the last section. A vertical line is drawn at a distance S from the origin. Suppose it is at AB in Fig. 13.22*a*. Then the ray directions are given by the perpendiculars to the refractive index surfaces where the line cuts them. They are now always in the plane of incidence, so there is no lateral deviation. Suppose that a wave-packet travels upwards in an ionosphere in which the electron density increases monotonically. Eventually a level

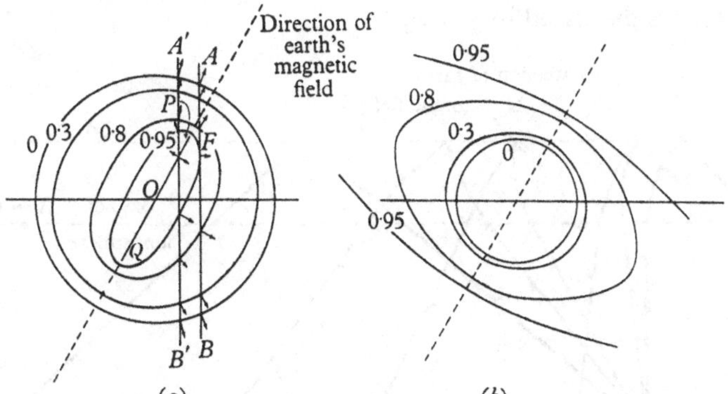

Fig. 13.22. (*a*) shows a cross-section of the refractive index surfaces for the ordinary ray by the magnetic meridian plane. The figures by the curves are values of X. (*b*) shows the corresponding ray surfaces. In (*a*) the curve marked 0 is a circle of unit radius, and in (*b*) it is a circle of radius c. In this example $Y = \frac{1}{2}$, but the curves have a similar form for frequencies both above and below the gyro-frequency.

is reached where the line AB just touches a refractive index surface, as at F in Fig. 13.22*a*. Here the ray is horizontal and this is the level of reflection. The phenomenon is similar to that described in the last section. The surface which the line just touches gives the value of X at the level of reflection. If the angle of incidence, arc sin S, increases, the line AB moves further from the origin and the surface which is touched corresponds to a smaller value of X. The reflection level therefore decreases as the angle of incidence increases.

Suppose, however, that the vertical line is at $A'B'$ in Fig. 13.22*a*. The ray directions are given as before by the perpendiculars to the refractive index surfaces where the line cuts them, but there is now no surface which is touched by the line. Instead, the line $A'B'$ cuts the line PQ, which refers to the level where $X = 1$. Above PQ the outward perpendicular to the refractive index surface is directed obliquely upwards; below PQ

the perpendicular is obliquely downwards. On the line PQ the ray therefore reverses its direction abruptly, and this is the level of reflection. Here the ray path is perpendicular to the line PQ, that is to the earth's magnetic field. The ray path never becomes horizontal, but has a cusp at the level $X = 1$, called by Poeverlein the 'Spitze'. Fig. 13.23 shows some typical ray paths for various angles of incidence. It is similar to a diagram given by Poeverlein (1950). As long as the angle of incidence is such that the line $A'B'$ cuts the line PQ, there is always a Spitze, and the reflection level is where $X = 1$ for all angles of incidence. In the limiting cases the line $A'B'$ passes through one of the points P or Q. This is discussed in § 13.23.

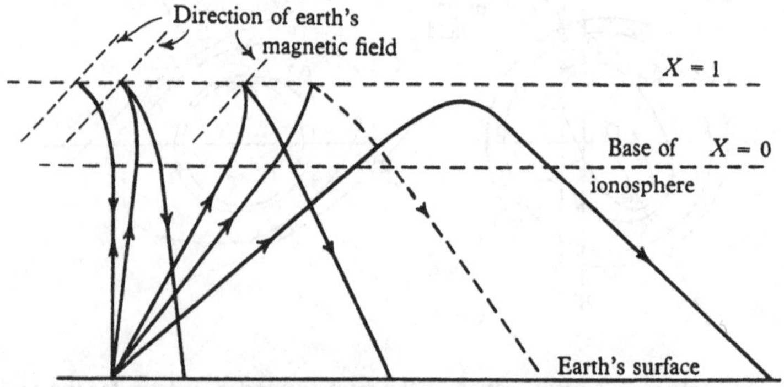

Fig. 13.23. Typical ray paths for the ordinary ray in the magnetic meridian, showing the 'Spitze'.

If the plane of incidence is turned very slightly from the magnetic meridian, the line PQ moves out of the plane of the diagram in Fig. 13.22*a*. The line $A'B'$ then touches one of the refractive index surfaces, for X slightly less than unity. This is a long cigar-shaped surface close to PQ. The perpendicular to it is a horizontal line almost perpendicular to the plane of the diagram. The ray path is then a twisted curve in three dimensions, and Figs. 13.24*a* and *b* show how it would appear when seen from the west and the north respectively. The receiver is not in the plane of incidence through the transmitter, for the reasons explained in § 13.15. A ray path with a Spitze is a limiting case, when the width of the curve of Fig. 13.24*b*, in the east–west direction becomes infinitesimally small.

13.23 The refractive index surfaces for the extraordinary ray when $Y < 1$

When $Y < 1$ so that the frequency is above the gyro-frequency, one set of refractive index surfaces, for the extraordinary ray, extends from $X = 0$ to

$X = 1 - Y$. The cross-section of these by the magnetic meridian is shown in Fig. 13.25*a* for the case $Y = \frac{1}{2}$. The outermost curve for $X = 0$ is a circle of unit radius. The curves get smaller as X increases and eventually shrink to a point when $X = 1 - Y$. They never shrink to a line as in Fig. 13.22*a* so that there is no phenomenon of the Spitze for the extraordinary ray. The value of X at reflection decreases as S increases, for the whole range $0 \leqslant S \leqslant 1$.

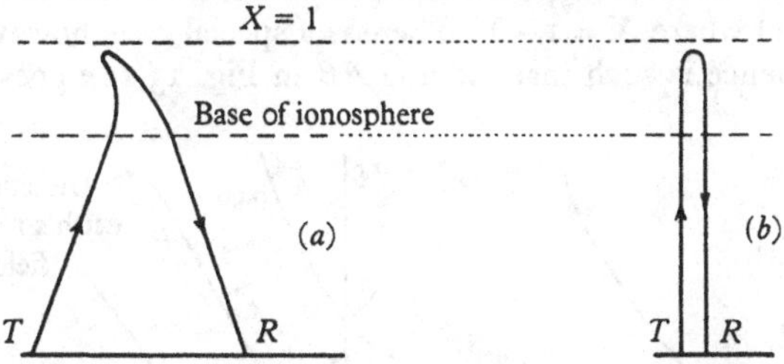

Fig. 13.24. Projections of the ray path for the ordinary ray when the plane of incidence is slightly inclined to the magnetic meridian. (*a*) is projected on to the plane of incidence, and (*b*) on to a plane at right angles to this.

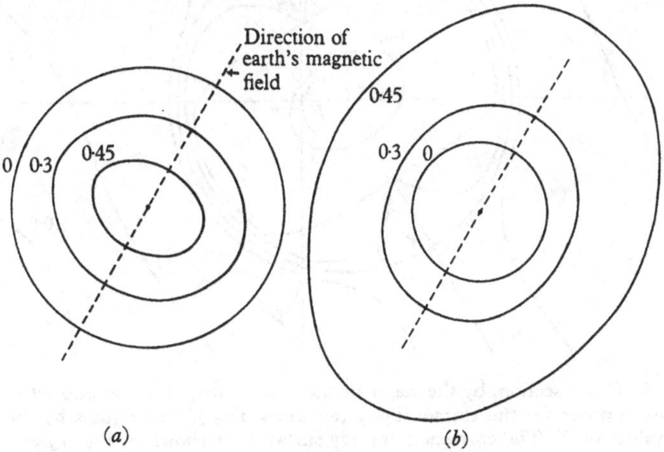

Fig. 13.25. (*a*) shows a cross-section of the refractive index surfaces for the extraordinary ray by the magnetic meridian plane when $Y = \frac{1}{2}$. The figures by the curves are values of X. (*b*) shows the corresponding ray surfaces. In (*a*) the curve marked 0 is a circle of unit radius, and in (*b*) it is a circle of radius c. Curves for higher values of X are shown in Fig. 13.26.

The refractive index for the extraordinary ray is imaginary when

$$1 - Y < X < 1 - Y^2,$$

but for $X > 1 - Y^2$ it is again real, and another set of refractive index surfaces can be constructed. The cross-section of these by the magnetic meridian is shown in Fig. 13.26. Now some values of μ are greater than unity, so that there are some curves outside the unit circle. There is no curve for $X = 0$. The curves shrink to zero when $X = 1 + Y$. When $X = 1$, $\mu = 1$ (see §6.2), so the curve for $X = 1$ includes the unit circle, but it also includes the two straight line segments PP_1 and QQ_1 which are along the direction of the earth's magnetic field. This is because for purely longitudinal propagation there is an ambiguity in the value of μ when $X = 1$. The points P and Q in Fig. 13.26 are the same as those in Fig. 13.22*a*. The two sets of surfaces have these points in common.

In general for a wave-packet incident from below, the extraordinary ray could not reach the levels depicted in Fig. 13.26, for it would be reflected at or below the level where $X = 1 - Y$. There is a special case, however, when the angle of incidence is such that the line AB in Fig. 13.22*a* goes through the

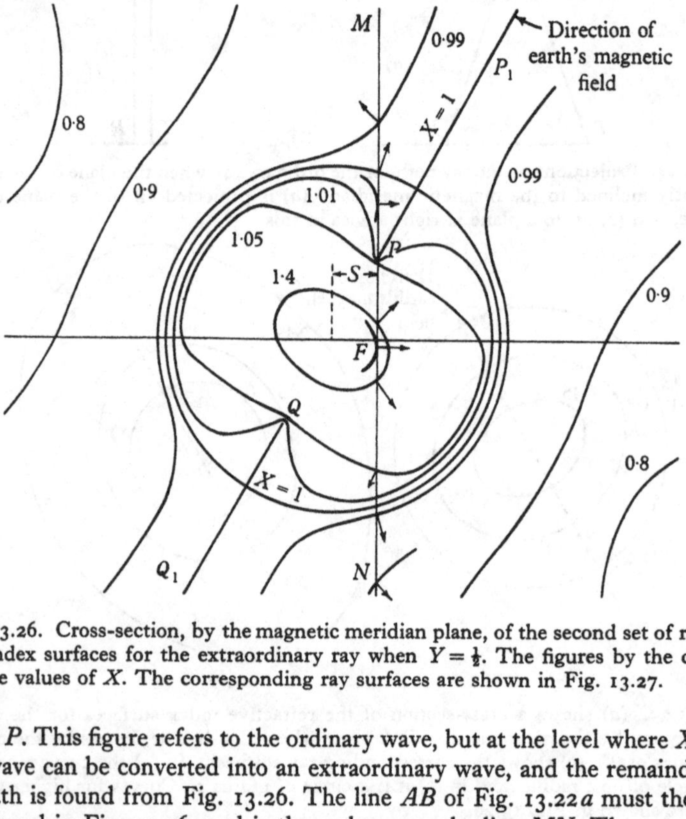

Fig. 13.26. Cross-section, by the magnetic meridian plane, of the second set of refractive index surfaces for the extraordinary ray when $Y = \frac{1}{2}$. The figures by the curves are the values of X. The corresponding ray surfaces are shown in Fig. 13.27.

point P. This figure refers to the ordinary wave, but at the level where $X = 1$ the wave can be converted into an extraordinary wave, and the remainder of its path is found from Fig. 13.26. The line AB of Fig. 13.22*a* must then be continued in Fig. 13.26, and is there shown as the line MN. The two parts PM and PN correspond to different rays. The part PM refers to a ray travelling downwards and arriving at the level $X = 1$ from above. This ray cannot be present since there is no source of energy above the level $X = 1$. Thus the incident ordinary ray splits into two parts at the level corresponding to P. One part is the reflected ordinary ray whose path is found from Fig. 13.22*a* as explained in § 13.22. The other is an extraordinary ray whose path is found from the line PN in Fig. 13.26. The ray travels on up to the level where PN just touches a refractive index surface at F. Here the ray is horizontal so this is a level of reflection. The ray then comes down again and crosses the level where $X = 1$, but can never leave the ionosphere because the level $X = 0$ does not appear in Fig. 13.26. The line PN successively cuts curves of increasing radius and the ray becomes more and more horizontal. The wave-packet therefore travels nearly horizontally just above the level where $X = 1 - Y^2$. In fact its energy would ultimately be absorbed because of electron collisions, which

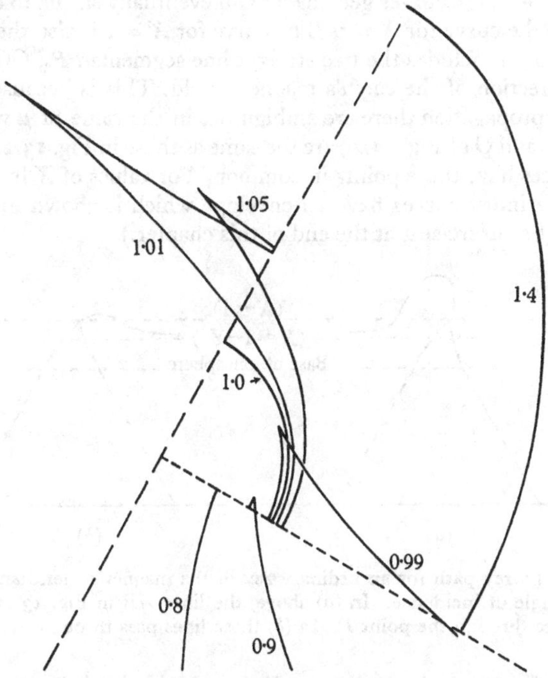

Fig. 13.27. Ray surfaces for the extraordinary ray when $Y = \frac{1}{2}$. These correspond to the parts of refractive index surfaces (Fig. 13.26) in the quadrant clockwise from the line PP_1. Note that the ray is not necessarily in the same quadrant as the wave-normal. The figures by the curves are the values of X.

cannot be neglected in this case. The ray path is shown in Fig. 13.28*a*. The behaviour is similar when the line AB in Figs. 13.22 and 13.26 goes through the point Q. The ray path for this case is shown in Fig. 13.28*b*.

In the absence of collisions the conversion of an ordinary to an extraordinary wave would only occur when the incident ray was exactly in the magnetic meridian and then only when the angle of incidence had the precise value $\arcsin \pm l[Y/(1 \pm Y)]^{\frac{1}{2}}$ (+ for the point P, and − for Q in Fig. 13.26; see § 13.9 and (13.31)). If collisions were allowed for, there would be a small cone of angles of incidence near the critical value, for which the ray would split into two components. The theory of this splitting has not been worked out very fully. The effect of collisions on the refractive index surfaces has been discussed by Forsgren (1951).

13.24 The refractive index surfaces for the extraordinary ray when $Y > 1$

When $Y > 1$, the cross-sections by the magnetic meridian plane of the refractive index surfaces for the extraordinary ray are as shown in Fig. 13.29. These correspond to the branch b of the curves in Figs. 6.7 and 6.8. As X approaches $1+Y$, the curves get smaller and eventually shrink to a point. The unit circle is the curve for $X = 0$. The curve for $X = 1$ is also the unit circle, but in addition it includes the two straight line segments PP_2, QQ_2 which are along the direction of the earth's magnetic field. This is because for purely longitudinal propagation there are ambiguities in the value of μ when $X = 1$. The points P and Q of Fig. 13.29 are the same as those in Fig. 13.22*a*. The two sets of surfaces have these points in common. For values of X less than unity the refractive index curves have an envelope which is shown in the inset of Fig. 13.29. (See exercise 2 at the end of this chapter.)

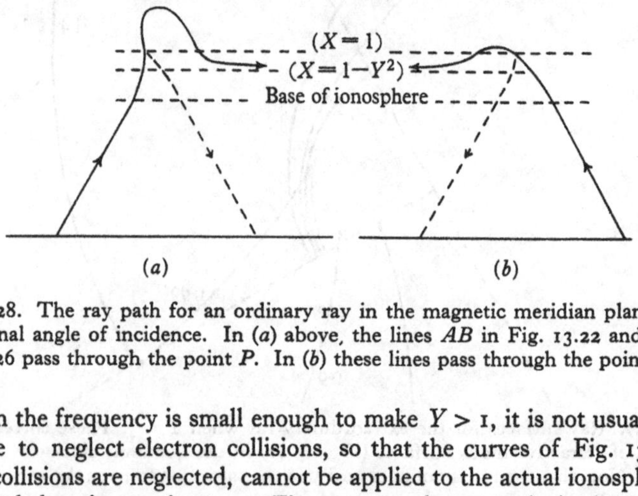

Fig. 13.28. The ray path for an ordinary ray in the magnetic meridian plane at the transitional angle of incidence. In (*a*) above, the lines AB in Fig. 13.22 and MN in Fig. 13.26 pass through the point P. In (*b*) these lines pass through the point Q.

When the frequency is small enough to make $Y > 1$, it is not usually permissible to neglect electron collisions, so that the curves of Fig. 13.29, in which collisions are neglected, cannot be applied to the actual ionosphere. It is, nevertheless, instructive to use Fig. 13.29 to plot ray paths in the magnetic meridian, for a fictitious ionosphere in which collisions are absent. This is done by Poeverlein's construction exactly as in §§ 13.21 and 13.22. A vertical line AB is drawn at a distance S from the origin. Six examples are shown in Fig. 13.30. For the line A_1B_1 (Fig. 13.29) the ray path is as in Fig. 13.30*a*, and for the line A_2B_2 it is as in Fig. 13.30*b*. In both these cases when the upgoing ray passes the level where $X = 1$, its direction is parallel to the incident ray, and when the downgoing ray passes this level its direction is parallel to the emergent ray below the ionosphere. Fig. 13.30*c* shows the ray path for the line A_3B_3 in Fig. 13.29. This crosses the straight segment PP_2 to the right of the point R and to the left of the right branch of the envelope, so that it does not intersect this branch. When the ray first reaches the level where $X = 1$, the corresponding point is S in Fig. 13.30*c*. The ray direction is here suddenly reversed, in exactly the way described in § 13.22 for the ordinary ray, and there is a 'Spitze' in the ray path. The ray next goes down to a level where it is horizontal, corresponding to the point T in the figures. Then it rises again, passes through the level where $X = 1$ (point U, Figs. 13.29 and 13.30c), and again becomes horizontal at the level corresponding to point V. Thereafter it travels downwards, again passing the level where $X = 1$, and eventually reaches the ground.

Fig. 13.30d shows the ray path when A_3B_3 passes between P and R in Fig. 13.29, and Fig. 13.30e is the path when A_3B_3 passes through R. The detailed

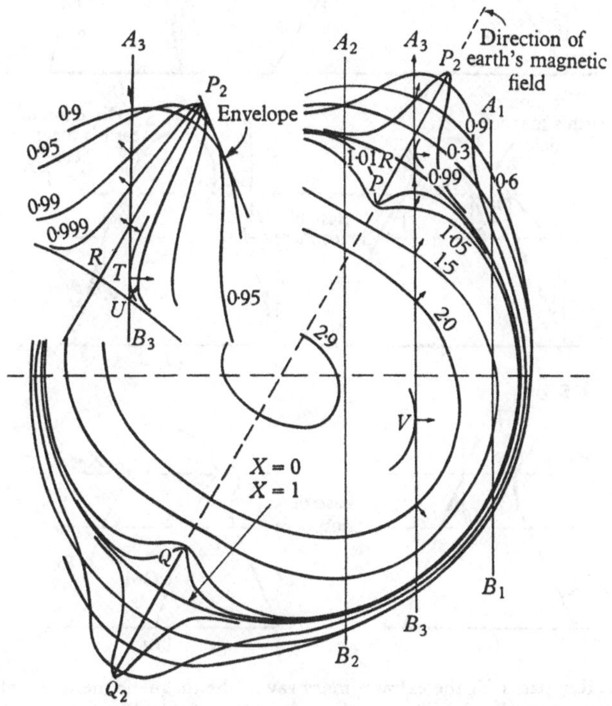

Fig. 13.29. Cross-section, by the magnetic meridian plane, of the refractive index surfaces for the extraordinary ray when $Y = 2$. The figures by the curves are the values of X. Inset is an enlarged view of part of the diagram.

tracing of these rays may be left as an exercise for the reader, who should verify that at the level where $X = 1$, corresponding to the point R, the ray is inclined to the vertical at an angle $\frac{1}{2}\pi - 2\theta_I$ where θ_I is the angle of incidence.

When the line AB in Fig. 13.29 passes through the point P the ray path can be traced as before until the representative point reaches P. The extraordinary wave can here be converted into an ordinary wave, for the point P is common to the surfaces in Figs. 13.29 and 13.22. The rest of the ray path is then found from Fig. 13.22, and the ray returns to earth as an ordinary wave although it started out as an extraordinary wave. The complete ray path in this case is as shown in Fig. 13.30*f*. The electron collisions would make the wave split into an ordinary and extraordinary wave for a cone of angles of incidence near the critical value, as explained at the end of § 13.23.

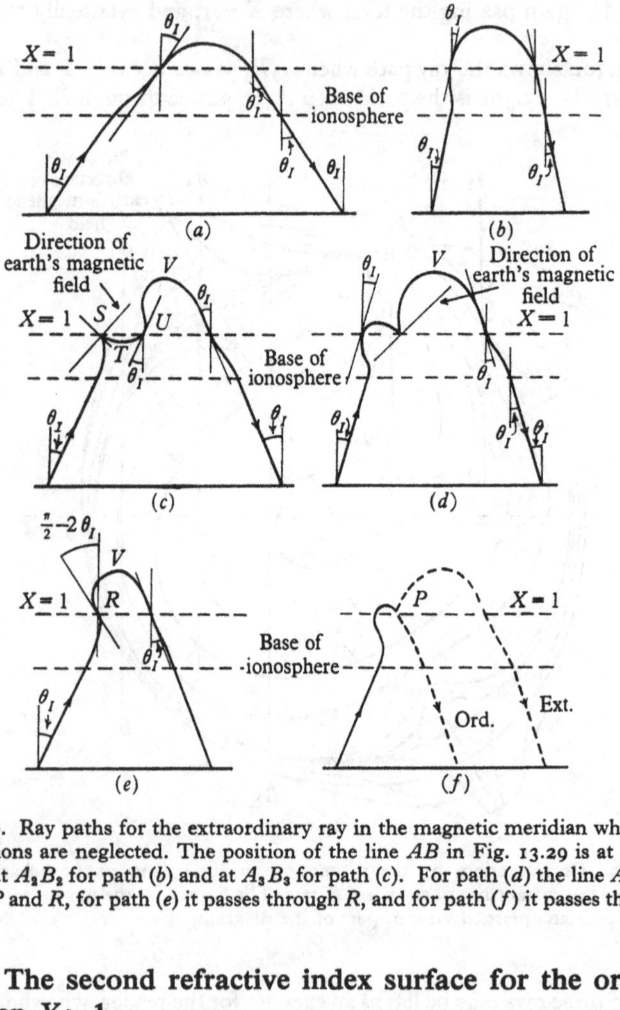

Fig. 13.30. Ray paths for the extraordinary ray in the magnetic meridian when $Y > 1$ and collisions are neglected. The position of the line AB in Fig. 13.29 is at A_1B_1 for path (*a*), at A_2B_2 for path (*b*) and at A_3B_3 for path (*c*). For path (*d*) the line AB passes between P and R, for path (*e*) it passes through R, and for path (*f*) it passes through P.

13.25 The second refractive index surface for the ordinary ray when $Y > 1$

The refractive index surfaces for the ordinary and extraordinary rays, when the plane of incidence is the magnetic meridian and when $Y > 1$, have been described in §§ 13.22 and 13.24 respectively. The refractive index for the ordinary ray is imaginary when $1 < X < (Y^2-1)/(Y_L^2-1)$, but for $X > (Y^2-1)/(Y_L^2-1)$ it is again real, and a third set of refractive index surfaces can be constructed. The cross-section of these by the magnetic meridian is shown in Fig. 13.31. Now all values of μ are greater than unity, and the unit circle is not one of the curves. There is no curve for $X = 0$. The curve for $X = 1$ is the two straight line segments from P_2 and Q_2 out to infinity, along the direction of the earth's magnetic field. The points $P_2 Q_2$ in Fig. 13.31 are the same as those in Fig. 13.29, and the two sets of surfaces have these points in common. There are no surfaces in this diagram when $Y_L < 1$, and this sets an upper limit to the angle ϑ between the wave normal and the earth's magnetic field, namely $\cos\vartheta = 1/Y$. In the case illustrated $Y = 2$, and the upper limit is 60°. When Y is very large, this upper limit approaches 90°.

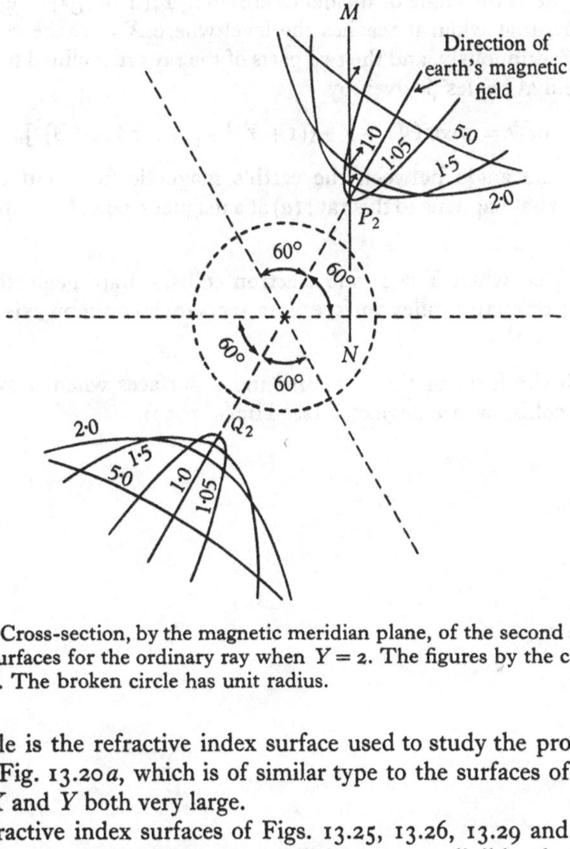

Fig. 13.31. Cross-section, by the magnetic meridian plane, of the second set of refractive index surfaces for the ordinary ray when $Y = 2$. The figures by the curves are the values of X. The broken circle has unit radius.

An example is the refractive index surface used to study the propagation of whistlers, Fig. 13.20*a*, which is of similar type to the surfaces of Fig. 13.31, but with X and Y both very large.

The refractive index surfaces of Figs. 13.25, 13.26, 13.29 and 13.31 have been drawn assuming that electron collisions are negligible. At frequencies small enough to make $Y > 1$ it is rarely permissible to neglect collisions, so that great care is needed in using Figs. 13.29 and 13.31 to draw conclusions about actual propagation problems. If, nevertheless, collisions are neglected, then for a wave-packet incident from below, the ordinary ray could not, in general, reach the levels depicted in Fig. 13.31, for it would be reflected at or below the level where $X = 1$. There is a special case, however, when the angle of incidence is such that the line AB in Fig. 13.29 goes through the point P_2. Fig. 13.29 refers to the extraordinary ray but, if the representative point reaches P_2, the ray can be converted to an ordinary ray at the level where $X = 1$, and the remainder of its path is found from Fig. 13.31 (compare § 13.23). The line AB of Fig. 13.29 must then be continued in Fig. 13.31 and is there shown as the line MN. The ray would continue to travel upwards indefinitely, provided that X continues to increase as the height increases.

Examples

1. An ordinary ray is incident on the ionosphere obliquely in the magnetic meridian plane at the angle of incidence arc sin$\{l[Y/(Y+1)]^{\frac{1}{2}}\}$ (see §§ 13.9 and 13.23). Show that when it reaches the level where $X = 1$, the ray direction changes discontinuously, and the two parts of the ray are inclined to the earth's magnetic field at angles ψ given by

$$\cot\psi = 2\cot\Theta[1+Y\pm\{(1+Y)^2+\tfrac{1}{2}(1+Y)\tan^2\Theta\}^{\frac{1}{2}}],$$

where Θ is the angle between the earth's magnetic field and the vertical ($l = \sin\Theta$). What happens to this ray: (*a*) at a magnetic pole, (*b*) at the magnetic equator?

2. Prove that when $Y > 1$, and electron collisions are neglected, the envelope of the refractive index surfaces intersects their common axis at an angle arc tan$\{2(Y-1)\}^{\frac{1}{2}}$.

3. Sketch the form of the refractive index surfaces when heavy ions are present but collisions are neglected (see Hines, 1957).

CHAPTER 14

THE GENERAL PROBLEM OF RAY TRACING

14.1 Introduction

In the last chapter the properties of the refractive index surfaces have been discussed, for a homogeneous medium. They could be used to find the ray direction when the direction of the wave-normal was known, and thus the general shape of the ray path could be found for a slowly varying, horizontally stratified ionosphere. The exact ray paths were not found, however, and in the preceding sections the paths were described only qualitatively. To find the ray paths exactly (13.17) and (13.20) can be used. These are differential equations relating the coordinates x, y, z of a point on the ray path. Thus, by integration, the path can be found. The expressions (13.20) use the Booker quartic, which can only be applied to a horizontally stratified ionosphere, and the earth must be assumed to be flat. In this chapter, therefore, an alternative method will be described to which these severe restrictions do not apply. The method uses the properties of the refractive index surface and the ray surface at each point. It is a mathematical expression of the ideas of the last part of ch. 13.

The object of the following sections is to show how the differential equations for a ray path may be found for any coordinate system and for any slowly varying magnetoionic medium. The derivation of the equation is a useful illustration of the ideas of the last chapter, but discussion of their solutions is beyond the scope of this book (see references in § 14.4).

Haselgrove (1954) has shown how the differential equations for a very general coordinate system can be derived from Fermat's principle of stationary time. This states that the time of travel of a wave-front along the ray path between two points is a maximum or minimum for small variations of the path. A wave-front means some particular feature of the wave, such as a wave-crest. Fermat's principle does not apply to the time of travel of a wave-packet.

A wave-front travels with the ray velocity V_R (see § 13.19). Hence if ds is an element of the ray path, the time of travel is $\int \frac{1}{V_R}\, ds$ which is equal to $\frac{1}{c}\int \mathscr{M} ds$, from (13.61), where $\mathscr{M}$ is the 'ray refractive index', and Fermat's principle may be stated, in the notation of the calculus of variations, thus:

$$\delta \int \mathscr{M} ds = 0, \tag{14.1}$$

where the limits of the integral are at the end-points of the path.

The truth of Fermat's principle for an anisotropic medium will be established by working in Cartesian coordinates. But (14.1) is independent of the particular coordinate system used, and once it is proved it can be applied to more general coordinates. For the proof the idea of the Eikonal function (§ 14.3) is used to derive the canonical equations for a ray path in Cartesian coordinates. But first the equations of the refractive index surface and the ray surface must be found.

In most of this chapter electron collisions are neglected and the refractive index n is the same as its real part μ. The effect of a small collision frequency is considered in § 14.7.

14.2 Equations of the refractive index surface and the ray surface

To construct the refractive index surface for the point (x, y, z) in the ionosphere we consider a three-dimensional space which will be called the 'refractive index space'. In it we use Cartesian coordinates p_x, p_y, p_z, whose axes are parallel to the x, y, z axes, respectively, of ordinary space. Consider a wave-front at the point x, y, z, and draw a line from the origin in the refractive index space parallel to the wave-normal and of length equal to the refractive index μ. Let p_x, p_y, p_z be the coordinates of its end-point. Then $(p_x^2+p_y^2+p_z^2)^{\frac{1}{2}} = \mu$, and the direction cosines of the wave-normal are

$$\frac{p_x}{(p_x^2+p_y^2+p_z^2)^{\frac{1}{2}}}, \quad \frac{p_y}{(p_x^2+p_y^2+p_z^2)^{\frac{1}{2}}}, \quad \frac{p_z}{(p_x^2+p_y^2+p_z^2)^{\frac{1}{2}}}. \tag{14.2}$$

Now let the wave-normal successively take all possible directions. Then the locus of the point p_x, p_y, p_z is the refractive index surface. The refractive index μ is given by the Appleton–Hartree formula and is a function of position (x, y, z) and of the direction of the wave-normal. This will be indicated by writing it $\mu(x, y, z; p_x, p_y, p_z)$ where p_x, p_y, p_z occur only in the combinations (14.2). Then the equation of the refractive index surface may be written

$$G(x, y, z; p_x, p_y, p_z) \equiv \frac{(p_x^2+p_y^2+p_z^2)^{\frac{1}{2}}}{\mu(x, y, z; p_x, p_y, p_z)} = 1. \tag{14.3}$$

The ray direction is the normal to the refractive index surface (§ 13.18) and has direction cosines proportional to $\partial G/\partial p_x$, $\partial G/\partial p_y$, $\partial G/\partial p_z$. Now x, y, z are the coordinates of the point where a wave-front intersects the ray, so that the components of the ray velocity V_R are $\dot{x}$, $\dot{y}$, $\dot{z}$, where a dot represents d/dt and t is time. Hence $\dot{x}$, $\dot{y}$, $\dot{z}$ are proportional to $\partial G/\partial p_x$, $\partial G/\partial p_y$, $\partial G/\partial p_z$ respectively. To find the constant of proportionality choose the x-axis to be parallel to the wave-normal of any particular wave-packet. Then $p_y = p_z = 0$, $p_x = \mu$, and the partial derivative of each of the quantities (14.2) with respect to p_x is zero, so that $(\partial/\partial p_x)(1/\mu) = 0$, and $\partial G/\partial p_x = 1/\mu$. Now $\dot{x}$ is the component of the ray velocity in the direction of the wave-normal, that is c/μ. Hence $\partial G/\partial p_x = \dot{x}/c$ and the required constant of proportionality is $1/c$. The components of the ray velocity are therefore given by

$$\dot{x} = c\frac{\partial G}{\partial p_x}, \quad \dot{y} = c\frac{\partial G}{\partial p_y}, \quad \dot{z} = c\frac{\partial G}{\partial p_z}. \tag{14.4}$$

An equation for the ray surface may be found in a similar way as follows. Consider a three-dimensional space, to be called the 'ray space', in which the Cartesian coordinates are $\dot{x}$, $\dot{y}$, $\dot{z}$, and whose axes are parallel to the x, y, z axes of ordinary space. Consider a wave-front at the point x, y, z and draw a line from the origin in the ray space, parallel to the ray direction and of length equal to the ray velocity V_R. Let $\dot{x}$, $\dot{y}$, $\dot{z}$ be the coordinates of its end-point. Then $(\dot{x}^2+\dot{y}^2+\dot{z}^2)^{\frac{1}{2}} = V_R$, and the direction cosines of the ray are

$$\frac{\dot{x}}{(\dot{x}^2+\dot{y}^2+\dot{z}^2)^{\frac{1}{2}}}, \quad \frac{\dot{y}}{(\dot{x}^2+\dot{y}^2+\dot{z}^2)^{\frac{1}{2}}}, \quad \frac{\dot{z}}{(\dot{x}^2+\dot{y}^2+\dot{z}^2)^{\frac{1}{2}}}. \tag{14.5}$$

Let the wave-normal at x, y, z successively take all possible directions. Then the locus of the point $\dot{x}$, $\dot{y}$, $\dot{z}$ in the ray space, is the ray surface. Now $V_R = c/\mathcal{M}$ where $\mathcal{M}$ is the ray refractive index (13.61), and is a function of position (x, y, z) and of the direction of the ray. This will be indicated by writing it $\mathcal{M}(x, y, z; \dot{x}, \dot{y}, \dot{z})$, where $\dot{x}$, $\dot{y}$, $\dot{z}$ occur only in the combinations (14.5). Then the equation of the ray surface may be written

$$F(x, y, z; \dot{x}, \dot{y}, \dot{z}) \equiv \frac{1}{c}(\dot{x}^2+\dot{y}^2+\dot{z}^2)^{\frac{1}{2}}\,\mathcal{M}(x, y, z; \dot{x}, \dot{y}, \dot{z}) = 1. \tag{14.6}$$

The direction of the wave-normal is the normal to the ray surface (§ 13.19) and therefore has direction cosines proportional to $\partial F/\partial \dot{x}$, $\partial F/\partial \dot{y}$, $\partial F/\partial \dot{z}$. Hence p_x, p_y, p_z are proportional to $\partial F/\partial \dot{x}$, $\partial F/\partial \dot{y}$, $\partial F/\partial \dot{z}$ respectively. The constant of proportionality may be shown to be c, by a similar method to that used for (14.4), that is by choosing the x-axis to be parallel to the ray. Hence

$$p_x = c\frac{\partial F}{\partial \dot{x}}, \quad p_y = c\frac{\partial F}{\partial \dot{y}}, \quad p_z = c\frac{\partial F}{\partial \dot{z}}. \tag{14.7}$$

The equations (14.4) and (14.7) express the reciprocal properties of the refractive index surface and the ray surface. The function G, (14.3), is homogeneous in the variables p_x, p_y, p_z and of degree 1. Hence by Euler's theorem for homogeneous functions (see, for example, Gibson, 1929, p. 412)

$$p_x\frac{\partial G}{\partial p_x}+p_y\frac{\partial G}{\partial p_y}+p_z\frac{\partial G}{\partial p_z} = 1, \tag{14.8}$$

whence, from (14.4)

$$\dot{x}p_x+\dot{y}p_y+\dot{z}p_z = c. \tag{14.9}$$

Similarly, the function F, (14.6), is homogeneous in $\dot{x}$, $\dot{y}$, $\dot{z}$, and Euler's theorem again gives (14.9), which is an alternative way of expressing the reciprocal properties of the two surfaces.

For a given point in the ionosphere the refractive index surface is given by (14.3) and could be plotted using the Appleton–Hartree formula. Some examples have already been given (Figs. 13.22*a*, 13.25*a*, 13.26, 13.29, 13.31). The equation (14.3) is of the fourth degree in p_x, p_y, p_z and is a kind of generalized form of the Booker quartic.

The plotting of the ray surface is less easy. Its equation could be obtained formally by eliminating p_x, p_y, p_z from the four equations (14.3) and (14.4). The result is an equation of the twelfth degree in $\dot{x}$, $\dot{y}$, z, which is too complicated to give specifically, and ray surfaces must usually be found by numerical means. Some examples are shown in Figs. 13.22*b*, 13.25*b* and 13.27.

14.3 The Eikonal function

The properties derived so far for the refractive index surface and the ray surface apply only to a single point in the ionosphere, or rather to a fictitious homogeneous medium with the properties of the ionosphere at the point considered. They tell us nothing yet about the path of a ray in an inhomogeneous ionosphere. In the present section, therefore, we consider how a wave travels in an inhomogeneous but slowly varying ionosphere.

In a homogeneous medium we know that one possible solution of Maxwell's equations represents a plane wave with its wave-normal in a given direction. This problem was discussed fully in ch. 5. There the

z-axis was chosen to be parallel to the wave-normal, and it was shown that the electric field $\mathbf{E}$ of the wave is given by $\mathbf{E} = \mathbf{E}_0 \exp(-ik\mu z)$, where $\mathbf{E}_0$ is a constant vector and μ is the refractive index, given by the Appleton–Hartree formula. For a different set of Cartesian axes this expression for $\mathbf{E}$ becomes

$$\mathbf{E} = \mathbf{E}_0 \exp\{-ik(xp_x + yp_y + zp_z)\}, \tag{14.10}$$

where p_x, p_y, p_z are proportional to the direction cosines of the wave-normal with respect to the new axes, and

$$p_x^2 + p_y^2 + p_z^2 = \mu^2. \tag{14.11}$$

Thus p_x, p_y, p_z are the same as the quantities defined in § 14.2. We now consider whether a solution similar to (14.10) is possible in an inhomogeneous but slowly varying medium.

The problem of the transition from a homogeneous to a slowly varying medium was discussed in § 9.5 where the 'phase memory' concept was introduced, and a similar argument may be used here. Thus in a small distance δx the phase of the wave (14.10) changes by $kp_x\delta x$. If p_x is a slowly varying function of position, then the change of phase in traversing a distance x would be

$$k\int_0^x p_x dx.$$

More generally the difference in the phase of the wave at the origin and at the point (x, y, z) is

$$k\left(\int_0^x p_x dx + \int_0^y p_y dy + \int_0^z p_z dz\right).$$

This is nothing more than an extension to three dimensions of the phase-memory concept.

We therefore assume that there exists a function

$$\mathscr{E}(x, y, z; p_x, p_y, p_z) = k\left(\int^x p_x dx + \int^y p_y dy + \int^z p_z dz\right), \tag{14.12}$$

such that $\mathbf{E} = \mathbf{E}_0 \exp(-i\mathscr{E})$ is a solution of Maxwell's equations at each point of the medium. If this $\mathbf{E}$ were substituted in Maxwell's equations it would give terms containing p_x, p_y, p_z, which are the same as for a homogeneous medium and must therefore cancel. In addition there are terms involving spatial derivatives of p_x, p_y, p_z, and X. In a slowly varying medium these are very small. The assumption made here is that they are small enough to be neglected. It is this assumption which makes possible the use of 'ray theory'. For a very full discussion see Suchy (1952, 1953, 1954).

The function $\mathscr{E}$ is sometimes called the Eikonal function and may be thought of as the spatial part of the phase of the wave. An example of it for an isotropic medium was mentioned in § 9.5.

From (14.12) it is clear that:

$$p_x = \frac{1}{k}\frac{\partial \mathscr{E}}{\partial x}, \quad p_y = \frac{1}{k}\frac{\partial \mathscr{E}}{\partial y}, \quad p_z = \frac{1}{k}\frac{\partial \mathscr{E}}{\partial z}, \tag{14.13}$$

whence

$$\mathbf{p} = \frac{1}{k}\operatorname{grad}\mathscr{E}, \tag{14.14}$$

where $\mathbf{p}$ is the vector whose components are p_x, p_y, p_z. Hence

$$\frac{\partial p_z}{\partial y} = \frac{\partial p_y}{\partial z}, \quad \frac{\partial p_x}{\partial z} = \frac{\partial p_z}{\partial x}, \quad \frac{\partial p_y}{\partial x} = \frac{\partial p_x}{\partial y}, \tag{14.15}$$

or $\operatorname{curl}\mathbf{p} = 0$. This equation may be considered as a result of applying Maxwell's equations at each point in the ionosphere.

14.4 The canonical equations for a ray, and the generalisation of Snell's law

Now let x, y, z be the coordinates of the point where a ray intersects a wave-front. For example, x, y, z may be the coordinates of a wave-crest as it travels along a ray. At every point, p_x, p_y, p_z must satisfy the equation $G(x, y, z; p_x, p_y, p_z) = 1$ (14.3) for the refractive index surfaces. Hence $dG/dx = 0$, that is, on a ray:

$$\frac{\partial G}{\partial x} + \frac{\partial G}{\partial p_x}\frac{\partial p_x}{\partial x} + \frac{\partial G}{\partial p_y}\frac{\partial p_y}{\partial x} + \frac{\partial G}{\partial p_z}\frac{\partial p_z}{\partial x} = 0. \tag{14.16}$$

Now use (14.15) and (14.4). Then

$$\frac{\partial G}{\partial x} + \frac{1}{c}\left(\frac{\partial p_x}{\partial x}\dot{x} + \frac{\partial p_x}{\partial y}\dot{y} + \frac{\partial p_x}{\partial z}\dot{z}\right) = 0, \tag{14.17}$$

which gives $\partial G/\partial x = -(1/c)(dp_x/dt)$ where d/dt applies to the moving point (x, y, z). We shall write $dp_x/dt = \dot{p}_x$. The conditions $dG/dy = 0$, $dG/dz = 0$ give similar relations. Hence

$$\dot{p}_x = -c\frac{\partial G}{\partial x}, \quad \dot{p}_y = -c\frac{\partial G}{\partial y}, \quad \dot{p}_z = -c\frac{\partial G}{\partial z}. \tag{14.18}$$

The equations (14.4) and (14.18) together are called the canonical equations for a ray (in Cartesian coordinates). Their resemblance to Hamilton's canonical equations in dynamics is obvious.

Equations (14.18) are a kind of generalization of Snell's law. For example, consider a horizontally stratified ionosphere in which μ is a function of z only. Then with the notation of § 13.4 $p_x = S_1$, $p_y = S_2$, $p_z = q$. Now (14.3) shows that $\partial G/\partial x = \partial G/\partial y = 0$, whence (14.18) gives $dS_1/dt = dS_2/dt = 0$, so that S_1 and S_2 are constant along a ray, which is Snell's law.

The canonical equations (14.4) and (14.18) have been used by Haselgrove (1957) for tracing the ray path in the magnetic meridian when the ionosphere is horizontally stratified and the earth is assumed to be flat. Then μ is independent of the horizontal coordinates x and y. We take the x–z-plane to be the plane of incidence so that p_y is initially zero and the second equation (14.18) shows that it is always zero. The expression (14.3) for G then becomes

$$G(z; p_x, p_z) = \frac{(p_x^2+p_z^2)^{\frac{1}{2}}}{\mu(z; p_x, p_z)}. \tag{14.19}$$

It is convenient to introduce the angle θ between the wave-normal and the vertical as a new variable. Then

$$p_x = \mu\sin\theta, \quad p_z = \mu\cos\theta, \quad \tan\theta = p_x/p_z. \tag{14.20}$$

The refractive index μ depends on p_x and p_z only in the combinations $p_x(p_x^2+p_z^2)^{-\frac{1}{2}} = \sin\theta$, and $p_z(p_x^2+p_z^2)^{-\frac{1}{2}} = \cos\theta$. It is therefore a function only of z and θ and may be written $\mu(z, \theta)$. Then

$$\frac{\partial G}{\partial p_x} = \frac{p_x}{\mu(p_x^2+p_z^2)^{\frac{1}{2}}} - \frac{(p_x^2+p_z^2)^{\frac{1}{2}}}{\mu^2}\frac{\partial\mu}{\partial\theta}\frac{\partial\theta}{\partial p_x} = \frac{\sin\theta}{\mu} - \frac{\cos\theta}{\mu^2}\frac{\partial\mu}{\partial\theta} \tag{14.21}$$

and

$$\frac{\partial G}{\partial p_z} = \frac{p_z}{\mu(p_x^2+p_z^2)^{\frac{1}{2}}} - \frac{(p_x^2+p_z^2)^{\frac{1}{2}}}{\mu^2}\frac{\partial\mu}{\partial\theta}\frac{\partial\theta}{\partial p_z} = \frac{\cos\theta}{\mu} + \frac{\sin\theta}{\mu^2}\frac{\partial\mu}{\partial\theta}. \tag{14.22}$$

For propagation in the magnetic meridian plane $\partial\mu/\partial p_y$ and therefore $\partial G/\partial p_y$ are zero. Further $\partial G/\partial x$ is zero and the first equation (14.18) shows that p_x is constant on a ray (this is Snell's law). Hence

$$\frac{dp_x}{dt} = \sin\theta\frac{d\mu}{dt} + \mu\cos\theta\frac{d\theta}{dt} = 0, \tag{14.23}$$

so that

$$\frac{dp_z}{dt} = \cos\theta\frac{d\mu}{dt} - \mu\sin\theta\frac{d\theta}{dt} = -\frac{d\theta}{dt}\left(\mu\sin\theta + \frac{\mu\cos^2\theta}{\sin\theta}\right) = -\frac{\mu}{\sin\theta}\frac{d\theta}{dt}. \tag{14.24}$$

Equations (14.21), (14.22) and (14.24) are substituted in the first and third equations (14.4) and the last equation (14.18) respectively, and lead to

$$\left.\begin{aligned}\frac{dx}{dt} &= \frac{c}{\mu^2}\left(\mu\sin\theta - \cos\theta\frac{\partial\mu}{\partial\theta}\right),\\ \frac{dz}{dt} &= \frac{c}{\mu^2}\left(\mu\cos\theta + \sin\theta\frac{\partial\mu}{\partial\theta}\right),\\ \frac{d\theta}{dt} &= -\frac{c}{\mu^2}\sin\theta\frac{\partial\mu}{\partial z},\end{aligned}\right\} \tag{14.25}$$

which are the equations used by Haselgrove (1957) (with different notation) for computing ray paths on a digital computer.

When the earth cannot be regarded as flat, it is convenient to use spherical polar coordinates with the centre of the earth as origin. Equations for this case have also been given by Haselgrove (1954) (see Examples 2 and 3 at the end of this chapter) and by Maeda and Kimura (1956) who used them, with an ingenious approximation, to calculate the paths of whistlers.

The problem of ray tracing in the ionosphere when the earth's magnetic field is allowed for, has been discussed by Marcou, Pfister and Ulwick (1958) using a slightly different but essentially equivalent method. See also Al'pert (1948).

14.5 Other relations between the equations for the ray surface and the refractive index surface

To establish Fermat's principle we need a set of relations between the functions G, (14.3) and F, (14.6). Since $F = 1$ at each point of a ray, $dF/dx = 0$ so that

$$\begin{aligned}\frac{\partial F}{\partial x} &= -\left(\frac{\partial F}{\partial \dot{x}}\frac{\partial \dot{x}}{\partial x} + \frac{\partial F}{\partial \dot{y}}\frac{\partial \dot{y}}{\partial x} + \frac{\partial F}{\partial \dot{z}}\frac{\partial \dot{z}}{\partial x}\right)\\ &= -\frac{1}{c}\left(p_x\frac{\partial \dot{x}}{\partial x} + p_y\frac{\partial \dot{y}}{\partial x} + p_z\frac{\partial \dot{z}}{\partial x}\right)\end{aligned} \tag{14.26}$$

from equation (14.7). Similarly, $dG/dx = 0$ which leads to (14.17). Now add (14.26) and (14.17) and use (14.9). Then

$$\frac{\partial F}{\partial x} + \frac{\partial G}{\partial x} = -\frac{1}{c}\frac{\partial}{\partial x}(\dot{x}p_x + \dot{y}p_y + \dot{z}p_z) = 0. \tag{14.27}$$

Similar results hold for the y- and z-derivatives. Hence

$$\frac{\partial F}{\partial x} = -\frac{\partial G}{\partial x}, \quad \frac{\partial F}{\partial y} = -\frac{\partial G}{\partial y}, \quad \frac{\partial F}{\partial z} = -\frac{\partial G}{\partial z}. \tag{14.28}$$

14.6 Fermat's principle

From (14.28), (14.18) and (14.7) we have, for the moving point where a wave-front intersects a ray,

$$\left.\begin{aligned} \frac{\partial F}{\partial x} &= -\frac{1}{c}\frac{dp_x}{dt} = -\frac{d}{dt}\left(\frac{\partial F}{\partial \dot{x}}\right), \\ \text{and similarly} \qquad \frac{\partial F}{\partial y} &= -\frac{d}{dt}\left(\frac{\partial F}{\partial \dot{y}}\right), \\ \frac{\partial F}{\partial z} &= -\frac{d}{dt}\left(\frac{\partial F}{\partial \dot{z}}\right). \end{aligned}\right\} \tag{14.29}$$

Now these are Euler's equations in the calculus of variations, and they show that

$$\delta \int_{(A)}^{(B)} F\,dt = 0, \tag{14.30}$$

where the integral is taken along a ray joining the fixed points A and B. Equation (14.30) means that if the integral is evaluated for a number of different paths between A and B its value is a maximum or minimum for the actual ray path. The function F is given by (14.6) which is the equation of the ray surface, and must hold at all points of the true ray path (although it need not hold at points on the varied paths). Let s be distance measured along the ray. Then $(ds/dt)^2 = \dot{x}^2 + \dot{y}^2 + \dot{z}^2$, and substitution of (14.6) in (14.30) gives

$$\delta \int_{(A)}^{(B)} \mathscr{M} ds = 0 \tag{14.31}$$

for a ray path, where $\mathscr{M}$ is the 'ray refractive index', defined by (13.65). Equations (14.31) or (14.30) show that the time of travel of a wave-front from A to B along a ray is stationary with respect to small variations of the path. This is Fermat's principle of stationary time.

It should again be stressed that this time of travel is not the time that it takes for a wave-packet to go from A to B, but is the time of travel of some feature of the wave such as a wave-crest. The time of travel of a wave-packet depends on the equivalent path, and is discussed in the next section.

Fermat's principle has here been derived using Cartesian coordinates. But the result (14.31) is independent of any particular coordinate system, and can therefore be used with other more general systems. The canonical equations for a ray in a very general coordinate system have been given by Haselgrove (1954).

14.7 Equivalent path and absorption

It was shown in § 13.18 that a wave-packet travels in the direction of the ray with the group velocity U given by (13.59). The time of travel over a given path is equal to P'/c where P' is called the equivalent path. Hence

$$\frac{P'}{c} = \int_{(A)}^{(B)} \frac{ds}{U}$$

and (13.59) shows that $$P' = \int_{(A)}^{(B)} \mu' \cos\alpha\, ds, \qquad (14.32)$$

where μ' is the group refractive index (§ 12.3), and α is the angle between the ray and the wave-normal. Now

$$ds = V_R dt = \frac{c}{\mathscr{M}} dt = \frac{c}{\mu\cos\alpha} dt.$$

Hence (14.32) becomes

$$P' = c\int_{(A)}^{(B)} \frac{\mu'}{\mu} dt = c\int_{(A)}^{(B)} \left(1 + \frac{f}{\mu}\frac{\partial\mu}{\partial f}\right) dt \qquad (14.33)$$

from (10.6) (where f is the frequency).

If electron collisions are allowed for, the refractive index is complex, and the wave is attenuated as it travels. Provided that the collision-frequency is small, however, the preceding theory can be applied to the real part of the refractive index and the ray path can be found by ignoring the imaginary part. Once the path is known, the attenuation can be found as follows. Let $-\chi$ be the imaginary part of the refractive index. This determines the attenuation in the direction of the wave-normal. The attenuation along the ray path is therefore

$$A = \exp\left\{k\int_{(A)}^{(B)} \chi\cos\alpha\, ds\right\}, \qquad (14.34)$$

where the integral is along the ray path, and A is the factor by which the amplitude is reduced. Now $ds\cos\alpha = (c dt/\mu)$ where μ is the real part of the refractive index. Hence (14.34) may be written

$$\log A = kc\int_{(A)}^{(B)} \frac{\chi}{\mu} dt. \qquad (14.35)$$

The six canonical differential equations (such as (14.4) and (14.18) in Cartesian coordinates or a similar set in other coordinates) can be used to find the ray path, by a step-by-step integration process in a digital computer (see Haselgrove, 1954, and § 14.4). Here the independent variable is t, the time of travel of a wave-front along a ray. At the same time it is convenient to find the equivalent path and the attenuation. This may be done by expressing (14.33) and (14.35) as additional differential equations, thus:

$$\frac{dP'}{dt} = c\left(1 + \frac{f}{\mu}\frac{\partial \mu}{\partial f}\right), \quad \frac{d(\log A)}{dt} = kc\frac{\chi}{\mu}. \tag{14.36}$$

The six canonical equations and the two equations (14.36) can then be integrated at the same time.

14.8 The problem of finding the maximum usable frequency

The need for predicting the M.U.F. for given transmitting and receiving stations has already been mentioned in ch. 11 and methods of estimating it were described in §§ 11.13 and 11.15. There the effect of the earth's magnetic field was neglected, and it was shown that by making measurements of the $h'(f)$ curve at vertical incidence it is then possible to find the M.U.F. for a given path exactly, for example by Newbern Smith's method (§ 11.15). The methods made use of Martyn's theorem and Breit and Tuve's theorem (§ 11.9), but these are not valid when the earth's magnetic field is allowed for, and the methods of ch. 11 were therefore only approximate.

There is no simple method of computing the M.U.F. using the full theory. The problem could be solved, in principle, using the ray-tracing methods described in the preceding sections, but the work would be exceedingly laborious. It would be necessary, first, to adopt some model for the ionosphere. This could be predicted from measurements at vertical incidence. Then for a given frequency and bearing a number of rays could be traced for various angles of incidence and the horizontal ranges found. There would in general be a minimum range or 'skip distance'. By repeating this at other bearings a polar diagram could be plotted showing how the skip distance depends on bearing. Such diagrams would have to be constructed for a number of frequencies. Then for a given receiving station the M.U.F. is the greatest frequency for which the station is beyond the skip distance. A programme of this kind would, however, be exacting even for the fastest digital computer at present available.

A possible alternative approach is to find the factor by which Martyn's theorem is in error. The extent to which this depends on frequency and on the electron density profile is not yet known, but if the dependence is slight it may still be possible to use the methods of ch. 11 with suitable correcting factors. Haselgrove (1957) has begun a study of this problem. Let P' be the true equivalent path of an oblique ray and let P'_V be its value as predicted by Martyn's theorem. Then the percentage error is $100(P' - P'_V)/P'_V$. Haselgrove gives curves showing how this error depends on angle of incidence for rays in the magnetic meridian only, but for various frequencies. She uses the ordinary ray and assumes that the electron density profile is a parabola, and that the earth is flat. This work establishes an important method which could be extended to other electron density profiles, other azimuths and to a curved earth.

Examples

1. Show that when the earth's magnetic field is neglected, the canonical equations for a ray in Cartesian coordinates become

$$\frac{\dot{x}}{c}=\frac{p_x}{\mu^2},\quad \frac{\dot{y}}{c}=\frac{p_y}{\mu^2},\quad \frac{\dot{z}}{c}=\frac{p_z}{\mu^2},$$

$$\frac{\dot{p}_x}{c}=\frac{1}{\mu}\frac{\partial\mu}{\partial x},\quad \frac{\dot{p}_y}{c}=\frac{1}{\mu}\frac{\partial\mu}{\partial y},\quad \frac{\dot{p}_z}{c}=\frac{1}{\mu}\frac{\partial\mu}{\partial z}.$$

2. Let r, θ, ϕ be the spherical polar coordinates of a point on a ray. Let the wave-normal at this point make an angle ψ with the radius and let the plane containing the wave-normal and the radius make an angle η with the plane ϕ = constant. Thus ψ, η are the polar angles of the wave-normal referred to axes whose origin is at the point r, θ, ϕ. Let

$$p_r=\mu\cos\psi,\quad p_\theta=\mu\sin\psi\cos\eta,\quad p_\phi=\mu\sin\psi\sin\eta.$$

Show that when the earth's magnetic field is neglected, the canonical equations for a ray may be written:

$$\frac{\dot{r}}{c}=\frac{p_r}{\mu^2},\qquad \frac{\dot{p}_r}{c}=\frac{1}{\mu}\frac{\partial\mu}{\partial r}+\frac{p_\theta^2+p_\phi^2}{r\mu^2};$$

$$\frac{\dot{\theta}}{c}=\frac{p_\theta}{r\mu^2},\qquad \frac{\dot{p}_\theta}{c}=\frac{1}{r\mu}\frac{\partial\mu}{\partial\theta}+\frac{p_\phi^2\cos\theta-p_r p_\theta}{r\mu^2};$$

$$\frac{\dot{\phi}}{c}=\frac{p_\phi}{r\sin\theta\,\mu^2},\qquad \frac{\dot{p}_\phi}{c}=\frac{1}{r\mu\sin\theta}\frac{\partial\mu}{\partial\phi}-\frac{p_\phi(p_r+p_\theta\cot\theta)}{r\mu^2}.$$

3. In the general case when the earth's magnetic field is allowed for, the refractive index μ is a function of r, θ, ϕ and a homogeneous function of p_r, p_θ, p_ϕ (notation as in the preceding example), and may be written $\mu(r,\theta,\phi;p_r,p_\theta,p_\phi)$. Let

$$G(r,\theta,\phi;p_r,p_\theta,p_\phi)=\frac{(p_r^2+p_\theta^2+p_\phi^2)^{\frac{1}{2}}}{\mu(r,\theta,\phi;p_r,p_\theta,p_\phi)}.$$

Show that the canonical equations for a ray in spherical polar coordinates are

$$\frac{\dot{r}}{c}=\frac{\partial G}{\partial p_r},\qquad \frac{\dot{p}_r}{c}=\frac{1}{\mu}\frac{\partial\mu}{\partial r}+\frac{p_\theta}{r}\frac{\partial G}{\partial p_\theta}+\frac{p_\phi}{r}\frac{\partial G}{\partial p_\phi};$$

$$\frac{\dot{\theta}}{c}=\frac{1}{r}\frac{\partial G}{\partial p_\theta},\qquad \frac{\dot{p}_\theta}{c}=\frac{1}{r\mu}\frac{\partial\mu}{\partial\theta}-\frac{p_\theta}{r}\frac{\partial G}{\partial p_r}+\frac{p_\theta\cot\theta}{r}\frac{\partial G}{\partial p_\phi};$$

$$\frac{\dot{\phi}}{c}=\frac{1}{r\sin\theta}\frac{\partial G}{\partial p_\phi},\qquad \frac{\dot{p}_\phi}{c}=\frac{1}{r\mu\sin\theta}\frac{\partial\mu}{\partial\phi}-\frac{p_\phi}{r}\frac{\partial G}{\partial p_r}-\frac{p_\phi\cot\theta}{r}\frac{\partial G}{\partial p_\theta}.$$

CHAPTER 15

THE AIRY INTEGRAL FUNCTION, AND THE STOKES PHENOMENON

15.1 Introduction

In ch. 9 it was shown that at most levels in a slowly varying ionosphere the propagation of a radio wave is described by approximate solutions of the differential equations, known as the W.K.B. solutions. The earth's magnetic field was neglected, and it was shown that the W.K.B. solutions are good approximations to the true solutions when the conditions (9.29) (for vertical incidence) or (9.61) and (9.72) (for oblique incidence) are fulfilled. These conditions fail, however, near levels where $\mathfrak{n} = 0$ (vertical incidence) or $q = 0$ (oblique incidence), which are the levels of reflection. It was stated in § 9.8 that at the reflection level the upgoing W.K.B. solution is converted into the downgoing W.K.B. solution with the same amplitude factor, and this led to the expressions (9.37) (vertical incidence) or (9.62) (oblique incidence) for the reflection coefficient. The justification for this statement is examined in ch. 16 and it is shown to require only a small modification. For this purpose the solutions of the differential equations must be examined in more detail for levels near a zero of $\mathfrak{n}$ or q. Throughout this chapter the earth's magnetic field is neglected.

The detailed study of the differential equations near levels where $q = 0$ also leads to a better understanding of the process of reflection, which is closely connected with the 'Stokes phenomenon'.

15.2 Linear gradient of electron density associated with an isolated zero of q

If electron collisions are neglected, q^2 is related to the electron number density N thus

$$q^2 = C^2 - X, \tag{15.1}$$

where X is given by (3.5) and is directly proportional to N, and $C = \cos\theta_I$ where θ_I is the angle between the incident wave-normal and the vertical. Thus q depends on z through X, and is zero when $X = C^2$.

The simplest example of a zero of q occurs when X is a slowly increasing monotonic function of z, as shown in Fig. 15.1a. Then q^2 is a decreasing monotonic function of z as shown in Fig. 15.1b. It is zero where $z = z_0$, and to a first approximation the variation of q^2 with z may be taken as linear near z_0, so that

$$q^2 \approx -\mathfrak{a}(z-z_0), \tag{15.2}$$

where $\mathfrak{a}$ is a constant. The right side of (15.2) may be regarded as the first term in the Taylor expansion for q^2 about the point z_0. It was shown in § 9.6 that the W.K.B. solutions in this case are good approximations provided that $|z-z_0|$ exceeds a certain minimum value, M. The argument was there given for vertical incidence but may be applied to oblique incidence when $\mathfrak{n}$ is replaced by q. In the present chapter it is assumed that (15.2) may be used for q^2 for all values of $|z-z_0|$ from zero up to M.

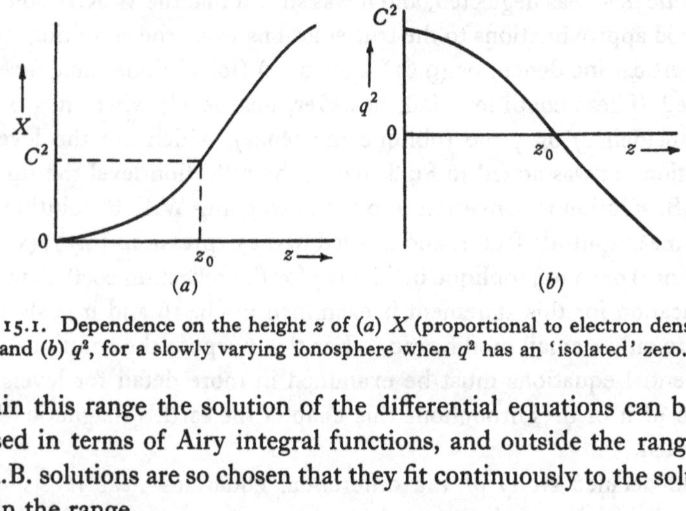

Fig. 15.1. Dependence on the height z of (a) X (proportional to electron density), and (b) q^2, for a slowly varying ionosphere when q^2 has an 'isolated' zero.

Within this range the solution of the differential equations can be expressed in terms of Airy integral functions, and outside the range the W.K.B. solutions are so chosen that they fit continuously to the solution within the range.

If q^2 is given exactly by (15.2) there is only one value of z, namely z_0 which makes $q = 0$. If, however, the line in Fig. 15.1*b* is slightly curved where $z = z_0$ then (15.2) may be replaced by

$$q^2 \approx -\mathfrak{a}(z-z_0)+\mathfrak{b}(z-z_0)^2, \tag{15.3}$$

where $\mathfrak{b}$ determines the curvature at $z = z_0$. Now q is zero both at $z = z_0$ and at $z = z_0+\mathfrak{a}/\mathfrak{b}$. This case is illustrated in Figs. 15.2*a* and *b*, and could occur, for a frequency less than the penetration-frequency, when the electron density has a maximum value. If $\mathfrak{b}$ is small enough to be neglected in (15.3), then the second zero is at a great distance $\mathfrak{a}/\mathfrak{b}$ from the first, provided that $\mathfrak{a}$ is not small.

Hence the condition that the variation of q^2 with z shall be nearly linear, near a zero of z, is equivalent to saying that this zero is at a great distance from the next nearest zero. In other words, the zero of q must be isolated. A more exact statement of this condition is given in § 16.8.

If $\mathfrak{a}$ is very small, the two zeros of (15.3) can be close together even when $\mathfrak{b}$ is small. This case could occur when the frequency is only just below the penetration-frequency. Here the W.K.B. solutions fail for a range of z which includes both zeros. It is then necessary to study the differential equations when q^2 varies according to the parabolic law. This leads to the phenomena of partial penetration and reflection, and is discussed in § 17.5. For frequencies just above the penetration-frequency q has no zeros at any real value of z, but there are zeros in the complex z plane. This case also is discussed in § 17.5.

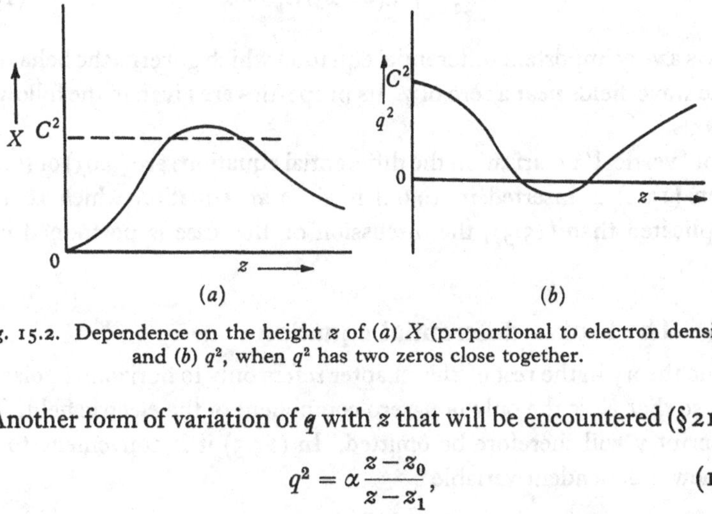

Fig. 15.2. Dependence on the height z of (a) X (proportional to electron density), and (b) q^2, when q^2 has two zeros close together.

Another form of variation of q with z that will be encountered (§ 21.15) is

$$q^2 = \alpha\frac{z-z_0}{z-z_1}, \tag{15.4}$$

where α, z_0, z_1 are constants. Here q^2 has only one zero where $z = z_0$, but it has an infinity where $z = z_1$. The curvature of the curve of q^2 versus z at $z = z_0$ is $-2\alpha/(z_0-z_1)^2$. If this is to be small, $|z_0-z_1|$ must be large, so that the infinity and the zero must be well separated. When the expression (15.2) is used for q^2, therefore, it implies that the zero of q at $z = z_0$ is at a great distance both from other zeros and from infinities.

15.3 The differential equation for horizontal polarisation and oblique incidence

The differential equations to be satisfied by the wave-fields for oblique incidence were derived in § 9.11 for the case when the earth's magnetic field is neglected. It was shown that they separate into two sets, one for fields in which the electric vector is everywhere horizontal, and the other for fields in which the electric vector is everywhere in the plane of incidence. These two cases are usually said to apply to horizontal and vertical polarisation respectively. For horizontal polarisation the equations are (9.57) or (9.58), where q is given by (9.56). If the effect of electron collisions is neglected, so that $Z = 0$, then (9.56) reduces to (15.1). If $S^2 = 0$ so that $C^2 = 1$, (9.56) gives $q^2 = \mathfrak{n}^2$, and the differential equations are the same as those for vertical incidence (9.8) or (9.2) and (9.6). The solution for vertical incidence is therefore a special case of the solution for oblique incidence with horizontal polarisation.

When (15.2) is inserted in (9.58), it gives

$$\frac{d^2E_y}{dz^2} - k^2\mathfrak{a}(z-z_0)\,E_y = 0. \tag{15.5}$$

This is a very important differential equation which governs the behaviour of the wave-fields near a zero of q. Its properties are given in the following sections.

For 'vertical' polarisation the differential equations are (9.63) or (9.64). When (15.2) is inserted in (9.64) it gives an equation which is more complicated than (15.5); the discussion of this case is postponed until § 16.12.

15.4 The Stokes differential equation

The theory in the rest of this chapter refers only to horizontal polarisation, so that E_y is the only non-zero component of the electric field. The subscript y will therefore be omitted. In (15.5) it is convenient to use the new independent variable

$$\zeta = (k^2\mathfrak{a})^{\frac{1}{3}}(z-z_0), \tag{15.6}$$

where the value of $(k^2\mathfrak{a})^{\frac{1}{3}}$ is taken to be real and positive. ζ is thus a measure of the height. Then (15.5) becomes

$$\frac{d^2E}{d\zeta^2} = \zeta E. \tag{15.7}$$

This is known as the Stokes differential equation.† It has no singularities when ζ is finite and its solution must therefore be finite and single valued, except possibly at $\zeta = \infty$. It is necessary to study the properties of these solutions for both real and complex values of ζ.

† The name 'Stokes differential equation' is sometimes given to the equation $(d^2E/dx^2)+xE = 0$ which is easily converted to the form (15.7) by the substitution $x = (-1)^{\frac{1}{3}}\zeta$.

Solutions of (15.7) can be found as series in ascending powers of ζ, by the standard method. Assume that a solution is $E = a_0 + a_1\zeta + a_2\zeta^2 + \dots$. Substitute this in (15.7) and equate powers of ζ. This gives relations between the constants a_0, a_1, a_2, etc., and leads finally to

$$E = a_0\left\{1 + \frac{\zeta^3}{3.2} + \frac{\zeta^6}{6.5.3.2} + \frac{\zeta^9}{9.8.6.5.3.2} + \dots\right\} + a_1\left\{\zeta + \frac{\zeta^4}{4.3} + \frac{\zeta^7}{7.6.4.3} + \frac{\zeta^{10}}{10.9.7.6.4.3} + \dots\right\}, \quad (15.8)$$

which contains the two arbitrary constants a_0 and a_1, and is therefore the most general solution. The series are convergent for all ζ, which confirms that every solution of (15.7) is finite, continuous and single valued. The two series separately have no particular physical significance. The values of the constants a_0 and a_1 for the functions Ai(ζ) and Bi(ζ) (see § 15.6) are as follows (see Jeffreys and Jeffreys, 1956; or Miller, 1946):

$$\left.\begin{aligned} &\text{For Ai}(\zeta): \quad a_0 = 3^{-\frac{2}{3}}/(-\tfrac{1}{3})!, \quad a_1 = -3^{-\frac{1}{3}}/(-\tfrac{2}{3})!,\\ &\text{For Bi}(\zeta): \quad a_0 = 3^{-\frac{1}{6}}/(-\tfrac{1}{3})!, \quad a_1 = 3^{\frac{1}{6}}/(-\tfrac{2}{3})!. \end{aligned}\right\} \quad (15.9)$$

15.5 Qualitative discussion of the solutions of the Stokes equation

Equation (15.7) shows that if E and ζ are real, $d^2E/d\zeta^2$ is real. If $dE/d\zeta$ is also real, then E must be real for all real values of ζ. This is also apparent from (15.8). It is of interest to trace the curve of E versus ζ when ζ and E are both real. If ζ is positive, (15.7) shows that the curvature of the curve has the same sign as E. Hence the curve is convex towards the line $E = 0$. If the curve is traced step by step from $\zeta = 0$ upwards, there are three possibilities which are illustrated in Fig. 15.3. First, if the initial slope is sufficiently negative, the curve can cross the line $E = 0$ (curve A). When it does so, the sign of the curvature changes, and for higher values of ζ the magnitude of the slope must increase indefinitely. Hence the curve moves indefinitely further from the line $E = 0$ and can never cross it again. Secondly, if the initial slope is positive or only slightly negative, the slope can become zero before the curve reaches the line $E = 0$ (curve B). Thereafter the curve moves indefinitely further from this line and can never cross it. In both these cases E ultimately becomes indefinitely large as ζ increases. The third possibility occurs for one particular negative value of the initial slope. The curve then approaches the line $E = 0$, and never actually reaches it, but gets closer and closer to it (curve C). The slope must always have the opposite sign to E, and E must become smaller and smaller as ζ increases. This last case is of particular importance, and the solution Ai(ζ), described later, has this property.

When ζ is negative, the curvature has the opposite sign to E. Hence the curve is concave towards the line $E = 0$. If the curve is traced step by step from $\zeta = 0$ towards increasingly negative values of ζ, it must always curve towards the line $E = 0$ and eventually cross it. The sign of the curvature then changes so that the curve again bends towards the line $E = 0$, and crosses it again. In all cases therefore the function E is oscillatory and as ζ becomes more negative the curvature for a given $|E|$ increases, so that the oscillation gets more rapid and its amplitude gets smaller. This is illustrated for all three curves in Fig. 15.3.

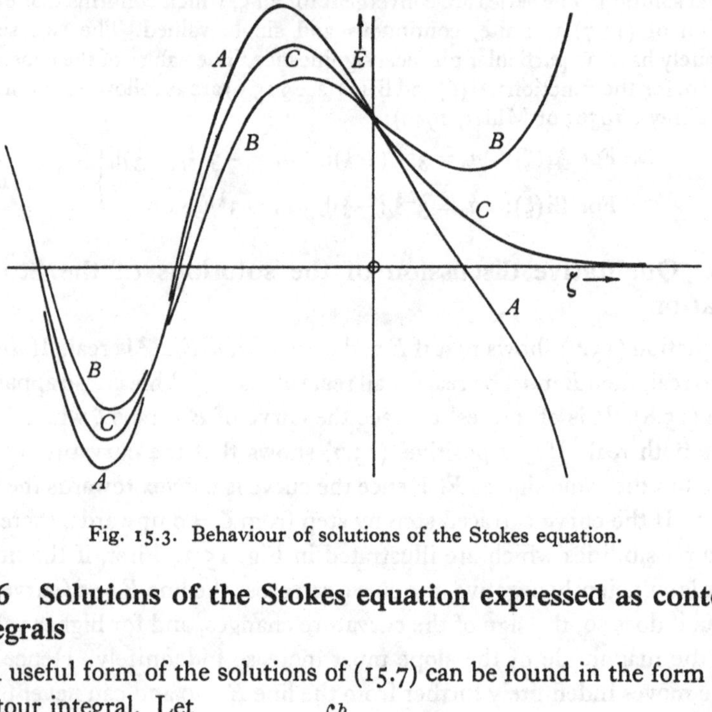

Fig. 15.3. Behaviour of solutions of the Stokes equation.

15.6 Solutions of the Stokes equation expressed as contour integrals

A useful form of the solutions of (15.7) can be found in the form of a contour integral. Let

$$E = \int_a^b e^{\zeta t} f(t)\, dt, \tag{15.10}$$

where t is a complex variable and the integral is evaluated along some contour in the complex t plane, whose end-points a and b are to be specified later. Since (15.10) must satisfy (15.7), it is necessary that

$$\int_a^b (t^2 - \zeta) f(t)\, e^{\zeta t}\, dt = 0.$$

The second term can be integrated by parts which gives

$$-e^{\zeta t} f(t)\Big|_a^b + \int_a^b \left\{ t^2 f(t) + \frac{df(t)}{dt} \right\} e^{\zeta t}\, dt = 0. \tag{15.11}$$

The limits a, b are to be chosen so that the first term vanishes at both limits. Then (15.11) is satisfied if

$$\frac{df(t)}{dt} + t^2 f(t) = 0, \tag{15.12}$$

that is if
$$f(t) = A e^{-\frac{1}{3}t^3}, \tag{15.13}$$

where A is a constant. The limits a and b must therefore be chosen so that $e^{-\frac{1}{3}t^3+\zeta t}$ is zero for both. This is only possible if

$$|t| \to \infty \quad \text{and} \quad 2\pi n - \tfrac{1}{2}\pi < 3 \arg t < 2\pi n + \tfrac{1}{2}\pi,$$

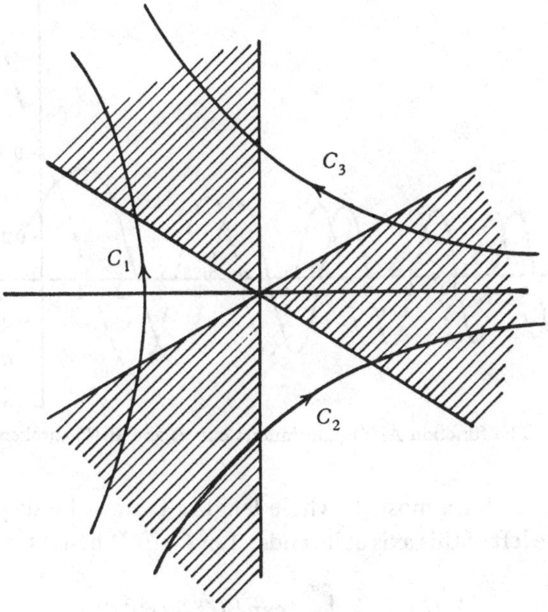

Fig. 15.4. The complex t-plane.

where n is an integer. Fig. 15.4 is a diagram of the complex t-plane, and a and b must each be at ∞ in one of the shaded sectors. They cannot both be in the same sector, for then the integral (15.10) would be zero. Hence the contour may be chosen in three ways, as shown by the three curves C_1, C_2, C_3. This might appear at first to give three independent solutions of (15.7). But the contour C_1 can be distorted so as to coincide with the two contours $C_2 + C_3$, so that

$$\int_{C_1} = \int_{C_2} + \int_{C_3},$$

and therefore there are only two independent solutions.

Jeffreys and Jeffreys (1956) define the two functions Ai (ζ) and Bi (ζ) as follows:

$$\mathrm{Ai}(\zeta) = \frac{1}{2\pi i}\int_{C_1} \exp(-\tfrac{1}{3}t^3 + \zeta t)\, dt, \tag{15.14}$$

$$\mathrm{Bi}(\zeta) = \frac{1}{2\pi}\int_{C_2} \exp(-\tfrac{1}{3}t^3 + \zeta t)\, dt - \frac{1}{2\pi}\int_{C_3} \exp(-\tfrac{1}{3}t^3 + \zeta t)\, dt. \tag{15.15}$$

In (15.14) the contour C_1 can be distorted so as to coincide with the

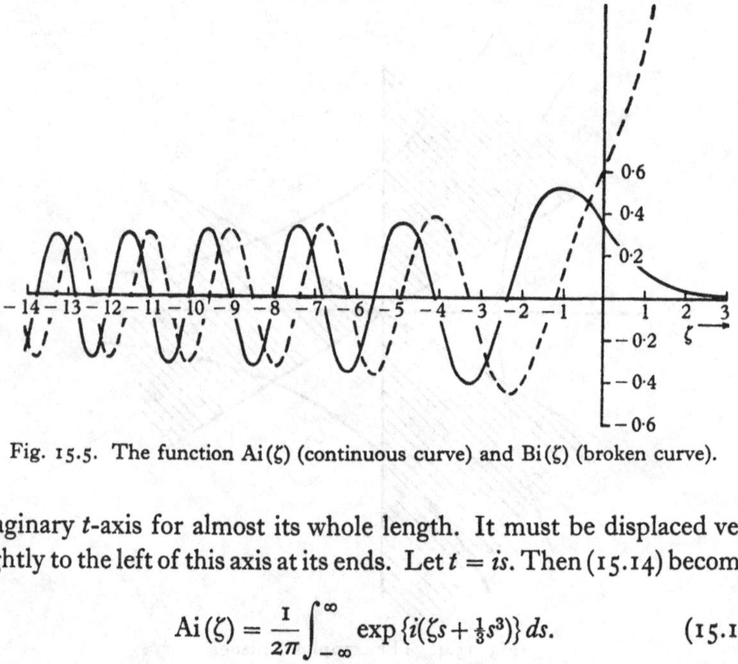

Fig. 15.5. The function Ai (ζ) (continuous curve) and Bi (ζ) (broken curve).

imaginary t-axis for almost its whole length. It must be displaced very slightly to the left of this axis at its ends. Let $t = is$. Then (15.14) becomes

$$\mathrm{Ai}(\zeta) = \frac{1}{2\pi}\int_{-\infty}^{\infty} \exp\{i(\zeta s + \tfrac{1}{3}s^3)\}\, ds. \tag{15.16}$$

Here the imaginary part of the integrand is an odd function of s, and contributes nothing to the integral, which may therefore be written

$$\mathrm{Ai}(\zeta) = \frac{1}{\pi}\int_{0}^{\infty} \cos(\zeta s + \tfrac{1}{3}s^3)\, ds. \tag{15.17}$$

Apart from a constant, this is the same as the expression used by Airy (1838, 1849). It is known as the Airy integral, and Ai (ζ) is called the Airy integral function.

The functions Ai (ζ) and Bi (ζ) are shown in Fig. 15.5.

15.7 Solutions of the Stokes equation expressed as Bessel functions

The Stokes equation (15.7) can be converted into Bessel's equation by changing both the dependent and independent variables. First let

$$\tfrac{2}{3}\zeta^{\frac{3}{2}} = i\xi. \tag{15.18}$$

Then (15.17) becomes

$$\frac{d^2E}{d\xi^2}+\frac{1}{3\xi}\frac{dE}{d\xi}+E=0. \tag{15.19}$$

Next let

$$E = \xi^{\frac{1}{3}}\mathscr{E} = (-\tfrac{2}{3}i)^{\frac{1}{3}}\zeta^{\frac{1}{2}}\mathscr{E}. \tag{15.20}$$

Then (15.19) becomes

$$\frac{d^2\mathscr{E}}{d\xi^2}+\frac{1}{\xi}\frac{d\mathscr{E}}{d\xi}+\left(1-\frac{1}{9\xi^2}\right)\mathscr{E}=0. \tag{15.21}$$

This is Bessel's equation of order one-third. The transformations (15.18) and (15.20) are complicated and introduce ambiguities. For example, (15.18) does not define ξ completely, since there is an ambiguity of the sign of the term on the left. Consequently the use of the Bessel function solutions is liable to lead to errors. Moreover, one solution of Bessel's equation (15.21) has a singularity where $\xi = 0$, whereas the solutions of the Stokes equation have no singularities. The transformations (15.18) and (15.20) in fact introduce compensating singularities. The relations between the Bessel functions and $\mathrm{Ai}(\zeta)$, $\mathrm{Bi}(\zeta)$ are given by Watson (1944) and Miller (1946).

15.8 Tables of the Airy integral functions

Some useful tables of functions closely related to $\mathrm{Ai}(\zeta)$, $\mathrm{Bi}(\zeta)$, have been prepared by the Staff of the Computation Laboratory at Cambridge, Mass. (1945). These bear the misleading title 'Tables of Modified Hankel functions of order one-third, and of their derivatives'. The functions tabulated, $h_1(z)$ and $h_2(z)$ and their derivatives, are not, strictly speaking, Hankel functions (the term Hankel function normally means a solution of Bessel's equation), but are solutions of the Stokes equation (15.7) with ζ replaced by $-\zeta$. They are related to Ai and Bi as follows:

$$h_1(z) = (12)^{\frac{1}{6}}e^{-\frac{1}{6}\pi i}\{\mathrm{Ai}(-z)-i\,\mathrm{Bi}(-z)\}, \tag{15.22}$$

$$h_2(z) = (12)^{\frac{1}{6}}e^{\frac{1}{6}\pi i}\{\mathrm{Ai}(-z)+i\,\mathrm{Bi}(-z)\}, \tag{15.23}$$

$$\mathrm{Ai}(z) = \tfrac{1}{2}(12)^{-\frac{1}{6}}\{e^{\frac{1}{6}\pi i}h_1(-z)+e^{-\frac{1}{6}\pi i}h_2(-z)\}, \tag{15.24}$$

$$\mathrm{Bi}(z) = \tfrac{1}{2}i(12)^{-\frac{1}{6}}\{e^{\frac{1}{6}\pi i}h_1(-z)-e^{-\frac{1}{6}\pi i}h_2(-z)\}. \tag{15.25}$$

The tables of $h_1(z)$, $h_2(z)$ are for both real and complex values of z, and may readily be used to find $\mathrm{Ai}(z)$, $\mathrm{Bi}(z)$. They cover the range $|z| \leqslant 6$ (approx.) at intervals of 0·1 in $\mathscr{R}(z)$ and $\mathscr{I}(z)$

Tables of $\mathrm{Ai}(z)$, $\mathrm{Bi}(z)$ and their derivatives and other related functions for real values of z are given by Miller (1946), who also gives a most useful summary of the properties of these functions. Tables of $\mathrm{Ai}(z)$, $\mathrm{Bi}(z)$ and their derivatives for complex values of z are given by Woodward and Woodward (1946). These are for a coarser interval (0·2 in $\mathscr{R}(z)$ and $\mathscr{I}(z)$), and cover a smaller range than the tables of $h_1(z)$, $h_2(z)$ mentioned above, but full instructions are given for bivariate interpolation.

15.9 The W.K.B. solutions of the Stokes equation

Approximate solutions of the Stokes equation (15.7) can be found by the W.K.B. method of § 9.5. Comparison of (15.7) with (9.8) shows that we may put $k^2\mathfrak{n}^2 = -\zeta$, and the two W.K.B. solutions (9.26) then lead to

$$E = \zeta^{-\frac{1}{4}} \exp(-\tfrac{2}{3}\zeta^{\frac{3}{2}}) \tag{15.26}$$

and

$$E = \zeta^{-\frac{1}{4}} \exp(\tfrac{2}{3}\zeta^{\frac{3}{2}}). \tag{15.27}$$

The condition that these are good approximations to solutions of (15.7) is given by (9.29), which becomes in this case

$$|\zeta| \gg \tfrac{1}{2}. \tag{15.28}$$

The approximations therefore fail when $|\zeta|$ is small, but provided it is large enough, any solution of (15.7) can be expressed with good accuracy as a linear combination of (15.26) and (15.27).

15.10 The Stokes phenomenon of the 'discontinuity of the constants'

The expressions (15.26) and (15.27), are multiple-valued functions, whereas any solution of (15.7) is single valued. Hence a solution of (15.7) cannot be represented by the same combination of (15.26) and (15.27) for all values of ζ. To illustrate this, consider the function $\mathrm{Ai}(\zeta)$. It was mentioned in § 15.5 that when ζ is real and positive, this function decreases steadily as ζ increases (Fig. 15.3, curve C), but never becomes zero. Hence the W.K.B. approximation for $\mathrm{Ai}(\zeta)$ when ζ is real and positive must be given by

$$\mathrm{Ai}(\zeta) \sim A\zeta^{-\frac{1}{4}} \exp(-\tfrac{2}{3}\zeta^{\frac{3}{2}}) \quad (\text{for } \arg\zeta = 0), \tag{15.29}$$

where A is a constant and the positive values of $\zeta^{-\frac{1}{4}}$, $\zeta^{\frac{3}{2}}$ are used. It cannot include a multiple of (15.27) for this would make the function increase indefinitely for large ζ. Now let $|\zeta|$ be kept constant, and let $\arg\zeta$ increase continuously from 0 to π, so that ζ becomes real and negative. Then the expression (15.29) becomes

$$Ae^{-\frac{1}{4}i\pi}\,|\zeta^{-\frac{1}{4}}| \exp\{\tfrac{2}{3}i|\zeta^{\frac{3}{2}}|\} \quad (\text{for } \arg\zeta = \pi), \tag{15.30}$$

which is a complex function. But $\mathrm{Ai}(\zeta)$ is real for all real values of ζ, so that (15.30) cannot represent $\mathrm{Ai}(\zeta)$ when ζ is real and negative, and the correct representation must include multiples of both (15.26) and (15.27). It is shown later than an additional term should be added to (15.29) when $\arg\zeta > \frac{2}{3}\pi$.

The W.K.B. approximation to the most general solution of the Stokes equation is

$$A\zeta^{-\frac{1}{4}}\exp\left(-\tfrac{2}{3}\zeta^{\frac{3}{2}}\right)+B\zeta^{-\frac{1}{4}}\exp\left(\tfrac{2}{3}\zeta^{\frac{3}{2}}\right), \tag{15.31}$$

but this can apply only to a part of the complex ζ-plane. If ζ moves out of this part, one of the arbitrary constants A or B must be changed. This phenomenon was discovered by Stokes (1858), and is called the 'Stokes phenomenon of the discontinuity of the arbitrary constants'.

15.11 Stokes lines and anti-Stokes lines

The exponents of both the exponentials in (15.31) are real if

$$\arg\zeta = 0, \quad \tfrac{2}{3}\pi, \quad \text{or} \quad \tfrac{4}{3}\pi, \tag{15.32}$$

and then if $|\zeta|$ becomes indefinitely large, one exponential becomes indefinitely small, and the other indefinitely large. Moreover, if $|\zeta|$ is kept constant, and $\arg\zeta$ is varied, both exponentials have maximum or minimum values when $\arg\zeta$ is given by (15.32), which defines three lines radiating from the origin of the complex ζ-plane. These are known as the 'Stokes lines'.

The exponents have equal moduli if

$$\arg\zeta = \tfrac{1}{3}\pi, \quad \pi, \quad \text{or} \quad \tfrac{5}{3}\pi, \tag{15.33}$$

and then if $|\zeta|$ becomes indefinitely large, the moduli of both exponentials remain equal to unity, but the terms oscillate more and more rapidly. The radial lines defined by (15.33) are called the 'anti-Stokes lines'. For all values of $\arg\zeta$ except those in (15.33) one exponential must have modulus greater than unity, and the other must have modulus less than unity. The larger exponential remains so, as long as ζ lies in the 120° sector between two anti-Stokes lines, and the corresponding term in (15.31) is called the 'Dominant' term. The other term, containing the smaller exponential, is called the 'Subdominant' term. Each term changes from dominant to subdominant, or the reverse, when ζ crosses an anti-Stokes line.

It was shown in § 15.10 that one of the constants A and B must change when ζ crosses some line in the complex ζ-plane. It is fairly clear that the constant in the dominant term cannot change, for this would give a detectable discontinuity in the function (15.31), which would mean that it would not even approximately satisfy the differential equation. Hence the constant which changes must be that in the subdominant term, and the most likely place for the change to occur is on a Stokes line, for there the ratio of the subdominant to the dominant term is smallest.

The solution (15.31) is approximate. It was shown by Stokes (1858) that when the constant in the subdominant term changes on a Stokes line, the size of the discontinuity is less than the error involved in the approximation.

In the following two sections it will be assumed that the arbitrary constant in the subdominant term may change on a Stokes line. These sections are descriptive and are intended to help towards an understanding of the Stokes phenomenon. A more formal mathematical proof of the results is given in § 15.18.

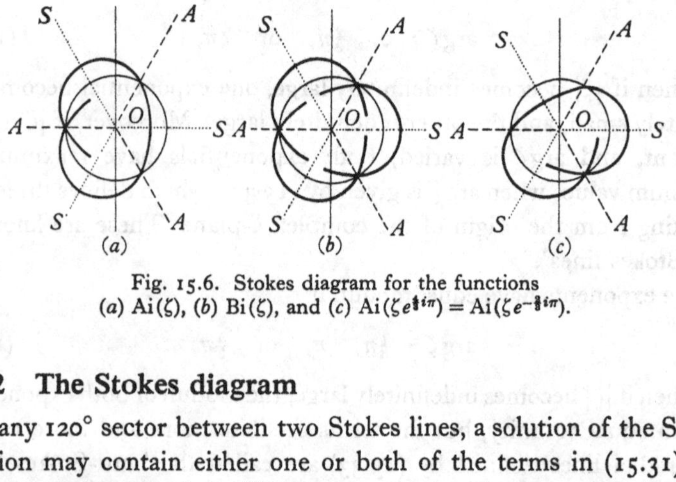

Fig. 15.6. Stokes diagram for the functions (*a*) $\mathrm{Ai}(\zeta)$, (*b*) $\mathrm{Bi}(\zeta)$, and (*c*) $\mathrm{Ai}(\zeta e^{\frac{2}{3}i\pi}) = \mathrm{Ai}(\zeta e^{-\frac{2}{3}i\pi})$.

15.12 The Stokes diagram

In any 120° sector between two Stokes lines, a solution of the Stokes equation may contain either one or both of the terms in (15.31). To indicate the nature of the W.K.B. or asymptotic approximation to a particular solution, Stokes (1858) used a diagram constructed as follows. Radial lines are drawn from a fixed point O (Fig. 15.6) in the directions of the Stokes and anti-Stokes lines, and labelled S or A respectively. A circle is drawn with centre O. In this diagram radial directions indicate values of $\arg \zeta$ but radial distances do not indicate values of $|\zeta|$. The plane of the diagram, therefore, is not the complex ζ-plane. Two heavy lines are drawn in the diagram, one inside and one outside the circle, and both cross the circle on the anti-Stokes lines. For some value of $\arg \zeta$ a radial line is drawn in the corresponding direction. If it crosses a heavy line inside the circle, this shows that there is a subdominant term in the asymptotic approximation. If it crosses a heavy line outside the circle, this shows that there is a dominant term.

Where the heavy line crosses a Stokes line inside the circle, the constant multiplying the associated subdominant term may change, and this is indicated by a break in the heavy line. The heavy line crosses the circle on the anti-Stokes lines because there the associated terms change from dominant to subdominant or the reverse. The heavy line must remain unbroken except where it meets a Stokes line inside the circle, for only there can the associated constant change. Moreover, this change can only occur if the dominant term is present. If it were absent, a change in the constant of the subdominant term would give a detectable discontinuity, since there is now no dominant term to mask it.

These properties are illustrated in Fig. 15.6*a*, which is the Stokes diagram for the function $\mathrm{Ai}(\zeta)$. In the sector $-\frac{2}{3}\pi < \arg\zeta < \frac{2}{3}\pi$ there is only one term in the asymptotic approximation, and this is subdominant in the sector $-\frac{1}{3}\pi < \arg\zeta < \frac{1}{3}\pi$. The constant cannot change on the Stokes line at $\arg\zeta = 0$ because there is no dominant term. The same term is dominant on the Stokes line at $\arg\zeta = \frac{2}{3}\pi$, and for greater values of $\arg\zeta$ there is also a subdominant term, which becomes dominant when $\arg\zeta = \pi$, and here the original term again becomes subdominant. This term disappears beyond the Stokes line at $\arg\zeta = \frac{4}{3}\pi$ or $-\frac{2}{3}\pi$.

Another example is given in Fig. 15.6*b*, which is the Stokes diagram for the function $\mathrm{Bi}(\zeta)$. Here there are both dominant and subdominant terms for all values of $\arg\zeta$, and the constant in the subdominant term changes on all three Stokes lines.

15.13 Definition of the Stokes constant

It is now necessary to determine by how much the constant in the subdominant term changes when a Stokes line is crossed. Suppose that in the sector $0 < \arg\zeta < \frac{2}{3}\pi$ (sector I of Fig. 15.7) a given solution of the Stokes equation has the asymptotic approximation

$$\zeta^{-\frac{1}{4}}[A_1 \exp(-\tfrac{2}{3}\zeta^{\frac{3}{2}}) + B_1 \exp(\tfrac{2}{3}\zeta^{\frac{3}{2}})]. \tag{15.34}$$

On the Stokes line at $\arg\zeta = \frac{2}{3}\pi$ (S_2 in Fig. 15.7) the first term is dominant, and hence for the sector $\frac{2}{3}\pi < \arg\zeta < \frac{4}{3}\pi$ (sector II) the asymptotic approximation is

$$\zeta^{-\frac{1}{4}}[A_1 \exp(-\tfrac{2}{3}\zeta^{\frac{3}{2}}) + B_2 \exp(\tfrac{2}{3}\zeta^{\frac{3}{2}})]. \tag{15.35}$$

The constant in the subdominant term has changed by $B_2 - B_1$. Now this change is zero if A_1 is zero. It cannot depend on B_1, for it would be unaltered if we added to (15.34) any multiple of the solution in which $A_1 = 0$. Since the differential equation is linear, $B_2 - B_1$ must be proportional to A_1, so that

$$B_2 - B_1 = \lambda_2 A_1, \tag{15.36}$$

where λ_2 is a constant called the 'Stokes constant' for the Stokes line at $\arg\zeta = \frac{2}{3}\pi$. It gives the change in the constant for the subdominant term when the Stokes line is crossed in a counter-clockwise direction. If the crossing is clockwise, the Stokes constant has the opposite sign. Stokes constants can be defined in a similar way for the other two Stokes lines. It will be shown in §§ 15.14 and 15.18 that for the Stokes equation (15.7) all three Stokes constants are equal to i.

15.14 Furry's derivation of the Stokes constants for the Stokes equation

The following derivation of the Stokes constants seems to have been used first by Furry (1947). The two terms in (15.34) are multiple-valued functions of ζ with a branch point at the origin. Hence we introduce a cut in the complex ζ-plane from 0 to $-\infty$ along the real axis, and take $-\pi \leqslant \arg\zeta \leqslant \pi$. Consider a solution whose asymptotic approximation is (15.34) in sector I of Fig. 15.7. Then in the top part of sector II it is

$$\zeta^{-\frac{1}{4}}[A_1 \exp(-\tfrac{2}{3}\zeta^{\frac{3}{2}}) + (B_1 + \lambda_2 A_1)\exp(\tfrac{2}{3}\zeta^{\frac{3}{2}})]. \tag{15.37}$$

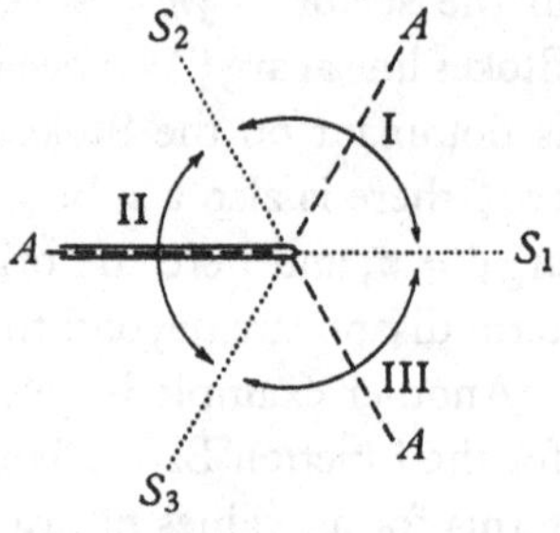

Fig. 15.7. The complex ζ-plane.

On the Stokes line S_1 the second term of (15.34) is dominant, and the constant in the first term changes so that in sector III the asymptotic approximation is

$$\zeta^{-\frac{1}{4}}[(A_1 - \lambda_1 B_1)\exp(-\tfrac{2}{3}\zeta^{\frac{3}{2}}) + B_1 \exp(\tfrac{2}{3}\zeta^{\frac{3}{2}})], \tag{15.38}$$

where λ_1 is the Stokes constant for the Stokes line S_1. On the Stokes line S_3 the first term of (15.38) is dominant, and the constant in the second term changes so that in the lower part of sector II the asymptotic approximation is

$$\zeta^{-\frac{1}{4}}[(A_1 - \lambda_1 B_1)\exp(-\tfrac{2}{3}\zeta^{\frac{3}{2}}) + \{B_1 - \lambda_3(A_1 - \lambda_1 B_1)\}\exp(\tfrac{2}{3}\zeta^{\frac{3}{2}})], \tag{15.39}$$

where λ_3 is the Stokes constant for the Stokes line 3. Now the cut passes through the middle of sector II. The solution must be continuous across the cut, and hence (15.39) and (15.37) must agree. On crossing the cut from top to bottom, $\zeta^{-\frac{1}{4}}$ changes by a factor $e^{\frac{1}{2}i\pi}$, and $\zeta^{\frac{3}{2}}$ changes sign. Hence by equating coefficients of the exponentials in (15.37) and (15.39) we obtain

$$i(A_1 - \lambda_1 B_1) = B_1 + \lambda_2 A_1, \tag{15.40}$$

$$i\{B_1 - \lambda_3(A_1 - \lambda_1 B_1)\} = A_1. \tag{15.41}$$

Now this argument must apply whatever the values of A_1 and B_1. If $A_1 = 0$, (15.40) shows that $\lambda_1 = i$. If $B_1 = 0$, (15.40) gives $\lambda_2 = i$, and (15.41) gives $\lambda_3 = i$. Hence

$$\lambda_1 = \lambda_2 = \lambda_3 = i. \tag{15.42}$$

From the differential equation (15.7) it could have been predicted that all three Stokes constants are the same, for this equation is unaltered if ζ is replaced by $\zeta e^{\frac{2}{3}\pi i}$, and it therefore has the same properties on each Stokes line.

15.15 Asymptotic approximations obtained from the contour integrals

Although the method of the preceding section gives the correct values of the Stokes constants, it is perhaps not entirely satisfactory mathematically. An alternative method of determining the asymptotic approximations makes use of the contour integrals (15.14) and (15.15) which can be evaluated approximately by the method of steepest descents. An outline of this method is given in the following sections. For full mathematical details the reader should consult a standard treatise such as Jeffreys and Jeffreys (1956).

15.16 Summary of some important properties of complex variables

A function $f(t)$ of a complex variable t may be expressed as the sum of its real and imaginary parts, thus:

$$f(t) = \phi(t) + i\psi(t), \tag{15.43}$$

and similarly

$$t = u + iv, \tag{15.44}$$

where ϕ, ψ, u, v are real. Then ϕ and ψ satisfy the Cauchy relations

$$\frac{\partial\phi}{\partial u} = \frac{\partial\psi}{\partial v}, \quad \frac{\partial\phi}{\partial v} = -\frac{\partial\psi}{\partial u}. \tag{15.45}$$

Let grad ϕ be the vector whose components are $\partial\phi/\partial u$, $\partial\phi/\partial v$ in the directions of the real and imaginary t-axes respectively. Then (15.45) shows that

$$\operatorname{grad}\phi \,.\, \operatorname{grad}\psi = 0, \tag{15.46}$$

so that the vectors grad ϕ and grad ψ are at right angles, which means that lines in the complex t-plane for which ϕ is constant are also lines along which ψ changes most rapidly, and vice versa. The function $\exp f(t)$ has constant modulus on lines of constant ϕ, which are called 'level lines' for this function. The same function has constant argument or phase on lines for which ψ is constant. On these lines ϕ is changing most rapidly, so that they are called lines of 'steepest descent' (or 'steepest ascent'). Equation (15.46) shows that the level lines and lines of steepest descent are at right angles. In a contour map of $|\exp\{f(t)\}|$ the contours are level lines, and their orthogonal trajectories are the lines of steepest descent or ascent.

The Cauchy relations (15.45) arise from the requirement that $f(t)$ shall have a unique derivative, which is given by

$$\frac{df(t)}{dt} = \frac{\partial\phi}{\partial u} + i\frac{\partial\psi}{\partial u} = \frac{\partial\psi}{\partial v} - i\frac{\partial\phi}{\partial v}.$$

At a point where df/dt is zero, both grad ϕ and grad ψ are zero, so that both the modulus and argument of $\exp f(t)$ are stationary. Such a point is called a 'saddle point'. Let $f(t)$ have a saddle point where $t = t_0$, and let $f(t_0) = f_0$, $d^2f/dt^2 = A\,e^{i\alpha}$ at the saddle point, where A and α are real, but f_0 is in general complex. Then Taylor's theorem gives

$$f(t) = f_0 + \tfrac{1}{2}(t-t_0)^2 A\,e^{i\alpha} + O\{(t-t_0)^3\}. \tag{15.47}$$

Let $t - t_0 = s\,e^{i\theta}$ where s and θ are real. Then

$$|\exp\{f(t)\}| = |e^{f_0}|\exp\{\tfrac{1}{2}As^2\cos(2\theta+\alpha) + O(s^3)\},$$

which shows that at the saddle point there are four directions (two crossing lines) for which $|\exp\{f(t)\}|$ is independent of s. These are given by $2\theta+\alpha = \pm\frac{1}{2}\pi, \pm\frac{3}{2}\pi$. Hence through every saddle point there are two level lines crossing at right angles. Further

$$\arg[\exp\{f(t)\}] = \arg f_0 + \tfrac{1}{2}As^2\sin(2\theta+\alpha) + O(s^3), \tag{15.48}$$

which shows there are four directions (two crossing lines) for which the argument or phase of $\exp\{f(t)\}$ is independent of s. These are given by $2\theta+\alpha = 0, \pm\pi, 2\pi$, and if one of these directions is followed, in moving away from the saddle point, it becomes a line of steepest descent (or steepest ascent). Hence through every saddle point there are two such lines crossing at right angles, and their directions are at 45° to the level lines.

If any straight line through a saddle point is traversed, $|\exp\{f(t)\}|$ has a maximum or minimum value at the saddle point, according as $\cos(2\theta+\alpha)$ is negative or positive. If the line is a line of steepest descent or ascent, the rate of change of $|\exp\{f(t)\}|$ is greatest. In a contour map of $|\exp\{f(t)\}|$ a saddle point would appear as a mountain pass, or col, shaped like a saddle.

If the second derivative d^2f/dt^2 is also zero at a saddle point, it can be shown that two saddle points coalesce, and the point is called a 'double saddle point' or 'triple point'. To study the behaviour of the functions it is then necessary to include the third derivative of $f(t)$, but the results are not required in this book.

If the two equations in (15.45) are differentiated with respect to u, v (or v, u) respectively, and added, the result is

$$\frac{\partial^2\phi}{\partial u^2}+\frac{\partial^2\phi}{\partial v^2}=0, \quad \frac{\partial^2\psi}{\partial u^2}+\frac{\partial^2\psi}{\partial v^2}=0. \tag{15.49}$$

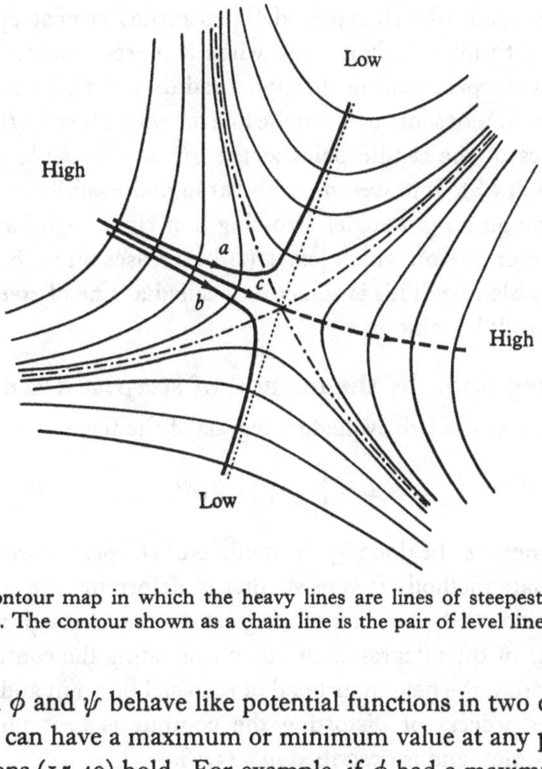

Fig. 15.8. Contour map in which the heavy lines are lines of steepest descent near a saddle point. The contour shown as a chain line is the pair of level lines through the saddle point.

Hence both ϕ and ψ behave like potential functions in two dimensions, and neither can have a maximum or minimum value at any point where the conditions (15.49) hold. For example, if ϕ had a maximum at some point, this would require that $\partial^2\phi/\partial u^2$ and $\partial^2\phi/\partial v^2$ are both negative, which would violate (15.49). A point where the conditions (15.49) and the Cauchy relations (15.45) do not hold is called a 'singular point' of the function $f(t)$. A line of steepest descent is a line on which ϕ decreases or increases at the greater possible rate. It is now clear that ϕ must continue to decrease or increase until the line meets a singular point, for there is no other way in which ϕ can have a maximum or minimum value.

Fig. 15.8 shows part of a contour map for $|\exp\{f(t)\}|$ near a saddle point. The line a is a line of steepest descent which passes very close to the saddle point but not through it. It must turn quite quickly through a right angle away from the saddle point, as shown. The line b is also a line of steepest descent which passes slightly to the other side of the saddle point; it also turns sharply through a right angle but in the opposite direction. The line c is a line of steepest descent which reaches the saddle point. To continue the descent it is necessary to turn through a right angle either to the right or left. If, instead, the line is continued straight through the saddle point, the value of $|\exp\{f(t)\}|$ is a minimum at the saddle point, and increases again when it is passed. Thus a true line of steepest descent turns sharply through a right angle when it meets a saddle point. The term 'line of steepest descent through a saddle point', however, is used in a slightly different sense to mean the line for which $|\exp\{f(t)\}|$ decreases on both sides of the saddle point at the greatest possible rate (shown dotted in Fig. 15.8). It passes smoothly through the saddle point and does not turn through a right angle. Crossing it at right angles at the saddle point is another line on which $|\exp\{f(t)\}|$ increases on both sides at the greatest possible rate. This is sometimes called a 'line of steepest ascent through the saddle point'.

15.17 Integration by the method of steepest descents

It is often necessary to evaluate integrals of the form

$$I = \int_C \exp\{f(t)\}\,dt. \qquad (15.50)$$

This can sometimes be done by the method of steepest descents which is an approximate method. It consists first in deforming the contour C so that it coincides with the lines of steepest descent through one or more saddle points of the integrand, and then evaluating the contributions to the integral from the neighbourhood of each saddle point and adding the results. The process of distorting the contour is best illustrated by specific examples, and is described in § 15.18.

The contribution from the neighbourhood of one saddle point is found as follows. The saddle point is assumed to be where $t = t_0$, and here $df/dt = 0$. The exponent $f(t)$ is expanded in the Taylor series (15.47), so that the integrand becomes

$$\exp\{f(t)\} = \exp f_0 \exp\left\{\frac{1}{2!}(t-t_0)^2 f_0'' + \frac{1}{3!}(t-t_0)^3 f_0''' + \dots\right\}, \qquad (15.51)$$

where f_0, f_0'', f_0''', are the values of f, d^2f/dt^2, d^3f/dt^3 at the saddle point. On the line of steepest descent through the saddle point the last exponent in curly brackets is real. It is zero at the saddle point and negative elsewhere. Hence we take

$$\frac{1}{2!}(t-t_0)^2 f_0'' + \frac{1}{3!}(t-t_0)^3 f_0''' + \dots = -\sigma^2, \qquad (15.52)$$

where σ is real, and is taken to be negative where the contour begins, and positive where it ends. Then the integral (15.50) becomes

$$I = e^{f_0} \int_{(-)}^{(+)} e^{-\sigma^2} \frac{dt}{d\sigma} d\sigma, \tag{15.53}$$

where the limits must be found from a study of the contour. Now the line of steepest descent must end either at infinity or at a singularity of $f(t)$. In the cases of interest here $f(t)$ has no singularities when t is finite, and the correct limits are $\pm\infty$. The term $e^{-\sigma^2}$ is largest where $\sigma = 0$, and becomes small very rapidly as $|\sigma|$ increases. Hence the largest contribution to the integral comes from near $\sigma = 0$, that is from near the saddle point, and even when the contour does not extend to infinity the limits in (15.53) may usually be taken as $\pm\infty$ with very small error.

The term $dt/d\sigma$ must now be found. In (15.52) σ is expressed as a series in power of $(t-t_0)$. This is a Taylor series and is convergent when $|t-t_0|$ is within some radius of convergence R (in the cases of interest here it is convergent for all $|t-t_0|$). From it $t-t_0$ can be expressed as a series in powers of σ, by the process known as 'reversion' (see, for example, Gibson, 1931, § 78). The first term of this series is obtained from (15.52) simply by neglecting the third and higher powers of $t-t_0$. Hence

$$t-t_0 = \pm\frac{i\sqrt{2}\,\sigma}{(f_0'')^{\frac{1}{2}}}\{1+\tfrac{1}{2}C_1\sigma+\tfrac{1}{3}C_2\sigma^2+\ldots\}. \tag{15.54}$$

Let $$f_0'' = Ae^{i\alpha} \quad (0 \leqslant \alpha < 2\pi). \tag{15.55}$$

Then the constant term in (15.54) is

$$\pm\frac{\sqrt{2}}{|A^{\frac{1}{2}}|}e^{\frac{1}{2}i(\pi-\alpha)}. \tag{15.56}$$

The sign depends on the direction Θ of the contour at the saddle point. It is positive if $-\frac{1}{2}\pi < \Theta \leqslant \frac{1}{2}\pi$.

The first two coefficients C_1, C_2 in the series (15.54) are given by

$$C_1 = \mp i\frac{2\sqrt{2}}{3}\frac{f_0'''}{(f_0'')^{\frac{3}{2}}},$$

$$C_2 = \{\tfrac{3}{4}f_0''f_0^{\text{iv}} - \tfrac{7}{4}(f_0''')^2\}/(f_0'')^3. \tag{15.57}$$

It can be shown that the series has a non-zero radius of convergence, although it may diverge when $|\sigma|$ exceeds a certain minimum value. The possibility of divergence is ignored in this section, but is discussed in § 15.21. The coefficients C_1, C_2, etc., contain descending powers of $(f_0'')^{\frac{1}{2}}$. If $|f_0''|$ is large, the terms of the series decrease rapidly at first, and in many cases it is only the first term that is of interest. Differentiation of (15.54) gives

$$\frac{dt}{d\sigma} = \pm\frac{\sqrt{2}}{|A^{\frac{1}{2}}|}e^{\frac{1}{2}i(\pi-\alpha)}\{1+C_1\sigma+C_2\sigma^2+\ldots\}, \tag{15.58}$$

and when this is substituted in the integral (15.53) with limits $\pm\infty$, it gives

$$I = \pm e^{f_0}\frac{\sqrt{2}}{|A^{\frac{1}{2}}|}e^{\frac{1}{2}i(\pi-\alpha)}\int_{-\infty}^{\infty}\{1+C_2\sigma^2+\ldots\}e^{-\sigma^2}d\sigma. \tag{15.59}$$

Odd powers of σ are omitted since the integrals are obviously zero. The remaining integrals can be evaluated, and the result is

$$I = \pm\frac{(2\pi)^{\frac{1}{2}}}{|A^{\frac{1}{2}}|}e^{f_0}e^{\frac{1}{2}i(\pi-\alpha)}\{1+\tfrac{1}{2}C_2+\tfrac{3}{4}C_4+\ldots\}. \tag{15.60}$$

The first term alone would have been obtained if the third and higher powers of $t-t_0$ had been neglected in (15.51).

15.18 Application of the method of steepest descents to solutions of the Stokes equation

The foregoing method will now be used to find approximate values of the contour integral (15.14) for Ai(ζ), where the contour C_1 is as shown in Fig. 15.4. The exponent $\zeta t-\frac{1}{3}t^3$ has two saddle points where $t = t_0 = \pm\zeta^{\frac{1}{2}}$, and the lines of steepest descent through these must first be found, and the contour must then be distorted to coincide with one or both of them. The exponent at either saddle point has the value

$$f_0 = \tfrac{2}{3}\zeta t_0, \tag{15.61}$$

and its second derivative is

$$f_0'' = -2t_0. \tag{15.62}$$

The configuration of the lines of steepest descent depends on the value of $\arg\zeta$, and it will be shown that sometimes one and sometimes both of these lines must be used. The transition from one case to the other gives the Stokes phenomenon.

Let $t = u+iv$ where u and v are real, and let $\zeta = m^2e^{i\theta}$ where m and θ are real and positive. Then the two saddle points are where

$$t = t_0 = \pm me^{\frac{1}{2}i\theta}. \tag{15.63}$$

On the lines of steepest descent and ascent through a saddle point, the imaginary part of the exponent $\zeta t-\frac{1}{3}t^3$ is constant. Hence the equations of these lines are given by

$$\mathscr{I}(t\zeta-\tfrac{1}{3}t^3) = \mathscr{I}(t_0\zeta-3t_0^3), \tag{15.64}$$

which leads to

$$\tfrac{1}{3}v^3-u^2v+vm^2\cos\theta+um^2\sin\theta = \pm\tfrac{2}{3}m^3\sin\tfrac{3}{2}\theta. \tag{15.65}$$

Suppose first that ζ is real and positive, so that $\theta = 0$. Then the saddle points lie on the real t-axis at $\pm m$ and (15.65) becomes

$$v(\tfrac{1}{3}v^2-u^2+m^2) = 0. \tag{15.66}$$

The curves are shown in Fig. 15.9. The line $v = 0$ is a line of steepest ascent through the saddle point S_1 and a line of steepest descent through the saddle point S_2. The other factor is the equation of a hyperbola whose asymptotes are given by $v = \pm\sqrt{3}\,u$. The left branch is a line of steepest descent through the saddle point S_1, and it begins and ends where $\arg t \sim \mp \frac{2}{3}\pi$, that is within the sectors where C_1 must begin and end. Hence the contour may be distorted to coincide with the left branch of the hyperbola as indicated by arrows in Fig. 15.9. The right branch of the hyperbola is a line of steepest ascent through the other saddle point S_2, and is not part of the contour, which therefore passes through only the one saddle point S_1 at $t_0 = -m$. Equations (15.61) and (15.62) then show that $f_0 = -\frac{2}{3}|\zeta^{\frac{3}{2}}|$, $f_0'' = 2|m^{\frac{1}{2}}| = 2|\zeta^{\frac{1}{2}}|$, so that $\alpha = 0$, $A = 2m$. The formula (15.60) may now be used to give the value of the integral (15.14). The

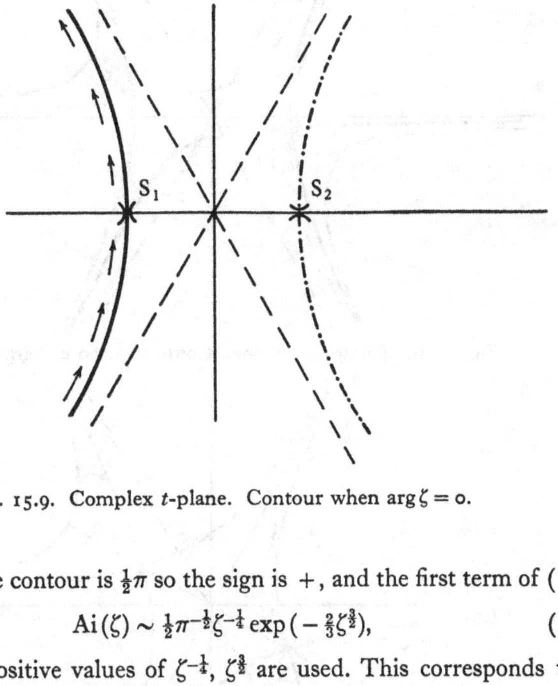

Fig. 15.9. Complex t-plane. Contour when $\arg\zeta = 0$.

direction Θ of the contour is $\frac{1}{2}\pi$ so the sign is $+$, and the first term of (15.60) gives

$$\mathrm{Ai}(\zeta) \sim \tfrac{1}{2}\pi^{-\frac{1}{2}}\zeta^{-\frac{1}{4}}\exp(-\tfrac{2}{3}\zeta^{\frac{3}{2}}), \tag{15.67}$$

where the real positive values of $\zeta^{-\frac{1}{4}}$, $\zeta^{\frac{3}{2}}$ are used. This corresponds to the approximate W.K.B. solution (15.26).

Next suppose that θ ($= \arg\zeta$) is in the range $0 < \theta < \frac{1}{3}\pi$. The expression (15.65) now gives two separate equations. The plus sign gives the lines of steepest ascent and descent through the saddle point S_2 at $t_0 = m e^{\frac{1}{2}i\theta}$, shown as a chain line in Fig. 15.10, and the minus sign gives those through the other saddle point S_1, shown as full lines. Each equation is now a cubic. The curve for the minus sign has two intersecting branches which are the lines of steepest descent and ascent through the saddle point S_1 at $t_0 = -m e^{\frac{1}{2}i\theta}$, and a third branch which does not pass through either saddle point. The asymptotes are the lines $v = 0$, $v = \pm\sqrt{3}\,u$. The line of steepest descent begins and ends in the correct sectors as before, and is used as the contour C_1, which is indicated by arrows in Fig. 15.10, and which therefore passes through only the one saddle point at $t_0 = -m e^{\frac{1}{2}i\theta}$. Equations (15.61) and (15.62) show that $f_0 = -\frac{2}{3}\zeta^{\frac{3}{2}}$ (where $\arg\zeta$ is $\frac{3}{2}\theta$) and $f_0'' = 2m e^{\frac{1}{2}i\theta}$, so that $\alpha = \frac{1}{2}\theta$, $A = 2m$. The direction Θ

of the contour at the saddle point is $\frac{1}{2}\pi - \frac{1}{4}\theta$, so the + sign is used in (15.60) whose first term gives

$$\mathrm{Ai}(\zeta) = \tfrac{1}{2}\pi^{-\frac{1}{2}}\,|m^{-\frac{1}{2}}|\,e^{-\frac{1}{4}i\theta}\exp(-\tfrac{2}{3}\zeta^{\frac{3}{2}}), \tag{15.68}$$

which is the same as (15.67) if $\zeta^{-\frac{1}{4}}$ takes the value which goes continuously into $|\zeta^{-\frac{1}{4}}|$ as θ decreases to zero.

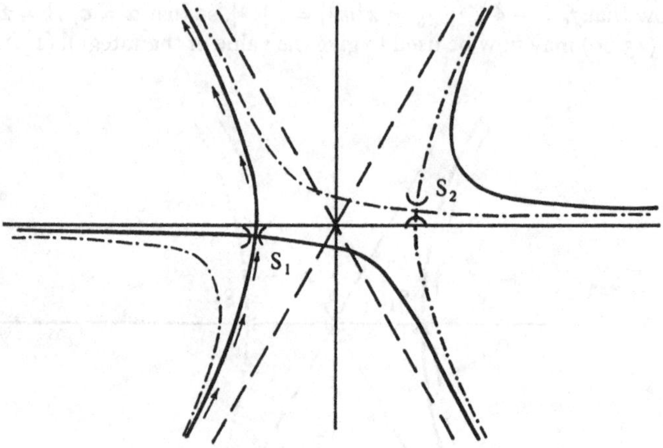

Fig. 15.10. Complex t-plane. Contour when $0 < \arg\zeta < \frac{1}{3}\pi$.

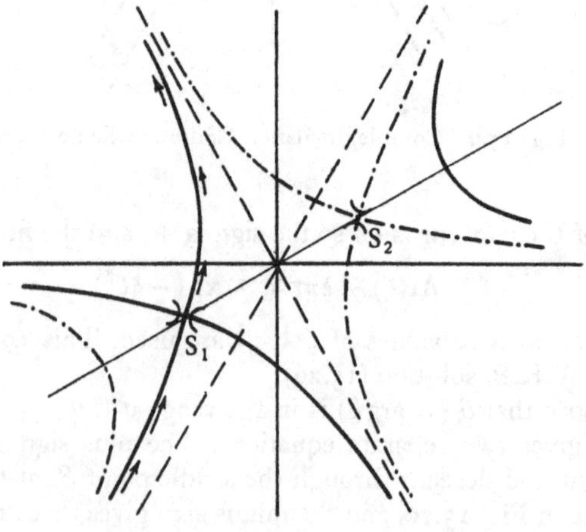

Fig. 15.11. Complex t-plane. Contour when $\arg\zeta = \frac{1}{3}\pi$.

The curves given by (15.65) can be drawn in a similar way for other values of $\theta = \arg\zeta$. Fig. 15.11 shows them for $\theta = \frac{1}{3}\pi$, which is an anti-Stokes line, and Fig. 15.12 shows them for $\frac{1}{3}\pi < \theta < \frac{2}{3}\pi$. In both these cases the contour is similar to those in Fig. 15.9 and 15.10, and the approximate value of the integral is given by (15.67) as before. In Fig. 15.12 the contour approaches the saddle point S_2 at $t_0 = me^{\frac{1}{2}i\theta}$, but turns fairly sharply through nearly a right angle to the left and ends in the correct sector.

Now consider the curves shown in Fig. 15.13 for $\theta = \frac{2}{3}\pi$ which is a Stokes line. They are like those for $\theta = 0$, but rotated through 60°. The line of steepest descent through the saddle point S_1 at $t_0 = -me^{\frac{1}{2}i\theta}$ is a straight line which begins in the correct sector (at $\arg t = -\frac{2}{3}\pi$), passes through the lower saddle

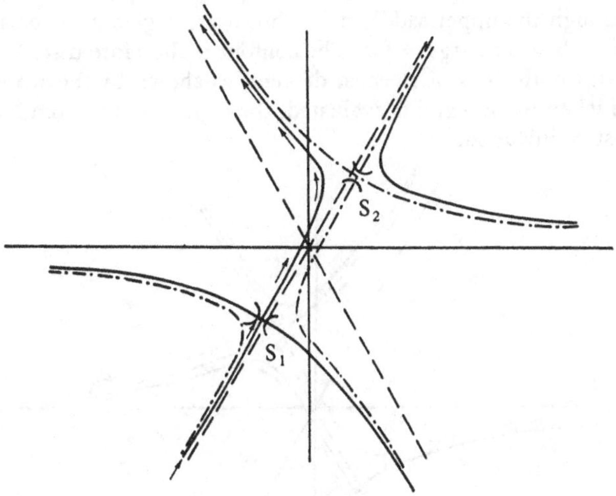

Fig. 15.12. Complex t-plane. Contour when $\frac{1}{3}\pi < \arg \zeta < \frac{2}{3}\pi$.

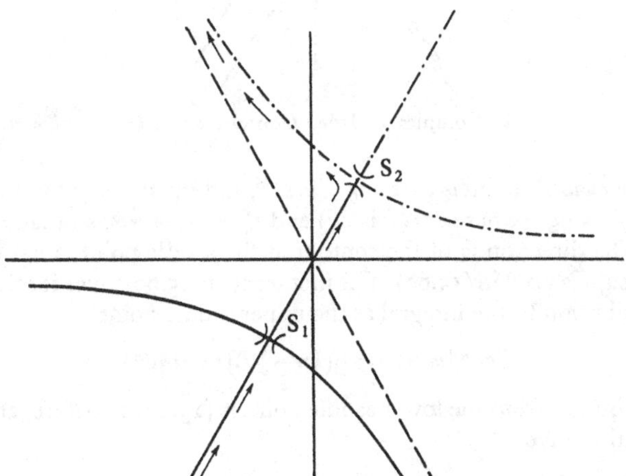

Fig. 15.13. Complex t-plane. Contour when $\arg \zeta = \frac{2}{3}\pi$. (Stokes line.)

point and then reaches the other saddle point S_2. To continue the descent, it is necessary to turn through a right angle either to the right or to the left. For the contour we chose a left turn as indicated by the arrows, so that the contour ends in the correct sector ($\arg t = \frac{2}{3}\pi$), and the approximate value of the integral is (15.67) as before.

Fig. 15.14 shows the curves for $\frac{2}{3}\pi < \theta < \pi$. The line of steepest descent through the lower saddle point begins in the correct sector, passes through the saddle point, but when it approaches the other saddle point it now turns fairly sharply through a right angle to the right, and ends near the asymptote $v = 0$, which is not in the correct sector for the end of the contour. The line of steepest descent through the upper saddle point, however, begins near the asymptote $v = 0$, and ends where $\arg t = \frac{2}{3}\pi$. The contour is therefore distorted so as to coincide with both lines of steepest descent, as shown by the arrows in Fig. 15.14, and when the integral is evaluated, the contributions from both saddle points must be included.

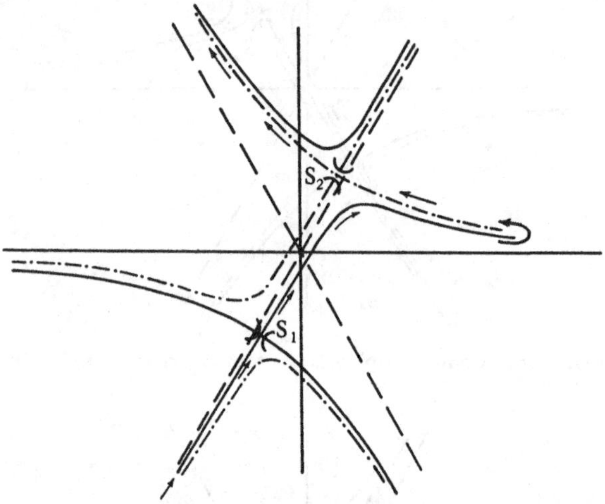

Fig. 15.14. Complex t-plane. Contour when $\frac{2}{3}\pi < \arg\zeta < \pi$.

The upper saddle point is where $t_0 = m e^{\frac{1}{2}i\theta}$, and equations (15.61) and (15.62) show that $f_0 = \frac{2}{3}\zeta^{\frac{3}{2}}$ (where $\arg\zeta^{\frac{3}{2}}$ is $\frac{3}{2}\theta$) and $f_0'' = -2m e^{\frac{1}{2}i\theta}$, so that $\alpha = \pi + \frac{1}{2}\theta$, $A = 2m$. The direction Θ of the contour at the saddle point is $\pi - \frac{1}{4}\theta$, so that the minus sign is used in (15.60). The first term of (15.60) gives for the approximate contribution to the integral at the upper saddle point

$$\tfrac{1}{2}\pi^{-\frac{1}{2}}\,|m^{-\frac{1}{2}}|\,\exp\left[i\left(\tfrac{1}{2}\pi - \tfrac{1}{4}\theta\right)\right]\exp\left(\tfrac{2}{3}\zeta^{\frac{3}{2}}\right). \qquad (15.69)$$

The contribution from the lower saddle point is (15.67) as before, and the two terms together give

$$\mathrm{Ai}(\zeta) \sim \tfrac{1}{2}\pi^{-\frac{1}{2}}\zeta^{-\frac{1}{4}}\left\{\exp\left(-\tfrac{2}{3}\zeta^{\frac{3}{2}}\right) + i\exp\left(\tfrac{2}{3}\zeta^{\frac{3}{2}}\right)\right\} \quad \left(\tfrac{2}{3}\pi \leqslant \arg\zeta \leqslant \tfrac{4}{3}\pi\right), \qquad (15.70)$$

where $\arg(\zeta^{\frac{3}{2}}) = \frac{3}{2}\theta$; $\arg(\zeta^{-\frac{1}{4}}) = -\frac{1}{4}\theta$. The factor i in (15.70) is the Stokes constant.

The above argument shows how the second term comes in discontinuously at the Stokes line $\theta = \frac{2}{3}\pi$. When $\theta - \frac{2}{3}\pi$ is positive but small, the value of the integrand $\exp(\zeta t - \frac{1}{3}t^3)$ is much smaller at the upper saddle point than at the lower saddle point, and the second term in (15.70) is small—it is the subdominant term. As $\arg\zeta$ approaches π, however, the two terms become more nearly equal, and on the anti-Stokes line at $\theta = \pi$ they have equal moduli. The curves for $\theta = \pi$ are shown in Fig. 15.15. They are like those for $\theta = \frac{1}{3}\pi$, but rotated through 60°. Here ζ is real and negative, and the two saddle points are on the imaginary axis. The curves could be traced in a similar way for other values of θ. For the Stokes line at $\theta = \frac{4}{3}\pi$ they would be like those for $\theta = \frac{2}{3}\pi$, but rotated through 60°. For $\frac{4}{3}\pi < \theta < 2\pi$ it is again possible to choose the contour so that it coincides with a line of steepest descent through only one saddle point, and in this range there is only one term in the asymptotic approximation. When $\theta = 2\pi$, the curves again become as in Fig. 15.9. When θ

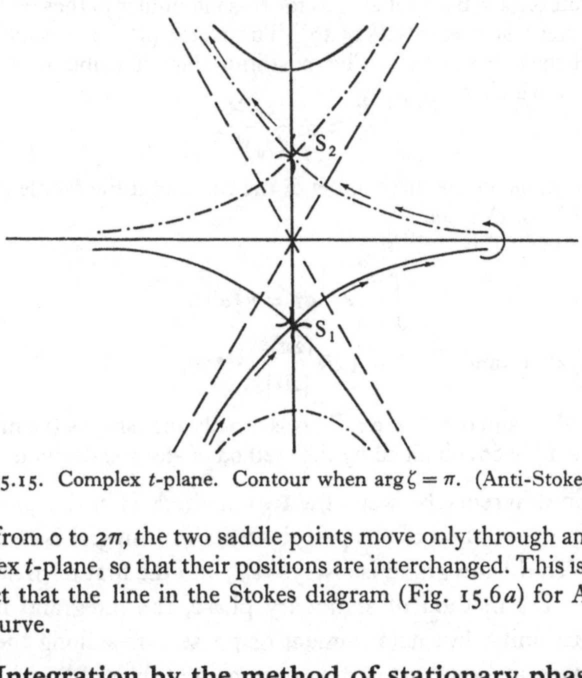

Fig. 15.15. Complex t-plane. Contour when $\arg\zeta = \pi$. (Anti-Stokes line.)

increases from 0 to 2π, the two saddle points move only through an angle π in the complex t-plane, so that their positions are interchanged. This is illustrated by the fact that the line in the Stokes diagram (Fig. 15.6*a*) for $\mathrm{Ai}(\zeta)$ is not a closed curve.

15.19 Integration by the method of stationary phase

Instead of using the method of steepest descents, it is sometimes more convenient to evaluate integrals of the type (15.50) by the method of stationary phase. This consists first in deforming the contour C so that it coincides with a level line through one or more saddle points of the integrand, and then evaluating the contributions to the integral from the neighbourhood of each saddle point and adding the results.

The contribution from the neighbourhood of one saddle point, at $t = t_0$, is found as follows. The exponent $f(t)$ is expanded in a Taylor series about $t = t_0$, and since df/dt is zero, the result is given by (15.51). On a level line through the saddle point the last exponent in curly brackets in (15.51) is purely imaginary, and is zero at the saddle point. There are two level lines through each saddle point, and for one the phase is a minimum, and for the other it is a maximum. Suppose that we have chosen the one on which the phase is a minimum. Then we take

$$\frac{1}{2!}(t-t_0)^2 f_0'' + \frac{1}{3!}(t-t_0)^3 f_0''' + \ldots = i\tau^2, \qquad (15.71)$$

where τ is real, and is taken to be negative where the contour begins, and positive where it ends. Then the integral (15.50) becomes

$$I = e^{f_0}\int_{-\infty}^{\infty} e^{i\tau^2}\left(\frac{dt}{d\tau}\right)d\tau, \qquad (15.72)$$

where the limits have been set at $\pm\infty$ for reasons similar to those given for the method of steepest descents (§ 15.17). The factor $(dt/d\tau)$ may be found by reversion of the series (15.71). The most important contribution comes from the first term, which is

$$\frac{dt}{d\tau} \approx \pm\frac{\sqrt{2}\,e^{i\frac{1}{4}\pi}}{(f_0'')^{\frac{1}{2}}}. \qquad (15.73)$$

The sign depends on the direction Θ of the contour at the saddle point. It is positive if $-\frac{3}{4}\pi < \Theta \leqslant \frac{1}{4}\pi$.

The integral for this term is

$$\int_{-\infty}^{\infty} e^{i\tau^2}\,d\tau = \pi^{\frac{1}{2}}\,e^{i\frac{1}{4}\pi}, \qquad (15.74)$$

so that (15.72) becomes

$$I \sim \pm e^{f_0}\frac{(2\pi)^{\frac{1}{2}}}{|A^{\frac{1}{2}}|}\,e^{\frac{1}{2}i(\pi-\alpha)}, \qquad (15.75)$$

where $f_0'' = A\,e^{i\alpha}$ and $0 \leqslant \alpha < 2\pi$. This is exactly the same as the first term of the expression (15.60) obtained by the method of steepest descents.

The main difference between the two methods is in the choice of the contour. In the method of steepest descents the integral is reduced to a form in which the integrand is always real, and the first term is the error integral. In the method of stationary phase, the integrand in (15.74) has modulus unity, but its argument or phase varies along the contour and is proportional to τ^2. It is thus stationary at the saddle point, where $\tau = 0$. In this case the integral is reduced to the complex Fresnel integral (15.74). Its value may be regarded as the resultant of a Cornu spiral. This is illustrated in Fig. 15.16, which is like a vector diagram of the kind used in the study of optical diffraction. It shows the contribution to the integral (15.50) from a level line through one saddle point. The length and direction of an element δs of the curve show the magnitude and phase respectively of the integrand for some element δt of the level line, which is the contour. The elements δs all have equal lengths for equal values of $|\delta t|$, because the integrand has constant modulus. Contributions δs from near the saddle point have nearly the same direction since the phase is stationary there. On receding from the saddle point the phase changes more and more rapidly, so that the curve spirals in more and more quickly. There are two spiral arms for the two sides of the saddle point. The resultant of the whole spiral is the value of the integral.

In physical optics the complex Fresnel integral (15.74) is often used with finite limits and gives the Fresnel diffraction pattern of some finite aperture. Its value is the resultant of some limited part of the Cornu spiral. In the method of stationary phase, however, the resultant of the whole spiral is used. Its value depends on the rate at which the curve spirals in, and this in turn depends on $|f_0''|$. If this is large, the phase

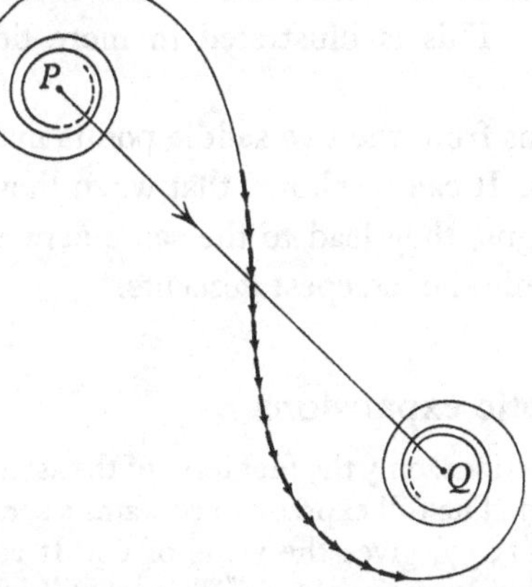

Fig. 15.16. Illustrating integration by the method of stationary phase. The diagram is the complex plane, and the arrows are small complex contributions to the integral. Near the middle the phase is stationary so that the arrows have nearly constant direction. The line PQ is the resultant, a complex number equal to the contribution to the integral from near one saddle point.

changes rapidly as we recede from the saddle point and the resultant of the spiral is small. The formula (15.75) shows that the resultant is proportional to $|f_0''|^{-\frac{1}{2}}$.

If the level line through the saddle point were the one for which the phase is a maximum, the right-hand side of (15.71) would be $-i\tau^2$. This gives a Cornu spiral whose curvature is opposite to that of Fig. 15.16. It can be shown that the resultant is given by (15.75) as before, where the sign is positive if $-\frac{1}{4}\pi \leqslant \Theta < \frac{3}{4}\pi$ (where Θ is the direction of the contour at the saddle point).

15.20 Method of stationary phase applied to the Airy integral function

A good illustration of the use of the method of stationary phase is provided by the contour integral (15.14) for $\mathrm{Ai}(\zeta)$ when ζ is real and negative. Then the saddle points $t_0 = \pm\zeta^{\frac{1}{2}}$ lie on the imaginary axis, which is a level line through both saddle points. The contour may be distorted to coincide with the imaginary axis, and it is convenient to let $t = is$. This was done in § 15.6 and was shown to lead to Airy's expression (15.17). For the method of stationary phase, however, we use the form (15.16). The saddle points are where $s = \pm|\zeta^{\frac{1}{2}}|$. When $s = +|\zeta^{\frac{1}{2}}|$ the phase $\zeta s + \frac{1}{3}s^3$ is a minimum, and the Cornu spiral is as in Fig. 15.16. When $s = -|\zeta^{\frac{1}{2}}|$ the phase is a maximum, and the Cornu spiral has the opposite curvature to that in Fig. 15.16. The asymptotic approximation to Ai (ζ) is therefore the sum of the resultants of two equal Cornu spirals with opposite curvature. This is illustrated in more detail in § 15.23, and Fig. 15.20.

The contributions from the two saddle points may each be expressed in the form (15.75). It can be shown that when they are added together with the correct signs, they lead to the same expression (15.70) as was obtained by the method of steepest descents.

15.21 Asymptotic expansions

In the last three sections, only the first term of the asymptotic approximation (15.60) has been used. The full expression contains a series $1 + \frac{1}{2}C_2 + \frac{3}{4}C_4 + \ldots$, and the expression (15.57) gives the value of C_2. It can be shown that the denominator of C_{2n} is proportional to $(f_0'')^{3n}$, so that if $|f_0''|$ is large enough, the terms can be made to decrease rapidly at first. A series of this kind is called an asymptotic expansion. It can be shown (Jeffreys and Jeffreys, § 17.023) that for any function which exhibits the Stokes phenomenon the series is ultimately divergent. It can further be shown that when only a finite number of consecutive terms is used, the error is less than the last term included. To obtain the best accuracy the series should therefore be used as far as the smallest term.

A formal definition of an asymptotic expansion was given by Poincaré (1886). The proof that the series (15.60) obtained by the method of steepest descents is an asymptotic expansion in the Poincaré sense is known as Watson's lemma. For a full discussion see Jeffreys and Jeffreys (1956). A most illuminating treatment of asymptotic expansions has been given recently by Dingle (1958, 1959).

Where asymptotic approximations are used in this book, it is seldom necessary to go beyond the first term. The asymptotic expansions for the functions Ai (ζ), Bi (ζ) are given in full by Miller (1946).

15.22 The range of validity of asymptotic approximations

Poincaré's definition of an asymptotic approximation is equivalent to the following. Suppose that $f(\zeta), g(\zeta)$ are two functions such that $\lim_{|\zeta|\to\infty} f(\zeta)/g(\zeta) = 1$.

Then $f(\zeta)$ is an asymptotic approximation to $g(\zeta)$ in the Poincaré sense.† Now equation (15.70) shows that for the range $\frac{2}{3}\pi \leqslant \arg\zeta \leqslant \frac{4}{3}\pi$ the asymptotic approximation for $\mathrm{Ai}(\zeta)$ contains two exponential terms. For the range $\frac{2}{3}\pi \leqslant \arg\zeta < \pi$ the first term is dominant, and the second subdominant. For a fixed $\arg\zeta$ in this range, let $|\zeta|$ increase indefinitely. Then the ratio of the dominant to the subdominant term becomes indefinitely large, and the ratio of $\mathrm{Ai}(\zeta)$ to the dominant term tends to unity as $|\zeta|$ tends to ∞. Hence the asymptotic approximation to $\mathrm{Ai}(\zeta)$ in the Poincaré sense need only contain the dominant term. This is the first exponential term of equation (15.70) for $\frac{2}{3}\pi < \arg\zeta < \pi$, and the second exponential term for $\pi < \arg\zeta < \frac{4}{3}\pi$. It is only on the anti-Stokes line, where $\arg\zeta = \pi$ exactly, that it is necessary to include both terms, according to the Poincaré criterion.

The Poincaré definition thus applies only to the limiting behaviour when $|\zeta|$ tends to infinity. When $|\zeta|$ is finite, it may clearly be insufficient to include only the dominant term. For example, when $\arg\zeta$ is very close to π, the two terms are nearly equal, and in computing $\mathrm{Ai}(\zeta)$ it would be necessary to include both. Care is therefore needed when specifying the range of $\arg\zeta$ over which a given asymptotic approximation is valid. For $\mathrm{Ai}(\zeta)$ the best way of doing this is as follows:

$$\mathrm{Ai}(\zeta) \sim \tfrac{1}{2}\pi^{-\frac{1}{2}}\zeta^{-\frac{1}{4}}\exp(-\tfrac{2}{3}\zeta^{\frac{3}{2}}) \quad \text{for} \quad -\tfrac{2}{3}\pi \leqslant \arg\zeta \leqslant \tfrac{2}{3}\pi, \tag{15.76}$$

$$\mathrm{Ai}(\zeta) \sim \tfrac{1}{2}\pi^{-\frac{1}{2}}\zeta^{-\frac{1}{4}}\{\exp(-\tfrac{2}{3}\zeta^{\frac{3}{2}}) + i\exp(\tfrac{2}{3}\zeta^{\frac{3}{2}})\} \quad \text{for} \quad \tfrac{2}{3}\pi \leqslant \arg\zeta \leqslant \tfrac{4}{3}\pi, \tag{15.77}$$

where a fractional power ζ^f always means $|\zeta^f|\exp\{if\arg\zeta\}$. Here the ranges of $\arg\zeta$ end on Stokes lines, and this convention is used in the present book. According to the Poincaré definition the range in (15.76) could be written $-\pi < \arg\zeta < \pi$, and in (15.77) it could be written $\frac{1}{3}\pi < \arg\zeta < \frac{5}{3}\pi$. These ranges end just short of anti-Stokes lines, and the inequality signs are $<$ and not $\leqslant$, so that two anti-Stokes lines are just outside the range at the ends. The two ranges have a common part where there are both dominant and subdominant terms, since in the Poincaré sense it does not matter whether the subdominant term is included or not.

In most text-books of mathematics which give asymptotic approximations, the Poincaré definition is used, and this can sometimes lead to errors. An example of this is given in § 17.4, for the parabolic cylinder function. There are some further comments on this topic in § 17.2.

The asymptotic approximations for $\mathrm{Ai}'(\zeta)$ will also be needed later. They can be found from those for $\mathrm{Ai}(\zeta)$ by differentiating the asymptotic expansion, which gives new expansions for $\mathrm{Ai}'(\zeta)$. The first terms of these are as follows:

$$\mathrm{Ai}'(\zeta) \sim -\tfrac{1}{2}\pi^{-\frac{1}{2}}\zeta^{\frac{1}{4}}\exp(-\tfrac{2}{3}\zeta^{\frac{3}{2}}) \quad \text{for} \quad -\tfrac{2}{3}\pi \leqslant \arg\zeta \leqslant \tfrac{2}{3}\pi, \tag{15.78}$$

$$\mathrm{Ai}'(\zeta) \sim \tfrac{1}{2}\pi^{-\frac{1}{2}}\zeta^{\frac{1}{4}}\{-\exp(-\tfrac{2}{3}\zeta^{\frac{3}{2}}) + i\exp(\tfrac{2}{3}\zeta^{\frac{3}{2}})\} \quad \text{for} \quad \tfrac{2}{3}\pi \leqslant \arg\zeta \leqslant \tfrac{4}{3}\pi. \tag{15.79}$$

† The full Poincaré definition is as follows. Let $S_n(\zeta)$ be the sum up to the term $A_n\zeta^{-n}$ of the series $S(\zeta) = 1 + A_1/\zeta + \ldots + A_n/\zeta^n + \ldots$ and let $R_n(\zeta) = f(\zeta)/g(\zeta) - S_n(\zeta)$. Then $g(\zeta)\,S(\zeta)$ is an asymptotic approximation to $f(\zeta)$ if $\lim_{|\zeta|\to\infty} \zeta^n R_n(\zeta) = 0$ for all n, and we write $f(\zeta) \sim g(\zeta)\,S(\zeta)$. In the special case when $n = 0$, only the first term of the series $S(\zeta)$ is used, and this leads to the simplified definition given above.

15.23 The choice of a fundamental system of solutions of the Stokes equation

Since the Stokes differential equation (15.7) is of the second order, it has two independent solutions which may be chosen in various ways, and these may be called the fundamental solutions. Any other solution is then a linear combination of them. There is no absolute criterion for these solutions and their choice is a matter of convenience. An obvious choice for one fundamental solution is the Airy integral function $\mathrm{Ai}(\zeta)$. As a second solution Jeffreys and Jeffreys use the function $\mathrm{Bi}(\zeta)$ defined in § 15.6. The choice of $\mathrm{Ai}(\zeta)$ and $\mathrm{Bi}(\zeta)$ has the advantage that both functions are real when ζ is real. The asymptotic approximation for $\mathrm{Bi}(\zeta)$ may be found as in § 15.18, from the contour integral (15.15), and is as follows:

$$\mathrm{Bi}(\zeta) \sim \tfrac{1}{2}\pi^{-\frac{1}{2}}\zeta^{-\frac{1}{4}}\{i\exp(-\tfrac{2}{3}\zeta^{\frac{3}{2}})+2\exp(\tfrac{2}{3}\zeta^{\frac{3}{2}})\} \quad \text{for} \quad 0 \leqslant \arg\zeta \leqslant \tfrac{2}{3}\pi, \qquad (15.80)$$

$$\mathrm{Bi}(\zeta) \sim \tfrac{1}{2}\pi^{-\frac{1}{2}}\zeta^{-\frac{1}{4}}\{i\exp(-\tfrac{2}{3}\zeta^{\frac{3}{2}})+\exp(\tfrac{2}{3}\zeta^{\frac{3}{2}})\} \quad \text{for} \quad \tfrac{2}{3}\pi \leqslant \arg\zeta \leqslant \tfrac{4}{3}\pi, \qquad (15.81)$$

$$\mathrm{Bi}(\zeta) \sim \tfrac{1}{2}\pi^{-\frac{1}{2}}\zeta^{-\frac{1}{4}}\{2i\exp(-\tfrac{2}{3}\zeta^{\frac{3}{2}})+\exp(\tfrac{2}{3}\zeta^{\frac{3}{2}})\} \quad \text{for} \quad \tfrac{4}{3}\pi \leqslant \arg\zeta \leqslant 2\pi, \qquad (15.82)$$

where a fractional power of ζ has the meaning given in the preceding section. An alternative form of (15.82) is

$$\mathrm{Bi}(\zeta) \sim \tfrac{1}{2}\pi^{-\frac{1}{2}}\zeta^{-\frac{1}{4}}\{-i\exp(-\tfrac{2}{3}\zeta^{\frac{3}{2}})+2\exp(\tfrac{2}{3}\zeta^{\frac{3}{2}})\} \quad \text{for} \quad -\tfrac{2}{3}\pi \leqslant \arg\zeta \leqslant 0. \qquad (15.83)$$

In these formulae the ranges of $\arg\zeta$ end on Stokes lines, as explained in the preceding section. There are both dominant and subdominant terms in all three ranges of $\arg\zeta$ and the asymptotic behaviour of $\mathrm{Bi}(\zeta)$ is therefore considerably more complicated than for $\mathrm{Ai}(\zeta)$. This is illustrated by the Stokes diagram for $\mathrm{Bi}(\zeta)$ given in Fig. 15.6*b*.

It is easily shown that the Stokes equation (15.7) is unaltered when ζ is replaced by $\zeta\exp(\frac{2}{3}i\pi)$, or $\zeta\exp(\frac{4}{3}i\pi)$. Hence $\mathrm{Ai}(\zeta e^{\frac{2}{3}i\pi})$ and $\mathrm{Ai}(\zeta e^{\frac{4}{3}i\pi})$ are solutions of the Stokes equation. For some purposes it is convenient to use $\mathrm{Ai}(\zeta e^{\frac{2}{3}i\pi})$ as the second fundamental solution, instead of $\mathrm{Bi}(\zeta)$. It readily follows from (15.76), (15.77) that the asymptotic approximation for $\mathrm{Ai}(\zeta e^{\frac{2}{3}i\pi})$ is

$$\mathrm{Ai}(\zeta e^{\frac{2}{3}i\pi}) \sim \tfrac{1}{2}\pi^{-\frac{1}{2}}\zeta^{-\frac{1}{4}}e^{-\frac{1}{6}\pi i}\exp(\tfrac{2}{3}\zeta^{\frac{3}{2}}) \quad \text{for} \quad -\tfrac{4}{3}\pi \leqslant \arg\zeta \leqslant 0, \qquad (15.84)$$

$$\mathrm{Ai}(\zeta e^{\frac{2}{3}i\pi}) \sim \tfrac{1}{2}\pi^{-\frac{1}{2}}\zeta^{-\frac{1}{4}}e^{-\frac{1}{6}\pi i}\{\exp(\tfrac{2}{3}\zeta^{\frac{3}{2}})+i\exp(-\tfrac{2}{3}\zeta^{\frac{3}{2}})\} \quad \text{for} \quad 0 \leqslant \arg\zeta \leqslant \tfrac{2}{3}\pi. \qquad (15.85)$$

The Stokes diagram for $\mathrm{Ai}(\zeta e^{\frac{4}{3}i\pi})$ is shown in Fig. 15.6*c* and is obtained from that for $\mathrm{Ai}(\zeta)$ (Fig. 15.6*a*) by rotation through 120° clockwise.

It must be possible to express $\mathrm{Ai}(\zeta e^{\frac{2}{3}i\pi})$ and $\mathrm{Ai}(\zeta e^{\frac{4}{3}i\pi})$ as linear combinations of $\mathrm{Ai}(\zeta)$ and $\mathrm{Bi}(\zeta)$. It can be shown (see, for example, Miller, 1946), that

$$\mathrm{Ai}(\zeta e^{\frac{2}{3}i\pi}) = -\tfrac{1}{2}e^{\frac{1}{3}i\pi}\{\mathrm{Ai}(\zeta)-i\,\mathrm{Bi}(\zeta)\}, \qquad (15.86)$$

$$\mathrm{Ai}(\zeta e^{\frac{4}{3}i\pi}) = -\tfrac{1}{2}e^{\frac{2}{3}i\pi}\{\mathrm{Ai}(\zeta)+i\,\mathrm{Bi}(\zeta)\}. \qquad (15.87)$$

If S_1 and S_2 are any two solutions of a second-order differential equation then $S_1S_2'-S_2S_1'$ is called the Wronskian for these solutions, and is written $W(S_1, S_2)$. It can be shown that its derivative is zero for a differential equation which has no first derivative term, and therefore W is a constant. This applies in particular to the Stokes equation.

The Wronskian for $\mathrm{Ai}(\zeta)$ and $\mathrm{Bi}(\zeta)$ can be found for very large values of ζ by using the asymptotic approximations (15.76) for $\mathrm{Ai}(\zeta)$ and (15.80) for $\mathrm{Bi}(\zeta)$ in which the first (subdominant) term can be neglected. It is then easily shown that

$$\mathrm{Ai}(\zeta)\,\mathrm{Bi}'(\zeta) - \mathrm{Bi}(\zeta)\,\mathrm{Ai}'(\zeta) = 1/\pi, \tag{15.88}$$

and since the Wronskian is constant, this must be true for all ζ. Similarly, the Wronskian for $\mathrm{Ai}(\zeta)$ and $\mathrm{Ai}(\zeta e^{\frac{2}{3}i\pi})$ can be shown to be

$$\mathrm{Ai}(\zeta)\frac{d}{d\zeta}\mathrm{Ai}(\zeta e^{\frac{2}{3}i\pi}) - \mathrm{Ai}'(\zeta)\,\mathrm{Ai}(\zeta e^{\frac{2}{3}i\pi}) = \frac{e^{-\frac{1}{6}\pi i}}{2\pi}. \tag{15.89}$$

15.24 Connection formulae, or circuit relations

Equations (15.76) and (15.77) show that in the special case when ζ is real, the asymptotic approximations for $\mathrm{Ai}(\zeta)$ are different according as ζ is positive or negative. A formula which gives one asymptotic approximation when the other is known, is sometimes called a 'connection formula' or 'circuit relation'. For example, the connection formula for $\mathrm{Ai}(\zeta)$ is

$$\zeta^{-\frac{1}{4}}\{\exp(-\tfrac{2}{3}\zeta^{\frac{3}{2}}) + i\exp(\tfrac{3}{2}\zeta^{\frac{3}{2}})\} \leftrightarrow \zeta^{-\frac{1}{4}}\exp(-\tfrac{2}{3}\zeta^{\frac{3}{2}}) \tag{15.90}$$

($\arg\zeta = \pi$ on left, and 0 on right). The use of the double arrow was introduced by Jeffreys (1923). The terms on the left are for negative ζ and those on the right for positive ζ. Clearly, in order to give a connection formula it is necessary to know the value of the Stokes constant.

The connection formula for $\mathrm{Bi}(\zeta)$ is written slightly differently thus

$$\zeta^{-\frac{1}{4}}\{i\exp(-\tfrac{2}{3}\zeta^{\frac{3}{2}}) + \exp(\tfrac{2}{3}\zeta^{\frac{3}{2}})\} \rightarrow 2\zeta^{-\frac{1}{4}}\exp(\tfrac{2}{3}\zeta^{\frac{3}{2}}). \tag{15.91}$$

Since the line $\arg\zeta = 0$ is a Stokes line it does not matter whether the subdominant term is present or not. But the dominant term alone, for positive ζ, is insufficient to determine the asymptotic behaviour for negative ζ. On the other hand, the asymptotic behaviour for negative ζ determines completely the behaviour for positive ζ. Hence, following Langer (1934), we use only a single arrow in (15.91).

15.25 The intensity of light near a caustic

The integral (15.17) was originally derived by Airy (1838, 1849) in a study of the variation of the intensity of light near a caustic. Caustics can also be formed by radio waves, and it is therefore useful to give the theory. Huyghens's principle is used to construct 'amplitude-phase' diagrams for a series of neighbouring points. Each diagram is a spiral, somewhat analogous to Cornu's spiral, and its vector resultant gives the amplitude and phase of the light at the point considered.

Consider a parallel beam of light incident from the left on a convex lens (Fig. 15.17). For simplicity the lens is assumed to be cylindrical, and the problem is treated as though it were in two dimensions only. It is well known that the rays of the beam are tangents to a caustic curve XY. It is required to find the intensity of the light at the points of a line AB perpendicular to the caustic. Let Q be a typical point of this line. Consider a plane wave-front of the beam (RS in Fig. 15.17), just before it reaches the lens. Imagine this wave-front to be divided into infinitesimal strips of equal width δs, which all radiate cylindrical waves of the same intensity. On arrival at Q these waves all have the same amplitude, but their phases depend on the position, s, of the strip from which each originates. If Q is inside the caustic, as shown in Fig. 15.17, two geometrical rays pass through it. Assume that one of these, RQ, is included in the narrow pencil between rays 1 and 2, and the other, SQ, in the pencil between rays 4 and 5.

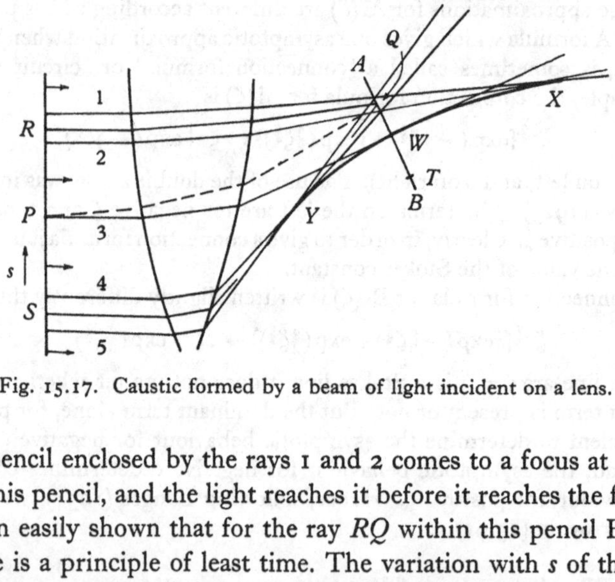

Fig. 15.17. Caustic formed by a beam of light incident on a lens.

The pencil enclosed by the rays 1 and 2 comes to a focus at X. Q is within this pencil, and the light reaches it before it reaches the focus X. It is then easily shown that for the ray RQ within this pencil Fermat's principle is a principle of least time. The variation with s of the phase $\phi(s)$ therefore has a minimum for the Huyghens wavelets which originate near R. The pencil enclosed by the rays 4 and 5 comes to a focus at Y. Q is within the pencil and beyond the focus, and it can then similarly be shown that for the ray SQ Fermat's principle is a principle of greatest time, so that the phase $\phi(s)$ has a maximum for Huyghens wavelets which originate near S. $\phi(s)$ has no other turning points (Fig. 15.18, curve Q). If Q is outside the caustic, at T say, no geometrical rays pass through it. Then $\phi(s)$ has no turning points, and varies as shown in Fig. 15.18, curve T. If Q is on the caustic, at W say, the two turning points coincide (Fig. 15.18, curve W).

Let the zero of s be chosen midway between the two turning points, or at the point of least slope when there are no turning points. Then the form of the curves in Fig. 15.18 is given by the cubic law

$$\phi = K(\zeta s + \tfrac{1}{3}s^3), \tag{15.92}$$

where ζ depends on the distance of Q from the caustic, and K is a constant. Clearly the scale of s can be chosen so that $K = 1$. ζ is negative when Q is on the illuminated side of the caustic as in Fig. 15.17, and positive when Q is outside the caustic.

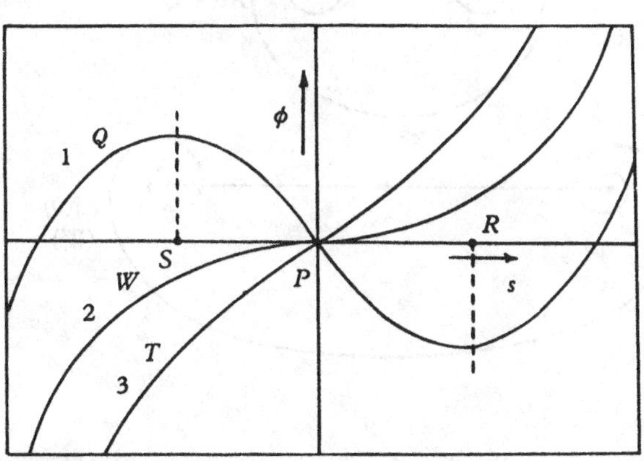

Fig. 15.18. Variation of phase $\phi(s)$ of Huyghens wavelet with position, s, of source on initial wave-front.

The value of ζ can be found as follows. Let the distance $QW = u$, and let the radius of curvature of the caustic be R. For curve 1, in Fig. 15.18, ζ is negative, and the difference between the maximum and minimum values of ϕ is $\frac{4}{3}\zeta^{\frac{3}{2}}$ (with $K = 1$). This is the difference between the phases of the light arriving at Q via the ray pencils 4–5 and 1–2. Now any two rays which are close together, such as 4 and 5, have the same phase where they meet on the caustic. Hence the phase difference between the waves arriving at X and Y is just the phase difference that would arise from the curved path YWX along the caustic, namely $4\pi R\theta/\lambda$ where 2θ is the angle subtended by the arc XY at its centre of curvature. Thus the difference between the phases of the light arriving at Q via the two ray pencils is

$$\frac{2\pi}{\lambda}\{YQ + QX - \text{arc}(YWX)\} = 4\pi R(\tan\theta - \theta)/\lambda \approx \frac{4\pi R\theta^3}{3\lambda},$$

so that $\zeta^3 \approx \pi^2R^2\theta^6/\lambda^2$. Now $QW = u \doteqdot \frac{1}{2}R\theta^2$, so that when u is small

$$\zeta = 2u\left(\frac{\pi^2}{R\lambda^2}\right)^{\frac{1}{3}}. \tag{15.93}$$

Clearly ζ could be expressed by an ascending power series in $u = QW$, and if ζ is small enough, powers higher than the first can be neglected.

For each position of the point Q an amplitude-phase diagram may now be constructed. The curvature of this at a given point is proportional to the slope at the corresponding point in the (ϕ, s)-curve. Since this slope eventually increases indefinitely both for s increasing and for s decreasing, the amplitude-phase diagrams end in spirals. Typical diagrams are shown in Fig. 15.19, corresponding to the three curves of Fig. 15.18.

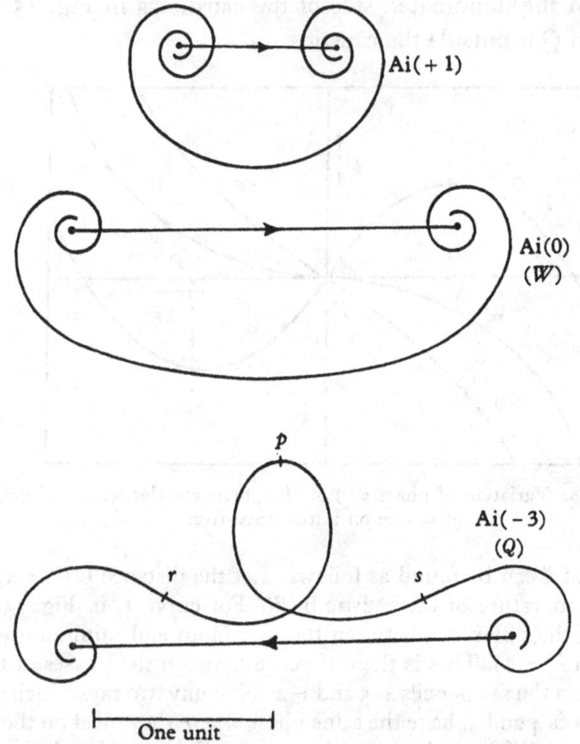

Fig. 15.19. Spirals associated with the Airy integral function. In the third spiral the points p, r, s correspond roughly to the points P, R, S in Figs. 15.17 and 15.18.

There is an obvious analogy between these diagrams and Cornu's spiral, for which the (ϕ, s)-curve is a parabola. Cornu's spiral always has the same form. In diffraction problems whose solution involves a Cornu spiral, variations of intensity occur, because different fractions of the spiral are used to determine the intensity reaching different points. In the present problem, however, the shape of the spiral is different for different values of ζ. For all points of the line AB, the resultant of the whole of the spiral is found, but the intensity varies because the shape of the spiral varies from point to point.

In Fig. 15.19, let the x-axis be chosen parallel to the element corresponding to $s = 0$. Then the diagrams are symmetrical about the y-axis, and the resultant vector is always parallel to the x-axis. When ζ is positive (first diagram of Fig. 15.19), the curvature of the spiral always has the same sign and the length of the resultant decreases monotonically as ζ increases. When ζ is negative (third diagram of Fig. 15.19), the middle section of the spiral has opposite curvature from the rest, and as ζ decreases the limiting points of the spiral repeatedly move completely round the origin. The resultant thus alternates in sign, and in the third diagram of Fig. 15.19 it is negative.

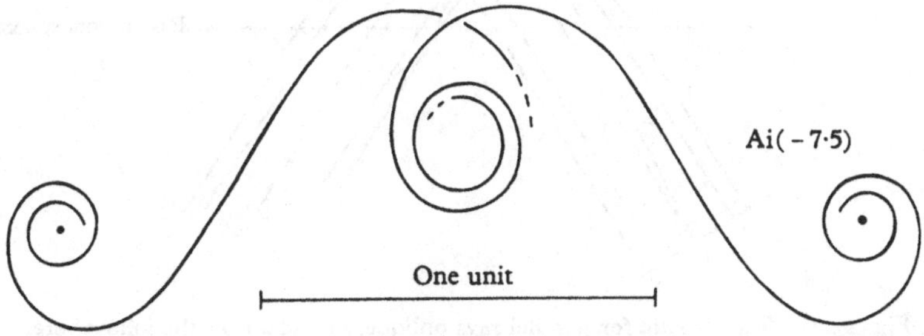

Fig. 15.20. Spiral associated with Airy integral function of large negative argument, showing resemblance to two Cornu spirals. The part of the left half of the diagram where overlapping occurs is not shown.

The contribution of an element δs of the wave-front to the resultant vector is proportional to $\delta s \cos\phi$, and hence the resultant vector is proportional to

$$\int_{-\infty}^{\infty} \cos\phi \, ds = \int_{-\infty}^{\infty} \cos\left(\zeta s + \tfrac{1}{3}s^3\right) ds. \tag{15.94}$$

Apart from a constant, this is the standard expression (15.17) for the Airy integral function, $\mathrm{Ai}(\zeta)$.

When Q is well beyond the caustic on the illuminated side (ζ large and negative), the turning points of the (ϕ, s)-curve are widely separated, and near each turning point the curve is approximately a parabola. Hence the portion of the amplitude-phase diagram corresponding to the neighbourhood of each turning point is approximately a Cornu spiral. The two Cornu spirals curve opposite ways, and together constitute the more complex spiral whose resultant is the Airy integral function (see Fig. 15.20). When ζ is large and negative, the Airy integral function $\mathrm{Ai}(\zeta)$ is given by (15.70), and it was shown in § 15.20 that the two terms may be regarded as the resultants of the two Cornu spirals.

It was shown in § 11.7 that when a radio wave is reflected from the ionosphere there may sometimes be a 'skip' distance, which is a point where a caustic curve meets the ground. The caustic is the envelope of a series of reflected rays, and the law of variation of the signal over the ground is given by the Airy integral function. For details see §§ 11.7 and 11.12.

In Fig. 15.21 the line RS represents a wave-front of a plane wave which is obliquely incident on the ionosphere. The lines 1, 2, 3, ..., are rays which are refracted by the ionosphere, and they all become horizontal

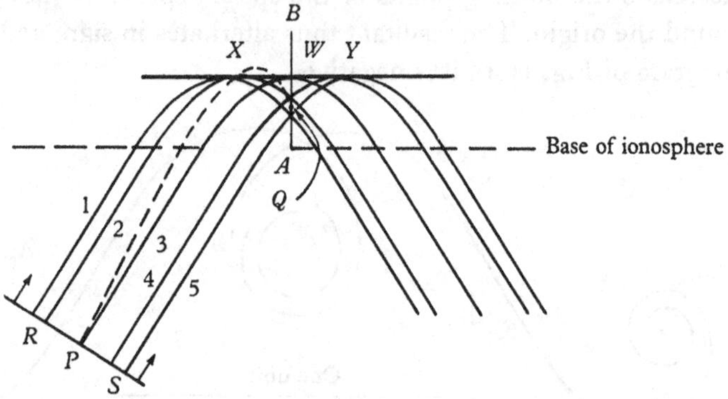

Fig. 15.21. The caustic for parallel rays obliquely incident on the ionosphere.

at the same level. The straight line XWY is the envelope of the rays and may be considered to be a caustic 'curve' for this case. Now Huyghens principle could be used to find how the electric field varies along a vertical line AB. The Huyghens wavelets arriving at Q from various points of the wave-front RS would travel along curved paths, but if Q were below the caustic the phase would be stationary for the two ray pencils 1–2 and 4–5, and the amplitude-phase diagram would be a spiral similar to the third diagram of Fig. 15.19. If Q were above the caustic no part of the wave-front RS would give Huyghens wavelets with stationary phase, and the amplitude-phase diagram would be a spiral similar to the first diagram of Fig. 15.19. Thus we might expect that the variation of electric field with height is given by the Airy integral function, and this is shown to be the case by other methods, in § 16.2, when the variation of electron density with height is linear. This treatment by Huyghens's principle is of only limited application, however. For example, it cannot be used at vertical incidence, because no Huyghens wavelets could go above the level where $\mu = 0$, whereas it can be shown that there is an appreciable electric field above this level.

CHAPTER 16

LINEAR GRADIENT OF ELECTRON DENSITY

16.1 Introduction

The results of ch. 15 will now be used to find the reflection coefficient of the ionosphere for profiles in which the variation of electron number density N with height z is linear over some part of the range. The earth's magnetic field is neglected, and in the first part of this chapter the waves are assumed to be horizontally polarised at oblique incidence. The reflection of waves polarised with the electric vector in the plane of incidence ('vertical' polarisation) is discussed in § 16.12 onwards.

In previous chapters it was assumed that the reflection coefficient of a slowly varying ionosphere was given by (9.37), which was based on 'ray theory' arguments. We must now examine the justification for using this expression, which will be found to require a small modification. In this chapter we begin the study of the 'full wave' theory, to which most of the rest of this book is devoted.

16.2 Purely linear profile. Electron collisions neglected

Suppose that the base of the ionosphere is at a height $z = h_0$ from the ground, and above this the electron number density N increases linearly with height. Then $X = a(z-h_0)$ for $z > h_0$, since X is proportional to N. Here a depends on frequency, but is constant at a fixed frequency. If electron collisions are neglected, $Z = 0$ and (9.56) then shows that

$$q^2 = \begin{cases} C^2 - a(z-h_0) & \text{when} \quad z \geqslant h_0, \\ C^2 & \text{when} \quad z \leqslant h_0. \end{cases} \tag{16.1}$$

The total electric field of the wave is assumed to have only a horizontal component E_y, which must satisfy the differential equation (9.58). This becomes

$$\frac{d^2E_y}{dz^2} + k^2\{C^2 - a(z-h_0)\}E_y = 0 \quad \text{for} \quad z \geqslant h_0, \tag{16.2}$$

$$\frac{d^2E_y}{dz^2} + k^2C^2E_y = 0 \quad \text{for} \quad z \leqslant h_0. \tag{16.3}$$

Equation (16.2) is the Stokes equation. It may be reduced to the standard form (15.7) by the substitution

$$\zeta = (k^2\mathfrak{a})^{\frac{1}{3}}(z - h_0 - C^2/\mathfrak{a}),$$

where the value of $(k^2\mathfrak{a})^{\frac{1}{3}}$ is taken to be real and positive. The variable ζ is thus a new measure of the height, in which the origin is taken at the 'reflection level' $z = h_0 + C^2/\mathfrak{a}$ where $q = 0$, and the scale is changed by the factor $(k^2\mathfrak{a})^{\frac{1}{3}}$. Two independent solutions of (16.2) are then $\mathrm{Ai}(\zeta)$, and $\mathrm{Bi}(\zeta)$ (see § 15.23), and the required solution of (16.2) must be a linear combination of these. Now when z is very large and positive, ζ is also large and positive and $\mathrm{Ai}(\zeta)$ and $\mathrm{Bi}(\zeta)$ may then be replaced by their asymptotic approximations. That for $\mathrm{Ai}(\zeta)$ is given by (15.76) and includes only the subdominant term $\zeta^{-\frac{1}{4}}\exp(-\frac{2}{3}\zeta^{\frac{3}{2}})$ which gets indefinitely smaller as ζ increases. That for $\mathrm{Bi}(\zeta)$ is given by (15.80) or (15.83) and includes the dominant term $\zeta^{-\frac{1}{4}}\exp(\frac{2}{3}\zeta^{\frac{3}{2}})$ which gets indefinitely larger as ζ increases. If the solution included any multiple of $\mathrm{Bi}(\zeta)$ the electric field would get larger and larger as the height z increased, and the energy stored in the field would be extremely large at very great heights. This obviously could not happen in a field produced by radio waves incident on the ionosphere from below. Hence the solution cannot contain any multiple of $\mathrm{Bi}(\zeta)$ and so the physical conditions of the problem show that in the required solution the electric field E_y is some multiple of $\mathrm{Ai}(\zeta)$. Further discussions of how to choose the correct solution are given in §§ 16.3, 16.5 and 16.6.

The horizontal component $\mathscr{H}_x$ of the magnetic field is found from the Maxwell equation (9.57). Above the level $z = h_0$ the fields are therefore given by

$$\left.\begin{aligned} E_y &= K\,\mathrm{Ai}(\zeta), \\ \mathscr{H}_x &= -iK(\mathfrak{a}/k)^{\frac{1}{3}}\,\mathrm{Ai}'(\zeta), \end{aligned}\right\} \tag{16.4}$$

where K is a constant.

Below the level $z = h_0$ the medium is free space and two independent solutions of the differential equation (16.3) may be taken to be e^{-ikCz} and e^{+ikCz}. The first represents an upgoing wave, the incident wave, and the second represents a downgoing wave, the reflected wave. Suppose that the electric field of the incident wave has unit amplitude and the reflection coefficient is R. Then the solution of (16.3) below the level $z = h_0$ is

$$\left.\begin{aligned} E_y &= e^{-ikCz} + R\,e^{ikCz}, \\ \mathscr{H}_x &= -C\,e^{-ikCz} + RC\,e^{ikCz}. \end{aligned}\right\} \tag{16.5}$$

At the base of the ionosphere where $z = h_0$ there is a discontinuity of gradient of electron density and the two solutions (16.4) and (16.5) must here be fitted together correctly. The process is very similar to that discussed in §8.2 for reflection at a sharp boundary, and the arguments of that section show that the boundary conditions (8.1) and (8.2) must hold also in the present problem. Hence

$$e^{-ikCh_0} + R\,e^{ikCh_0} = K\,\mathrm{Ai}\{-(k^2\mathfrak{a})^{\frac{1}{3}}C^2/\mathfrak{a}\}, \tag{16.6}$$

$$-C\,e^{-ikCh_0} + RC\,e^{ikCh_0} = -iK(\mathfrak{a}/k)^{\frac{1}{3}}\,\mathrm{Ai}'\{-(k^2\mathfrak{a})^{\frac{1}{3}}C^2/\mathfrak{a}\}. \tag{16.7}$$

By dividing these two equations, the constant K is cancelled, and a rearrangement then gives

$$R = e^{-2ikCh_0}\frac{C\,\mathrm{Ai}(\zeta_0) - i(\mathfrak{a}/k)^{\frac{1}{3}}\,\mathrm{Ai}'(\zeta_0)}{C\,\mathrm{Ai}(\zeta_0) + i(\mathfrak{a}/k)^{\frac{1}{3}}\,\mathrm{Ai}'(\zeta_0)}, \tag{16.8}$$

where ζ_0 is written for $-C^2(k/\mathfrak{a})^{\frac{2}{3}}$. Since ζ_0 is real, the numerator and denominator of the last term of (16.8) are complex conjugates, and hence $|R| = 1$. The reflection is therefore complete, which was to be expected since electron collisions are neglected and all the energy in the incident wave should be reflected. The phase difference at the ground between the incident and reflected waves is equal to $\arg R$.

The functions Ai and Ai′ in (16.8) may be replaced by their asymptotic approximations (15.77), (15.79) (for $\arg\zeta_0 = \pi$) if $|\zeta_0|$ is large enough, that is if the gradient $\mathfrak{a}$ is small and C is not too small. This is equivalent to the condition that the W.K.B. solutions of (16.2) must be good approximations at the boundary $z = h_0$, and the upper limit to the value of $\mathfrak{a}$ may therefore be found from (9.61). When this condition is satisfied, (16.8) leads to

$$R = i\exp\{-2ikCh_0 - \tfrac{4}{3}iC^3k/\mathfrak{a}\}. \tag{16.9}$$

Fig. 16.1 shows how the electric field E_y and the horizontal component $\mathscr{H}_x$ of the magnetic field vary with height z in a typical case. Below the boundary at $z = h_0$ the upgoing and downgoing waves have equal amplitudes and form a standing wave-system in which the zeros are spaced λ/C apart. This is continued into the ionosphere, but here the spacing of the zeros and the amplitudes of the maxima vary with height. Above the 'reflection' level, where $q = 0$, there are no more oscillations of the fields in space. The amplitude here decreases as the height z increases but the field has the same phase everywhere, and (16.4) shows that E_y and $\mathscr{H}_x$ are in quadrature. The disturbance is very similar to the 'evanescent wave' described in §4.6.

16.3 Application to a slowly varying profile

We now discuss the reflection of radio waves from an ionosphere in which the electron number density increases continuously and monotonically with height z. Electron collisions are neglected and the waves are assumed to be obliquely incident and horizontally polarised. The medium is assumed to be slowly varying, as explained below.

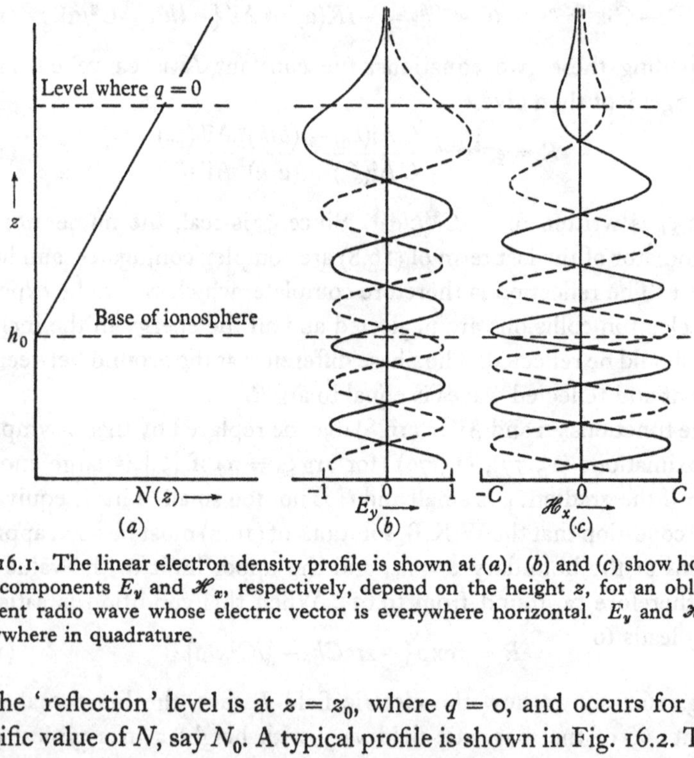

Fig. 16.1. The linear electron density profile is shown at (a). (b) and (c) show how the field components E_y and $\mathscr{H}_x$, respectively, depend on the height z, for an obliquely incident radio wave whose electric vector is everywhere horizontal. E_y and $\mathscr{H}_x$ are everywhere in quadrature.

The 'reflection' level is at $z = z_0$, where $q = 0$, and occurs for some specific value of N, say N_0. A typical profile is shown in Fig. 16.2. There is a certain range of height, suppose it is cd, in Fig. 16.2, where $|q|$ is so small that the condition (9.61) does not apply, and the W.K.B. solutions cannot be used. Outside this range the W.K.B. solutions are good approximations because the medium is slowly varying. It is now assumed that N varies linearly with z over a range be which includes the smaller range cd. Thus in proceeding away from the reflection level z_0, the profile $N(z)$ remains linear until $|z - z_0|$ is large enough for the asymptotic approximations to be used. Thereafter the W.K.B. solutions are valid and it does not matter if the $N(z)$ profile departs from linearity. The question as to how close $N(z)$ must be to linearity is considered in § 16.8.

The electric field E_y of the wave must satisfy the differential equation (9.58), viz.

$$\frac{d^2E_y}{dz^2}+k^2q^2E_y=0.$$

Within the range *be* q^2 varies linearly with z so that

$$q^2=-\mathfrak{a}(z-z_0), \tag{16.10}$$

where $\mathfrak{a}$ is a constant. The differential equation is then

$$\frac{d^2E_y}{dz^2}=\mathfrak{a}k^2(z-z_0)\,E_y, \tag{16.11}$$

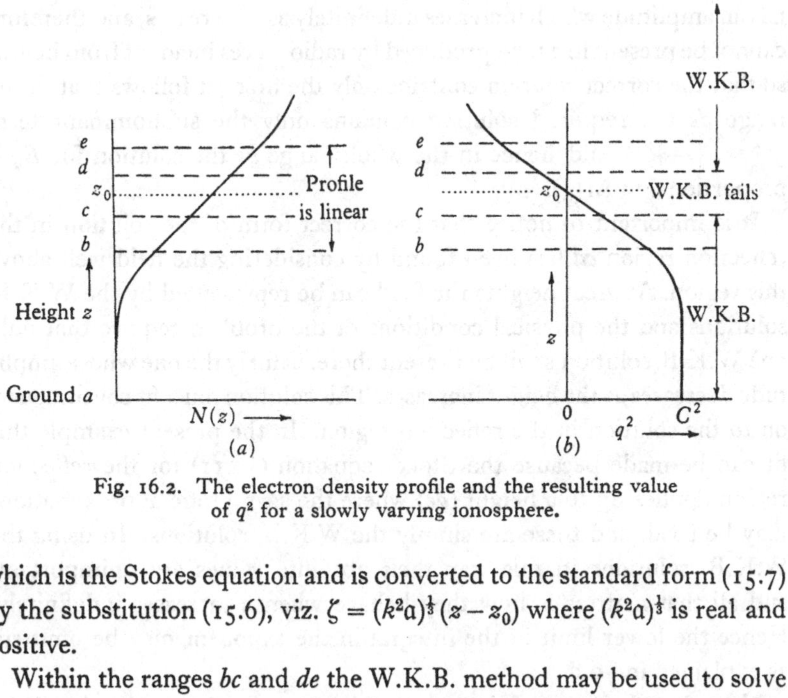

Fig. 16.2. The electron density profile and the resulting value of q^2 for a slowly varying ionosphere.

which is the Stokes equation and is converted to the standard form (15.7) by the substitution (15.6), viz. $\zeta=(k^2\mathfrak{a})^{\frac{1}{3}}(z-z_0)$ where $(k^2\mathfrak{a})^{\frac{1}{3}}$ is real and positive.

Within the ranges *bc* and *de* the W.K.B. method may be used to solve the Stokes equation, as in § 15.9, and the result is the two asymptotic approximations (15.26) and (15.27). Alternatively the W.K.B. method may be applied directly to the differential equation (9.58), which gives the two solutions (9.59), in which q is found from (9.56), and is given by $\{C^2-X/(1-iZ)\}^{\frac{1}{2}}$, where the value of the square root in the fourth quadrant is taken. In the present example $Z=0$, so that q is either real and positive, or negative imaginary. Consider first the range *de* where q is negative imaginary. The solutions $Aq^{-\frac{1}{2}}\exp\left\{-ik\int^z q\,dz\right\}$ from (9.59) and $\zeta^{-\frac{1}{4}}\exp\{-\frac{2}{3}\zeta^{\frac{3}{2}}\}$ from (15.26) are identical apart from a multiplying constant, provided that the positive value of $\zeta^{\frac{3}{2}}$ is used. Similarly, the solutions

$$Aq^{-\frac{1}{2}}\exp\left\{ik\int^{z} q\,dz\right\} \quad\text{and}\quad \zeta^{-\frac{1}{4}}\exp\{\tfrac{2}{3}\zeta^{\frac{3}{2}}\}$$

are identical within the range *de*. Above the level *e* this equality may disappear if the profile departs from linearity. The possible solutions are now

$$Aq^{-\frac{1}{2}}\exp\left\{-ik\int^{z} q\,dz\right\} \quad\text{and}\quad Aq^{-\frac{1}{2}}\exp\left\{ik\int^{z} q\,dz\right\},$$

and the required solution is a linear combination of them. The second has an amplitude which increases indefinitely as z increases, and therefore cannot be present in a field produced by radio waves incident from below, so that the correct solution contains only the first. It follows that in the range *de* the required solution contains only the subdominant term $\zeta^{-\frac{1}{4}}\exp\{-\frac{2}{3}\zeta^{\frac{3}{2}}\}$, and hence in the whole range *be* the solution for E_y is proportional to $\mathrm{Ai}(\zeta)$.

It is important to notice that the correct form of the solution in the reflection region *cd* has been found by considering the field well above this region. At great heights the field can be represented by the W.K.B. solutions and the physical conditions of the problem require that only one W.K.B. solution shall be present there, namely the one whose amplitude decreases as the height increases. This solution must fit continuously on to the solution in the reflection region. In the present example this fit can be made because the Stokes equation (16.11) for the reflection region applies up to a height (*de*) where the asymptotic approximations may be used, and these are simply the W.K.B. solutions. In using the W.K.B. solutions in this way their absolute values are unimportant, and all that matters is how they behave when z increases indefinitely. Hence the lower limit of the integral in the exponent may be omitted, as explained in § 9.5.

Now that the form of the solution in the reflection region has been found, its asymptotic approximation *below* the reflection level may be fitted to the W.K.B. solutions there, and thence the reflection coefficient is found. Within the range *ac* (Fig. 16.2) the two W.K.B. solutions are

$$q^{-\frac{1}{2}}\exp\left\{\mp ik\int_0^{z} q\,dz\right\},$$

and the required solution is proportional to

$$q^{-\frac{1}{2}}\left\{\exp\left(-ik\int_0^{z} q\,dz\right) + R\exp\left(ik\int_0^{z} q\,dz\right)\right\}, \qquad (16.12)$$

where R is the reflection coefficient and is the ratio of the amplitudes of the downgoing wave to the upgoing wave at the ground, where $z = 0$. Notice that the lower limits of the integrals are now specified. The expression (16.12) may be multiplied by any constant. Let it be multiplied by

$$\exp\left(ik\int_0^{z_0} q\,dz\right).$$

It then becomes

$$q^{-\frac{1}{2}}\left\{\exp\left(-ik\int_{z_0}^{z} q\,dz\right)+R\exp\left(2ik\int_0^{z_0} q\,dz\right)\exp\left(ik\int_{z_0}^{z} q\,dz\right)\right\}. \tag{16.13}$$

This is a good approximation to the required solution provided that z is below the level c. The fixed lower limit z_0 of the integrals is above this level, but this does not matter, provided that the variable upper limit z is below c. In particular (16.13) holds in the range bc, where q is given by (16.10) so that

$$\int_{z_0}^{z} q\,dz = -\int_{z}^{z_0} \mathfrak{a}^{\frac{1}{3}}(z_0-z)^{\frac{1}{2}}\,dz = -\tfrac{2}{3}\mathfrak{a}^{\frac{1}{3}}(z_0-z)^{\frac{3}{2}}.$$

Now ζ is given by (15.6) in this range, and is negative. Let $\arg\zeta = \pi$ and $\arg(\zeta^{\frac{3}{2}}) = \frac{3}{2}\pi$. Then

$$ik\int_{z_0}^{z} q\,dz = \tfrac{2}{3}\zeta^{\frac{3}{2}}. \tag{16.14}$$

Now $q^{-\frac{1}{2}}$ is proportional to $\zeta^{-\frac{1}{4}}$. Hence (16.13) is proportional to

$$\zeta^{-\frac{1}{4}}\left\{\exp(-\tfrac{2}{3}\zeta^{\frac{3}{2}})+R\exp\left(2ik\int_0^{z_0} q\,dz\right)\exp(\tfrac{2}{3}\zeta^{\frac{3}{2}})\right\}. \tag{16.15}$$

But we have already seen that the correct solution in the range *be* is proportional to $\mathrm{Ai}(\zeta)$. Thus for the range bc the expression (16.15) must agree with the asymptotic approximation for $\mathrm{Ai}(\zeta)$ when $\arg\zeta = \pi$, given by (15.77). Hence

$$R\exp\left(2ik\int_0^{z_0} q\,dz\right) = i. \tag{16.16}$$

This is simply the Stokes constant of the Stokes equation (see §15.13). Hence the reflection coefficient is

$$R = i\exp\left(-2ik\int_0^{z_0} q\,dz\right). \tag{16.17}$$

This should be compared with (9.37), which was based on an intuitive 'ray theory' argument. The 'full wave' theory of the present section shows that the ray-theory result (9.37) must be multiplied by a factor i. In other words there is a phase advance of $\frac{1}{2}\pi$ associated with reflection, which would not be predicted from the approximate ray theory. In many problems this phase advance is unimportant, and the expression (9.37) may then be used. This has been done throughout chs. 10 to 14.

16.4 The effect of electron collisions. The height z as a complex variable

When the effect of electron collisions is included the variable q is given by (9.56), viz. $q^2 = C^2 - [X/(1-iZ)]$. Hence q is complex and since X and Z are real and positive for all real heights z, q cannot be zero at any real height, and it would seem that there is no 'reflection' level. The W.K.B. solution which represents the upgoing wave is

$$q^{-\frac{1}{2}} \exp\left\{-ik\int_0^z q\,dz\right\}.$$

The real part of q is always positive and hence the planes of constant phase travel upwards at all levels. The imaginary part of q is negative (see § 4.5; the argument there applies to n but can be used also for q), so that the waves are attenuated as they travel. This would seem to show that a wave incident from below simply travels upwards indefinitely and its energy is absorbed by the medium. There may, however, be a range of height where the W.K.B. solutions fail and then reflection occurs. Although q cannot be zero at real heights, it may become small enough for the condition (9.61) to be violated, and this determines the level of reflection.

The condition for q to be zero is

$$X = C^2(1-iZ). \qquad (16.18)$$

In many problems Z and X are known analytic functions of z. It is then often possible to find a complex value of z which satisfies (16.18). For example, if Z is a constant and $X = \alpha(z-h_0)$, then q has a zero where $z = h_0 + (1-iZ)C^2/\alpha = z_0$. This is a complex height which therefore does not correspond to any real level in the ionosphere, but it is nevertheless of great importance mathematically. Fig. 16.3 shows a diagram of the complex z-plane for this case, and the point $z = z_0$ is below the real z-axis. Surrounding it there is a region (shaded in the figure) where q is so small that the condition (9.61) is violated. This region may intersect a short range of the real z-axis, which is then the range of height where reflection occurs.

The variable q is now treated as a function of the complex variable z. Equation (9.56) shows that q is a two-valued function of z with a zero at $z = z_0$, which is also a branch point of q. Similarly, the electric field E_y is a complex function of the complex variable z. The idea of the height z as a complex variable will be used extensively throughout this book.

16.5 Constant collision-frequency. Purely linear profile of electron density

Suppose that the electron number density $N(z)$ varies with height z as described in § 16.2, but that in addition the electron collision-frequency ν is constant, so that Z is constant and not zero. Now q^2 is given by (9.56), and hence (16.1) must be replaced by

$$q^2 = \begin{cases} C^2 - \dfrac{a}{1-iZ}(z-h_0) & \text{when} \quad z \geqslant h_0, \\ C^2 & \text{when} \quad z \leqslant h_0. \end{cases} \tag{16.19}$$

Below the ionosphere this is the same as before and leads to the differential equation (16.3). Above the lower boundary of the ionosphere at $z = h_0$, however, the differential equation becomes (compare (16.2)):

$$\frac{d^2E_y}{dz^2} + k^2\left\{C^2 - \frac{a}{1-iZ}(z-h_0)\right\}E_y = 0. \tag{16.20}$$

This again is the Stokes equation. To convert it to the standard form we take

$$\zeta = \left(\frac{k^2 a}{1-iZ}\right)^{\frac{1}{3}} \{z - h_0 - (1-iZ)\,C^2/a\}. \tag{16.21}$$

The variable ζ is thus a new measure of height, in which the origin is taken where $q = 0$, that is at the complex height $z = z_0$, where

$$z_0 = h_0 + (1-iZ)\,C^2/a. \tag{16.22}$$

The scale is changed by the complex factor

$$g = \left(\frac{k^2 a}{1-iZ}\right)^{\frac{1}{3}}, \tag{16.23}$$

and since this is a cube root it may be chosen in three different ways. When Z was zero (§ 16.2) we chose the value which was real and positive. We now choose the value which goes continuously into a real positive value as $Z \to 0$. Hence

$$\arg g = \tfrac{1}{3}\arctan Z. \tag{16.24}$$

This change of scale includes a rotation in the complex z-plane through an angle $\arg g$, which must be less than $\frac{1}{6}\pi$. Fig. 16.3 is a diagram of the complex z-plane showing the positions of the real and imaginary ζ-axes.

A solution of the equation (16.20) must now be chosen which satisfies the physical conditions of the problem at great heights, that is, when z is real, positive and large. Then $|\zeta|$ is large, and $\arg \zeta \to \arg g$, so that $\arg \zeta$ is in the range $0 \leqslant \arg \zeta \leqslant \frac{2}{3}\pi$. The asymptotic approximation of the solution $\mathrm{Ai}(\zeta)$ is given by (15.76) which contains the factor $\exp(-\frac{2}{3}\zeta^{\frac{3}{2}})$, where $\zeta^{\frac{3}{2}}$ has a negative imaginary part and a positive real part. Hence it represents a wave whose amplitude decreases as $|\zeta|$ or $|z|$ increases, and whose phase is propagated upwards. This is therefore an acceptable solution. The asymptotic approximation for $\mathrm{Bi}(\zeta)$ is given by (15.80) and includes a term with the factor $\exp(\frac{2}{3}\zeta^{\frac{3}{2}})$, which represents a wave whose amplitude increases indefinitely as $|\zeta|$ or $|z|$ increases, and whose phase is propagated downwards. It can be shown that the energy in this wave flows downwards, and it could not therefore be produced by a wave incident on the ionosphere from below. Hence the correct solution is proportional to $\mathrm{Ai}(\zeta)$. Notice that when collisions were neglected (§ 16.2) the asymptotic approximation to $\mathrm{Ai}(\zeta)$ at great heights resembled an evanescent wave, since there was no variation of phase with height. When collisions are included there is some upward phase propagation since ζ is no longer real.

The horizontal component $\mathscr{H}_x$ of the magnetic field is found, as before, from the Maxwell equation (9.57). Above the level $z = h_0$ the fields are therefore given (compare (16.4)) by

$$\left.\begin{aligned} E_y &= K\,\mathrm{Ai}(\zeta), \\ \mathscr{H}_x &= -iK\{\alpha/(1-iZ)\,k\}^{\frac{1}{3}}\,\mathrm{Ai}'(\zeta), \end{aligned}\right\} \tag{16.25}$$

where K is constant. This solution must be fitted to that below the boundary at $z = h_0$, which may be done exactly as in § 16.2. Hence it can be shown that the reflection coefficient R is given by

$$R = e^{-2ikCh_0}\frac{C\,\mathrm{Ai}(\zeta_0) - i\{\alpha/(1-iZ)\,k\}^{\frac{1}{3}}\,\mathrm{Ai}'(\zeta_0)}{C\,\mathrm{Ai}(\zeta_0) + i\{\alpha/(1-iZ)\,k\}^{\frac{1}{3}}\,\mathrm{Ai}'(\zeta_0)}, \tag{16.26}$$

where $\zeta_0 = -C^2\{k(1-iZ)/\alpha\}^{\frac{2}{3}}$ and the value of $(1-iZ)^{\frac{1}{3}}$ is in the fourth quadrant.

If $|\zeta_0|$ is large enough, the functions Ai and Ai$'$ may be replaced by their asymptotic approximation (15.77) and (15.79), since now

$$\pi \leqslant \arg \zeta \leqslant \tfrac{5}{6}\pi.$$

This leads to $$R = i\exp\{-2ikCh_0 - \tfrac{4}{3}iC^3k(1-iZ)/\alpha\}. \tag{16.27}$$

When collisions are neglected $Z = 0$, and (16.27) then reduces to (16.9) for which $|R| = 1$. When collisions are present, the exponent of (16.27) has a negative real part, and hence

$$|R| = \exp\{-\tfrac{4}{3}C^3kZ/\alpha\}. \qquad (16.28)$$

This is less than unity because some of the incident energy is absorbed.

16.6 The slowly varying profile when collisions are included. Derivation of the phase integral formula

In § 16.3 the expression (16.17) was derived for the reflection coefficient of a slowly varying ionosphere when electron collisions were neglected. The corresponding result when collisions are present will now be derived.

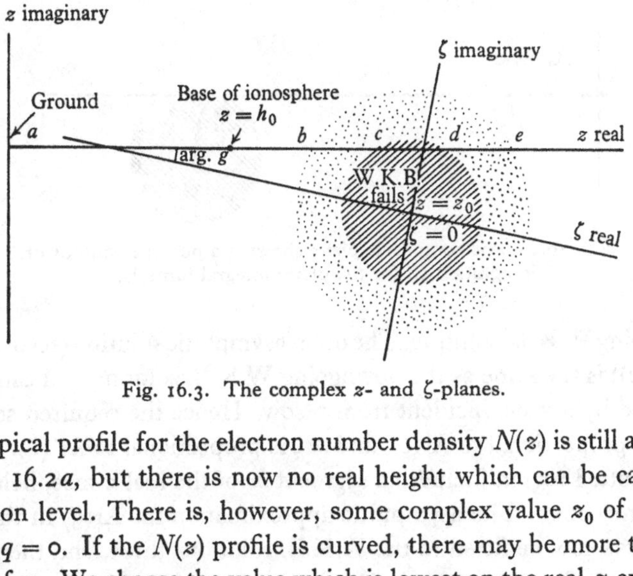

Fig. 16.3. The complex z- and ζ-planes.

A typical profile for the electron number density $N(z)$ is still as shown in Fig. 16.2a, but there is now no real height which can be called the reflection level. There is, however, some complex value z_0 of z which makes $q = 0$. If the $N(z)$ profile is curved, there may be more than one value of z_0. We choose the value which is lowest on the real z-axis when $Z = 0$. Then when $Z \neq 0$ it can be shown that this value moves below the real z-axis as in Figs. 16.3 or 16.4.

Near $z = z_0$ there is a region of the complex z-plane (shaded in Figs. 16.3 and 16.4) where $|q|$ is so small that the condition (9.61) does not apply and the W.K.B. solutions cannot be used. Outside this region the W.K.B. solutions are good approximations because the medium is slowly varying. It is now assumed that Z is constant, and that $N(z)$ varies linearly with z over a region of the complex z-plane (dotted in Fig. 16.3) which completely includes the smaller shaded region. Thus in proceeding away from z_0 the function $q^2(z)$ remains linear until $|z-z_0|$ is large enough for the asymptotic approximations to be used. Beyond this the W.K.B. solutions are valid and it does not matter if $q^2(z)$ departs from linearity. The question as to how close $q^2(z)$ must be to linearity is considered in § 16.8. Departures from linearity can occur either because $N(z)$ is not linear, or because Z is not constant.

Within the dotted region, therefore, the electric field E_y must satisfy the Stokes equation (16.20), and the variable ζ of equation (16.21) may be used as before. The region includes a segment *de* of the real z-axis where the asymptotic approximations may be used. At the upper end of this segment (just below *e*) $\arg\zeta$ must be in the range $0 \leqslant \arg\zeta \leqslant \frac{2}{3}\pi$. The only acceptable asymptotic solution is (15.76), for this is the same as

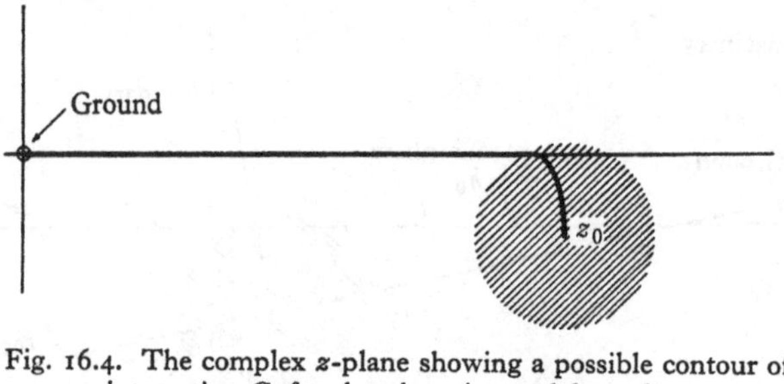

Fig. 16.4. The complex z-plane showing a possible contour of integration C, for the phase integral formula.

the upgoing W.K.B. solution. The other asymptotic solution (second term of (15.80)) is the same as the downgoing W.K.B. solution and cannot be generated by a wave incident from below. Hence the required solution for E_y is proportional to $\mathrm{Ai}(\zeta)$, and $\mathscr{H}_x$ is proportional to $\mathrm{Ai}'(\zeta)$.

The dotted region includes a segment *bc* of the real z-axis, below the level of reflection. The asymptotic approximation for $\mathrm{Ai}(\zeta)$ in this segment must now be fitted to the W.K.B. solutions there, and thence the reflection coefficient is found. It is now assumed that at the lower end of this segment (just above *b*) $\arg\zeta > \frac{2}{3}\pi$ (this condition is violated only if the profile $q^2(z)$ departs too much from linearity). The required approximation is then (15.77).

Within the range *ab* the required solution is (16.12). As before this may be multiplied by the constant $\exp\left(ik\int_0^{z_0} q\,dz\right)$. Here the upper limit of the integral is complex so that z is a complex variable and the integral is a contour integral. The contour is assumed to lie along the real axis as far as the variable limit z in (16.12). Thereafter it must leave the real axis to reach the point z_0 (Fig. 16.4). Then (16.12) is converted to (16.13) where the integrals are now contour integrals, and the contours are on the thick line of Fig. 16.4. The rest of the argument of § 16.3 may now be used to find the reflection coefficient and we obtain finally

$$R = i\exp\left(-2ik\int_0^{z_0} q\,dz\right), \tag{16.29}$$

which is the Phase Integral Formula for the reflection coefficient. The value of the integral in (16.29) is independent of the precise position of the contour, which may therefore be deformed to other positions provided that the end-points remain fixed and that no singularities of the integrand q are crossed.

16.7 Discussion of the phase integral formula

The complete expression for the W.K.B. solution which represents the upgoing wave is $E_y = q^{-\frac{1}{2}}\exp\left\{ik\left(ct - Sx - \int_0^z q\,dz\right)\right\}$. The exponent, without the factor i, is the phase, ϕ, of the wave, and is in general complex since q is complex. This idea of a complex phase is very convenient and will be used extensively. The variations with x and z of the real part of ϕ give the change of phase of the wave in the ordinary sense. The variations of the imaginary part of ϕ are associated with changes of amplitude. The integral $k\int_0^z q\,dz$ is the part of the complex phase which depends on z, and is called the 'phase integral'.

In the expression (16.29) for the reflection coefficient, the exponent is $2i$ times the phase integral, evaluated for some contour from the ground ($z = 0$) to the complex point of reflection ($z = z_0$). It may be called the total change of phase of a wave which goes from the ground to the reflection point and back, but it must be remembered that the reflection point is in general at a complex height. Equation (16.29) is called the 'phase integral' formula for the reflection coefficient. It is an example of Eckersley's 'Phase Integral Method' to which ch. 20 is devoted.

The value of the phase integral is unaffected by the precise position of the contour C, which may therefore be chosen so as to make the evaluation of the integral as easy as possible. Different possible contours are shown in Fig. 16.5 ($\Gamma_1, \Gamma_2, \Gamma_3$). It often proves convenient to use the contour Γ_1 which consists of two straight lines AB and BC at right angles. For the part AB the variable z is real, and the real part of the phase integral, $2k\int_0^{\mathscr{R}(z_0)} \mathscr{R}(q)\,dz$ is simply the change of phase when the wave travels from the ground to the height $\mathscr{R}(z_0)$ and down again. The imaginary part is $2k\int_0^{\mathscr{R}(z_0)} \mathscr{I}(q)\,dz$ which shows that there is a reduction in amplitude by a factor $\exp\left\{2k\int_0^{\mathscr{R}(z_0)} \mathscr{I}(q)\,dz\right\}$ associated with the part AB of the contour. ($\mathscr{I}(q)$ is always negative when z is real.) This occurs because some energy is absorbed from the wave on both the upward and downward paths. These changes of phase and amplitude would be expected on a simple ray theory.

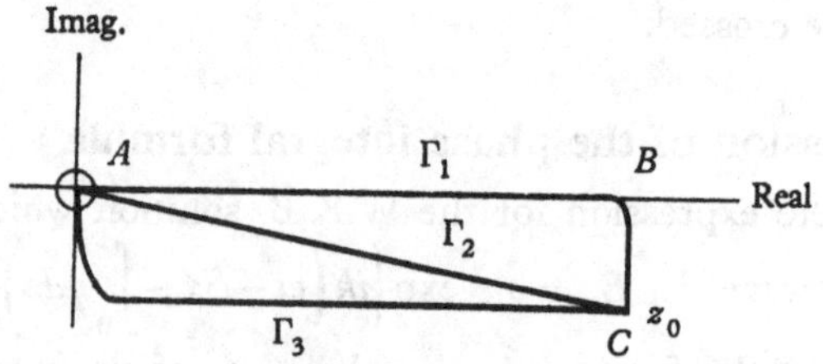

Fig. 16.5. Complex z-plane showing various possible contours for the phase integral formula.

On the part BC of the contour we may write $z = \mathscr{R}(z_0) - i\mathfrak{z}$, where $\mathfrak{z}$ is real and positive, and the contributions to the phase integral are:

$$\text{Real part} \quad 2k\int_0^{-\mathscr{I}(z_0)} \mathscr{I}(q)\,d\mathfrak{z}; \tag{16.30}$$

$$\text{Imaginary part} \quad -2k\int_0^{-\mathscr{I}(z_0)} \mathscr{R}(q)\,d\mathfrak{z}. \tag{16.31}$$

Hence there is an additional phase shift (16.30) and a further reduction of amplitude by a factor equal to the exponential of (16.31). These would not have been predicted from the simple ray theory and may be thought of as associated with the processes occurring very near the level of the reflection.

For the phase integral formula to be valid, two conditions are necessary:

(i) The square of the refractive index, $\mathrm{n}^2(z)$, must vary linearly with height z over a range near the reflection level where $q = 0$, the degree of linearity being specified later, equations (16.36) or (16.38).

(ii) There must be a range of height z where the W.K.B. solutions are good approximations, that is where (9.61) holds, and this must extend from the ground into the region where (i) applies.

The left side of (9.61) is inversely proportional to the square of the frequency and is therefore largest at low frequencies. Hence the condition is most easily violated at low frequencies. Similarly, the condition (16.36) is less likely to be fulfilled at low frequencies.

The phase integral formula (16.29) is discussed further in § 20.3, and some applications of it are given in §§ 17.3 and 17.7.

16.8 Effect of curvature of the electron density profile

We must now investigate how far $N(z)$ may depart from linearity at the reflection level without seriously affecting the conclusions of the last section. A slight curvature of the $N(z)$ profile means that squares and higher powers of $(z-z_0)$ may appear in (16.10). Suppose that

$$q^2 = -\mathfrak{A}(z-z_0)+\mathfrak{B}(z-z_0)^2. \tag{16.32}$$

If this is substituted in the differential equation (9.58) and the variable ζ given by (15.6) is used, we obtain

$$\frac{d^2E_y}{d\zeta^2}-\{\zeta-\mathfrak{B}(k\mathfrak{A}^2)^{-\frac{4}{3}}\zeta^2\}E_y = 0. \tag{16.33}$$

If $E_y = \mathrm{Ai}(\zeta)$ is substituted in (16.33) the left side becomes

$$\mathfrak{B}(k\mathfrak{A}^2)^{-\frac{4}{3}}\zeta^2\,\mathrm{Ai}(\zeta), \tag{16.34}$$

which arises from the last term. If $\mathrm{Ai}(\zeta)$ is to be a good approximation to the solution, then (16.34) must be small compared with either of the remaining terms. This requires that

$$|\mathfrak{B}(k\mathfrak{A}^2)^{-\frac{4}{3}}\zeta| \ll 1, \tag{16.35}$$

and this must hold for values of $|\zeta|$ large enough for the asymptotic approximations (15.26) and (15.27) to be used, that is for $|\zeta| \gg \frac{1}{2}$ from (15.28). If we take $|\zeta| \approx 1$, then (16.35) gives

$$|\mathfrak{B}| \ll |(k\mathfrak{A}^2)^{\frac{4}{3}}|. \tag{16.36}$$

If N_0 is the electron number density at the reflection level, we have

$$\mathfrak{A} = \frac{C^2}{N_0}\frac{dN}{dz}, \quad \mathfrak{B} = -\frac{C^2}{2N_0}\frac{d^2N}{dz^2}, \tag{16.37}$$

where the gradient dN/dz and curvature d^2N/dz^2 of the electron density profile are measured at the reflection level. Then (16.36) becomes

$$\left|\frac{1}{k^2}\frac{d^2N}{dz^2}\right| \ll 2\left|\left(\frac{C^2}{N_0}\right)^{\frac{1}{3}}\left(\frac{1}{k}\frac{dN}{dz}\right)^{\frac{4}{3}}\right|. \tag{16.38}$$

The conditions (16.36) or (16.38) are most likely to fail at low frequencies since k is then small. They will also fail at the maximum of an ionospheric layer, for then dN/dz is very small.

When the electron density profile is curved, the asymptotic approximations

(15.26) and (15.27) are no longer equal to the W.K.B. solutions (9.59). This difficulty may be overcome by using the variable

$$\eta = \left[\tfrac{3}{2}ik\int_0^\zeta q\,d\zeta\right]^{\frac{2}{3}} \tag{16.39}$$

instead of ζ, and taking Ai(η) as the solution of (9.58). It can be shown that $\eta = \zeta + O(\zeta^2)$, and in particular if q^2 is given by (16.10) then $\eta \equiv \zeta$. The arguments of § 16.3 may then be used with very little change. The variable η has been used by Langer (1937).

The derivation of the phase integral formula (16.29) depended on the properties of the solution near the zero of q at $z = z_0$ in a linear profile $N(z)$. If this profile is curved, as in the expression (16.32) for example, q still has a zero at $z = z_0$ but it has another zero where $z = z_0 + \mathfrak{A}/\mathfrak{B}$. If, however, $\mathfrak{B}$ is small enough to satisfy (16.36), this second zero is at a great distance from z_0 so that the phase integral formula is not appreciably affected. Both zeros are surrounded by regions where the asymptotic approximations (15.26) and (15.27) cannot be used. If (16.36) is not satisfied, then $\mathfrak{B}$ is large enough to bring the second zero too near to that at $z = z_0$, so that these regions partly overlap. Then the phase integral formula is not accurate, because the Airy function Ai(ζ) (see § 16.6) is not a sufficiently accurate solution near $z = z_0$.

More generally we may say that the phase integral formula (16.29) is accurate as long as the branch point of q at $z = z_0$ is isolated, but fails when other singularities of q approach too close to it. In some cases it is possible to modify the phase integral formula to allow for the other singularities, by replacing the Stokes constant i in (16.29) by a different constant. Examples are given in §§ 16.17 and 20.4. But it is more usual in these cases to abandon the attempt to treat the ionosphere like a slowly varying medium, and to use a 'full wave' solution (see ch. 17) or to solve the equations numerically (ch. 22).

16.9 Reflection at a discontinuity of gradient

Electromagnetic waves are reflected at the bounding surface between two media with different refractive indices, and ch. 8 was devoted to this topic. There is then a discontinuity in the refractive index. It is now of interest to ask whether reflection can occur at a surface where the refractive index is continuous, but the gradient of the refractive index is discontinuous.

A problem where there is such a discontinuity of gradient has already been discussed in § 16.5, where it was assumed that Z is constant and the profile of electron number density is linear, above the base of the ionosphere, so that there is a discontinuity of $dN(z)/dz$ where $z = h_0$. The expression (16.26) was deduced for the reflection coefficient. It has been shown that radio waves are reflected near the level where $z = z_0$. If there is also some reflection from the discontinuity at $z = h_0$, then (16.26) must be the result of combining these two reflections. Hence we must now determine whether (16.26) includes any contribution from the level $z = h_0$.

The field immediately above $z = h_0$ is given by (16.25), with $\zeta = \zeta_0$. If $|\zeta_0|$ is large enough the function Ai (ζ) may be replaced by its asymptotic approximation (15.77), which contains two terms of which one is an upgoing wave and the other a downgoing wave. Some (or all) of this downgoing wave is transmitted through the discontinuity and gives a downgoing wave at the ground, but this is not part of the reflection at $z = h_0$ and must be excluded. We therefore need to consider a different field configuration in which there is only an upgoing wave just above the level $z = h_0$.

First suppose that the gradients of $N(z)$ and of $\mathfrak{n}(z)$ and $q(z)$ above the discontinuity at $z = h_0$ are so small that the W.K.B. solutions may be used. Then the field is given by the W.K.B. solution representing an upgoing wave, namely

$$E_y = Aq^{-\frac{1}{2}} \exp\left\{-ik\int_{h_0}^{z} q\,dz\right\}, \tag{16.40}$$

$$\mathscr{H}_x = -Aq^{\frac{1}{2}} \exp\left\{-ik\int_{h_0}^{z} q\,dz\right\} \tag{16.41}$$

(from (9.59) and (9.60)), where A is a constant and the lower limit of the phase integral has been chosen to be at $z = h_0$. Immediately above the boundary $q = C$, $z = h_0$ and the expressions become

$$E_y = AC^{-\frac{1}{2}}, \tag{16.42}$$

$$\mathscr{H}_x = -AC^{\frac{1}{2}}. \tag{16.43}$$

Below the discontinuity the fields are given by

$$E_y = e^{-ikC(z-h_0)} + R\,e^{ikC(z-h_0)}, \tag{16.44}$$

$$\mathscr{H}_x = -C\,e^{-ikC(z-h_0)} + RC\,e^{ikC(z-h_0)}, \tag{16.45}$$

where R is the reflection coefficient, that is, the ratio of the amplitudes of the downgoing to the upgoing wave. Immediately below the discontinuity (where $z = h_0$) these expressions become

$$E_y = 1 + R, \tag{16.46}$$

$$\mathscr{H}_x = -C + CR, \tag{16.47}$$

and the boundary conditions (8.1) and (8.2) require that these are equated to (16.42) and (16.43) respectively. This gives at once

$$R = 0. \tag{16.48}$$

In this derivation the approximate W.K.B. solutions have been used.

The same result may be derived more simply as follows. For an upgoing W.K.B. solution (9.59) and (9.60) show that

$$\mathscr{H}_x = -qE_y. \tag{16.49}$$

Now q is the same on the two sides of a discontinuity of gradient, so that the ratio $E_y/\mathscr{H}_x$ (the wave impedance) is the same for upgoing waves on the two sides of the discontinuity. Hence there is no reflection.

We conclude that if the W.K.B. solutions are good approximations on both sides of a discontinuity of gradient of N or q, the reflection coefficient is zero to the same degree of approximation.

This result is illustrated by equation (16.27) which was obtained by using the asymptotic approximations, that is, the W.K.B. approximations, in (16.26). It is easy to show that (16.27) is exactly the phase integral formula (16.29), which gives the reflection at $z = z_0$, but does not involve the gradient of q and therefore takes no account of its discontinuity at $z = h_0$.

Next suppose that the gradients of $N(z)$ and of $\mathfrak{n}(z)$ and $q(z)$ above the discontinuity are so large that the W.K.B. solutions cannot be used. Then it is not possible to define a purely upgoing wave in this region and the problem of finding the reflection coefficient of a discontinuity has no meaning. The reflection coefficient of the whole ionosphere is still given by (16.26) but this does not now reduce to the phase integral formula, and the effects of the discontinuity at $z = h_0$ and of the zero of q at $z = z_0$ cannot be separated.

16.10 Linear gradient between two homogeneous regions

Consider an ionosphere in which the electron density profile is as shown in Fig. 16.6. The electron density N is zero below the base of the ionosphere at $z = h_0$. Above the level $z = h_1$, $N = N_1$, a constant, so that the ionosphere is there homogeneous. Between these levels $N(z)$ varies linearly with z. It is required to find the reflection coefficient. This problem was solved by Hartree (1929). The solution is given here only for the case of horizontal polarisation, when electron collisions are neglected. The effect of a constant collision-frequency could be included by an extension of the method of § 16.5. The treatment of 'vertical' polarisation at oblique incidence is more difficult, and in general could only be tackled by a numerical solution of the differential equations.

When $z \geqslant h_1$, let $X = X_1 = N_1 e^2/(\epsilon_0 m\omega^2)$, a constant. Then for the range $h_0 \leqslant z \leqslant h_1$, $X = X_1(z-h_0)/(h_1-h_0)$ and (9.56) shows that

$$q^2 = \begin{cases} C^2 & \text{for } z \leqslant h_0, \\ C^2 - X_1\dfrac{z-h_0}{h_1-h_0} & \text{for } h_0 \leqslant z \leqslant h_1, \\ C^2 - X_1 & \text{for } h_1 \leqslant z. \end{cases} \tag{16.50}$$

The variable q is everywhere either real and positive, or negative imaginary. When $z \geqslant h_1$, let $q = q_1$ where

$$q_1 = (C^2 - X_1)^{\frac{1}{2}}. \tag{16.51}$$

The electric field E_y satisfies the differential equation $d^2E_y/dz^2 + k^2q^2E_y = 0$. Above the level $z = h_1$ this has the two solutions e^{-ikq_1z} and e^{+ikq_1z}. Only the first is acceptable since the second represents either a downcoming wave (if q_1 is real and positive) or an evanescent wave whose amplitude increases indefinitely as z increases (if q_1 is negative imaginary). The field component $\mathscr{H}_x$ may be found from the Maxwell equation (9.57). Hence just above the level $z = h_1$ we have

$$\left.\begin{aligned} E_y &= D e^{-ikq_1z}, \\ \mathscr{H}_x &= -q_1 D e^{-ikq_1z}, \end{aligned}\right\} \tag{16.52}$$

where D is a constant.

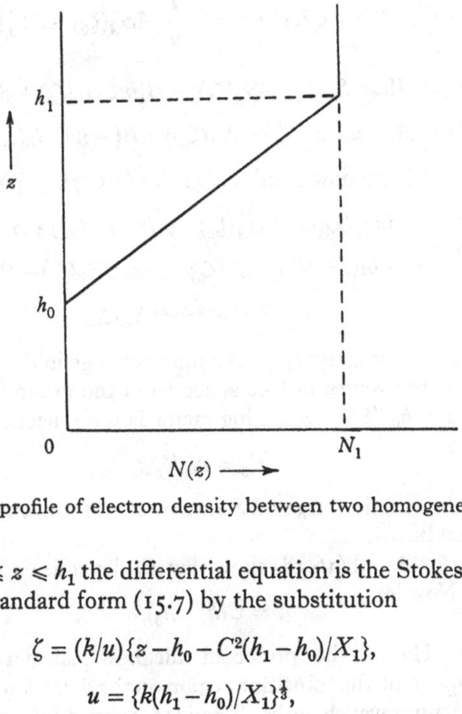

Fig. 16.6. Linear profile of electron density between two homogeneous regions.

In the range $h_0 \leqslant z \leqslant h_1$ the differential equation is the Stokes equation and is reduced to the standard form (15.7) by the substitution

$$\zeta = (k/u)\{z - h_0 - C^2(h_1 - h_0)/X_1\}, \tag{16.53}$$

where
$$u = \{k(h_1 - h_0)/X_1\}^{\frac{1}{3}}, \tag{16.54}$$

and the real positive value of the cube root is used. Let $S_A(\zeta)$, $S_B(\zeta)$ be any two independent solutions of the Stokes equation. Then E_y is a linear combination of them, and the field component $\mathscr{H}_x$ may again be found from the Maxwell equation (9.57). Hence in this range

$$\left.\begin{aligned} E_y &= AS_A(\zeta) + BS_B(\zeta), \\ \mathscr{H}_x &= -(i/u)\{AS'_A(\zeta) + BS'_B(\zeta)\}, \end{aligned}\right\} \tag{16.55}$$

where A and B are constants. At $z = h_1$ let $\zeta = \zeta_1$ so that

$$\zeta_1 = -(C^2 - X_1)\{k(h_1 - h_0)/X_1\}^{\frac{2}{3}} = -q_1^2 u^2. \tag{16.56}$$

Here there is a discontinuity of gradient of q, but E_y and $\mathscr{H}_x$ must be continuous. By equating (16.55) and (16.52) at $z = h_1$ we obtain

$$A\{iuq_1 S_A(\zeta_1) + S_A'(\zeta_1)\} + B\{iuq_1 S_B(\zeta_1) + S_B'(\zeta_1)\} = 0. \tag{16.57}$$

In the range $z \leqslant h_0$, that is below the ionosphere, the fields are given by (16.5). At $z = h_0$ let $\zeta = \zeta_0$ so that

$$\zeta_0 = -C^2 u^2. \tag{16.58}$$

The fields must be continuous at $z = h_0$, and by equating (16.5) and (16.55) at this level we obtain:

$$\left.\begin{aligned} e^{-ikCh_0} + R\,e^{ikCh_0} &= AS_A(\zeta_0) + BS_B(\zeta_0), \\ -C\,e^{-ikCh_0} + CR\,e^{ikCh_0} &= -\frac{i}{u}\{AS_A'(\zeta_0) + BS_B'(\zeta_0)\}, \end{aligned}\right\} \tag{16.59}$$

whence

$$2iuCR\,e^{ikCh_0} = A\{iuCS_A(\zeta_0) + S_A'(\zeta_0)\} + B\{iuCS_B(\zeta_0) + S_B'(\zeta_0)\}, \tag{16.60}$$

$$2iuC\,e^{-ikCh_0} = A\{-iuCS_A(\zeta_0) + S_A'(\zeta_0)\} + B\{-iuCS_B(\zeta_0) + S_B'(\zeta_0)\}. \tag{16.61}$$

To find R, A and B are now eliminated from (16.57), (16.60) and (16.61). Let

$$\Delta_\pm = \begin{vmatrix} iuq_1 S_A(\zeta_1) + S_A'(\zeta_1) & iuq_1 S_B(\zeta_1) + S_B'(\zeta_1) \\ \pm iuCS_A(\zeta_0) + S_A'(\zeta_0) & \pm iuCS_B(\zeta_0) + S_B'(\zeta_0) \end{vmatrix}. \tag{16.62}$$

Then

$$R = e^{-2ikCh_0}\Delta_+/\Delta_-. \tag{16.63}$$

The exponential factor arises from the phase change in the upward and downward passage of the waves in free space from the ground to the base of the ionosphere at $z = h_0$. The remaining factor is the reflection coefficient of the ionosphere, viz.

$$R_0 = \Delta_+/\Delta_-. \tag{16.64}$$

The choice of the functions S_A and S_B is a matter of convenience in calculation, and is discussed below.

Equations (16.56) and (16.58) show that R_0 depends on the two parameters C/q_1 and $q_1 u$. Now let

$$d = C(h_1 - h_0)/\lambda. \tag{16.65}$$

This is called by Hartree the 'projected thickness', since it is the projection of the thickness $h_1 - h_0$ of the transition region on the direction of the incident ray, measured in vacuum wavelengths. Then (16.51) and (16.53) give

$$(q_1 u)^3 = 2\pi d \bigg/ \left\{\left(\frac{C}{q_1}\right)^3 - \frac{C}{q_1}\right\}. \tag{16.66}$$

Hence we may equally well take C/q_1 and d as the two parameters on which R_0 depends.

Before discussing the formula (16.64) some properties of the reflection coefficient R_0 can be deduced by simpler arguments based on the results of earlier sections. Suppose that in the transition region where $h_0 \leqslant z \leqslant h_1$, the gradient of electron density is so low that the W.K.B. approximations can be used in some part of the range. Then there are two possibilities.

(*a*) *The W.K.B. approximations are good throughout the range* $h_0 \leqslant z \leqslant h_1$. This could only happen if q is everywhere positive, that is if $X_1 < C^2$. Then the W.K.B. approximations are good on both sides of the discontinuities in the gradient of q at $z = h_0$ and $z = h_1$. It was shown in § 16.9 that there is no appreciable reflection in these conditions. Hence the reflection coefficient of the ionosphere is then negligibly small.

(*b*) *W.K.B. approximations are good at* $z = h_1$, *but* q *has a zero below this.* In this case q_1 is negative imaginary so that the wave in the region where $z > h_1$ is evanescent and its amplitude decreases as z increases. Just below $z = h$, the W.K.B. solution is the corresponding evanescent approximation to the solution of the Stokes equation, that is the subdominant term (15.76) (see § 16.2). Hence the solution in the transition region is simply $\mathrm{Ai}(\zeta)$ which is the same as if the discontinuity at $z = h_1$ did not exist. The problem thus reduces to that solved in § 16.2, and the reflection coefficient is given by (16.8). The 'reflection' level is then where $q = 0$ which must be within the range $h_0 < z < h_1$. If the W.K.B. approximations are good also at $z = h_0$, then the reflection coefficient can be further simplified and is given by (16.9).

When these approximations cannot be made, the formulae (16.62) and (16.63) must be used, with a suitable choice of the functions S_A and S_B. One possible choice is to take $S_A(\zeta) = \mathrm{Ai}(\zeta)$, $S_B(\zeta) = \mathrm{Bi}(\zeta)$. The determinants $\Delta_\pm$ can then be computed by using the tables of $\mathrm{Ai}(\zeta)$, $\mathrm{Bi}(\zeta)$ and their derivatives (Miller, 1946). An alternative to this is available when u is small. Then S_A and S_B may be taken as the two series in (15.8), so that

$$\left.\begin{aligned} S_A(\zeta) &= 1 + \frac{\zeta^3}{3.2} + \frac{\zeta^6}{6.5.3.2} + O(\zeta^9), \\ S_B(\zeta) &= \zeta + \frac{\zeta^4}{4.3} + \frac{\zeta^7}{7.6.4.3} + O(\zeta^{10}), \end{aligned}\right\} \tag{16.67}$$

and

$$\left.\begin{aligned} iuq_1 S_A(\zeta_1) + S_A'(\zeta_1) &= iuq_1 + \frac{(iuq_1)^4}{2} + \frac{(iuq_1)^7}{3.2} + \frac{(iuq_1)^{10}}{5.3.2} + \frac{(iuq_1)^{13}}{6.5.3.2} + O(u^{16}), \\ iuq_1 S_B(\zeta_1) + S_B'(\zeta_1) &= 1 + (iuq_1)^3 + \frac{(iuq_1)^6}{3} + \frac{(iuq_1)^9}{4.3} + \frac{(iuq_1)^{12}}{6.4.3} + O(u^{15}). \end{aligned}\right\} \tag{16.68}$$

The terms of the last rows of the determinants (16.62) are obtained from (16.68) on replacing q_1 by $\pm C$. The determinants may now be evaluated. They have a common factor iu which may be cancelled and (16.64) becomes

$$R_0 = \frac{C-q_1}{C+q_1} \frac{1 - \tfrac{1}{2}i\dfrac{2\pi d}{C}(C-q_1) + \dfrac{1}{6}\left(\dfrac{2\pi d}{C}\right)^2 (C^2 - Cq_1 + q_1^2) + O(d^3)}{1 + \tfrac{1}{2}i\dfrac{2\pi d}{C}(C-q_1) - \dfrac{1}{6}\left(\dfrac{2\pi d}{C}\right)^2 (C^2 - Cq_1 + q_1^2) + O(d^3)}. \tag{16.69}$$

When the projected thickness d (from (16.65)) tends to zero, the transition layer becomes indefinitely thin, so that the problem is that of reflection at the sharp boundary between two homogeneous media. The reflection coefficient (16.69) is then the same as the Fresnel formula (8.22), namely

$$R_1 = \frac{C - q_1}{C + q_1}. \tag{16.70}$$

Hartree has given curves showing how $|R_0/R_1|^2$ depends on d for various values of C/q_1.

Fig. 16.7. The double linear profile of electron density.

16.11 Symmetrical ionosphere with double linear profile

As a rough approximation to an ionospheric layer with a maximum of electron density at its centre, we may take the profile shown in Fig. 16.7, in which the electron density is zero above and below the ionosphere, and increases linearly to a maximum at the centre, the profile being symmetrical. As before in this chapter, electron collisions are neglected, and we consider only horizontal polarisation. This case was also discussed by Hartree (1929). It is convenient to take the origin of the height z at the centre of the ionosphere, so that

$$X = X_0(1 - |z/z_1|) \quad \text{for} \quad |z| < z_1, \tag{16.71}$$

where X_0 is the maximum value of X, and the top and bottom of the ionosphere are at $z = \pm z_1$. Hence

$$q^2 = \begin{cases} C^2 & |z| \geqslant z_1, \\ C^2 - X_0 + X_0(z/z_1) & 0 \leqslant z \leqslant z_1, \\ C^2 - X_0 - X_0(z/z_1) & -z_1 \leqslant z \leqslant 0. \end{cases} \tag{16.72}$$

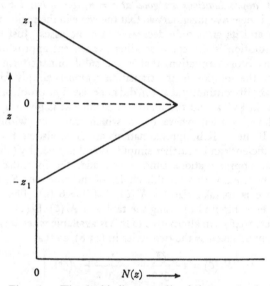

Just above the top of the ionosphere there is only an upgoing wave, and hence

$$\left.\begin{aligned} E_y &= D e^{-ikCz}, \\ \mathscr{H}_x &= -CD e^{-ikCz}, \end{aligned}\right\} \tag{16.73}$$

where D is a constant. Now consider the upper half of the ionosphere for which $0 \leqslant z \leqslant z_1$, and let

$$\left.\begin{aligned} u &= \left(\frac{kz_1}{X_0}\right)^{\frac{1}{3}}, \quad \zeta = -\frac{k}{u}\{z - z_1 + C^2 z_1/X_0\}, \\ \zeta_1 &= -\frac{k}{u}\frac{C^2 z_1}{X_0}, \quad \zeta_0 = -\frac{kz_1}{u}\left(\frac{C^2}{X_0} - 1\right). \end{aligned}\right\} \tag{16.74}$$

Then the equation satisfied by E_y reduces to the Stokes equation. Let $S_A(\zeta)$, $S_B(\zeta)$ be any two independent solutions. Then

$$\left.\begin{aligned} g(\zeta) &= S_A(\zeta)S_B(\zeta_0) - S_B(\zeta)S_A(\zeta_0), \\ h(\zeta) &= -S_A(\zeta)S_B'(\zeta_0) + S_B(\zeta)S_A'(\zeta_0) \end{aligned}\right\} \tag{16.75}$$

are also solutions such that

$$g(\zeta_0) = 0, \quad h'(\zeta_0) = 0. \tag{16.76}$$

The solution for the upper half of the ionosphere may be written

$$\left.\begin{aligned} E_y &= Ag(\zeta) + Bh(\zeta), \\ \mathscr{H}_x &= \frac{i}{u}\{Ag'(\zeta) + Bh'(\zeta)\}, \end{aligned}\right\} \tag{16.77}$$

where A and B are constants. (This differs slightly from (16.55) because of the minus sign in (16.74).) The fields must be continuous at $z = z_1$ and hence (16.77) and (16.73) are equated, which gives

$$A\{iuCg(\zeta_1) - g'(\zeta_1)\} + B\{iuCh(\zeta_1) - h'(\zeta_1)\} = 0. \tag{16.78}$$

In the lower half of the ionosphere, for which $-z_1 \leqslant z \leqslant 0$, let

$$\xi = \frac{k}{u}\left\{z + z_1 - \frac{C^2 z_1}{X_0}\right\}, \quad \xi_1 = -\frac{k}{u}\frac{C^2 z_1}{X_0} = \zeta_1, \quad \xi_0 = -\frac{kz_1}{u}\left(\frac{C^2}{X_0} - 1\right) = \zeta_0. \tag{16.79}$$

Then E_y again satisfies the Stokes equation, and we take as the solution

$$\left.\begin{aligned} E_y &= -A_1 g(\xi) + B_1 h(\xi), \\ \mathscr{H}_x &= \frac{i}{u}\{A_1 g'(\xi) - B_1 h'(\xi)\}, \end{aligned}\right\} \tag{16.80}$$

where A_1 and B_1 are constants. Continuity of the fields at $z = 0$ requires that (16.77) and (16.80) are equal when $\zeta = \xi = \zeta_0$. Then (16.76) leads to

$$A_1 = A, \quad B_1 = B. \tag{16.81}$$

Notice how the functions (16.75) were deliberately chosen to give this simple result.

Below the ionosphere the field components are

$$\left.\begin{aligned} E_y &= e^{-ikC(z+z_1)} + R_0 e^{ikC(z+z_1)}, \\ \mathscr{H}_x &= -Ce^{-ikC(z+z_1)} + R_0 C e^{ikC(z+z_1)}, \end{aligned}\right\} \tag{16.82}$$

where R_0 is the reflection coefficient measured at the base of the ionosphere. At this level, $z = -z_1$, the two components (16.82) are equal to the pair (16.80) with $\xi = \xi_1 = \zeta_1$, which leads (compare (16.60) and (16.61)) to

$$\left.\begin{aligned} 2iuCR_0 &= -A\{iuCg(\zeta_1)+g'(\zeta_1)\}+B\{iuCh(\zeta_1)+h'(\zeta_1)\}, \\ 2iuC &= -A\{iuCg(\zeta_1)-g'(\zeta_1)\}+B\{iuCh(\zeta_1)-h'(\zeta_1)\}. \end{aligned}\right\} \tag{16.83}$$

The constants A and B are now eliminated from (16.78) and (16.83). Let

$$\Delta_{\pm} = \begin{vmatrix} iuCg(\zeta_1)-g'(\zeta_1) & iuCh(\zeta_1)-h'(\zeta_1) \\ -iuCg(\zeta_1)\mp g'(\zeta_1) & iuCh(\zeta_1)\pm h'(\zeta_1) \end{vmatrix}. \tag{16.84}$$

Then

$$R_0 = \Delta_+/\Delta_-. \tag{16.85}$$

On evaluating the determinants, this gives

$$R_0 = \frac{u^2C^2g(\zeta_1)\,h(\zeta_1)+g'(\zeta_1)\,h'(\zeta_1)}{u^2C^2g(\zeta_1)\,h(\zeta_1)+iuC\{g(\zeta_1)\,h'(\zeta_1)+g'(\zeta_1)\,h(\zeta_1)\}-g'(\zeta_1)\,h'(\zeta_1)}. \tag{16.86}$$

If $g(\zeta)$, $h(\zeta)$ are chosen to be real functions, then

$$|R_0|^2 = \frac{\{u^2C^2g(\zeta_1)\,h(\zeta_1)+g'(\zeta_1)\,h'(\zeta_1)\}^2}{\{u^2C^2g(\zeta_1)\,h(\zeta_1)+g'(\zeta_1)\,h'(\zeta_1)\}^2+u^2C^2\{g(\zeta_1)\,h'(\zeta_1)-g'(\zeta_1)\,h(\zeta_1)\}^2}. \tag{16.87}$$

Here the last term of the denominator is the Wronskian and is constant. It is convenient to choose the functions g and h so that it is unity. Hartree (1929) has given curves showing how $|R_0|^2$ depends on the projected thickness $2z_1C/\lambda$ for various values of X_0/C^2.

The transmission coefficient of the ionosphere is the ratio of the amplitude of the upgoing wave above the ionosphere (16.73) to the amplitude of the incident wave below the ionosphere. It is therefore simply the constant D in (16.73) and could be found by an extension of the foregoing analysis. But $|D|^2$ can be found more simply, for since collisions are neglected there is no loss of energy from the waves and hence $|D|^2+|R|^2 = 1$.

For an ionospheric layer of the kind shown in Fig. 16.7, we may write $X_0 = f_p^2/f^2$, where f is the frequency and f_p is called the penetration-frequency of the layer. On a simple ray theory it would be expected that waves normally incident would be completely reflected if $f < f_p$, and completely transmitted if $f > f_p$. The full wave theory shows, however, that there is a small range of frequencies near f_p for which partial penetration and reflection can occur. The size of the range depends on the layer thickness $2z_1$; it is large for thin layers and small for thick layers. Curves showing how $|R_0|$ varies with frequency for normal incidence are given in Fig. 16.8 for various values of the thickness $2z_1$. Most of these curves were calculated by the methods of ch. 22, but a few points were checked using the formula (16.87). The curves should be compared with Figs. 8.9, 17.3 and 17.5, which give similar curves for other models of the ionosphere.

16.12 The differential equation for oblique incidence applicable when the electric vector is parallel to the plane of incidence

So far in this chapter we have discussed the reflection of horizontally polarised waves at oblique incidence. The problem must now be considered when the electric vector is parallel to the plane of incidence, for

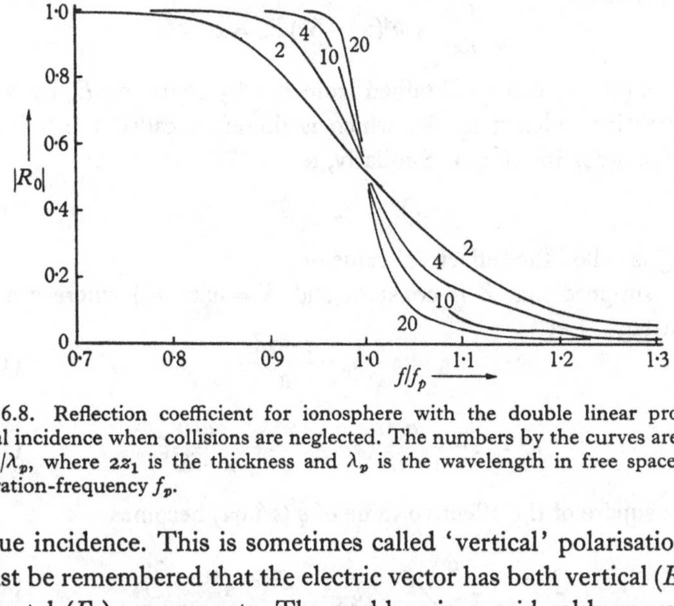

Fig. 16.8. Reflection coefficient for ionosphere with the double linear profile, at vertical incidence when collisions are neglected. The numbers by the curves are values of $2z_1/\lambda_p$, where $2z_1$ is the thickness and λ_p is the wavelength in free space at the penetration-frequency f_p.

oblique incidence. This is sometimes called 'vertical' polarisation, but it must be remembered that the electric vector has both vertical (E_z) and horizontal (E_x) components. The problem is considerably more complicated, and the solution is less complete.

The differential equations for 'vertical' polarisation were formulated in §9.13. One form is (9.64) which shows how the magnetic field $\mathscr{H}_y$ of the wave depends on the height z. Any linear second-order differential equation with a term containing the first derivative ($d\mathscr{H}_y/dz$ in (9.64)) can be reduced to its 'normal form', that is, a form without a first derivative term, by a change of the dependent variable. We take

$$V = \mathscr{H}_y/\mathfrak{n} \qquad (16.88)$$

and (9.64) becomes

$$\frac{d^2V}{dz^2} + k^2\left[q^2 + \frac{1}{2k^2\mathfrak{n}^2}\frac{d^2(\mathfrak{n}^2)}{dz^2} - \frac{3}{4k^2\mathfrak{n}^4}\left\{\frac{d(\mathfrak{n}^2)}{dz}\right\}^2\right]V = 0. \qquad (16.89)$$

Since the medium is slowly varying, the terms $d(\mathfrak{n}^2)/dz$ and $d^2(\mathfrak{n}^2)/dz^2$ are small, but they will nevertheless be of importance near the level of reflection where $q = 0$. Now let

$$\mathfrak{n}^2 + \frac{1}{2k^2\mathfrak{n}^2}\frac{d^2(\mathfrak{n}^2)}{dz^2} - \frac{3}{4k^2\mathfrak{n}^4}\left\{\frac{d(\mathfrak{n}^2)}{dz}\right\}^2 = M^2. \tag{16.90}$$

Then (16.89) becomes

$$\frac{d^2V}{dz^2} + k^2(M^2 - S^2)V = 0. \tag{16.91}$$

This should be compared with (9.58), for horizontal polarisation, which may be written

$$\frac{d^2E_y}{dz^2} + k^2(\mathfrak{n}^2 - S^2)E_y = 0.$$

Equation (16.91) can be obtained from this by replacing E_y by V, and the refractive index $\mathfrak{n}$ by M, which is therefore called the 'effective' refractive index in (16.91). Similarly, let

$$Q^2 = M^2 - S^2. \tag{16.92}$$

Then Q is called the 'effective' value of q.

Now suppose that Z is constant and $X = \alpha(z - h_0)$ where α and h_0 are constant. Let

$$\xi = z - h_0 - \frac{1 - iZ}{\alpha}. \tag{16.93}$$

Then
$$\mathfrak{n}^2 = 1 - \frac{\alpha}{1 - iZ}(z - h_0) = -\frac{\alpha}{1 - iZ}\xi, \tag{16.94}$$

and the square of the effective value of q (16.92) becomes

$$Q^2 = -\frac{\alpha}{1 - iZ}\xi - \frac{3}{4k^2\xi^2} - S^2 = q^2 - \frac{3}{4k^2\xi^2}. \tag{16.95}$$

Evidently Q is infinite where $\xi = 0$, and this is a singular point of the differential equation (16.91). It may be close to the point where $Q = 0$, and then the function Q^2 is not even approximately linear there. Hence the presence of the singularity means that at the reflection level (16.91) does not approximate to the Stokes equation, and the reflection process is more complicated for vertical than for horizontal polarisation.

Equation (16.90) is not the only possible form for the square of the 'effective' refractive index. By using other variables (for example $\mathfrak{n}E_x/q$) a different form is obtained. But for 'vertical' polarisation it is never possible to find a variable which makes $M = \mathfrak{n}$ except in the special case $S = 0$, that is, for vertical incidence.

With the substitution (16.94), the differential equation (9.64) becomes

$$\frac{d^2\mathscr{H}_y}{d\xi^2} - \frac{1}{\xi}\frac{d\mathscr{H}_y}{d\xi} + k^2\left\{-\frac{\mathfrak{a}}{1-iZ}\xi - S^2\right\}\mathscr{H}_y = 0, \tag{16.96}$$

where (9.56) has been used. Now let

$$\mathfrak{z} = \left(\frac{k^2\mathfrak{a}}{1-iZ}\right)^{\frac{1}{3}}\xi, \tag{16.97}$$

where the cube root is chosen to lie in the first quadrant (and is real and positive when $Z = 0$). Then (16.96) gives

$$\frac{d^2\mathscr{H}_y}{d\mathfrak{z}^2} - \frac{1}{\mathfrak{z}}\frac{d\mathscr{H}_y}{d\mathfrak{z}} - (\mathfrak{z}+B)\mathscr{H}_y = 0, \tag{16.98}$$

where
$$B = S^2\{(1-iZ)k/\mathfrak{a}\}^{\frac{2}{3}}. \tag{16.99}$$

In the special case of vertical incidence, $B = 0$ and equation (16.98) becomes

$$\frac{d^2\mathscr{H}_y}{d\mathfrak{z}^2} - \frac{1}{\mathfrak{z}}\frac{d\mathscr{H}_y}{d\mathfrak{z}} - \mathfrak{z}\mathscr{H}_y = 0. \tag{16.100}$$

Now let $\mathfrak{S}(\mathfrak{z})$ be any solution of the Stokes equation so that $\mathfrak{S}'' = \mathfrak{z}\mathfrak{S}$. Then it is easy to show that $\mathscr{H}_y = \mathfrak{S}'(\mathfrak{z})$ is a solution of (16.100). For example one solution of (16.100) is $\mathrm{Ai}'(\mathfrak{z})$. Reflection at vertical incidence could equally well be studied for waves polarised with the electric vector parallel to the y-axis. The solution would then be as in §§16.5 or 16.6, where it was shown that near the level of reflection the magnetic field is proportional to $\mathrm{Ai}'(\mathfrak{z})$ (16.25). Hence the solution $\mathrm{Ai}'(\mathfrak{z})$ of (16.100) satisfies the physical conditions for waves incident from below, at vertical incidence.

It can be shown by the method of §15.14 that the Stokes constant of (16.100) is $-i$. Now for 'vertical' polarisation the reflection coefficient of the ionosphere is ${}_{\parallel}R_{\parallel}$ and (7.16) shows that it is the ratio of the magnetic fields of the upgoing and downgoing waves. Thus the phase integral formula must now be written

$${}_{\parallel}R_{\parallel} = -i\exp\left\{-2ik\int_0^{z_0}\mathfrak{n}\,dz\right\}.$$

If a reflection coefficient R were defined as the ratio of the electric fields, a further change of sign would be necessary which would lead to the same phase integral formula, (16.29), as was derived for horizontal polarisation in §16.6.

16.13 The behaviour of the fields near a zero of the refractive index for 'vertical' polarisation at oblique incidence

The study of the reflection of vertically polarised waves really requires a full treatment of the differential equation (16.98), similar to that given in ch. 15 for the Stokes equation. The account given in this and the following four sections is much shorter and less complete. Part of it is based on a treatment given by Försterling and Wüster (1951).

A solution of (16.98) can be found as a series of ascending powers of $\mathfrak{z}$. Let

$$\mathscr{H}_y = \mathfrak{z}^\beta\{a_0 + a_1\mathfrak{z} + a_2\mathfrak{z}^2 + \dots\}. \tag{16.101}$$

This is substituted in (16.98) and coefficients of successive powers of $\mathfrak{z}$ are equated to zero, which gives:

$$\left.\begin{aligned} \beta(\beta-2) &= 0,\\ a_1(\beta+1)(\beta-1) &= 0,\\ a_2(\beta+2)\beta &= a_0 B,\\ a_3(\beta+3)(\beta+1) &= a_1 B + a_0,\\ a_4(\beta+4)(\beta+2) &= a_2 B + a_1,\\ &\dots\dots\dots\dots \end{aligned}\right\} \tag{16.102}$$

The first is the indicial equation and gives $\beta = 0$, or 2. The third equation shows that $\beta = 0$ is impossible if $B \neq 0$. Hence $\beta = 2$, and

$$a_1 = 0, \quad a_n = \frac{a_{n-3} + Ba_{n-2}}{n(n+2)} \quad \text{for} \quad n = 2, 3, 4, \dots, \tag{16.103}$$

where $a_{-1} = 0$. The method therefore gives only one solution, namely

$$v_1(\mathfrak{z}) = \mathfrak{z}^2 + a_2\mathfrak{z}^4 + a_3\mathfrak{z}^5 + \dots, \tag{16.104}$$

where the arbitrary constant a_0 has been taken equal to unity. A second solution may be found (see, for example, Whittaker and Watson, 1935), as follows. Let

$$v_2(\mathfrak{z}) = Kv_1(\mathfrak{z})\log\mathfrak{z} + 1 + b_1\mathfrak{z} + b_2\mathfrak{z}^2 + \dots. \tag{16.105}$$

Substitute in the differential equation (16.98) and equate coefficients of powers of $\mathfrak{z}$ to zero. For $\mathfrak{z}^{-1}$, $\mathfrak{z}^0$ and $\mathfrak{z}$ this gives respectively

$$b_1 = 0, \quad K = \tfrac{1}{2}B, \quad b_3 = \tfrac{1}{3}. \tag{16.106}$$

The value of b_2 is arbitrary and may be taken as zero. If it is not zero, the effect is simply to add a multiple of $v_1(\mathfrak{z})$ on to $v_2(\mathfrak{z})$. The higher powers of $\mathfrak{z}$ then give

$$b_n = \frac{Bb_{n-2} + b_{n-3} - B(n-1)\,a_{n-2}}{n(n-2)}. \tag{16.107}$$

Notice that if B is zero (vertical incidence), then K is zero and the logarithmic term disappears from (16.105). The solutions $v_1(\mathfrak{z})$ and $v_2(\mathfrak{z})$ are then simply derivatives of the two series in (15.8).

The general solution of (16.98) is

$$\mathscr{H}_y = A_1 v_1(\mathfrak{z}) + A_2 v_2(\mathfrak{z}), \tag{16.108}$$

where A_1 and A_2 are constants. Now for vertical incidence the required solution is proportional to $\mathrm{Ai}'(\mathfrak{z})$ and (15.8) and (15.9) show that

$$\mathrm{Ai}'(\mathfrak{z}) = \{\tfrac{1}{2}.3^{-\frac{2}{3}}/(-\tfrac{1}{3})!\}\, v_1(\mathfrak{z}) - \{3^{-\frac{1}{3}}/(-\tfrac{2}{3})!\}\, v_2(\mathfrak{z}) \quad \text{for} \quad B = 0. \tag{16.109}$$

Hence A_2 is not zero when $B = 0$, and it cannot drop discontinuously to zero when $B \neq 0$. Thus the solution $\mathscr{H}_y$ for oblique incidence must contain a multiple of $v_2(\mathfrak{z})$, that is, it must contain the logarithmic term in (16.105).

The field components E_x and E_z may be found from the Maxwell equations (9.52) and (9.54) respectively and are as follows:

$$E_x = -ik\left(\frac{1-iZ}{k^2\mathfrak{a}}\right)^{\frac{1}{3}} [(A_1 + \tfrac{1}{2}A_2 B \log \mathfrak{z})(2 + 4a_2\mathfrak{z}^2 + 5a_3\mathfrak{z}^3 + \ldots)$$

$$+ \tfrac{1}{2}A_2 B(1 + a_2\mathfrak{z}^2 + a_3\mathfrak{z}^3 + \ldots) + A_3(3b_3\mathfrak{z} + 4b_4\mathfrak{z}^2 + \ldots)], \tag{16.110}$$

$$E_z = k^2 S\left(\frac{1-iZ}{k^2\mathfrak{a}}\right)^{\frac{2}{3}} [(A_1 + \tfrac{1}{2}A_2 B \log \mathfrak{z})(\mathfrak{z} + a_2\mathfrak{z}^3 + a_3\mathfrak{z}^4 + \ldots)$$

$$+ A_2(1/\mathfrak{z} + b_3\mathfrak{z}^2 + b_4\mathfrak{z}^3 + \ldots)]. \tag{16.111}$$

Thus E_x contains a term $A_2 B \log \mathfrak{z}$, and E_z contains a term $A_2/\mathfrak{z}$, which are both infinite where $\mathfrak{z} = 0$, that is where $\mathfrak{n} = 0$. This is usually above the reflection level where $q = 0$.

In the ionosphere the electrons always make some collisions so that Z is never exactly zero. Consequently we cannot have $\mathfrak{n} = 0$ at any real height z. For example, (16.94) shows that z is complex when $\mathfrak{n}^2$ is zero. But if Z is small, the condition $\mathfrak{n} = 0$ may hold at a point very close to the real axis in the complex z-plane. At real heights near this point the terms $\log \mathfrak{z}$ and $1/\mathfrak{z}$ in E_x and E_z respectively may be very large. Thus the electric field may become large at this level for vertical polarisation at oblique incidence. This phenomenon does not occur for horizontal polarisation.

16.14 The generation of harmonics in the ionosphere

The field component E_z imparts vertical motions to the electrons. When E_z is large, these vertical motions are large, and within one cycle of oscillation of the field any one electron moves to different levels where E_z is different. Thus one electron encounters very different fields E_z within one oscillation. The vertical force to which it is subjected does not, therefore, vary harmonically with time, and the electron motion is not simple harmonic. The motion could be Fourier analysed into a series of frequencies equal to the wave-frequency and its harmonics. The theory of this process has been developed by Feinstein (1950) and Försterling and Wüster (1951) who showed that harmonics of the wave-frequency can be generated in this way near a level where $\mathfrak{n} = 0$. This process would be most marked when Z is very small, for then the point where $\mathfrak{n} = 0$ is near the real axis in the complex z-plane. Some energy must go into the harmonics generated, and this comes from the original wave which is therefore attenuated. The effect is thus similar to that of damping of the electron motions. Even if the collision-frequency were zero, some energy could be removed from the wave as harmonics, and this would lead to a reflection coefficient with modulus less than unity.

16.15 The phase integral formula for 'vertical' polarisation at oblique incidence

In ch. 10 it has been assumed that at frequencies which are great enough the 'ray' theory holds, so that the reflection coefficient of the ionosphere is given by (9.62), for any state of polarisation of the wave. This is the same as the phase integral formula (16.29), except for the omission of a factor i, which is unimportant at high frequencies. The justification for using these formulae has been given, for horizontal polarisation only, in § 16.6, and was based on the fact that near the level of reflection, where $q = 0$, the function $q^2(z)$ could be assumed to be linear, and therefore one component of the field was given by the Stokes equation. The factor i in (16.29) is the Stokes constant of the Stokes equation (§ 15.13).

The justification for using (9.62) or (16.29) for 'vertical' polarisation at oblique incidence must now be examined. The differential equation near the level of reflection may be reduced to the form (16.98) but this is less simple than the Stokes equation, and its Stokes constant is no longer in general equal to i, but depends on the angle of incidence and on the gradient of electron density at the reflection level.

One possible approach is to apply the method of § 16.6 to the differential equation (16.91) but this is only possible if the variation of Q^2 with z is nearly linear in a region close to the zero of Q. Now (16.94) shows that q^2 is zero when $\xi = \xi_0$ where

$$\xi_0 = -\frac{1-iZ}{\mathfrak{a}} S^2. \tag{16.112}$$

Here Q^2 differs from q^2 by the term $3/(4k^2\xi_0^2)$ in (16.95) and if ξ_0 is large enough, this difference is small. Hence if S is large enough, q and Q have zeros at the same point $\xi = \xi_0$. Q^2 can now be expanded in a Taylor series about this point. This gives approximately

$$Q^2 = -\frac{\mathfrak{a}}{1-iZ}(\xi-\xi_0) - \frac{9\mathfrak{a}^4}{4k^2(1-iZ)^4 S^8}(\xi-\xi_0)^2 + \ldots.$$

The criterion of linearity (16.36) may now be applied, and gives

$$S^3 \gg \left(\frac{3}{2}\right)^{\frac{3}{4}} \left|\frac{\mathfrak{a}}{k(1-iZ)}\right|. \tag{16.113}$$

In terms of B, equation (16.99), this becomes

$$|B| \gg 1{\cdot}36.$$

Notice that a large value of S means that the zero and infinity of Q^2 are well separated as discussed in § 15.2.

If the condition (16.113) holds, the method of § 16.6 can be used and gives for the reflection coefficient

$$R = i\exp\left\{-2ik\int_{0\atop C}^{z_0} Q\,dz\right\}. \tag{16.114}$$

Now Q only differs appreciably from q near the point $\xi = 0$. The contour C may be chosen to pass well away from this point. Hence Q may be replaced by q in (16.114) which is then the same as the phase integral formula (16.29). It should be stressed, however, that in (16.114) the reflection coefficient R is $_{\parallel}R_{\parallel}$, defined in (7.16) as the ratio of the *magnetic* fields of the reflected and incident waves, whereas in (16.29) R is $_{\perp}R_{\perp}$, the ratio of the electric fields.

We conclude that the phase integral formula can be used to give $_{\parallel}R_{\parallel}$ for vertical polarisation if the angle of incidence is large enough for (16.113) to be satisfied. If the angle of incidence is zero, the Stokes constant i in (16.114) must be replaced by $-i$ (see end of § 16.12). For intermediate values of the angle of incidence the Stokes constant makes a transition from $-i$ to $+i$. This is discussed in the next section.

16.16 Asymptotic approximations for the solutions of the differential equation for 'vertical' polarisation

The asymptotic approximations for the solutions of (16.98) may be found as follows. The equation is first reduced to normal form by setting $\mathscr{H}_y = G\mathfrak{z}^{\frac{1}{2}}$. This gives

$$\frac{d^2G}{d\mathfrak{z}^2} = G\{\mathfrak{z} + B + \tfrac{3}{4}\mathfrak{z}^{-2}\}. \tag{16.115}$$

The W.K.B. method of §9.5 is now used and gives

$$\mathscr{H}_y \sim \mathfrak{z}^{\frac{1}{2}}\{\mathfrak{z}+B+\tfrac{3}{4}\mathfrak{z}^{-2}\}^{-\frac{1}{4}}\exp\left\{\pm\int(\mathfrak{z}+B+\tfrac{3}{4}\mathfrak{z}^{-2})^{\frac{1}{2}}\,d\mathfrak{z}\right\}. \tag{16.116}$$

When $\mathfrak{z}$ is large compared with B and with unity, the integrand may be expanded by the binomial theorem and integrated term by term. Only positive powers of $\mathfrak{z}$ need be retained in the exponent. Hence

$$\mathscr{H}_y \sim h(\mathfrak{z})\exp\{\pm(\tfrac{2}{3}\mathfrak{z}^{\frac{3}{2}}+B\mathfrak{z}^{\frac{1}{2}})\}, \tag{16.117}$$

where the abbreviation $h(\mathfrak{z})$ is used for the factor preceding the exponential. When $|\mathfrak{z}|$ is very large, the factor $\exp(\pm\frac{2}{3}\mathfrak{z}^{\frac{3}{2}})$ predominates. It occurs also in the asymptotic approximations to the solutions of the Stokes equation. Hence the Stokes lines and the anti-Stokes lines of the equations (16.98) or (16.115) are the same as for the Stokes equation.

At great heights in the ionosphere $|\mathfrak{z}|$ is large, and (16.97) shows that $0 \leqslant \arg\mathfrak{z} < \frac{1}{6}\pi$. In particular $\mathfrak{z}$ is real and positive if there are no collisions so that $Z = 0$. Here the required solution contains only the subdominant term, representing either an upward travelling wave at great heights, or a disturbance whose amplitude decreases as z increases. Well below the level of reflection $\frac{2}{3}\pi < \arg\mathfrak{z} \leqslant \pi$. In going from great to small heights, therefore, the Stokes line, at $\arg\mathfrak{z} = \frac{2}{3}\pi$, is crossed and there are two terms in the asymptotic approximation below the level of reflection. We now require a connection formula between the two asymptotic approximations. This is of the form (15.90) and may be written

$$h(\mathfrak{z})\{\exp(-\tfrac{2}{3}\mathfrak{z}^{\frac{3}{2}}-B\mathfrak{z}^{\frac{1}{2}})+\mathscr{S}\exp(\tfrac{2}{3}\mathfrak{z}^{\frac{3}{2}}+B\mathfrak{z}^{\frac{1}{2}})\} \leftrightarrow h(\mathfrak{z})\exp(-\tfrac{2}{3}\mathfrak{z}^{\frac{3}{2}}-B\mathfrak{z}^{\frac{1}{2}}), \tag{16.118}$$

where $\mathscr{S}$ is the Stokes constant for (16.98).

When $B = 0$, the solutions of (16.98) are simply derivatives of solutions of the Stokes equation, and then (16.118) is similar to (15.90) and the Stokes constant $\mathscr{S} = -i$. When $B \neq 0$, however, the differential equation is much more complicated and cannot be reduced to a form for which the asymptotic behaviour of the solutions is completely known. The Stokes constant $\mathscr{S}$ must then be found by numerical computation. It is a function of B denoted by $\mathscr{S}(B)$. The method of computing it has not been published and is therefore given in an Appendix. The results are shown in Fig. 16.9, where the values of $\mathscr{S}$ are plotted in the complex plane for various values of B. The diagram shows that $\mathscr{S} \to -i$ when $B \to 0$ and $\mathscr{S} \to +i$ when B is large, as we should expect.

16.17 Application of the phase integral formula

It is evident from (16.21), (16.93), (16.97) and (16.99) that

$$\zeta = \mathfrak{z} + B. \tag{16.119}$$

It was shown in §16.3 that when $|\zeta|$ is large enough, the exponentials $\exp\{\mp\frac{2}{3}\zeta^{\frac{3}{2}}\}$ are proportional to the exponentials $\exp\left\{\mp ik\int^{z} q\,dz\right\}$ in the W.K.B. solutions.

Now when $|\zeta|$ is large, $\frac{2}{3}\zeta^{\frac{3}{2}} = \frac{2}{3}(\mathfrak{z}+B)^{\frac{3}{2}} = \frac{2}{3}\mathfrak{z}^{\frac{3}{2}} + B\mathfrak{z}^{\frac{1}{2}} + O(\mathfrak{z}^{-\frac{1}{2}})$. If negative powers of $\mathfrak{z}$ are neglected, this is the same as the exponent in the asymptotic approximations (16.117), which are therefore proportional to the W.K.B. solutions. Further, the factor $h(\mathfrak{z})$ is proportional to $q^{-\frac{1}{2}}$ when $|\zeta|$ is large. Hence the arguments of § 16.6 may be applied to the differential equation (16.98), and the only difference is that the Stokes constant $\mathscr{S}(B)$ appears in (16.118) instead of the Stokes constant i in (15.90). Thus the reflection coefficient is

$$R = \mathscr{S}(B)\exp\left\{-2ik\underset{C}{\int_0^{z_0}} q\,dz\right\}, \qquad (16.120)$$

which replaces (16.29).

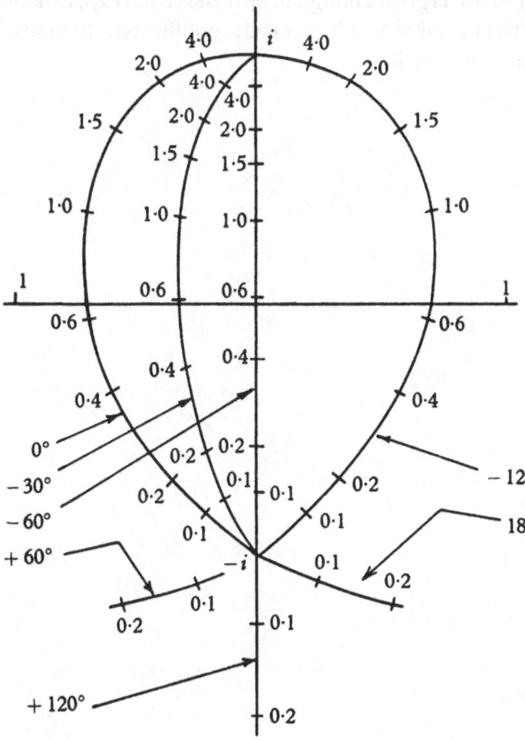

Fig. 16.9. The complex plane showing values of the Stokes constant $\mathscr{S}$ for the differential equation (16.98). Numbers by the curves show the values of $|B|$ and arg B. Only values of arg B from $-\frac{1}{3}\pi$ to 0 are of interest when finding reflection coefficients, but curves for arg $B = \pm\frac{2}{3}\pi$ are also included for mathematical interest.

In the special case when electron collisions are neglected, B is purely real, and the integral in (16.114) is also real, so that the exponential has modulus unity. But Fig. 16.9 shows that $|\mathscr{S}(B)|$ is less than unity, and hence $|R| < 1$. The smallest value is about 0·72 and occurs when $B \approx 0{\cdot}45$. This means that not all the incident energy is reflected from the ionosphere, even when there is no physical mechanism for converting the energy of the waves into heat.

This may be because some energy is converted into waves of the harmonic frequencies mentioned in § 16.14, since harmonics would be most easily generated when the collision-frequency is small.

In the lower ionosphere a typical value of α is about $0{\cdot}6\,\text{km}^{-1}$. This gives the following values of B, (16.99), for various frequencies:

Frequency	16 kc/s	2 Mc/s	6 Mc/s
Approximate value of B	$1{\cdot}6S^2$	$40S^2$	$80S^2$

The value $B = 0{\cdot}45$ therefore corresponds to an angle of incidence of about 30° at 16 kc/s, or 6° at 2 Mc/s.

It is doubtful whether the difference between (16.120) and the phase integral formula (16.29) is great enough to be observed in experiments. Moreover, the theory for the actual ionosphere needs modification to allow for the effect of the earth's magnetic field.

CHAPTER 17

VARIOUS ELECTRON DENSITY PROFILES WHEN THE EARTH'S MAGNETIC FIELD IS NEGLECTED

17.1 Introduction

Chapters 10 to 14 were devoted to some results of applying 'ray theory' to the reflecting properties of the ionosphere. They were based on the use of the reflection coefficient as given by the formula (9.62), whose validity was examined in ch. 16, and it was found to require a small modification, namely the inclusion of a factor i. This led to the phase integral formula (16.29), which is the basis of much of the theory of the reflection of radio waves from the ionosphere, for high frequencies (above 1 Mc/s). The conditions under which it can be used were listed in § 16.7.

In the present chapter we investigate reflection from the ionosphere when these conditions fail, and the results are therefore mainly applicable to low frequencies. Even at high frequencies, however, the conditions may fail for an ionospheric layer with a maximum of electron density, when the frequency is close to the penetration frequency. There can then be partial penetration and reflection, which is not predicted by a pure ray theory, and this phenomenon is also investigated.

The number of profiles of electron density that can be investigated by the methods of this chapter is small, since only those profiles can be used which lead to differential equations whose solutions have properties which are well enough understood. For other profiles the differential equations must be solved numerically as explained in ch. 22.

The process of finding the reflection coefficient of the ionosphere may be divided into four steps:

(*a*) Formulate the differential equation satisfied by some field component of the wave.

(*b*) Find a solution which satisfies the physical conditions at a great height.

(*c*) Use the connection formula or circuit relations to get the form of the solution below the ionosphere.

(*d*) Separate this solution into upgoing and downgoing waves, and thus find the reflection coefficient.

The above process has been used in ch. 16 to find the reflection coefficient for a linear profile of electron density. It will now be used for other profiles.

In this chapter the earth's magnetic field is neglected and the waves are assumed to be horizontally polarised at oblique incidence. Some related problems, with the effect of the earth's magnetic field included, are discussed in ch. 21.

17.2 Exponential profile. Constant collision-frequency

Radio waves of very low frequency (less than 100 kc/s) are reflected low down in the ionosphere, usually between 70 and 100 km. The profile $N(z)$ of electron density at these levels is not very well known, but one possibility is that reflection occurs in the lower part of the E-layer, whose maximum is near 110 km. If the E-layer is a Chapman layer (see § 1.5), then the variation of $N(z)$ in its lower part is roughly exponential. The exponential profile is therefore very important in the study of the reflection of low-frequency radio waves in the lower part of an ionospheric layer.

Assume that the electron density N varies exponentially with height. Since X is proportional to N

$$X = e^{\alpha z}, \tag{17.1}$$

where α is a constant, and the origin of the height z is chosen where $X = 1$. It is important to notice that this choice of origin depends on the frequency f, since X is inversely proportional to f^2 (3.6). The electron collision-frequency is assumed to be constant and therefore Z is constant, and (9.56) gives

$$q^2 = C^2 - \frac{e^{\alpha z}}{1 - iZ}. \tag{17.2}$$

For horizontally polarised waves at oblique incidence the differential equation satisfied by the electric field E_y is (9.58), which becomes

$$\frac{d^2E_y}{dz^2} + k^2\left(C^2 - \frac{e^{\alpha z}}{1 - iZ}\right)E_y = 0. \tag{17.3}$$

Now let

$$\zeta = \frac{2k}{\alpha} e^{\frac{1}{2}\alpha z} e^{i(\frac{1}{2}\pi)} (1 - iZ)^{-\frac{1}{2}}, \tag{17.4}$$

where the root $(1 - iZ)^{-\frac{1}{2}}$ is chosen to have a positive real part, and let

$$\nu = 2ikC/\alpha \tag{17.5}$$

(throughout this section ν has the meaning (17.5) and must not be confused with the ν used earlier for collision-frequency). Then (17.3) becomes

$$\frac{d^2E_y}{d\zeta^2}+\frac{1}{\zeta}\frac{dE_y}{d\zeta}+\left(1-\frac{\mathsf{v}^2}{\zeta^2}\right)E_y=0, \tag{17.6}$$

which is Bessel's equation of order v. A solution must now be found which represents an upgoing wave only, at great heights. Equation (17.2) shows that when z is very large

$$q^2\approx-\frac{e^{az}}{1-iZ}, \tag{17.7}$$

whence

$$q\approx-ie^{\frac{1}{2}az}(1-iZ)^{-\frac{1}{2}}, \tag{17.8}$$

where the sign is minus because q has a positive real part and a negative imaginary part. Hence

$$k\int_0^z q\,dz\approx-\zeta. \tag{17.9}$$

At a great height the upgoing W.K.B. solution contains the factor

$$\exp\left\{-ik\int_0^z q\,dz\right\},$$

that is $e^{i\zeta}$. Similarly, the downgoing W.K.B. solution contains the factor $e^{-i\zeta}$ and this cannot appear in the required solution.

Two independent solutions of (17.6) are $H^{(1)}_\mathsf{v}(\zeta)$ and $H^{(2)}_\mathsf{v}(\zeta)$, where Watson's (1944) notation is used for Bessel functions of the third kind, sometimes called Hankel functions. The order v of these functions is purely imaginary. The appearance of a complex order is perhaps unusual in physical problems, but it does not affect the definition of the Hankel functions as given by Watson. Now (17.4) shows that when z is positive

$$\tfrac{1}{2}\pi\leqslant\arg\zeta<\pi, \tag{17.10}$$

and for this range the asymptotic approximations (first terms) given by Watson (1944, p. 201) are as follows:†

$$H^{(1)}_\mathsf{v}(\zeta)\sim(2/\pi\zeta)^{\frac{1}{2}}\exp\{i(\zeta-\tfrac{1}{2}\mathsf{v}\pi-\tfrac{1}{4}\pi)\}\quad\text{for}\quad-\pi<\arg\zeta<2\pi, \tag{17.11}$$

$$H^{(2)}_\mathsf{v}(\zeta)\sim(2/\pi\zeta)^{\frac{1}{2}}\exp\{-i(\zeta-\tfrac{1}{2}\mathsf{v}\pi-\tfrac{1}{4}\pi)\}\quad\text{for}\quad-2\pi<\arg\zeta<\pi. \tag{17.12}$$

† Bessel's equation (17.6) has anti-Stokes lines where $\arg\zeta=0$, $\pm\pi$, and Stokes lines where $\arg\zeta=\pm\frac{1}{2}\pi$. Watson gives asymptotic approximations in the Poincaré sense, and the range of $\arg\zeta$ therefore extends nearly to anti-Stokes lines. It was explained in § 15.22 how this may lead to errors when $|\zeta|$ is not indefinitely large. It might be better to use a range of $\arg\zeta$ which ends on Stokes lines, so that in (17.11) the range should be $-\frac{1}{2}\pi\leqslant\arg\zeta\leqslant\frac{3}{2}\pi$, and only the subdominant term is present when

Hence the required solution is $H_\nu^{(1)}(\zeta)$. This function must now be examined at very low levels, where it is to be separated into upgoing and downgoing waves.

Bessel's equation (17.6) may be solved by assuming that there are solutions of the form $\zeta^\beta(a_0+a_1\zeta+\ldots)$. This is substituted in the differential equation and coefficients of successive powers of ζ are equated to zero. In this way two solutions can be found, which are independent when ν is not an integer. In the present problem ν is never an integer, and so we can use these solutions as alternatives to $H_\nu^{(1)}(\zeta)$ and $H_\nu^{(2)}(\zeta)$. They are written

$$\left.\begin{aligned} J_\nu(\zeta) &= \frac{(\frac{1}{2}\zeta)^\nu}{\nu!}\{1+a_2\zeta^2+\ldots\},\\ J_{-\nu}(\zeta) &= \frac{(\frac{1}{2}\zeta)^{-\nu}}{(-\nu)!}\{1+b_2\zeta^2+\ldots\}.\end{aligned}\right\} \tag{17.13}$$

The values of a_2, b_2, etc., can be found but are of no interest in the present problem. The solution $H_\nu^{(1)}(\zeta)$ must be a linear combination of (17.13) and it is shown by Watson (1944, p. 74) that

$$H_\nu^{(1)}(\zeta) = \frac{J_{-\nu}(\zeta)-e^{-\nu\pi i}J_\nu(\zeta)}{i\sin\nu\pi}. \tag{17.14}$$

This provides the connection-formula in the present problem.

At the ground z is very large and negative, so that $|\zeta|$ is very small. Hence only the first terms of the series (17.13) need be retained, and (17.14) then shows that the solution near the ground is proportional to

$$\frac{(\frac{1}{2}\zeta)^{-\nu}}{(-\nu)!}-e^{-\nu\pi i}\frac{(\frac{1}{2}\zeta)^\nu}{\nu!}, \tag{17.15}$$

since the denominator in (17.14) is a constant which is never zero, and can therefore be omitted. The factor $\zeta^{-\nu}$ in the first term is proportional to e^{-ikCz} (from (17.4)), which is the upgoing wave. Similarly, the second term contains a factor e^{ikCz} which is the downgoing wave. The ratio of the two terms is therefore the reflection coefficient R, given by

$$R = -\left(\frac{k}{\alpha}\right)^{2\nu}(1-iZ)^{-\nu}\frac{(-\nu)!}{\nu!}e^{-2ikCh_1}, \tag{17.16}$$

$0<\arg\zeta<\pi$. This includes the range (17.10). Similarly, in (17.12) the range should be $-\frac{3}{2}\pi\leqslant\arg\zeta\leqslant\frac{1}{2}\pi$, and only the subdominant term is present when $-\pi<\arg\zeta<0$. The reader will find it instructive to draw the Stokes diagram for Bessel's equation, and to prove that the Stokes constant is $2i\cos\nu\pi$. Since Bessel functions are not in general single valued, the form of Stokes diagram used in Fig. 15.6 is not suitable. An alternative form suggested by Heading (1957) may be used, in which arg ζ is plotted as abscissa and the circle of Fig. 15.6 is replaced by a horizontal line.

where h_1 is the height above the ground of the level where $X = 1$. Since ν is purely imaginary, all factors in (17.16) except $(1-iZ)^{-\nu}$ have modulus unity and hence

$$|R| = \exp\left\{-\frac{2kC}{\alpha}\arctan Z\right\}, \tag{17.17}$$

$$\arg R = \pi + \frac{4kC}{\alpha}\log\left(\frac{k}{\alpha}\right) - \frac{kC}{\alpha}\log(1+Z^2) + 2\arg\{(-2ikC/\alpha)!\} - 2kCh_1. \tag{17.18}$$

Equation (17.17) shows that when electron collisions are neglected, so that $Z = 0$, the reflection coefficient has modulus unity. This was to be expected, since no energy can be absorbed from the wave.

17.3 The phase integral formula applied to the exponential layer

It is interesting to use the phase integral formula (16.29) to find the reflection coefficient of the exponential profile of § 17.2. Let z_0 be the value of z which makes $q = 0$ in (17.2). Then the phase integral is

$$2k\int_{-h_1}^{z_0} q\,dz = 2kCh_1 + \frac{4kC}{\alpha}\log\left[C(1-iZ)^{\frac{1}{2}} + \{C^2(1-iZ) - e^{-\alpha h_1}\}^{\frac{1}{2}}\right]$$
$$-\frac{4kC}{\alpha}\{1 - e^{-\alpha h_1}C^{-2}(1-iZ)^{-1}\}^{\frac{1}{2}}. \tag{17.19}$$

Now $e^{-\alpha h_1}$ is very much less than unity, and may be neglected, so that (17.19) becomes

$$2kCh_1 + \frac{4kC}{\alpha}(\log 2C - 1) + \frac{kC}{\alpha}\log(1+Z^2) - i\frac{2kC}{\alpha}\arctan Z, \tag{17.20}$$

and the phase integral formula (16.29) then gives

$$|R| = \exp\left\{-\frac{2kC}{\alpha}\arctan Z\right\}, \tag{17.21}$$

$$\arg R = \frac{\pi}{2} - \frac{4kC}{\alpha}(\log 2C - 1) - \frac{kC}{\alpha}\log(1+Z^2) - 2kCh_1. \tag{17.22}$$

Equation (17.21) is the same as (17.17) so that the phase integral formula gives the correct value of $|R|$ for all values of α. The difference between the values (17.18) and (17.22) for $\arg R$ is

$$\frac{\pi}{2} + \frac{4kC}{\alpha}\left\{\log\left(\frac{2kC}{\alpha}\right) - 1\right\} + 2\arg\{(-2ikC/\alpha)!\}. \tag{17.23}$$

When the ionosphere is 'slowly varying' α is small and $2kC/\alpha$ is large. The first term of Stirling's formula may then be used for the factorial function, and this makes (17.23) zero. Some values of (17.23) for smaller values of kC/α are given in Table 17.1.

Table 17.1. *Values obtained by* (17.23)

$\frac{kC}{\alpha}$	Frequency for normal incidence when $\alpha = 0{\cdot}6$ km^{-1}	Value of phase difference (17.23)
0	0	1·57 (90°)
0·05	1·4 kc/s	0·75 (43°)
0·1	2·86 kc/s	0·45 (26°)
1·0	28·6 kc/s	0·17 (10°)
4·0	1·14 Mc/s	0·04 (2·3°)

The value $\alpha = 0{\cdot}6\,\text{km}^{-1}$ used in the second column is roughly typical for the lower ionosphere. Hence, in this example the phase integral formula is remarkably accurate, since it always gives the correct value of $|R|$ and the error in phase never exceeds 90°.

17.4 The parabolic layer

Suppose that the electron number density N is given by the parabolic law (10.27). One curve in Fig. 10.4 shows how N varies with height z in this case. Some results of applying ray theory to this profile have been given in §§ 10.7, 11.6 and 11.11, in which the electron collision-frequency was neglected. The problem will now be examined using full wave theory, for horizontally polarised waves at oblique incidence, when the electron collision-frequency is constant. It will be shown that the ray theory needs some modification for frequencies near the penetration-frequency. This profile is particularly important in the study of layers with a maximum of electron density, since it provides one way of studying the phenomenon of partial penetration and reflection.

Since X is proportional to N we may write

$$X = X_m\{1-(z-z_m)^2/a^2\} \quad \text{for} \quad |z-z_m| \leqslant a, \tag{17.24}$$

where X_m is the maximum value of X, corresponding to $N = N_m$. Then (9.56) gives

$$q^2 = C^2 - \frac{X_m}{1-iZ}\left\{\frac{1-(z-z_m)^2}{a^2}\right\}, \tag{17.25}$$

where Z is taken as constant, and the differential equation satisfied by the electric field E_y is (9.58) which becomes

$$\frac{d^2E_y}{dz^2}+\left\{k^2\left(C^2-\frac{X_m}{1-iZ}\right)+\frac{k^2X_m(z-z_m)^2}{(1-iZ)a^2}\right\}E_y=0. \tag{17.26}$$

Now let
$$\zeta=\left\{\frac{-4k^2X_m}{(1-iZ)a^2}\right\}^{\frac{1}{4}}(z-z_m) \tag{17.27}$$

and
$$n+\tfrac{1}{2}=\left\{\frac{-4k^2X_m}{(1-iZ)a^2}\right\}^{-\frac{1}{2}}k^2\left(C^2-\frac{X_m}{1-iZ}\right). \tag{17.28}$$

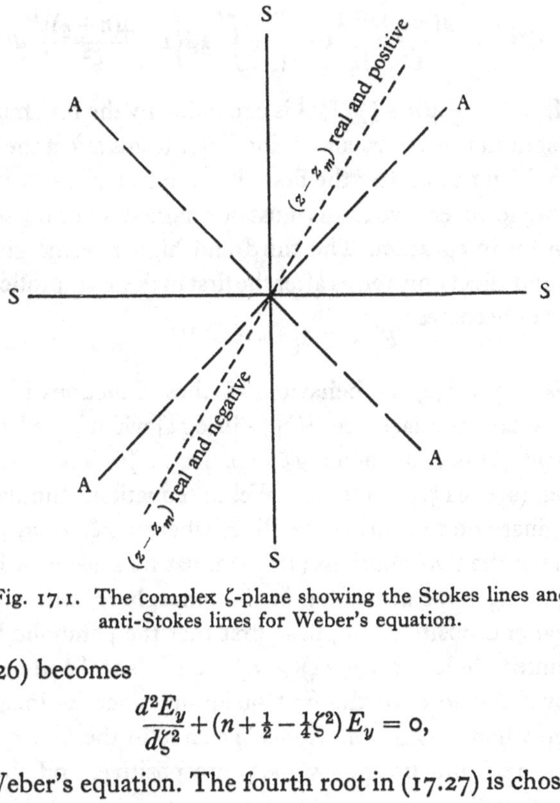

Fig. 17.1. The complex ζ-plane showing the Stokes lines and anti-Stokes lines for Weber's equation.

Then (17.26) becomes

$$\frac{d^2E_y}{d\zeta^2}+(n+\tfrac{1}{2}-\tfrac{1}{4}\zeta^2)E_y=0, \tag{17.29}$$

which is Weber's equation. The fourth root in (17.27) is chosen so that

$$\tfrac{1}{4}\pi<\arg\zeta<\tfrac{3}{8}\pi \quad \text{when } z-z_m \text{ is real and positive,} \tag{17.30}$$

which makes $\arg\zeta\to\frac{1}{4}\pi$ as $Z\to 0$, and $\arg\zeta\to\frac{3}{8}\pi$ as $Z\to\infty$. This choice ensures that the function $D_n(\zeta)$ (see p. 361) is the required solution. A different choice of the fourth root would lead to the use of a less convenient function. From this choice it also follows that the reciprocal square root in (17.28) must be negative imaginary when $Z\to 0$. Fig. 17.1 is a diagram of the complex ζ-plane showing the line for which z is real.

Weber's equation (17.29) has no singularities when ζ is finite, so that its solutions are finite, continuous and single valued for all ζ except ∞. Two independent solutions can be found by the standard method of solution in series of ascending powers of ζ.

Each of the two asymptotic approximations to the solutions of Weber's equation consists of a function of the form $\zeta^r \exp \phi(\zeta)$ multiplying a series in descending powers of ζ, that is an asymptotic expansion (see § 15.21). Only the first term of this series is needed in the present problem, and this can be found by the W.K.B. method of § 9.5, which gives

$$E_y \sim \zeta^{-\frac{1}{2}} \left\{ 1 - \frac{4(n+\frac{1}{2})}{\zeta^2} \right\}^{-\frac{1}{4}} \exp \left[\pm \int^{\zeta} \tfrac{1}{2}\zeta \left\{ 1 - \frac{4(n+\frac{1}{2})}{\zeta^2} \right\}^{\frac{1}{2}} d\zeta \right]. \quad (17.31)$$

When the factor $\{1 - 4(n+\frac{1}{2})/\zeta^2\}^{-\frac{1}{4}}$ is expanded by the binomial theorem, only the first term need be retained since later terms affect the asymptotic expansion only in terms after the first. For the factor $\{1-4(n+\frac{1}{2})/\zeta^2\}^{\frac{1}{2}}$ in the exponent, however, two terms must be retained, since the second gives a logarithm on integration. The third and higher terms give negative powers of ζ and affect only terms after the first in the asymptotic expansion. Hence (17.31) becomes

$$E_y \sim \zeta^{-\frac{1}{2}} \zeta^{\pm(n+\frac{1}{2})} e^{\mp\frac{1}{4}\zeta^2}. \quad (17.32)$$

When $|\zeta|$ is very large, the behaviour of these functions is determined mainly by the exponential since $|\frac{1}{4}\zeta^2| \gg |n \log \zeta|$ when $|\zeta|$ is large enough. The exponent $\frac{1}{4}\zeta^2$ is real when $\arg \zeta = 0, \frac{1}{2}\pi, \pi, \frac{3}{2}\pi$. These are therefore Stokes lines (see § 15.11) for the Weber equation. Similarly, $\frac{1}{4}\zeta^2$ is purely imaginary on the anti-Stokes lines where $\arg \zeta = \frac{1}{4}\pi, \frac{3}{4}\pi, \frac{5}{4}\pi, \frac{7}{4}\pi$.

Next, one of the two solutions (17.32) must be selected which represents an upgoing wave at the top of the ionosphere, where $z - z_m = a$, which is real and positive. Suppose first that the parabolic law (17.24) could continue to hold for $(z - z_m) > a$. Then X is real and negative, and (17.25) shows that q^2 is in the first quadrant, since its imaginary part changes sign when $z - z_m = a$. Hence q also is in the first quadrant, so that both its real and imaginary parts are positive, and therefore the upgoing wave, with

$$E_y = q^{-\frac{1}{2}} \exp \left\{ -\int^z q\, dz \right\},$$

increases in amplitude as z increases. Now the range where the W.K.B. solutions are good approximations includes very large values of z, and we shall assume that it extends down to below the level $z - z_m = a$. In the whole of this range the upgoing W.K.B. solution can be identified. Below $z - z_m = a$ its amplitude decreases as z increases and it is therefore the required solution. When $|\zeta|$ is very large, however, that is, well above the level $z - z_m = a$, the amplitude increases as $|\zeta|$ increases, and it is therefore the *dominant* term of the pair (17.32), namely

$$E_y \sim \zeta^n e^{-\frac{1}{4}\zeta^2} \quad \text{for} \quad \tfrac{1}{4}\pi < \arg\zeta < \tfrac{3}{8}\pi. \tag{17.33}$$

The dominancy of this term is determined by the exponential, but the solution is used only for comparatively small values of $|\zeta|$, where the factor ζ^n can have an appreciable effect. It is important, however, that $|\zeta|$ is still large enough for the W.K.B. solutions to be good approximations when $z - z_m = a$.

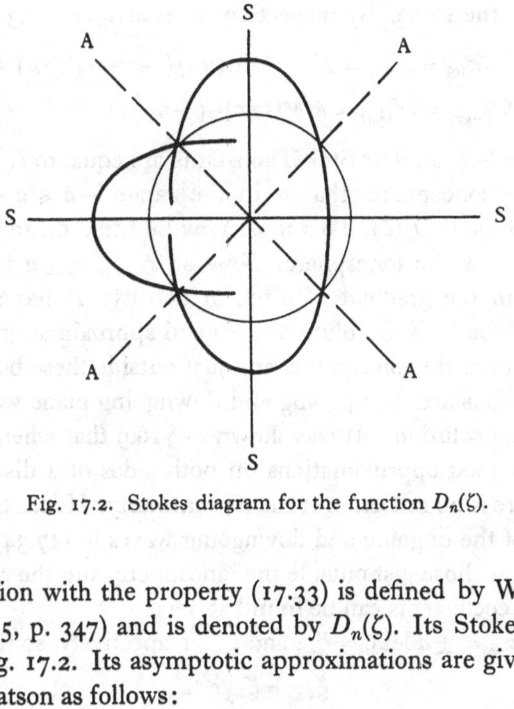

Fig. 17.2. Stokes diagram for the function $D_n(\zeta)$.

The function with the property (17.33) is defined by Whittaker and Watson (1935, p. 347) and is denoted by $D_n(\zeta)$. Its Stokes diagram is shown in Fig. 17.2. Its asymptotic approximations are given by Whittaker and Watson as follows:

$$D_n(\zeta) \sim \begin{cases} \zeta^n e^{-\frac{1}{4}\zeta^2} \quad \text{for} \quad -\frac{3}{4}\pi < \arg\zeta < \frac{3}{4}\pi, \\ \zeta^n e^{-\frac{1}{4}\zeta^2} - \dfrac{(2\pi)^{\frac{1}{2}}}{(-n-1)!} e^{-n\pi i} \zeta^{-(n+1)} e^{\frac{1}{4}\zeta^2} \quad \text{for} \quad -\frac{5}{4}\pi < \arg\zeta < -\frac{1}{4}\pi, \end{cases}$$

but the ranges of $\arg\zeta$ extend nearly to anti-Stokes lines. Although this satisfies Poincaré's criterion for an asymptotic approximation, it was explained in § 15.22 how it may lead to errors when $|\zeta|$ is not indefinitely large. We therefore use ranges of $\arg\zeta$ which end on Stokes lines, thus:

$$D_n(\zeta) \sim \begin{cases} \zeta^n e^{-\frac{1}{4}\zeta^2} \quad \text{for} \quad -\frac{1}{2}\pi \leqslant \arg\zeta \leqslant \frac{1}{2}\pi, & (17.34) \\ \zeta^n e^{-\frac{1}{4}\zeta^2} - \dfrac{(2\pi)^{\frac{1}{2}}}{(-n-1)!} e^{-n\pi i} \zeta^{-(n+1)} e^{\frac{1}{4}\zeta^2} \quad \text{for} \quad -\pi \leqslant \arg\zeta \leqslant -\frac{1}{2}\pi. & (17.35) \end{cases}$$

Weber's equation (17.29) is unaltered if ζ is replaced by $-\zeta$. Hence the Stokes constants $\mathscr{S}_{(0)}$ and $\mathscr{S}_{(\pi)}$ on the lines $\arg\zeta = 0, \pi$ are the same, and similarly those on the lines $\arg\zeta = \pm\frac{1}{2}\pi$, viz. $\mathscr{S}_{(-\frac{1}{2}\pi)}$, $\mathscr{S}_{(\frac{1}{2}\pi)}$ are the same, but in general $\mathscr{S}_{(0)} \neq \mathscr{S}_{(\frac{1}{2}\pi)}$. When $n = -\frac{1}{2}$, however, Weber's equation is unaltered when ζ is replaced by $i\zeta$, and then all four Stokes constants are the same. By inspection of equations (17.34) it is easily seen that

$$\left.\begin{aligned} \mathscr{S}_{(0)} = \mathscr{S}_{(\pi)} &= 2i\, e^{2\pi n i} \sin(n\pi)(-n-1)!\,(2\pi)^{-\frac{1}{2}}, \\ \mathscr{S}_{(-\frac{1}{2}\pi)} = \mathscr{S}_{(\frac{1}{2}\pi)} &= e^{-n\pi i}(2\pi)^{\frac{1}{2}}/(-n-1)!, \end{aligned}\right\} \quad (17.36)$$

and when $n = -\frac{1}{2}$, all four Stokes constants are equal to $i\sqrt{2}$.

Within the ionosphere, that is in the range $-a \leqslant z - z_m \leqslant a$, the required solution is $D_n(\zeta)$. This must now be fitted on to the solutions above and below the ionosphere. Now at $z - z_m = \pm a$ there are discontinuities in the gradient of electron density. It has already been assumed that the W.K.B. solutions are good approximations just within the boundaries of the ionosphere, and just outside these boundaries the W.K.B. solutions are the upgoing and downgoing plane waves, and are therefore exact solutions. It was shown in § 16.9 that when the W.K.B. solutions are good approximations on both sides of a discontinuity of gradient, there is no reflection at the discontinuity. Hence the (complex) amplitudes of the upgoing and downgoing waves in (17.34) and (17.35) are the same as those just outside the ionosphere, and the reflection and transmission coefficients can be found at once.

When $z - z_m = \pm a$ let $\zeta = \zeta_{+a}$ and ζ_{-a}, respectively, so that

$$\zeta_{+a} = \zeta_{-a} e^{i\pi}. \quad (17.37)$$

In (17.35) the first term is the downgoing wave and the second is the upgoing wave and their ratio is the reflection coefficient:

$$R = -(-n-1)! \exp(n\pi i) \exp(-\tfrac{1}{2}\zeta_{-a}^2)\, \zeta_{-a}^{(2n+1)} (2\pi)^{-\frac{1}{2}}. \quad (17.38)$$

Similarly (17.34) is the upgoing wave at the top of the ionosphere and its ratio to the upgoing wave in (17.35) gives the transmission coefficient

$$T = R e^{n\pi i}, \quad (17.39)$$

where (17.37) has been used.

For (17.38) and (17.39) to be valid it is necessary that the W.K.B. solutions shall be good approximations just within the boundaries of the ionosphere, so that the condition (9.61) must hold there. This leads to

$$X_m \left| \frac{5}{2C^2} \frac{X_m}{1-iZ} - 1 \right| \ll 2k^2a^2C^4 |1-iZ|. \qquad (17.40)$$

For normal incidence ($C = 1$), at the penetration frequency ($X_m = 1$), when collisions are neglected ($Z = 0$), this gives

$$a/\lambda \gg 0.14, \qquad (17.41)$$

where λ is the wavelength *in vacuo*. Hence the theory of this section applies only when the thickness of the ionosphere is several free-space wavelengths. This is usually true for the F-layer, but it is possible that the E-layer is sometimes so thin that this theory fails.

The reflecting properties of a thin parabolic layer may thus be affected by the discontinuities in the gradient of electron density at the boundaries. This problem has been discussed by Rydbeck (1943) who used more accurate formulae for $D_n(\zeta)$ and its derivative.

The application of the phase integral method to the parabolic ionosphere is discussed in §20.4.

17.5 Partial penetration and reflection

The formulae (17.38) and (17.39) are very complicated when the effect of electron collisions is included. If collisions are neglected, however, a simpler treatment is possible. Hence we take $Z = 0$. Then no energy can be lost from the waves so that

$$|R|^2 + |T|^2 = 1. \qquad (17.42)$$

Equation (17.28) shows that $(n+\frac{1}{2})$ is purely imaginary. Hence let

$$n + \tfrac{1}{2} = -\tfrac{1}{2}iD, \qquad (17.43)$$

where D is real and is given by

$$D = ak(C^2 - X_m)X_m^{-\frac{1}{2}}. \qquad (17.44)$$

Thus D is negative for frequencies below penetration, and positive for frequencies above it. Equation (17.39) shows that

$$|T|^2 = |R|^2 e^{\pi D}, \qquad (17.45)$$

whence

$$|R|^2 = \frac{1}{1+e^{\pi D}}, \quad |T|^2 = \frac{1}{1+e^{-\pi D}}. \qquad (17.46)$$

On a simple ray theory the waves would just penetrate the layer when $X_m = C^2$. Then $D = 0$ and the full wave theory shows that

$$|R|^2 = |T|^2 = \tfrac{1}{2}.$$

Fig. 17.3 shows how $|R|$ and $|T|$ depend on D. It should be compared with Figs. 8.9, 16.8 and 17.5, which give similar curves for other models of the ionosphere.

In the special case of normal incidence, $C = 1$. Further, $X_m = f_p^2/f^2$ where f_p is the penetration-frequency. Let

$$\Delta f = f - f_p, \quad a/\lambda_p = m, \tag{17.47}$$

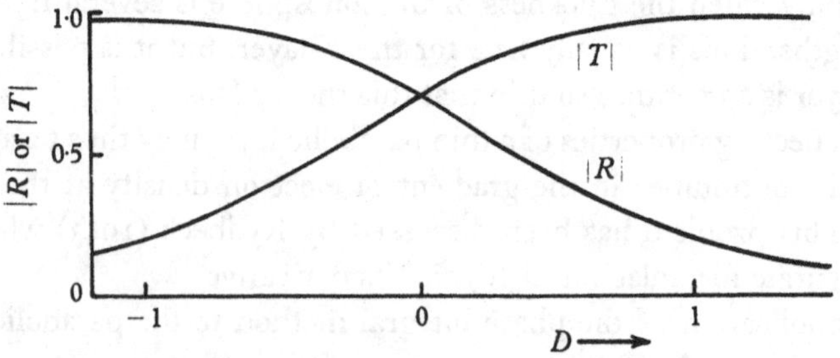

Fig. 17.3. Reflection coefficient $|R|$ and transmission coefficient $|T|$ for a parabolic layer when electron collisions and the earth's magnetic field are neglected. The abscissa D is zero at the penetration-frequency and varies nearly linearly with frequency.

where λ_p is the free-space wavelength at the penetration-frequency, and consider frequencies close to penetration so that $|\Delta f| \ll f_p$. Then

$$D = \frac{2a\pi}{\lambda_p}\frac{f^2 - f_p^2}{f_p^2} \approx \frac{4\pi m \Delta f}{f_p}. \tag{17.48}$$

Equation (17.45) shows that when $|D| = 0{\cdot}7$, $|T^2|/|R|^2 \approx 10$ or $\frac{1}{10}$. It is therefore probable that partial penetration and reflection could only be observed over a frequency range $2\Delta f$ small enough to make $|D| < 0{\cdot}7$. Now for the F-layer m is of the order of 1000 at the penetration-frequency and hence $2\Delta f$ must be less than $0{\cdot}00006 f_p$, that is of the order of 1 kc. This is too small to observe, and we conclude that for a layer as thick as the F-layer the transition from total reflection to total transmission occurs quite sharply at the penetration-frequency, as predicted by ray theory. For the E-layer, however, partial penetration and reflection is often observed over a frequency range of the order $2\Delta f \approx 0{\cdot}04 f_p$. If this is explained by the theory of the present section then m must be of the order of 3.

17.6 The equivalent height of reflection for a parabolic layer

In § 10.7 the 'ray theory' was used to deduce the equivalent height of reflection of radio waves in a parabolic layer. This theory may fail for frequencies near the penetration-frequency, and the 'full wave' theory will therefore now be used. Only vertical incidence is discussed ($C = 1$) and electron collisions are neglected ($Z = 0$). The theory taking account of electron collisions has been given by Rydbeck (1943).

The equivalent height of reflection $h'(f)$ is given in terms of the phase height $h(f)$ by the formula (10.15). When the phase of the reflected wave varies with frequency through some mechanism other than propagation through a medium, there would still be an effective group retardation which would result in an observed equivalent height. At the bottom of the parabolic layer the phase difference between the incident and reflected waves is $\arg R$ so that

$$h(f) = z_m - a - (\arg R)\,\lambda/4\pi = z_m - a - \tfrac{1}{2}(\arg R)/k$$

and
$$h'(f) = z_m - a - \frac{1}{2}\frac{d(\arg R)}{dk} = z_m - a - \frac{1}{2}\frac{d(\arg R)}{dD}\frac{dD}{dk}$$

$$= z_m - a - a\frac{k}{k_p}\frac{d(\arg R)}{dD}, \qquad (17.49)$$

where $k_p = 2\pi/\lambda_p$. It has here been assumed that $h'(f)$ depends only on the variation of phase with frequency, and we have ignored the variation of amplitude $|R|$ with frequency. This is satisfactory as long as $d|R|/df$ is not too large. Equation (10.15) was based on the idea of 'group' propagation, which fails when $d|R|/df$ is large (see, for example, Stratton, 1941, ch. V).

Let k_p be the value of k at the penetration-frequency. Then (17.27) gives

$$\zeta_{-a} = (2ak_p)^{\frac{1}{2}}\, e^{-i(\frac{3}{4}\pi)}, \qquad (17.50)$$

and (17.38) may be written

$$R = -(2\pi)^{-\frac{1}{2}}(-\tfrac{1}{2}+\tfrac{1}{2}iD)!\exp\{-\tfrac{1}{2}iD\log(2ak_p) - i(\tfrac{1}{2}\pi) - iak_p - \tfrac{1}{4}\pi D\}. \qquad (17.51)$$

Stirling's formula may be used for the factorial function. The quantity $|-\frac{1}{2}+\frac{1}{2}iD|$ is smallest when $D = 0$ and is then equal to $\frac{1}{2}$. For this value it is not accurate enough to use only the first term of the series in Stirling's formula, and two terms must be used. Hence we take

$$(2\pi)^{-\frac{1}{2}}(-\tfrac{1}{2}+\tfrac{1}{2}iD)! \sim \left\{1 - \frac{1}{6(1-iD)}\right\}\exp\{\tfrac{1}{2} - \tfrac{1}{2}iD + \tfrac{1}{2}iD\log(-\tfrac{1}{2}+\tfrac{1}{2}iD)\}, \qquad (17.52)$$

and (17.51) then gives

$$\arg R \sim \tfrac{1}{2}\pi - ak_p - \tfrac{1}{2}D[1+\log(2ak_p) - \log\{\tfrac{1}{2}(1+D^2)^{\frac{1}{2}}\}] \\ - \arctan\{D/(6D^2+5)\}, \quad (17.53)$$

whence (17.49) gives

$$h'(f) \sim z_m - a + a\left(\frac{k}{k_p}\right)\left[\tfrac{1}{2} - \tfrac{1}{2}\log\frac{(1+D^2)^{\frac{1}{2}}}{4ak_p} - \frac{\frac{1}{2}D^2+1}{D^2+1} + \frac{\frac{6}{5}}{1+(6D/5)^2}\right]. \quad (17.54)$$

This is nearly symmetrical about the penetration-frequency where $k = k_p$, $D = 0$. It has a maximum value there equal to

$$z_m - \tfrac{3}{10}a + \tfrac{1}{2}a\log(4ak_p).$$

For waves which travel right through the parabolic layer we may calculate the contribution to the equivalent height from the region between the ground and the top of the layer, as was done by 'ray theory' in § 10.7. The contribution is

$$h'(f) = z_m - a - \frac{d(\arg T)}{dk}, \quad (17.55)$$

and when (17.39) is used

$$h'(f) \sim z_m - a + 2a\left(\frac{k}{k_p}\right)\left[\tfrac{1}{2} - \tfrac{1}{2}\log\left\{\frac{(1+D^2)^{\frac{1}{2}}}{4ak_p}\right\} - \frac{\frac{1}{2}D^2+1}{D^2+1} + \frac{\frac{6}{5}}{1+(6D/5)^2}\right]. \quad (17.56)$$

The expressions (17.54) and (17.56) are shown plotted as functions of f/f_p in Fig. 17.4, and the corresponding curves for the 'ray' theory, given by (10.33) and (10.37), are also shown for comparison. The curves are very nearly the same when $|D| \gg 0$,† but near the penetration-frequency the ray theory gives values of $h'(f)$ which approach infinity, whereas the full wave theory shows that it remains finite. The difference between the two is most marked when m is small, that is for thin layers.

17.7 Electron density with square law increase

Suppose that the electron number density is proportional to the square of the height above the base of the ionosphere, so that

$$X = \begin{cases} \beta(z-z_1)^2 & \text{for } z \geq z_1, \\ 0 & \text{for } z \leq z_1, \end{cases} \quad (17.57)$$

where β is a constant. The reflection coefficient in this case was found by

† See note 1 on p. 529.

Rydbeck (1944), and a closely related problem was discussed by Wilkes (1940). The problem is interesting because there is a discontinuity in curvature of the profile at the base of the ionosphere. Equation (9.56) gives

$$q^2 = C^2 - \beta(z-z_1)^2/(1-iZ) \quad \text{for} \quad z \geqslant z_1 \tag{17.58}$$

and the electric field E_y must satisfy the differential equation (9.58) which becomes

$$\frac{d^2E_y}{dz^2} + k^2\{C^2 - \beta(z-z_1)^2/(1-iZ)\}E_y = 0. \tag{17.59}$$

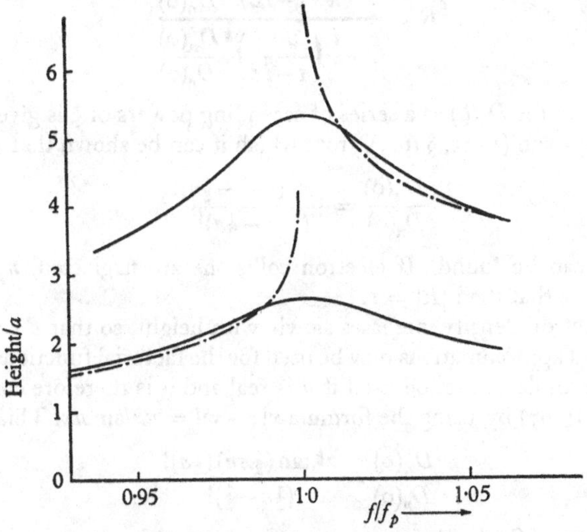

Fig. 17.4. The contribution to the equivalent height of reflection from a parabolic layer of half-thickness a, according to 'full wave' theory when electron collisions and the earth's magnetic field are neglected. In this example $m = a/\lambda_p = 1{\cdot}99$. The lower curve is for reflection, and the upper curve for waves which have penetrated the layer. The chain curves show the corresponding results for 'ray theory'.

As before it will be assumed that Z is constant (constant collision-frequency). Now let

$$\zeta = \left(\frac{4k^2\beta}{1-iZ}\right)^{\frac{1}{4}}(z-z_1), \tag{17.60}$$

$$n+\tfrac{1}{2} = \left(\frac{4k^2\beta}{1-iZ}\right)^{-\frac{1}{2}} k^2C^2. \tag{17.61}$$

Then (17.59) is converted to the Weber equation (17.29). The fourth root in (17.60) is chosen to be in the first quadrant so that, when $z-z_1$ is positive, $0 < \arg\zeta < \frac{1}{8}\pi$. One solution of (17.29) is $D_n(\zeta)$, whose asymptotic approximation is (17.34) in this case. This represents an upgoing wave at great heights and is, therefore, the required solution.

At the base of the ionosphere, where $z = z_1$,

$$E_y = D_n(0), \tag{17.62}$$

and from (9.49)

$$\mathscr{H}_x = -\frac{i}{k}\left(\frac{4k^2\beta}{1-iZ}\right)^{\frac{1}{4}} D_n'(0). \tag{17.63}$$

Just below the base of the ionosphere the field consists of an upgoing and a downgoing wave so that

$$E_y = A(R+1), \tag{17.64}$$

$$\mathcal{H}_x = AC(R-1), \tag{17.65}$$

where A is the amplitude of the incident wave, and R is the reflection coefficient. These are equated to (17.62) and (17.63), respectively, which gives

$$R = \frac{C - \dfrac{i}{k}\left(\dfrac{4k^2\beta}{1-iZ}\right)^{\frac{1}{4}} \dfrac{D_n'(0)}{D_n(0)}}{C + \dfrac{i}{k}\left(\dfrac{4k^2\beta}{1-iZ}\right)^{\frac{1}{4}} \dfrac{D_n'(0)}{D_n(0)}}. \tag{17.66}$$

An expression for $D_n(\zeta)$ as a series of ascending powers of ζ is given by Whittaker and Watson (1935, § 16.5), from which it can be shown that

$$\frac{D_n'(0)}{D_n(0)} = \frac{-2^{\frac{1}{2}}(-\frac{1}{2}-\frac{1}{2}n)!}{(-1-\frac{1}{2}n)!}, \tag{17.67}$$

whence R can be found. If electron collisions are neglected, n is real and (17.66) shows that then $|R| = 1$.

If the electron density increases slowly with height, so that β is small, then n is large and approximations may be used for the factorial functions in (17.67). Stirling's formula cannot be used if n is real and it is therefore convenient to transform (17.67) by using the formula $x!\,(-x)! = \pi x/\sin \pi x$. This leads to

$$\frac{D_n'(0)}{D_n(0)} = \frac{2^{\frac{1}{2}} \tan(\frac{1}{2}\pi n)\,(\frac{1}{2}n)!}{(\frac{1}{2}n-\frac{1}{2})!}. \tag{17.68}$$

The first term of Stirling's formula may now be used for the factorial functions, and if terms of order $1/n^2$ are neglected, this gives

$$\frac{D_n'(0)}{D_n(0)} \sim n^{\frac{1}{2}} \tan(\frac{1}{2}\pi n), \tag{17.69}$$

which may now be inserted in (17.66). If (17.61) is used and terms of order $1/n^2$ are again neglected

$$R \approx e^{-i\pi n}. \tag{17.70}$$

This is the same as the result given by the phase integral formula (16.29).

17.8 The sinusoidal layer

The parabolic profile of electron density discussed in §§ 17.4 to 17.6 is not entirely satisfactory as a model of the ionosphere, because it has discontinuities in the gradient of electron density at the top and bottom boundaries. A profile which does not have this disadvantage is the sinusoidal profile, in which

$$X = \tfrac{1}{2}X_m\left\{1 + \cos\left(\pi \frac{z-z_m}{a}\right)\right\} \quad \text{for} \quad |z-z_m| \leqslant a,$$

$$X = 0 \quad \text{for} \quad |z-z_m| \geqslant a.$$

The top and bottom of the ionosphere are at $z = z_m \pm a$, and a is called the half-thickness.

This expression for X may be inserted in the formula (9.56) for q^2, which in turn is substituted in the differential equation (9.58). If Z is assumed to be constant, the differential equation can be converted into one of the standard forms of Mathieu's equation (see, for example, Whittaker and Watson, 1935, ch. XIX, or Brainerd and Weygandt, 1940) by a change of the independent variable. It is possible to choose solutions which are Mathieu functions and which satisfy the boundary conditions at the top of the ionosphere. The reflection coefficient at the bottom can then be found using tables of these functions. Tables of suitable Mathieu functions have been given by Gray, Mervin and Brainerd (1948), but the range they cover is only suitable for frequencies below penetration when electron collisions are neglected.

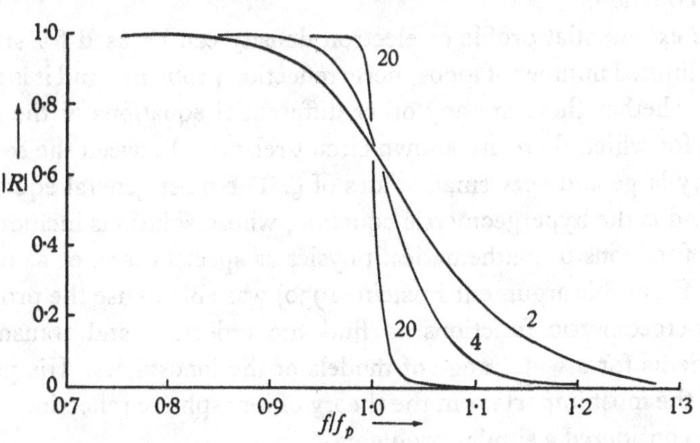

Fig. 17.5. Reflection coefficient $|R|$ for ionosphere with sinusoidal electron density profile, at vertical incidence. The numbers by the curves are the values of $2a/\lambda_p$ where a is the 'half thickness' and λ_p is the wavelength in free space at the penetration-frequency f_p.

This problem can, however, be solved very easily by one of the numerical methods described in ch. 22. Some results for vertical incidence when $Z = 0$ are shown in Fig. 17.5. They were calculated by integration of the second equation in example 1 of ch. 22. The curves show only the reflection coefficient $|R|$, but in this case there is no loss of energy so that $|R|^2+|T|^2 = 1$, and the transmission coefficient $|T|$ could easily be found. The curves display the phenomenon of partial penetration and reflection and should be compared with Figs. 8.9, 16.8 and 17.3, which show corresponding results for other electron density profiles.

17.9 Circuit relations. Introduction to Epstein's theory

In § 17.2 the reflection coefficient was found for an ionosphere in which the electron density varied exponentially with height z, and the electric field was found to be represented by a Bessel function $H_\nu^{(1)}(Ke^{\frac{1}{2}\alpha z})$ (where K and α are constants), which was chosen to satisfy the physical conditions at great heights. Below the ionosphere z is large and negative and $|Ke^{\frac{1}{2}\alpha z}|$ is very small. The form of the function $H_\nu^{(1)}$ is known in these circumstances (17.14), and it could be expressed as the sum of terms representing upgoing and downgoing waves. The use of the exponential variable $\zeta = Ke^{\frac{1}{2}\alpha z}$ is especially convenient here because of its very small value below the ionosphere. The success of the method depended on knowing the circuit relation between the function (17.11), representing an upgoing wave when ζ is very large, and the two functions (17.13) representing downgoing and upgoing waves, respectively, when ζ is very small. This circuit relation is a known property of Bessel's differential equation.

The exponential profile of electron density can be used for studying only a limited number of ionospheric reflection problems, and it is natural to ask whether there are any other differential equations of the second order, for which there are known circuit relations between the solutions for very large and very small values of ζ. The most general equation of this kind is the hypergeometric equation, whose solutions include many of the functions of mathematical physics as special cases, or as limiting cases. From this argument Epstein (1930) was able to use the properties of hypergeometric functions to find the reflection and transmission coefficients for a wide range of models of the ionosphere. His paper is one of the most important in the theory of ionospheric reflection. Eckart (1930) considered a similar problem.

17.10 The hypergeometric differential equation

The hypergeometric equation may be written:†

$$\zeta(1-\zeta)\frac{d^2u}{d\zeta^2}+\{c-(a+b+1)\zeta\}\frac{du}{d\zeta}-abu=0. \tag{17.71}$$

It has regular singularities at $\zeta = 0$, 1 and ∞. To find a solution we try first a series of ascending powers of ζ. Hence insert

$$u = \zeta^\beta(1+a_1\zeta+a_2\zeta^2+\dots) \tag{17.72}$$

in (17.71) and equate coefficients of successive increasing powers of ζ. This gives

$$\left.\begin{aligned} \beta(\beta-1+c) &= 0,\\ a_1(\beta+1)(\beta+c) &= \beta(\beta+a+b)+ab,\\ a_2(\beta+2)(\beta+c+1) &= a_1\{(\beta+1)(\beta+a+b+1)+ab\},\\ &\dots\dots\dots\dots\\ a_n(\beta+n)(\beta+c+n-1) &= a_{n-1}\{(\beta+n-1)(\beta+a+b+n-1)+ab\}.\end{aligned}\right\} \tag{17.73}$$

† This form is used by Jeffreys and Jeffreys (1956) and by Whittaker and Watson (1935). Epstein (1930) used a different form, and the meaning of his symbols a, b are different. The symbol c here is a new constant, not the velocity of light.

The first of these is the indicial equation and gives $\beta = 0$ or $1-c$. For $\beta = 0$ we obtain

$$u = 1 + \frac{ab}{c.1!}\zeta + \frac{a(a+1)b(b+1)}{c(c+1)2!}\zeta^2 + \dots$$
$$+ \frac{a(a+1)\dots(a+n-1)b(b+1)\dots(b+n-1)}{c(c+1)\dots(a+n-1)n!}\zeta^n + \dots$$
$$= F(a,b;c;\zeta), \tag{17.74}$$

which is called the hypergeometric series and can easily be shown to be convergent when $|\zeta| < 1$. To find a second solution take $\beta = 1-c$. If c is a positive integer, one of the equations (17.73) cannot be satisfied, and there is then no second solution of the form (17.72). A second solution containing a logarithm can be found, but this special case does not arise in the present problem. Apart from this, (17.73) gives

$$u = (-\zeta)^{1-c}\left\{1 + \frac{(a-c+1)(b-c+1)}{(2-c)1!}\zeta\right.$$
$$\left. + \frac{(a-c+1)(a-c+2)(b-c+1)(b-c+2)}{(2-c)(3-c)2!}\zeta^2 + \dots\right\}$$
$$= (-\zeta)^{1-c}F(a-c+1, b-c+1; 2-c; \zeta) \tag{17.75}$$

(it is convenient to introduce the constant factor $(-1)^{1-c}$ since it appears in later formulae).

Alternatively, we may take as a trial solution a series in descending powers of ζ. This is possible because the singularity of (17.71) at infinity is regular. (It is not possible, for example, with the Stokes equation (15.7) or Bessel's equation (17.6), which both have essential singularities at infinity.) Thus insert

$$u = \zeta^\beta\left(1 + \frac{b_1}{\zeta} + \frac{b_2}{\zeta^2} + \dots\right)$$

in (17.71) and equate coefficients of successive decreasing powers of ζ. Then

$$(\beta+a)(\beta+b) = 0,$$
$$b_1\{(\beta-1)(\beta+a+b-1)+ab\} = \beta(\beta-1+c),$$
$$b_2\{(\beta-2)(\beta+a+b-2)+ab\} = b_1(\beta-1)(\beta-2+c),$$
$$\dots\dots\dots\dots\dots\dots\dots\dots\dots\dots\dots\dots$$
$$b_n\{(\beta-n)(\beta+a+b-n)+ab\} = b_{n-1}(\beta-n+1)(\beta-n+c).$$

The first of these is the indicial equation and gives $\beta = -a$, or $-b$. If a and b are not negative integers, which is true for the problems considered here, we obtain the two solutions:

$$u = (-\zeta)^{-a}F(b, b-c+1; b-a+1; \zeta^{-1}), \tag{17.76}$$
$$u = (-\zeta)^{-b}F(a, a-c+1; a-b+1; \zeta^{-1}), \tag{17.77}$$

where constant factors $(-1)^{-a}(-1)^{-b}$ have again been introduced for convenience later. These series are convergent provided that $|\zeta| > 1$. They are therefore slightly different from the asymptotic expansions considered in § 15.21. The functions (17.76) and (17.77) have branch points at infinity because of the factors ζ^{-a}, ζ^{-b}, but they give solutions valid for any range 2π of $\arg\zeta$, and they do not display the Stokes phenomenon (see Jeffreys and Jeffreys, § 17.023).

Since the differential equation (17.71) is of the second order, each of the solutions (17.76) and (17.77) must be expressible as a linear combination of the solutions (17.74) and (17.75). The expressions are the required circuit relations, and are derived in the next section.

The solution (17.75) may have a branch point at the origin, and (17.76) and (17.77) may have branch points at infinity. It can be shown that all four solutions (17.74), (17.75), (17.76) and (17.77) or their analytic continuations have branch points at $\zeta = 1$. Hence a cut is introduced in the complex ζ-plane extending along the whole of the positive real axis. The solutions are then single valued provided that the cut is not crossed, and in future the values used will be those for which

$$|\arg(-\zeta)| < \pi. \tag{17.78}$$

17.11 The circuit relations for the hypergeometric function

The series (17.74) converges only when $|\zeta| < 1$ and cannot therefore be considered to be a solution of (17.71) when $|\zeta| > 1$. Similarly, the series (17.76) and (17.77) can only be considered solutions when $|\zeta| > 1$. It is possible, however, by the process of analytic continuation (Whittaker and Watson, §§ 5.5 and 14.51) to express (17.74) in a form which is valid both inside and outside the domain $|\zeta| < 1$. This form is a contour integral which is equal to (17.61) when $|\zeta| < 1$, and can be expressed as a linear combination of (17.76) and (17.77) when $|\zeta| > 1$. In this way the circuit relations are derived.

To express the series (17.74) as a contour integral we arrange that the terms are the residues of a series of poles of the integrand. The factorial function $(-s)!$ has poles with residue $(-1)^{s-1}/(s-1)!$ where $s = 1, 2, 3, \ldots$ (see, for example, Copson, 1935, p. 207). Hence we use the integral

$$\frac{1}{2\pi i}\int_C \frac{(a+s-2)!\,(b+s-2)!\,(-s)!}{(c+s-2)!}(-\zeta)^{s-1}\,ds. \tag{17.79}$$

Provided that the contour is suitably chosen, the residue of the pole at $s = n$ contributes

$$-\frac{(a+n-2)!\,(b+n-2)!}{(c+n-2)!\,(n-1)!}\zeta^{n-1}$$

which is

$$-\frac{(a-1)!\,(b-1)!}{(c-1)!}$$

times the nth term of the series (17.74). The sign would be + if the pole were encircled clockwise. The contour must therefore be chosen so that all the poles of $(-s)!$ are included and all the poles of $(a+s-2)!$ and $(b+s-2)!$ are excluded when $|\zeta| < 1$. The poles of the factorial functions must be distinct and the poles of $(-s)!$ must not be annulled by zeros of $1/(c+n-2)!$, so that neither a nor b nor c can be a negative integer or zero, but these special cases do not arise in the present problem. Part of the contour is shown in Fig. 17.6 as a solid line. It begins and ends on the imaginary axis of the s-plane and is curved to separate the poles. Suppose that it extends from $-i(N+\frac{1}{2})$ to $+i(N+\frac{1}{2})$ where N is a large integer, and is closed by a semicircle S_1 of radius $N+\frac{1}{2}$ centred on the origin. This passes midway between two poles. Now it

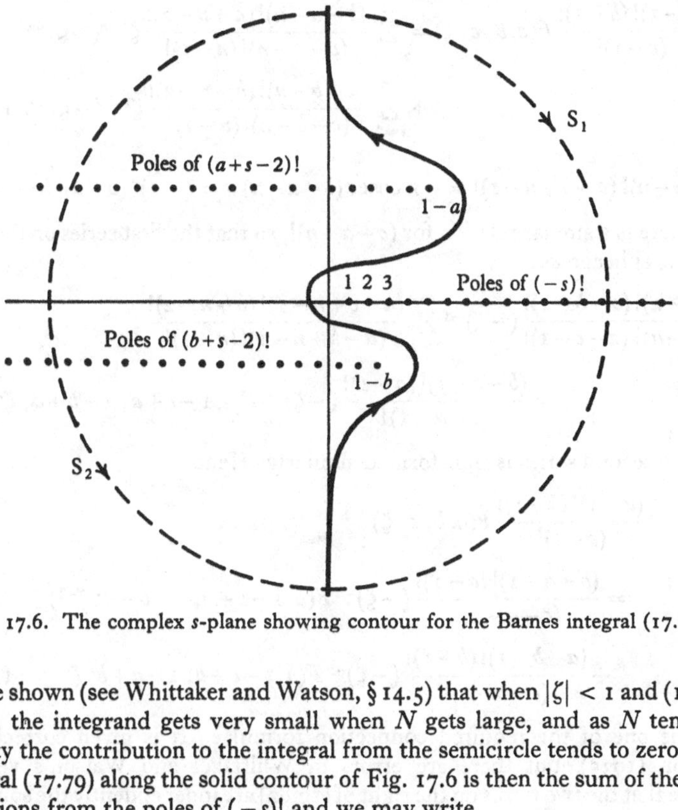

Fig. 17.6. The complex s-plane showing contour for the Barnes integral (17.69).

can be shown (see Whittaker and Watson, § 14.5) that when $|\zeta| < 1$ and (17.78) holds, the integrand gets very small when N gets large, and as N tends to infinity the contribution to the integral from the semicircle tends to zero. The integral (17.79) along the solid contour of Fig. 17.6 is then the sum of the contributions from the poles of $(-s)!$ and we may write

$$\frac{(a-1)!\,(b-1)!}{(c-1)!} F(a,b;c;\zeta) = \frac{1}{2\pi i}\int_{-i\infty}^{i\infty} \frac{(a+s-2)!\,(b+s-2)!\,(-s)!}{(c+s-2)!}(-\zeta)^{s-1}\,ds. \tag{17.80}$$

Integrals of this kind were used by Barnes (1908). Equation (17.80) is true when $|\zeta| < 1$. But the integral exists also when $|\zeta| > 1$, and therefore provides the required analytic continuation of the function on the left side. Previously $F(a, b; c; \zeta)$ was called the hypergeometric series (17.74) and was defined only when $|\zeta| < 1$. It is now defined by (17.80) for other values of ζ and is called the hypergeometric function.

When $|\zeta| > 1$ the contour cannot be closed by the semicircle S_1 because now the integrand becomes very large as $N \to \infty$. It can, however, be closed by the semicircle S_2 to the left, provided that the radius is chosen so that S_2 passes between poles of $(a+s-2)!$ and $(b+s-2)!$. The radius is then increased in steps and it can be shown that the contribution to the integral from the semicircle tends to zero as the radius tends to infinity. The integral is therefore the sum of the contributions from the two sets of poles on the left of the contour, so that

$$\frac{(a-1)!(b-1)!}{(c-1)!} F(a,b;c;\zeta) = \sum_{n=1}^{\infty} \frac{(b-a-n)!(a+n-2)!}{(c-a-n)!(n-1)!} \zeta^{1-n}(-\zeta)^{-a} + \sum_{n=1}^{\infty} \frac{(a-b-n)!(b+n-2)!}{(c-b-n)!(n-1)!} \zeta^{1-n}(-\zeta)^{-b}. \quad (17.81)$$

Now

$$(b-a-n)!(a-b+n-1)! = -\pi \operatorname{cosec}(b-a-n)\pi = (-1)^n (b-a)!(a-b-1)!$$

and there is a similar relation for $(c-a-n)!$, so that the first series on the right of (17.81) becomes

$$\frac{(b-a)!(a-b-1)!}{(c-a)!(a-c-1)!} (-\zeta)^{-a} \sum_{n=1}^{\infty} \frac{(a-c+n-1)!(a+n-2)!}{(a-b+n-1)!(n-1)!} \zeta^{1-n}$$
$$= \frac{(b-a-1)!(a-1)!}{(c-a-1)!} (-\zeta)^{-a} F(a, 1-c+a; 1-b+a; \zeta^{-1}),$$

and the second series is transformed similarly. Hence

$$\frac{(a-1)!(b-1)!}{(c-1)!} F(a,b;c;\zeta)$$
$$= \frac{(b-a-1)!(a-1)!}{(c-a-1)!} (-\zeta)^{-a} F(a, 1-c+a; 1-b+a; \zeta^{-1})$$
$$+ \frac{(a-b-1)!(b-1)!}{(c-b-1)!} (-\zeta)^{-b} F(b, 1-c+b; 1-a+b; \zeta^{-1}). \quad (17.82)$$

This is one of the required connection formulae. It is given correctly by Copson (1935), but there are errors in Whittaker and Watson's version. Notice that the two terms on the right of (17.82) are independently the solutions (17.76) and (17.77) of the hypergeometric equation.

The connection formula for the solution (17.75) can be found by replacing a by $a-c+1$, b by $b-c+1$, and c by $2-c$ in (17.81). It is

$$(-\zeta)^{1-c} \frac{(a-c)!(b-c)!}{(1-c)!} F(a-c+1, b-c+1; 2-c; \zeta)$$
$$= \frac{(b-a-1)!(a-c)!}{(-a)!} (-\zeta)^{-a} F(a, 1-c+a; 1-b+a; \zeta^{-1})$$
$$+ \frac{(a-b-1)!(b-c)!}{(-b)!} (-\zeta)^{-b} F(b, 1-c+b; 1-a+b; \zeta^{-1}). \quad (17.83)$$

By suitably combining (17.82) and (17.83) the following two formulae can be deduced:

$$\frac{(-c)!}{(-b)!\,(a-c)!}F(a,b;c;\zeta)$$

$$+\frac{(c-2)!}{(c-b-1)!\,(a-1)!}(-\zeta)^{1-c}F(a-c+1,b-c+1;2-c;\zeta)$$

$$=\frac{1}{(a-b)!}(-\zeta)^{-a}F(a,1-c+a;1-b+a;\zeta^{-1}), \quad (17.84)$$

$$\frac{(-c)!}{(-a)!\,(b-c)!}F(a,b;c;\zeta)$$

$$+\frac{(c-2)!}{(c-a-1)!\,(b-1)!}(-\zeta)^{1-c}F(a-c+1;b-c+1;2-c;\zeta)$$

$$=\frac{1}{(b-a)!}(-\zeta)^{-b}F(b,1-c+b;1-a+b;\zeta^{-1}). \quad (17.85)$$

Either of these two formulae could equally well be deduced by forming an integral of Barnes's type for the functions on the right, giving the series when $|\zeta| > 1$, and evaluating the integral when $|\zeta| < 1$. Different series of poles must be used according as $|\zeta| > 1$ or < 1, exactly as for the integral (17.80).

For the work of the present chapter the hypergeometric functions have now fulfilled their purpose and need not be used again. Those on the left of (17.82) to (17.85) will only be used when $|\zeta|$ is very small, so that only the first term (unity) of the series (17.74) is appreciable. Similarly, those on the right will only be used for very large $|\zeta|$ and again only the first term (unity) is appreciable. Hence in the remainder of this chapter the hypergeometric functions will be set equal to unity.

17.12 Application to the wave-equation

In § 17.9 it was indicated that the circuit relations can be applied to reflection from the ionosphere when the variable ζ is an exponential function of the height z. Hence in the hypergeometric equation (17.71) let†

$$-\zeta = e^{\mathfrak{z}}, \quad (17.86)$$

$$\mathfrak{z} = (z/\sigma)+\mathfrak{b}, \quad (17.87)$$

where σ is a constant which determines the scale of the vertical structure of the ionosphere. The constants σ and $\mathfrak{b}$ are chosen later to suit particular problems. In terms of $\mathfrak{z}$ as independent variable (17.71) becomes

$$(1+e^{\mathfrak{z}})\frac{d^2u}{d\mathfrak{z}^2}+\{c-1+(a+b)\,e^{\mathfrak{z}}\}\frac{du}{d\mathfrak{z}}+ab\,e^{\mathfrak{z}}u = 0. \quad (17.88)$$

† Rawer (1939) has shown how transformations more general than (17.86) may be used. In this way the range of Epstein profiles may be greatly extended.

This is now transformed to its normal form (without a first derivative term), by changing the dependent variable thus:

$$u = E \exp\{\tfrac{1}{2}(1-c)\,\mathfrak{z}\}\,(1+e^{\mathfrak{z}})^{\frac{1}{2}(c-1-a-b)}, \tag{17.89}$$

which converts (17.88) to

$$\frac{d^2E}{dz^2} + k^2q^2E = 0, \tag{17.90}$$

where
$$q^2 = \epsilon_1 + \frac{e^{\mathfrak{z}}}{(e^{\mathfrak{z}}+1)^2}\{(\epsilon_2-\epsilon_1)(e^{\mathfrak{z}}+1)+\epsilon_3\}, \tag{17.91}$$

and
$$\epsilon_1 = -\tfrac{1}{4}(c-1)^2/\sigma^2k^2, \tag{17.92}$$

$$\epsilon_2 = -\tfrac{1}{4}(a-b)^2/\sigma^2k^2, \tag{17.93}$$

$$\epsilon_3 = \tfrac{1}{4}(a+b-c+1)(a+b-c-1)/\sigma^2k^2. \tag{17.94}$$

Equation (17.90) is the same as (9.58) and E may be regarded as the electric field of a horizontally polarised wave obliquely incident on the ionosphere. The expression (17.91) therefore determines what profiles of electron density can be studied using this theory. We shall see that by suitably assigning the constants σ, $\mathfrak{b}$, ϵ_1, ϵ_2 and ϵ_3, a very wide range of models of the ionosphere can be investigated.

Below the ionosphere z is large and negative and (17.91) becomes $q^2 \approx \epsilon_1$. But (9.56) shows that here $q = C$. Hence $\epsilon_1 = C^2$ (where $C = \cos\theta_I$; θ_I is the angle of incidence) and

$$c-1 = -2ik\sigma C, \tag{17.95}$$

where the choice of sign made in taking the square root of (17.92) is arbitrary. (The other sign would lead to the same final result, but some of the intervening formulae including (17.106) to (17.109) would be different.) At a great height in the ionosphere z is large and positive and (17.91) becomes $q^2 \approx \epsilon_2$. Hence q tends to a constant value, q_2 say, and

$$a-b = -2ik\sigma q_2, \tag{17.96}$$

where the sign must be chosen to make $a-b = c-1$ when $q_2 = C$. At intermediate heights the variation of q with z is determined by ϵ_3, and (17.94) shows that

$$a+b-c = \pm(4k^2\sigma^2\epsilon_3+1)^{\frac{1}{2}}, \tag{17.97}$$

where either sign may be chosen.

17.13 The reflection and transmission coefficients of an Epstein layer

In § 17.10 four independent solutions—(17.74), (17.75), (17.76) and (17.77)—were found for the hypergeometric equation. The field variable E associated with each of these must now be examined. First, when z is large and negative, $e^{\mathfrak{z}}$ is very small compared to unity, and the transformation (17.89) becomes

$$E = u \exp\{\tfrac{1}{2}(c-1)\,\mathfrak{z}\}. \tag{17.98}$$

In (17.74) and (17.75) the hypergeometric functions are unity when z is large and negative, and when (17.86), (17.87) and (17.95) are used, these solutions lead to the following:

Equation (17.74): $u = 1$,

$$E = \exp\{\tfrac{1}{2}(c-1)\,\mathfrak{z}\} = \exp(-ik\sigma C\mathfrak{b})\exp(-ikCz). \tag{17.99}$$

Equation (17.75): $u = (-\zeta)^{1-c}$,

$$E = \exp\{\tfrac{1}{2}(1-c)\,\mathfrak{z}\} = \exp(ik\sigma C\mathfrak{b})\exp(ikCz). \tag{17.100}$$

Hence (17.74) and (17.99) represent the upgoing wave or incident wave below the ionosphere, and (17.75) and (17.100) represent the downgoing or reflected wave.

Next, when z is large and positive, $e^{\mathfrak{z}}$ is very large compared to unity and the transformation (17.89) becomes

$$E = u \exp\{\tfrac{1}{2}(b-a)\,\mathfrak{z}\}. \tag{17.101}$$

In (17.76) and (17.77) the hypergeometric functions are unity when z is large and positive and when (17.86) is used they lead to:

Equation (17.76): $u = (-\zeta)^{-a}$,

$$E = \exp\{\tfrac{1}{2}(b-a)\,\mathfrak{z}\} = \exp(ikq_2\sigma\mathfrak{b})\exp(ikq_2 z). \tag{17.102}$$

Equation (17.77): $u = (-\zeta)^{-b}$,

$$E = \exp\{\tfrac{1}{2}(a-b)\,\mathfrak{z}\} = \exp(-ik\sigma q_2\mathfrak{b})\exp(-ikq_2 z). \tag{17.103}$$

Hence (17.77) and (17.103) represent the upgoing wave or transmitted wave above the ionosphere, and (17.76) and (17.102) represent a downgoing wave above the ionosphere, which cannot be present when the only source of energy is below the ionosphere.

To find the reflection and transmission coefficients we select a solution in which (17.76) or (17.102) is absent. The required solution is (17.77), which is a linear combination of the solutions (17.74) and (17.75) given by the circuit relation (17.85). Hence the solution below the ionosphere, from (17.85) and (17.100), is

$$E = \frac{(-c)!}{(-a)!(b-c)!} e^{-ik\sigma C\mathfrak{b}} e^{-ikCz} + \frac{(c-2)!}{(c-a-1)!(b-1)!} e^{ik\sigma C\mathfrak{b}} e^{ikCz}, \tag{17.104}$$

and above the ionosphere it is

$$E = \frac{1}{(b-a)!} e^{-ik\sigma q_2\mathfrak{b}} e^{-ikq_2 z}. \tag{17.105}$$

The reflection coefficient R is the ratio of the second to the first term in (17.104) and for some level $z = z_1$ it is

$$R = \frac{(c-2)!(-a)!(b-c)!}{(c-a-1)!(b-1)!(-c)!} e^{2ik\sigma C\mathfrak{b}} e^{2ikCz_1}. \tag{17.106}$$

The transmission coefficient T is the ratio of (17.105) to the first term of (17.104). If the transmitted wave is observed at $z = z_2$ and the incident wave at $z = z_1$, then

$$T = \frac{(-a)!(b-c)!}{(b-a)!(-c)!} e^{ik\sigma\mathfrak{b}(C-q_2)} e^{-ik(q_2 z_2 - Cz_1)}. \tag{17.107}$$

To find the reflection coefficient R' and transmission coefficient T' when the incident wave comes from above the ionosphere, we select a solution in which the upgoing wave (17.74) or (17.99) is absent. The required solution is (17.83) and

$$R' = \frac{(a-b-1)!(b-c)!(-a)!}{(b-a-1)!(a-c)!(-b)!} e^{-2ik\sigma q_2\mathfrak{b}} e^{-2ikq_2 z_2}, \tag{17.108}$$

$$T' = \frac{(b-c)!(-a)!}{(1-c)!(b-a-1)!} e^{ik\sigma\mathfrak{b}(C-q_2)} e^{ik(Cz_1 - q_2 z_2)}. \tag{17.109}$$

17.14 Epstein profiles

Equation (17.91) gives the most general profile that can be studied by Epstein's method. Now $\mathfrak{n}^2 = q^2 + S^2$ ($S = \sin\theta_I$, where θ_I is the angle of incidence) so that the refractive index $\mathfrak{n}$ is given by

$$\mathfrak{n}^2 = \epsilon_1 + S^2 + \frac{e^{\mathfrak{z}}}{(e^{\mathfrak{z}}+1)^2}\{(\epsilon_2 - \epsilon_1)(e^{\mathfrak{z}}+1) + \epsilon_3\}. \tag{17.110}$$

Suppose that $\mathfrak{b} = z_m/\sigma$ and σ is real and positive. Then $\mathfrak{z}$ varies linearly with the height z. Some typical curves of $\mathfrak{n}^2$ are shown in Fig. 17.7 for the case when ϵ_1, ϵ_2, ϵ_3 are real, which would occur if electron collisions were neglected. In general they may be complex, although in many ionosphere problems ϵ_1 is real and equal to C^2. Two important special cases are considered in later sections. The first is for $\epsilon_3 = 0$, when $\mathfrak{n}^2$ makes a continuous monotonic change from unity below the ionosphere to some other constant value at great heights (§ 17.15). The second case is when $\epsilon_1 = \epsilon_2 = C^2$ so that there is free space above and below the ionosphere. This is the 'sech²' layer, discussed in §§ 10.8 and 17.16.

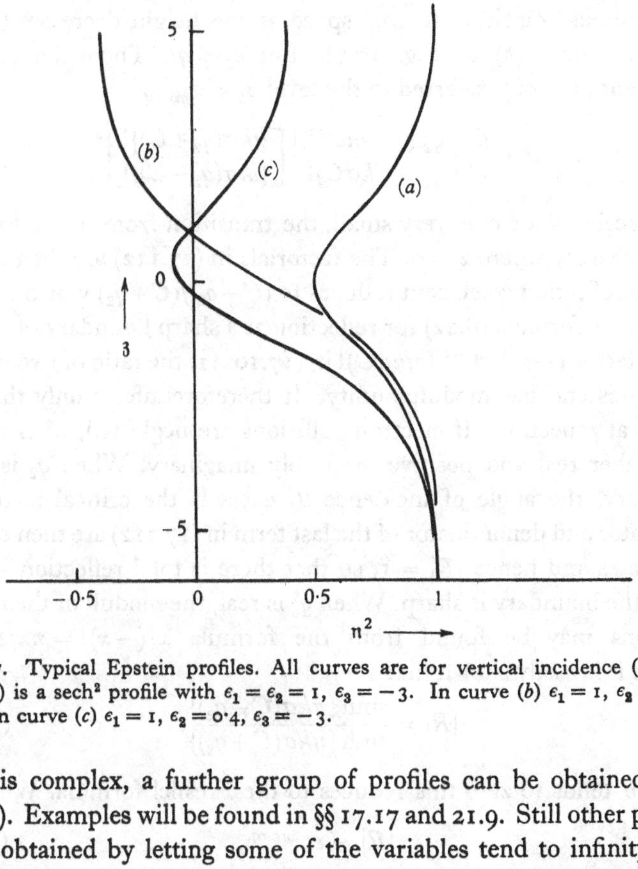

Fig. 17.7. Typical Epstein profiles. All curves are for vertical incidence ($S^2 = 0$). Curve (a) is a sech² profile with $\epsilon_1 = \epsilon_2 = 1$, $\epsilon_3 = -3$. In curve (b) $\epsilon_1 = 1$, $\epsilon_2 = -0{\cdot}4$, $\epsilon_3 = 0$. In curve (c) $\epsilon_1 = 1$, $\epsilon_2 = 0{\cdot}4$, $\epsilon_3 = -3$.

If $\mathfrak{b}$ is complex, a further group of profiles can be obtained from (17.110). Examples will be found in §§ 17.17 and 21.9. Still other profiles can be obtained by letting some of the variables tend to infinity. For example, let

$$\mathfrak{b} = i\pi - \log \epsilon_2 - \log(1 - iZ), \quad \epsilon_1 = C^2, \quad \epsilon_3 = 0, \quad \sigma = 1/\alpha$$

and let ϵ_2 tend to infinity. Then (17.110) becomes $\mathfrak{n}^2 = 1 - e^{\alpha z}/(1 - iZ)$ which is simply the exponential profile (17.2). The reader should verify that the reflection coefficient (17.106) is the same as (17.16) in this case.

17.15 Ionosphere with gradual boundary

If, in (17.110) we set $\mathfrak{b} = z_m/\sigma$, $\epsilon_1 = C^2$, $\epsilon_3 = 0$, it becomes

$$\mathfrak{n}^2 = S^2 + \tfrac{1}{2}(\epsilon_2 + C^2) + \tfrac{1}{2}(\epsilon_2 - C^2)\tanh\{\tfrac{1}{2}(z - z_m)/\sigma\}. \quad (17.111)$$

Hence below the ionosphere where $z - z_m$ is large and negative $\mathfrak{n}^2 = 1$. At great heights where $z - z_m$ is large and positive, $\mathfrak{n}^2 = \epsilon_2 + S^2 = \mathfrak{n}_2^2$ say, a constant, so that the medium is ultimately homogeneous, and there is a continuous transition to free space as the height decreases (see, for example, curve (*b*) of Fig. 17.7). Let $\epsilon_2 = q_2^2$. Then the reflection coefficient (17.106), referred to the level $z_1 = z_m$, is

$$R = \frac{C - q_2}{C + q_2} \frac{(-2ik\sigma C)!}{(2ik\sigma C)!} \left[\frac{\{ik\sigma(q_2 + C)\}!}{\{ik\sigma(q_2 - C)\}!}\right]^2. \quad (17.112)$$

If the scale factor σ is very small, the transition from $\mathfrak{n} = 1$ to $\mathfrak{n} = \mathfrak{n}_2$ occurs sharply where $z = 0$. The factorials in (17.112) are then all unity and the reflection coefficient reduces to $(C - q_2)/(C + q_2)$ which is simply the Fresnel formula (8.22) for reflection at a sharp boundary.

The factor $(-2ik\sigma C)!/(2ik\sigma C)!$ in (17.102) is the ratio of two complex conjugates and has modulus unity. It therefore affects only the phase change at reflection. If electron collisions are neglected, $\mathfrak{n}^2$ is real and q_2 is either real and positive or purely imaginary. When q_2 is purely imaginary, the angle of incidence θ_I exceeds the critical angle. The numerator and denominator of the last term in (17.112) are then complex conjugates and hence $|R| = 1$, so that there is total reflection, whether or not the boundary is sharp. When q^2 is real, the moduli of the factorial functions may be found from the formula $x!(-x)! = \pi x/\sin(\pi x)$, whence it is easily shown that

$$|R| = \frac{\sinh\{\pi k\sigma(C - q_2)\}}{\sinh\{\pi k\sigma(C + q_2)\}}. \quad (17.113)$$

When σ tends to zero this reduces to the Fresnel formula. When σ is large it gives

$$|R| \sim e^{-2\pi k\sigma q_2}, \quad (17.114)$$

which tends to zero as $\sigma \to \infty$.

17.16 The 'sech²' profile

If in (17.110) we set $\mathfrak{b} = z_m/\sigma$, $\epsilon_1 = \epsilon_2 = C^2$, it becomes

$$\mathfrak{n}^2 = 1 + \tfrac{1}{4}\epsilon_3 \operatorname{sech}^2\{\tfrac{1}{2}(z - z_m)/\sigma\}. \quad (17.115)$$

Hence there is free space, $\mathfrak{n}^2 = 1$, both below and above the ionosphere where $|z-z_m|$ is large. The ionosphere is a symmetrical layer with its centre at $z = z_m$, where $n^2 = 1+\frac{1}{4}\epsilon_3$. This profile has been used by Rawer (1939) to study the partial penetration and reflection from a layer with a maximum of electron density. It is in some respects better than the parabolic layer studied in §§ 17.4 to 17.6 since there is no discontinuity in the gradient of electron density at the top and bottom of the ionosphere, and hence the 'sech²' profile can be used to study very thin layers and is not subject to the limitation mentioned at the end of § 17.4.

Let

$$\epsilon_3 = -\frac{4X_m}{1-iZ}, \tag{17.116}$$

so that the collision-frequency is constant and the electron density is given by

$$X = X_m \operatorname{sech}^2\{\tfrac{1}{2}(z-z_m)/\sigma\}. \tag{17.117}$$

From (17.95) and (17.96) it follows that $a-b = c-1 = -2ik\sigma C$. Let $4k^2\sigma^2\epsilon_3+1 = 4\gamma^2$ so that $a+b-c = 2\gamma$ from (17.97). Then the reflection coefficient (17.106), referred to the level $z_1 = z_m$, is

$$R = -\frac{(-2ik\sigma C)!\,(2ik\sigma C-\gamma-\frac{1}{2})!\,(2ik\sigma C+\gamma-\frac{1}{2})!}{(2ik\sigma C)!\,(-\gamma-\frac{1}{2})!\,(\gamma-\frac{1}{2})!}, \tag{17.118}$$

and the transmission coefficient (17.107) when both incident and transmitted waves are referred to the same level ($z_1 = z_2$) is

$$T = 2ik\sigma C\frac{(2ik\sigma C-\gamma-\frac{1}{2})!\,(2ik\sigma C+\gamma-\frac{1}{2})!}{\{(2ik\sigma C)!\}^2}. \tag{17.119}$$

It can be verified that $R \to 0$ when $\sigma \to 0$, that is when the layer becomes indefinitely thin.

If electron collisions are neglected, ϵ_3 is equal to $-4k_p^2/k^2$ where $k_p c/2\pi$ is the penetration-frequency (c here is the velocity of light *in vacuo*), and

$$4\gamma^2 = 1-16\sigma^2 k_p^2. \tag{17.120}$$

Thus γ is independent of frequency, and is either real or purely imaginary. In both these cases the moduli of the factorial functions can be found from the formula $x!\,(-x)! = \pi x/\sin(\pi x)$ and (17.118) and (17.119) give

$$|R|^2 = \frac{\cos 2\pi\gamma+1}{\cos 2\pi\gamma+\cosh 4\pi k\sigma C}, \tag{17.121}$$

$$|T|^2 = \frac{\cosh 4\pi k\sigma C-1}{\cos 2\pi\gamma+\cosh 4\pi k\sigma C}. \tag{17.122}$$

These expressions can be used to study the partial penetration and reflection near the penetration-frequency of the layer.

It is interesting to compare the 'sech²' profile with a parabolic profile having the same penetration-frequency and the same curvature at its maximum. It can be shown that the parabola then has half-thickness $a = 2\sigma$. Now for frequencies above about 2 Mc/s, most ionospheric layers have $a/\lambda \geqslant 2$, so that in the cases of greatest interest $16\sigma^2 k_p^2 \gg 1$ and $i\gamma \approx 2\sigma k_p \gg 1$. Thus $\cos 2\pi\gamma \gg 1$, so that the 1 may be neglected in the numerator of (17.121), and we may take $\cos 2\pi\gamma \approx \frac{1}{2}\exp(4\pi\sigma k_p)$. It can then be shown that (17.121) reduces to the formula (17.46) for the parabolic profile.

When the effect of electron collisions is included the formulae (17.118) and (17.119) cannot be so easily simplified and it is necessary to know the values of the factorial functions with complex argument. Rawer (1939) has computed $|R|^2$ and $|T|^2$ for various values of Z and curves are given in his paper.

Equations (17.121) and (17.122) show that

$$|R|/|T| = \frac{\cosh 2\pi k\sigma C}{\cos 2\pi\gamma}.$$

At high enough frequencies $2\pi k\sigma C \gg 1$ so that

$$\cosh 2\pi k\sigma C \approx \tfrac{1}{2}e^{2\pi k\sigma C}$$

and
$$\log|R/T| \approx 2\pi k\sigma C + \text{constant}.$$

Thus if $\log|R/T|$ is plotted against frequency (proportional to k) a straight line is obtained whose slope gives σ which is a measure of the thickness of the ionospheric layer. This method has been applied to observations at vertical incidence by Briggs (1951 *a*, *b*). Even if electron collisions are included, it may be assumed that $Z \ll 1$ at high frequencies, so that γ is still nearly constant and the method can still be used.

The equivalent height of reflection for radio waves vertically incident on a 'sech²' layer can be found by the method used in § 17.6 for the parabolic layer. Since the centre of the 'sech²' layer is at a height z_m above the ground

$$h'(f) = z_m - \frac{1}{2}\frac{d}{dk}(\arg R), \tag{17.123}$$

where R is given by (17.118) with $C = 1$. Now in cases of practical interest σ is of order 1 km or more, and k is of order 20 km⁻¹ (at 1 Mc/s), so that the factorial functions all contain large numbers and the first term of Stirling's formula may be used for each. When electron collisions are neglected, γ is either real or purely imaginary, and this approximation leads to

$$h'(f) = z_m + \tfrac{1}{2}\sigma \frac{\tfrac{1}{4} - \gamma^2 + 4k^2\sigma^2}{\{4k^2\sigma^2 + (\tfrac{1}{2}+\gamma)^2\}\{4k^2\sigma^2 + (\tfrac{1}{2}-\gamma)^2\}} - \tfrac{1}{2}\sigma \log\left[\left\{1 + \frac{(\tfrac{1}{2}+\gamma)^2}{4k^2\sigma^2}\right\}\left\{1 + \frac{(\tfrac{1}{2}-\gamma)^2}{4k^2\sigma^2}\right\}\right]. \quad (17.124)$$

This is shown as a function of f/f_p in Fig. 17.8 for various values of σ, and the corresponding curve for the ray theory, given by (10.42), is also shown for comparison. These results should be compared with Fig. 17.4 which gives similar curves for a parabolic layer.

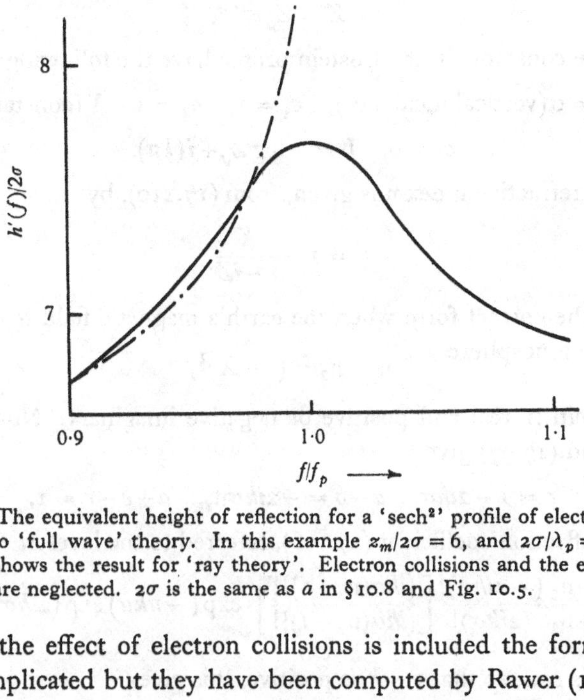

Fig. 17.8. The equivalent height of reflection for a 'sech²' profile of electron density according to 'full wave' theory. In this example $z_m/2\sigma = 6$, and $2\sigma/\lambda_p = 1{\cdot}99$. The chain line shows the result for 'ray theory'. Electron collisions and the earth's magnetic field are neglected. 2σ is the same as a in § 10.8 and Fig. 10.5.

When the effect of electron collisions is included the formulae are more complicated but they have been computed by Rawer (1939) who gives curves of $h'(f)$ versus f for various values of Z and σ.

17.17 Fixed electron density and varying collision-frequency

In the preceding sections dealing with various models of the ionosphere, the electron collision-frequency has been assumed to be constant, and the electron number density varied in such a way that it was zero below the ionosphere, thus giving a refractive index of unity. The electron density is never exactly zero since there are small numbers of charges in the air even near the ground. At a frequency of 16 kc/s an electron number density of only about 3 cm^{-3} would be enough to make $X = 1$ and if there were no collisions this would mean that one refractive index is zero. It is because the collision-frequency, and therefore Z, is large at this frequency that the refractive index is near to unity even when $X > 1$. Thus the increase of Z as the height decreases must play an important part in the behaviour of the refractive index at low levels. To study this effect it is interesting to assume that X is constant and to suppose that variation of the refractive index occurs only because the collision frequency varies. The theory is given here for vertical incidence. The extension to oblique incidence for horizontal polarisation is not difficult.

Let
$$Z = Z_g e^{-z/\sigma}, \tag{17.125}$$
and let the constants in the Epstein profile have the following values:
$$S = 0 \,(\text{vertical incidence}), \quad \epsilon_1 = 1, \quad \epsilon_2 = 1 - X\,(\text{constant}),$$
$$\epsilon_3 = 0, \quad \mathfrak{b} = -\log Z_g + i(\tfrac{1}{2}\pi). \tag{17.126}$$
Then the refractive index $\mathfrak{n}$ is given, from (17.110), by
$$\mathfrak{n}^2 = 1 - \frac{X}{1 - iZ}, \tag{17.127}$$
which is the correct form when the earth's magnetic field is neglected. Above the ionosphere
$$\mathfrak{n} = \mathfrak{n}_2 = (1 - X)^{\frac{1}{2}}, \tag{17.128}$$
which is either real and positive or negative imaginary. Now (17.95), (17.96) and (17.97) give
$$c = 1 - 2ik\sigma, \quad a - b = -2ik\sigma\mathfrak{n}_2, \quad a + b - c = 1, \tag{17.129}$$
and the reflection coefficient (17.106), referred to the level $z_1 = 0$, is
$$R = \frac{1 - \mathfrak{n}_2}{1 + \mathfrak{n}_2} \frac{(-2ik\sigma)!}{(2ik\sigma)!} \left[\frac{\{ik\sigma(\mathfrak{n}_2 + 1)\}!}{\{ik\sigma(\mathfrak{n}_2 - 1)\}!}\right]^2 \exp(-\pi k\sigma) \exp(2ik\sigma \log Z_g). \tag{17.130}$$
When $\mathfrak{n}_2$ is real, so that $1 - X$ is positive, this gives
$$|R| = \frac{\sinh\{\pi k\sigma\,|\mathfrak{n}_2 - 1|\}}{\sinh\{\pi k\sigma(\mathfrak{n}_2 + 1)\}} e^{-\pi k\sigma}, \tag{17.131}$$
and when $\mathfrak{n}_2$ is negative imaginary, so that X is negative, it gives
$$|R| = e^{-\pi k\sigma}. \tag{17.132}$$
This method can be applied to the case of vertical incidence when the earth's magnetic field is assumed to be vertical.

A further application of Epstein's results is given in § 21.9.

CHAPTER 18

ANISOTROPIC MEDIA. COUPLED WAVE-EQUATIONS AND W.K.B. SOLUTIONS

18.1 Introduction

Many of the problems of radio wave-propagation in the ionosphere can be solved by the methods of 'ray' theory. This is equivalent to using the W.K.B. solutions of the differential equations, and a formal justification was given in chs. 9, 15 and 16 for the case when the earth's magnetic field was neglected. In chs. 12 to 14 ray theory was applied also to problems in which the earth's magnetic field was included, and the validity of this must now be established.

In the present chapter, the differential equations are derived for an inhomogeneous, anisotropic ionosphere, and their W.K.B. solutions are then discussed. For this purpose the equations are expressed in coupled form. This chapter is therefore an extension of ch. 9 to anisotropic media. The W.K.B. solutions fail near levels of reflection or coupling.

18.2 The differential equations

Cartesian coordinates are used with the z-axis directed vertically upwards, so that the electron number density and collision-frequency are functions only of z. A plane wave is incident on the ionosphere from below with its normal in the x–z-plane at an angle θ_I to the vertical. Let $S = \sin\theta_I$, $C = \cos\theta_I$. As in §9.10 we may imagine the ionosphere to be replaced by a number of thin discrete strata in each of which the medium is homogeneous. Then for the waves in each stratum, (9.48) gives $\partial/\partial x \equiv -ikS$; $\partial/\partial y \equiv 0$, and these are true no matter how thin the strata. They may be expected to hold in the limit for the continuously varying ionosphere. They are therefore used in the Maxwell equations (2.31). They show that all field quantities contain a factor e^{-ikSx}. This is assumed to be omitted, in the same way as the time factor $e^{i\omega t}$ is omitted. Then the terms which remain are functions of z only and the partial differentiation sign $\partial/\partial z$ may be replaced by the total differentiation sign d/dz. Then (2.31) become:

$$-\frac{dE_y}{dz} = -ik\mathscr{H}_x, \tag{18.1}$$

$$\frac{dE_x}{dz} + ikSE_z = -ik\mathscr{H}_y, \tag{18.2}$$

$$-ikSE_y = -ik\mathscr{H}_z, \tag{18.3}$$

$$-\frac{d\mathscr{H}_y}{dz} = \frac{ik}{\epsilon_0}D_x, \tag{18.4}$$

$$\frac{d\mathscr{H}_x}{dz} + ikS\mathscr{H}_z = \frac{ik}{\epsilon_0}D_y, \tag{18.5}$$

$$-ikS\mathscr{H}_y = \frac{ik}{\epsilon_0}D_z. \tag{18.6}$$

When the earth's magnetic field was neglected, the electric displacement **D** was given by the simple relation $\mathbf{D} = \epsilon_0 n^2\mathbf{E}$ where n^2 has the value (4.8). With the earth's field included, however, the expression for **D** is more complicated. Equation (2.6) shows that $\mathbf{D} = \epsilon_0\mathbf{E} + \mathbf{P}$ and the electric polarisation **P** is given by (3.24). If these are substituted in (18.4), (18.5) and (18.6), and if (18.3) is used to eliminate $\mathscr{H}_z$, then:

$$\frac{d\mathscr{H}_y}{dz} = -ik[(1+M_{xx})E_x + M_{xy}E_y + M_{xz}E_z], \tag{18.7}$$

$$\frac{d\mathscr{H}_x}{dz} = ik[M_{yx}E_x + (1 - S^2 + M_{yy})E_y + M_{yz}E_z], \tag{18.8}$$

$$-S\mathscr{H}_y = M_{zx}E_x + M_{zy}E_y + (1 + M_{zz})E_z, \tag{18.9}$$

where the M_{ij} are the elements of the susceptibility matrix **M** and are functions of z. From these equations and (18.1) and (18.2) we can eliminate E_z, which gives:

$$-\frac{1}{ik}\frac{dE_x}{dz} = -\frac{SM_{zx}}{1+M_{zz}}E_x - \frac{SM_{zy}}{1+M_{zz}}E_y + \frac{C^2+M_{zz}}{1+M_{zz}}\mathscr{H}_y, \tag{18.10}$$

$$-\frac{1}{ik}\frac{dE_y}{dz} = -\mathscr{H}_x, \tag{18.11}$$

$$\begin{aligned}-\frac{1}{ik}\frac{d\mathscr{H}_x}{dz} &= \left(-M_{yx} + \frac{M_{yz}M_{zx}}{1+M_{zz}}\right)E_x \\ &\quad - \left(C^2 + M_{yy} - \frac{M_{yz}M_{zy}}{1+M_{zz}}\right)E_y + \frac{SM_{yz}}{1+M_{zz}}\mathscr{H}_y,\end{aligned} \tag{18.12}$$

$$\begin{aligned}-\frac{1}{ik}\frac{d\mathscr{H}_y}{dz} &= \left(1 + M_{xx} - \frac{M_{xz}M_{zx}}{1+M_{zz}}\right)E_x \\ &\quad + \left(M_{xy} - \frac{M_{xz}M_{zy}}{1+M_{zz}}\right)E_y - \frac{SM_{xz}}{1+M_{zz}}\mathscr{H}_y.\end{aligned} \tag{18.13}$$

The equations in this form are the starting-point for much of the later work in this book.

18.3 The four characteristic waves

The equations (18.10) to (18.13) are four simultaneous linear differential equations of the first order, with the four dependent variables E_x, E_y, $\mathscr{H}_x$ and $\mathscr{H}_y$. Three of these could be eliminated to give a single linear differential equation of the fourth order, which in general has four independent solutions. It is shown later that in many cases of interest there are parts of the ionosphere where these four solutions may be regarded as the upgoing ordinary, upgoing extraordinary, downgoing ordinary and downgoing extraordinary waves, propagated independently of each other. At some levels the independence breaks down, and these are levels of reflection or coupling.

In the special case when the ionosphere is homogeneous, the coefficients in (18.10) to (18.13) are constants independent of z, and the fourth-order equation is then a linear equation with constant coefficients. A standard method of solving such an equation is to set the variable equal to $e^{\lambda z}$ (where λ is a constant to be found), which gives a quartic equation in λ, called the 'characteristic equation', whose four roots give four solutions. Clearly each of the variables E_x, E_y, $\mathscr{H}_x$, $\mathscr{H}_y$ is then a multiple of $e^{\lambda z}$ for each solution, and $d/dz \equiv \lambda$. There is therefore no need to form the fourth-order equation. It is here convenient to take $\lambda = -ikq$, so that $-(1/ik)(d/dz) \equiv q$. If this is substituted in (18.10) to (18.13) and the variables E_x, E_y, $\mathscr{H}_x$, $\mathscr{H}_y$ are eliminated, we obtain the determinantal equation for q:

$$\begin{vmatrix} -\dfrac{SM_{zx}}{1+M_{zz}}-q & -\dfrac{SM_{zy}}{1+M_{zz}} & 0 & \dfrac{C^2+M_{zz}}{1+M_{zz}} \\ 0 & -q & -1 & 0 \\ -M_{yx}+\dfrac{M_{yz}M_{zx}}{1+M_{zz}} & -C^2-M_{yy}+\dfrac{M_{zy}M_{yz}}{1+M_{zz}} & -q & \dfrac{SM_{yz}}{1+M_{zz}} \\ 1+M_{xx}-\dfrac{M_{xz}M_{zx}}{1+M_{zz}} & M_{xy}-\dfrac{M_{xz}M_{zy}}{1+M_{zz}} & 0 & -\dfrac{SM_{xz}}{1+M_{zz}}-q \end{vmatrix} = 0. \tag{18.14}$$

This is the Booker quartic, which may therefore be regarded as the characteristic equation of the differential equations when the ionosphere is homogeneous. It can be shown that (18.14) reduces to the form (13.12) with $S_2 = 0$, $S_1 = S$, and thence to (13.13) and (13.14).

For a homogeneous ionosphere, therefore, each solution of the quartic

(18.14) gives a wave in which all the field variables contain a factor e^{-ikqz}. Thus a typical solution is

$$E_x = E_x^{(1)} e^{-ikq_1 z}, \quad E_y = E_y^{(1)} e^{-ikq_1 z}, \quad \mathscr{H}_x = \mathscr{H}_x^{(1)} e^{-ikq_1 z}, \quad \mathscr{H}_y = \mathscr{H}_y^{(1)} e^{-ikq_1 z}, \tag{18.15}$$

where q_1 is one root of the quartic. This will be called a 'characteristic wave'. The four quantities $E_x^{(1)}$, $E_y^{(1)}$, $\mathscr{H}_x^{(1)}$, $\mathscr{H}_y^{(1)}$, are constants whose ratios could be found by substituting $d/dz = -ikq_1$ in (18.10) to (18.13) and then solving them. These ratios describe the wave-polarisation and wave-impedance, and are given later for some special cases (§§ 18.6, 18.12, 18.14 and 18.15). Expressions for the ratios $E_x:E_y:E_z$ have been given by Booker (1936) for the general case, but they are rather complicated and are not needed here.

In ch. 13 the Booker quartic was used to find the path of a wave-packet in the ionosphere, when electron collisions are neglected. Then the coefficients in the quartic are all real and the roots are either (*a*) all real, or (*b*) two are real and two form a conjugate complex pair, or (*c*) the four roots are in two conjugate complex pairs. By considering the motion of wave-packets it was shown in § 13.7 that in case (*a*) the four solutions give two upgoing and two downgoing rays, and in case (*b*) one of the two real roots gives an upgoing ray and the other a downgoing ray. It was further shown in § 13.13 that a wave-packet moves in the same direction as the energy flow as given by the complex Poynting vector.

When the effect of electron collisions is included, it can similarly be shown that two roots of the quartic correspond to upgoing waves, and two to downgoing waves. This is best done by considering the complex Poynting vector $\overline{\Pi}$ (2.40). It is easily shown that for any solution of (18.10) to (18.13) corresponding to one value of q, that is for a single characteristic wave, $\overline{\Pi}$ is independent of x and y, and depends on z only through a factor $\exp\{2kz\mathscr{I}(q)\}$. Now electromagnetic energy is converted into heat in the medium. Hence if energy is travelling upwards, $\overline{\Pi}_z$ is positive, and must decrease as z increases, so that $\mathscr{I}(q)$ is negative. Similarly, for a downgoing wave $\overline{\Pi}_z$ is negative and $\mathscr{I}(q)$ must be positive. Now it was proved in § 8.18 that two roots of the quartic have positive $\mathscr{I}(q)$ and the other two have negative $\mathscr{I}(q)$. Hence two solutions correspond to upgoing waves and two to downgoing waves.

For any level in the ionosphere the four characteristic waves are solutions of (18.10) to (18.13) for a fictitious ionosphere which is homogeneous, with the same values of X and Z as the actual ionosphere at that level. The characteristic waves are not, therefore, solutions for the actual ionosphere, but they play an important part in deriving valid solutions, and their properties must be well understood. This is particularly true for a 'slowly varying' ionosphere, for which it can be shown that there are approximate W.K.B. solutions resembling the characteristic waves in many respects. For ionospheric profiles which are not slowly varying, and for very low frequencies, the characteristic waves are less important and a more detailed study of the differential equations is necessary.

18.4 Matrix form of the equations

The equations (18.10) to (18.13) may be written concisely in matrix notation. Let $\mathbf{e}$ denote the column matrix

$$\mathbf{e} = \begin{pmatrix} E_x \\ -E_y \\ \mathscr{H}_x \\ \mathscr{H}_y \end{pmatrix}, \tag{18.16}$$

and let $\mathbf{T}$ denote the 4×4 matrix:

$$\mathbf{T} = \begin{pmatrix} -\dfrac{SM_{zx}}{1+M_{zz}} & \dfrac{SM_{zy}}{1+M_{zz}} & 0 & \dfrac{C^2+M_{zz}}{1+M_{zz}} \\ 0 & 0 & 1 & 0 \\ \dfrac{M_{yz}M_{zx}}{1+M_{zz}} - M_{yx} & C^2+M_{yy}-\dfrac{M_{yz}M_{zy}}{1+M_{zz}} & 0 & \dfrac{SM_{yz}}{1+M_{zz}} \\ 1+M_{xx}-\dfrac{M_{xz}M_{zx}}{1+M_{zz}} & \dfrac{M_{xz}M_{zy}}{1+M_{zz}}-M_{xy} & 0 & \dfrac{-SM_{xz}}{1+M_{zz}} \end{pmatrix}, \tag{18.17}$$

where the order and signs of the elements of $\mathbf{e}$ have been chosen to give some symmetry in $\mathbf{T}$. Then (18.10) to (18.13) become

$$\frac{d\mathbf{e}}{dz} = -ik\mathbf{T}\mathbf{e}. \tag{18.18}$$

The matrix $\mathbf{T}$ has four characteristic roots or eigenvalues q_i ($i = 1, 2, 3, 4$) which satisfy the characteristic equation

$$\det(\mathbf{T}-q\mathbf{I}) = 0, \tag{18.19}$$

where $\mathbf{I}$ is the unit 4×4 matrix. From the preceding section it is clear that (18.19) is the Booker quartic (18.14). Thus at any level in the ionosphere the four values of q are the four characteristic roots of the matrix $\mathbf{T}$.

Corresponding to any root q_i there is a column matrix $\mathbf{e} = \mathbf{s}^{(i)}$ which satisfies

$$\mathbf{T}\mathbf{s}^{(i)} = q_i\mathbf{s}^{(i)} \quad \text{(not summed)}, \tag{18.20}$$

and $\mathbf{s}^{(i)}$ is called a characteristic column matrix or eigenvector of the matrix $\mathbf{T}$. Only the ratios of the elements of $\mathbf{s}^{(i)}$ are determined by (18.20). It can easily be shown that $\mathbf{s}^{(i)}$ is proportional to any column of the adjoint† of the matrix $\mathbf{T}-q_j\mathbf{I}$.

In a loss-free medium the susceptibility matrix $\mathbf{M}$ is Hermitian, that is $M_{ij}=M_{ji}^*$ (see § 3.6), and (18.17) shows that then

$$T_{5-j,5-i}=T_{ij}^*, \tag{18.21}$$

so that $\mathbf{T}$ may be described as 'Hermitian with respect to the trailing diagonal'. In a lossy medium this is no longer true, but still $T_{5-j,5-i}$ is the same as T_{ij} except for a change of sign of i where it appears explicitly, that is not in $U=1-iZ$, but everywhere else.

To understand the physical significance of the $\mathbf{s}^{(i)}$, consider a fictitious homogeneous ionosphere having the properties of the actual ionosphere at the height considered. In such a medium $\mathbf{T}$ would be independent of the height z, and the $\mathbf{s}^{(i)}$ could also be chosen to be independent of z. Then one solution of (18.18) is $\mathbf{e}=\mathbf{s}^{(i)}\exp(-ikq_i z)$. Hence the elements of $\mathbf{s}^{(i)}$ are proportional to the field quantities (18.16) for the ith characteristic wave; let a superscript (i) be used to distinguish field quantities in this wave. Then, for example, $-s_2^{(i)}/s_1^{(i)}=E_y^{(i)}/E_x^{(i)}$. For a vertically incident wave this ratio is the wave-polarisation. Similarly,

$$\frac{s_1^{(i)}}{s_4^{(i)}}=\frac{E_x^{(i)}}{\mathscr{H}_y^{(i)}}=\frac{1}{Z_0}\frac{E_x^{(i)}}{H_y^{(i)}}.$$

For a vertically incident wave

$$\frac{E_x^{(i)}}{H_y^{(i)}}=-\frac{E_y^{(i)}}{H_x^{(i)}}$$

and this ratio is sometimes called the 'intrinsic impedance' of the medium. Hence for vertical incidence the ratios of the elements of $\mathbf{s}^{(i)}$ give the wave-polarisation and intrinsic impedance for the ith characteristic wave.

For oblique incidence the wave-polarisation and intrinsic impedance may still be expressed in terms of the horizontal components of the fields. Thus each eigenvector $\mathbf{s}^{(i)}$ may still conveniently be thought of as giving these for the ith characteristic wave.

† Let $\mathbf{a}$ be a square matrix with elements a_{ij} and let A_{ij} be the cofactor of a_{ij} in det $\mathbf{a}$. Then the adjoint of $\mathbf{a}$ is denoted by adj $\mathbf{a}$ and its elements are given by $(\text{adj }\mathbf{a})_{ij}=A_{ji}$.

18.5 The differential equations for vertical incidence

For vertical incidence $S = 0$, $C = 1$ and the equations (18.1) to (18.6) take the simpler form

$$\frac{dE_y}{dz} = ik\mathscr{H}_x, \quad \frac{dE_x}{dz} = -ik\mathscr{H}_y, \quad \mathscr{H}_z = 0; \tag{18.22}$$

$$\frac{d\mathscr{H}_y}{dz} = -\frac{ik}{\epsilon_0} D_x, \quad \frac{d\mathscr{H}_x}{dz} = \frac{ik}{\epsilon_0} D_y, \quad D_z = 0. \tag{18.23}$$

If $\mathscr{H}_x$ and $\mathscr{H}_y$ are eliminated

$$\frac{d^2E_x}{dz^2} + \frac{k^2}{\epsilon_0} D_x = 0, \tag{18.24}$$

$$\frac{d^2E_y}{dz^2} + \frac{k^2}{\epsilon_0} D_y = 0. \tag{18.25}$$

Now D_x, D_y, D_z can be expressed in terms of E_x, E_y, E_z from (2.6) and (3.24). The last of equations (18.23) gives

$$M_{zx}E_x + M_{zy}E_y + (1 + M_{zz})E_z = 0, \tag{18.26}$$

whence E_z may be eliminated from the expressions for D_x, D_y. Hence (18.24) and (18.25) become

$$\frac{1}{k^2}\frac{d^2E_x}{dz^2} + \left(1 + M_{xx} - \frac{M_{xz}M_{zx}}{1 + M_{zz}}\right) E_x + \left(M_{xy} - \frac{M_{xz}M_{zy}}{1 + M_{zz}}\right) E_y = 0, \tag{18.27}$$

$$\frac{1}{k^2}\frac{d^2E_y}{dz^2} + \left(1 + M_{yy} - \frac{M_{yz}M_{zy}}{1 + M_{zz}}\right) E_y + \left(M_{yx} - \frac{M_{yz}M_{zx}}{1 + M_{zz}}\right) E_x = 0. \tag{18.28}$$

Let $\mathfrak{n}_o$, $\mathfrak{n}_x$ be the two refractive indices for waves travelling upwards in a fictitious homogeneous medium with the properties of the actual ionosphere, at each level, and let ρ_o, ρ_x be the corresponding values of the wave-polarisation E_y/E_x. Then the four quantities $\mathfrak{n}_o$, $\mathfrak{n}_x$, ρ_o, ρ_x are functions of z and are expressible in terms of the M_{ij}. They provide an alternative way of specifying the properties of the medium at each level, and the coefficients in (18.27) and (18.28) can be expressed in terms of them. It can be shown that the equations then become

$$\left.\begin{aligned} &\frac{1}{k^2}\frac{d^2E_x}{dz^2} + \frac{\rho_o\mathfrak{n}_x^2 - \rho_x\mathfrak{n}_o^2}{\rho_o - \rho_x} E_x + \frac{\mathfrak{n}_o^2 - \mathfrak{n}_x^2}{\rho_o - \rho_x} E_y = 0, \\ &\frac{1}{k^2}\frac{d^2E_y}{dz^2} + \frac{\rho_o\mathfrak{n}_o^2 - \rho_x\mathfrak{n}_x^2}{\rho_o - \rho_x} E_y - \frac{\mathfrak{n}_o^2 - \mathfrak{n}_x^2}{\rho_o - \rho_x} E_x = 0. \end{aligned}\right\} \tag{18.29}$$

The proof is left as an exercise for the reader, who should also verify that the equations reduce to the correct form in special cases, e.g. at the magnetic equator and at the magnetic poles.

18.6 The W.K.B. solution for vertical incidence on a loss-free medium

Suppose that the electron density varies so slowly with height z that a range of z can be found in which the refractive indices $\mathfrak{n}_o$, $\mathfrak{n}_x$ are nearly constant. Consider an upgoing progressive wave, say an 'ordinary' wave, in this range with its wave-normal vertically upwards. Then its field components would be given by

$$\left.\begin{aligned} E_x &= A\exp(-ik\mathfrak{n}_o z), & E_y &= \rho_0 A\exp(-ik\mathfrak{n}_o z);\\ \mathscr{H}_x &= -\mathfrak{n}_o\rho_o A\exp(-ik\mathfrak{n}_o z), & \mathscr{H}_y &= \mathfrak{n}_o A\exp(-ik\mathfrak{n}_o z);\end{aligned}\right\} \quad (18.30)$$

where A is a constant to a first approximation, and ρ_o is the polarisation for the ordinary wave, given by the magnetoionic theory (§ 5.3). These ought to satisfy the differential equations (18.22) and (18.23) but the fit is not very good unless $d\mathfrak{n}_o/dz$, $d^2\mathfrak{n}_o/dz^2$, $d\rho_o/dz$ are small.

When the wave (18.30) passes through a small thickness δz of the ionosphere, the change of phase is $k\mathfrak{n}_o\delta z$. When it passes through a finite thickness z, the change of phase would be

$$k\int_0^z \mathfrak{n}_o\,dz.$$

This suggests that a better solution than (18.30) might be

$$\left.\begin{aligned} E_x &= A\exp\left(-ik\int_0^z \mathfrak{n}_o\,dz\right), & E_y &= \rho_o A\exp\left(-ik\int_0^z \mathfrak{n}_o\,dz\right);\\ \mathscr{H}_x &= -\mathfrak{n}_o\rho_o A\exp\left(-ik\int_0^z \mathfrak{n}_o\,dz\right), & \mathscr{H}_y &= \mathfrak{n}_o A\exp\left(-ik\int_0^z \mathfrak{n}_o\,dz\right).\end{aligned}\right\} \quad (18.31)$$

This satisfies the second equation (18.22) exactly, but the other equations in (18.22) and (18.23) are approximately satisfied only when $d\mathfrak{n}_0/dz$, $d\rho_0/dz$ are small. However, the agreement is better than for (18.30). The integral in (18.31) constitutes the 'phase memory', and (18.31) is an extension of § 9.3 to the anisotropic ionosphere. A similar (approximate) solution could be found for the upgoing extraordinary wave, with $\mathfrak{n}_x$, ρ_x, instead of $\mathfrak{n}_o$, ρ_o.

To obtain a solution which is more accurate than (18.31), A may be treated as a function of z, and if the medium is loss-free the method of §9.4 may be used to find A. In a progressive wave the upward flow of energy must be the same for all z, and hence the z-component of the complex Poynting vector (§2.13) must be constant. Thus

$$\mathscr{R}(E_x \mathscr{H}_y^* - E_y \mathscr{H}_x^*)$$

is constant, and since $\mathfrak{n}_o$ is real for a progressive wave, (18.31) gives

$$\mathfrak{n}_o AA^*(1 + \rho_o \rho_o^*) = \text{constant}.$$

Now ρ_o is purely imaginary in a loss-free medium (see §5.4), so that $\rho_o \rho_o^* = -\rho_o^2$. Then (18.31) suggests that

$$A = A_0 \mathfrak{n}_o^{-\frac{1}{2}}(1 - \rho_o^2)^{-\frac{1}{2}}, \tag{18.32}$$

where A_0 is a constant, and (18.31) becomes

$$\left.\begin{aligned} E_x &= \mathfrak{n}_o^{-\frac{1}{2}}(1-\rho_o^2)^{-\frac{1}{2}} A_0 \exp\left(-ik\int_0^z \mathfrak{n}_o\,dz\right), \\ E_y &= \rho_o \mathfrak{n}_o^{-\frac{1}{2}}(1-\rho_o^2)^{-\frac{1}{2}} A_0 \exp\left(-ik\int_0^z \mathfrak{n}_o\,dz\right), \\ \mathscr{H}_x &= -\mathfrak{n}_o^{\frac{1}{2}} \rho_o (1-\rho_o^2)^{-\frac{1}{2}} A_0 \exp\left(-ik\int_0^z \mathfrak{n}_o\,dz\right), \\ \mathscr{H}_y &= \mathfrak{n}_o^{\frac{1}{2}}(1-\rho_o^2)^{-\frac{1}{2}} A_0 \exp\left(-ik\int_0^z \mathfrak{n}_o\,dz\right). \end{aligned}\right\} \tag{18.33}$$

This is the W.K.B. solution for the upgoing ordinary wave. Although it is deduced here only for a region where $\mathfrak{n}_o$ is real, the more rigorous derivations given later (§18.13) show that it is the same when $\mathfrak{n}_o$ is complex. For the downgoing ordinary wave the corresponding solution is

$$\left.\begin{aligned} E_x &= \mathfrak{n}_o^{-\frac{1}{2}}(1-\rho_o^2)^{-\frac{1}{2}} A_0 \exp\left(ik\int_0^z \mathfrak{n}_o\,dz\right), \\ E_y &= \rho_o \mathfrak{n}_o^{-\frac{1}{2}}(1-\rho_o^2)^{-\frac{1}{2}} A_0 \exp\left(ik\int_0^z \mathfrak{n}_o\,dz\right), \\ \mathscr{H}_x &= \mathfrak{n}_o^{\frac{1}{2}} \rho_o (1-\rho_o^2)^{-\frac{1}{2}} A_0 \exp\left(ik\int_0^z \mathfrak{n}_o\,dz\right), \\ \mathscr{H}_y &= -\mathfrak{n}_o^{\frac{1}{2}}(1-\rho_o^2)^{-\frac{1}{2}} A_0 \exp\left(ik\int_0^z \mathfrak{n}_o\,dz\right), \end{aligned}\right\} \tag{18.34}$$

where ρ_o is the *same* as for the upgoing ordinary wave (see end of §5.4). There are two similar W.K.B. solutions for the upgoing and downgoing extraordinary waves, with $\mathfrak{n}_x$ for $\mathfrak{n}_o$ and ρ_x for ρ_o.

18.7 Introduction to W.K.B. solutions in the general case

In ch. 9 two different methods were given for deriving the W.K.B. solutions when the earth's magnetic field is neglected. In the first, § 9.5, the differential equation for E_y was formulated, (9.8), and the change of variable $E_y = A \exp\{i\phi(z)\}$, (9.17), was then made. This led to a Ricatti-type equation, (9.22), for ϕ. To use the same method in the general case it would be necessary to formulate the differential equation for one field variable. We have already seen that this is of the fourth order and it is too complicated to be worth deriving specifically, although in principle the method could be used.

It is better, however, to extend the second method, § 9.9, in which the fields were expressed as the sum of partial fields for the upgoing and downgoing waves. In this way the differential equations were expressed in the coupled form (9.42). Neglecting the coupling terms on the right-hand side gave two separate differential equations, whose solutions were the W.K.B. solutions for the upgoing and downgoing waves. The extension of this method to the general case is given in §§ 18.10 and 18.11 where the fields are expressed as the sums of the partial fields for the four characteristic waves, and the differential equations are expressed in the general coupled form. Neglect of the coupling terms gives four separate equations whose solutions are the four W.K.B. solutions.

Coupled forms of the differential equations have been used to solve many problems in radio-wave propagation, and the subject is therefore most important. A review of it is given in § 18.8 and an account of Försterling's form of coupled equations in § 18.9. The rest of this chapter is concerned with other forms of coupled equations and their relation to the W.K.B. solutions. Ch. 19 is devoted to various applications of coupled equations.

18.8 Introduction to coupled wave-equations

The term 'coupled equations' is usually given to a set of simultaneous ordinary differential equations with the following properties:

(1) There is one independent variable which in this book is the height z or a linear function of it.

(2) The number of equations is the same as the number of dependent variables.

(3) In each equation one dependent variable appears in derivatives up to a higher order than any other. The terms in this variable are called 'principal' terms and the remaining terms are called 'coupling' terms.

(4) The principal terms contain a different dependent variable in each equation, so that each dependent variable appears in the principal terms of one and only one equation.

It is often possible to choose the dependent variables so that the coupling terms are small over some range of z. Then the equations may be solved by successive approximations. As a first approximation the coupling terms are neglected, and the resulting equations can then be solved. The values thus obtained for the dependent variables are substituted in the coupling terms and the resulting equations are solved to give a better approximation. Some examples of this process are given in §§ 19.5 and 19.8. Coupled equations are extremely important, mainly because they enable solutions to be found by successive approximation.

As an example of coupled equations consider (18.27) and (18.28). In (18.27) the terms in E_x are principal terms and the last term (in E_y) is the coupling term; similarly, in (18.28) the principal terms contain E_y and the last term (in E_x) is the coupling term. In these equations, however, the coupling terms are in general of comparable magnitude to the principal terms, even in a homogeneous medium, so that we could not use a method of successive approximation in which the coupling terms are at first neglected. An exception occurs when the earth's magnetic field is in the x–z-plane and nearly horizontal, for then it can be shown that the coefficients of E_y in (18.27) and of E_x in (18.28) are small. If the earth's field is exactly horizontal (parallel to the x-axis) these coefficients are zero so that the coupling terms vanish and the equations separate into two independent second-order equations, one for E_x and the other for E_y. This is a special case of equations (18.50) and (18.51).

It has been assumed earlier, especially in chs. 12 to 14, that the ordinary and extraordinary waves can be considered to be propagated independently. This suggests that the dependent variables should be chosen so that one refers to the ordinary wave only, and the other to the extraordinary wave. The coupling terms should then be very small if the assumption is justified. Equations of this kind were given first by Försterling (1942) for the case of vertical incidence (see § 18.9). One of Försterling's variables, $\mathfrak{F}_o$, gave the field of the total ordinary† wave, including both upgoing and downgoing ordinary waves. The other, $\mathfrak{F}_x$, similarly gave the total extraordinary† wave. Thus Försterling did not separate the four characteristic waves completely. If the coupling terms in his equations are neglected, the resulting equations are two second-

† The terms 'ordinary' and 'extraordinary' are used here only for convenience of description. They may be ambiguous as explained in § 6.15.

order equations governing the independent propagation of the ordinary and extraordinary waves.

The theory was extended to the general case by Clemmow and Heading (1954), who used four independent variables for the four characteristic waves (§ 18.10). If the coupling terms are neglected in their equations, the resulting four equations lead to the four W.K.B. solutions. Where the coupling terms are not small there is interaction between the characteristic waves, which may constitute reflection, or may give strong coupling between characteristic waves travelling in the same direction.

Where a coupling term is large, the method of successive approximations cannot be used, and other methods of solving the equations must be found.

18.9 Försterling's coupled equations for vertical incidence

In the differential equations (18.24) and (18.25) the two variables E_x, E_y refer to the total fields. Each of these is now to be expressed as the sum of the fields for the ordinary and extraordinary waves, and hence

$$E_x = E_x^{(o)} + E_x^{(x)}, \tag{18.35}$$

$$E_y = E_y^{(o)} + E_y^{(x)}. \tag{18.36}$$

The four variables $E_x^{(o)}$, $E_x^{(x)}$, $E_y^{(o)}$ and $E_y^{(x)}$ may be made to satisfy two further relations. Now for waves in a homogeneous medium they satisfy

$$\frac{E_y^{(o)}}{E_x^{(o)}} = \rho_o, \quad \frac{E_y^{(x)}}{E_x^{(x)}} = \rho_x, \tag{18.37}$$

where ρ_o, ρ_x are the two wave-polarisations as given by the magnetoionic theory (§§ 5.3, 5.4). These equations apply for both upgoing and downgoing waves, since the same system of axes is used for both (see end of § 5.4). Equations (18.37) are the two further relations and are used to *define* the ordinary and extraordinary waves in the variable medium. Now (5.19) gives $\rho_o \rho_x = 1$, so that $E_x^{(x)} = \rho_o E_y^{(x)}$, and hence (18.35) and (18.36) become

$$E_x = E_x^{(o)} + \rho_o E_y^{(x)}, \tag{18.38}$$

$$E_y = \rho_o E_x^{(o)} + E_y^{(x)}. \tag{18.39}$$

The electric displacement **D** is derived from the electric field by the constitutive relations, which are the same for a homogeneous and a variable medium (see § 5.6). For the ordinary wave, (5.31) and (5.32) show that

$$D_x^{(o)} = \epsilon_0 \mathfrak{n}_o^2 E_x^{(o)}, \quad D_y^{(o)} = \epsilon_0 \mathfrak{n}_o^2 E_y^{(o)}, \tag{18.40}$$

and similarly for the extraordinary wave

$$D_x^{(x)} = \epsilon_0 \mathfrak{n}_x^2 E_x^{(x)}, \quad D_y^{(x)} = \epsilon_0 \mathfrak{n}_x^2 E_y^{(x)}, \tag{18.41}$$

where $\mathfrak{n}_o$, $\mathfrak{n}_x$ are the two refractive indices as given by the Appleton–Hartree formula. The total electric displacement $\mathbf{D}$ must be the sum of these components and hence

$$\frac{1}{\epsilon_0} D_x = \mathfrak{n}_o^2 E_x^{(o)} + \mathfrak{n}_x^2 \rho_o E_y^{(x)}, \tag{18.42}$$

$$\frac{1}{\epsilon_0} D_y = \mathfrak{n}_o^2 \rho_o E_x^{(o)} + \mathfrak{n}_x^2 E_y^{(x)}. \tag{18.43}$$

Equations (18.38), (18.39), (18.42) and (18.43) are now substituted in (18.24) and (18.25). A dash ′ is used to denote $(1/k)(d/dz)$. Thus

$$E_x^{(o)''} + \rho_o E_y^{(x)''} + 2\rho_o' E_y^{(x)'} + \rho_o'' E_y^{(x)} + \mathfrak{n}_o^2 E_x^{(o)} + \mathfrak{n}_x^2 \rho_o E_y^{(x)} = 0, \tag{18.44}$$

$$\rho_o E_x^{(o)''} + E_y^{(x)''} + 2\rho_o' E_x^{(o)'} + \rho_o'' E_x^{(o)} + \mathfrak{n}_o^2 \rho_o E_x^{(o)} + \mathfrak{n}_x^2 E_y^{(x)} = 0. \tag{18.45}$$

Now (18.45) is multiplied by ρ_o and (18.44) is subtracted; similarly (18.44) is multiplied by ρ_o and (18.45) is subtracted. This gives

$$(\rho_o^2 - 1)(E_x^{(o)''} + \mathfrak{n}_o^2 E_x^{(o)}) + \rho_o \rho_o'' E_x^{(o)} + 2\rho_o \rho_o' E_x^{(o)'} = 2\rho_o' E_y^{(x)'} + \rho_x'' E_y^{(x)}, \tag{18.46}$$

$$(\rho_o^2 - 1)(E_y^{(x)''} + \mathfrak{n}_x^2 E_y^{(x)}) + \rho_o \rho_o'' E_y^{(x)} + 2\rho_o \rho_o' E_y^{(x)'} = 2\rho_o' E_x^{(o)'} + \rho_o'' E_x^{(o)}. \tag{18.47}$$

The equations are now in the required coupled form, but some further simplifications are possible. New dependent variables are chosen which remove the first-order derivative of the principal terms. Thus let

$$E_x^{(o)} = \mathfrak{F}_o(\rho_o^2 - 1)^{-\frac{1}{2}}, \quad E_y^{(x)} = \mathfrak{F}_x(\rho_o^2 - 1)^{-\frac{1}{2}}. \tag{18.48}$$

Further let

$$\psi = \frac{\rho_o'}{\rho_o^2 - 1} = \frac{1}{2k} \frac{d}{dz} \log\left(\frac{\rho_o - 1}{\rho_o + 1}\right). \tag{18.49}$$

Then the equations become

$$\mathfrak{F}_o'' + \mathfrak{F}_o(\mathfrak{n}_o^2 + \psi^2) = \psi' \mathfrak{F}_x + 2\psi \mathfrak{F}_x', \tag{18.50}$$

$$\mathfrak{F}_x'' + \mathfrak{F}_x(\mathfrak{n}_x^2 + \psi^2) = \psi' \mathfrak{F}_o + 2\psi \mathfrak{F}_o', \tag{18.51}$$

which are Försterling's coupled equations, arranged with the principal terms on the left and the coupling terms on the right. The variable ψ which appears in the coupling terms is called the 'coupling parameter' and plays an important part in all the theory of coupled equations. It may be regarded as a function of the state of the ionosphere at each level, and its properties are discussed fully in § 19.2. Notice that it is a derivative with respect to the height z, and is therefore zero in a homogeneous medium.

If the coupling parameter ψ is everywhere negligibly small, the equations become

$$\mathfrak{F}_o'' + \mathfrak{n}_o^2 \mathfrak{F}_o = 0, \tag{18.52}$$

$$\mathfrak{F}_x'' + \mathfrak{n}_x^2 \mathfrak{F}_x = 0, \tag{18.53}$$

which shows that the ordinary and extraordinary waves are then propagated and reflected independently. These could be treated in exactly the same way as (9.8) for the isotropic case, and much of the theory of chs. 9, 15 and 16 could be applied to them.

Some applications of Försterling's equations are given in §§ 19.8, 21.13 and 22.6.

18.10 Coupled equations in the general case, in matrix form

The coupled equations for the general case were first given by Clemmow and Heading (1954). They are most conveniently derived from the matrix form (18.18) of the differential equations, which are to be transformed by using four new dependent variables f_1, f_2, f_3, f_4. These may be written as a column matrix $\mathbf{f}$, thus

$$\mathbf{f} = \begin{pmatrix} f_1 \\ f_2 \\ f_3 \\ f_4 \end{pmatrix}. \tag{18.54}$$

These are linear functions of the old variables $\mathbf{e}$ (18.16), and the relation may be written in matrix form

$$\mathbf{e} = \mathbf{S}\mathbf{f}, \tag{18.55}$$

where $\mathbf{S}$ is a 4×4 matrix which is to be chosen so as to convert the equations to the required coupled form. It depends on the electron density and collision-frequency, and therefore on the height z, and expresses a property of the ionosphere at each level. In a homogeneous ionosphere it is a constant matrix. Provided that $\mathbf{S}$ is non-singular

$$\mathbf{f} = \mathbf{S}^{-1}\mathbf{e}. \tag{18.56}$$

As before a dash ′ is used to denote $(1/k)(d/dz)$. The substitution of (18.56) in (18.18) gives

$$\mathbf{f}' + i\mathbf{S}^{-1}\mathbf{T}\mathbf{S}\mathbf{f} = -\mathbf{S}^{-1}\mathbf{S}'\mathbf{f}. \tag{18.57}$$

The right side of (18.57) vanishes in a homogeneous medium because of the factor $\mathbf{S}'$. The equations are therefore in the required coupled form if $\mathbf{S}$ is chosen so that f_i is the only element of $\mathbf{f}$ in the left side of the ith equation, that is $\mathbf{S}^{-1}\mathbf{T}\mathbf{S}$ is a diagonal matrix. This can be done if the characteristic roots q_i of $\mathbf{T}$ are all distinct.

Let $\mathbf{s}^{(i)}$ be an eigenvector of $\mathbf{T}$ which satisfies (18.20). Then there are four eigenvectors $\mathbf{s}^{(i)}$ corresponding to the four characteristic roots q_i. Let $\mathbf{S}$ be the matrix whose ith column is $\mathbf{s}^{(i)}$. Then it is easily shown that

$$\mathbf{S}^{-1}\mathbf{T}\mathbf{S} = \begin{pmatrix} q_1 & 0 & 0 & 0 \\ 0 & q_2 & 0 & 0 \\ 0 & 0 & q_3 & 0 \\ 0 & 0 & 0 & q_4 \end{pmatrix} = \mathbf{\Delta}, \quad \text{say,} \tag{18.58}$$

and the equations (18.57) become

$$\mathbf{f}' + i\mathbf{\Delta}\mathbf{f} = -\mathbf{S}^{-1}\mathbf{S}'\mathbf{f}. \tag{18.59}$$

For any eigenvector $\mathbf{s}^{(i)}$ only the ratios of the elements are determined, so that each column of $\mathbf{S}$ may be multiplied by any number without affecting the above arguments. Hence further conditions can be imposed upon $\mathbf{S}$, and it is shown later than $\mathbf{S}^{-1}\mathbf{S}'$ can sometimes be made antisymmetric, so that its diagonal terms vanish. Then the equations (18.59) are the required coupled equations with the principal terms on the left and the coupling terms on the right.

18.11 Expressions for the elements of S, S⁻¹, and −S⁻¹S′

Equation (18.58) gives

$$\mathbf{TS} = \mathbf{S\Delta}, \tag{18.60}$$

which may be solved for the ratios of the elements of the jth column of $\mathbf{S}$, thus:†

$$S_{1j}:S_{2j}:S_{3j}:S_{4j} = a_3 q_j + a_4 : A_j : q_j A_j : a_5 q_j + a_6 \quad (j = 1, 2, 3, 4), \tag{18.61}$$

where

$$A_i = q_j^2 + a_1 q_j + a_2, \tag{18.62}$$

$$\left.\begin{aligned} a_1 &= -(T_{11} + T_{44}) &&= 2SY^2 lnD, \\ a_2 &= T_{11}T_{44} - T_{14}T_{41} &&= -C^2 + \{U(U-X) - l^2Y^2C^2 - n^2Y^2S^2\}D, \\ a_3 &= T_{12} &&= S(mnY^2 - ilUY)D, \\ a_4 &= T_{14}T_{42} - T_{12}T_{44} &&= -\{C^2 lmY^2 + inY(UC^2 - X)\}D, \\ a_5 &= T_{42} &&= -\{lmY^2 + inY(U-X)\}D, \\ a_6 &= T_{41}T_{12} - T_{11}T_{42} &&= S\{mnY^2 - ilY(U-X)\}D; \end{aligned}\right\} \tag{18.63}$$

$$D = X/\{U(U^2 - Y^2) - X(U^2 - n^2Y^2)\}. \tag{18.64}$$

Similarly, (18.58) leads to

$$\tilde{\mathbf{T}}\tilde{\mathbf{S}}^{-1} = \tilde{\mathbf{S}}^{-1}\mathbf{\Delta}, \tag{18.65}$$

where the symbol ~ over a matrix indicates its transpose. Now $\mathbf{T}$ has the property (see § 18.4) that T_{ij} is obtained from $T_{5-j,5-i}$ by changing the sign of $i\,(=\sqrt{-1})$ everywhere except in $U = 1 - iZ$. Thus it follows that

$$(S^{-1})_{i1}:(S^{-1})_{i2}:(S^{-1})_{i3}:(S^{-1})_{i4} = b_5 q_i + b_6 : q_i A_i : A_i : b_3 q_i + b_4 \quad (i = 1, 2, 3, 4), \tag{18.66}$$

† In this and the next section the summation convention is *not* implied when a repeated suffix appears.

where $(S^{-1})_{ij}$ denotes the element of $\mathbf{S}^{-1}$ (not the reciprocal of S_{ij}), and b_3, b_4, b_5, b_6, are obtained from a_3, a_4, a_5, a_6, respectively, by changing the sign of i everywhere except in $U = 1 - iZ$.

The determination of $\mathbf{S}$ and $\mathbf{S}^{-1}$ can now be completed by using the condition that the diagonal terms of $\mathbf{S}^{-1}\mathbf{S}$ are unity. Let

$$\left.\begin{aligned} F_1 &= (q_1-q_2)(q_1-q_3)(q_1-q_4),\\ F_2 &= (q_2-q_1)(q_2-q_3)(q_2-q_4),\\ F_3 &= (q_3-q_1)(q_3-q_2)(q_3-q_4),\\ F_4 &= (q_4-q_1)(q_4-q_2)(q_4-q_3).\end{aligned}\right\} \qquad (18.67)$$

With the notation of this section the quartic equation (18.19) is

$$(q^2-T_{32})(q^2+a_1q+a_2)-(a_3b_5+b_3a_5)q-(a_4b_5+a_6b_3) = 0, \qquad (18.68)$$

from which F_i may be obtained by differentiating the left side with respect to q and then setting $q = q_i$. This gives

$$F_i = 2q_iA_i+(q_i^2-T_{32})(2q_i+a_1)-(a_3b_5+b_3a_5), \qquad (18.69)$$

whence

$$\begin{aligned} A_iF_i &= 2q_iA_i^2+(2q_i+a_1)\{(a_3b_5+b_3a_5)q_i+a_4b_5+a_6b_3\}-(a_3b_5+b_3a_5)A_i\\ &= 2q_iA_i^2+q_i^2(a_3b_5+b_3a_5)+2q_i(a_4b_5+a_6b_3)+a_1(a_4b_5+a_6b_3)\\ &\qquad -a_2(a_3b_5+b_3a_5). \end{aligned} \qquad (18.70)$$

Now it can be shown that

$$a_4b_5+a_6b_3 = a_3b_6+a_5b_4, \qquad (18.71)$$

$$a_1(a_4b_5+a_6b_3)-a_2(a_3b_5+b_3a_5) = a_4b_6+b_4a_6, \qquad (18.72)$$

whence from (18.70)

$$A_iF_i = (a_3q_i+a_4)(b_5q_i+b_6)+2q_iA_i^2+(a_5q_i+a_6)(b_3q_i+b_4). \qquad (18.73)$$

Now the right side of (18.73) is the product of the row matrix whose elements are the right side of (18.66), and the column matrix whose elements are the right side of (18.61). Hence we may take

$$(S_{1j}, S_{2j}, S_{3j}, S_{4j}) = (A_jF_j)^{-\frac{1}{2}}(a_3q_j+a_4, A_j, q_jA_j, a_5q_j+a_6), \qquad (18.74)$$

$$\{(S^{-1})_{i1}:(S^{-1})_{i2}:(S^{-1})_{i3}:(S^{-1})_{i4}\} = (A_iF_i)^{-\frac{1}{2}}(b_5q_i+b_6, q_iA_i, A_i, b_3q_i+b_4). \qquad (18.75)$$

These are convenient symmetrical forms which make $\mathbf{S}^{-1}\mathbf{S}'$ antisymmetric in some special cases.

Let

$$\boldsymbol{\Gamma} = -\mathbf{S}^{-1}\mathbf{S}'. \qquad (18.76)$$

Now $\mathbf{S}^{-1}\mathbf{S}$ is a constant (unity), so that $\mathbf{S}^{-1}\mathbf{S}'+(\mathbf{S}^{-1})'\mathbf{S} = 0$. Hence

$$\boldsymbol{\Gamma} = \tfrac{1}{2}\{(\mathbf{S}^{-1})'\mathbf{S}-\mathbf{S}^{-1}\mathbf{S}'\}. \qquad (18.77)$$

Substitution from (18.74) and (18.75) in (18.77) now gives Γ_{ij}. The relation

$$[\{(A_iF_i)^{-\frac{1}{2}}\}'(A_jF_j)^{-\frac{1}{2}}-(A_iF_i)^{-\frac{1}{2}}\{(A_jF_j)^{-\frac{1}{2}}\}']$$
$$\times[(b_5q_i+b_6)(a_3q_j+a_4)+A_iA_j(q_i+q_j)+(b_3q_i+b_4)(a_5q_j+a_6)]=0 \quad (18.78)$$

holds for all i, j since the first bracket vanishes when $i = j$, and the second when $i \neq j$ because $\mathbf{S}^{-1}\mathbf{S}$ is diagonal. The result is therefore

$$\begin{aligned}2\Gamma_{ij} = (A_iF_iA_jF_j)^{-\frac{1}{2}}\{&(b_5q_i+b_6)'(a_3q_j+a_4)-(b_5q_i+b_6)(a_3q_j+a_4)'\\ &+(q_iA_i)'A_j-q_iA_iA_j'+q_jA_jA_i'-(q_jA_j)'A_i\\ &+(b_3q_i+b_4)'(a_5q_j+a_6)-(b_3q_i+b_4)(a_5q_j+a_6)'\}.\end{aligned} \quad (18.79)$$

Finally, with the help of (18.71) a rearrangement of (18.79) gives

$$\begin{aligned}2\Gamma_{ii} = (A_iF_i)^{-1}\{&q_i^2(a_3b_5'-a_3'b_5+a_5b_3'-a_5'b_3)+a_4b_6'-a_4'b_6+a_6b_4'-a_6'b_4\\ &+q_i(a_3b_6'-a_3'b_6+a_4b_5'-a_4'b_5+a_5b_4'-a_5'b_4+a_6b_3'-a_6'b_3)\},\end{aligned} \quad (18.80)$$

$$\begin{aligned}2\Gamma_{ij} = (A_iF_iA_jF_j)^{-\frac{1}{2}}\{&(q_i'q_j-q_j'q_i)(a_3b_5+a_5b_3)\\ &+q_iq_j(a_3b_5'-a_3'b_5+a_5b_3'-a_5'b_3)+q_i(a_4b_5'-a_4'b_5+a_6b_3'-a_6'b_3)\\ &+q_j(a_3b_6'-a_3'b_6+a_5b_4'-a_5'b_4)+(q_i+q_j)(A_i'A_j-A_iA_j')\\ &+(q_i'-q_j')(a_4b_5+a_6b_3+A_iA_j)+a_4b_6'-a_4'b_6+a_6b_4'-a_6'b_4\}.\end{aligned} \quad (18.81)$$

The coupled equations (18.59) may now be written more explicitly thus

$$f_i'+iq_if_i=\Gamma_{i1}f_1+\Gamma_{i2}f_2+\Gamma_{i3}f_3+\Gamma_{i4}f_4 \quad (i=1,2,3,4), \quad (18.82)$$

where Γ_{ij} is given by (18.80), (18.81), and from (18.56) and (18.75)

$$f_i=(A_iF_i)^{-\frac{1}{2}}\{(b_5q_i+b_6)E_x-q_iA_iE_y+A_i\mathscr{H}_x+(b_3q_i+b_4)\mathscr{H}_y\}. \quad (18.83)$$

Each coupling coefficient Γ_{ij} contains a factor $(F_iF_j)^{\frac{1}{2}}$ in its denominator. Some of these factors are zero at a point in the complex z-plane where two roots of the Booker quartic become equal, and near this point the coupling coefficients are very large. Such a point may be a point of reflection, or a point where coupling between two upgoing (or downgoing) waves is very strong. In the general treatment of this section there is no distinction between reflection and coupling, and mathematically they are the same phenomena.

In the following sections, where the W.K.B. solutions are derived, it is assumed that the coupling terms Γ_{ij} $(i \neq j)$ are small. Thus the W.K.B. solutions will cease to be good approximations near points in the complex z-plane where two roots of the Booker quartic become equal.

18.12 The W.K.B. solutions in the general case

Every term of the elements of $\mathbf{\Gamma}$ contains a derivative with respect to the height z of some function of the parameters X, Z of the ionosphere. Where these vary slowly enough with height, and where no two of the q_i become nearly equal, the terms of $\mathbf{\Gamma}$ are small quantities of the first order. Then there is an approximate solution in which coupling is ignored, and this solution is associated with a single characteristic wave. This means that all the f_i except one, say f_j, are small quantities of the first order. Then if terms of the second order are omitted but terms of the first order are retained,

$$f_j' + iq_j f_j = \Gamma_{jj} f_j, \tag{18.84}$$

with the solution $$f_j = \exp\left(-ik\int^z q_j\,dz + k\int^z \Gamma_{jj}\,dz\right). \tag{18.85}$$

The corresponding field components are, from (18.55) and (18.74),

$$(E_x, E_y, \mathscr{H}_x, \mathscr{H}_y) = (A_j F_j)^{-\frac{1}{2}}(a_3 q_j + a_4, -A_j, q_j A_j, a_5 q_j + a_6)$$
$$\times \exp\left(-ik\int^z q_j\,dz + k\int^z \Gamma_{jj}\,dz\right), \tag{18.86}$$

which is a generalisation, for the fourth-order system, of the W.K.B. solutions described in ch. 9 for equations of the second order.

It has been shown (Budden and Clemmow, 1957) that when there are no collisions ($U = 1$) the energy flow is as expected in the wave given by (18.86). In this case the q_j are either real or in conjugate complex pairs. Where q_j is real the solution represents a progressive wave and the vertical component $\overline{\Pi}_z$ of the complex Poynting vector is independent of z. Where q_j is complex the wave is evanescent and $\overline{\Pi}_z$ is zero.

18.13 The first-order coupled equations for vertical incidence

When $\theta_I = 0$, so that $S = 0$, $C = 1$, the axes may be chosen so that the magnetic meridian is in the x–z-plane, which makes $m = 0$. Then

$$\left.\begin{aligned} a_1 = a_3 = b_3 = a_6 = b_6 = 0,\\ a_2 = -1 + \{U(U-X) - l^2Y^2\}D,\\ a_4 = a_5 = -b_4 = -b_5 = -inY(U-X)D.\end{aligned}\right\} \tag{18.87}$$

The quartic (18.68) becomes

$$q^4 + q^2(a_2 - T_{32}) + a_4^2 - a_2 T_{32} = 0, \tag{18.88}$$

which is the Appleton–Hartree formula. Its solutions may be written

$$q_1 = -q_2 = \mathfrak{n}_o, \quad q_3 = -q_4 = \mathfrak{n}_x, \tag{18.89}$$

where $\mathfrak{n}_o$, $\mathfrak{n}_x$ are the refractive indices for the ordinary and extraordinary waves. Further

$$A_1 = A_2 = \mathfrak{n}_o^2 + a_2, \quad A_3 = A_4 = \mathfrak{n}_x^2 + a_2, \quad A_1 A_3 = a_4^2, \tag{18.90}$$

$$F_1 = -F_2 = 2\mathfrak{n}_o(\mathfrak{n}_o^2 - \mathfrak{n}_x^2), \quad F_3 = -F_4 = -2\mathfrak{n}_x(\mathfrak{n}_o^2 - \mathfrak{n}_x^2). \tag{18.91}$$

The expression (18.74) for **S** may now be written down. In particular

$$(S_{11}, S_{21}, S_{31}, S_{41}) = (A_1F_1)^{-\frac{1}{2}}\{a_4, \mathfrak{n}_o^2+a_2, \mathfrak{n}_o(\mathfrak{n}_o^2+a_2), a_4\mathfrak{n}_o\}, \quad (18.92)$$

which refers to the upgoing ordinary wave, and shows that for this wave

$$\frac{E_y}{E_x} = -\frac{\mathscr{H}_x}{\mathscr{H}_y} = -\frac{\mathfrak{n}_o^2+a_2}{a_4} = \rho_o \quad (18.93)$$

from the definition (5.1) of ρ for the ordinary wave. The same result is obtained for the downgoing ordinary wave (see end of § 5.4 for remarks on the sign of ρ_o). Similarly, for the upgoing or downgoing extraordinary waves

$$\frac{E_y}{E_x} = -\frac{\mathfrak{n}_x^2+a_2}{a_4} = \rho_x. \quad (18.94)$$

Equations (18.90) and (18.88) give

$$\rho_o\rho_x = 1, \quad \rho_o+\rho_x = -(T_{32}+a_2)/a_4 = \frac{iYl^2}{(U-X)n}, \quad (18.95)$$

which shows that ρ_o, ρ_x are roots of the quadratic (5.11).

From (18.80), $\Gamma_{ii} = 0$, and from (18.81)

$$2\Gamma_{ij} = (A_iF_iA_jF_j)^{-\frac{1}{2}}\{(A_i'A_j - A_iA_j')(q_i+q_j) + (A_iA_j - a_4^2)(q_i'-q_j')\}, \quad (18.96)$$

which is antisymmetric. Consider, first, the case $i = 1$, $j = 3$. In the reduction of the square root a sign convention must be adopted. Hence in the denominator of (18.96) take

$$(-\mathfrak{n}_o)^{\frac{1}{2}} = i\mathfrak{n}_o^{\frac{1}{2}}, \quad (-\mathfrak{n}_x)^{\frac{1}{2}} = i\mathfrak{n}_x^{\frac{1}{2}}, \quad (-\rho_o)^{\frac{1}{2}} = i\rho_o^{\frac{1}{2}}, \quad (-\rho_x)^{\frac{1}{2}} = i\rho_x^{\frac{1}{2}}, \quad (18.97)$$

where the real parts of $\mathfrak{n}_o^{\frac{1}{2}}$, $\mathfrak{n}_x^{\frac{1}{2}}$, $\rho_o^{\frac{1}{2}}$, $\rho_x^{\frac{1}{2}}$ are positive. Now

$$A_1 = A_2 = -a_4\rho_o, \quad A_3 = A_4 = -a_4\rho_x; \quad (18.98)$$

$$A_1'A_3 - A_3'A_1 = a_4^2(\rho_o'\rho_x - \rho_o\rho_x') = 2a_4^2\rho_o'\rho_x. \quad (18.99)$$

Further, from (18.91), (18.93) and (18.94)

$$-F_1 = F_2 = 2a_4\mathfrak{n}_o(\rho_o-\rho_x), \quad F_3 = -F_4 = 2a_4\mathfrak{n}_x(\rho_o-\rho_x), \quad (18.100)$$

so that
$$(A_1F_1F_3A_3)^{\frac{1}{2}} = -2ia_4^2(\rho_o-\rho_x)(\mathfrak{n}_o\mathfrak{n}_x)^{\frac{1}{2}}. \quad (18.101)$$

Hence
$$(A_1'A_3 - A_3'A_1)(A_1F_1A_3F_3)^{-\frac{1}{2}} = i\psi(\mathfrak{n}_o\mathfrak{n}_x)^{-\frac{1}{2}}, \quad (18.102)$$

where ψ is the coupling parameter introduced by Försterling and given by (18.49). Because of the last equation (18.90) the second term of

(18.96) vanishes when $i = 1, j = 3$. The other terms may similarly be found, and when they are inserted in the equations (18.82) they lead to the following four first-order coupled equations:

$$\left.\begin{aligned}
f_1' + i\mathfrak{n}_o f_1 &= -(\mathfrak{n}_o'/2i\mathfrak{n}_o)f_2 \\
&\qquad + \tfrac{1}{2}i\psi(\mathfrak{n}_o+\mathfrak{n}_x)(\mathfrak{n}_o\mathfrak{n}_x)^{-\frac{1}{2}}f_3 + \tfrac{1}{2}\psi(\mathfrak{n}_o-\mathfrak{n}_x)(\mathfrak{n}_o\mathfrak{n}_x)^{-\frac{1}{2}}f_4, \\
f_2' - i\mathfrak{n}_o f_2 &= (\mathfrak{n}_o'/2i\mathfrak{n}_o)f_1 \\
&\qquad + \tfrac{1}{2}\psi(\mathfrak{n}_o-\mathfrak{n}_x)(\mathfrak{n}_o\mathfrak{n}_x)^{-\frac{1}{2}}f_3 - \tfrac{1}{2}i\psi(\mathfrak{n}_o+\mathfrak{n}_x)(\mathfrak{n}_o\mathfrak{n}_x)^{-\frac{1}{2}}f_4, \\
f_3' + i\mathfrak{n}_x f_3 &= -\tfrac{1}{2}i\psi(\mathfrak{n}_o+\mathfrak{n}_x)(\mathfrak{n}_o\mathfrak{n}_x)^{-\frac{1}{2}}f_1 \\
&\qquad - \tfrac{1}{2}\psi(\mathfrak{n}_o-\mathfrak{n}_x)(\mathfrak{n}_o\mathfrak{n}_x)^{-\frac{1}{2}}f_2 + (\mathfrak{n}_x'/2i\mathfrak{n}_x)f_4, \\
f_4' - i\mathfrak{n}_x f_4 &= -\tfrac{1}{2}\psi(\mathfrak{n}_o-\mathfrak{n}_x)(\mathfrak{n}_o\mathfrak{n}_x)^{-\frac{1}{2}}f_1 \\
&\qquad + \tfrac{1}{2}i\psi(\mathfrak{n}_o+\mathfrak{n}_x)(\mathfrak{n}_o\mathfrak{n}_x)^{-\frac{1}{2}}f_2 - (\mathfrak{n}_x'/2i\mathfrak{n}_x)f_3,
\end{aligned}\right\} \tag{18.103}$$

where

$$\left.\begin{aligned}
f_1 &= \{2\mathfrak{n}_o(\rho_o^2-1)\}^{-\frac{1}{2}}\{\mathfrak{n}_o E_x - \mathfrak{n}_o\rho_o E_y + \rho_o\mathscr{H}_x + \mathscr{H}_y\}, \\
f_2 &= -i\{2\mathfrak{n}_o(\rho_o^2-1)\}^{-\frac{1}{2}}\{\mathfrak{n}_o E_x - \mathfrak{n}_o\rho_o E_y - \rho_o\mathscr{H}_x - \mathscr{H}_y\}, \\
f_3 &= -i\{2\mathfrak{n}_x(1-\rho_x^2)\}^{-\frac{1}{2}}\{-\mathfrak{n}_x E_x + \mathfrak{n}_x\rho_x E_y - \rho_x\mathscr{H}_x - \mathscr{H}_y\}, \\
f_4 &= \{2\mathfrak{n}_x(1-\rho_x^2)\}^{-\frac{1}{2}}\{-\mathfrak{n}_x E_x + \mathfrak{n}_x\rho_x E_y + \rho_x\mathscr{H}_x + \mathscr{H}_y\};
\end{aligned}\right\} \tag{18.104}$$

and

$$\left.\begin{aligned}
E_x &= -\{2(\rho_o^2-1)\}^{-\frac{1}{2}}\{\mathfrak{n}_o^{-\frac{1}{2}}f_1 + i\mathfrak{n}_o^{-\frac{1}{2}}f_2 + i\rho_o\mathfrak{n}_x^{-\frac{1}{2}}f_3 + \rho_o\mathfrak{n}_x^{-\frac{1}{2}}f_4\}, \\
E_y &= -\{2(\rho_o^2-1)\}^{-\frac{1}{2}}\{\rho_o\mathfrak{n}_o^{-\frac{1}{2}}f_1 + i\rho_o\mathfrak{n}_o^{-\frac{1}{2}}f_2 + i\mathfrak{n}_x^{-\frac{1}{2}}f_3 + \mathfrak{n}_x^{-\frac{1}{2}}f_4\}, \\
\mathscr{H}_x &= \{2(\rho_o^2-1)\}^{-\frac{1}{2}}\{\rho_o\mathfrak{n}_o^{\frac{1}{2}}f_1 - i\rho_o\mathfrak{n}_o^{\frac{1}{2}}f_2 + i\mathfrak{n}_x^{\frac{1}{2}}f_3 - \mathfrak{n}_x^{\frac{1}{2}}f_4\}, \\
\mathscr{H}_y &= \{2(\rho_o^2-1)\}^{-\frac{1}{2}}\{-\mathfrak{n}_o^{\frac{1}{2}}f_1 + i\mathfrak{n}_o^{\frac{1}{2}}f_2 - i\rho_o\mathfrak{n}_x^{\frac{1}{2}}f_3 + \rho_o\mathfrak{n}_x^{\frac{1}{2}}f_4\}.
\end{aligned}\right\} \tag{18.105}$$

The elements Γ_{12} and Γ_{21} on the right of (18.103) are large near levels where $\mathfrak{n} = 0$. They are associated with coupling between the upgoing and downgoing ordinary waves, that is, with reflection of the ordinary wave. Similarly, Γ_{34} and Γ_{43} are associated with reflection of the extraordinary wave. The elements Γ_{13} and Γ_{31} are associated with coupling between the upgoing ordinary and extraordinary waves, and similarly Γ_{24} and Γ_{42} are associated with the two downgoing waves. It is these two kinds of coupling which may be responsible for the 'Z-trace' sometimes observed in pulsed reflections from the ionosphere at high latitudes (see § 19.7). These terms become large when ψ is large, that is when $\mathfrak{n}_o \approx \mathfrak{n}_x$, for it can be shown that ψ has a factor $(\mathfrak{n}_o - \mathfrak{n}_x)^{-2}$ (see § 19.3). Thus Γ_{13}, Γ_{31}, Γ_{24} and Γ_{42} have infinities of this order.

The elements Γ_{14} and Γ_{41} determine coupling between the upgoing ordinary wave and the downgoing extraordinary wave. These terms are also large when ψ is large, but they have infinities only of the lower order $(\mathfrak{n}_o - \mathfrak{n}_x)^{-1}$. A similar argument applies to Γ_{23} and Γ_{32}. This kind of coupling is less important at high frequencies, but at lower frequencies it is responsible for one kind of reflection, the 'coupling echo', observed with pulsed signals at 150 kc/s (Kelso, Nearhoof, Nertney and Waynick, 1951; Lindquist, 1953) (see § 19.9).

Notice that when the earth's magnetic field is neglected so that ψ is zero, the equations (18.103) separate into two pairs, one pair containing only f_1 and f_2, and the other pair containing only f_3 and f_4. By the substitutions $f_1 = \mathfrak{n}_o^{\frac{1}{2}} E_y^{(1)}$, $f_2 = \mathfrak{n}_o^{\frac{1}{2}} E_y^{(2)}$, the first pair can be converted to the form (9.42). Similarly, the second pair can be converted to the same form. The method used in § 9.9 to derive (9.42) is essentially the same as that used in this chapter.

18.14 The W.K.B. solutions for vertical incidence

At levels where all the coupling terms in (18.103) can be neglected the W.K.B. solutions can be found as indicated in § 18.12. For the upgoing ordinary wave take $f_2 = f_3 = f_4 = 0$. Then the first equation (18.103) has the solution

$$f_1 = \exp\left(-ik\int^z \mathfrak{n}_o dz\right)$$

and the corresponding field components are, from (18.105):

$$\left.\begin{aligned} E_x &= -\{2(\rho_o^2 - 1)\mathfrak{n}_o\}^{-\frac{1}{2}} \exp\left(-ik\int^z \mathfrak{n}_o dz\right), \\ E_y &= -\rho_o\{2(\rho_o^2 - 1)\mathfrak{n}_o\}^{-\frac{1}{2}} \exp\left(-ik\int^z \mathfrak{n}_o dz\right), \\ \mathscr{H}_x &= \rho_o \mathfrak{n}_o^{\frac{1}{2}}\{2(\rho_o^2 - 1)\}^{-\frac{1}{2}} \exp\left(-ik\int^z \mathfrak{n}_o dz\right), \\ \mathscr{H}_y &= -\mathfrak{n}_o^{\frac{1}{2}}\{2(\rho_o^2 - 1)\}^{-\frac{1}{2}} \exp\left(-ik\int^z \mathfrak{n}_o dz\right). \end{aligned}\right\} \quad (18.106)$$

This is the W.K.B. solution for the upgoing ordinary wave at vertical incidence for the most general case. Apart from an unimportant constant factor, it agrees with (18.34) which were derived for the special case of a loss-free medium.

The other three W.K.B. solutions for vertical incidence are as follows. For the downgoing ordinary wave:

$$\left.\begin{aligned}
E_x &= -i\{2(\rho_o^2-1)\mathfrak{n}_o\}^{-\frac{1}{2}}\exp\left(ik\int^z \mathfrak{n}_o\,dz\right),\\
E_y &= -i\rho_o\{2(\rho_o^2-1)\mathfrak{n}_o\}^{-\frac{1}{2}}\exp\left(ik\int^z \mathfrak{n}_o\,dz\right),\\
\mathscr{H}_x &= -i\rho_o\mathfrak{n}_o^{\frac{1}{2}}\{2(\rho_o^2-1)\}^{-\frac{1}{2}}\exp\left(ik\int^z \mathfrak{n}_o\,dz\right),\\
\mathscr{H}_y &= i\mathfrak{n}_o^{\frac{1}{2}}\{2(\rho_o^2-1)\}^{-\frac{1}{2}}\exp\left(ik\int^z \mathfrak{n}_o\,dz\right).
\end{aligned}\right\} \quad (18.107)$$

For the upgoing extraordinary wave:

$$\left.\begin{aligned}
E_x &= -i\rho_o\{2(\rho_o^2-1)\mathfrak{n}_x\}^{-\frac{1}{2}}\exp\left(-ik\int^z \mathfrak{n}_x\,dz\right),\\
E_y &= -i\{2(\rho_o^2-1)\mathfrak{n}_x\}^{-\frac{1}{2}}\exp\left(-ik\int^z \mathfrak{n}_x\,dz\right),\\
\mathscr{H}_x &= i\mathfrak{n}_x^{\frac{1}{2}}\{2(\rho_o^2-1)\}^{-\frac{1}{2}}\exp\left(-ik\int^z \mathfrak{n}_x\,dz\right),\\
\mathscr{H}_y &= -i\rho_o\mathfrak{n}_x^{\frac{1}{2}}\{2(\rho_o^2-1)\}^{-\frac{1}{2}}\exp\left(-ik\int^z \mathfrak{n}_x\,dz\right).
\end{aligned}\right\} \quad (18.108)$$

For the downgoing extraordinary wave:

$$\left.\begin{aligned}
E_x &= \rho_o\{2(\rho_o^2-1)\mathfrak{n}_x\}^{-\frac{1}{2}}\exp\left(ik\int^z \mathfrak{n}_x\,dz\right),\\
E_y &= \{2(\rho_o^2-1)\mathfrak{n}_x\}^{-\frac{1}{2}}\exp\left(ik\int^z \mathfrak{n}_x\,dz\right),\\
\mathscr{H}_x &= \mathfrak{n}_x^{\frac{1}{2}}\{2(\rho_o^2-1)\}^{-\frac{1}{2}}\exp\left(ik\int^z \mathfrak{n}_x\,dz\right),\\
\mathscr{H}_y &= -\rho_o\mathfrak{n}_x^{\frac{1}{2}}\{2(\rho_o^2-1)\}^{-\frac{1}{2}}\exp\left(ik\int^z \mathfrak{n}_x\,dz\right).
\end{aligned}\right\} \quad (18.109)$$

18.15 The first-order equations in other special cases

The first-order coupled equations (18.82) for the general case of oblique incidence are too complicated to be of much use for computation. There are two cases, however, where they can be expressed in a simpler form. These are (*a*) when the earth's magnetic field is assumed to be vertical, and (*b*) for propagation in the magnetic meridian at the magnetic equator.

In case (*a*), $l = m = 0$, $n = 1$, so that (18.63) and (18.64) give

$$\left.\begin{aligned} a_1 &= a_3 = a_6 = b_3 = b_6 = 0,\\ a_2 &= -C^2+\{U(U-X)-Y^2S^2\}D,\\ a_4 &= -b_4 = -iY(UC^2-X)D,\\ a_5 &= -b_5 = -iY(U-X)D. \end{aligned}\right\} \quad (18.110)$$

$$D = X/\{(U-X)(U^2-Y^2)\}, \quad (18.111)$$

and the quartic (18.68) becomes

$$q^4+q^2(a_2-T_{32})+a_4a_5-a_2T_{32} = 0. \quad (18.112)$$

Its solutions may be written (compare (18.89))

$$q_1 = -q_2 = q_o, \quad q_3 = -q_4 = q_x. \quad (18.113)$$

Further, from (18.62)

$$A_1 = A_2 = q_o^2+a_2, \quad A_3 = A_4 = q_x^2+a_2, \quad A_1A_3 = a_4a_5; \quad (18.114)$$

$$F_1 = -F_2 = 2q_o(q_o^2-q_x^2), \quad F_3 = -F_4 = -2q_x(q_o^2-q_x^2). \quad (18.115)$$

Now let $\quad \rho_o = -(q_o^2+a_2)(a_4a_5)^{-\frac{1}{2}}, \quad \rho_x = -(q_x^2+a_2)(a_4a_5)^{-\frac{1}{2}}, \quad$ (18.116)

so that $\rho_o\rho_x = 1$ and ρ_o, ρ_x are the roots of

$$\rho^2+\rho\frac{iYS^2}{\{(U-X)(UC^2-X)\}^{\frac{1}{2}}}+1 = 0, \quad (18.117)$$

and $\quad A_1 = A_2 = -(a_4a_5)^{\frac{1}{2}}\rho_o, \quad A_3 = A_4 = -(a_4a_5)^{\frac{1}{2}}\rho_x; \quad$ (18.118)

$$-F_1 = F_2 = 2(a_4a_5)^{\frac{1}{2}}q_o(\rho_o-\rho_x), \quad F_3 = -F_4 = 2(a_4a_5)^{\frac{1}{2}}q_x(\rho_o-\rho_x). \quad (18.119)$$

The expression (18.81) for Γ_{ij} becomes

$$2\Gamma_{ij} = \frac{(a_4'a_5-a_5'a_4)(q_i-q_j)+(A_i'A_j-A_iA_j')(q_i+q_j)+(A_iA_j-a_4a_5)(q_i'-q_j')}{(A_iF_iA_jF_j)^{\frac{1}{2}}}, \quad (18.120)$$

which is antisymmetric, and its form differs from the corresponding result (18.96) for vertical incidence, only by the inclusion of the additional first term in the numerator. Now let

$$2\chi = \frac{a_4'a_5-a_5'a_4}{a_4a_5(\rho_o-\rho_x)} = (\rho_o-\rho_x)^{-1}\frac{1}{k}\frac{d}{dz}\log\left(\frac{UC^2-X}{U-X}\right), \quad (18.121)$$

and $\quad \psi = \dfrac{\rho_o'\rho_x}{\rho_o-\rho_x} = \dfrac{\rho_o'}{\rho_o^2-1}. \quad$ (18.122)

Then

$$\left.\begin{aligned}\Gamma_{12} &= -\Gamma_{21} = i\chi/\rho_o - q_o'/2iq_o,\\ \Gamma_{13} &= -\Gamma_{31} = -\Gamma_{24} = \Gamma_{42} = \tfrac{1}{2}\chi(q_o - q_x)(q_o q_x)^{-\frac{1}{2}} + \tfrac{1}{2}i\psi(q_o + q_x)(q_o q_x)^{-\frac{1}{2}},\\ \Gamma_{14} &= -\Gamma_{41} = \Gamma_{23} = -\Gamma_{32} = \tfrac{1}{2}\chi(q_o + q_x)(q_o q_x)^{-\frac{1}{2}} + \tfrac{1}{2}i\psi(q_o - q_x)(q_o q_x)^{-\frac{1}{2}},\\ \Gamma_{34} &= -\Gamma_{43} = i\chi/\rho_x + q_x'/2iq_x,\end{aligned}\right\} \tag{18.123}$$

and these values may be substituted in the first-order equations (18.82). The variables f_i in this case are

$$\left.\begin{aligned}f_1 &= \{2q_o(\rho_o^2-1)\}^{-\frac{1}{2}}\{(a_5/a_4)^{\frac{1}{2}}q_o E_x - q_o\rho_o E_y + \rho_o\mathscr{H}_x + (a_4/a_5)^{\frac{1}{2}}\mathscr{H}_y\},\\ f_2 &= -i\{2q_o(\rho_o^2-1)\}^{-\frac{1}{2}}\{(a_5/a_4)^{\frac{1}{2}}q_o E_x - q_o\rho_o E_y - \rho_o\mathscr{H}_x - (a_4/a_5)^{\frac{1}{2}}\mathscr{H}_y\},\\ f_3 &= -i\{2q_x(1-\rho_x^2)\}^{-\frac{1}{2}}\{-(a_5/a_4)^{\frac{1}{2}}q_x E_x + q_x\rho_x E_y - \rho_x\mathscr{H}_x - (a_4/a_5)^{\frac{1}{2}}\mathscr{H}_y\},\\ f_4 &= \{2q_x(1-\rho_x^2)\}^{-\frac{1}{2}}\{-(a_5/a_4)^{\frac{1}{2}}q_x E_x + q_x\rho_x E_y + \rho_x\mathscr{H}_x + (a_4/a_5)^{\frac{1}{2}}\mathscr{H}_y\}.\end{aligned}\right\} \tag{18.124}$$

In case (*b*) (propagation in the magnetic meridian at the magnetic equator) $m = n = 0$, $l = 1$, so that (18.63) gives

$$a_1 = a_4 = a_5 = b_4 = b_5 = 0, \tag{18.125}$$

and it follows from (18.81) that the expression for Γ_{ij} is obtained from (18.120) on replacing a_4 by a_6 and a_5 by a_3. Thus the analysis is the same as for case (*a*) and the details need not be given.

18.16 Second-order coupled equations

The second-order coupled equations (18.50) and (18.51) for vertical incidence were first given by Försterling (1942) and may be derived directly from Maxwell's equations and the constitutive relations of the ionosphere, as in § 18.9, without first formulating the first-order coupled equations. Clemmow and Heading (1954) gave an alternative derivation of the second-order coupled equations in which they started from the first-order equations (18.82), and recombined the four variables f_j so as to give two new dependent variables, corresponding to $\mathfrak{F}_o$, $\mathfrak{F}_x$ in (18.50) and (18.51). Their method can be applied not only to vertical incidence but to any case where the Booker quartic (18.14) or (18.68) reduces to a quadratic in q^2. Their mathematics was given only in generalised matrix form. It has been applied by Budden and Clemmow (1957) to the special cases of (i) vertical incidence (giving an alternative derivation of Försterling's equations) and (ii) oblique incidence when the earth's magnetic field is assumed to be vertical† (case (*a*) of § 18.15).

† Second-order equations for this case, but derived by a different method, were also given by Heading (1953).

In the past, considerable use has been made of Försterling's equations (18.50) and (18.51) for numerical computation (see, for example, Gibbons and Nertney, 1951, 1952; Rydbeck, 1950), since they were published long before the first-order equations. It is probable, however, that where coupled equations can be used, the first-order equations are more convenient than the second order. Some examples of the use of both types are given in ch. 19.

The commonest way of using coupled equations is in a method of successive approximations. If the coupling terms are small they are neglected to a first approximation and the resulting equations, containing only the principal terms, can be solved. The solutions are inserted in the coupling terms and the equations are again solved to give a better approximation. This process is obviously impossible when any of the coupling terms becomes large.

As an example of strong coupling consider the reflection of waves at vertical incidence in an isotropic medium. One form of the coupled equations for this case was given in (9.42), where the variables are $E_y^{(1)}$ (upgoing wave) and $E_y^{(2)}$ (downgoing wave), and the coupling coefficient ($\Gamma_{12} = \Gamma_{21}$) is $(1/2\mathfrak{n}k)(d\mathfrak{n}/dz)$. As long as this is small, the coupling terms may be neglected and the solutions of the resulting equations are the W.K.B. solutions for the upgoing and downgoing waves. In the region of the complex z-plane near a zero of $\mathfrak{n}$, however, the coupling terms are very large, even when the medium is slowly varying. It is then no longer convenient to use the coupled equations, and instead the second-order equation (9.8) must be used. It was shown in ch. 16 that for a slowly varying medium this reduces to the Stokes equation (16.11) near the reflection level, and the reflection coefficient depends on the properties of the Airy integral function. This led to the very important phase integral formula (16.29) for the reflection coefficient.

Similarly in the general case, two coupling coefficients, Γ_{ij} and Γ_{ji} say, become very large near the point in the complex z-plane where the two roots q_i and q_j of the Booker quartic are equal. This is because of the factor $(F_iF_j)^{-\frac{1}{2}}$ in the expression (18.81) for Γ_{ij}. Then the two first-order coupled equations which contain these terms are useless for computation by successive approximations and the equations are therefore recombined into a single second-order equation, which, in some important special cases, can be reduced to the Stokes equation. (This may be true in the general case, although it has never been proved, as far as the author is aware.) An example, for coupling between the ordinary and extraordinary waves, is given in § 19.4. This has the important result that there is a phase integral formula for determining coupling. It was used by Eckersley (1950) and is described in §§ 19.6 and 20.7.

18.17 Condition for the validity of the W.K.B. solutions

The W.K.B. solutions for oblique incidence on an anisotropic ionosphere in the general case were given by (18.85). These are approximate solutions, and to test whether the approximation is good, the expressions could be substituted back into the differential equations (18.10) to (18.13) or (18.18). This should give a condition analogous to (9.29) which was derived for the special case of vertical incidence on an isotropic ionosphere. In the general case, however, the algebra is too complicated to give a useful result. A better test is to consider a solution consisting of two W.K.B. solutions of equal amplitude, and to examine the coupling terms in (18.82). If these are small compared with either of the principal terms on the left-hand side, then it is usually safe to assume that the W.K.B. solutions are good approximations. A discussion of the order of magnitude of the coupling terms in some special cases is given in ch. 19.

It has already been shown that some coupling terms become large, and therefore the W.K.B. solutions fail, in regions of the complex z-plane near points where two roots of the Booker quartic become equal. The solution can then be discussed in terms of the Airy integral function as indicated in § 18.16 and developed in § 19.4. Another case of failure of the W.K.B. solutions occurs when $1+M_{zz}$ is zero. This term is the denominator in some of the coefficients in the differential equations (18.10) to (18.13), and is proportional to the coefficient α in the Booker quartic (13.13). When it is zero, one of the roots of the quartic is infinite. Some solutions in special cases of this kind are discussed in §§ 21.14 and 21.15.

It is possible that there are other cases of failure of the W.K.B. solutions even in a slowly varying medium. For the coupling coefficients (18.81) contain factors $A_i^{\frac{1}{2}}$, $A_j^{\frac{1}{2}}$ in their denominators, and these might be zero at points where the roots of the quartic are all distinct. Such cases do not occur for vertical incidence. No very detailed study of the coupled equations for oblique incidence has yet been made, and it is possible that they have many interesting properties as yet undiscovered.

Example

Show that for oblique incidence when the earth's magnetic field is assumed to be vertical, the two second-order coupled equations may be written:

$$h_o'' + \left\{k^2 q_o^2 + \psi^2 + \frac{\eta''}{\eta(\rho_o^2 - 1)}\right\} h_o = \left\{-\psi' + \frac{\eta'' \rho_o}{\eta(\rho_o^2 - 1)}\right\} h_x - 2\psi h_x',$$

$$h_x'' + \left\{k^2 q_x^2 + \psi^2 + \frac{\eta''}{\eta(\rho_x^2 - 1)}\right\} h^x = \left\{-\psi' + \frac{\eta'' \rho_x}{\eta(\rho_x^2 - 1)}\right\} h_o - 2\psi h_o',$$

where q_o, q_x are roots of the Booker quartic, ρ_o, ρ_x are given by (18.116), a_4, a_5 are given by (18.110), ψ is given by (18.122), and

$$h_o = (\rho_o^2 - 1)^{-\frac{1}{2}} \left\{\left(\frac{a_5}{a_4}\right)^{\frac{1}{2}} E_x - \rho_o E_y\right\},$$

$$h_x = (1 - \rho_x^2)^{-\frac{1}{2}} \left\{\left(\frac{a_5}{a_4}\right)^{\frac{1}{2}} E_x - \rho_x E_y\right\},$$

$$\eta^2 = \frac{U - X}{UC^2 - X}.$$

(See Budden and Clemmow, 1957.)

CHAPTER 19
APPLICATIONS OF COUPLED WAVE-EQUATIONS

19.1 Introduction

The coupled wave-equations derived in the last chapter are important because they can be used to derive the W.K.B. solutions for a slowly varying anisotropic ionosphere. The coupled equations for the most general case of oblique incidence (given in first-order form in § 18.11) are, however, very complicated and have never been used for numerical computation, as far as the author is aware. Only for vertical incidence are the equations simple enough to be used for computing, and the problems discussed in the present chapter are entirely for vertical incidence. Most published work uses the second-order equations (18.50) and (18.51) as given by Försterling (1942), but it is often simpler to use the first-order equations (18.103). Both forms contain the coupling parameter ψ, whose properties must therefore be well understood. They are discussed in the next section.

At vertical incidence coupling is important in two distinct ways. First, there may be a level in the ionosphere where the coupling parameter ψ is large. This is called the coupling level and is usually near the level where $X = 1$. When this coupling is strong it can give rise at high frequencies to the phenomenon of the Z-trace (§ 19.7), and at lower frequencies to the so-called 'coupling echo' (§ 19.9). Secondly, coupling is important in the lowest part of the ionosphere. Here, although ψ may be small, it can have a cumulative effect over a large range of height. Thus it is important in determining the 'limiting polarisation' of a downcoming wave (§§ 19.11 to 19.13).

19.2 Properties of the coupling parameter ψ

The coupling parameter ψ is given by (18.49) which, in view of the relation $\rho_o \rho_x = 1$, (5.19) can be written

$$\psi = \frac{\rho_o'}{\rho_o^2 - 1} = \frac{\rho_x'}{\rho_x^2 - 1} = \frac{1}{2k} \frac{d}{dz} \log\left(\frac{\rho_o - 1}{\rho_o + 1}\right). \tag{19.1}$$

Substitution of the value of ρ_o (or ρ_x) from (5.13) gives

$$\psi = \frac{\frac{1}{4} i Y_T^2 Y_L}{(1 - X - iZ)^2 \, Y_L^2 + \frac{1}{4} Y_T^4} (-X' - iZ'). \tag{19.2}$$

It thus depends on the rates of variation of X (proportional to electron number density N), and Z (proportional to electron collision-frequency ν) with the height z. It is zero in a homogeneous medium.

It will sometimes be convenient to separate ψ into factors thus:

$$\psi = \Psi\epsilon, \quad \epsilon = X' + iZ', \quad \Psi = \frac{-\frac{1}{4}iY_T^2 Y_L}{(1-X-iZ)^2\, Y_L^2 + \frac{1}{4}Y_T^4}, \tag{19.3}$$

where ϵ is a factor determined by the gradients of electron number density and collision-frequency, and is small in a slowly varying medium.

The denominator in (19.2) may be zero for some values of X and Z. Then Ψ and ψ are infinite even when ϵ is very small. This is called 'critical coupling'. At real heights z, X and Z are real and for critical coupling it is necessary that simultaneously

$$X = 1, \quad Z = \frac{1}{2}\left|\frac{Y_T^2}{Y_L}\right| = Z_c. \tag{19.4}$$

These relations also give the critical cases discussed in § 5.4 and illustrated in Figs. 6.13 to 6.16. The last condition is independent of frequency and gives $\nu = \omega_c$ where ω_c is the critical value of the collision-frequency, given by equation (5.17). The function $\nu(z)$ is believed to be a monotonically decreasing function of z, so that it attains the value ω_c at some fixed height (about 90 km in temperate latitudes), to be called the 'critical height'. The other condition $X = 1$ depends on the frequency, and in general the relations (19.4) are not simultaneously true at any real height. There is always one frequency, however, which makes $X = 1$ at the critical height, and this will be called the 'critical coupling frequency'. At this frequency ψ becomes infinite at the critical height.

It is often convenient to treat X and Z as analytic (complex) functions of z. Then for complex values of z, X and Z can become complex, and the condition for the vanishing of the denominator in (19.2) is no longer restricted to (19.4). A point in the complex z-plane which makes the denominator of (19.2) zero is called a coupling point. Thus at the critical coupling-frequency (but at no other frequency) there is a coupling point on the real z-axis. For a further discussion of coupling points see §§ 19.4 and 20.6.

Below the ionosphere X is zero, but Z and Z' are not zero and ψ becomes

$$\psi = \frac{\frac{1}{4}iY_T^2 Y_L Z'}{(1-iZ)^2\, Y_L^2 + \frac{1}{4}Y_T^4}. \tag{19.5}$$

In this case there are no electrons and the medium behaves like free space. Yet (19.5) shows that the coupling parameter is not zero. This has important consequences in the theory of limiting polarisation and is discussed in § 19.12.

Some typical curves showing how $|\psi|$ depends on height z are given in Figs. 19.1 to 19.3. Here a model has been assumed for the ionosphere

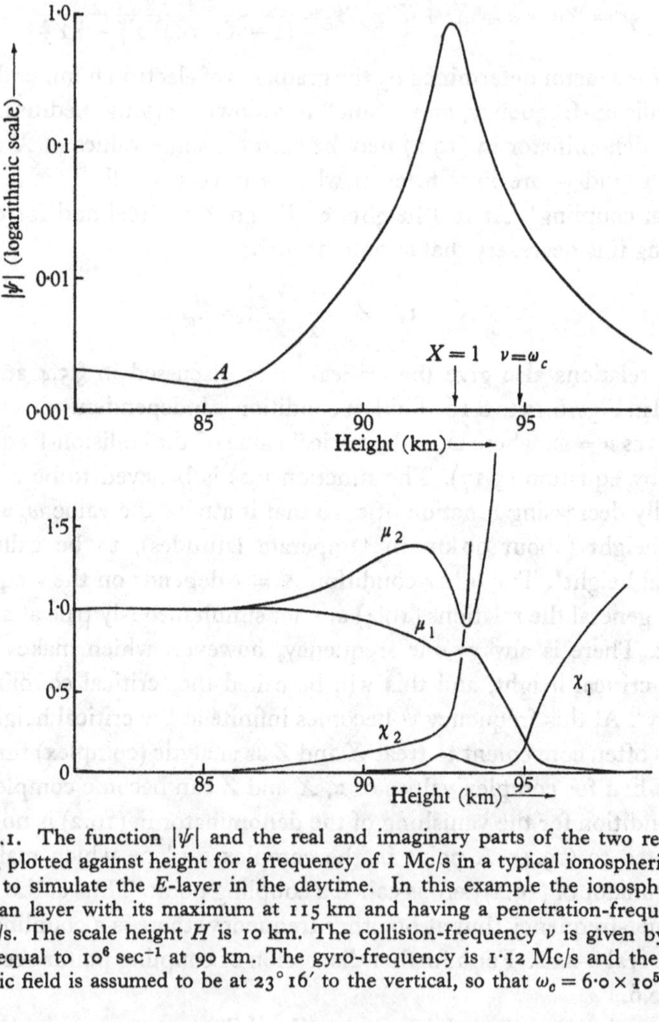

Fig. 19.1. The function $|\psi|$ and the real and imaginary parts of the two refractive indices, plotted against height for a frequency of 1 Mc/s in a typical ionospheric layer, chosen to simulate the E-layer in the daytime. In this example the ionosphere is a Chapman layer with its maximum at 115 km and having a penetration-frequency of 4·4 Mc/s. The scale height H is 10 km. The collision-frequency ν is given by (1.11) and is equal to 10^6 sec^{-1} at 90 km. The gyro-frequency is 1·12 Mc/s and the earth's magnetic field is assumed to be at 23° 16′ to the vertical, so that $\omega_c = 6{\cdot}0 \times 10^5$ sec^{-1}.

in which both X and Z vary with height z. Curves of the real and imaginary parts of the two refractive indices are also shown for comparison. The levels where $\nu = \omega_c$ and $X = 1$ are indicated. All the curves of $|\psi|$ show a bend marked A in the figures. At levels below this, the electron density is negligible and the value of ϵ in (19.3) is determind by Z'.

Thus at these low levels it is the variation of collision-frequency with height which ensures that ψ is not zero. At higher levels the electron density is appreciable and varies much more rapidly than the collision-frequency. Hence at levels above A, Z' is negligible compared with X', and ϵ is determined predominantly by X'. At higher levels the curves show a maximum. This is contributed by the factor, Ψ (19.3), and is

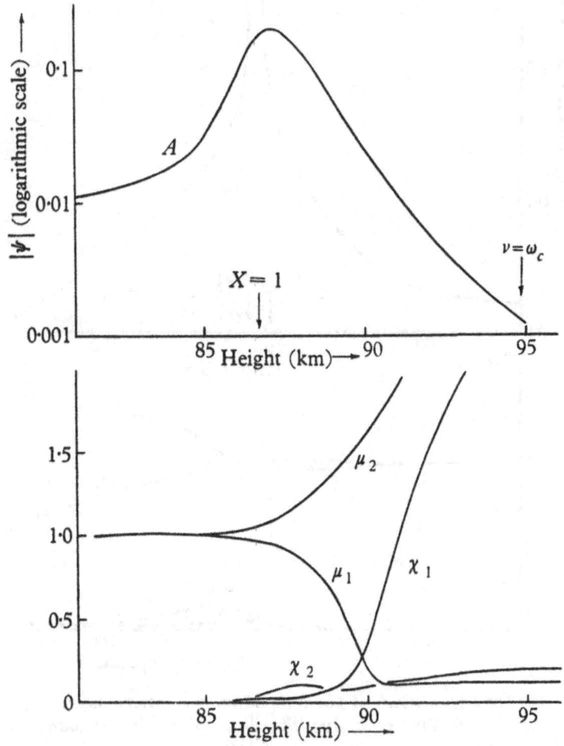

Fig. 19.2. Curves for a frequency of 160 kc/s and for the same ionosphere as in Fig. 19.1.

very marked when the levels of $X = 1$ and $\nu = \omega_c$ are close together (Fig. 19.3), that is near critical coupling. It is less marked when these two levels are well separated (Fig. 19.2).

At greater frequencies (above about 2 Mc/s) the level where $X = 1$ is nearly always well above the level where $\nu = \omega_c$, and it is then permissible to neglect electron collisions, as was done in most of chs. 12 to 14. Then ψ is given by

$$\psi = \frac{-\frac{1}{4}iY_T^2 Y_L X'}{(1-X)^2 Y_L^2 + \frac{1}{4}Y_T^4}. \tag{19.6}$$

If typical numerical values are inserted, it can be shown that at high frequencies $|\psi|$ remains small, so that there is no close approach to critical coupling. If ψ is neglected completely, then Försterling's coupled equations (18.50), (18.51) become

$$\mathfrak{F}_o'' + \mathfrak{n}_o^2 \mathfrak{F}_o = 0, \quad \mathfrak{F}_x'' + \mathfrak{n}_x^2 \mathfrak{F}_x = 0. \tag{19.7}$$

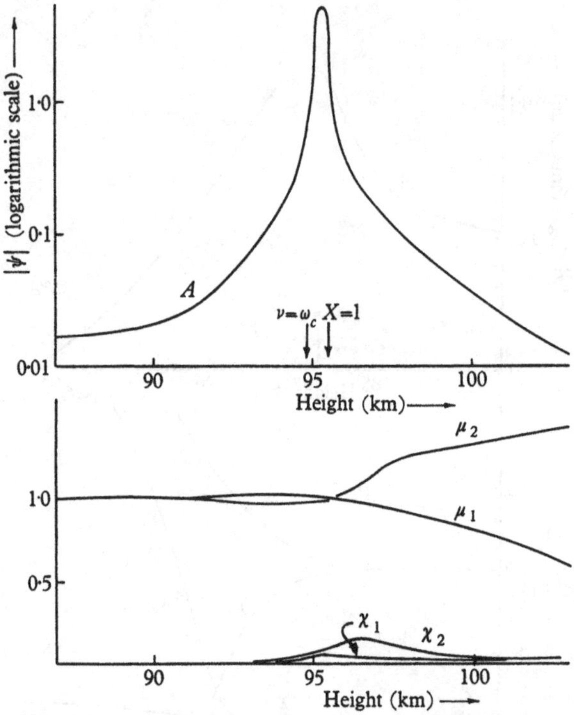

Fig. 19.3. Curves for a frequency of 160 kc/s in an ionospheric layer chosen to simulate the lower part of the E-layer at night. The ionosphere is the same as in Fig. 19.1 except that the penetration-frequency is 0·5 Mc/s.

These are two independent second-order equations, each of the same form as that used when the earth's magnetic field was neglected (9.58). They imply that the ordinary and extraordinary waves are propagated and reflected independently. In a slowly varying medium each equation may be treated by the method of § 16.6, and a phase integral formula may be derived for the reflection coefficient.

It is nearly always permissible to neglect coupling for high frequencies at vertical incidence, except in the lowest part of the ionosphere where the limiting polarisation is determined (§§ 19.11 to 19.13). Another possible exception is in the phenomenon of the Z-trace (§ 19.7).

19.3 Behaviour of the coefficients near a coupling point

To continue the mathematical study of the first-order coupled wave-equations (18.103) it is necessary to examine the behaviour of the coefficients near a coupling point in the complex z-plane.

A coupling point is where the denominator in (19.2) or (19.3) vanishes, that is where

$$(1-X-iZ)Y_L = \pm\tfrac{1}{2}iY_T^2. \tag{19.8}$$

This is also the condition that the square root vanishes in the Appleton–Hartree formula (6.1), and that the quadratic equation (5.12) for ρ_o, ρ_x shall have both roots equal to unity. Hence at a coupling point

$$\mathfrak{n}_o = \mathfrak{n}_x, \quad \rho_o = \rho_x = 1. \tag{19.9}$$

Let (19.8) be satisfied where $z = z_P$ and let $k(z-z_P) = \zeta$. We choose the minus sign; the analysis for the plus sign is similar. Assume that $X(z)$ and $Z(z)$ are analytic near z_P. Then (19.8) may be expanded by Taylor's theorem and gives

$$(1-X-iZ)Y_L+\tfrac{1}{2}iY_T^2 = -Y_L\epsilon\zeta+O(\zeta^2), \tag{19.10}$$

where ϵ is given by (19.3), that is $(1/k)\,(d/dz)\,(X+iZ)$. Now (19.10) is one factor of the denominator of ψ or Ψ (19.3). The other factor is

$$(1-X-iZ)Y_L-\tfrac{1}{2}iY_T^2 = -iY_T^2+O(\zeta). \tag{19.11}$$

Hence the coupling parameter ψ becomes

$$\psi = \frac{-\tfrac{1}{4}iY_T^2Y_L\epsilon}{Y_L\epsilon\zeta iY_T^2}\{1+O(\zeta)\} = -\frac{1}{4\zeta}\{1+O(\zeta)\}. \tag{19.12}$$

Thus at the coupling point ψ has a simple pole whose residue is independent of ϵ, that is, it does not depend on whether or not the medium is slowly varying. In the special case when Z is constant and X varies linearly with z, ψ has two simple poles at the values of X which satisfy (19.8), and has no other singularities.

Let $\mathfrak{n}_c$ be the common value of the two refractive indices at the coupling point. Then it can similarly be shown that

$$\left.\begin{aligned}\mathfrak{n}_o &= \mathfrak{n}_c+\tfrac{1}{2}M_1(\epsilon\zeta)^{\frac{1}{2}}+M_2\epsilon\zeta+O\{(\epsilon\zeta)^{\frac{3}{2}}\},\\ \mathfrak{n}_x &= \mathfrak{n}_c-\tfrac{1}{2}M_1(\epsilon\zeta)^{\frac{1}{2}}+M_2\epsilon\zeta+O\{(\epsilon\zeta)^{\frac{3}{2}}\},\end{aligned}\right\} \tag{19.13}$$

$$\mathfrak{n}_o-\mathfrak{n}_x = M_1(\epsilon\zeta)^{\frac{1}{2}}\{1+O(\epsilon\zeta)\}, \tag{19.14}$$

$$\tfrac{1}{2}(\mathfrak{n}_o+\mathfrak{n}_x)/(\mathfrak{n}_o\mathfrak{n}_x)^{\frac{1}{2}} = \eta\,(\text{say}) = 1+O(\epsilon\zeta), \tag{19.15}$$

$$\left.\begin{matrix}\rho_o\\ \rho_x\end{matrix}\right\} = 1\pm\tfrac{1}{2}N_1(\epsilon\zeta)^{\frac{1}{2}}+O(\epsilon\zeta), \tag{19.16}$$

$$\rho_o-\rho_x = N_1(\epsilon\zeta)^{\frac{1}{2}}\{1+O(\epsilon\zeta)\}, \tag{19.17}$$

$$\frac{\rho_o+1}{\rho_o-1} = (U-X)^{\frac{1}{2}}(\epsilon\zeta)^{-\frac{1}{2}}\{1+O(\epsilon\zeta)\}, \tag{19.18}$$

where M_1, M_2, N_1 are constants. These results are needed in the next section, and in ch. 20.

19.4 Properties of the coupled differential equations near a reflection point and near a coupling point

It has already been mentioned (e.g. in §§9.9 and 18.11) that 'reflection' and 'coupling' are very similar phenomena, each being associated with a point in the complex z-plane where two roots of the Booker quartic become equal. At such a point the two associated waves are strongly coupled, but it would be expected that the remaining two waves continue to be propagated independently in a medium which is sufficiently slowly varying. For example, consider the reflection point for the ordinary wave, that is the point where $\mathfrak{n}_o = 0$. For a slowly varying medium it would be at a great distance from other reflection points and from coupling points in the complex z-plane, so that in its neighbourhood ψ is very small, and the coupling terms in the Försterling equations (18.50) and (18.51) may be neglected, and the equations have the form (19.7). The second of these refers to the extraordinary wave which is propagated independently, and since $\mathfrak{n}_x$ is not small, the two W.K.B. solutions for the extraordinary wave may be used. Thus the Försterling equations show that strong coupling between two of the waves leaves the other two unaffected. Another proof of this result was given by Booker (1936).

This result is not immediately apparent when the first-order coupled equations (18.103) are examined. For example, at the point where $\mathfrak{n}_o = 0$, the coefficient of f_3 in the first of equations (18.103) is infinite. This would seem to imply that the upgoing extraordinary wave is critically coupled to the upgoing ordinary wave at this reflection point, whereas we have just shown from Försterling equations that it is independently propagated. Hence care is needed when using the first-order coupled equations near a reflection or coupling point.

Now consider the coupling point described in the last section. Here the two upgoing waves, ordinary and extraordinary, are strongly coupled, but the two downgoing waves are also strongly coupled. Thus all four W.K.B. solutions fail to be good approximations. It is proved below that in a slowly varying medium the pair of upgoing waves, though closely coupled together, is propagated independently of the pair of downgoing waves. Thus it ought to be possible to express the equations as a pair of coupled second-order equations with two new dependent variables, one referring to upgoing waves only and the other to downgoing waves only. This, however, does not seem to have been done.

To prove the independence of the upgoing pair and the downgoing pair of waves near a coupling point, it is convenient to use the differential equations in the form (18.29). The coefficients in these equations can be found from (5.13) and the Appleton–Hartree formula in the form (5.35), and can be shown to be

$$\left.\begin{aligned}\frac{\mathfrak{n}_o^2-\mathfrak{n}_x^2}{\rho_o-\rho_x} &= \frac{iY_L X(U-X)}{(U^2-Y_L^2)(U-X)-UY_T^2} = A(\zeta) \quad \text{say},\\ \frac{\rho_o\mathfrak{n}_x^2-\rho_x\mathfrak{n}_o^2}{\rho_o-\rho_x} &= \frac{-X\{U(U-X)-Y_T^2\}}{(U^2-Y_L^2)(U-X)-UY_T^2} = B(\zeta),\\ \frac{\rho_o\mathfrak{n}_o^2-\rho_x\mathfrak{n}_x^2}{\rho_o-\rho_x} &= \frac{-XU(U-X)}{(U^2-Y_L^2)(U-X)-UY_T^2} = C(\zeta).\end{aligned}\right\} \qquad (19.19)$$

Thus the coefficients are analytic at the coupling point, which is therefore an ordinary point of the differential equations. (It will be recalled that for an isotropic medium a reflection point is similarly an ordinary point; see § 15.4.) The coefficients may be expanded in Taylor series thus:

$$A(\zeta) = A_0 + A_1\epsilon\zeta + \dots, \quad B(\zeta) = B_0 + B_1\epsilon\zeta + \dots, \quad C(\zeta) = C_0 + C_1\epsilon\zeta + \dots, \tag{19.20}$$

where A_0, A_1, B_0, B_1, C_0, C_1 are constants, and ϵ is defined by (19.3) and is small in a slowly varying medium.

Now take as solutions

$$\left.\begin{aligned} E_x &= a_0 + a_1\zeta + a_2\zeta^2 + \dots, \\ E_y &= b_0 + b_1\zeta + b_2\zeta^2 + \dots. \end{aligned}\right\} \tag{19.21}$$

These are substituted in the differential equations and coefficients of successive powers of ζ are equated to zero. This gives a_2, b_2 and higher coefficients in terms of a_0, b_0, a_1, b_1, and these four may be assigned arbitrarily. This was to be expected since there must be four independent solutions. From (19.21) the Maxwell equations (18.22) give

$$\left.\begin{aligned} \mathscr{H}_x &= -(b_1 + 2b_2\zeta + \dots), \\ \mathscr{H}_y &= i(a_1 + 2a_2\zeta + \dots). \end{aligned}\right\} \tag{19.22}$$

We now inquire whether a_0, a_1, b_0, b_1 can be chosen so that in a slowly varying medium the two downgoing waves are absent. This requires that $f_2 = f_4 = 0$, that is, from (18.104):

$$\left.\begin{aligned} \mathfrak{n}_o E_x - \mathfrak{n}_o\rho_o E_y - \rho_o\mathscr{H}_x - \mathscr{H}_y &= 0, \\ -\mathfrak{n}_x E_x + \mathfrak{n}_x\rho_x E_y + \rho_x\mathscr{H}_x + \mathscr{H}_y &= 0. \end{aligned}\right\} \tag{19.23}$$

First add these and divide by $(\rho_o - \rho_x)$; next multiply the first by ρ_x and the second by ρ_o, add, and divide by $(\rho_o - \rho_x)$. This gives the pair of equations:

$$\left.\begin{aligned} \frac{\mathfrak{n}_o - \mathfrak{n}_x}{\rho_o - \rho_x} E_x - \frac{\rho_o\mathfrak{n}_o - \rho_x\mathfrak{n}_x}{\rho_o - \rho_x} E_y + \mathscr{H}_x &= 0, \\ \frac{\rho_x\mathfrak{n}_o - \rho_o\mathfrak{n}_x}{\rho_o - \rho_x} E_x - \frac{\mathfrak{n}_o - \mathfrak{n}_x}{\rho_o - \rho_x} E_y + \mathscr{H}_y &= 0. \end{aligned}\right\} \tag{19.24}$$

Now (19.13) to (19.17) show that the coefficients are analytic near $\zeta = 0$, and we may write

$$\frac{\mathfrak{n}_o - \mathfrak{n}_x}{\rho_o - \rho_x} = \alpha_0 + \alpha_1\epsilon\zeta + \dots, \quad \frac{\rho_o\mathfrak{n}_o - \rho_x\mathfrak{n}_x}{\rho_o - \rho_x} = \beta_0 + \beta_1\epsilon\zeta + \dots,$$

$$\frac{\rho_x\mathfrak{n}_o - \rho_o\mathfrak{n}_x}{\rho_o - \rho_x} = \gamma_0 + \gamma_1\epsilon\zeta + \dots. \tag{19.25}$$

Substitute for E_x, E_y, $\mathscr{H}_x$, $\mathscr{H}_y$ from (19.21) and (19.22), and equate coefficients of ζ^0, ζ. This gives four equations, homogeneous in a_0, b_0, a_1, b_1, which can be satisfied in two independent ways. The coefficients of ζ^2, ζ^3, etc., in (19.25) all contain ϵ^2 or terms in X'', Z'', and they are therefore very small in a slowly varying medium. Thus two independent solutions exist which, up to terms in ϵ, represent only upgoing waves. Similarly, two solutions can be found which represent downgoing waves only. This establishes the required result that near a coupling point in a slowly varying medium the pair of upgoing waves is propagated independently of the pair of downgoing waves.

This same method can also be used to show that it is impossible to find a solution which, near the coupling point, represents, say, the two ordinary waves only, that is a solution in which $f_3 = f_4 = 0$. The proof may be left as an exercise for the reader.

For a slowly varying medium it has been shown (§§ 16.6 and 19.2) that near a reflection point the differential equations can be reduced to the Stokes equation, which makes it possible to use the phase integral method for finding reflection coefficients. It will now be shown that a similar reduction is possible at a coupling point.

Consider a solution in which downgoing waves are absent, so that $f_2 = f_4 = 0$. Then (18.103) become

$$\left.\begin{aligned} f_1' + i\mathfrak{n}_o f_1 &= i\psi\eta f_3, \\ f_3' + i\mathfrak{n}_x f_3 &= -i\psi\eta f_1. \end{aligned}\right\} \tag{19.26}$$

Here η is given by (19.15), which shows that it differs from unity only by a small quantity of order $\epsilon\zeta$. Hence we shall take $\eta = 1$, which simplifies the algebra. The proof can be formulated without this simplification but is then much longer. Now let

$$f_1 = g_1 \exp\left\{-\tfrac{1}{2}ik\int^z (\mathfrak{n}_o + \mathfrak{n}_x)\,dz\right\}, \quad f_3 = g_3 \exp\left\{-\tfrac{1}{2}ik\int^z (\mathfrak{n}_o + \mathfrak{n}_x)\,dz\right\}. \tag{19.27}$$

Then (19.26) gives

$$\left.\begin{aligned} g_1' + \tfrac{1}{2}i(\mathfrak{n}_o - \mathfrak{n}_x)g_1 &= i\psi g_3, \\ g_3' - \tfrac{1}{2}i(\mathfrak{n}_o - \mathfrak{n}_x)g_3 &= -i\psi g_1. \end{aligned}\right\} \tag{19.28}$$

Next let $$g_+ = (g_1 + ig_3)\left(\frac{\rho_o + 1}{\rho_o - 1}\right)^{\frac{1}{2}}, \quad g_- = (g_1 - ig_3)\left(\frac{\rho_o - 1}{\rho_o + 1}\right)^{\frac{1}{2}}. \tag{19.29}$$

Then the equations become

$$g_+' = -\tfrac{1}{2}i(\mathfrak{n}_o - \mathfrak{n}_x)\frac{\rho_o + 1}{\rho_o - 1}g_-, \tag{19.30}$$

$$g_-' = -\tfrac{1}{2}i(\mathfrak{n}_o - \mathfrak{n}_x)\frac{\rho_o - 1}{\rho_o + 1}g_+, \tag{19.31}$$

where the expression (18.49) for ψ has been used. Now (19.14) and (19.18) show that the coefficients on the right are analytic at $\zeta = 0$, and the coefficient in the second equation contains a factor ζ. Hence the equations may be written

$$\frac{dg_+}{d\zeta} = K_1(\epsilon\zeta)g_-, \quad \frac{dg_-}{d\zeta} = \epsilon\zeta K_2(\epsilon\zeta)g_+,$$

where $K_1(\epsilon\zeta)$ and $K_2(\epsilon\zeta)$ are analytic and non-zero at $\zeta = 0$. If ϵ is very small, K_1 and K_2 may be taken as constants and the first equation (19.31) may be differentiated again with respect to ζ and the second used. This gives

$$\frac{d^2g_+}{d\zeta^2} = \epsilon K_1 K_2 \zeta g_+$$

which is the Stokes equation. Alternatively, let

$$\xi = \int_0^\zeta K_1(\epsilon\zeta)\,d\zeta = \zeta\{K_1(0) + O(\epsilon\zeta)\}. \tag{19.32}$$

Then the equations (19.31) become

$$\frac{dg_-}{d\xi} = \xi\{K_3 + O(\epsilon\xi)\}g_+, \quad \frac{dg_+}{d\xi} = g_-, \tag{19.33}$$

where K_3 is a constant, and these lead at once to

$$\frac{d^2g_+}{d\xi^2} = \xi\{K_3 + O(\epsilon\xi)\}g_+. \tag{19.34}$$

When the terms indicated by $O(\epsilon\xi)$ are small, which applies when the ionosphere is slowly varying, the equation (19.34) behaves like the Stokes equation for small ξ. This has the very important result that the phase integral method can be used for finding coupling coefficients (see § 20.7).

19.5 The use of successive approximations

The use of the method of successive approximations with the first-order coupled equations (18.103) may be illustrated by considering a very important example where the frequency is above the gyro-frequency, so that $Y < 1$, and the electron number density N increases monotonically with the height z. We shall also assume that the electron collision-frequency is small so that $Z \ll 1$. The coupling parameter ψ will then be so small that it can be neglected except near the level where $X = 1$. A wave travelling upwards into the ionosphere splits into ordinary and extraordinary components. The extraordinary wave is reflected near the level where $X = 1 - Y$. Just above this level, therefore, the only upgoing wave is the ordinary wave. This enters the coupling region and can now generate another extraordinary wave. Our problem is to find the amplitude of this wave above the coupling region. First, therefore, we ignore the downgoing waves, and take as the 'zero order' solution an expression which represents an upgoing ordinary wave alone. Thus f_1 must satisfy $f_1' + ik\mathrm{n}_o f_1 = 0$. The solution is

$$\text{zero-order approximation} \quad \left\{\begin{aligned} f_1 &= \exp\left(-ik\int_0^z \mathrm{n}_o\,dz\right), \\ f_2 &= f_3 = f_4 = 0. \end{aligned}\right\} \tag{19.35}$$

This is now inserted in the right side of (18.103) and the equations are solved to give a better approximation. The first of equations (18.103) does not contain f_1 on the right, and is unaffected. The second shows that a downgoing ordinary wave (symbol f_2) is generated near the level where $\mathfrak{n}_o = 0$; this is simply the reflection of the ordinary wave. The third equation (18.103) becomes

$$f_3' + i\mathfrak{n}_x f_3 = -\tfrac{1}{2}i\psi(\mathfrak{n}_o + \mathfrak{n}_x)(\mathfrak{n}_o \mathfrak{n}_x)^{-\frac{1}{2}} \exp\left(-ik\int_0^z \mathfrak{n}_o\, dz\right), \quad (19.36)$$

whose solution is

$$f_3 = \exp\left(-ik\int_0^z \mathfrak{n}_x\, dz\right)$$
$$\times \int_a^z -\tfrac{1}{2}ik\psi(\mathfrak{n}_o + \mathfrak{n}_x)(\mathfrak{n}_o \mathfrak{n}_x)^{-\frac{1}{2}} \exp\left\{-ik\int_0^z (\mathfrak{n}_o - \mathfrak{n}_x)\, dz\right\} dz, \quad (19.37)$$

where the lower limit a of the integral must be chosen to satisfy the 'boundary conditions'. Now ψ is very small, so that the integrand of the second integral is also very small, except in the coupling region. Hence a may refer to any level below the coupling region, for then if z is also below the coupling region (19.37) gives $f_3 = 0$, which correctly shows that no extraordinary wave enters the coupling region from below. If z is above the coupling region the second integral in (19.37) is a constant, equal to the amplitude of the extraordinary wave f_3 generated by coupling. In fact (19.37) shows that the coupling region may be thought of as a distributed source of upgoing extraordinary wave, the amplitude of the source at any level being equal to the integrand.

Equation (19.37) is a first-order approximation to the solution for the extraordinary wave when $X > 1$. This could now be inserted in the right side of the first equation (18.103) and a new solution obtained for f_1. This would be a second-order approximation, but the algebra is so complicated that it is not worth proceeding, and in fact the method of successive approximations is rarely useful beyond the first order. If greater accuracy is required it is usually better to solve (18.10) to (18.13) numerically.

When the zero-order approximation (19.35) is inserted in the fourth equation (18.103) it gives

$$f_4' - i\mathfrak{n}_x f_4 = -\tfrac{1}{2}\psi(\mathfrak{n}_o - \mathfrak{n}_x)(\mathfrak{n}_o \mathfrak{n}_x)^{-\frac{1}{2}} \exp\left(-ik\int_0^z \mathfrak{n}_o\, dz\right), \quad (19.38)$$

whose solution is

$$f_4 = \exp\left(ik\int_0^z \mathfrak{n}_x\, dz\right)$$
$$\times \int_z^b -\tfrac{1}{2}k\psi(\mathfrak{n}_o - \mathfrak{n}_x)(\mathfrak{n}_o \mathfrak{n}_x)^{-\frac{1}{2}} \exp\left\{-ik\int_0^z (\mathfrak{n}_o + \mathfrak{n}_x)\, dz\right\} dz, \quad (19.39)$$

where the upper limit b may be any level above the coupling region, for then if z is also above the coupling region, (19.39) is zero which shows correctly that no extraordinary wave enters the coupling region from above. The result (19.39) is discussed later, in § 19.9.

19.6 The phase integral formula for coupling

When the ionosphere varies slowly enough with height z, the phase integral method may be used to derive an expression for the extraordinary wave generated by coupling. The theory is given later in § 20.7. Let $z = z_P$ be the position of the coupling point in the complex z-plane,

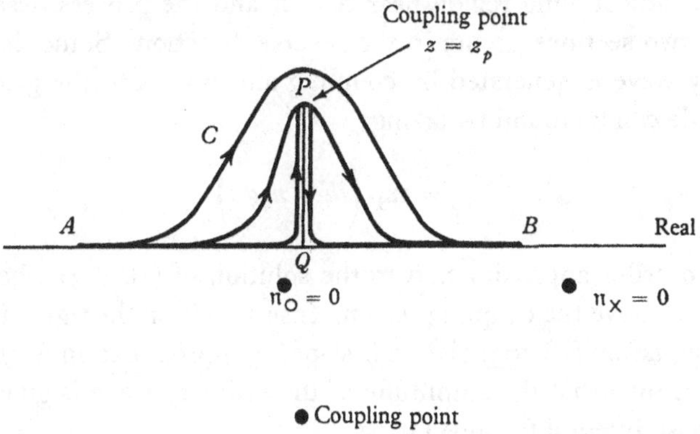

Fig. 19.4. The complex z-plane showing possible contours for use with the phase integral formula for coupling.

given by (19.8) with the minus sign. Then it is shown in § 20.7 that the formula (19.37) may be replaced by

$$f_3 = \exp\left(-ik\int_0^{z_P} \mathfrak{n}_o\,dz\right)\exp\left(-ik\int_{z_P}^{z} \mathfrak{n}_x\,dz\right), \qquad (19.40)$$

where the integrals are contour integrals in the complex z-plane. This result implies that the upgoing ordinary wave proceeds to the coupling point z_P (integral in first exponential) and then becomes an extraordinary wave. The function $\mathfrak{n}$ has a branch point at P and the transition from ordinary to extraordinary would occur automatically if the contour passed above P. Since $\mathfrak{n}$ is not infinite at P the contour may be distorted away from P as shown at C in Fig. 19.4, and (19.40) then becomes:

$$f_3 = \exp\left(-ik\int_{0\ (C)}^{z} \mathfrak{n}\,dz\right), \qquad (19.41)$$

where $\mathfrak{n}$ refers to the ordinary wave when z is below the coupling region. Equation (19.41) is the 'phase integral' formula for coupling.

19.7 The Z-trace

The extraordinary wave generated in the coupling region travels upwards in the region where $1 < X < 1 + Y$. Here the refractive index n_x is purely real when Z is zero (see Fig. 6.3), and is almost real when Z is small. The wave is reflected near where $X = 1 + Y$, and then becomes a downgoing extraordinary wave approaching the coupling region from above. The refractive index n_x is infinite near where $X = (1 - Y^2)/(1 - Y_L^2)$ and the wave is absorbed as it approaches this level (see § 21.16) and cannot reach the ground as an extraordinary wave. But it first passes through the coupling region near $X = 1$, and the process described in the last two sections occurs in the reverse direction. Some downgoing ordinary wave is generated by coupling and travels to the ground. Its amplitude can be found by taking

$$f_4 = \exp\left(ik\int_b^z n_x dz\right)$$

as a zero-order approximation to the solution of (18.103) where b is a value of z above the coupling region. Inserting it on the right side of the second equation (18.103) gives a first-order approximation for f_2. It can thus be shown that the amplitude of the ordinary wave is given by the same phase integral formula (19.41).

When pulses of radio waves are reflected at normal incidence and observed at the ground using a cathode-ray oscillograph, as described in chs. 10 and 12, records of the function $h'(f)$ are obtained which normally consist of two curves, one for the ordinary and one for the extraordinary wave (see, for example, Figs. 12.6 and 12.7). Sometimes, especially at high latitudes, a third curve is observed in which $h'(f)$ for the F-layer has a greater value than for either of the other curves (Harang, 1936; Meek, 1948; Newstead, 1948; Toshniwal, 1935). The form of the $h'(f)$ record near the penetration-frequency of the F-layer is then as sketched in Fig. 19.5. The third echo is usually known as the Z-trace. Its polarisation shows that when it leaves the ionosphere it is an ordinary wave, but its equivalent height shows that it must have been reflected near the level where $X = 1 + Y$. It has been suggested (e.g. Rydbeck, 1950) that this echo is produced by the twofold coupling process described in this and the preceding sections.

It is instructive to estimate the order of magnitude of the Z-trace echo assuming that this is the correct explanation. For simplicity we take $Z = 0$ so that n_o is real for $X < 1$ and n_x is real for $X > 1$. The contour

C (Fig. 19.4), can be distorted to coincide with the path $AQPQB$, and the real part of the exponent in (19.41) is determined entirely by the two sides of the line PQ. Assume that the electron density varies linearly with height z, so that $X = \alpha z$. Let $\mathscr{G}$ be the modulus of the expression (19.41). Then

$$\mathscr{G} = \exp \mathscr{R}\left\{\frac{-ik}{\alpha}\int_{(Q)}^{(P)} (\mathfrak{n}_o - \mathfrak{n}_x)\, dX\right\}. \tag{19.42}$$

At Q: $X = 1$ and at P: $X = 1 + \frac{1}{2}iY_T^2/Y_L = 1 + iZ_c$. At intermediate points let $X = 1 + i\zeta$. Then (19.42) becomes

$$\mathscr{G} = \exp\left\{\frac{k}{\alpha}\int_0^{Z_c} \mathscr{R}(\mathfrak{n}_o - \mathfrak{n}_x)\, d\zeta\right\}. \tag{19.43}$$

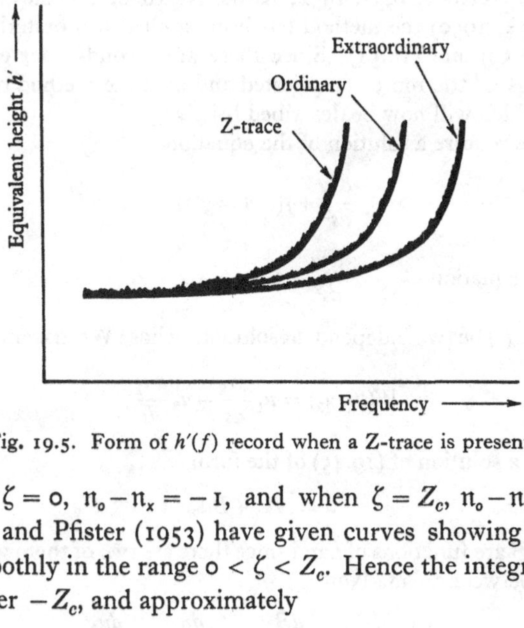

Fig. 19.5. Form of $h'(f)$ record when a Z-trace is present.

Now when $\zeta = 0$, $\mathfrak{n}_o - \mathfrak{n}_x = -1$, and when $\zeta = Z_c$, $\mathfrak{n}_o - \mathfrak{n}_x = 0$. Rydbeck (1950) and Pfister (1953) have given curves showing that $\mathfrak{n}_o - \mathfrak{n}_x$ behaves smoothly in the range $0 < \zeta < Z_c$. Hence the integral in (19.43) is of the order $-Z_c$, and approximately

$$\mathscr{G} \approx \exp\left(-\frac{k}{\alpha}Z_c\right) = \exp\left(-\frac{\omega_c}{\alpha c}\right). \tag{19.44}$$

For temperate latitudes, take $\omega_c \approx 10^6\,\text{sec}^{-1}$, $\alpha = \frac{1}{10}\,\text{km}^{-1}$. Then $\mathscr{G} \approx e^{-33} = 4{\cdot}5 \times 10^{-15}$. To give the amplitude of the Z-trace this figure must be squared since the coupling process occurs twice. Hence the amplitude of the Z-echo would be far too weak to be detected. But since it has often been observed even in low latitudes (e.g. India, see Toshniwal, 1935) there must be some other explanation. One possibility is that occasionally there can be a very steep gradient of electron density near the coupling level, so that α has a much larger value than was assumed above. A more likely explanation (Ellis, 1956) is that the Z-trace arises from waves which travel obliquely so that in the coupling region the wave-normal is nearly parallel to the earth's magnetic field and ω_c is very small. Such waves would not be reflected back to a receiver near the transmitter, but if there were irregularities near the reflection level ($X = 1 + Y$) some energy would be scattered so as to return along the same path.

19.8 The method of 'variation of parameters'

In § 19.5 the method of successive approximations was illustrated using the first-order coupled equations (18.103). In a number of important papers (Gibbons and Nertney, 1951, 1952; Kelso, Nearhoof, Nertney and Waynick, 1951; Rydbeck, 1950) the method has been applied to Försterling's coupled equations (18.50) and (18.51). Since these are second-order equations, the mathematics is a little more complicated and uses the method of variation of parameters which will now be described briefly.

Suppose we require a solution of the equation

$$\frac{d^2u}{ds^2} + f(s)\,u = g(s). \tag{19.45}$$

Consider the equation
$$\frac{d^2u}{ds^2} + f(s)\,u = 0, \tag{19.46}$$

and let $u_1(s)$, $u_2(s)$ be two independent solutions whose Wronskian is denoted by

$$W(u_1, u_2) = u_1\frac{du_2}{ds} - u_2\frac{du_1}{ds}. \tag{19.47}$$

We now seek a solution of (19.45) of the form

$$u = Au_1 + Bu_2, \tag{19.48}$$

where A and B are functions of s and since there are two of them we can impose one relation between them. Now

$$\frac{du}{ds} = \frac{dA}{ds}u_1 + \frac{dB}{ds}u_2 + A\frac{du_1}{ds} + B\frac{du_2}{ds}. \tag{19.49}$$

Take
$$u_1\frac{dA}{ds} + u_2\frac{dB}{ds} = 0 \tag{19.50}$$

as the imposed relation. Then

$$\frac{d^2u}{ds^2} = \frac{du_1}{ds}\frac{dA}{ds} + \frac{du_2}{ds}\frac{dB}{ds} + A\frac{d^2u_1}{ds^2} + B\frac{d^2u_2}{ds^2}. \tag{19.51}$$

Substitute in (19.45) and use (19.46) with $u = u_1, u_2$. Then

$$\frac{du_1}{ds}\frac{dA}{ds} + \frac{du_2}{ds}\frac{dB}{ds} = g(s). \tag{19.52}$$

Now (19.50) and (19.52) may be solved to give

$$\frac{dA}{ds} = -\frac{u_2 g}{W}, \quad \frac{dB}{ds} = \frac{u_1 g}{W} \tag{19.53}$$

and the solution (19.48) becomes

$$u = -u_1 \int_a^s \frac{u_2 g}{W} ds - u_2 \int_s^b \frac{u_1 g}{W} ds, \tag{19.54}$$

where a and b are arbitrary constants, to be determined from the boundary conditions. In the cases of interest here, the differential equations are in the normal form, that is, like (19.45) they have no first derivative term, so that the Wronskian W is constant and may be taken outside the integrals (19.54).

The application of this result to the Försterling equations (18.50) and (18.51) will now be illustrated by considering a specific problem in which an ordinary wave is incident from below upon the coupling region. The waves which go through the coupling region may be reflected at a higher level, but the reflected waves will at first be ignored since the effect of coupling on waves coming down from above can be treated separately.

The coupling parameter ψ is assumed to be negligible except in the coupling region, and is there assumed to be so small that the right sides may be neglected to a first approximation. The factors ψ^2 on the left will be neglected completely; they could be allowed for by replacing $\mathfrak{n}_o$, $\mathfrak{n}_x$ by $(\mathfrak{n}_o^2+\psi^2)^{\frac{1}{2}}$, $(\mathfrak{n}_x^2+\psi^2)^{\frac{1}{2}}$ respectively in the following argument. It is also assumed that $\mathfrak{n}_o$ and $\mathfrak{n}_x$ are not near zero in the coupling region, so that the reflection levels are not close to the coupling region. Take $s = kz$ so that the equations become

$$\frac{d^2\mathfrak{F}_o}{ds^2} + \mathfrak{n}_o^2 \mathfrak{F}_o = 0 \quad \text{and} \quad \frac{d^2\mathfrak{F}_x}{ds^2} + \mathfrak{n}_x^2 \mathfrak{F}_x = 0$$

(compare (19.46)), and their W.K.B. solutions are

$$\left.\begin{aligned} \mathfrak{F}_o^{(1)} &= \mathfrak{n}_o^{-\frac{1}{2}} \exp\left(-i\int_0^s \mathfrak{n}_o\, ds\right), & \mathfrak{F}_o^{(2)} &= \mathfrak{n}_o^{-\frac{1}{2}} \exp\left(i\int_0^s \mathfrak{n}_o\, ds\right), \\ \mathfrak{F}_x^{(1)} &= \mathfrak{n}_x^{-\frac{1}{2}} \exp\left(-i\int_0^s \mathfrak{n}_x\, ds\right), & \mathfrak{F}_x^{(2)} &= \mathfrak{n}_x^{-\frac{1}{2}} \exp\left(i\int_0^s \mathfrak{n}_x\, ds\right). \end{aligned}\right\} \tag{19.55}$$

These may be used in the zero-order approximation. There is no upgoing extraordinary wave below the coupling region, and no downgoing waves above it, so that at all levels the zero-order approximation is

$$\mathfrak{F}_o = \mathfrak{F}_o^{(1)}, \quad \mathfrak{F}_x = 0. \tag{19.56}$$

To obtain a more accurate solution this result is substituted in the right sides of (18.50) and (18.51). This leaves (18.50) unaffected so that to the first order $\mathfrak{F}_o$ is unchanged. Equation (18.51), however, becomes (neglecting ψ^2 on the left):

$$\frac{d^2\mathfrak{F}_x}{ds^2} + \mathfrak{n}_x^2 \mathfrak{F}_x = g(s), \tag{19.57}$$

where $$g(s) = \psi'\mathfrak{F}_o^{(1)} + 2\psi\mathfrak{F}_o^{(1)\prime}. \qquad (19.58)$$

The formula (19.54) is now applied with $u_1 = \mathfrak{F}_x^{(1)}$, $u_2 = \mathfrak{F}_x^{(2)}$ so that $W = 2i$. Thus to the first order

$$\mathfrak{F}_x = \tfrac{1}{2}i\mathfrak{F}_x^{(1)}\int_a^s \mathfrak{F}_x^{(2)} g(s)\,ds + \tfrac{1}{2}i\mathfrak{F}_x^{(2)}\int_s^b \mathfrak{F}_x^{(1)} g(s)\,ds. \qquad (19.59)$$

The limits a, b must be chosen to satisfy the 'boundary' conditions. When s is outside the coupling region the integrands in (19.59) are zero, and the integrals are constant. When s is well below the coupling region, the first integral must be zero since $\mathfrak{F}_x^{(1)}$ is an upgoing extraordinary wave which is absent there. Similarly, the second integral must be zero when s is above the coupling region since $\mathfrak{F}_x^{(2)}$ is a downgoing extraordinary wave. Thus a is any value of s well below the coupling region, and b is any value well above it. Equation (19.59) shows that above the coupling region there is an upgoing extraordinary wave (first term), and below it there is a downgoing extraordinary wave (second term). These are generated by coupling from the incident upgoing ordinary wave. In fact (19.59) shows that the coupling region may be thought of as two distributed sources of these waves, the amplitudes of the sources at any level being equal to the integrands.

The Försterling equations can similarly be used to find the amplitudes of the waves generated by coupling from a downgoing ordinary wave or from upgoing or downgoing extraordinary waves.

19.9 The 'coupling echo'

Equations (19.39) and (19.59) show that an upgoing ordinary wave gives rise to a downgoing extraordinary wave from the coupling region. Although (19.39) was derived for the special case when $Y < 1$ and Z is small, it still applies more generally, and in particular, when $Y > 1$. Then both extraordinary and ordinary waves reach the coupling region, and it can similarly be shown that an upgoing extraordinary wave gives a downgoing ordinary wave. These waves have been observed in measurements with pulses at vertical incidence on a frequency of 150 kc/s (Kelso *et al.* 1951). The downgoing waves reach the ground as a pulse known as the 'coupling echo'. Measurements of the equivalent height of reflection confirm that it is returned from the coupling region. Some authors use the term 'backscatter' for the coupling process which generates this echo, but it is important to remember that it is really a reflection process and not quite the same as the kind of scattering that would occur from irregularities of electron density.

The two expressions in (19.59) and (19.39) for the amplitude of the coupling echo can be shown to be the same to the order of approximation used. (In (19.59), integrate by parts and use the property that ψ is zero outside the coupling region, and that $\mathfrak{n}_o'$ is small compared with $\mathfrak{n}_o$.)

The derivation in § 19.5 from the first-order equations is to be preferred since it is simpler and fewer approximations are made.

It was shown in § 19.4 that in a slowly varying medium the coupling between upgoing and downgoing waves is very weak in the coupling region. The presence of a coupling echo therefore shows that the ionosphere is not slowly varying, but must change appreciably within one wavelength. This is most likely to occur when the wavelength is long and explains why coupling echos are only observed at frequencies below about 500 kc/s.

19.10 The transition through critical coupling

The critical coupling-frequency has already been defined (§ 19.2) as that frequency which makes $X = 1$ at the level where $\nu = \omega_c$. For smaller frequencies we shall say that conditions are 'less than critical', and for greater frequencies that they are 'greater than critical'. (Some authors use the terms 'quasi-longitudinal' and 'quasi-transverse' respectively, but these have misleading implications and are best avoided.) At high enough frequencies, therefore, conditions are greater than critical. This is the usual case encountered for vertical incidence at high frequencies in temperate latitudes, and we can then speak without ambiguity of an ordinary and an extraordinary wave. For frequencies which are less than critical, however, there is an ambiguity in these terms, already mentioned in §§ 6.13 and 6.15. In temperate latitudes the critical coupling-frequency is usually in the range 200 kc/s to 1 Mc/s, and is rarely smaller than 150 kc/s. For very low frequencies, therefore, conditions are always less than critical. In the examples of Figs. 19.1 and 19.2, conditions are less than critical, and in Fig. 19.3 they are very slightly greater than critical.

It is now interesting to inquire what would be observed if conditions changed continuously from greater than to less than critical. This might happen if the transmitter-frequency was continuously decreased, or if the frequency was fixed and the electron density was increasing. In describing the phenomenon we shall suppose that the transmitter-frequency is being slowly decreased, and that the critical coupling-frequency has some fixed value less than the gyro-frequency. It is also assumed that the electron density, and therefore X, is a monotonically increasing function of the height z. The discussion applies only for vertical incidence.

Suppose that curves are plotted showing how the real and imaginary parts of the two refractive indices depend on the height z. For frequencies greater than critical these curves have the form of Fig. 6.16*a*, *b*. At the critical coupling-frequency they are as in Fig. 6.16*c* and at frequencies less than critical they are as in Figs. 6.16*d*, *e*.

Consider, first, Figs. 6.16*a*, *b* for conditions greater than critical. If there were no collisions we should expect the ordinary wave to be reflected near where $X = 1$. The effect of collisions, however, is to attenuate this wave heavily. This is partly because, for $Y > 1$, the refractive index for the ordinary wave has an infinity, and it can be seen in Fig. 6.16*a* that $\mathfrak{n}_o$ does not approach zero near $X = 1$, and that its imaginary part becomes large above this level. Thus the ordinary wave is almost completely absorbed and its reflection is not observed. The only observed reflection is the extraordinary wave, which is returned from near the level where $X = 1 + Y$ (beyond the range covered in Fig. 6.16). Its refractive index is given by the chain curves in Figs. 6.16*a*, *b*. The polarisation of this reflected wave is difficult to calculate but it is shown in § 19.13 that it is given approximately by the value of ρ_x at some point in the lowest part of the ionosphere. Here X is very small, and if it is neglected, (5.13) and (5.16) give for the polarisation of the downcoming wave

$$\rho_x = \frac{iZ_c}{1 - iZ} + i\left\{1 + \frac{Z_c^2}{(1 - iZ)^2}\right\}^{\frac{1}{2}},$$

where the real part of the square root is positive. This represents an elliptically polarised wave with a right-handed sense.

We might also expect that the upgoing ordinary wave could generate some extraordinary wave by coupling. This would be reflected where $X = 1 + Y$ and on its downward path could generate some ordinary wave, exactly as for the Z-trace. The amplitude of the resulting reflection could be found by the phase integral formula (19.40), and is equal to the square of the expression (19.42). For frequencies much greater than critical it is extremely small because the range of integration PQ in (19.42) is large. But as the critical coupling-frequency is approached the point P in Fig. 19.4 moves on to the real z-axis and the range of integration PQ approaches zero. Thus at the critical coupling-frequency this Z-reflection cannot be ignored, and we should expect it then to have roughly the same amplitude as the extraordinary reflected wave.

Below the level $X = 1$ this Z-wave is an ordinary wave so that the associated refractive index is given by the solid curves in Figs. 6.16*a*, *b*. Its polarisation is given approximately by

$$\rho_o = \frac{iZ_c}{1 - iZ} - i\left\{1 + \frac{Z_c^2}{(1 - iZ)^2}\right\}^{\frac{1}{2}}.$$

This is markedly different from ρ_x and represents an elliptically polarised wave with a left-handed sense.

Now consider Figs. 6.16*d*, *e*, which apply to frequencies less than critical. Here the wave reflected where $X = 1 + Y$ is represented by a chain curve when $X > 1$ but by a solid curve when $X < 1$. On emerging from the ionosphere, therefore, it is what we previously called an ordinary wave, and its polarisation is given by ρ_o. The refractive index curves for a wave which starts as an upgoing extraordinary wave (chain curves in the left halves of Fig. 6.16*d*, *e*) now never go near zero, and so there is no reflection of this wave. It could give some of the other wave by coupling near the level $X = 1$. The amplitude could be found from a phase integral formula similar to (19.40). The point P in Fig. 19.4 is now below the real axis, and the contour would have to go below it. We should expect this amplitude to be very small for frequencies less than critical, just as the Z-reflection was small for frequencies above critical. Thus as the frequency passes down through the critical coupling-frequency, the Z-reflection becomes relatively bigger and bigger. At frequencies less than critical it is the main reflection.

For a range of frequencies near critical, when the two reflections have comparable amplitudes, the polarisation of the resultant reflected waves could be found by combining the two components with the correct phases and amplitudes. This transition range of frequency is probably very small.

The most important effect to be expected is thus the change in polarisation from

$$\rho_x = \frac{iZ_c}{1 - iZ} + i\left\{1 + \frac{Z_c^2}{(1 - iZ)^2}\right\}^{\frac{1}{2}}$$

to

$$\rho_o = \frac{iZ_c}{1 - iZ} - i\left\{1 + \frac{Z_c^2}{(1 - iZ)^2}\right\}^{\frac{1}{2}}.$$

These are markedly different, and we should therefore expect that the transition through the critical coupling-frequency gives a readily observable change in polarisation. The author has heard references to observations of this phenomenon but does not know of any published account of it.

This topic is discussed further in § 20.8. A related topic is discussed by Lepechinsky (1956), and Landmark and Lied (1957).

19.11 Introduction to limiting polarisation

In the earlier chapters it has been assumed that when a characteristic wave, ordinary or extraordinary, travels through the ionosphere, its polarisation is given by the magnetoionic theory of §§ 5.2 and 5.3, and changes as X and Z change. The justification for this assumption must now be examined.

The polarisation ρ is given by (5.13). When $X = 0$ this becomes

$$\rho = \frac{iY_T^2}{2Y_L(1-iZ)} \pm i\left\{\frac{Y_T^4}{4Y_L^2(1-iZ)^2}+1\right\}^{\frac{1}{2}}, \qquad (19.60)$$

so that even when there are no electrons, ρ has a value which depends on Z. In the free space below the ionosphere it is possible even when electrons are absent, to speak of the electron collision-frequency ν, and therefore of Z. These would be roughly proportional to the pressure and would therefore change markedly within the height range of about 60 km below the ionosphere. Thus a pure ordinary wave travelling downwards would show a marked change of polarisation in this range. But clearly the actual polarisation could not change, since the medium behaves just like free space. Here the assumption that a characteristic wave retains its correct polarisation must fail. It will be shown in the next section that some of the other characteristic wave is generated by coupling, and the amount is just enough to keep the polarisation constant. This coupling is only effective when the electron density is small. In the ionosphere itself where the electrons are numerous, the ordinary and extraordinary waves retain their correct polarisations and are independently propagated (except in the coupling region discussed in earlier sections). The following problems must therefore be examined:

(*a*) What conditions are necessary for a characteristic wave to retain its polarisation as given by equation (5.13)?

(*b*) When one characteristic wave travels down through the ionosphere, what determines its 'limiting polarisation' when it reaches the ground?

These problems have been discussed by several authors, for example, Eckersley and Millington (1939), Booker (1936).

Consider an ordinary wave travelling down through the lower ionosphere into the free space below. Here the coupling parameter ψ is extremely small, and in a small height range the amount of extraordinary wave generated by coupling is also small. But the two refractive indices $\mathfrak{n}_o$, $\mathfrak{n}_x$ are both equal to unity, so that the ordinary wave and the small extraordinary wave both travel with the same phase velocity. Thus as more of the extraordinary wave is generated, it adds to the amplitude of that already present, because the small contributions arriving from higher levels all have the same phase. In other words, although the coupling is small, it is cumulative, and in the next section it is shown that this maintains the polarisation constant. At higher levels where the electrons make $\mathfrak{n}_o \neq \mathfrak{n}_x$, the ordinary and extraordinary waves have different phase velocities, so that the small contributions to the extraordinary wave are not phase coherent, and interfere destructively. When this happens there is no cumulative generation of extraordinary wave and the ordinary wave then retains its correct polarisation. It will be shown that the transition occurs near the level where $|\frac{1}{4}(\mathfrak{n}_o-\mathfrak{n}_x)^2-\frac{1}{2}i(\mathfrak{n}_o-\mathfrak{n}_x)'| \approx |\psi|^2$ and it is here that the limiting polarisation is determined. It is convenient to call this the 'limiting region'. This is quite different from the 'coupling region' discussed earlier. In the coupling region $X \approx 1$ and ψ is large. In the limiting region X and ψ are both very small.

19.12 The free space below the ionosphere

To study downgoing waves in the lower ionosphere we shall use the first-order coupled equations (18.103), where f_2, f_4 refer to the downgoing ordinary and extraordinary waves, respectively. These are propagated independently of the upgoing waves so that we may take $f_1 = f_3 = 0$ (in any case the coefficients of f_1, f_3 in the second and fourth equations (18.103) are negligible). Where the electron density is zero, $\mathfrak{n}_o = \mathfrak{n}_x = 1$. Let $s = kz$. A dash $'$ denotes $(1/k)(d/dz) = d/ds$ and the equations become

$$\left.\begin{aligned} f_2'-if_2 &= -i\psi f_4, \\ f_4'-if_4 &= i\psi f_2. \end{aligned}\right\} \tag{19.61}$$

Two independent exact solutions of these are

$$f_2 = e^{is}\exp\int_p^s \psi\, ds, \quad f_4 = i\,e^{is}\exp\int_p^s \psi\, ds, \tag{19.62}$$

and

$$f_2 = e^{is}\exp\int_p^s -\psi\, ds, \quad f_4 = -i\,e^{is}\exp\int_p^s -\psi\, ds, \tag{19.63}$$

where the lower limit p is some fixed point in the complex s-plane. The most general solution is a linear combination of (19.62) and (19.63), and may be written

$$\left.\begin{aligned} f_2 &= e^{is}\left\{a\exp\int_p^s \psi\, ds + b\exp\int_p^s -\psi\, ds\right\}, \\ f_4 &= i\,e^{is}\left\{a\exp\int_p^s \psi\, ds - b\exp\int_p^s -\psi\, ds\right\}, \end{aligned}\right\} \tag{19.64}$$

where a and b are constants which are to be found from the initial conditions. Their values depend on the choice of p. Now ψ is given by (19.1), whence

$$\int_p^s \psi\, ds = \tfrac{1}{2}\log\left\{\frac{\rho_o-1}{\rho_o+1}\frac{\rho_p+1}{\rho_p-1}\right\}, \tag{19.65}$$

where ρ_o is the polarisation of a pure ordinary wave, and ρ_p is the value of ρ_o at the point $s = p$. It is convenient to write

$$\sigma = \frac{E_y-E_x}{E_y+E_x} = \frac{\rho-1}{\rho+1}, \quad \sigma_o = \frac{\rho_o-1}{\rho_o+1}, \quad \sigma_p = \frac{\rho_p-1}{\rho_p+1}, \tag{19.66}$$

so that σ is the polarisation referred to new axes formed by a rotation of 45° about the z-axis, and σ_o is its value for a pure ordinary wave. Then σ_p is the value of σ_o at the point $s = p$. Now (19.65) is equal to $\tfrac{1}{2}\log(\sigma_o/\sigma_p)$ and substitution in (19.64) gives

$$f_2 = e^{is}\left\{a\left(\frac{\sigma_o}{\sigma_p}\right)^{\frac{1}{2}} + b\left(\frac{\sigma_p}{\sigma_o}\right)^{\frac{1}{2}}\right\}, \quad f_4 = i\,e^{is}\left\{a\left(\frac{\sigma_o}{\sigma_p}\right)^{\frac{1}{2}} - b\left(\frac{\sigma_p}{\sigma_o}\right)^{\frac{1}{2}}\right\}. \tag{19.67}$$

Consider (18.105) with $f_1 = f_3 = 0$, $\mathfrak{n}_o = \mathfrak{n}_x = 1$. With (19.67) they give

$$\sigma = \frac{E_y-E_x}{E_y+E_x} = \sigma_o\frac{if_2-f_4}{if_2+f_4} = \frac{b}{a}\sigma_p. \tag{19.68}$$

This gives the polarisation of the resultant wave below the ionosphere, and shows that it is independent of s and z. Hence we have proved that the coupling is just sufficient to keep the polarisation constant.

19.13 The differential equation for the study of limiting polarisation

In the lowest part of the ionosphere, including the limiting region, the two refractive indices $\mathfrak{n}_o$, $\mathfrak{n}_x$ are still close to unity but their difference is comparable with ψ, and is no longer negligible. In the coupling terms, however, where ψ is a factor, it is permissible to take $\tfrac{1}{2}(\mathfrak{n}_o+\mathfrak{n}_x)(\mathfrak{n}_o\mathfrak{n}_x)^{-\frac{1}{2}} = 1$. Then the second and fourth coupled equations (18.103) become

$$\left.\begin{aligned} f_2' - i\mathfrak{n}_o f_2 &= -i\psi f_4, \\ f_4' - i\mathfrak{n}_x f_4 &= i\psi f_2. \end{aligned}\right\} \tag{19.69}$$

Now introduce the new variable

$$u = f_4/f_2 \quad \text{so that} \quad u' = \frac{f_4'}{f_2} - \frac{u^2 f_2'}{f_4}. \tag{19.70}$$

Then u satisfies the differential equation

$$u' + iu(\mathfrak{n}_o - \mathfrak{n}_x) - i\psi(1+u^2) = 0, \tag{19.71}$$

which is a non-linear equation of Riccati type. If u could be found below the ionosphere, the limiting polarisation σ_l could easily be found from (19.68) which shows that $\sigma_l = \sigma_o(i-u)/(i+u)$. The wave above the limiting region is a pure ordinary wave so that there $u = 0$. The equation (19.71) is in a form suitable for integration using a step-by-step process. It has been used in this way by Barron (1960). The integration proceeds downwards through the limiting region starting with the value $u = 0$.

The Riccati differential equation (19.71) can be converted into a second-order linear differential equation. For let

$$\psi u = i\frac{v'}{v} + \tfrac{1}{2}i\frac{\psi'}{\psi} + \tfrac{1}{2}(\mathfrak{n}_o - \mathfrak{n}_x). \qquad (19.72)$$

Then the new variable v satisfies

$$v'' + v\left\{\tfrac{1}{4}(\mathfrak{n}_o - \mathfrak{n}_x)^2 - \tfrac{1}{2}i(\mathfrak{n}_o - \mathfrak{n}_x)' - \psi^2 + \frac{1}{2}\frac{\psi''}{\psi} - \frac{3}{4}\left(\frac{\psi'}{\psi}\right)^2 + \tfrac{1}{2}i(\mathfrak{n}_o - \mathfrak{n}_x)\frac{\psi'}{\psi}\right\} = 0. \qquad (19.73)$$

It is probable that ψ'/ψ and ψ''/ψ are negligible in the limiting region, in which case

$$v'' + v\{\tfrac{1}{4}(\mathfrak{n}_o - \mathfrak{n}_x)^2 - \tfrac{1}{2}i(\mathfrak{n}_o - \mathfrak{n}_x)' - \psi^2\} = 0. \qquad (19.74)$$

Some discussion of a solution of this equation has been given (Budden, 1952*a*), but only for conditions which are not likely to hold in the real ionosphere.

Below the limiting region $\mathfrak{n}_o - \mathfrak{n}_x$ is negligible, and it can be verified that any solution of (19.73) gives a constant polarisation. Above the limiting region ψ^2 is negligible in (19.74), and one solution is then

$$v = \exp\left\{\tfrac{1}{2}ik\int^z (\mathfrak{n}_o - \mathfrak{n}_x)\,dz\right\}.$$

This gives $u = 0$, which is correct for a downcoming pure ordinary wave. Hence the limiting region is where

$$|\tfrac{1}{4}(\mathfrak{n}_o - \mathfrak{n}_x)^2 - \tfrac{1}{2}i(\mathfrak{n}_o - \mathfrak{n}_x)'| \approx |\psi^2|. \qquad (19.75)$$

We can now say that if $|\psi^2|$ is negligible compared with the left side of (19.75), then each characteristic wave retains the polarisation given by the magneto-ionic theory. Further, if the left side of (19.75) is negligible compared with $|\psi^2|$ then any wave is propagated without change of polarisation. This answers the problem (*a*) of § 19.11.

To answer the problem (*b*), that is the determination of limiting polarisation, the differential equation (19.71) must be studied in the intervening limiting region, and this work is now in progress (Barron, 1960). It might be expected that the limiting polarisation is equal to ρ for some level near the limiting region. Here X is very small and may be taken as zero, so that ρ is given by (19.60). In the lower ionosphere ν is about $10^6\,\mathrm{sec}^{-1}$. For a frequency of 2 Mc/s this gives $Z \approx 1/4\pi$ so that $Z \ll 1$ for high frequencies, and to a first approximation (19.60) is independent of Z, and becomes (since $Z_c = |\tfrac{1}{2}Y_T^2/Y_L|$):

$$\rho = iZ_c \pm i(1 + Z_c^2)^{\frac{1}{2}}. \qquad (19.76)$$

Thus to obtain information about the value of Z in the limiting region it would be necessary to measure the polarisation very accurately. But this cannot be done, and the observed polarisations are sometimes found to be variable (Landmark, 1955; Morgan and Johnson, 1954), probably because of moving irregularities in the ionosphere.

Examples

1. Estimate the maximum value attained by the coupling parameter $|\psi|$ (19.2) for a Chapman layer in the high-frequency range (above about 2 Mc/s).

[Answer: $$\frac{1+\sqrt{3}}{4\pi}\frac{\cos\Theta}{\sin^2\Theta}\frac{\lambda_H}{H}$$ approximately,

where H is the scale height, λ_H is the wavelength in free space at the gyro-frequency, and Θ is the inclination of the earth's magnetic field to the vertical.]

2. Sketch the form of the $h'(f)$ curves you would expect if the critical coupling-frequency were above the gyro-frequency. How would the polarisation of the reflected components depend on the frequency? (A full solution of this problem would probably involve extensive numerical computation.)

CHAPTER 20

THE PHASE INTEGRAL METHOD

20.1 Introduction

When the ionosphere is slowly varying and electron collisions are neglected, a radio wave vertically incident from below travels up to the level where the refractive index $\mathfrak{n}$ is zero. There it is reflected and returns to the ground. Since $\mathfrak{n}$ is everywhere real on the path, the reflection coefficient R has modulus unity and argument equal to the total change of phase

$$-2k\int_0^{z_0}\mathfrak{n}\,dz+\tfrac{1}{2}\pi,$$

where $z = 0$ at the ground and $z = z_0$ where $\mathfrak{n} = 0$. This expression (without the $\frac{1}{2}\pi$) is the phase integral and is the result of applying a simple ray theory. The $\frac{1}{2}\pi$ is a correction mentioned below. When electron collisions are included, $\mathfrak{n}$ is no longer real and is never zero for any real value of z, but it was shown in ch. 16 that the reflection coefficient R for a slowly varying isotropic medium is still given by

$$R = i\exp\left\{-2ik\int_0^{z_0}\mathfrak{n}\,dz\right\}, \tag{20.1}$$

where the integral now extends along a contour in the complex z-plane from the ground at $z = 0$ to the complex value z_0 which makes $\mathfrak{n}(z) = 0$. This is the phase integral formula and may be thought of as a generalisation of ray theory. The exponent in (20.1) is now complex so that $|R|$ is no longer unity, and (20.1) gives both the amplitude and phase of the reflection coefficient. The factor i is a Stokes constant associated with the zero of $\mathfrak{n}$ as shown in ch. 16 and is not predicted by ray theory. It was further shown in § 19.2 that the phase integral formula (20.1) may be used to find the reflection coefficient for either the ordinary or extraordinary wave in an anisotropic ionosphere, provided it is slowly varying, so that the coupling is negligible. A phase integral formula for finding the coupling coefficient was mentioned in § 19.6.

These formulae are examples of the application of the Phase Integral Method, which is widely used in the study of wave-propagation in many branches of Physics. It was applied to the ionosphere by Eckersley (1931, 1932 *a–c*, 1950) in a series of brilliant but difficult papers. The power of the method does not seem to have been fully appreciated and consequently some of the results given in this chapter are tentative and need further study. A full and masterly treatment was given by Heading (1953), but unfortunately this has not been published.

One important application of the method is to the 'normal mode' theory of propagation of radio waves to great distances. In effect, this treats the space between the earth and the ionosphere as a wave-guide, and the phase integral method is used to find the propagation constants of the normal modes. The subject is beyond the scope of this book, but see Watson (1919*a*, *b*), Eckersley (1931, 1932*b*, *c*), Bremmer (1949*a*), Budden (1951*a*, 1952*b*, 1953, 1957), Booker and Walkinshaw (1946).

The phase integral method is closely connected with the general theory of wave-propagation and has consequently been discussed in some detail in earlier chapters. The object of the present chapter is to collect together and extend the results already derived, and to formulate rules for using the method. Before proceeding with the present chapter the reader should be thoroughly familiar with the material in ch. 9 and §§16.1 to 16.8.

20.2 The Riemann surface for the refractive index

Every phase integral formula uses a phase integral of the form $k\int \mathfrak{n}(z)\,dz$ for vertical incidence, or $k\int q(z)\,dz$ for oblique incidence, where q is one root of the Booker quartic. For vertical incidence q is the same thing as the refractive index $\mathfrak{n}$ given by the Appleton–Hartree formula. In most of this chapter we shall speak of vertical incidence and use $\mathfrak{n}$, but it must be remembered that the theory could be extended to oblique incidence.

For an isotropic medium the function $\mathfrak{n}(z)$ is given by

$$\mathfrak{n}(z) = \left\{1 - \frac{X(z)}{1 - iZ(z)}\right\}^{\frac{1}{2}}. \tag{20.2}$$

It is two-valued and has branch points, which are also zeros, where $X = 1 - iZ$. The number and position of these branch points in the complex z-plane depends on the functions $X(z)$ and $Z(z)$. Thus $\mathfrak{n}(z)$ is represented by a Riemann surface with two sheets which touch only at the branch points. In this example the two values of $\mathfrak{n}$ are zero at the branch points and are the same with opposite signs at all other points. The formula (20.1) may now be written in the alternative form

$$R = i\exp\left\{-ik\int_C \mathfrak{n}\,dz\right\}, \tag{20.3}$$

where the contour C extends from $z = z_1$, to the branch point $z = z_0$ with $\mathfrak{n}$ on one sheet of its Riemann surface, and then from $z = z_0$ back to $z = z_1$ with $\mathfrak{n}$ on the other sheet (contour C_1 in Fig. 20.1). Alternatively the contour may be distorted away from the branch point into a small circle as shown in contour C_2, since if the circle is very small, it contributes nothing to the integral. Now if the point representing $\mathfrak{n}(z)$ always remains in the Riemann surface it automatically goes from one sheet to the other when the branch point is encircled. Finally, the contour may be distorted away from the branch point altogether as shown at C_3. This does not change the value of the integral provided that no singularities are crossed. We then obtain the most general form of the phase integral formula (20.3), in which the contour C may be any contour which encircles the branch point, provided that it does not enclose any other singularities.

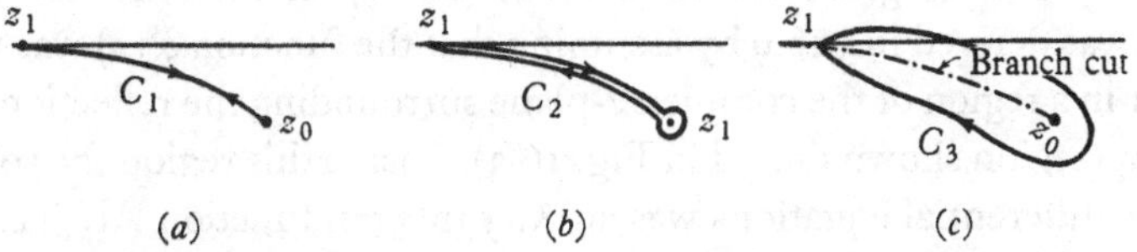

Fig. 20.1. Diagrams of the complex z-plane showing possible contours when the phase integral formula is used for finding the reflection coefficient.

Since the point which represents $\mathfrak{n}(z)$ always remains in the Riemann surface, it follows that the values of $\mathfrak{n}$ on the outward and return paths of C_3 are not in the same sheet, and the contour is not a closed contour. Strictly speaking this should be indicated by drawing a branch cut in the complex z-plane as shown in Fig. 20.1*c*, so that in the cut plane $\mathfrak{n}(z)$ can be treated as a single-valued function. The presence of this cut is implied in all phase integral formulae, though usually it is not shown in diagrams of the complex z-plane.

For an anisotropic medium with oblique incidence $\mathfrak{n}(z)$ is replaced by $q(z)$ which is a four-valued function and is therefore represented by a Riemann surface with four sheets. Two of these sheets touch at a point where two roots of the Booker quartic become equal, and this is a branch point of the function $q(z)$. By following a contour which encircles such a branch point we pass to another sheet. If from a point on the real z-axis we go in this way from a value of q which refers to an upgoing wave to another value which refers to a downgoing wave, then the branch point is associated with reflection. Alternatively, if q on the two sheets refers to waves both travelling upwards, or both downwards, then the branch point is associated with coupling. In either case the value of q at the branch point need not be zero, although it happened that in the examples discussed in earlier chapters the reflection branch points were also zeros of q or $\mathfrak{n}$ (§§16.6 to 16.8).

For an anisotropic medium with vertical incidence the function $\mathfrak{n}(z)$ is given by the Appleton–Hartree formula and it also has four sheets. The reflection branch points are also zeros of $\mathfrak{n}$, but there are coupling branch points where $\mathfrak{n}$ is not zero. The Riemann surface for $\mathfrak{n}$ is described in more detail in §20.6.

20.3 The linear electron density profile

This section is a continuation of §§16.2 to 16.8 and the same notation is used. The earth's magnetic field is neglected and the electric vector of the waves is assumed to be horizontal.

The phase integral formula (20.3) or (16.29) for the reflection coefficient was derived in §16.6 by assuming that the function $q^2(z)$ was nearly linear in a region of the complex z-plane surrounding the reflection point $z = z_0$ (region shown dotted in Fig. 16.3). Inside this region the solution of the differential equations was an Airy integral function $\mathrm{Ai}(\zeta)$ and the region had to be large enough for the asymptotic forms of this function to be good approximations on its boundary. Then it does not matter if the profile $q^2(z)$ departs from linearity outside the region. The first factor i in the phase integral formula is the Stokes constant of the function $\mathrm{Ai}(\zeta)$. The Stokes phenomenon plays an important part in the theory of the phase integral method, and it is useful to draw the Stokes diagram associated with the various branch points.

The two asymptotic forms of the solution of the differential equations are

$$E = q^{-\frac{1}{2}} \exp\left(\pm ik \int_{z_0}^{z} q\,dz\right). \tag{20.4}$$

On anti-Stokes lines the exponent must be purely imaginary so that $\int_{z_0}^{z} q\,dz$ is purely real. Similarly on the Stokes lines $\int_{z_0}^{z} q\,dz$ is purely imaginary. Fig. 20.2a is a diagram of the complex z-plane for a profile which is nearly linear, that is, it satisfies the criteria (16.36) or (16.38) within a region (dotted in the figure) surrounding the reflection point z_0. The Stokes lines and anti-Stokes lines are nearly straight within this region, but outside it they may be curved.

The left end of the real z-axis represents the free space below the ionosphere, and here $q = C$, a real constant. Hence the imaginary part of

$\int_{z_0}^{z} q\,dz$ is constant, which suggests that one anti-Stokes line is close to the negative end of the real z-axis, as indicated in Fig. 20.2*a*. If z_1 is some real height below the ionosphere, then the equation of this anti-Stokes line is

$$\mathscr{I}\left\{\int_{z_1}^{z} q\,dz + \int_{z_0}^{z_1} q\,dz\right\} = 0.$$

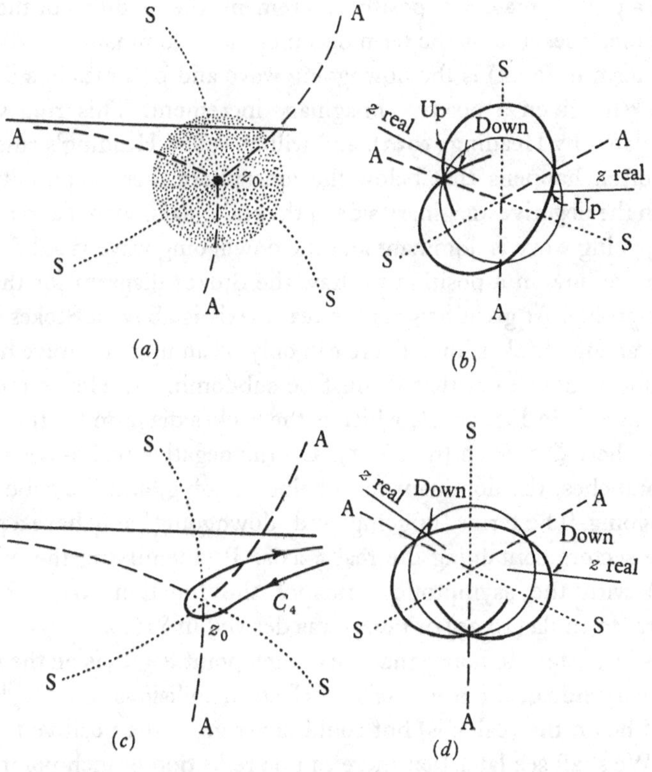

Fig. 20.2. (*a*) is the complex z-plane showing how the Stokes lines (S) and anti-Stokes lines (A) can become curved outside the region (shown dotted) where q^2 is linear. (*b*) and (*d*) are the Stokes diagrams when the incident wave comes from below and above, respectively. (*c*) shows the position of the contour C_4 in the complex z-plane.

For example, suppose that in the lowest ionosphere $q \sim C - e^{\gamma z}$, where γ is a real constant (this happens in the example of § 17.3). Then the anti-Stokes line is asymptotic to the line

$$\mathscr{I}(z) = -\frac{1}{C}\mathscr{I}\left\{\int_{z_0}^{z_1} q\,dz\right\}.$$

At points which are not on anti-Stokes lines, one of the terms (20.4) is dominant and the other subdominant. Consider the term

$$q^{-\frac{1}{2}} \exp\left(-ik\int_{z_0}^{z} q\,dz\right),$$

and suppose that on the real z-axis $\mathscr{R}(q)$ is positive, so that the wave is upgoing. (This is always true for the isotropic medium considered here, but may not be true for an anisotropic medium; see § 13.7.) If z is now given a purely imaginary positive increment, the modulus ot the exponential increases, that is the term becomes more dominant. Similarly, the other term in (20.4) is the downgoing wave and becomes less dominant when z is given a positive imaginary increment. This rule was first formulated by Heading (1953), and will be called Heading's rule.

It often happens that below the ionosphere there is an anti-Stokes line on the negative imaginary side of the real z-axis, so that at real heights the upgoing wave is dominant and the downgoing wave is subdominant.

We are now in a position to draw the Stokes diagram for the nearly linear profile. At great heights the real z-axis is above a Stokes line and below an anti-Stokes line. There can only be an upgoing wave here, and Heading's rule shows that it must be subdominant. Hence the Stokes diagram is as in Fig. 20.2*b*, which is the Stokes diagram for the function $\mathrm{Ai}(\zeta)$ where ζ is given by (16.21). On the negative real z-axis there are two branches, the dominant term being upgoing, and the subdominant downgoing. The terms 'upgoing' and 'downgoing' only have a meaning in the sectors containing the real z-axis. By identifying the two terms (20.4) with the asymptotic forms of the function $\mathrm{Ai}(\zeta)$ the phase integral formula (16.29) or (20.3) was derived in § 16.6.

It is important to notice that the branch point $z = z_0$ is on the negative imaginary side of the real z-axis. If electron collisions were neglected, it would be on the real axis, but could never go to the positive imaginary side. We shall see later that there can be reflection branch points on the positive imaginary side of the real axis, but the phase integral formula for *reflection* cannot be applied to them. On the other hand, the phase integral formula for *coupling* can be applied to coupling branch points on either side (see § 20.6).†

It is instructive to study the nearly linear profile when the incident wave comes from above the reflection level. Here there is a region where $\mathscr{I}(q)$ is very large so that both upgoing and downgoing waves are heavily

† Throughout this book it is assumed that the time-factor is $e^{+i\omega t}$. If a time-factor $e^{-i\omega t}$ were used it would be necessary to replace i by $-i$ everywhere, and then the useful reflection branch points would be on the positive imaginary side of the real z-axis.

attenuated as they travel. In this case there can be no upgoing wave below the ionosphere so that, in the Stokes diagram, the sector containing the negative real z-axis can only have a downgoing wave and Heading's rule shows that this is subdominant. Hence the Stokes diagram must be as in Fig. 20.2*d*, which is for the function $\mathrm{Ai}(\zeta e^{-\frac{2}{3}\pi i})$. At great heights there is no upgoing wave and there can be no reflection. If, in spite of this, the phase integral were evaluated using the contour C_4 of Fig. 20.2*c* it would give an extremely small value of the reflection coefficient, because of the heavy attenuation of both downgoing and upgoing waves. In this example the incident wave is travelling in the direction of z decreasing. It is more usual to take the direction of the incident wave as that of increasing z, which would necessitate changing the sign of z, and rotating the diagram of the complex z-plane (Fig. 20.2*c*) through 180°. The reflection branch point would then be on the positive imaginary side of the real axis. This confirms the statement made earlier that the phase integral formula for *reflection* cannot be used for a reflection branch point on the positive imaginary side of the real z-axis.

20.4 The parabolic electron density profile

The problem of reflection from a parabolic ionosphere, when the earth's magnetic field is neglected and the electric vector of the waves is horizontal, has already been discussed in §§ 17.4 to 17.6, where it was shown that it can be solved exactly using the parabolic cylinder functions. The same problem will now be investigated by the phase integral method, using the same notation. The electron density is a maximum at the real height $z = z_m$.

The function $q(z)$ is given by (17.25). It has two branch points, which are also zeros, where

$$z = z_m \pm a\{C^2(1-iZ)/X_m - 1\}^{\frac{1}{2}}. \tag{20.5}$$

These are on a straight line through the point $z = z_m$, and equidistant from it. They are marked z_1, z_2 in Fig. 20.3, which is a diagram of the complex z-plane. The point z_1 has the smaller real part and must lie on the negative imaginary side of the real z-axis. If $X_m > 1$ and $Z = 0$, the point z_1 would be on the real axis, and would be the level of reflection for a wave incident from below. Thus z_1 is the correct branch point to use when applying the phase integral method, and z_2 must not be used since it is on the positive imaginary side of the real axis. Let

$$k_p = k\{X_m/(1-iZ)\}^{\frac{1}{2}}. \tag{20.6}$$

This symbol was used in § 17.6 for the special case $Z = 0$ and was then the value of k at the penetration-frequency. It is now used in a more general sense and is complex. We also use the symbol D defined by (17.43) and (17.28). Then (20.5) shows that the distance between the branch points of q is

$$|z_2 - z_1| = 2\,|(aD/k_p)^{\frac{1}{2}}|. \tag{20.7}$$

The reflection coefficient R is to be found for the lower edge of the ionosphere, where $z = z_m - a$, as was done in § 17.4. It must not be found

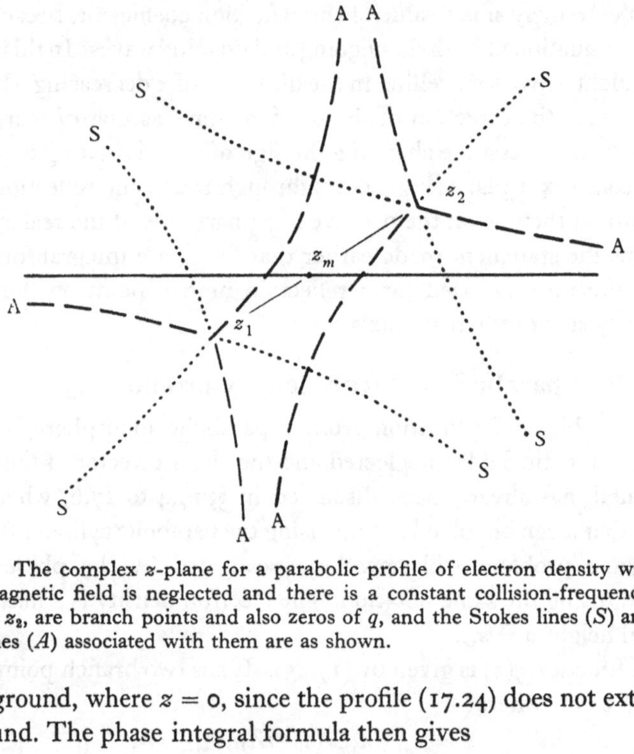

Fig. 20.3. The complex z-plane for a parabolic profile of electron density when the earth's magnetic field is neglected and there is a constant collision-frequency. The points z_1, z_2, are branch points and also zeros of q, and the Stokes lines (S) and anti-Stokes lines (A) associated with them are as shown.

for the ground, where $z = 0$, since the profile (17.24) does not extend to the ground. The phase integral formula then gives

$$R = i \exp\left\{-2ik\int_{-a}^{z_1} q\,dz\right\}, \tag{20.8}$$

where $q(z)$ is given by (17.25). The integral can be evaluated by elementary methods, and gives

$$\log R = \tfrac{1}{2}\pi i - ikaC - \tfrac{1}{2}\pi D - \tfrac{1}{2}iD \log\frac{Ck+k_p}{Ck-k_p}. \tag{20.9}$$

This should be compared with the result (17.38) obtained by the methods of 'full wave theory', which gives

$$\log R = \tfrac{1}{2}\pi i - iak_p - \tfrac{1}{4}\pi D - \tfrac{1}{2}iD\log(2ak_p)$$
$$+ \log\{(2\pi)^{-\frac{1}{2}}(-\tfrac{1}{2}+\tfrac{1}{2}iD)!\}. \tag{20.10}$$

It is not immediately apparent that (20.9) and (20.10) are approximately the same when $|D|$ is large, although in the special case discussed later it is shown that they are not very different (Fig. 20.4).

For frequencies near penetration, or for a thin ionosphere, however, D is not large and there is then a considerable difference between (20.8) and the correct value of R as given by (17.38). The phase integral formula

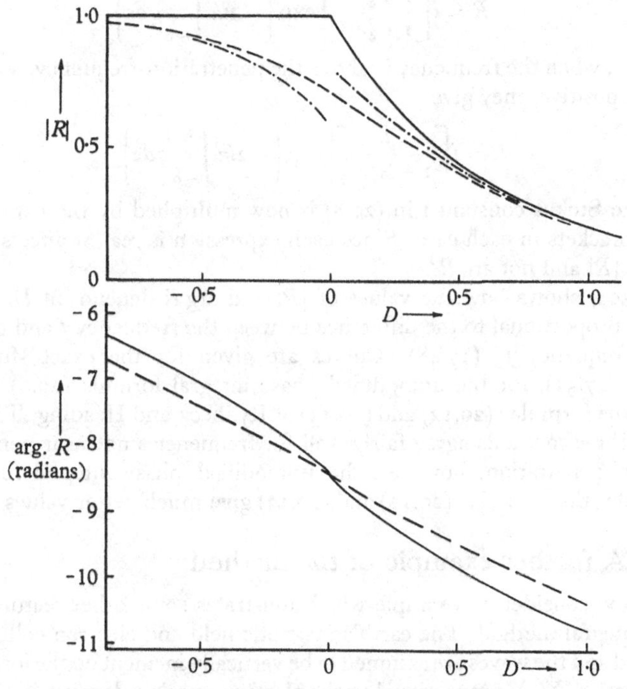

Fig. 20.4. The reflection coefficient R for vertical incidence on a parabolic layer according to the phase integral formula (solid curves), the formulae of Heading and Rydbeck (chain curves), and the correct full wave theory (broken curves). The phase integral formula and the Heading–Rydbeck formulae give the same values of arg R but not of $|R|$. In this example $2\pi a/\lambda_p = ak_p = 10$. (This affects the values of arg R but not of $|R|$.) Electron collisions are neglected.

is unreliable because the branch points at z_1 and z_2 are so close together that there is no region between them where the W.K.B. solutions are good approximations.

In this case it is usual to abandon the phase integral method and use the more exact methods of ch. 17. It is, however, interesting to inquire whether the formula can be amended to allow for the proximity of the second branch point. Methods of doing this have been suggested by Rydbeck (1948), and Heading (1953), who derived the same formulae but by different methods. Their results are, briefly, as follows. Suppose that electron collisions are negligible,† so that $Z = 0$. Let

$$J = -ik\int_{z_1}^{z_2} q\,dz = \frac{\pi}{2}D, \tag{20.11}$$

where $\mathscr{R}(q)$ is positive or zero, and $\mathscr{I}(q)$ is negative or zero, and D is defined by equation (17.44). Now let the frequency be below the penetration-frequency so that J is real and negative. Then Rydbeck and Heading give for the reflection coefficient

$$R = i\left[\frac{1-\frac{1}{4}e^{2J}}{1+\frac{1}{4}e^{2J}}\right]\exp\left\{-2ik\int_{-a}^{z} q\,dz\right\}. \tag{20.12}$$

Similarly, when the frequency is above the penetration-frequency, so that J is real and positive, they give

$$R = i\left[\frac{1}{1+\frac{1}{4}e^{-2J}}\right]\exp\left\{-2ik\int_{-a}^{z_1} q\,dz\right\}. \tag{20.13}$$

Thus the Stokes constant i in (20.8) is now multiplied by the expression in square brackets in each case. Since each expression is real, it affects only the value of $|R|$ and not $\arg R$.

Fig. 20.4 shows how the values of $|R|$ and $\arg R$ depend on D, which is roughly proportional to the difference between the frequency f and the penetration-frequency f_p (17.48). Curves are given for the exact 'full wave' formula (17.51), for the unmodified phase integral formula (20.8) or (20.9) and for the formulae (20.12) and (20.13) of Rydbeck and Heading. They show that all three methods agree fairly well for frequencies not near penetration. Close to penetration, however, the unmodified phase integral formula is unreliable; the formulae (20.12) and (20.13) give much better values of $|R|$.

20.5 A further example of the method

We now consider an example which illustrates some other features of the phase integral method. The earth's magnetic field and electron collisions are neglected and the waves are assumed to be vertically incident on the ionosphere. The function $X(z)$, proportional to the electron number density, is given by

$$X = \tfrac{1}{2}X_0[1+\tanh\{\tfrac{1}{2}(z-z_m)/\sigma\}], \tag{20.14}$$

so that

$$\mathfrak{n} = \left(1-\tfrac{1}{2}X_0[1+\tanh\{\tfrac{1}{2}(z-z_m)/\sigma\}]\right)^{\frac{1}{2}}, \tag{20.15}$$

where σ and X_0 are constant, and $X_0 < 1$. Fig. 20.5 shows how $\mathfrak{n}$ depends on z in this case. It changes monotonically from a constant value unity below the ionosphere to another constant positive value $\mathfrak{n}_2 = (1-X_0)^{\frac{1}{2}}$ at great heights. On a simple ray theory there would be no reflection from this ionosphere since, for real z, $\mathfrak{n}$ is always real and never becomes zero. The full wave theory, using Epstein's method, was given for this example in § 17.15.

The reflection coefficient R will now be found by the phase integral method. The function $\mathfrak{n}$ has branch points, which are zeros, where

$$\tanh\{\tfrac{1}{2}(z-z_m)/\sigma\} = \frac{2}{X_0} - 1,$$

† Heading also considered a case where electron collisions were not negligible.

that is, where $$z = z_m + 2\sigma\left\{\coth\left(\frac{2}{X_0} - 1\right) + (2n+1)i\frac{\pi}{2}\right\},\tag{20.16}$$

where n is a positive or negative integer. There are infinities of $\mathfrak{n}$, which are also branch points, where

$$z = z_m + 2\sigma(2n+1)i\frac{\pi}{2}.\tag{20.17}$$

Fig. 20.6 is a diagram of the complex z-plane showing the positions of these singularities. The phase integral formula is most accurate when the ionosphere is slowly varying, that is when σ is large, and (20.16) and (20.17) show that the singularities are then well separated.

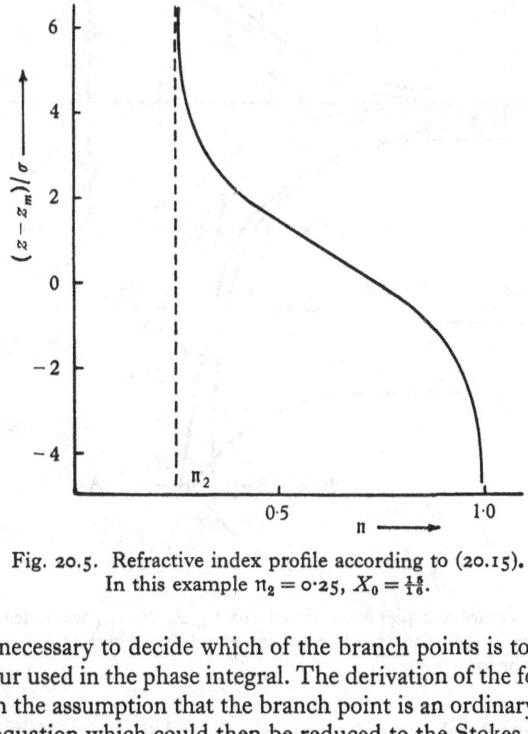

Fig. 20.5. Refractive index profile according to (20.15). In this example $\mathfrak{n}_2 = 0{\cdot}25$, $X_0 = \frac{15}{16}$.

It is now necessary to decide which of the branch points is to be encircled by the contour used in the phase integral. The derivation of the formula (20.1) was based on the assumption that the branch point is an ordinary point of the differential equation which could then be reduced to the Stokes equation (see § 16.6). This shows that one of the zeros must be used, and not one of the infinities. It is shown later that the correct branch point is the zero B_1, below the real axis and nearest to it, and the contour is therefore as shown at C_1 in Fig. 20.6.

In the phase integral $k\int_C \mathfrak{n}\,dz$ the contour starts at the ground, $z = 0$, and here $\mathfrak{n}$ is very slightly less than unity. Let it be $1-\epsilon$. Then

$$(1-\epsilon)^2 = 1 - \tfrac{1}{2}X_0\{1 - \tanh(\tfrac{1}{2}z_m/\sigma)\},\tag{20.18}$$

and since z_m/σ is assumed to be large

$$\epsilon \approx \tfrac{1}{2}X_0 e^{-z_m/\sigma}. \tag{20.19}$$

The contour ends where $\mathfrak{n}$ is equal to $-1+\epsilon$. The integral may be evaluated by changing to $\mathfrak{n}$ as the variable of integration. After some reduction it becomes

$$-2k\sigma\int_{+1-\epsilon}^{-1+\epsilon}\left\{\frac{1-X_0}{\mathfrak{n}^2+X_0-1}-\frac{1}{\mathfrak{n}^2-1}\right\}d\mathfrak{n}, \tag{20.20}$$

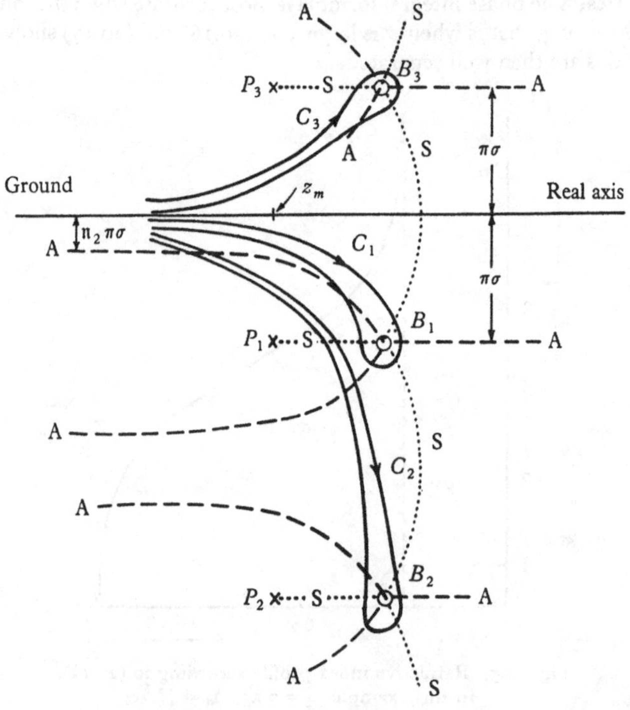

Fig. 20.6. The complex z-plane showing zeros B_1, B_2, B_3, ..., and poles P_1, P_2, P_3, ..., of the refractive index $\mathfrak{n}$ (20.15). The Stokes lines (S) and anti-Stokes lines (A) for the zeros are also shown.

so that the integrand has poles at $\mathfrak{n} = \pm\mathfrak{n}_2$ and $\mathfrak{n} = \pm 1$. Now consider the contour C_1 to be traversed in the direction of the arrow in Fig. 20.6. At first the real part of $\mathfrak{n}$ decreases, and when $\mathscr{I}(z)$ becomes negative $\mathscr{I}(\mathfrak{n})$ becomes positive. Then as the branch point is encircled, both $\mathscr{R}(\mathfrak{n})$ and $\mathscr{I}(\mathfrak{n})$ change sign. Thus the contour to be used in the complex $\mathfrak{n}$-plane is as shown at C_1 in Fig. 20.7. It passes above the pole at $+\mathfrak{n}_2$ and below that at $-\mathfrak{n}_2$. If (20.20) is evaluated directly, and ϵ is neglected in comparison with 1 and $\mathfrak{n}_2$, it gives

$$2k\sigma\mathfrak{n}_2\log\frac{1-\mathfrak{n}_2}{1+\mathfrak{n}_2}-2k\sigma\log\left(\tfrac{1}{2}\epsilon\right),$$

but this is its value when the contour passes entirely above both poles. To it must be added $2\pi i$ times the residue at the pole at $\mathfrak{n} = -\mathfrak{n}_2$. Hence finally

$$\phi \equiv k \int_C \mathfrak{n}\, dz = 2k\sigma\mathfrak{n}_2 \log \frac{1-\mathfrak{n}_2}{1+\mathfrak{n}_2} - 2k\sigma \log(\tfrac{1}{2}\epsilon) - 2\pi i k\sigma\mathfrak{n}_2, \quad (20.21)$$

and (20.3) gives

$$R = ie^{-i\phi} = i\left(\frac{1-\mathfrak{n}_2}{1+\mathfrak{n}_2}\right)^{-2ik\sigma\mathfrak{n}_2} \left(\frac{\epsilon}{2}\right)^{2ik\sigma} e^{-2\pi k\sigma\mathfrak{n}_2}. \quad (20.22)$$

All factors except the last have modulus unity. Hence $|R| = e^{-2\pi k\sigma\mathfrak{n}_2}$, which agrees with the result (17.114) given by Epstein's method when σ is large. The factor $(\frac{1}{2}\epsilon)^{2ik\sigma}$ is equal to $(\frac{1}{4}X_0)^{2ik\sigma} e^{-2ikz_m}$. Now (20.22) gives R as measured at the ground. If R is to be referred to the level $z = z_m$, as it was in § 17.14, the factor e^{-2ikz_m} must be omitted. Then the remaining factors agree with the expression (17.112) when Stirling's approximation is used for the factorial functions. This confirms that the choice of the branch point B_1 (Fig. 20.6) was correct.

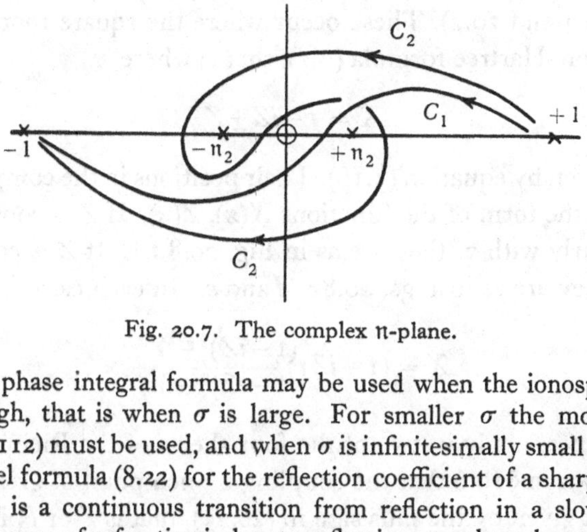

Fig. 20.7. The complex $\mathfrak{n}$-plane.

Thus the phase integral formula may be used when the ionosphere varies slowly enough, that is when σ is large. For smaller σ the more accurate formula (17.112) must be used, and when σ is infinitesimally small this reduces to the Fresnel formula (8.22) for the reflection coefficient of a sharp boundary. Hence there is a continuous transition from reflection in a slowly varying medium at a zero of refractive index, to reflection at a sharp boundary.

Suppose now that the branch point B_2 had been chosen and the contour C_2 used in Fig. 20.6. By studying the changes of $\mathfrak{n}$ it can be shown that the corresponding contour C_2 in the complex $\mathfrak{n}$-plane (Fig. 20.7) has loops round each pole at $\mathfrak{n} = \pm\mathfrak{n}_2$ as shown, and the last term in (20.11) would have to be replaced by $6\pi i k\sigma\mathfrak{n}_2$, which would clearly give the wrong value of R. Alternatively, suppose that the branch point B_3 had been chosen, and the contour C_3 used in Fig. 20.6. Inspection shows that for each element of the path C_3, the increment $-ik\mathfrak{n}\delta z$ in the exponent of the phase integral formula (20.3) has a positive real part so that this contour gives $|R| > 1$ which is obviously wrong. This confirms the result given in § 20.3 that the reflection branch point must not be on the positive imaginary side of the real z-axis. Similar arguments for the other branch points show that B_1 is the only branch point which gives the right value for R.

20.6 Coupling branch points and their Stokes lines and anti-Stokes lines

In the rest of this chapter we shall discuss the propagation of radio waves at vertical incidence when the earth's magnetic field is allowed for. The application of the phase integral method to reflection in this case is very similar to the treatment used in the preceding sections for an isotropic medium, except that there are now two independent waves, ordinary and extraordinary, which must be treated separately. The reflection branch points are also zeros of the two refractive indices $\mathfrak{n}_o$ and $\mathfrak{n}_x$.

In addition, however, there may be coupling between the ordinary and extraordinary waves travelling in the same direction, and to study it we need to investigate the coupling branch points of the function $\mathfrak{n}(z)$ (§§ 19.2, 19.3 and 20.2). These occur where the square root vanishes in the Appleton–Hartree formula (6.1), that is where

$$X = 1 - iZ \pm Z_c, \tag{20.23}$$

Z_c being given by equation (5.16). Their positions in the complex z-plane depend on the form of the functions $X(z)$, $Z(z)$. If Z is constant and X varies linearly with z, they are as in Fig. 20.8 *a*, *b*. If Z is constant, and $X = e^{\alpha z}$, they are as in Figs. 20.8 *c*, *d* and *e*. In each case there is a pole where

$$X = (1 - iZ)\frac{(1 - iZ)^2 - Y^2}{(1 - iZ)^2 - Y_L^2},$$

but this appears only in two of the four sheets of the Riemann surface. It was shown in § 19.6 that the coupling branch point of greater importance is that given by the plus sign in (20.23), that is P or P_1 in Figs. 19.4 and 20.8.

In § 19.4 it was shown that near a coupling point in a slowly varying medium the differential equations can be separated into two sets, one for upgoing waves only and the other for downgoing waves only, and each set can be reduced to the Stokes equation. Here we shall consider upgoing waves; the discussion for downgoing waves is very similar. The variable which satisfies the Stokes equation (19.34) is g_+, defined by (19.29). This in turn is related to the field variables E_x, E_y, $\mathscr{H}_x$, $\mathscr{H}_y$, through the transformations (19.27) and (18.104). Of these the most important is (19.27), for it shows that to obtain E_x, E_y, $\mathscr{H}_x$, $\mathscr{H}_y$ from g_+, the transformations include a multiplication by

$$\exp\left\{-\tfrac{1}{2}ik\int^z (\mathfrak{n}_o + \mathfrak{n}_x)\,dz\right\}.$$

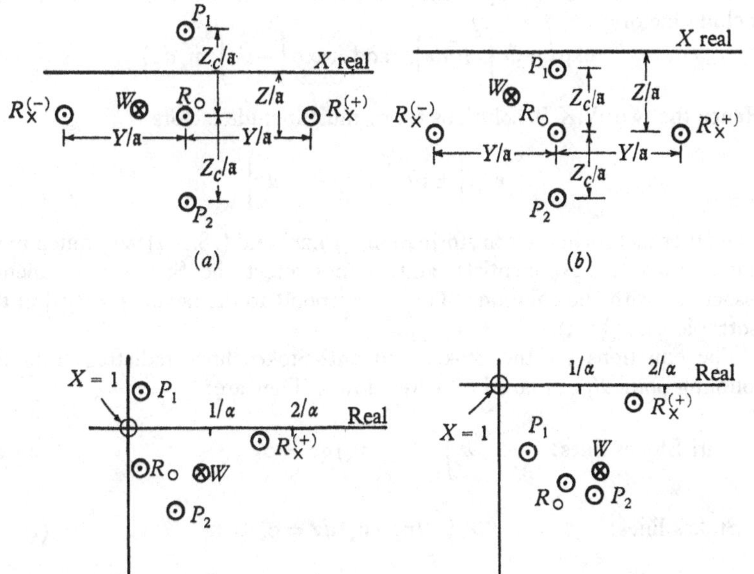

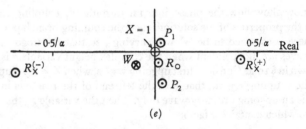

Fig. 20.8. Positions of the singularities of $\mathfrak{n}(z)$ in the complex z-plane, when Z is constant. In (a) and (b) $X = \mathfrak{a}(z-h_0)$. In (c), (d) and (e) $X = e^{\alpha z}$. P_1 and P_2 are coupling branch points, R_0 is the reflection branch point for the ordinary ray, where $X = 1 - iZ$, and $R_X^{(+)}$, $R_X^{(-)}$ are the two reflection branch points for the extraordinary ray where $X = 1 + Y - iZ$ and $1 - Y - iZ$ respectively. W is a pole which appears in only two of the four sheets of the Riemann surface. In (c), (d), (e), each point occurs again repeatedly at intervals 2π in the direction of the imaginary axis. In (a), (c) and (e) $Z < Z_c$, and in (b) and (d) $Z > Z_c$. In (c) $Y = 4$, $Z = \frac{1}{2}Z_c \approx \frac{1}{2}$. In (d) $Y = 4$, $Z = 2Z_c \approx 2$. In (e) $Y = \frac{1}{2}$, $Z = \frac{1}{2}Z_c \approx \frac{1}{2}$.

Now at points which are far enough away from the singularities of n, the W.K.B. solutions must be good approximations, and for the two upgoing waves these include factors

$$\exp\left\{-ik\int^z n_o\,dz\right\} \quad\text{and}\quad \exp\left\{-ik\int^z n_x\,dz\right\}.$$

Hence the two W.K.B. solutions for g_+ must include factors

$$\exp\left\{\pm\tfrac{1}{2}ik\int^z (n_o - n_x)\,dz\right\}.$$

The other factors in the transformations (19.29) and (18.104) vary much more slowly than the exponentials, and do not affect the Stokes phenomenon associated with the solutions. They correspond to the factor $n^{-\frac{1}{2}}$ used in the isotropic case (§9.5).

The equations for the Stokes and anti-Stokes lines radiating from the coupling point P_1 can now be written down. They are:

Anti-Stokes lines: $$\mathscr{R}\int_{z_P}^{z}(n_o - n_x)\,dz = 0, \qquad (20.24)$$

Stokes lines: $$\mathscr{I}\int_{z_P}^{z}(n_o - n_x)\,dz = 0, \qquad (20.25)$$

where z_P is used for the value of z at the coupling point P. From the properties of the Stokes equation it is obvious that there must be three of each.

The positions of these lines have been computed (on EDSAC) for several cases when Z is constant, and some examples are shown in Figs. 20.9*a*, 20.10 and 20.11.

20.7 The phase integral method for coupling

We shall now show how the phase integral formula for coupling can be derived from the properties of the solutions near the coupling branch point P_1. The ionosphere is assumed to be 'slowly varying', so that the upgoing and downgoing waves are independent when the complex height variable z is near P, as was shown in §19.4. Consider first upgoing waves, when $Z < Z_c$. Suppose that below the coupling region, that is at the left end of the real axis in Fig. 20.9*a*, there is an upgoing ordinary wave only. Then the variable g_+ there has only one term which contains a factor

$$\exp\left\{-\tfrac{1}{2}ik\int^z (n_o - n_x)\,dz\right\}.$$

Now at the point of intersection of the Stokes line S_1 with the real x-axis, n_o is almost purely real and n_x is almost negative imaginary whence it easily follows that this term is subdominant. Hence the Stokes diagram can be drawn, and is as shown in Fig. 20.9*b*. It is the Stokes diagram of a function $g_+ = \mathrm{Ai}(\xi)$ where ξ is measured from P_1 and is real for the direction of the Stokes line S_1. Above the ionosphere where X is real, positive and large, we must use the sector between the lines A_3 and S_3 where g_+ has two terms. In this sector g_+ must contain a factor

$$\exp\left\{-\tfrac{1}{2}ik\int_{z_P}^{z}(\mathfrak{n}_o-\mathfrak{n}_x)\,dz\right\}+i\exp\left\{\tfrac{1}{2}ik\int_{z_P}^{z}(\mathfrak{n}_o-\mathfrak{n}_x)\,dz\right\}, \qquad (20.26)$$

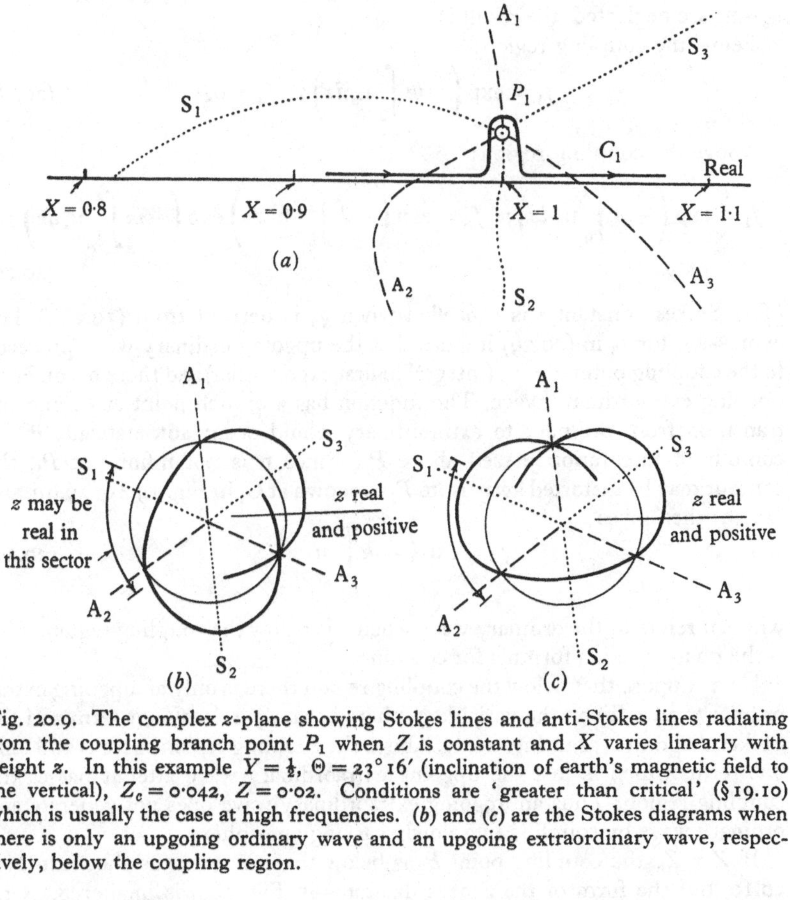

Fig. 20.9. The complex z-plane showing Stokes lines and anti-Stokes lines radiating from the coupling branch point P_1 when Z is constant and X varies linearly with height z. In this example $Y=\frac{1}{2}$, $\Theta=23°16'$ (inclination of earth's magnetic field to the vertical), $Z_c=0{\cdot}042$, $Z=0{\cdot}02$. Conditions are 'greater than critical' (§ 19.10) which is usually the case at high frequencies. (b) and (c) are the Stokes diagrams when there is only an upgoing ordinary wave and an upgoing extraordinary wave, respectively, below the coupling region.

where the Stokes constant i multiplies the term which becomes more dominant on rotating anticlockwise. This result may be expressed in the form of a connection formula (§ 15.24) for g_+, thus:

$$\exp\left\{-\tfrac{1}{2}ik\int_{z_P}^{z}(\mathfrak{n}_o-\mathfrak{n}_x)\,dz\right\}\leftrightarrow\exp\left\{-\tfrac{1}{2}ik\int_{z_P}^{z}(\mathfrak{n}_o-\mathfrak{n}_x)\,dz\right\}$$
$$+\,i\exp\left\{\tfrac{1}{2}ik\int_{z_P}^{z}(\mathfrak{n}_o-\mathfrak{n}_x)\,dz\right\}, \qquad (20.27)$$

Below coupling region Above coupling region

where the same slowly varying factor $[(\rho_o+1)/(\rho_o-1)]^{\frac{1}{2}}$ is omitted from each term. On the right the first term is the upgoing ordinary wave, and the second term is an upgoing extraordinary wave generated by coupling.

Equation (20.27) may be used with the transformations (19.29) and (19.27) to derive expressions for the variables f_1 and f_3 below and above the coupling region. If derivatives of the slowly varying terms $[(\rho_o+1)/(\rho_o-1)]^{\frac{1}{2}}$ and of $\mathfrak{n}_o-\mathfrak{n}_x$ are neglected, the result is

Below the coupling region:

$$f_1 = \exp\left(-ik\int_0^z \mathfrak{n}_o\,dz\right); \quad f_3 = 0. \tag{20.28}$$

Above the coupling region:

$$f_1 = \exp\left(-ik\int_0^z \mathfrak{n}_o\,dz\right); \quad f_3 = \exp\left(-ik\int_0^{z_P} \mathfrak{n}_o\,dz\right)\exp\left(-ik\int_{z_P}^z \mathfrak{n}_x\,dz\right). \tag{20.29}$$

(The Stokes constant i is cancelled when g_3 is derived from (19.29).) The expression for f_3 in (20.29) implies that the upgoing ordinary wave proceeds to the coupling point $z = z_P$ (integral in first exponential) and there becomes an upgoing extraordinary wave. The function has a branch point at P_1 and the transition from ordinary to extraordinary would occur automatically if the contour of integration passed above P_1. Since $\mathfrak{n}$ is not infinite at P_1, the contour may be distorted away from P_1 as shown at C_1 in Fig. 19.4 or 20.9a, and (20.29) then gives

$$f_3 = \exp\left(-ik\int_{0\;(C_1)}^z \mathfrak{n}\,dz\right), \tag{20.30}$$

where $\mathfrak{n}$ refers to the ordinary wave when z is below the coupling region. This is the phase integral formula for coupling.

Next suppose that below the coupling region there is only an upgoing extraordinary wave. Then the variable g_+ there has only a dominant term and the Stokes diagram is as in Fig. 20.9c. Above the coupling region there is still only one term, which is just the upgoing extraordinary wave after it passes the coupling region. Thus an upgoing extraordinary wave does not generate any ordinary wave by coupling in a slowly varying ionosphere.

If $Z > Z_c$ the coupling point P_1 is below the real z-axis as shown in Fig. 20.10, but the form of the Stokes diagrams in Fig. 20.9 is unaltered. One magnetoionic component of the incident wave travels upwards until it is reflected near the level where $X = 1 + Y$. This wave should be called 'ordinary' when $X < 1$ and 'extraordinary' when $X > 1$ and is an example of the ambiguity of nomenclature mentioned in § 6.15. In the Stokes diagram of Fig. 20.9b it is subdominant below the coupling region. The contour C now passes entirely along the real z-axis, and the formula (20.30) simply expresses the change of the (complex) phase of the wave at real heights. It can generate some of the other magnetoionic component by coupling, and the amplitude could be found by using the contour C_2 of Fig. 20.10, but this wave is of little interest since it is like an evanescent wave if $f > f_H$, and if $f < f_H$ it would never be reflected at any level above the coupling region, but would travel upwards until it is absorbed or penetrates right through the ionosphere.

For downcoming waves the treatment is very similar. When $Z < Z_c$ a downcoming extraordinary wave generates some ordinary wave by coupling, but a downcoming ordinary wave does not generate any extraordinary wave. The reader should construct the Stokes diagrams for these cases and show that when there is a downcoming extraordinary wave above the coupling region, given by

$$f_4 = \exp\left(ik \int_a^z \mathfrak{n}_x dz\right)$$

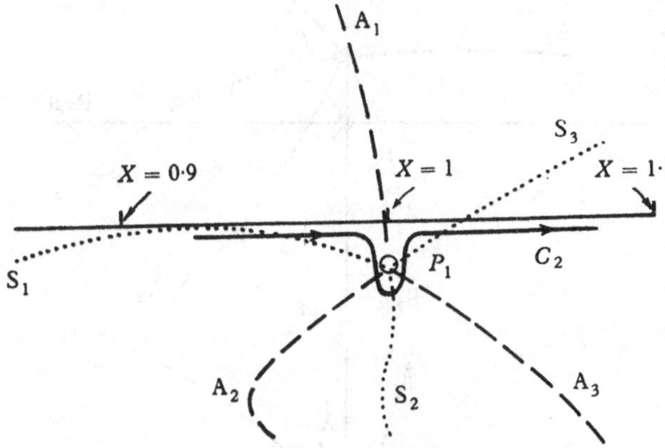

Fig. 20.10. The complex z-plane showing Stokes lines and anti-Stokes lines radiating from the coupling branch point P_1 when Z is constant and X varies linearly with height z. In this example $Y = \frac{1}{2}$, $\Theta = 23° 16'$, $Z_c = 0{\cdot}042$, $Z = 0{\cdot}06$. Conditions are 'less than critical' (§ 19.10).

(a and z are above the coupling region), it generates an ordinary wave given by

$$f_2 = \exp\left(ik \int_{a}^{z} \mathfrak{n}\, dz\right),$$
$$C$$

where z is now below the coupling region, the contour C is as in Fig. 20.9a, and $\mathfrak{n}$ means $\mathfrak{n}_x$ above the coupling region and therefore $\mathfrak{n}_o$ below it.

20.8 Further discussion of the transition through critical coupling (continued from § 19.10)

Fig. 20.11 shows the Stokes lines and anti-Stokes lines radiating from the coupling branch point P_1, when $Y > 1$ and the electron density increases exponentially with height. These curves are therefore typical of what might be expected in the lowest part of the ionosphere for frequencies well below the gyro-frequency. In Fig. 20.11a conditions are greater than critical and in Fig. 20.11b they are less than critical. Suppose now that above the coupling region there is a downgoing wave which was reflected near the level where

$X = 1 + Y$. It can be shown that the variable g_- (§ 19.4) for this wave gives a subdominant term. Below the coupling region there are two downgoing waves which will eventually reach the ground. These are associated with the two refractive indices and have markedly different polarisations, as explained in § 19.10.

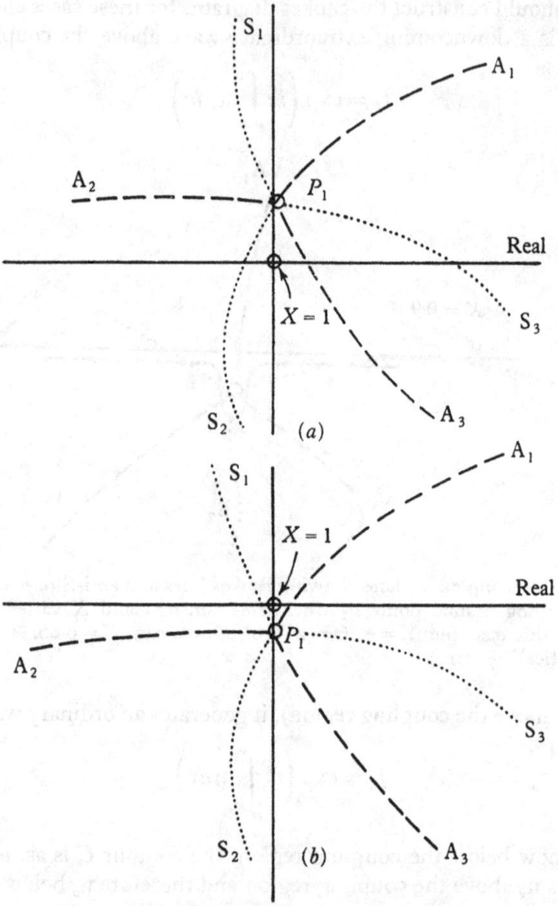

Fig. 20.11. The complex z-plane showing Stokes lines and anti-Stokes lines radiating from the coupling branch point P_1 when Z is constant and $X = e^{\alpha z}$. In this example $Y = 2$, $\Theta = 23°16'$, $Z_c = 0{\cdot}17$. In (*a*) $Z = 0{\cdot}1$, so that conditions are 'greater than critical', and in (*b*) $Z = 0{\cdot}2$ so that conditions are 'less than critical' which is usually the case at very low frequencies.

The Stokes diagram associated with the coupling point P_1 for this case is as in Fig. 20.12. It shows that when conditions change from greater than critical to less than critical, the dominancy of two downgoing waves below the ionosphere is interchanged. This is because the real z-axis in Fig. 20.11 moves across the anti-Stokes line A_2 during the transition through critical coupling.

Hence the wave reaching the ground with greater amplitude must have a polarisation which is markedly different for conditions greater than and less than critical. This confirms the conclusions of § 19.10.

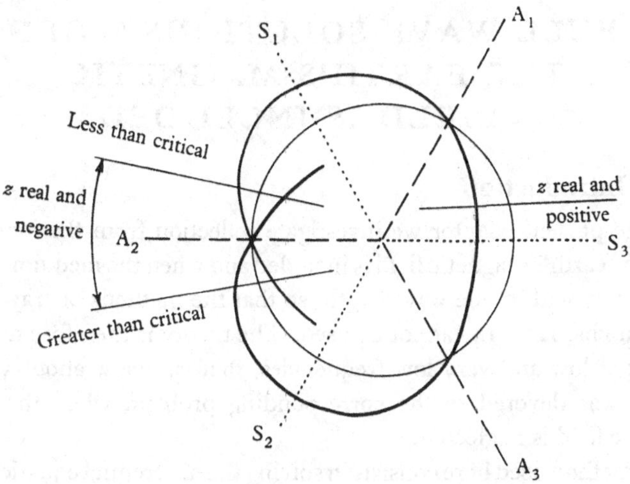

Fig. 20.12. Stokes diagram for coupling when the frequency is less than the gyro-frequency, and a downgoing wave (reflected near where $X = 1 + Y$) is approaching the coupling region from above.

Example

Use the phase integral method to find the reflection coefficient at vertical incidence for an ionosphere in which the earth's magnetic field is neglected, the electron density is constant, and the electron collision-frequency decreases exponentially as the height increases. Compare your result with the 'full wave' theory of § 17.17.

CHAPTER 21

FULL WAVE SOLUTIONS WHEN THE EARTH'S MAGNETIC FIELD IS INCLUDED

21.1 Introduction

In the present chapter we investigate reflection from the ionosphere when the earth's magnetic field is included and when the medium changes appreciably within one wavelength, so that the methods of 'ray theory' given in chs. 12 to 14 cannot be used. The theory is therefore mainly of interest at low and very low frequencies, that is, below about 500 kc/s. Ch. 17 was devoted to the corresponding problem when the earth's magnetic field is neglected.

The method used here consists in solving the differential equations with suitable 'boundary conditions' above the reflection levels, and using the solution below the ionosphere to find the reflection coefficient. In this process there are four steps as described in § 17.1.

The reflection coefficient of the ionosphere depends on the state of polarisation of the incident wave, and in general the reflected and incident waves do not have the same polarisation. It was shown in §§ 7.4 and 7.5 that the reflecting properties of the ionosphere can be completely specified by a set of four (complex) quantities, and the most commonly used set is ${}_{\parallel}R_{\parallel}$, ${}_{\parallel}R_{\perp}$, ${}_{\perp}R_{\parallel}$, ${}_{\perp}R_{\perp}$ defined in § 7.4. In the present chapter expressions for these quantities will be found in some special cases.

Because of the complexity of the differential equations it is necessary to make drastic simplifying assumptions before they can be solved in terms of known functions. The range of ionospheric models that can be studied is therefore much more restricted than for the isotropic ionosphere (ch. 17), but the results illustrate some important general principles and are useful in checking the numerical solutions obtained by the methods of ch. 22.

21.2 The differential equations

The differential equations for the most general horizontally stratified ionosphere have been given in §§ 18.2 and 18.4. In nearly all of the present chapter it will be assumed that the earth's magnetic field is vertical so that the direction cosines of the vector **Y** (§ 3.4) are 0, 0, 1 (northern hemisphere) and the susceptibility matrix **M** (3.24) becomes:

$$\mathbf{M} = -X\begin{pmatrix} b_1 & -ib_2 & 0 \\ ib_2 & b_1 & 0 \\ 0 & 0 & b_4 \end{pmatrix}, \tag{21.1}$$

where $$b_1 = \frac{U}{U^2-Y^2}, \quad b_2 = \frac{Y}{U^2-Y^2}, \quad b_4 = \frac{1}{U} \tag{21.2}$$

(the symbols b_1 to b_4 have been used by several authors—Budden, 1955*a*, *b*; Heading, 1955: b_3 is not used here). Now (21.1) is inserted in (18.10) to (18.13). It is convenient to take $s = kz$ as the independent variable, proportional to height, and to use a dash ′ to denote d/ds. Then the equations become

$$\left.\begin{aligned} E'_x &= i\left(\frac{S^2}{1-Xb_4}-1\right)\mathscr{H}_y, \\ E'_y &= i\mathscr{H}_x, \\ \mathscr{H}'_x &= Xb_2E_x + i(C^2 - Xb_1)E_y, \\ \mathscr{H}'_y &= -i(1-Xb_1)E_x + Xb_2E_y. \end{aligned}\right\} \tag{21.3}$$

They were given substantially in this form by Hartree (1931*b*). The equations apply to oblique incidence when the earth's magnetic field is vertical. By making further assumptions about the functions $X(z)$ and $Z(z)$ Heading and Whipple (1952) were able to solve them in terms of known functions. Before discussing this solution, we shall consider the simpler case of vertical incidence.

21.3 Vertical incidence and vertical magnetic field

For vertical incidence $S = 0$, $C = 1$, and (21.3) become

$$\left.\begin{aligned} E'_x &= -i\mathscr{H}_y, & E'_y &= i\mathscr{H}_x, \\ \mathscr{H}'_x &= Xb_2E_x + i(1-Xb_1)E_y, & \mathscr{H}'_y &= -i(1-Xb_1)E_x + Xb_2E_y, \end{aligned}\right\} \tag{21.4}$$

which can be rearranged thus:

$$\left.\begin{aligned} (E_x + iE_y)' &= -(\mathscr{H}_x + i\mathscr{H}_y), \\ (\mathscr{H}_x + i\mathscr{H}_y)' &= (1 - Xb_1 + Xb_2)(E_x + iE_y), \end{aligned}\right\} \tag{21.5}$$

and $$\left.\begin{aligned} (E_x - iE_y)' &= (\mathscr{H}_x - i\mathscr{H}_y), \\ (\mathscr{H}_x - i\mathscr{H}_y)' &= -(1 - Xb_1 - Xb_2)(E_x - iE_y). \end{aligned}\right\} \tag{21.6}$$

Hence the equations separate into the two independent pairs (21.5) and (21.6). By eliminating $\mathscr{H}_x \pm i\mathscr{H}_y$ the equations can be converted to the two independent second-order equations

$$(E_x+iE_y)''+\{1-Xb_1+Xb_2\}(E_x+iE_y)=0 \qquad (21.7)$$

and

$$(E_x-iE_y)''+\{1-Xb_1-Xb_2\}(E_x-iE_y)=0. \qquad (21.8)$$

This separation is important for it shows that there are two waves which are propagated and reflected independently. For example, suppose that at some level both E_x-iE_y and $\mathscr{H}_x-i\mathscr{H}_y$ are zero. Then (21.6) shows that they remain zero at all levels so that the wave is circularly polarised and the rotation is clockwise to an observer looking in the direction of z increasing. Thus where it is possible to speak of upgoing and downgoing waves, for example below the ionosphere, the upgoing wave has a right-handed sense and the downgoing wave a left-handed sense (see §§ 5.4 and 8.14). The propagation is governed by (21.5) or (21.7) which shows that there is reflection near where $1-Xb_1+Xb_2=0$. Similarly, if E_x+iE_y and $\mathscr{H}_x+i\mathscr{H}_y$ are zero at some level, the wave is again circularly polarised but with the opposite sense. Its propagation is now governed by (21.8) and it is reflected near where $1-Xb_1-Xb_2=0$. In fact the expressions in brackets {} in (21.7) and (21.8) are the two values of $\mathfrak{n}^2$ as given by the Appleton–Hartree formula for purely longitudinal propagation ((6.19); $\mathfrak{n}$ is the refractive index).

In practice the incident wave is often linearly polarised. It must then be resolved into two circularly polarised components whose reflection coefficients are found separately. The reflected waves are then recombined below the ionosphere. An example of this process is given in § 21.5.

21.4 Exponential profile of electron density. Constant collision-frequency

The exponential profile of electron density is important at very low frequencies for the reasons given in § 17.2. As in that section we now take Z to be constant, and $X=e^{\alpha z}$. Then in terms of the height variable $s=kz$

$$X=e^{\beta s}, \quad \text{where} \quad \beta=\alpha/k. \qquad (21.9)$$

The equations (21.7) and (21.8) may be considered together. Let

$$F_r=E_x+iE_y, \quad F_l=E_x-iE_y, \qquad (21.10)$$

$$b_1-b_2=\frac{1}{U+Y}=b_r, \quad b_1+b_2=\frac{1}{U-Y}=b_l, \qquad (21.11)$$

where the subscripts r and l indicate that the incident wave has right- or left-handed circular polarisation, respectively. Then both (21.7) and (21.8) may be written

$$F'' + (1 - b\,e^{\beta s})F = 0, \tag{21.12}$$

where F and b are to have the same subscript, r or l. This is of the same form as (17.3) and may be solved in the same way, in terms of Bessel functions. The process is the same as in § 17.2 and need only be given briefly here. Let

$$\zeta = \frac{2i}{\beta} b^{\frac{1}{2}} e^{\frac{1}{2}\beta s}, \quad \nu = 2i/\beta, \tag{21.13}$$

where $b^{\frac{1}{2}}$ is chosen so that its imaginary part is positive. Then (21.12) reduces to Bessel's equation (17.6) and a solution must be found which represents an upgoing wave only, at great heights. The correct solution is

$$F = H_\nu^{(1)}(\zeta), \tag{21.14}$$

whose asymptotic form contains a factor $\exp\{-(2/\beta) b^{\frac{1}{2}} e^{\frac{1}{2}\beta s}\}$ which clearly represents an upgoing wave since $\mathscr{I}(b^{\frac{1}{2}})$ is positive.

Below the ionosphere the function (21.14) is separated into upgoing and downgoing waves by using the formulae (17.14) and (17.15), which, with (21.13) show that when s is large and negative F is proportional to

$$e^{-is} - b^\nu \beta^{-2\nu} \frac{(-\nu)!}{\nu!} e^{is}, \tag{21.15}$$

and the reflection coefficient for one circularly polarised component wave, referred to the level $s = 0$, is

$$R = -b^\nu \beta^{-2\nu} \frac{(-\nu)!}{\nu!}. \tag{21.16}$$

Since ν is purely imaginary, the two factorial functions are complex conjugates and their ratio has modulus unity. The factor $\beta^{-2\nu}$ also has modulus unity. Hence

$$|R| = |b^{2i/\beta}| = \exp\left(-\frac{2}{\beta}\arg b\right). \tag{21.17}$$

Now let b take the value b_r, (21.11). Then the reflection coefficient R becomes ${}_rR_l$, where the first subscript r denotes that the incident wave has right-handed circular polarisation and the second subscript l denotes that the reflected wave has left-handed circular polarisation, as explained in § 7.7, the absolute direction of rotation being the same for both waves. Now $\arg b_r = \arctan Z/(1 + Y)$ and (21.17) gives

$$|{}_rR_l| = \exp\left(-\frac{2}{\beta}\arctan\frac{Z}{1+Y}\right). \tag{21.18}$$

If electron collisions are neglected so that $Z = 0$, then $|{}_rR_l| = 1$, so that the wave is totally reflected. For this case the refractive index is zero where $X = 1 + Y$, and according to the 'ray theory' of ch. 12 there should be total reflection at this level in a slowly varying medium. We have now shown that for an exponential profile when collisions are neglected there is total reflection of this component even when the medium is not slowly varying.

Next let b take the value b_l, (21.11), so that R becomes ${}_lR_r$. Then $\arg b_l = \arctan Z/(1-Y)$, and (21.17) gives

$$|{}_lR_r| = \exp\left(-\frac{2}{\beta}\arctan\frac{Z}{1-Y}\right). \tag{21.19}$$

For frequencies above the gyro-frequency $Y < 1$, and if electron collisions are again neglected then $|{}_lR_r| = 1$ for this component also. Here the refractive index becomes zero where $X = 1 - Y$.

These formulae are more interesting, however, for low frequencies so that $Y > 1$. The condition $\mathscr{I}(b^{\frac{1}{2}}) > 0$ implies that $\arg b$ is in the range 0 to π so that (21.19) now gives

$$|{}_lR_r| = \exp\left(-\frac{2\pi}{\beta} + \frac{2}{\beta}\arctan\frac{Z}{Y-1}\right) \quad \text{for} \quad Y > 1. \tag{21.20}$$

If $Z = 0$, $|{}_lR_r| = e^{-2\pi/\beta}$. In a slowly varying medium β is small so that the reflection coefficient for this component is small. Now the condition $X = 1 - Y$ is not obeyed at any real height and according to the 'ray theory' of ch. 12 there should be no reflection of this component in a slowly varying medium. The full wave theory shows that for an exponential profile there is some reflection, which is small in a slowly varying medium but becomes larger when the medium varies more quickly.

21.5 Exponential profile (continued). Incident wave linearly polarised

Equation (21.16) gives the two reflection coefficients ${}_rR_l$ and ${}_lR_r$ when the incident wave is circularly polarised. It is also evident that ${}_rR_r = {}_lR_l = 0$ so that the matrix $\mathbf{R}_0$, (7.21), is

$$\mathbf{R}_0 = \begin{pmatrix} 0 & {}_lR_r \\ {}_rR_l & 0 \end{pmatrix}. \tag{21.21}$$

If the incident wave is linearly polarised, the reflecting properties of the ionosphere are conveniently specified by ${}_{\parallel}R_{\parallel}$, ${}_{\parallel}R_{\perp}$, ${}_{\perp}R_{\parallel}$, ${}_{\perp}R_{\perp}$, that is by the matrix **R**, (7.19). This is easily found from the transformation (7.29), which gives

$$\mathbf{R} = \mathbf{U}\mathbf{R}_0\mathbf{U}^{-1} = \frac{1}{2}\begin{pmatrix} {}_rR_l + {}_lR_r & i\,{}_lR_r - i\,{}_rR_l \\ i\,{}_lR_r - i\,{}_rR_l & -{}_rR_l - {}_lR_r \end{pmatrix}, \tag{21.22}$$

so that

$$\left.\begin{aligned} {}_{\parallel}R_{\parallel} &= -{}_{\perp}R_{\perp} = \tfrac{1}{2}({}_rR_l + {}_lR_r), \\ {}_{\parallel}R_{\perp} &= {}_{\perp}R_{\parallel} = \tfrac{1}{2}i({}_lR_r - {}_rR_l). \end{aligned}\right\} \tag{21.23}$$

When the two results (21.16) are used this gives

$$\left.\begin{aligned} {}_{\parallel}R_{\parallel} &= -{}_{\perp}R_{\perp} = -\tfrac{1}{2}\beta^{-2\nu}\frac{(-\nu)!}{\nu!}(b_r^{\nu} + b_l^{\nu}), \\ {}_{\perp}R_{\parallel} &= {}_{\parallel}R_{\perp} = \tfrac{1}{2}i\beta^{-2\nu}\frac{(-\nu)!}{\nu!}(b_r^{\nu} - b_l^{\nu}). \end{aligned}\right\} \tag{21.24}$$

These formulae are of interest in the special case of very low frequencies when $Y \gg 1$ (for example, at 16 kc/s, $Y \approx 80$). If 1 is neglected compared with Y in (21.11)

$$b_r = \frac{1}{Y - iZ}, \quad b_l = \frac{1}{-Y - iZ}, \tag{21.25}$$

so that $|b_r| = |b_l|$ and b_r^{ν}, b_l^{ν} have the same argument (since ν is purely imaginary). Then (21.24) shows that ${}_{\parallel}R_{\parallel}$ and ${}_{\parallel}R_{\perp}$ are in quadrature, so that when the incident wave is linearly polarised, the polarisation ellipse of the reflected wave has its major axis in the plane of polarisation of the incident wave. It can be shown that in the northern hemisphere the rotation in the reflected wave has a left-handed sense. Further

$$\left.\begin{aligned} |{}_{\parallel}R_{\parallel}| &= \frac{1}{2}\left[\exp\left(-\frac{2}{\beta}\arctan\frac{Z}{Y}\right) + \exp\left\{-\frac{2}{\beta}\left(\pi - \arctan\frac{Z}{Y}\right)\right\}\right] \\ &= e^{-\pi/\beta}\cosh\left\{\left(\pi - 2\arctan\frac{Z}{Y}\right)\Big/\beta\right\}, \\ |{}_{\parallel}R_{\perp}| &= e^{-\pi/\beta}\sinh\left\{\left(\pi - 2\arctan\frac{Z}{Y}\right)\Big/\beta\right\}. \end{aligned}\right\} \tag{21.26}$$

These results were given by Stanley (1950) who derived them in a different way as a limiting case of Epstein's method (see end of § 17.14). They are important because in observations using a frequency of 16 kc/s at nearly vertical incidence (Bracewell *et al.* 1951) it was found that the reflected wave was elliptically polarised with a left-handed sense and with the major axis of the ellipse parallel to the plane of polarisation of the incident wave.

When β is very small (21.26) shows that $|_{\parallel}R_{\parallel}| \approx |_{\parallel}R_{\perp}|$ so that the reflected wave is nearly circularly polarised. This is the case of the very slowly varying medium in which one circularly polarised component of the incident wave is practically completely absorbed.

21.6 Other electron density profiles for vertical field and vertical incidence

The two differential equations (21.7) and (21.8) are of the same form as (9.58), which was solved for various electron density profiles, in ch. 17, where the earth's magnetic field was neglected. Thus all the profiles used in ch. 17 can also be used for vertical incidence when the earth's magnetic field is also vertical. For example, if the electron density has the parabolic profile of § 17.4, the differential equation (17.26) is replaced by two equations obtained by putting F_r, F_l for E_y, $1 \pm Y - iZ$ for $1 - iZ$, and $C^2 = 1$. The two reflection coefficients ${}_rR_l$ and ${}_lR_r$ are then found exactly as in § 17.4 and they can be combined to give ${}_{\parallel}R_{\parallel}$, ${}_{\parallel}R_{\perp}$, ${}_{\perp}R_{\parallel}$, ${}_{\perp}R_{\perp}$ by the method of the last section. Some results for this case were given by Pfister (1949).

Wilkes (1940) has discussed the case where the electron density is proportional to the square of the height, as in § 17.7. He also discussed the case where X is constant and $Z \propto 1/z$, which can be solved in terms of confluent hypergeometric functions. For a similar model see the example at the end of ch. 20.

21.7 Vertical magnetic field and oblique incidence. Introduction to Heading and Whipple's method

When the waves are incident obliquely on the ionosphere the differential equations (21.3) cannot in general be separated into two second-order equations, but are equivalent to a single fourth-order equation. There are very few cases where the properties of fourth-order equations have been studied; most of the functions of mathematical physics are solutions of second-order equations. Hence it is necessary to make some approximations before progress is possible.

A very important solution of (21.3) was given by Heading and Whipple (1952) who made the following assumptions:

(1) $|1 - iZ| \ll Y$. Thus the collision-frequency is very small compared with the gyro-frequency.

(2) X increases monotonically with the height z.

(3) At the level where $X \approx |1 - iZ|$, Z is constant and X is proportional to $e^{\alpha z}$ where α is constant.

(4) At the level where $X \approx Y$, X is proportional to $e^{\gamma z}$ where γ is another constant. Here Z is negligible because of (1).

The assumption (1) is probably not true for low frequencies in the actual ionosphere, so that Heading and Whipple's results are not immediately applicable to the observations, but they illustrate some important physical principles and they provide a most valuable check of results obtained by the numerical methods of ch. 22.

It is now convenient to divide the ionosphere into five regions, as shown in Fig. 21.1. The rather unusual numbering is used to agree with Heading and Whipple who speak only of regions I and II. In the lowest

Region II(*a*)	$X \gg Y$
Region II	$\lvert 1-iZ\rvert \ll X \approx Y$ (Partially reflecting)
Region I(*a*)	$\lvert 1-iZ\rvert \ll X \ll Y$
Region I	$X \approx \lvert 1-iZ\rvert$ (Partially reflecting)
Region O	$X \ll \lvert 1-iZ\rvert \ll Y$ (Like free space)
	Ground

Fig. 21.1. The regions of the ionosphere as used in Heading and Whipple's method.

region, O, $X \ll |1-iZ| \ll Y$ so that both refractive indices are very close to unity; this is therefore simply the free space below the ionosphere. In the highest region II(a), X is very large and both refractive indices are large, being roughly proportional to $X^{\frac{1}{2}}$, so that the W.K.B. solutions are good approximations. This is above all the reflecting levels so that the solution must contain only upgoing waves. Regions I and II are the reflecting regions and are discussed in later sections. In the intermediate region I(a) there is no reflection but the waves have some interesting features described in the next section.

21.8 Regions O, I and I(a)

In the three lowest regions the conditions $X \ll Y$, and $|1-iZ| \ll Y$ both hold, and (21.2) shows that b_1 and b_2 are then negligible. Then (21.3) thus become:

$$E'_x = i\left(\frac{S^2}{1-X/U}-1\right)\mathscr{H}_y, \quad \mathscr{H}'_y = -iE_x, \tag{21.27}$$

and

$$E'_y = i\mathscr{H}_x, \quad \mathscr{H}'_x = iC^2E_y. \tag{21.28}$$

They have separated into two independent pairs. The pair (21.28) refers to linearly polarised waves with the electric vector horizontal, and shows that these waves are propagated just as though regions O, I and I(*a*) were free space.

The pair (21.27) refers to linearly polarised waves with the electric vector in the plane of incidence. It is shown in the next section that these can be partially reflected from region I, where $X \approx |1-iZ|$.

We now examine the solutions of (21.27) in region I(*a*). Here $|X/U| \gg 1$ so that the term $S^2/[1-(X/U)]$ is negligible and the equations take the very simple form

$$E'_x = -i\mathscr{H}_y, \quad \mathscr{H}'_y = -iE_x. \tag{21.29}$$

In formulating the original equations it was assumed (§ 18.2) that all field quantities include a factor e^{-ikSx}. If this factor is restored, a solution of (21.29) is

$$E_x = \mathscr{H}_y = \exp\{-ik(Sx+z)\}, \tag{21.30}$$

which represents a wave travelling obliquely upwards with phase velocity $c(1+S^2)^{-\frac{1}{2}}$ so that the refractive index is $(1+S^2)^{\frac{1}{2}}$. Its wave-normal is at an angle θ to the vertical where $\tan\theta = S = \sin\theta_I$ (θ_I is the angle of incidence below the ionosphere). The vertical component E_z of the electric field can be found from (18.9) which, with (21.1) and (21.2), gives $E_z = -S\mathscr{H}_y/[1-(X/U)]$ which is negligible because $|X/U| \gg 1$. Hence the electric vector of the wave is horizontal and in the plane of incidence. There is a second solution of (21.29), namely

$$E_x = -\mathscr{H}_y = \exp\{-ik(Sx-z)\}, \tag{21.31}$$

which represents a wave travelling obliquely downwards. Provided that $|U| \ll X \ll Y$, the propagation of these waves is unaffected by the electron density.

The above results may be summarised in more physical language as follows. The assumption $Y \gg X$ means that the earth's magnetic field is extremely large so that electrons are prevented from moving across it. Hence electrons can only move vertically, and will not respond appreciably to forces tending to move them horizontally. This explains why a wave with its electric vector horizontal is unaffected by the electrons, so that the three lower regions behave like free space for this wave. In region II and above, however, X is comparable with Y and the electrons are now so numerous that they can begin to affect this wave.

For a wave with its electric vector in the plane of incidence there is a force eE_z tending to move the electrons vertically. But in region I(*a*), where $X \gg |1-iZ|$, the electrons are so numerous that the medium effectively has infinite conductivity in the z-direction. Hence it cannot sustain a vertical component of the electric field and the only electric field is horizontal as in (21.30) and (21.31). In region O there are so few electrons that the medium is like free space. The transition occurs in region I which is a region of finite conductivity increasing as the height increases.

21.9 Reflection and transmission coefficients of region I

In region I it is convenient to treat $\mathscr{H}_y$ as the dependent variable. Let $z = z_1$ be the level where $X = |1-iZ|$. Then the assumption (3), §21.7, gives $X = |1-iZ|\, e^{\alpha(z-z_1)}$. Further, let $\chi = \arctan Z = -\arg(1-iZ)$. Then

$$\frac{X}{1-iZ} = \exp\{\alpha(z-z_1)+i\chi\}$$

and, since $s = kz$, (21.27) become

$$\frac{d^2\mathscr{H}_y}{dz^2}+k^2\left(1-\frac{S^2}{1-\exp\{\alpha(z-z_1)+i\chi\}}\right)\mathscr{H}_y = 0. \tag{21.32}$$

This is of the form considered by Epstein and discussed in §§ 17.9 to 17.17 (compare (17.90)). The expression in brackets in (21.32) may be called the effective value of q^2 and should be compared with (17.91), with which it is identical provided that we take

$$\epsilon_1 = C^2,\quad \epsilon_2 = 1,\quad \epsilon_3 = 0,\quad \sigma = 1/\alpha,\quad \mathfrak{b} = -\alpha z_1+i\chi-i\pi. \tag{21.33}$$

In the last of these expressions no other multiple of π is permissible because of the condition (17.78). Equations (17.92), (17.93) and (17.94) can now be used to find a, b and c. Next, (17.106) to (17.109) are used to give the reflection and transmission coefficients. In the present problem these apply to the magnetic field $\mathscr{H}_y$ of the wave, whereas the conventional coefficients refer to the electric field. For the reflection coefficients this makes no difference, but for the transmission coefficients it must be remembered that below region I: $E_x = \pm C\mathscr{H}_y$, and above it $E_x = \pm\mathscr{H}_y$ ((21.30) and (21.31)). Hence the result (17.107) must be multiplied by C, and (17.109) by $1/C$. We shall use small letters r, t and r', t' to denote the reflection and transmission coefficients of region I referred to the level $X = |U|$, that is $z = z_1$, for the electric field of waves incident from below (r, t), and from above (r', t'). The sign convention used is given in §7.5. The results are

$$r = -\frac{1-C}{1+C}\frac{(-2ikC/\alpha)!}{(2ikC/\alpha)!}\left[\frac{\{ik(C+1)/\alpha\}!}{\{ik(1-C)/\alpha\}!}\right]^2\exp\{2kC(\pi-\chi)/\alpha\}, \tag{21.34}$$

$$t = \frac{2C}{1+C}\frac{[\{ik(C+1)/\alpha\}!]^2}{(2ikC/\alpha)!\,(2ik/\alpha)!}\exp\{-k(1-C)(\pi-\chi)/\alpha\}, \tag{21.35}$$

$$r' = \frac{1-C}{1+C}\frac{(-2ik/\alpha)!}{(2ik/\alpha)!}\left[\frac{\{ik(C+1)/\alpha\}!}{\{ik(C-1)/\alpha\}!}\right]^2\exp\{-2k(\pi-\chi)/\alpha\}, \tag{21.36}$$

$$t' = \frac{2}{C+1}\frac{[\{ik(C+1)/\alpha\}!]^2}{(2ikC/\alpha)!\,(2ik/\alpha)!}\exp\{-k(1-C)(\pi-\chi)/\alpha\}. \tag{21.37}$$

These show that when $C = 1$ (vertical incidence) there is complete transmission and no reflection, which confirms that region I behaves like free space when the electric field is entirely horizontal. Their moduli can be found by applying the formula $x!(-x)! = \pi x/\sin(\pi x)$, whence it can be shown that

$$\left.\begin{aligned} |r| &= \frac{\sinh\{\pi k(1-C)/\alpha\}}{\sinh\{\pi k(1+C)/\alpha\}}\exp\{2kC(\pi-\chi)/\alpha\}, \\ |r'| &= \frac{\sinh\{\pi k(1-C)/\alpha\}}{\sinh\{\pi k(1+C)/\alpha\}}\exp\{-2k(\pi-\chi)/\alpha\}, \\ C\,|t'| = |t| &= C^{-\frac{1}{2}}\frac{\{\sinh(2\pi k/\alpha)\sinh(2\pi kC/\alpha)\}^{\frac{1}{2}}}{\sinh\{\pi k(1+C)/\alpha\}}\exp\{-k(1-C)(\pi-\chi)/\alpha\}. \end{aligned}\right\} \tag{21.38}$$

If $C \neq 1$ and α is very small so that the medium is slowly varying, these reduce to

$$\left.\begin{aligned} |r| \approx \exp\{-2kC\chi/\alpha\}, \quad |r'| \approx \exp\{-2k(\pi C+\pi-\chi)/\alpha\}, \\ |t'| = C\,|t| = C^{-\frac{1}{2}}\exp\{-k(1-C)(\pi-\chi)/\alpha\}. \end{aligned}\right\} \tag{21.39}$$

If electron collisions are neglected, $\chi = 0$, and then $|r| \approx 1$ but $|r'|$, $|t|$ and $|t'|$ remain very small. Thus, when waves are incident from below region I, all their energy is reflected, but when they are incident from above it, nearly all the energy is absorbed. This is at first surprising since there is no physical mechanism for converting electrical energy into heat. The explanation is given later, §21.16. It is connected with the fact that the refractive index is infinite at one level in region I.

21.10 Regions II and II(*a*)

In the two upper regions X is comparable with or greater than Y, so that Xb_1, Xb_2 in (21.3) are no longer negligible. Since $|U| \ll Y$, we have from (21.2)

$$Xb_1 = -\frac{XU}{Y^2}, \quad Xb_2 = -\frac{X}{Y}, \tag{21.40}$$

and since $S^2/(1-Xb_4) \ll 1$ the differential equations (21.3) become:

$$\left.\begin{aligned} E_x'' + \left(1+\frac{XU}{Y^2}\right)E_x &= i\frac{X}{Y}E_y, \\ E_y'' + \left(C^2+\frac{XU}{Y^2}\right)E_y &= -i\frac{X}{Y}E_x. \end{aligned}\right\} \tag{21.41}$$

According to simple ray theory we expect some reflection to occur near the level where X and Y are comparable, that is in region II. Above this level, where $X \gg Y$, both refractive indices are of the order $(X/Y)^{\frac{1}{2}}$ and are large compared to unity and we expect the W.K.B. solutions to be good approximations. Thus there is no reflection above region II.

Since $|U| \ll Y$ in region II, the terms XU/Y^2 are neglible compared with 1 and C^2, and may be omitted. Then the equations become:

$$\left.\begin{aligned} E_x'' + E_x &= i\frac{X}{Y}E_y, \\ E_y'' + C^2 E_y &= -i\frac{X}{Y}E_x. \end{aligned}\right\} \tag{21.42}$$

It might be thought that there is a level just above region II where XU/Y^2 is comparable with 1 and must be retained. But this is already above the reflection level as can be verified by studying the second-order coupled equations for this case (see § 18.16).

21.11 The reflection coefficients of region II

The two equations (21.42) are equivalent to a single fourth-order equation. They were solved by Heading and Whipple (1952)† for the special case where X is proportional to $e^{\gamma z}$. The details of the solution are not given here since the mathematics is rather long and specialised, and is given fully and clearly in the original paper. In this section we give an outline of the method and then quote the results. The treatment is in many respects similar to the theory of §§ 17.10 to 17.13.

Suppose that $z = z_2$ is the level where $X/Y = 4\gamma^2/k^2$. Then at other levels

$$\frac{X}{Y} = \frac{4\gamma^2}{k^2}\exp\{\gamma(z-z_2)\}. \tag{21.43}$$

Introduce the new independent variable $w = \exp\{2\gamma(z-z_2)\}$ and eliminate one of the dependent variables, say E_x, from (21.42). Then they are converted to a single fourth-order differential equation for E_y, which can be solved as a series in ascending powers of w. In this way four independent series solutions are obtained for E_y, and with each is associated another similar series for E_x. In the lowest part of region II $|w|$ is very small so that only the first terms of these series are important, and these are shown in the following table:

$$\left.\begin{array}{ll} \quad E_x & \quad E_y \\ \text{Solution 1:} & \\ \exp\{ik(z-z_2)\}, & \exp\{\gamma(z-z_2)\}\exp\{ik(z-z_2)\}, \\ \text{Solution 2:} & \\ \exp\{-ik(z-z_2)\}, & \exp\{\gamma(z-z_2)\}\exp\{-ik(z-z_2)\}, \\ \text{Solution 3:} & \\ \exp\{\gamma(z-z_2)\}\exp\{ikC(z-z_2)\}, & \exp\{ikC(z-z_2)\}, \\ \text{Solution 4:} & \\ \exp\{\gamma(z-z_2)\}\exp\{-ikC(z-z_2)\}, & \exp\{-ikC(z-z_2)\}. \end{array}\right\} \tag{21.44}$$

† They were also studied by Wilkes (1947) for the special case when X is a linear function of z. He showed that the solution can then be expressed as the sum of a number of contour integrals.

Here the factor $\exp\{\gamma(z-z_2)\}$ is very small at the bottom of region II so that solutions 1 and 2 represent waves with the electric vector in the plane of incidence, and are to be identified with the waves in region I (*a*) given by equations (21.31) (downgoing) and (21.30) (upgoing) respectively. Similarly, solutions 3 and 4 represent linearly polarised waves with the electric vector horizontal, travelling as though in free space.

A linear combination of the above four solutions must now be taken in such a way that the physical conditions at great heights are satisfied. Here the four W.K.B. solutions are good approximations and two of them represent upgoing waves. These give two independent linear combinations of the above solutions, which may be written $E_x = J$, $E_x = K$. Then the most general solution which satisfies the physical conditions is $E_x = J + AK$, where A is an arbitrary constant.

To find the combinations J, K, Heading and Whipple convert the series solutions into integrals of Barnes's type (§ 17.11) whose asymptotic values for large $|w|$ can be found by the method of steepest descents. In this way they show that the ratios of the four solutions (21.44) must be

$$\begin{aligned}&(1+A\,e^{-2\pi a})(-2ia-1)!\,(-\tfrac{1}{2}+ib-ia)!\,(-\tfrac{1}{2}-ib-ia)!\\&:(1+A\,e^{2\pi a})(2ia-1)!\,(-\tfrac{1}{2}+ib+ia)!\,(-\tfrac{1}{2}-ib+ia)!\\&:i(1-A\,e^{-2\pi b})(-2ib-1)!\,(-\tfrac{1}{2}+ia-ib)!\,(-\tfrac{1}{2}-ia-ib)!\\&:i(1-A\,e^{2\pi b})(2ib-1)!\,(-\tfrac{1}{2}+ia+ib)!\,(-\tfrac{1}{2}-ia+ib)!,\end{aligned} \tag{21.45}$$

where
$$a = k/2\gamma, \quad b = kC/2\gamma. \tag{21.46}$$

Now the components of the reflection coefficient matrix for region II can be found. They are referred to the level $z = z_2$ and are denoted by ${}_\parallel\rho_\parallel$, ${}_\parallel\rho_\perp$, ${}_\perp\rho_\parallel$, ${}_\perp\rho_\perp$ to distinguish them from ${}_\parallel R_\parallel$, etc., which are used for the ionosphere as a whole.

For example, ${}_\parallel\rho_\parallel$ and ${}_\parallel\rho_\perp$ give the reflected wave when the incident wave has no electric vector perpendicular to the plane of incidence. Hence, in (21.45), solution 4 must have zero amplitude which gives $A = e^{-2\pi b}$. Then ${}_\parallel\rho_\parallel$ is the ratio of the values of E_x in solutions 1 and 2, and ${}_\parallel\rho_\perp$ is the ratio of the values of E_y in solution 3 to E_x in solution 2. Thus the four components of the reflection coefficient are: †

$$\left.\begin{aligned}{}_\parallel\rho_\parallel &= -\frac{1+\exp\{-2\pi(a+b)\}}{1+\exp\{2\pi(a-b)\}}\,\frac{(-2ia-1)!\,(-\tfrac{1}{2}-ia+ib)!\,(-\tfrac{1}{2}-ia-ib)!}{(2ia-1)!\,(-\tfrac{1}{2}+ia+ib)!\,(-\tfrac{1}{2}+ia-ib)!},\\ {}_\parallel\rho_\perp &= -i\frac{1-\exp(-4\pi b)}{1+\exp\{2\pi(a-b)\}}\,\frac{(-2ib-1)!\,(-\tfrac{1}{2}-ia-ib)!}{(2ia-1)!\,(-\tfrac{1}{2}+ia+ib)!},\\ {}_\perp\rho_\parallel &= -i\frac{1-\exp(-4\pi a)}{1+\exp\{2\pi(b-a)\}}\,\frac{(-2ia-1)!\,(-\tfrac{1}{2}-ia-ib)!}{(2ib-1)!\,(-\tfrac{1}{2}+ia+ib)!},\\ {}_\perp\rho_\perp &= \frac{1+\exp\{-2\pi(a+b)\}}{1+\exp\{2\pi(b-a)\}}\,\frac{(-2ib-1)!\,(-\tfrac{1}{2}+ia-ib)!\,(-\tfrac{1}{2}-ia-ib)!}{(2ib-1)!\,(-\tfrac{1}{2}+ia+ib)!\,(-\tfrac{1}{2}-ia+ib)!}.\end{aligned}\right\} \tag{21.47}$$

† Heading and Whipple adopt a sign convention for the reflection coefficients which is different from that used here (see §7.5), so that their symbols ρ_{xx}, etc., have the following meanings;

$$\rho_{xx} = -\,{}_\parallel\rho_\parallel, \quad \rho_{xy} = -\,{}_\parallel\rho_\perp, \quad \rho_{yx} = {}_\perp\rho_\parallel, \quad \rho_{yy} = {}_\perp\rho_\perp.$$

For vertical incidence the differential equations (21.42) separate into two independent second-order equations which can be solved by the method of §§21.4 and 21.5. In this case $a = b$ and it is easily shown that the formulae (21.47) are then the same as (21.24).

In deriving the formulae (21.47) it was assumed that the earth's magnetic field is vertical. Heading (1955) has shown that the equations for region II can also be solved by the same method for oblique incidence when the earth's magnetic field is oblique and in the plane of incidence, that is for propagation from (magnetic) north to south or south to north. The equations for region I, however, cannot be solved by the method of §21.9 for this more general case, and there it is necessary to assume that either the earth's magnetic field or the wave-normal is vertical. The solutions for vertical incidence with oblique magnetic field have been given in full by Heading (1955).

21.12 The combined effect of regions I and II

The reflection coefficients (21.47) of region II must now be combined with the reflection and transmission coefficients (21.34) to (21.37) of region I to give the resultant reflection coefficients of the whole ionosphere. It must be remembered that the coefficients (21.47) are referred to the level $z = z_2$, whereas those in (21.34) to (21.37) are referred to the level $z = z_1$. The reflection coefficients ${}_{\parallel}R_{\parallel}$, ${}_{\parallel}R_{\perp}$, ${}_{\perp}R_{\parallel}$, ${}_{\perp}R_{\perp}$ for the whole ionosphere will be referred to the level $z = z_1$.

Below the ionosphere let the field components E_x and E_y be

$$E_x = -A_1 \exp\{-ikC(z-z_1)\} + A_2 \exp\{ikC(z-z_1)\}, \tag{21.48}$$

$$E_y = A_3 \exp\{-ikC(z-z_1)\} + A_4 \exp\{ikC(z-z_1)\}. \tag{21.49}$$

The first term of (21.48) has a minus sign because of the sign convention used for the reflection coefficients (§7.5). The two waves in (21.49) are unaffected by region I so that this expression holds throughout regions O, I and I (*a*). Between regions I and II let the field components be

$$E_x = -B_1 \exp\{-ik(z-z_1)\} + B_2 \exp\{ik(z-z_1)\}, \tag{21.50}$$

$$E_y = A_3 \exp\{-ikC(z-z_1)\} + A_4 \exp\{ikC(z-z_1)\}. \tag{21.51}$$

Now the downgoing wave in (21.48) (with coefficient A_2) arises partly from reflection of the upgoing wave (coefficient $-A_1$) and partly from the downgoing wave in (21.50) (coefficient B_2), after partial transmission through region I. Hence

$$A_2 = rA_1 + t'B_2. \tag{21.52}$$

Similarly, the upgoing wave in (21.50) (coefficient $-B_1$) arises partly from reflection of the downgoing wave (coefficient B_2) and partly from the upgoing wave in (21.48) (coefficient $-A_1$), after partial transmission through region I. Hence

$$B_1 = r'B_2 + tA_1. \tag{21.53}$$

The reflection coefficients (21.47) of region II give the two further relations:

$$B_2 \exp\{ik(z_2-z_1)\} = {}_{\parallel}\rho_{\parallel} B_1 \exp\{-ik(z_2-z_1)\} + {}_{\perp}\rho_{\parallel} A_3 \exp\{-ikC(z_2-z_1)\}, \tag{21.54}$$

$$A_4 \exp\{ikC(z_2-z_1)\} = {}_{\parallel}\rho_{\perp} B_1 \exp\{-ik(z_2-z_1)\} + {}_{\perp}\rho_{\perp} A_3 \exp\{-ikC(z_2-z_1)\}, \tag{21.55}$$

where the exponential factors allow for the different reference levels used for regions I and II.

Now suppose that $A_3 = 0$ and $A_1 = 1$. Then (21.48) and (21.49) show that $A_2 = {}_{\parallel}R_{\parallel}$, $A_4 = {}_{\parallel}R_{\perp}$, and these can be found by solving (21.52) to (21.55). Similarly, if $A_1 = 0$ and $A_3 = 1$, then $A_2 = {}_{\perp}R_{\parallel}$, $A_4 = {}_{\perp}R_{\perp}$, and these can be found in the same way. The results are:

$$\left.\begin{aligned}
{}_{\parallel}R_{\parallel} &= r + \frac{tt'\,{}_{\parallel}\rho_{\parallel}}{\exp\{2ik(z_2-z_1)\} - r'\,{}_{\parallel}\rho_{\parallel}},\\
{}_{\parallel}R_{\perp} &= \frac{t\,{}_{\parallel}\rho_{\perp}\exp\{ik(z_2-z_1)(1-C)\}}{\exp\{2ik(z_2-z_1)\} - r'\,{}_{\parallel}\rho_{\parallel}},\\
{}_{\perp}R_{\parallel} &= \frac{t'\,{}_{\perp}\rho_{\parallel}\exp\{ik(z_2-z_1)(1-C)\}}{\exp\{2ik(z_2-z_1)\} - r'\,{}_{\parallel}\rho_{\parallel}},\\
{}_{\perp}R_{\perp} &= \left[{}_{\perp}\rho_{\perp} + \frac{r'\,{}_{\parallel}\rho_{\perp}\,{}_{\perp}\rho_{\parallel}}{\exp\{2ik(z_2-z_1)\} - r'\,{}_{\parallel}\rho_{\parallel}}\right]\exp\{-2ikC(z_2-z_1)\}.
\end{aligned}\right\} \tag{21.56}$$

Some curves showing how these quantities vary with angle of incidence in typical cases are given by Heading and Whipple (1952), who discuss the relation of their results to observations at very low frequencies. The formulae (21.56) have been used by Budden (1955*b*) to check numerical calculations.

21.13 The effect of an infinity in the refractive index

One refractive index $\mathfrak{n}$ of a magnetoionic medium, as given by the Appleton–Hartree formula (6.1), is infinite for a certain value of the electron density (6.5). It has often been stated or implied that when a vertically incident radio wave reaches a level in the ionosphere where this infinity occurs, it is reflected, and the expression for the electron density at this level has been called 'the fourth reflection condition'. In this and the following sections examples are given which show that in general there is no reflection at a level where the refractive index is infinite, but that the energy in the radio wave is absorbed. This applies even when the electrons make no collisions so that there is no physical mechanism for the absorption of energy. An explanation for this paradox is suggested in § 21.16.

In those cases where reflection has been thought to occur at the level of the 'fourth reflection condition', the reflection must in fact result from some other cause such as a zero or a sharp gradient of the refractive index.

The curves of Figs. 6.3, 6.4 and 6.8 show typical ways in which the infinity of refractive index can occur when electron collisions are neglected. In Figs. 6.3 and 6.4, $Y < 1$ and the infinity appears for the extraordinary wave, while in Fig. 6.8 $Y > 1$ and the infinity appears for the ordinary wave. In all three cases a wave travelling upwards into a region of increasing electron density would first encounter a zero of refractive index (at $X = 1 - Y$ in Figs. 6.3 and 6.4, or $X = 1$ in Fig. 6.8) and would be strongly reflected there. Above this zero the refractive index is purely imaginary and the wave is 'evanescent' (§4.6), so that in a slowly varying medium the amplitude at the level of the infinity would be very small, and this level could have little effect on the wave.

Suppose, however, that a wave could travel into a region of decreasing electron density in such a way that the refractive index is given by a point on branch c of the curves of Figs. 6.3, 6.4 or 6.8. Then it would encounter the infinity and it is important to know whether or not it is reflected. The following sections show that in a slowly varying medium it is completely absorbed and there is no reflection. If the medium is not slowly varying, the steep gradient of electron density is associated with both zeros and infinities of the refractive index in the complex z-plane, as illustrated by the example of §20.5. There may then be some reflection, but it should be associated with the zeros of refractive index and not with the infinities.

The theory given here is for vertical incidence only, so that Försterling's coupled wave-equations (18.50) and (18.51) can be used. These contain the coupling parameter ψ (19.2) which depends on the gradients of electron density and collision-frequency. If the medium is slowly varying these are small, and ψ is then small except possibly near the coupling level (§ 19.2), but this is not near the level of the infinity. In the present problem we are not concerned with coupling between the characteristic waves, and therefore ψ will be neglected. The two coupled wave-equations are then independent, and the ordinary and extraordinary waves are propagated independently. Each equation then reduces to

$$\frac{d^2\mathfrak{F}}{ds^2} + \mathfrak{n}^2\mathfrak{F} = 0, \tag{21.57}$$

where $s = kz$ is height measured in units of $\lambda/2\pi$. This is similar to (9.8) for an isotropic medium and may be discussed in a similar way. Approximate solutions are the two W.K.B. solutions

$$\mathfrak{F} = \mathfrak{n}^{-\frac{1}{2}} \exp\left\{\mp i \int^{s} \mathfrak{n}\, ds\right\}, \tag{21.58}$$

and the approximation is good provided that the condition (9.29) is satisfied.

In a slowly varying ionosphere it may be assumed, as a first approximation, that, for any given small range of s, X varies linearly with height s. Let electron collisions be neglected and let the origin of s be chosen where one refractive index is infinite, that is at $X = X_\infty$, where X_∞ is given by (6.8). Then

$$X - X_\infty = \alpha s, \tag{21.59}$$

where α is constant, and it is easily shown that for the branch of $\mathfrak{n}^2$ which has the infinity

$$\mathfrak{n}^2 = \frac{A}{\alpha s} + B + O(s), \tag{21.60}$$

where A and B are constants expressible in terms of Y, Y_L, Y_T. If $|s|$ is very small all terms except the first are negligible and substitution into (9.29) gives

$$|s| \gg 3\alpha/8A, \tag{21.61}$$

as the condition for the validity of the W.K.B. or 'ray theory' solutions (21.58). The condition fails near the infinity at $s = 0$, and a 'full wave' solution of the differential equation must be sought. The solutions for two important cases are considered in the next two sections.

21.14 Isolated infinity of refractive index

In this section it is assumed that α is so small that only the first term of (21.60) need be retained for values of $|s|$ up to that for which (21.61) and therefore (9.29) become valid. Then the differential equation is

$$\frac{d^2\mathfrak{F}}{ds^2} + \frac{\beta}{s}\mathfrak{F} = 0, \tag{21.62}$$

where β is a real positive constant. This equation represents waves propagated in a medium for which the profile of the square of the refractive index, $\mathfrak{n}^2$, is as shown in Fig. 21.2 (full curve). A solution is sought which gives the reflection coefficient for waves incident from the right. It must therefore fulfil the physical condition that when s is large and negative the solution represents a wave travelling to the left, or a disturbance whose amplitude decreases as s becomes more negative. If the electron collision-frequency is not negligible but is small and constant, the infinity of $\mathfrak{n}^2$ occurs for a value of X and therefore of s which has a small negative imaginary part. The zero of s is then chosen to be where X is equal to the real part of X_∞ and (21.62) becomes

$$\frac{d^2\mathfrak{F}}{ds^2}+\frac{\beta\mathfrak{F}}{s+i\gamma}=0, \tag{21.63}$$

where γ is real and positive. The real part of $\mathfrak{n}^2$ is then as shown by the broken curve in Fig. 21.2. Equations (21.62) and (21.63) are of standard form (see, for example, Watson, 1944, p. 97). A solution of (21.63) is

$$\mathfrak{F}=(s+i\gamma)^{\frac{1}{2}}\mathscr{C}\{2\beta^{\frac{1}{2}}(s+i\gamma)^{\frac{1}{2}}\}, \tag{21.64}$$

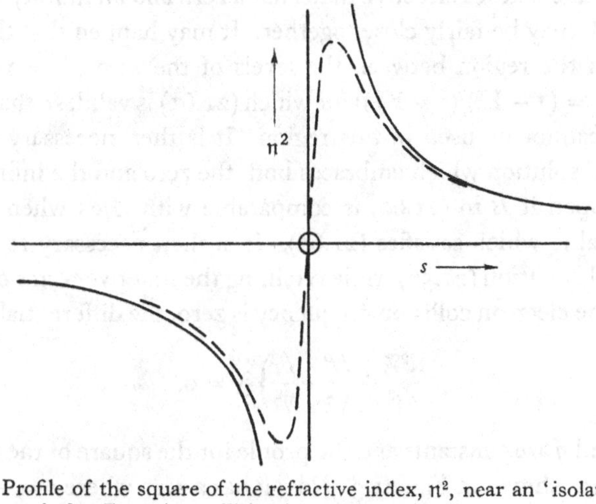

Fig. 21.2. Profile of the square of the refractive index, $\mathfrak{n}^2$, near an 'isolated' infinity. The full curve is for $Z=0$ (no collisions), and the broken curve is the real part of $\mathfrak{n}^2$ when there are collisions. The abscissa s is proportional to height.

where $\mathscr{C}$ denotes a Bessel function. If, for real s, the sign of the square root is chosen so that

$$0<\arg(s+i\gamma)^{\frac{1}{2}}<\pi, \tag{21.65}$$

then the Bessel function which satisfies the physical conditions is $\mathscr{C}=H_1^{(1)}$ where Watson's notation is used for Bessel functions of the third kind. When $|s|$ is large the asymptotic formulae for this function lead to the following results (Watson, p. 197). For s real and negative

$$\mathfrak{F}\sim(s+i\gamma)^{\frac{1}{4}}\exp[2\beta^{\frac{1}{2}}\{-|s|^{\frac{1}{2}}+\tfrac{1}{2}i\gamma/|s|^{\frac{1}{2}}\}], \tag{21.66}$$

and for s real and positive

$$\mathfrak{F}\sim(s+i\gamma)^{\frac{1}{4}}\exp\{2\beta^{\frac{1}{2}}i|s|^{\frac{1}{2}}\}. \tag{21.67}$$

A constant factor is omitted from (21.66) and (21.67). Equation (21.67) represents a wave travelling to the left in Fig. 21.2. This is the incident wave and may be identified with one W.K.B. solution. There is no other term and therefore no reflected wave. The argument applies in the limit when the collision-frequency, and therefore γ, is zero. The energy in the incident wave is still completely absorbed. The question as to what happens to this energy is discussed later, in §21.16.

21.15 Refractive index having infinity and zero

In the example illustrated in Fig. 6.8 the branches *a* and *c* refer to the ordinary wave whose refractive index has a zero and an infinity for values of X which may be fairly close together. It may happen that there is no height s in the region between the levels of the zero ($X = 1$) and the infinity ($X = (1-Y^2)/(1-Y_L^2)$) for which (21.61) is valid, so that W.K.B. solutions cannot be used in this region. It is then necessary to seek a 'full wave' solution which embraces both the zero and the infinity. This would happen if B in (21.60) is comparable with $A/\alpha s$ when s has the smallest value which satisfies (21.61). It is then necessary to solve the differential equation (21.57) while retaining the first two terms of (21.60).

When the electron collision-frequency is zero, the differential equation is

$$\frac{d^2\mathfrak{F}}{ds^2}+\left(\frac{\beta}{s}+\frac{\beta^2}{\eta^2}\right)\mathfrak{F}=0, \tag{21.68}$$

where β and η are constants and the profile for the square of the refractive index $\mathfrak{n}^2$ is as shown in Fig. 21.3. Now $\mathfrak{n}$ has a zero where $s = -\eta^2/\beta$ and an infinity where $s = 0$. If the collision-frequency is not zero but is small and constant, s in the second term of (21.68) is replaced by $s+i\gamma$ as in the example of §21.14 where γ is real and positive. Then the pole and zero of $\mathfrak{n}^2$ both lie just below the real s-axis.

Let

$$\zeta = 2i\beta s/\eta. \tag{21.69}$$

Then (21.68) becomes

$$\frac{d^2\mathfrak{F}}{d\zeta^2}+\{-\tfrac{1}{4}-\tfrac{1}{2}i\eta/\zeta\}\mathfrak{F}=0, \tag{21.70}$$

which is of the form given by Whittaker and Watson (1935, p. 337) for the confluent hypergeometric function.

Case (i). *Wave incident from below*

One solution of (21.70) is $\qquad \mathfrak{F} = W_{k,m}(\zeta), \qquad$ (21.71) where, in this case, $m = \pm\frac{1}{2}$, $k = -\frac{1}{2}i\eta$. When s is real and positive, $\arg\zeta = \frac{1}{2}\pi$ and for large $|s|$ (Whittaker and Watson, p. 343):

$$\mathfrak{F} \sim e^{-\frac{1}{2}\zeta}\zeta^k = \exp\{-is\beta/\eta - \tfrac{1}{2}i\eta\log s - \tfrac{1}{2}i\eta\log(2i\beta/\eta)\}. \qquad (21.72)$$

This represents an upgoing wave above the ionosphere. The solution (21.71) may therefore be used to find the reflection and transmission coefficients of the ionosphere for a wave incident from below. To do this, it is necessary to find the asymptotic form of (21.71) when s is real and negative. In this case $\arg\zeta$ is either $\frac{3}{2}\pi$ or $-\frac{1}{2}\pi$. Now it was shown earlier in this section that if a

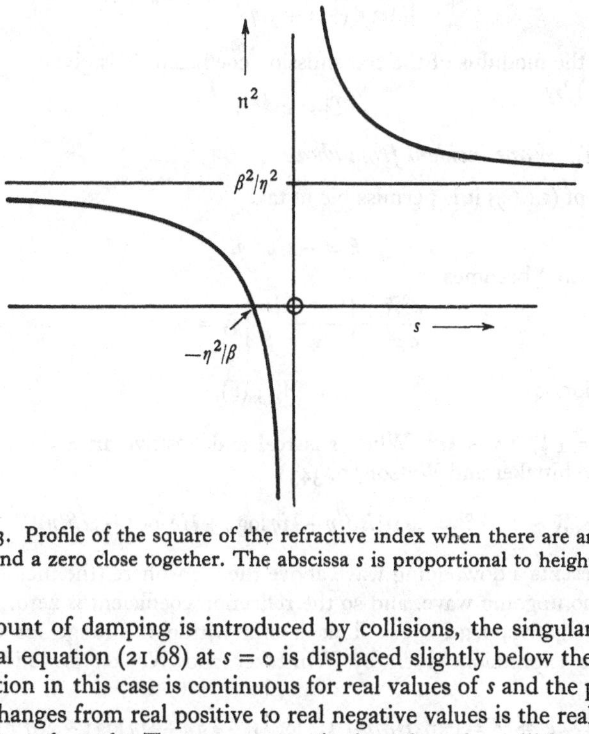

Fig. 21.3. Profile of the square of the refractive index when there are an infinity and a zero close together. The abscissa s is proportional to height.

small amount of damping is introduced by collisions, the singularity of the differential equation (21.68) at $s = 0$ is displaced slightly below the real axis. The solution in this case is continuous for real values of s and the path along which s changes from real positive to real negative values is the real axis, and passes above the pole. To preserve continuity when the collision-frequency tends to zero, the path along which s changes must still pass above the singularity at $s = 0$, and therefore $\arg s$ increases from 0 to π. Thus when s is real and negative, the correct value for $\arg\zeta$ is $\frac{3}{2}\pi$. But this is outside the range of $\arg\zeta$ for which the asymptotic form given for (21.71) by Whittaker and Watson is valid. The correct asymptotic form has been given by Heading (1953) and is

$$W_{k,m}(\zeta) \sim \zeta^k\exp(-\tfrac{1}{2}\zeta) + \frac{2\pi i\exp(2\pi ik)}{(-\frac{1}{2}-m-k)!\,(-\frac{1}{2}+m-k)!}\zeta^{-k}\exp(\tfrac{1}{2}\zeta) \qquad (21.73)$$

for $\frac{1}{2}\pi + \epsilon < \arg\zeta < \frac{5}{2}\pi - \epsilon$ where $\epsilon > 0$.

For the special values $m = \pm\frac{1}{2}$, $k = -\frac{1}{2}i\eta$ this becomes

$$\exp\{-is\beta/\eta - \tfrac{1}{2}i\eta\log|s| + \tfrac{1}{2}\pi\eta - \tfrac{1}{2}i\eta\log(2i\beta/\eta)\}$$

$$+\frac{2\pi i\exp(\pi\eta)}{(\frac{1}{2}i\eta)!\,(-1+\frac{1}{2}i\eta)!}\exp\{is\beta/\eta + \tfrac{1}{2}i\eta\log|s| - \tfrac{1}{2}\pi\eta + \tfrac{1}{2}i\eta\log(2i\beta/\eta)\}. \quad (21.74)$$

The first term represents an upgoing wave (the incident wave), and the second a downgoing wave (the reflected wave). The ratio of the moduli of these terms gives the modulus of the reflection coefficient R, which is

$$|R| = \frac{2\pi\exp(-\frac{1}{2}\pi\eta)}{|(\frac{1}{2}i\eta)!\,(-1+\frac{1}{2}i\eta)!|} = 1 - e^{-\pi\eta}. \quad (21.75)$$

Similarly, the modulus of the transmission coefficient T is given from (21.72) and (21.74) by

$$|T| = e^{-\frac{1}{2}\pi\eta}. \quad (21.76)$$

Case (ii). *Wave incident from above*

Instead of (21.69) it is permissible to take

$$\xi = -2i\beta s/\eta, \quad (21.77)$$

whence (21.68) becomes

$$\frac{d^2\mathfrak{F}}{d\xi^2} + \left\{-\frac{1}{4} + \frac{\frac{1}{2}i\eta}{\xi}\right\}\mathfrak{F} = 0. \quad (21.78)$$

One solution is

$$\mathfrak{F} = W_{k,m}(\xi), \quad (21.79)$$

where $m = \pm\frac{1}{2}$, $k = +\frac{1}{2}i\eta$. When s is real and positive $\arg\xi = -\frac{1}{2}\pi$ and for large $|s|$ (Whittaker and Watson, p. 343)

$$\mathfrak{F} \sim e^{-\frac{1}{2}\xi}\xi^k = \exp\{is\beta/\eta + \tfrac{1}{2}i\eta\log s + \tfrac{1}{2}i\eta\log(-2i\beta/\eta)\}. \quad (21.80)$$

This represents a downgoing wave above the ionosphere (the incident wave). There is no upgoing wave, and so the reflection coefficient is zero. When s is real and negative, $\arg\xi = \frac{1}{2}\pi$. This is still within the range for which the asymptotic expression given by Whittaker and Watson is valid, and for large $|s|$

$$\mathfrak{F} \sim e^{-\frac{1}{2}\xi}\xi^k = \exp\{is\beta/\eta + \tfrac{1}{2}i\eta\log|s| - \tfrac{1}{2}\pi\eta + \tfrac{1}{2}i\eta\log(-2i\beta/\eta)\}. \quad (21.81)$$

This represents a downgoing wave below the ionosphere (the transmitted wave). The modulus of the transmission coefficient T is given by

$$|T| = e^{-\frac{1}{2}\pi\eta}. \quad (21.82)$$

The results of this section may be summarised as follows:

For a wave incident from below the ionosphere

$$\left.\begin{aligned} |R| = 1 - e^{-\pi\eta}, \quad |T| = e^{-\frac{1}{2}\pi\eta},\\ |R|^2 + |T|^2 = 1 - e^{-\pi\eta} + e^{-2\pi\eta} < 1.\end{aligned}\right\} \quad (21.83)$$

For a wave incident from above the ionosphere

$$|R| = 0, \quad |T| = e^{-\frac{1}{2}\pi\eta}, \quad |R|^2+|T|^2 = e^{-\pi\eta} < 1. \qquad (21.84)$$

It is clear that in both cases the energy in the incident wave does not all reappear in the reflected and transmitted waves. Yet these results apply when the electrons make no collisions and there is apparently no mechanism in the ionosphere for absorbing energy. This paradox, in a slightly different context, was pointed out by Heading and Whipple (1952) (see § 21.9). A possible explanation is discussed in the next section.

21.16 The apparent loss of energy near an infinity of refractive index

Consider what happens when η is very large. This has the effect of making the ionosphere 'slowly varying' (except near $s = 0$). For a wave incident from below, the reflection coefficient is practically unity and the reflection may be thought of as occurring at the level $s = -\eta^2/\beta$, where $\mathfrak{n} = 0$. In the region between $s = -\eta^2/\beta$ and $s = 0$, the value of $\mathfrak{n}$ is purely imaginary and the disturbance is of the type usually called an evanescent wave. The amplitude of such a wave decreases rapidly when s increases above the value $-\eta^2/\beta$ and becomes negligibly small well below the level $s = 0$. The infinite value of $\mathfrak{n}$ at $s = 0$ therefore has little effect on the reflection process, and the transmission coefficient is negligibly small.

A wave incident from above, however, gives no reflected wave and the transmission coefficient is very small. This is not difficult to understand when there is a small amount of damping, for the wave enters a region where $|\mathfrak{n}|$ becomes very large, and if there are collisions, $\mathfrak{n}$ has an imaginary part which is large and results in the absorption of energy. However, if there are no collisions, $\mathfrak{n}$ is purely real above the level $s = 0$, and the energy cannot be absorbed. In ch. 13 it was shown that the velocity and direction of propagation of the energy in a radio wave are the same as those of a 'wave packet'. The direction does not, in general, coincide with the direction of the wave-normal. The method of § 13.13 is easily applied to the present problem in which a wave-packet travels down towards the level $s = 0$ where $\mathfrak{n}$ is infinite. If x, z are the horizontal and vertical coordinates of the wave-packet ($s = kz$), x being in the magnetic meridian, and if the principal direction of the wave-normal is vertical, then (13.44) shows that, when s is small, $dx/dz \approx K_1/s$ where K_1 is a constant. Hence, as the waves approach the level $s = 0$, the direction of the energy flow becomes more nearly horizontal. (Booker and Berkner,

1938, obtained a similar result near an infinity of $\mathfrak{n}$ which occurred because of the inclusion of the Lorentz polarisation term.) It was further shown in § 12.3, equation (12.18) that, just above the level $s = 0$, the vertical component U_z of the group velocity decreases as s decreases, and when s is small

$$U_z \approx K_2 s^{\frac{3}{2}}, \tag{21.85}$$

where K_2 is a constant. A wave-packet incident from above travels with approximately the group velocity. When it approaches the level $s = 0$, it is deviated horizontally. The time T that it takes to reach a level $s = \delta > 0$, is given approximately by

$$T = \int_{(+)}^{\delta} \frac{ds}{kU_z} \approx K_3 \delta^{-\frac{1}{2}} + \text{constant}, \tag{21.86}$$

where K_3 is a constant. This increases indefinitely as δ is made smaller. The wave-packet therefore steadily approaches the level $s = 0$, but never reaches it and instead moves horizontally in the magnetic meridian. The same applies to the energy in an infinite plane wave whose wave-normal is vertical. Then the energy density there increases indefinitely. In practice there must, of course, be a small amount of damping, which ensures that the electromagnetic energy is ultimately converted into heat.

In contrast to this result, consider the group velocity of an upgoing wave which approaches the level $s = s_0$, where the refractive index $\mathfrak{n}$ is zero. In this case, the vertical component U_z of the group velocity also tends to zero as s tends to s_0, but formulae (12.12) and (12.16) show that

$$U_z \approx K_4 (s_0 - s)^{\frac{1}{2}}, \tag{21.87}$$

where K_4 is constant. Hence the integral

$$T = \int^{s_0} \frac{ds}{kU_z} \tag{21.88}$$

converges, and in this case there is no indefinite accumulation of energy. Such integrals play an important part in the theory of equivalent heights of reflection of radio waves (see § 12.5).

The above crude argument, using group velocity, is strictly only applicable where the medium is 'slowly varying', and would fail near the level $s = 0$, where the 'full wave' solution must be used. It does, nevertheless, indicate that the level $s = 0$ where $\mathfrak{n}$ is infinite might act as a 'sink of energy' and the properties of the solution given in §§ 21.14 and 21.15 show that this is the case.

When the wave is incident from below, and η is not large, the region between $s = -\eta^2/\beta$ and $s = 0$ becomes narrower, and it is possible for some energy to penetrate this region. There is now an appreciable transmitted wave, but the result (21.83) shows that the reflected and transmitted waves do not contain all the energy in the incident wave, and, therefore, if there are no collisions, some energy must accumulate indefinitely near the level $s = 0$.

It is now clear that the condition of the ionosphere usually called the 'fourth reflection condition' is really not a reflection condition at all, and the use of the term should be discontinued.

The phenomenon described in the preceding sections is associated with an infinite value of the refractive index, and it has been shown that the level where this occurs might act as a 'sink of energy'. This is because energy can flow horizontally and accumulate indefinitely at this level. Hence, if the solution of any given propagation problem in the ionosphere is checked by considering the special case when there is no damping, it must not be assumed that the energy in the emerging waves should be the same as that in the incident wave.

In the examples discussed here, the infinite value of $\mathfrak{n}$ arose through the properties of the magnetoionic medium. There are other ways in which infinite values of the 'effective' refractive index can arise. An example was given in § 16.12 which applied to radio waves linearly polarised with the electric vector in the plane of incidence, obliquely incident on an isotropic ionosphere. The 'effective' refractive index M is given for this case by (16.90) and has an infinity where $\mathfrak{n} = 0$. In § 16.17 it was shown that even when electron collisions are neglected, the energy in the incident wave is not all returned, and this may be because some of it is converted into harmonics near the level where $\mathfrak{n} = 0$, as described in § 16.14.

Example

A linearly polarised wave is vertically incident on the ionosphere. The earth's magnetic field is vertical and $Y = \frac{3}{8}$. The electron density increases linearly with height z, so that $X = \alpha z$. Electron collisions are to be neglected. Prove that the reflected wave is always linearly polarised and show that its plane of polarisation is parallel to that of the incident wave when $\alpha = 1/n\lambda$, approximately, where n is an integer and λ is the wavelength in free space. What happens if Y is greater than 1?

CHAPTER 22

NUMERICAL METHODS FOR FINDING REFLECTION COEFFICIENTS

22.1 Introduction

The methods used for computing the reflecting properties of the ionosphere have changed markedly in recent years because of the advent of high-speed automatic digital computers. Some of the earlier papers on numerical methods were concerned with simplifying the algebra by approximations or by suitable choice of variables, so as to make the computations tractable on desk machines. Many valuable results were obtained in this way, but the same results can now be achieved much more quickly and without the need for approximations, on high-speed machines. The present chapter is not concerned with the detailed planning of the computations, nor with the program for any particular digital computer. Its object is to set out the general principles that must be followed.

At high frequencies, above about 1 Mc/s, it is usually possible to use the 'ray theory', and the reflection coefficient of the ionosphere can then be found by the methods given in detail in chs. 10 to 13. The numerical work usually involves simply the evaluation of an integral of the form $2k\int_0^{z_0} \mathfrak{n}\, dz$, where $\mathfrak{n}$ is the (complex) refractive index. Similarly, the equivalent height of reflection is found from $\int_0^{z_0} \mu'\, dz$ as explained in §§ 12.5 and 12.11 where μ' is the group refractive index. These integrals can be evaluated by using one of the standard integration methods such as the Gauss formula (see, for example, Hartree, *Numerical Analysis*, ch. VI).

At lower frequencies the 'ray theory' solutions may break down and it is then necessary to use more accurate solutions of the differential equations. Thus the problem of finding reflection coefficients reduces to solving the differential equations with suitable starting conditions, and most of the present chapter is devoted to this.

Two other problems involving numerical work have already been mentioned and will not be discussed further. These are: (1) finding the function $N(z)$ when the function $h'(f)$ is given. This involves inverting the integral equation (12.21)—see §§ 12.13 onwards. (2) Tracing ray paths in an inhomogeneous anisotropic ionosphere by solving the differential equations (14.4) and (14.18)—§ 14.4.

22.2 Methods of integrating differential equations

The differential equations governing the reflection of a plane radio wave from a horizontally stratified ionosphere have been given in several different forms in the preceding chapters, but always they consist of a set of simultaneous differential equations with several dependent variables but a single independent variable which is the height z or a multiple of it. They are equivalent to a single differential equation of the fourth order with only one dependent variable, but it is nearly always more convenient to use them in the form of several simultaneous equations of lower order. Such equations can be integrated using a step-by-step process. Suppose that the dependent variables are known at some value of the height z. Then the differential equations are used to estimate new values of the dependent variables at the height $z+\delta z$ where δz is the size of the step (in the problems discussed here, δz is negative). There are several standard methods for doing this, and most of them operate as follows. The differential equations are expressed as a set of n simultaneous first-order equations in which the dependent variables are $y_1, y_2, y_3, \ldots, y_n$. Then the equations may be written thus:

$$\left.\begin{aligned} \frac{dy_1}{dz} &= f_1(y_1, y_2, \ldots, y_n, z),\\ \frac{dy_2}{dz} &= f_2(y_1, y_2, \ldots, y_n, z),\\ &\cdots\cdots\cdots\cdots\\ \frac{dy_n}{dz} &= f_n(y_1, y_2, \ldots, y_n, z). \end{aligned}\right\} \quad (22.1)$$

The integration is started at some value of z, say z_1, at which $y_1, y_2, \ldots, y_n$ are all known. Then (22.1) is used to calculate the derivatives where $z = z_1$, whence estimates of $y_1, y_2, \ldots, y_n$ are made for a new value $z = z_2$. A first rough estimate would be $y_1(z_2) = y_1(z_1)+(z_2-z_1)f_1$, etc. This estimate is improved by recalculating the derivatives (22.1) at $z = z_2$ or at intermediate points. The final result is a corrected set of values of $y_1, y_2, \ldots, y_n$ at the new value $z = z_2$. The interval $z_2-z_1 = \delta z$ is called the step-size. Usually the process is repeated for successive intervals of the same size, though it is possible to change the step size if required.

One very successful method of this kind is the Runge–Kutta process as modified by Gill (1951) for use with a digital computer. Another method is known as the Milne three-point method (Milne, 1953) and has been used by Keitel (1955) for problems similar to those discussed here. In all these methods the right sides of (22.1) must be repeatedly computed, and in a digital computer this is done by an auxiliary sub-routine.

In this book it will be assumed that an integration routine of the above kind is available. To use it we must decide on: (i) the best dependent variables (§ 22.4), (ii) the method of choosing their initial values (§ 22.6), and (iii) the calculation of the reflection coefficients after the integration is complete (§ 22.7).

22.3 The size of the step

If one of the dependent variables, say y_1, were plotted against z, the result would be a continuous curve. The use of a finite step size means that the curve is being replaced by a series of straight line segments. This can lead to errors known as truncation errors, and to reduce these the step size should be as small as possible. It is inconvenient to use too small a step, however, for the derivatives (22.1) must be computed several times for each step and a very small step means that the total number of steps needed is large and the integration takes a very long time. Usually the best step size is found by trial.

Here the dependent variables represent a wave-motion, so that the wavelength λ in free space gives a guide to the size of step to use. The wavelength in the ionosphere is of the order $\lambda/|\mathfrak{n}|$ which may be small where $|\mathfrak{n}|$ is large. For many purposes a step size $\delta z = -\frac{1}{50}\lambda$ has proved satisfactory with the Runge–Kutta–Gill method in the author's experience. If, however, the electron density $N(z)$ varies very rapidly with height z, it may be necessary to use a smaller step.

22.4 The choice of dependent variable

The four equations (18.10) to (18.13) are the differential equations for the most general case of radio waves obliquely reflected from a horizontally stratified ionosphere. They were derived directly from Maxwell's equations together with the constitutive equations for the ionosphere. They are already in the form (22.1) of four first-order simultaneous differential equations so that they can be used without modification in a process like the Runge–Kutta–Gill process mentioned in § 22.2. The variables E_x, E_y, $\mathscr{H}_x$, and $\mathscr{H}_y$ are complex, so that the real and imaginary parts must be computed separately, and the equations are therefore equivalent to eight simultaneous first-order equations.

Although the variables E_x, E_y, $\mathscr{H}_x$ and $\mathscr{H}_y$ have been used successfully in this way (Budden, 1955) they contain more information than is required, since the reflection coefficients depend only on *ratios* of these variables. These ratios are known as wave-admittances, and by using them as the dependent variables the numerical work can be shortened considerably. This is the basis of the wave-admittance concept (§§ 22.8 and 22.9).

It is sometimes convenient to use other dependent variables. For example, for vertical incidence much important work has been done using Försterling's coupled wave-equations (18.50) and (18.51). These are useful in problems where the coupling parameter ψ is small in most of the ionosphere, for then the equations separate into the two independent second-order equations (18.52) and (18.53). For these the W.K.B. solutions may be good approximations in most of the ionosphere. Only in regions of reflection, where one of the refractive indices $\mathfrak{n}_o$ or $\mathfrak{n}_x$ approaches zero, must a full wave solution of the associated equation be used. In the coupling region where ψ is not negligible, the coupled equations can sometimes be solved by successive approximations, as explained in § 19.5. The disadvantages of the Försterling equations for numerical work are: (i) they can only be used for vertical incidence (coupled equations for oblique incidence are very complicated—see Budden and Clemmow, 1957); (ii) it is necessary to compute the functions $\mathfrak{n}_o^2$, $\mathfrak{n}_x^2$, ψ and ψ', which are much more complicated than the coefficients in (18.10) to (18.13), and ψ, ψ' involve first and second derivatives of the electron density and collision-frequency, whereas there are no such derivatives in (18.10) to (18.13); (iii) the method of successive approximations fails when ψ becomes very large, that is near critical coupling. In spite of these disadvantages the Försterling equations have been used very successfully (see, for example, Gibbons and Nertney, 1951, 1952; Kelso, Nearhoof, Nertney and Waynick, 1951).

To overcome the disadvantage (iii) near critical coupling a different form of second-order coupled equations has been suggested (Davids and Parkinson, 1955), in which the coupling terms remain small even in the coupling region near critical coupling. These make use of two new dependent variables which are simple functions of the 'principal axis components' of the electric field of the wave (§ 3.11). But the equations contain parameters which depend on first and second derivatives of the electron density so that the disadvantage (ii) still applies. The equations seem to have no advantage over the basic equations (18.10) to (18.13), which give no trouble even near critical coupling.

22.5 The three parts of the calculation of reflection coefficients

When reflection coefficients are calculated by integrating the differential equations, the calculation is divided into three parts as follows:

(1) A solution is found which satisfies the physical conditions above the ionosphere or at a great height within it. These conditions arise because the only influx of energy is in the incident wave below the ionosphere, and at a high enough level the solution must represent an upgoing wave. The differential equations are equivalent to a single equation of the fourth order which has four independent solutions. At a sufficiently high level, where the W.K.B. solutions may be used, two of these are upgoing and two are downgoing waves (see § 13.7). Hence there are two independent solutions which satisfy the physical conditions at a high level.

(2) Starting with one of the sets of dependent variables determined in 1, the step-by-step integration is performed proceeding downwards through the ionosphere, so that the step δz is negative. This is continued until the free space below the ionosphere is reached. It is repeated with the second set of starting values.

(3) The solution resulting from the first integration gives total values of E_x, E_y, $\mathscr{H}_x$ and $\mathscr{H}_y$, below the ionosphere. This set is now separated into an upgoing and a downgoing wave, each being in general elliptically polarised. The ratio of the amplitude of these two waves would give a reflection coefficient of the ionosphere, but this would apply only to an incident wave of a particular elliptical polarisation of no special interest. The process is therefore repeated with the solution resulting from the second integration, and the resulting upgoing and downgoing waves below the ionosphere in general have different polarisations from the first pair. A linear combination of the two solutions is now formed so that the incident wave of the combination is of unit amplitude and linearly polarised either in the plane of incidence or perpendicular to it. The associated reflected wave then gives two components of the reflection coefficient.

Stage 2 depends on the particular integration process used and needs no further description here. Stages 1 and 3 are dealt with in §§ 22.6 and 22.7, respectively.

When a wave-admittance variable is used, the two possible sets of starting values are combined, and only one integration is then necessary. The calculation of the reflection coefficient at the bottom of the ionosphere is also simplified. This method is described in §§ 22.8 and 22.9.

22.6 The starting solutions at a great height

It sometimes happens that the integration can be started at a level in the free space above the ionosphere, and the choice of the two starting solutions is then simple. Here there may be a transmitted wave travelling obliquely upwards. It can be resolved into two linearly polarised components with the electric vector, in, and perpendicular to, the x–z-plane (that is the plane of incidence) for which

$$E_x/C = \mathscr{H}_y = A, \quad E_y = \mathscr{H}_x = 0, \tag{22.2}$$

and
$$E_x = \mathscr{H}_y = 0, \quad E_y = -\mathscr{H}_x/C = B, \tag{22.3}$$

respectively. A and B are arbitrary constants which may be chosen to suit the selected integration method. Since reflection coefficients depend only on ratios of the field quantities E_x, E_y, $\mathscr{H}_x$ and $\mathscr{H}_y$ the values of A and B do not affect the final result. In fact the four dependent variables may be multiplied by the same constant, if necessary, during the integration, without affecting the results. This is permissible because the differential equations (18.10) to (18.13) are linear and homogeneous, and the device is sometimes used in digital computers to prevent the numbers from exceeding the capacity of the machine.

Radio waves of very low frequency are reflected at comparatively low levels in the ionosphere, certainly well below the height of maximum ionisation of the E-layer. It would then be tedious and unnecessary to carry out the integration over the whole ionosphere, and it is better to start at some level within the ionosphere, above the reflection levels. If there is such a level where X is of the order of 10^3, the starting values can be found as follows. For large X the two refractive indices are both large, being roughly proportional to $X^{\frac{1}{2}}$. The wavelength in the medium is therefore small and for sufficiently large X the medium can be treated as slowly varying, so that the W.K.B. solutions are good approximations. Moreover, Snell's law shows that the wave-normals must be nearly vertical, regardless of the value of the angle of incidence θ_I below the ionosphere. Hence let $\theta_I = 0$, $S = 0$, $C = 1$. The two polarisations ρ_o, ρ_x are nearly constant and equal to $-i$, $+i$, respectively, and the coupling parameter ψ is negligible. Then Försterling's coupled form of the differential equations reduces to

$$\frac{d^2\mathfrak{F}_o}{ds^2} + \mathfrak{n}_o^2\mathfrak{F}_o = 0, \quad \frac{d^2\mathfrak{F}_x}{ds^2} + \mathfrak{n}_x^2\mathfrak{F}_x = 0, \tag{22.4}$$

where $$2^{\frac{1}{2}}i\mathfrak{F}_o \approx E_x + iE_y, \quad 2^{\frac{1}{2}}\mathfrak{F}_x \approx E_x - iE_y, \tag{22.5}$$

$$\mathfrak{n}_o^2 \approx 1 - \frac{X}{U+Y_L}, \quad \mathfrak{n}_x^2 \approx 1 - \frac{X}{U-Y_L}. \tag{22.6}$$

The two equations (18.10) and (18.11) reduce to

$$\mathscr{H}_y = i\frac{dE_x}{ds}, \quad \mathscr{H}_x = -i\frac{dE_y}{ds}. \tag{22.7}$$

The two upgoing W.K.B. solutions of (22.4) can now be written down. For the first equation we take $\mathfrak{F}_x = 0$, and the four field quantities are in the ratios

$$E_x : E_y : \mathscr{H}_x : \mathscr{H}_y = 1 : -i : i\mathfrak{n}_o : \mathfrak{n}_o. \tag{22.8}$$

Similarly, for the second equation $\mathfrak{F}_o = 0$ and the four field quantities are in the ratios

$$E_x : E_y : \mathscr{H}_x : \mathscr{H}_y = 1 : i : -i\mathfrak{n}_x : \mathfrak{n}_x. \tag{22.9}$$

These results could also have been deduced from (18.106) and (18.108). Equations (22.8) and (22.9) are the required two sets of starting values.

To test the accuracy of these solutions the criterion (9.29) may be used. For both refractive indices $|\mathfrak{n}^2|$ is of the order of X/Y when X is large. Now the reflection level for very low frequencies is usually well below the maximum of the E-layer, and here the variation of X with height is roughly exponential so that X is proportional to $e^{\alpha z}$ where α is about 0·1 to 1·0 km^{-1}. Then (9.29) requires that $X \gg \alpha^2 Y/4k^2$ or $X \gg \alpha^2 Y\lambda^2/160$, where λ is the wavelength in free space. For a frequency of 16 kc/s (for which the method given here has been extensively used) $\lambda = 18{\cdot}75$ km, $Y \approx 80$. If $\alpha = 1$ km^{-1}, we require that $X \gg 200$. In practice, a value $X \approx 2000$ at the starting level was found to be satisfactory (Budden, 1955).

The initial solutions (22.8) and (22.9) represent waves whose amplitude increases as the integration proceeds downwards. Any error in an initial solution in general has the same effect as adding an unwanted downgoing wave. But such a wave decreases in amplitude as the integration proceeds downwards, and its effect at the bottom of the ionosphere is negligible provided that it was not too large at the beginning of the integration. Hence any small errors which arise from the approximations made in deriving the initial solutions are eliminated during the integrations.

The initial solutions (22.8) and (22.9) are the same for all angles of incidence, and are sufficiently accurate if the differential equations (18.10) to (18.13) are integrated as they stand. When wave-admittance variables are used, however, initial values derived from (22.8) and (22.9) are not accurate enough and must be corrected, for example, by the method of § 22.11; different initial solutions are then required for each angle of incidence.

22.7 Calculation of the components of the reflection coefficient

The two integrations give two sets of the four quantities E_x, E_y, $\mathscr{H}_x$ and $\mathscr{H}_y$, below the ionosphere. It is convenient to denote these by a column matrix **e** defined by (18.16) (note that E_y has a minus sign), so that two values $\mathbf{e}^{(1)}$ and $\mathbf{e}^{(2)}$ are found below the ionosphere. From them the four quantities $_{\parallel}R_{\parallel}$, $_{\parallel}R_{\perp}$, $_{\perp}R_{\parallel}$ and $_{\perp}R_{\perp}$ are to be calculated, that is the matrix **R** (§§ 7.4 to 7.6).

For each set **e** the four elements specify the total electromagnetic field of the wave. Below the ionosphere this may be resolved into upgoing and downgoing waves, each of which is, in general, elliptically polarised. Let the upgoing wave itself be resolved into linearly polarised components, with the electric field in the plane of propagation (often called the 'normal component') and perpendicular to it (called the 'abnormal component'). For the upgoing 'normal' component

$$E_x = C\mathscr{H}_y, \quad E_y = \mathscr{H}_x = 0, \tag{22.10}$$

and for the upgoing 'abnormal' component

$$\mathscr{H}_x = -CE_y, \quad E_x = \mathscr{H}_y = 0. \tag{22.11}$$

Similarly, the downgoing or reflected wave can be resolved into 'normal' and 'abnormal' components. For the downgoing 'normal' component

$$E_x = -C\mathscr{H}_y, \quad E_y = \mathscr{H}_x = 0, \tag{22.12}$$

and for the downgoing 'abnormal' component

$$\mathscr{H}_x = CE_y, \quad E_x = \mathscr{H}_y = 0. \tag{22.13}$$

The total field below the ionosphere is the sum of the fields of these four component waves, and is given by

$$\mathbf{e} = a\begin{pmatrix} C \\ 0 \\ 0 \\ 1 \end{pmatrix} + b\begin{pmatrix} 0 \\ -1 \\ -C \\ 0 \end{pmatrix} + c\begin{pmatrix} -C \\ 0 \\ 0 \\ 1 \end{pmatrix} + d\begin{pmatrix} 0 \\ -1 \\ C \\ 0 \end{pmatrix}, \tag{22.14}$$

where a, b, c, d are the (complex) amplitudes of the component waves and may be written as a column matrix **q**. a and b denote upgoing waves and each is proportional to $\exp(-iCs)$. Similarly, c and d denote downgoing waves and each is proportional to $\exp(iCs)$. Notice that when a and c have the same argument, that is, when ${}_{\parallel}R_{\parallel}$ is real and positive, the x components of the electric vectors of the upgoing and downgoing waves are in *opposite* directions, because of the sign convention (§7.5). The relation (22.14) can be written concisely in matrix form thus:

$$\mathbf{e} = \begin{pmatrix} C & 0 & -C & 0 \\ 0 & -1 & 0 & -1 \\ 0 & -C & 0 & C \\ 1 & 0 & 1 & 0 \end{pmatrix} \mathbf{q} = \mathbf{L}\mathbf{q}. \qquad (22.15)$$

Hence by inversion of the 4×4 matrix $\mathbf{L}$

$$\mathbf{q} = \mathbf{L}^{-1}\mathbf{e} = \tfrac{1}{2}\begin{pmatrix} 1/C & 0 & 0 & 1 \\ 0 & -1 & -1/C & 0 \\ -1/C & 0 & 0 & 1 \\ 0 & -1 & 1/C & 0 \end{pmatrix} \mathbf{e}. \qquad (22.16)$$

The matrix $\mathbf{q}$ may be partitioned into the two matrices

$$\mathbf{u} = \begin{pmatrix} a \\ b \end{pmatrix}, \quad \mathbf{d} = \begin{pmatrix} c \\ d \end{pmatrix}, \qquad (22.17)$$

thus separating the upgoing and downgoing components. Now from the definition of the reflection coefficient matrix $\mathbf{R}$, given in §7.6, it is clear that these components must satisfy the equation

$$\mathbf{d} = \mathbf{R}\mathbf{u}. \qquad (22.18)$$

The integration process gives two values $\mathbf{e}^{(1)}$ and $\mathbf{e}^{(2)}$ of $\mathbf{e}$, and therefore of $\mathbf{q}$ and (22.18) must apply to both. This enables a formal expression for $\mathbf{R}$ to be derived in matrix notation. The superscripts (1) and (2) are used to distinguish quantities derived from $\mathbf{e}^{(1)}$ and $\mathbf{e}^{(2)}$. Let

$$\mathbf{U} = \begin{pmatrix} a^{(1)} & a^{(2)} \\ b^{(1)} & b^{(2)} \end{pmatrix}, \quad \mathbf{D} = \begin{pmatrix} c^{(1)} & c^{(2)} \\ d^{(1)} & d^{(2)} \end{pmatrix}. \qquad (22.19)$$

Then, since $\mathbf{U}$ must be non-singular, it follows from (22.18) that

$$\mathbf{R} = \mathbf{D}\,.\,\mathbf{U}^{-1}. \qquad (22.20)$$

The elements of $\mathbf{R}$ could be written out in full in terms of the eight field quantities $E_x^{(1)}$, $E_y^{(1)}$, $\mathscr{H}_x^{(1)}$, $\mathscr{H}_y^{(1)}$, $E_x^{(2)}$, $E_y^{(2)}$, $\mathscr{H}_x^{(2)}$ and $\mathscr{H}_y^{(2)}$, but the expressions are very complicated. Equation (22.20) shows, however, that the elements of **R** depend only on ratios like $\mathscr{H}_i/E_j$ $(i, j = x, y)$ and not on the absolute values of the field quantities.

22.8 The wave-admittance in an isotropic ionosphere

The matrices **q**, **u**, **d**, **U** and **D** were defined in the preceding section for a level below the ionosphere. But these definitions may be used at any other level, although it is not then generally permissible to interpret these quantities in terms of upgoing and downgoing waves. Similarly, (22.20) may be used to define the matrix **R** at any level. Then **R** is a matrix which, below the ionosphere, becomes equal to the reflection coefficient matrix. **R** will in future be used in this more general sense. It is important to notice that the derivation of **R** from (22.20) requires the knowledge of two independent solutions of (18.10) to (18.13).

It is now natural to inquire whether **R** can be treated as the dependent variable of a differential equation, instead of **e**. This proves to be possible, and the differential equations for **R** have been given (Budden, 1955). It is, however, shorter to use as the dependent variable another 4×4 matrix **A** which is simply related to **R**. **A** is the generalised wave-admittance matrix defined in § 22.9. Before discussing it, the idea of a wave-admittance will be illustrated for the simpler case of an isotropic medium.

The use of wave-impedances in the solution of problems of wave-propagation was introduced by Schelkunoff (1938) and extended by Booker (1947). Equivalent methods have been applied to the problem of wave-propagation in an unbounded isotropic medium by several authors (for example, Hines, 1953; Bailey, 1954; G. Millington, private communication). A wave-admittance is the reciprocal of a wave-impedance and is preferred here because the algebra is slightly shorter.

For an isotropic medium equations (18.10) to (18.13) separate into the two independent pairs:

$$E'_x = -i(1 - S^2/\mathfrak{n}^2)\,\mathscr{H}_y, \quad \mathscr{H}'_y = -i\mathfrak{n}^2 E_x, \tag{22.21}$$

and

$$E'_y = i\mathscr{H}_x, \quad \mathscr{H}'_x = i(\mathfrak{n}^2 - S^2)\,E_y, \tag{22.22}$$

where $\mathfrak{n}^2 = 1 - X/(1 - iZ)$ is the square of the refractive index. Now let

$$A_1 = \frac{\mathscr{H}_y}{E_x} = Z_0\frac{H_y}{E_x}, \quad A_2 = \frac{\mathscr{H}_x}{E_y} = Z_0\frac{H_x}{E_y}. \tag{22.23}$$

The ratios H_y/E_x and H_x/E_y are called wave-admittances, and in m.k.s. units they are measured in ohm^{-1}. The ratios A_1 and A_2 are dimensionless numbers proportional to the wave-admittances, the constant of proportionality being Z_0, the characteristic impedance of free space (see § 2.9). The same constant Z_0 is used throughout this chapter, and it is convenient to speak of A_1 and A_2 as though they are themselves wave-admittances.

The differential equations satisfied by A_1 and A_2 may be found by differentiating (22.23) and substituting for E'_x, $\mathscr{H}'_y$, E'_y and $\mathscr{H}'_x$ from (22.21) and (22.22). The result is

$$iA'_1 = \mathfrak{n}^2 - (1 - S^2/\mathfrak{n}^2)\,A_1^2, \tag{22.24}$$

$$iA'_2 = A_2^2 - (\mathfrak{n}^2 - S^2). \tag{22.25}$$

These are non-linear first-order equations of Ricatti type, and are in a form suitable for integration on a digital computer by one of the methods mentioned in § 22.2.

Consider first equation (22.24) which applies to linearly polarised waves with the electric vector in the plane of incidence. If the integration is started in the free space above the ionosphere the starting value of A_1 must be that of an upgoing wave which satisfies (22.2). Hence the correct starting value is $A_1 = 1/C$. If, however, the starting level is within the ionosphere above the level of reflection, it must be where the upgoing W.K.B. solution (9.70) and (9.71) (with the upper sign) can be used, and then the starting value is

$$A_1 = q/\mathfrak{n}^2 = \left(C^2 - \frac{X}{1-iZ}\right)^{\frac{1}{2}} \Big/ \left(1 - \frac{X}{1-iZ}\right).$$

Notice that these starting values do not contain an arbitrary constant. This is in contrast to the starting values (22.2), (22.3) or (22.8), (22.9) for the field variables. The equation (22.24) can now be integrated proceeding downwards through the ionosphere, until the free space below it is reached. The reflection coefficient R can then be found as follows.

Below the ionosphere there is an upgoing and a downgoing wave. Let the field components of these be distinguished by superscripts (u) and (d) respectively. Then the fields of these waves satisfy

$$\frac{\mathscr{H}_y^{(u)}}{E_x^{(u)}} = \frac{1}{C}, \quad \frac{\mathscr{H}_y^{(d)}}{E_x^{(d)}} = -\frac{1}{C}, \tag{22.26}$$

and from the definition of the reflection coefficient R ($= {}_{\|}R_{\|}$, § 7.5):

$$E_x^{(d)} = -RE_x^{(u)}, \quad \mathscr{H}_y^{(d)} = R\mathscr{H}_y^{(u)}. \tag{22.27}$$

The total fields are

$$E_x = E_x^{(u)} + E_x^{(d)}, \quad \mathscr{H}_y = \mathscr{H}_y^{(u)} + \mathscr{H}_y^{(d)}, \tag{22.28}$$

and their ratio $\mathscr{H}_y/E_x$ is the value of A_1 obtained from the integration. Hence

$$A_1 = \frac{\mathscr{H}_y^{(u)}+\mathscr{H}_y^{(d)}}{E_x^{(u)}+E_x^{(d)}} = \frac{1+R}{1-R}\frac{1}{C}, \tag{22.29}$$

so that

$$R = \frac{CA_1-1}{CA_1+1}. \tag{22.30}$$

Next consider (22.25) which applies to linearly polarised waves with the electric vector horizontal. If the integration is started in free space, the starting value is $A_2 = -C$ (from (22.3)). If it is started where the upgoing W.K.B. solution (9.59) and (9.60) can be used, the starting value is

$$A_2 = -q = -\left(C^2 - \frac{X}{1-iZ}\right)^{\frac{1}{2}}.$$

When the integration is complete the reflection coefficient R (short for ${}_{\perp}R_{\perp}$) can be found from the value of A_2 below the ionosphere, and is easily shown to be

$$R = \frac{C+A_2}{C-A_2}. \tag{22.31}$$

22.9 The wave-admittance matrix A for an anisotropic ionosphere

As in §22.7 let the differential equations (18.10) to (18.13) have two independent solutions $\mathbf{e}^{(1)}$ and $\mathbf{e}^{(2)}$ whose elements are distinguished by the superscripts (1) and (2) respectively. The matrix admittance $\mathbf{A}$ is now defined thus:

$$\mathbf{A} = \begin{pmatrix} \mathscr{H}_y^{(1)} & \mathscr{H}_y^{(2)} \\ \mathscr{H}_x^{(1)} & \mathscr{H}_x^{(2)} \end{pmatrix} \begin{pmatrix} E_x^{(1)} & E_x^{(2)} \\ E_y^{(1)} & E_y^{(2)} \end{pmatrix}^{-1}. \tag{22.32}$$

The meaning of $\mathbf{A}$ may be illustrated by considering the isotropic ionosphere of the preceding section, for which we may take $E_y^{(1)} = \mathscr{H}_x^{(1)} = 0$, $E_x^{(2)} = \mathscr{H}_y^{(2)} = 0$, and $\mathbf{A}$ becomes simply

$$\begin{pmatrix} \mathscr{H}_y^{(1)}/E_x^{(1)} & 0 \\ 0 & \mathscr{H}_x^{(2)}/E_y^{(2)} \end{pmatrix}$$

so that it is a diagonal matrix with elements A_1 and A_2, the admittance variables of the isotropic ionosphere.

The value of $\mathbf{A}$ is independent of the way in which the solutions $\mathbf{e}^{(1)}$ and $\mathbf{e}^{(2)}$ are chosen, for it is easy to show that $\mathbf{A}$ is unaltered if $\mathbf{e}^{(2)}$ is changed by adding to it a multiple of $\mathbf{e}^{(1)}$.

The differential equation satisfied by **A** is found as follows. Equation (22.32) shows that for either superscript (1) or (2)

$$\mathbf{A}\begin{pmatrix}E_x\\E_y\end{pmatrix}=\begin{pmatrix}\mathscr{H}_y\\\mathscr{H}_x\end{pmatrix}. \tag{22.33}$$

This is differentiated to give

$$\mathbf{A}\begin{pmatrix}E'_x\\E'_y\end{pmatrix}+\mathbf{A}'\begin{pmatrix}E_x\\E_y\end{pmatrix}=\begin{pmatrix}\mathscr{H}'_y\\\mathscr{H}'_x\end{pmatrix}. \tag{22.34}$$

Now the matrix form (18.18) of the equations (18.10) to (18.13) may be written in full and partitioned as follows:

$$\begin{pmatrix}E'_x\\-E'_y\\\cdots\cdots\\\mathscr{H}'_x\\\mathscr{H}'_y\end{pmatrix}=-i\left(\begin{array}{cc:cc}T_{11}&T_{12}&0&T_{14}\\0&0&1&0\\\hdashline T_{31}&T_{32}&0&T_{34}\\T_{41}&T_{42}&0&T_{44}\end{array}\right)\begin{pmatrix}E_x\\-E_y\\\cdots\cdots\\\mathscr{H}_x\\\mathscr{H}_y\end{pmatrix}, \tag{22.35}$$

whence

$$\left.\begin{aligned}i\begin{pmatrix}E'_x\\E'_y\end{pmatrix}&=\begin{pmatrix}T_{11}&-T_{12}\\0&0\end{pmatrix}\begin{pmatrix}E_x\\E_y\end{pmatrix}+\begin{pmatrix}T_{14}&0\\0&-1\end{pmatrix}\begin{pmatrix}\mathscr{H}_y\\\mathscr{H}_x\end{pmatrix},\\i\begin{pmatrix}\mathscr{H}'_y\\\mathscr{H}'_x\end{pmatrix}&=\begin{pmatrix}T_{41}&-T_{42}\\T_{31}&-T_{32}\end{pmatrix}\begin{pmatrix}E_x\\E_y\end{pmatrix}+\begin{pmatrix}T_{44}&0\\T_{34}&0\end{pmatrix}\begin{pmatrix}\mathscr{H}_y\\\mathscr{H}_x\end{pmatrix}.\end{aligned}\right\} \tag{22.36}$$

These are combined with (22.34) and (22.33) to give

$$\begin{aligned}i\mathbf{A}'\begin{pmatrix}E_x\\E_y\end{pmatrix}=&\begin{pmatrix}T_{41}&-T_{42}\\T_{31}&-T_{32}\end{pmatrix}\begin{pmatrix}E_x\\E_y\end{pmatrix}+\begin{pmatrix}T_{44}&0\\T_{34}&0\end{pmatrix}\mathbf{A}\begin{pmatrix}E_x\\E_y\end{pmatrix}\\&-\mathbf{A}\begin{pmatrix}T_{11}&-T_{12}\\0&0\end{pmatrix}\begin{pmatrix}E_x\\E_y\end{pmatrix}-\mathbf{A}\begin{pmatrix}T_{14}&0\\0&-1\end{pmatrix}\mathbf{A}\begin{pmatrix}E_x\\E_y\end{pmatrix}.\end{aligned} \tag{22.37}$$

The column matrix $\begin{pmatrix}E_x\\E_y\end{pmatrix}$ appears on the right of every term in (22.37) which is valid when E_x and E_y have either superscript (1) or (2). Hence it is also valid if $\begin{pmatrix}E_x\\E_y\end{pmatrix}$ is replaced by the matrix $\mathbf{E}=\begin{pmatrix}E_x^{(1)}&E_x^{(2)}\\E_y^{(1)}&E_y^{(2)}\end{pmatrix}$ which is non-singular since the two superscripts refer to two independent solutions. Then (22.37) could be multiplied on the right by $\mathbf{E}^{-1}$. In this way the field components are eliminated and the differential equation satisfied by **A** is

$$i\mathbf{A}'=\mathbf{A}\begin{pmatrix}-T_{14}&0\\0&1\end{pmatrix}\mathbf{A}+\begin{pmatrix}T_{41}&-T_{42}\\T_{31}&-T_{32}\end{pmatrix}+\mathbf{A}\begin{pmatrix}-T_{11}&T_{12}\\0&0\end{pmatrix}+\begin{pmatrix}T_{44}&0\\T_{34}&0\end{pmatrix}\mathbf{A}. \tag{22.38}$$

This equation is in a suitable form for integration step-by-step downwards through the ionosphere. A suitable starting value of **A** can be found by the method given in the next section. When the integration is complete the value of **A** at the bottom of the ionosphere can be used to find the reflection coefficient matrix **R** (§ 22.11). **A** contains information about both the solutions $\mathbf{e}^{(1)}$ and $\mathbf{e}^{(2)}$ and so only one integration is necessary.

22.10 The starting value of A

If the integration is started in the free space above the ionosphere, the starting value $\mathbf{A}_0$ of **A** is easily found, for in (22.32) we may take

$$E_x^{(1)} = C, \quad \mathscr{H}_x^{(1)} = 0, \quad E_y^{(1)} = 0, \quad \mathscr{H}_y^{(1)} = 1,$$
$$E_x^{(2)} = 0, \quad \mathscr{H}_x^{(2)} = -C, \quad E_y^{(2)} = 1, \quad \mathscr{H}_y^{(2)} = 0,$$

whence
$$\mathbf{A}_0 = \begin{pmatrix} 1/C & 0 \\ 0 & -C \end{pmatrix}. \tag{22.39}$$

In free space all the elements T_{ij} (18.17) are zero except

$$T_{14} = T_{32} = C^2, \quad T_{23} = T_{41} = 1,$$

and with these values (22.38) and (22.39) give $\mathbf{A}' = 0$.

If the integration is started within the ionosphere at a level where the W.K.B. solutions are good approximations (see § 22.6) the two upgoing W.K.B. solutions in (18.86) could be used to find the starting value of **A**. But these are too complicated to be useful and we shall first consider vertical incidence, for which the upgoing W.K.B. solutions are given by (18.106) and (18.108). When these are substituted in (22.32) they give for the starting value

$$\mathbf{A}_0 = \frac{1}{\rho_o - \rho_x} \begin{pmatrix} \rho_o \mathfrak{n}_x - \rho_x \mathfrak{n}_o & \mathfrak{n}_o - \mathfrak{n}_x \\ \mathfrak{n}_o - \mathfrak{n}_x & \rho_x \mathfrak{n}_x - \rho_o \mathfrak{n}_o \end{pmatrix}. \tag{22.40}$$

Since X is very large at the starting level we may take $\rho_x \approx -\rho_o \approx i$ as in § 22.6, whence

$$\mathbf{A}_0 \approx \tfrac{1}{2} \begin{pmatrix} \mathfrak{n}_o + \mathfrak{n}_x & i(\mathfrak{n}_o - \mathfrak{n}_x) \\ i(\mathfrak{n}_o - \mathfrak{n}_x) & -(\mathfrak{n}_o + \mathfrak{n}_x) \end{pmatrix}. \tag{22.41}$$

An alternative way of finding the starting value $\mathbf{A}_0$ is based on the following argument. The starting values can be found from the upgoing W.K.B. solutions. But in a homogeneous medium the W.K.B. solutions satisfy the differential equations exactly, and represent plane waves travelling obliquely upwards. Consider, therefore a fictitious homogeneous medium with the same properties as the real ionosphere at the starting level. In such a medium the value of **A**, for upgoing waves only, is the same at all heights because the ratios $E_x : E_y : \mathscr{H}_x : \mathscr{H}_y$ are constants for both upgoing characteristic waves. If this value can be found and substituted in (22.38), it must give $\mathbf{A}' = 0$. If, however, an incorrect value of **A** is used, then $\mathbf{A}' \neq 0$. The incorrect **A** is associated with fields which contain both upgoing and downgoing waves. Now let this incorrect **A** be used as a starting value in (22.38) for an integration proceeding downwards through the fictitious homogeneous medium. The downgoing waves are attenuated and the upgoing waves are accentuated as we proceed downwards, and the field therefore becomes more nearly that for upgoing waves only. Thus **A** gets closer to the correct value for upgoing waves, so that **A**′ gets smaller and smaller. When **A**′ has become negligibly small, **A** must have the correct value for upgoing waves only, in the fictitious homogeneous ionosphere, and this is the same as for upgoing W.K.B. solutions at the starting level in the real ionosphere.

The method of finding the starting value $\mathbf{A}_0$ is therefore as follows. A rough estimate of $\mathbf{A}_0$ is first made; the value (22.41) is often useful for this. It is inserted in (22.38) which is then integrated step by step, proceeding downwards with X and Z held constant at their values at the starting level. After each step the elements of **A**′ are examined. When they are all less than a pre-assigned small quantity, **A** is very close to the correct starting value. The main integration through the real ionosphere may then be done.

The process just described can also be used to find the reflection coefficient matrix for a sharply bounded homogeneous medium, as explained in the following section.

22.11 Relation between the admittance matrix A and the reflection coefficient matrix R

The relation between **A** and **R** can be found by using (22.16), (22.19), (22.20) and (22.32). It is convenient to introduce the matrix notation

$$\mathbf{E} = \begin{pmatrix} E_x^{(1)} & E_x^{(2)} \\ E_y^{(1)} & E_y^{(2)} \end{pmatrix}, \quad \mathbf{H} = \begin{pmatrix} \mathscr{H}_y^{(1)} & \mathscr{H}_y^{(2)} \\ \mathscr{H}_x^{(1)} & \mathscr{H}_x^{(2)} \end{pmatrix}$$

so that

$$\mathbf{H} = \mathbf{AE}, \quad \mathbf{D} = \mathbf{RU}. \tag{22.42}$$

The matrices in (22.16) may be partitioned and re-arranged to give

$$\begin{pmatrix} a \\ b \end{pmatrix} = \tfrac{1}{2}\begin{pmatrix} 1/C & 0 \\ 0 & 1 \end{pmatrix}\begin{pmatrix} E_x \\ E_y \end{pmatrix} + \tfrac{1}{2}\begin{pmatrix} 1 & 0 \\ 0 & -1/C \end{pmatrix}\begin{pmatrix} \mathscr{H}_y \\ \mathscr{H}_x \end{pmatrix}, \tag{22.43}$$

$$\begin{pmatrix} c \\ d \end{pmatrix} = \tfrac{1}{2}\begin{pmatrix} -1/C & 0 \\ 0 & 1 \end{pmatrix}\begin{pmatrix} E_x \\ E_y \end{pmatrix} + \tfrac{1}{2}\begin{pmatrix} 1 & 0 \\ 0 & 1/C \end{pmatrix}\begin{pmatrix} \mathscr{H}_y \\ \mathscr{H}_x \end{pmatrix}. \tag{22.44}$$

This applies for both superscripts (1) and (2) so that

$$\mathbf{U} = \tfrac{1}{2}\begin{pmatrix} 1/C & 0 \\ 0 & 1 \end{pmatrix}\mathbf{E} + \tfrac{1}{2}\begin{pmatrix} 1 & 0 \\ 0 & -1/C \end{pmatrix}\mathbf{H}, \tag{22.45}$$

and

$$\mathbf{D} = \tfrac{1}{2}\begin{pmatrix} -1/C & 0 \\ 0 & 1 \end{pmatrix}\mathbf{E} + \tfrac{1}{2}\begin{pmatrix} 1 & 0 \\ 0 & 1/C \end{pmatrix}\mathbf{H}. \tag{22.46}$$

From (22.42), (22.45) and (22.46) **E** and **H** can now be eliminated, and the result is

$$\mathbf{R}\begin{pmatrix} 1/C & 0 \\ 0 & 1 \end{pmatrix} + \mathbf{R}\begin{pmatrix} 1 & 0 \\ 0 & -1/C \end{pmatrix}\mathbf{A} = \begin{pmatrix} -1/C & 0 \\ 0 & 1 \end{pmatrix} + \begin{pmatrix} 1 & 0 \\ 0 & 1/C \end{pmatrix}\mathbf{A}, \tag{22.47}$$

whence it is easily shown that

$$\mathbf{R} = \begin{pmatrix} 1 & 0 \\ 0 & -1 \end{pmatrix} + 2\begin{pmatrix} -CA_{11}-1 & A_{12} \\ A_{21} & 1-A_{22}/C \end{pmatrix}^{-1}. \tag{22.48}$$

This formula is used for calculating **R** in the free space below the ionosphere. Equation (22.48) is a definition of **R** at any level within the ionosphere and is equivalent to the definition (22.20).

There is a useful physical interpretation of **R** within the ionosphere. Suppose that the integration of (22.38) is stopped at a level within the ionosphere. This is equivalent to terminating the ionosphere by a sharp boundary at that level, with free space below, and if **R** is now calculated from (22.48) its value is the reflection coefficient for that boundary. Suppose now that (22.38) is integrated proceeding downwards for a very large distance through a homogeneous ionosphere. Then **A** acquires the value which applies to upgoing waves only and does not change thereafter. The integration is now stopped, which is equivalent to terminating the homogeneous ionosphere by a sharp boundary. **R** is calculated from (22.48), and is the reflection coefficient when there are only upgoing waves above the boundary. It is therefore the reflection coefficient matrix for a sharply bounded homogeneous medium, and will be denoted by $\mathbf{R}_1$. In particular if the medium is isotropic, $\mathbf{R}_1$ is diagonal and its components are the Fresnel reflection coefficients (8.14) and (8.22). In the general case it is convenient to call $\mathbf{R}_1$ the 'Fresnel reflection coefficient matrix'.

A method of calculating $\mathbf{R}_1$ was given in ch. 8, but was rather laborious because it involved solving the Booker quartic equation with complex coefficients. A much easier method, if a digital computer is available, is to integrate (22.38) for the homogeneous medium until all elements of $\mathbf{A}'$ are less than a pre-assigned small quantity, and then to calculate $\mathbf{R}_1$ from (22.48).

22.12 The differential equation for A

The differential equation (22.38) is very useful for computing and is therefore given in full:

$$\left.\begin{aligned} iA'_{11} &= -T_{14}A_{11}^2 + A_{12}A_{21} + (T_{44}-T_{11})A_{11} + T_{41}, \\ iA'_{12} &= -T_{14}A_{11}A_{12} + A_{12}A_{22} + T_{12}A_{11} + T_{44}A_{12} - T_{42}, \\ iA'_{21} &= -T_{14}A_{11}A_{21} + A_{21}A_{22} + T_{34}A_{11} - T_{11}A_{21} + T_{31}, \\ iA'_{22} &= -T_{14}A_{12}A_{21} + A_{22}^2 + T_{34}A_{12} + T_{12}A_{21} - T_{32}. \end{aligned}\right\} \tag{22.49}$$

(The independent variable is $s = kz$, and a dash $'$ means d/ds.) The elements of $\mathbf{T}$ are given by (18.17) and it is convenient to express them in terms of the auxiliary functions

$$b_1 = U/(U^2-Y^2), \quad b_2 = Y/(U^2-Y^2), \quad b_3 = Y^2/(U^2-Y^2)\,U, \tag{22.50}$$

$$T_{14} = W_1 = \frac{C^2 - b_1X + n^2b_3X}{1 - b_1X + n^2b_3X}, \quad W_2 = \frac{b_2X}{1 - b_1X + n^2b_3X},$$

$$W_3 = \frac{b_3X}{1 - b_1X + n^2b_3X}. \tag{22.51}$$

The equations (22.49) then become:

$$\left.\begin{aligned} iA'_{11} &= -W_1A_{11}^2 + A_{12}A_{21} + 2imSW_2A_{11} + 1 - Xb_1 + Xl^2b_3 \\ &\quad - X(l^2n^2b_3W_3 + m^2b_2W_2), \\ iA'_{12} &= -W_1A_{11}A_{12} + A_{12}A_{22} + S(mnW_3 - ilW_2)A_{11} + S(imW_2 - lnb_3)A_{12} \\ &\quad + X(inb_2 + lmb_3) - X\{lmn^2b_3W_3 - lmb_2W_2 - in(l^2+m^2)b_2W_3\}, \\ iA'_{21} &= -W_1A_{11}A_{21} + A_{21}A_{22} + S(mnW_3 + ilW_2)A_{11} + S(imW_2 + lnb_3)A_{21} \\ &\quad + X(inb_2 - lmb_3) + X\{lmn^2b_3W_3 - lmb_2W_2 + in(l^2+m^2)b_2W_3\}, \\ iA'_{22} &= -W_1A_{12}A_{21} + A_{22}^2 + S(mnW_3 + ilW_2)A_{12} + S(mnW_3 - ilW_2)A_{21} \\ &\quad - C^2 + Xb_1 - Xm^2b_3 + X(m^2n^2b_3W_3 + l^2b_2W_2). \end{aligned}\right\} \tag{22.52}$$

It is easily verified that when the earth's magnetic field is neglected these reduce to the two equations (22.24) and (22.25).

In the special case of propagation from (magnetic) east to west or west to east, $l = 0$, and inspection of (22.52) shows that $\mathbf{A}'$ is symmetric if $\mathbf{A}$

is symmetric. Now above the ionosphere the starting value of **A** is diagonal so that **A** is there symmetric, and the differential equation shows that it must remain symmetric at all levels. Hence only three (complex) variables are necessary in this case instead of four and the equations become

$$\left.\begin{aligned} iA'_{11} &= -W_1 A_{11}^2 + A_{12}^2 + 2imSW_2 A_{11} + 1 - Xb_1 - Xm^2 b_2 W_2, \\ iA'_{12} &= -W_1 A_{11} A_{12} + A_{12} A_{22} + SmnW_3 A_{11} + imSW_2 A_{12} \\ &\qquad + inXb_2 + inm^2 Xb_2 W_3, \\ iA'_{22} &= -W_1 A_{12}^2 + A_{22}^2 + 2SmnW_3 A_{12} - C^2 + Xb_1 \\ &\qquad - Xm^2 b_3 + Xm^2 n^2 b_2 W_2. \end{aligned}\right\} \quad (22.53)$$

Equation (22.48) shows that **R** is also symmetric in this case.

22.13 Symmetry properties of the differential equations

Inspection of (22.52) shows that if A_{12} and A_{21} are interchanged and the sign of l is reversed, the equations are unaltered. Now the equations in the two cases where l has opposite signs can be integrated using the same starting value (22.39) above the ionosphere. Hence the two values of **A** below the ionosphere would be transposes of each other, and the values of **R** derived from them would also be transposes.

This result can be stated more specifically as follows. Consider two directions of propagation for which the plane of incidence makes angles ψ_I, ψ_{II} with the magnetic meridian, and let superscripts (I) and (II) be used to distinguish the elements of **R** in these two cases. Let $\psi_I + \psi_{II} = \pi$, so that the two planes of incidence make equal angles with a plane at right angles to the magnetic meridian. Then if the angles of incidence are the same:

$$\left.\begin{aligned} {}_{\parallel}R_{\parallel}^{(\mathrm{I})} &= {}_{\parallel}R_{\parallel}^{(\mathrm{II})}, & {}_{\parallel}R_{\perp}^{(\mathrm{I})} &= {}_{\perp}R_{\parallel}^{(\mathrm{II})}, \\ {}_{\perp}R_{\parallel}^{(\mathrm{I})} &= {}_{\parallel}R_{\perp}^{(\mathrm{II})}, & {}_{\perp}R_{\perp}^{(\mathrm{I})} &= {}_{\perp}R_{\perp}^{(\mathrm{II})}. \end{aligned}\right\} \quad (22.54)$$

This property is used in the next chapter in the special case when $m = 0$, and then leads to a reciprocity theorem.

22.14 Equivalent height of reflection

Calculations made as in the preceding sections give the over-all reflection coefficient of the ionosphere. This may arise from waves reflected at several different levels in the ionosphere, adding together to give a resultant reflected wave. In measurements with 'Continuous Waves' at a single fixed frequency only this resultant reflection could be observed. But when pulsed signals are used, the reflections are often separated because the waves travel different distances and the times of travel are different. In other words, the equivalent heights of reflection (§§ 10.3 and 12.5) are different. It may be important to know whether a calculated reflection coefficient matrix **R** is the result of only one reflection or of several reflections, and in either case the equivalent heights of reflection need to be found. This problem has not been very fully studied, but certain general observations are possible.

The use of pulses is equivalent to using a range of frequencies instead of a single frequency. The frequencies spread over a 'band width' Δf which is of the order of $1/\tau$ where τ is the duration of the pulse. If a reflection coefficient is to be resolved into its components, it must clearly be studied over a range of frequencies of this order.

When the ionosphere is treated as an isotropic medium (chs. 10 and 17) it is usually possible to decide whether or not there is more than one component in the reflected wave, simply by examining the electron density profile. If two components are present there must be partial penetration and reflection in a region of maximum electron density below the highest reflection level. The discussions in §§ 17.5 and 17.16 show that this is unlikely to occur except for a very small range of frequency near the penetration-frequency. If two reflected waves are present they would interfere or reinforce depending on the relative phase. Thus $R(f)$ is an oscillatory function of frequency and if the reflections are to be resolved by radio pulses there must be at least one cycle of the oscillation within the band width Δf. This suggests that in doubtful cases $|R|$ and $\arg R$ should be plotted against frequency. The presence of two reflections could then be recognised and the components are easily separated. The equivalent height for a single reflection can be calculated by finding $(\partial/\partial f)(\arg R)$ (see § 17.6 for a worked example).

For an anisotropic ionosphere the problem is more difficult because the reflected wave may contain two components, over a wide range of frequency. These may be, for example, the ordinary and extraordinary waves reflected from different levels, or the reflected extraordinary wave together with the 'coupling echo' arising from backscatter in the coupling region (§ 19.9).

A simple test can be applied to find whether **R** arises from a single magneto-ionic component. For if it does, the reflected wave must always have the same polarisation appropriate to that component, regardless of the polarisation of the incident wave. Then it can be shown† (see example 1 at the end of ch. 7) that the matrix **R** is singular, that is ${}_{\parallel}R_{\parallel}\,{}_{\perp}R_{\perp} - {}_{\parallel}R_{\perp}\,{}_{\perp}R_{\parallel} = 0$. When this is true, it usually happens that the polarisation changes only very slowly with frequency. Then the equivalent height can be found by calculating $(\partial/\partial f)(\arg R)$, which has the same value for all components of **R**.

No general method seems to have been given for separating **R** into components when it is not singular, although empirical methods can sometimes be found in special cases. This problem needs further study. When resolution of the reflected wave into components is important, it may be advantageous to use the coupled wave-equations (see § 22.4 and the references given there).

† For the coupling echo, § 19.9, **R** is not singular, since the polarisation of the reflected wave depends markedly on that of the incident wave.

Example

Show that in an isotropic ionosphere the reflection coefficients ${}_{\parallel}R_{\parallel}$ and ${}_{\perp}R_{\perp}$, defined by (22.30) and (22.31), respectively, satisfy the differential equations:

$$2i\,{}_{\parallel}R'_{\parallel} = C\mathfrak{n}^2(1-{}_{\parallel}R_{\parallel})^2 - \frac{1}{C}\left(1-\frac{S^2}{\mathfrak{n}^2}\right)(1+{}_{\parallel}R_{\parallel})^2$$

and $$2i\,{}_{\perp}R'_{\perp} = C(1-{}_{\perp}R_{\perp})^2 - \frac{1}{C}(\mathfrak{n}^2-S^2)(1+{}_{\perp}R_{\perp})^2,$$

where $\mathfrak{n}$ is the refractive index.

(Equations similar to these have been given by Schelkunoff, 1951.)

CHAPTER 23

RECIPROCITY

23.1 Introduction

The reciprocity theorem for electrical systems is sometimes stated as follows (e.g. Carson, 1929): 'Let some closed electrical system have two pairs of terminals A and B. Let an e.m.f. V from a generator of zero internal impedance be applied to the terminal pair A, and let the current, i, across the terminals B be measured when B is short-circuited. Now let the e.mf. V be removed from A and applied to B. Then the current between the terminals A when they are short-circuited is equal to i.'

This theorem is true when the system is an electric circuit which is passive and linear and contains only resistances, inductances and capacities (lumped or distributed). There are some conditions when it is also true if the system includes a radiation link, so that it may then be applied if the terminal pairs A and B are those of two radio aerials. For this to hold, it is usually stated that the dielectric constants, conductivities and permeabilities of the media through which the radiation passes must be symmetric tensors (Sommerfeld, 1925; Dällenbach, 1942). This requirement is not fulfilled by a magnetoionic medium, so for two aerials joined by a radiation link which includes a reflection from, or transmission through, the ionosphere, the theorem does not necessarily hold, and it will in general be found that communication in one direction is easier than in the other. There are some special cases, however, where there is reciprocity for two aerials connected by a radiation link which traverses the ionosphere.

The proofs of the standard reciprocity theorems are not given here. (For the reciprocity theorem for circuits, see, for example, Guillemin, *Communication Networks*, vol. I, 1946, ch. IV, para. 7. For the reciprocity theorem for fields see, for example: Huxley, *Waveguides*, 1947, §7.17; and Schelkunoff and Friis, 1952.)

23.2 Aerials

The question whether a given radiation link via the ionosphere is reciprocal depends very much on the aerials used at the two ends of the path. The problem is really one of aerial design, rather than a propagation problem.

In this chapter we shall discuss the problem illustrated in Fig. 23.1. A and B are two points on the earth's surface. At A is placed an aerial which has a single pair of terminals. When a voltage of given frequency is applied to these, the aerial radiates and some energy from it is reflected from the ionosphere and reaches B where it is received by another aerial. The aerial at B may be of quite different design, but it also has a single pair of terminals which are short-circuited and the current is measured. Alternatively, the voltage is applied to the terminals of the aerial at B and the current across the short-circuited terminals at A is measured. The *same aerials* are used for transmission and reception at both places.

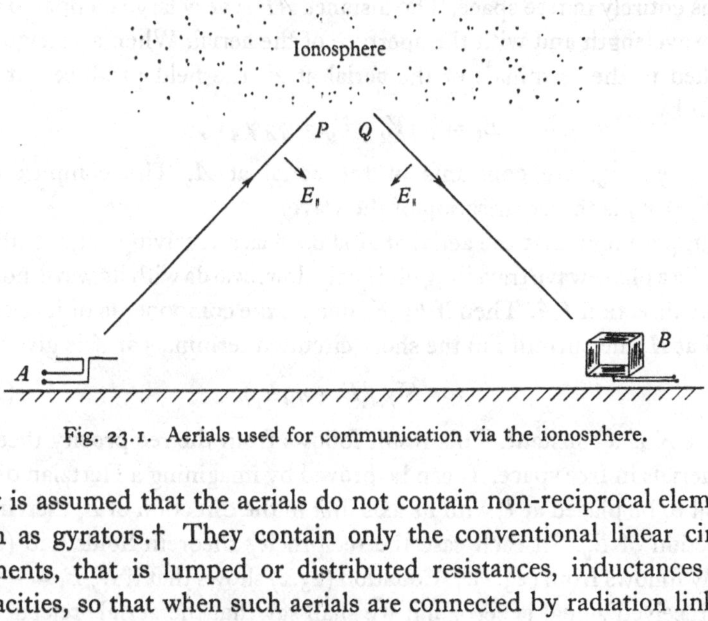

Fig. 23.1. Aerials used for communication via the ionosphere.

It is assumed that the aerials do not contain non-reciprocal elements such as gyrators.† They contain only the conventional linear circuit elements, that is lumped or distributed resistances, inductances and capacities, so that when such aerials are connected by radiation links in free space, the reciprocity theorem is obeyed.

When the aerial at A radiates, suppose that the part of a wave-front which reaches B travels via the point P just below the ionosphere. Then the signal at B is proportional to the strength of the field radiated from A in the direction AP. The radiated fields in other directions do not concern us since they do not reach B. Similarly, when the aerial at B radiates, assume that it is the field radiated in the direction BQ which reaches A.

† A gyrator is a device which usually contains a substance whose dielectric constant tensor or magnetic permeability tensor is not symmetric. The use of such devices is equivalent to using *different* aerials for transmission and reception.

A wave of given frequency travelling with its wave-normal in the direction AP or PA is in general elliptically polarised and may be specified by two components of its electric field denoted by E_y which is horizontal and perpendicular to AP, and $E_{\parallel}$ which is obliquely downwards in the plane of incidence, so that $E_{\parallel}$, E_y, and AP form a right-handed system. The ratio $E_y/E_{\parallel}$ will be called the polarisation of the wave, for waves in both directions (see end of § 5.4). Similarly, for a wave with its wave-normal in the direction QB the field is specified by E_y and $E_{\parallel}$ where E_y (but not $E_{\parallel}$) is the same direction as before and $E_{\parallel}$, E_y and QB form a right-handed system.

The point P is assumed to be just below the ionosphere so that the path AP is entirely in free space. The distance AP is very large compared with the wavelength and with the aperture of the aerial. When a voltage V is applied to the terminals of the aerial at A, the field produced at P is given by

$$E_{\parallel} = \gamma_A V, \quad E_y = \eta_A \gamma_A V, \tag{23.1}$$

where η_A, γ_A are constants of the aerial at A. The complex ratio $E_y/E_{\parallel} = \eta_A$ is the polarisation of the wave.

Suppose now that the aerial at A is used as a receiving aerial and that there is a plane wave travelling obliquely downwards with its wave-normal in the direction PA. Then if $E_{\parallel}$, E_y denote the components of its electric field at A, the current i in the short-circuited terminals at A is given by

$$i = K\gamma_A(E_{\parallel} + \eta_A E_y), \tag{23.2}$$

where K is a constant. This result follows from the reciprocity theorem for aerials in free space. It can be proved by imagining a Hertzian dipole aerial to be placed at P with its axis first in the direction of $E_{\parallel}$ then in the direction of E_y. In each case the reciprocity theorem holds and (23.2) easily follows from (23.1).† Equation (23.2) shows that if $E_y/E_{\parallel} = -1/\eta_A$ the received signal is zero, and we shall say that the aerial 'rejects' this signal.

Next consider the wave radiated in the direction BQ when the voltage V is applied to the terminals of the aerial at B, where Q is just below the ionosphere. For this wave

$$E_{\parallel} = \gamma_B V, \quad E_y = \eta_B \gamma_B V, \tag{23.3}$$

where η_B, γ_B are constants of the aerial at B, and $\eta_B = E_y/E_{\parallel}$ is the polarisation of the wave. If the aerial at B is used as a receiving aerial

† Equation (23.2) shows that there is reciprocity in phase as well as amplitude. The application of the theory of reciprocity to the phase has been discussed by Browne (1958).

and is placed in the field of a plane wave with its wave-normal in the direction QB and field components $E_{\|}, E_y$, then the current i in the short-circuited terminals is

$$i = K\gamma_B(E_{\|} + \eta_B E_y) \tag{23.4}$$

by the reciprocity theorem. Thus the aerial at B rejects a wave for which

$$E_y/E_{\|} = -1/\eta_B.$$

23.3 Goubau's reciprocity theorem

Goubau (1942) proved the equivalent of the following theorem: 'For any aerial system at A it is always possible to construct at least one aerial system at B so that there is reciprocity between the terminals of the two aerials.'

To prove it in the present context it is convenient to use the reflection coefficient matrix $\mathbf{R}$ of the ionosphere, defined in §7.6. This is different for waves going from A to B and from B to A and the symbols $\mathbf{R}^{(\mathrm{I})}$ and $\mathbf{R}^{(\mathrm{II})}$ respectively will be used.

Suppose that a voltage V is applied to the terminals of the aerial at A. Then the field at P is given by (23.1) and the field at B after reflection may be written in matrix form

$$\begin{pmatrix} E_{\|} \\ E_y \end{pmatrix} = K_1\gamma_A V\mathbf{R}^{(\mathrm{I})}\begin{pmatrix} 1 \\ \eta_A \end{pmatrix}, \tag{23.5}$$

where K_1 is a constant which allows for the extra distance QB travelled after reflection. Hence the current in the short-circuited terminals of the aerial at B is

$$\begin{aligned} i &= KK_1\gamma_A\gamma_B V(1, \eta_B)\mathbf{R}^{(\mathrm{I})}\begin{pmatrix} 1 \\ \eta_A \end{pmatrix} \\ &= KK_1\gamma_A\gamma_B V({}_{\|}R_{\|}^{(\mathrm{I})} + \eta_A\,{}_{\|}R_{\perp}^{(\mathrm{I})} + \eta_B\,{}_{\perp}R_{\|}^{(\mathrm{I})} + \eta_A\eta_B\,{}_{\perp}R_{\perp}^{(\mathrm{I})}). \end{aligned} \tag{23.6}$$

Similarly, when the voltage V is applied to the terminals of the aerial at B, the current in the short-circuited terminals of the aerial at A is

$$i = KK_1\gamma_A\gamma_B V({}_{\|}R_{\|}^{(\mathrm{II})} - \eta_B\,{}_{\|}R_{\perp}^{(\mathrm{II})} - \eta_A\,{}_{\perp}R_{\|}^{(\mathrm{II})} + \eta_A\eta_B\,{}_{\perp}R_{\perp}^{(\mathrm{II})}) \tag{23.7}$$

(the minus signs appear because of the sign convention used in the definition of $\mathbf{R}$ (§7.5)). For reciprocity the currents (23.6) and (23.7) must be equal, which requires that

$$-\eta_B = \frac{{}_{\|}R_{\|}^{(\mathrm{I})} - {}_{\|}R_{\|}^{(\mathrm{II})} + \eta_A({}_{\|}R_{\perp}^{(\mathrm{I})} + {}_{\perp}R_{\|}^{(\mathrm{II})})}{{}_{\perp}R_{\|}^{(\mathrm{I})} + {}_{\|}R_{\perp}^{(\mathrm{II})} + \eta_A({}_{\perp}R_{\perp}^{(\mathrm{I})} - {}_{\perp}R_{\perp}^{(\mathrm{II})})}. \tag{23.8}$$

Thus when η_A is given, an aerial is constructed at B which radiates the polarisation η_B given by (23.8). This is possible for any complex η_B and hence Goubau's theorem is proved.

If there is to be reciprocity for *any* value of η_B, it is necessary that the numerator and denominator of (23.8) shall both be zero. If there is to be reciprocity for any aerial systems whatever, that is for any η_A and η_B, then it is necessary that

$$_{\parallel}R_{\parallel}^{(\mathrm{I})} = {_{\parallel}R_{\parallel}^{(\mathrm{II})}}, \quad _{\perp}R_{\perp}^{(\mathrm{I})} = {_{\perp}R_{\perp}^{(\mathrm{II})}}, \quad _{\parallel}R_{\perp}^{(\mathrm{I})} = -{_{\perp}R_{\parallel}^{(\mathrm{II})}}, \quad _{\perp}R_{\parallel}^{(\mathrm{I})} = -{_{\parallel}R_{\perp}^{(\mathrm{II})}}. \tag{23.9}$$

This may be called 'complete reciprocity' and in general it can only be achieved when the ionosphere is isotropic.

23.4 One magnetoionic component

We now consider the case where the wave travelling though the ionosphere from A to B is a single magnetoionic component (ordinary or extraordinary wave). This problem often arises at high frequencies either because the other component is absorbed, or lost through penetration, or because the components are resolved in time by the use of pulses, so that observations can be made of only one component. We assume that the ionosphere can be treated as 'slowly varying', so that the ray theory of chs. 12 to 14 can be used.

The path of a wave-packet going from A to B can be found by the methods of ch. 13, using the Booker quartic. Such a path is for one magnetoionic component. It was shown in § 13.6 that if the wave-packet received at B is reversed in direction, it retraces the original path back to A. The polarisation, referred to fixed axes, is the same for both directions. It was further shown in § 13.16 that the attenuation is the same for both directions. It might therefore be thought that there is complete reciprocity in this case, but a simple example shows that this is not so. Suppose that at one end A of the path the two characteristic waves are circularly polarised with opposite senses. This might happen if the direction AP (Fig. 23.1) is parallel to the earth's magnetic field. Suppose that communication is to use the ordinary ray only, for which, at A, $E_y/E_\parallel = i$. The aerial at A is therefore constructed to radiate exactly this polarisation, so that no extraordinary wave is generated. The ordinary wave travels to B and can be received there on a suitable aerial. Now consider the reverse path. The ordinary wave arriving at A has the, polarisation $E_y/E_\parallel = i$, but this is completely rejected by the aerial at A.†

† It may seem surprising that an aerial can reject a signal with the same polarisation (referred to *fixed* axes) as it radiates. This is only true for circular polarisation.

Hence there cannot in general be reciprocity. Goubau's reciprocity theorem states that there must be reciprocity for *some* aerial at B, and in this case it must be an aerial which rejects the ordinary wave at B so that we have the trivial case where there is no communication in either direction. Communication with reciprocity would, of course, be possible if a different aerial were used at A.

It is now natural to ask what aerial systems must be used at A and B to give reciprocity with a single magnetoionic component. The answer is not simple but can be found as follows. We need expressions for the reflection coefficients $\mathbf{R}^{(\mathrm{I})}$ and $\mathbf{R}^{(\mathrm{II})}$. Suppose that communication is to use the ordinary wave whose polarisation is ρ_A at A, and ρ_B at B. Then $\mathbf{R}^{(\mathrm{I})}$ must have the following properties: (*a*) it gives a reflected wave of polarisation ρ_B no matter what the polarisation of the incident wave; (*b*) it gives zero reflection when the incident wave has the polarisation of the extraordinary wave at A.

It is therefore necessary to make some assumption about the extraordinary wave at A, and we assume that it has polarisation $1/\rho_A$.† (This is given by the magnetoionic theory (5.19), but may not be true below the ionosphere since the 'limiting polarisations' there do not necessarily obey the relation $\rho_o\rho_x = 1$.) Similarly, the extraordinary wave at B is assumed to have the polarisation $1/\rho_B$. Then $\mathbf{R}^{(\mathrm{I})}$ and $\mathbf{R}^{(\mathrm{II})}$ are given by

$$\mathbf{R}^{(\mathrm{I})} = {}_{\parallel}R_{\parallel}^{(\mathrm{I})}\begin{pmatrix} 1 & -\rho_A \\ \rho_B & -\rho_A\rho_B \end{pmatrix}, \quad \mathbf{R}^{(\mathrm{II})} = {}_{\parallel}R_{\parallel}^{(\mathrm{II})}\begin{pmatrix} 1 & \rho_B \\ \rho_A & \rho_A\rho_B \end{pmatrix} \tag{23.10}$$

(plus signs appear in the right-hand terms of $\mathbf{R}^{(\mathrm{II})}$ because of the sign convention used in the definition of $\mathbf{R}$ (§7.5)). The phase paths are equal for the two directions, which gives

$$\arg\{(1-\rho_A^2)\,{}_{\parallel}R_{\parallel}^{(\mathrm{I})}\} = \arg\{(1-\rho_B^2)\,{}_{\parallel}R_{\parallel}^{(\mathrm{II})}\}. \tag{23.11}$$

The attenuations are equal for the two directions, which means that if an ordinary wave with unit power flux is incident at one end, the power flux in the reflected wave is the same for both directions. This gives

$$\frac{|{}_{\parallel}R_{\parallel}^{(\mathrm{I})}|^2\,|1-\rho_A^2|^2(1+\rho_B\rho_B^*)}{1+\rho_A\rho_A^*} = \frac{|{}_{\parallel}R_{\parallel}^{(\mathrm{II})}|^2\,|1-\rho_B^2|^2(1+\rho_A\rho_A^*)}{1+\rho_B\rho_B^*} \tag{23.12}$$

(a star * denotes a complex conjugate).

Now combine (23.11) and (23.12), whence

$$\frac{{}_{\parallel}R_{\parallel}^{(\mathrm{I})}(1-\rho_A^2)}{1+\rho_A\rho_A^*} = \frac{{}_{\parallel}R_{\parallel}^{(\mathrm{II})}(1-\rho_B^2)}{1+\rho_B\rho_B^*}. \tag{23.13}$$

Suppose that the aerials at A and B have the constants defined in (23.1) to (23.4). Then if a voltage V is applied to the terminals of the aerial at A, the current in the short-circuited terminals of the aerial at B is

$$i = KK_1\gamma_A\gamma_B\,{}_{\parallel}R_{\parallel}^{(\mathrm{I})}(1-\rho_A\eta_A+\rho_B\eta_B-\rho_A\rho_B\eta_A\eta_B). \tag{23.14}$$

† See note 2 on p. 529.

Similarly, if V is applied to the terminals at B, the current at A is

$$i = KK_1\gamma_A\gamma_B\,{}_{\parallel}R_{\parallel}^{(\mathrm{II})}(1+\rho_A\eta_A-\rho_B\eta_B-\rho_A\rho_B\eta_A\eta_B). \qquad (23.15)$$

For reciprocity these must be equal, and when (23.13) is used, the condition is

$$\frac{1+\rho_A\rho_A^*}{1-\rho_A^2}(1-\rho_A\eta_A+\rho_B\eta_B-\rho_A\rho_B\eta_A\eta_B)$$

$$= \frac{1+\rho_B\rho_B^*}{1-\rho_B^2}(1+\rho_A\eta_A-\rho_B\eta_B-\rho_A\rho_B\eta_A\eta_B). \qquad (23.16)$$

This shows how to find η_B, for example, when η_A, ρ_A, ρ_B are known.

As an illustration of the use of (23.16) suppose that horizontal dipole aerials are used at both ends. Then $\eta_A = \eta_B \to \infty$† and (23.16) shows that for reciprocity

$$\frac{1+\rho_A\rho_A^*}{1-\rho_A^2} = \frac{1+\rho_B\rho_B^*}{1-\rho_B^2}.$$

This is satisfied if $\rho_A = \pm\rho_B$. In temperate latitudes a single magnetoionic component is often very nearly circularly polarised when it reaches the ground, so that this condition for reciprocity is fairly well satisfied.

23.5 Reciprocity with full wave solutions

At low frequencies the reflection coefficient matrix **R** must be found by using full wave theory, as in chs. 21 and 22. For communication in two opposite directions there is never complete reciprocity, but Goubau's reciprocity theorem is always true. There are some cases where the aerial systems which satisfy Goubau's theorem are particularly simple. These occur when the plane of incidence is the magnetic meridian, that is for communication from magnetic north to south or south to north.

In §22.13 it was shown that for two directions of propagation whose planes of incidence make angles ψ and $\pi-\psi$ with the magnetic meridian, the components of the reflection coefficient matrices satisfy (22.54). Now in the special case $\psi = 0$, the two planes of incidence coincide with the magnetic meridian and the directions of propagation are opposite.

Equation (22.54) gives

$${}_{\parallel}R_{\parallel}^{(\mathrm{I})} = {}_{\parallel}R_{\parallel}^{(\mathrm{II})}, \quad {}_{\perp}R_{\perp}^{(\mathrm{I})} = {}_{\perp}R_{\perp}^{(\mathrm{II})}, \qquad (23.17)$$

$${}_{\parallel}R_{\perp}^{(\mathrm{I})} = {}_{\perp}R_{\parallel}^{(\mathrm{II})}, \quad {}_{\perp}R_{\parallel}^{(\mathrm{I})} = {}_{\parallel}R_{\perp}^{(\mathrm{II})}. \qquad (23.18)$$

These are very different from the conditions (23.9) needed for complete reciprocity, because the last two relations in (23.9) have minus signs.

Equation (23.17) shows that for propagation from magnetic north to south or south to north, there is reciprocity if *both* aerials are vertical dipoles or if *both* are horizontal dipoles.

† γ_A and γ_B tend to zero, but $\gamma_A\eta_A$ and $\gamma_B\eta_B$ are finite and non-zero.

Suppose now, that there is a vertical dipole aerial at A and a horizontal dipole aerial at B. When A transmits, the current in the short-circuited terminals of the aerial at B is proportional to ${}_{\parallel}R_{\perp}^{(\mathrm{I})}$, but when B transmits, the current at A is proportional to $-{}_{\perp}R_{\parallel}^{(\mathrm{II})}$ (the minus sign is needed because of the sign convention used in the definition of $\mathbf{R}$; § 7.5). Hence there is reciprocity for the amplitude of the signals passed in the two directions, but the phase shifts differ by 180°. The same result applies for a horizontal dipole aerial at A and a vertical dipole aerial at B.

These results have been proved for the reflection of a single plane wave, but it can be shown (Budden, 1954) that they must also be true for the spherical wave from a small source, and also when allowance is made for the earth's curvature. They are based on the complete wave-equations, that is on a 'full wave' solution, and they therefore apply to all frequencies. Thus they apply both when there is almost specular reflection at very low frequencies, and when there is a gradual bending of the rays as in the return of high-frequency waves from the ionosphere. They also apply, for example, to waves which are returned from the F-layer and make a double passage through the E-layer, or to waves which make multiple reflections from the ionosphere and the surface of the earth. They are therefore applicable to radio transmission over any distance in the magnetic north–south direction.

APPENDIX

The Stokes constant for the differential equation (16.98) for 'vertical' polarisation

A radio wave is incident obliquely on the ionosphere with its wave-normal at an angle arc sin S to the vertical, and is linearly polarised with its electric vector in the plane of incidence. The earth's magnetic field is neglected, the electron collision-frequency is constant, and the electron density increases linearly with height z. The horizontal component $\mathscr{H}$ of the magnetic field of the wave then satisfies (16.98), namely

$$\frac{d^2\mathscr{H}}{d\mathfrak{z}^2} - \frac{1}{\mathfrak{z}}\frac{d\mathscr{H}}{d\mathfrak{z}} - (\mathfrak{z}+B)\mathscr{H} = 0, \tag{A. 1}$$

where $\mathfrak{z}$ is given by (16.97) and (16.93), and is a linear function of the height z, and B is a constant proportional to S^2, given by (16.99). This appendix describes a method of calculating the Stokes constant $\mathscr{S}$ of the equation (A. 1).

It is convenient to express (A. 1) as two simultaneous differential equations of the first order, thus

$$\frac{d\mathscr{H}}{d\mathfrak{z}} = F\mathfrak{z}, \quad \frac{dF}{d\mathfrak{z}} = \frac{\mathfrak{z}+B}{\mathfrak{z}}\mathscr{H}. \tag{A. 2}$$

It can be shown (see § 16.16) that, when $|\mathfrak{z}|$ is large enough, the two W.K.B. solutions of these equations give

$$F \sim \mathfrak{z}^{-\frac{1}{4}} \exp\{\pm \tfrac{2}{3}(\mathfrak{z}+B)^{\frac{3}{2}}\}. \tag{A. 3}$$

When $B = 0$ these are the same as for the Stokes equation, and F must then satisfy the Stokes equation exactly. Now (A. 3) shows that there is a Stokes line where $\arg \mathfrak{z} = \frac{2}{3}\pi$ and anti-Stokes lines where $\arg \mathfrak{z} = \frac{1}{3}\pi$ and π. A method will now be described for calculating the Stokes constant for the Stokes line $\arg \mathfrak{z} = \frac{2}{3}\pi$.

Fig. A. 1 is a diagram of the complex $\mathfrak{z}$-plane and P is a point on the anti-Stokes line $\arg \mathfrak{z} = \frac{1}{3}\pi$. Consider first a solution which, near P, satisfies

$$F = K\mathfrak{z}^{-\frac{1}{4}} \exp\{\tfrac{2}{3}(\mathfrak{z}+B)^{\frac{3}{2}}\}, \tag{A. 4}$$

where $\arg(\mathfrak{z}+B)^{\frac{3}{2}}$ is taken as near $+\frac{1}{2}\pi$. As $\arg \mathfrak{z}$ increases, this term becomes subdominant and so on crossing the Stokes line where $\arg \mathfrak{z} = \frac{2}{3}\pi$, the Stokes phenomenon does not occur, and when $\arg \mathfrak{z} = \pi$ the solution still satisfies (A. 4). Next consider the solution which at P satisfies

$$F = F_1 = K\mathfrak{z}^{-\frac{1}{4}} \exp\{-\tfrac{2}{3}(\mathfrak{z}+B)^{\frac{3}{2}}\}. \tag{A. 5}$$

Now as $\arg \mathfrak{z}$ increases, this term becomes dominant, and when $\arg \mathfrak{z}$ reaches the value π the solution (A. 5) must be replaced by

$$F = F_2 = K\mathfrak{z}^{-\frac{1}{4}}[\exp\{-\tfrac{2}{3}(\mathfrak{z}+B)^{\frac{3}{2}}\} - \mathscr{S} \exp\{\tfrac{2}{3}(\mathfrak{z}+B)^{\frac{3}{2}}\}], \tag{A. 6}$$

where $\mathscr{S}$ is the Stokes constant. The minus sign before $\mathscr{S}$ is needed because the dependent variable in (A. 1) is $\mathscr{H}$, whereas (A. 6) refers to F, which has a factor $d\mathscr{H}/d\mathfrak{z}$.

The constant $\mathscr{S}$ was calculated by integrating (A. 2) along a straight line PQ (see Fig. A. 1), where P and Q are at the same distance d from the origin, and are on the anti-Stokes lines $\arg \mathfrak{z} = \frac{1}{3}\pi$ and π, respectively. Thus, let r be distance along PQ measured from P. Then in terms of r the equations (A. 2) are

$$\frac{d\mathscr{H}}{dr} = -F\mathfrak{z}\exp(i\pi/6), \quad \frac{dF}{dr} = -\frac{\mathfrak{z}+B}{\mathfrak{z}}\mathscr{H}\exp(i\pi/6). \tag{A. 7}$$

Initially, at the point P, F is given by (A. 5), and (A. 2) show that the value of $\mathscr{H}$ is then

$$\mathscr{H} \approx -\mathfrak{z}^{\frac{1}{2}}F, \tag{A. 8}$$

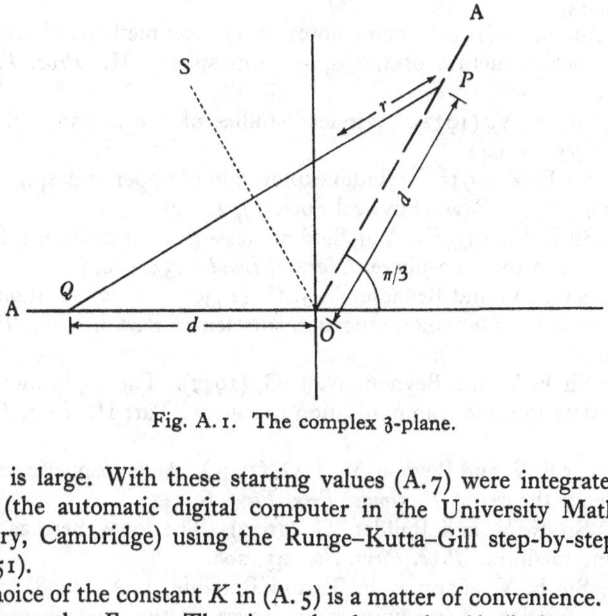

Fig. A. 1. The complex $\mathfrak{z}$-plane.

when $|\mathfrak{z}|$ is large. With these starting values (A. 7) were integrated on the EDSAC (the automatic digital computer in the University Mathematical Laboratory, Cambridge) using the Runge–Kutta–Gill step-by-step process (Gill, 1951).

The choice of the constant K in (A. 5) is a matter of convenience. It was in fact chosen so that $F_1 = 1$. Then it can be shown that (A. 6) gives

$$-\mathscr{S} = F_2\exp\{\tfrac{1}{6}i\pi - \tfrac{2}{3}(d\,e^{\frac{1}{3}i\pi}+B)^{\frac{3}{2}} - \tfrac{2}{3}(B-d)^{\frac{3}{2}}\} - \exp\{-\tfrac{4}{3}(B-d)^{\frac{3}{2}}\}. \tag{A. 9}$$

The integration was stopped when $\mathscr{I}(\mathfrak{z}) = 0$ and $\mathscr{S}$ was then calculated from (A. 9). In these calculations various values of d were used up to 16, and there were 200 steps or more in the integration. Some integrations were also performed with the starting value (A. 4) and it was confirmed that the Stokes phenomenon did not then occur.

Some results of these calculations are given in Fig. 16.9.

BIBLIOGRAPHY

Airy, Sir G. B. (1838). On the intensity of light in the neighbourhood of a caustic. *Trans. Camb. Phil. Soc.* **6**, 379.

Airy, Sir G. B. (1849). Supplement to a paper on the intensity of light in the neighbourhood of a caustic. *Trans. Camb. Phil. Soc.* **8**, 595.

Al'pert, Ya. L. (1948). On the trajectories of rays in a magneto-active ionised medium—the ionosphere. (In Russian.) *Izvestiya Akademii Nauk, SSSR. Seriya Fizicheskaya*, **12**, 241.

Appleton, Sir E. V. (1928). Some notes on wireless methods of investigating the electrical structure of the upper atmosphere. I. *Proc. Phys. Soc.* **41**, 43.

Appleton, Sir E. V. (1930). Some notes on wireless methods of investigating the electrical structure of the upper atmosphere. II. *Proc. Phys. Soc.* **42**, 321.

Appleton, Sir E. V. (1932). Wireless studies of the ionosphere. *J. Instn Elect. Engrs*, **71**, 642.

Appleton, Sir E. V. (1935*a*). Radio exploration of upper atmospheric ionisation. *Rep. Progr. Phys.* (Physical Society), **2**, 129.

Appleton, Sir E. V. (1935*b*). A method of measuring the collisional frequency of electrons in the ionosphere. *Nature, Lond.*, **135**, 618.

Appleton, Sir E. V. and Beynon, W. J. G. (1940). The application of ionospheric data to radio communication problems. Part I. *Proc. Phys. Soc.* **52**, 518.

Appleton, Sir E. V. and Beynon, W. J. G. (1947). The application of ionospheric data to radio communication problems. Part II. *Proc. Phys. Soc.* **59**, 58.

Appleton, Sir E. V. and Beynon, W. J. G. (1955). An ionospheric attenuation equivalence theorem. *J. Atmos. Terr. Phys.* **6**, 141.

Appleton, Sir E. V. and Builder, G. (1933). The ionosphere as a doubly refracting medium. *Proc. Phys. Soc.* **45**, 208.

Appleton, Sir E. V., Farmer, F. T. and Ratcliffe, J. A. (1938). Magnetic double refraction of medium radio waves in the ionosphere. *Nature, Lond.*, **141**, 409.

Bailey, V. A. (1954). Reflection of waves by an inhomogeneous medium. *Phys. Rev.* **96**, 865.

Banerjea, B. K. (1947). On the propagation of electromagnetic waves through the atmosphere. *Proc. Roy. Soc.* A, **190**, 67.

Barber, N. F. and Crombie, D. D. (1959). V.L.F. reflection from the ionosphere in the presence of a transverse magnetic field. *J. Atmos. Terr. Phys.* **16**, 37.

Barnes, E. W. (1908). A new development of the theory of hypergeometric functions. *Proc. Lond. Math. Soc.* (2), **6**, 141.

Barron, D. W. (1960). Numerical investigation of ionospheric problems. Ph.D. Thesis, Cambridge (unpublished).

Barron, D. W. and Budden, K. G. (1959). The numerical solution of differential equations governing reflection of long radio waves from the ionosphere. III. *Proc. Roy. Soc.* A, **249**, 387.

Becker, W. (1956). Tabellierung der vertikalen Gruppengeschwindigkeit ordentlicher Echos in der Ionosphäre. *Arch. Elektr. Übertr.* **11**, 166.

Berkner, L. V. *See* Booker.

Beynon, W. J. G. (1947). Oblique radio transmission in the ionosphere and the Lorentz polarisation term. *Proc. Phys. Soc.* **59**, 97.

Beynon, W. J. G. *See* Appleton.

Blair, J. C., Brown, J. N. and Watts, J. M. (1953). An ionosphere recorder for low frequencies. *J. Geophys. Res.* **58**, 99.

Boardman, E. M. *See* Burton.

Booker, H. G. (1934). Some general properties of the formulae of the magneto-ionic theory. *Proc. Roy. Soc.* A, **147**, 352.

Booker, H. G. (1935). The application of the magneto-ionic theory to the ionosphere. *Proc. Roy. Soc.* A, **150**, 267.

Booker, H. G. (1936). Oblique propagation of electromagnetic waves in a slowly varying non-isotropic medium. *Proc. Roy. Soc.* A, **155**, 235.

Booker, H. G. (1939). The propagation of wave packets incident obliquely on a stratified doubly refracting ionosphere. *Phil. Trans.* A, **237**, 411.

Booker, H. G. (1947). The elements of wave propagation using the impedance concept. *J. Instn Elect. Engrs*, **94**, part 3, 171.

Booker, H. G. (1949). The application of the magneto-ionic theory to radio waves incident obliquely upon a horizontally stratified ionosphere. *J. Geophys. Res.* **54**, 243.

Booker, H. G. and Berkner, L. V. (1938). An ionospheric investigation concerning the Lorentz polarisation corrections. *Terr. Magn. Atmos. Elect.* **43**, 427.

Booker, H. G. and Clemmow, P. C. (1950). The concept of an angular spectrum of plane waves and its relation to that of polar diagram and aperture distribution. *Proc. Instn Elect. Engrs*, **97**, part III, 11.

Booker, H. G. and Seaton, S. L. (1940). The relation between the actual and virtual ionospheric height. *Phys. Rev.* **57**, 87.

Booker, H. G. and Walkinshaw, W. (1946). The mode theory of tropospheric refraction and its relation to wave-guides and diffraction. *Report on Meteorological Factors in Radio Wave Propagation*, p. 80. London (Physical Society).

Bracewell, R. N., Budden, K. G., Ratcliffe, J. A., Straker, T. W. and Weekes, K. (1951). The ionospheric propagation of long and very long radio waves over distances less than 1000 km. *Proc. Instn Elect. Engrs*, **98**, part III, 221.

Bradbury, N. E. (1937). Fundamental mechanisms in the ionosphere. *J. Appl. Phys.* **8**, 709.

Bradbury, N. E. (1938). Ionisation, negative ion formation and recombination in the ionosphere. *Terr. Magn Atmos. Elect.* **43**, 55.

Brainerd, J. G. and Weygandt, C. N. (1940). Solutions of Mathieu's equation. *Phil. Mag.* **30**, 458.

Breit, G. and Tuve, M. A. (1926). A test of the existence of the conducting layer. *Phys. Rev.* **28**, 554.

Bremmer, H. (1949*a*). *Terrestrial Radio Waves*. New York: Elsevier Publ. Co., Inc.

Bremmer, H. (1949*b*). The propagation of electromagnetic waves through a stratified medium, and its W.K.B. approximation for oblique incidence. *Physica*, **15**, 593.

Briggs, B. H. (1951*a*). The determination of the collision frequency of electrons in the ionosphere from observations of the reflection coefficient of the abnormal *E*-layer. *J. Atmos. Terr. Phys.* **1**, 345.

Briggs, B. H. (1951*b*). An investigation of certain properties of the ionosphere by means of a rapid frequency-change experiment. *Proc. Phys. Soc.* B, **64**, 255.

Briggs, B. H. and Spencer, M. (1954). Horizontal movements in the ionosphere. A survey of experimental results. *Rep. Progr. Phys.* (Physical Society), **17**, 245.

Brown, J. N. *See* Blair.

Browne, J. (1958). A generalised form of the aerial reciprocity theorem. *Proc. Instn Elect. Engrs*, part C, **105**, 472.

Budden, K. G. (1951*a*). The propagation of a radio atmospheric. *Phil. Mag.* **42**, 1.

Budden, K. G. (1951*b*). The reflection of very low frequency radio waves at the surface of a sharply bounded ionosphere with superimposed magnetic field. *Phil. Mag.* **42**, 833.

Budden, K. G. (1952*a*). The theory of the limiting polarisation of radio waves reflected from the ionosphere. *Proc. Roy. Soc.* A, **215**, 215.

Budden, K. G. (1952*b*). The propagation of a radio atmospheric. II. *Phil. Mag.* **43**, 1179.

Budden, K. G. (1953). The propagation of very low frequency radio waves to great distances. *Phil. Mag.* **44**, 504.

Budden, K. G. (1954). A reciprocity theorem on the propagation of radio waves via the ionosphere. *Proc. Camb. Phil. Soc.* **50**, 604.

Budden, K. G. (1955*a*). The numerical solution of differential equations governing reflection of long radio waves from the ionosphere. *Proc. Roy. Soc.* A, **227**, 516.

Budden, K. G. (1955*b*). The numerical solution of the differential equations governing the reflection of long radio waves from the ionosphere. II. *Phil. Trans.* A, **248**, 45.

Budden, K. G. (1957). The 'waveguide mode' theory of the propagation of very low frequency radio waves. *Proc. Inst. Radio Engrs, N.Y.*, **45**, 772.

Budden, K. G. (1959). Effect of small irregularities on the constitutive relations for the ionosphere. *J. Res. Nat. Bur. Stand.* **63** D, 135.

Budden, K. G. and Clemmow, P. C. (1957). Coupled form of the differential equations governing radio propagation in the ionosphere. II. *Proc. Camb. Phil. Soc.* **53**, 669.

Budden, K. G. *See* Barron, Bracewell.

Builder, G. *See* Appleton.

Bullen, K. E. (1947). *Introduction to the Theory of Seismology*. Cambridge University Press.

Burton, E. T. and Boardman, E. M. (1933). Audio-frequency atmospherics. *Proc. Inst. Radio Engrs, N.Y.*, **21**, 1476.

Carson, J. R. (1929). Reciprocal theorems in radio communications. *Proc. Inst. Radio Engrs, N.Y.*, **17**, 952.

Chapman, S. (1931 *a*). The production of ionisation by monochromatic radiation incident upon a rotating atmosphere. Part I. *Proc. Phys. Soc.* **43**, 26.

Chapman, S. (1931 *b*). The production of ionisation by monochromatic radiation incident upon a rotating atmosphere. Part II. Grazing incidence. *Proc. Phys. Soc.* **43**, 483.

Chapman, S. (1939). The atmospheric height distribution of band absorbed solar radiation. *Proc. Phys. Soc.* **51**, 93.

Chatterjee, B. (1952). Effect of magnetic field in oblique propagation over equatorial regions. *Indian J. Phys.* **26**, 297.

Chatterjee, B. (1953). Oblique propagation of radio waves over a curved earth. *Indian J. Phys.* **27**, 257.

Clemmow, P. C. and Heading, J. (1954). Coupled forms of the differential equations governing radio propagation in the ionosphere. *Proc. Camb. Phil. Soc.* **50**, 319.

Clemmow, P. C. and Mullaly, R. F. (1955). The dependence of the refractive index in magnetoionic theory on the direction of the wave normal. *The Physics of the Ionosphere*, p. 340. The Physical Society.

Clemmow, P. C. *See* Booker, Budden.

Computation Laboratory, Cambridge, Mass. (1945). *Tables of Modified Hankel Functions of Order One-Third, and of their Derivatives*. Harvard University Press.

Copson, E. T. (1935). *Theory of Functions of a Complex Variable*. Oxford University Press.

Courant, R. and Hilbert, D. (1953). *Methods of Mathematical Physics*, vol. I (English translation). New York: Interscience Publishers.

Crary, J. H. *See* Helliwell.

Crombie, D. D. *See* Barber.

Crompton, R. W., Huxley, L. G. H. and Sutton, D. J. (1953). Experimental studies of the motions of slow electrons in air with application to the ionosphere. *Proc. Roy. Soc.* A, **218**, 507.

Dällenbach, W. (1942). Der Reziprozitätssatz des elektromagnetischen Feldes. *Arch. Elektrotech.* **36**, 153.

Darwin, C. G. (1924). The optical constants of matter. *Trans. Camb. Phil. Soc.* **23**, 137.

Darwin, C. G. (1934). The refractive index of an ionised medium. *Proc. Roy. Soc.* A, **146**, 17.

Darwin, C. G. (1943). The refractive index of an ionised medium. II. *Proc. Roy. Soc.* A, **182**, 152.

Davids, N. (1953). Optic axes and critical coupling in the ionosphere. *J. Geophys. Res.* **58**, 311.

Davids, N. and Parkinson, R. W. (1955). Wave solutions for critical and near-critical coupling conditions in the ionosphere. *J. Atmos. Terr. Phys.* **7**, 173.

De Groot, W. (1930). Some remarks on the analogy of certain cases of propagation of electromagnetic waves and the motion of a particle in a potential field. *Phil. Mag.* **10**, 521.

Dingle, R. B. (1958). Asymptotic expansions and converging factors. I to III. *Proc. Roy. Soc.* A, **244**, 456.

Dingle, R. B. (1959). Asymptotic expansions and converging factors. IV to VI. *Proc. Roy. Soc.* A, **249**, 270.

Eckart, C. (1930). The penetration of a potential barrier by electrons. *Phys. Rev.* **35**, 1303.

Eckersley, T. L. (1931). On the connection between the ray theory of electric waves and dynamics. *Proc. Roy. Soc.* A, **132**, 83.

Eckersley, T. L. (1932*a*). Studies in radio transmission. *J. Instn Elect. Engrs*, **71**, 405.

Eckersley, T. L. (1932*b*). Radio transmission problems treated by phase integral methods. *Proc. Roy. Soc.* A, **136**, 499.

Eckersley, T. L. (1932*c*). Long wave transmission, treated by phase integral methods. *Proc. Roy. Soc.* A, **137**, 158.

Eckersley, T. L. (1935). Musical atmospherics. *Nature, Lond.*, **135**, 104.

Eckersley, T. L. (1950). Coupling of the ordinary and extraordinary rays in the ionosphere. *Proc. Phys. Soc.* B, **63**, 49.

Eckersley, T. L. and Millington, G. (1939). The limiting polarisation of medium waves reflected from the ionosphere. *Proc. Phys. Soc.* **51**, 110.

Ellis, G. R. (1956). The *Z* propagation hole in the ionosphere. *J. Atmos. Terr. Phys.* **8**, 43.

Epstein, P. S. (1930). Reflection of waves in an inhomogeneous absorbing medium. *Proc. Nat. Acad. Sci., Wash.*, **16**, 627.

Farmer, F. T. and Ratcliffe, J. A. (1935). The absorption of wireless waves in the ionosphere. *Proc. Roy. Soc.* A, **151**, 370.

Farmer, F. T. *See* Appleton.

Feinstein, J. (1950). Higher order approximations in ionospheric wave propagation. *J. Geophys. Res.* **55**, 161.

Fejer, J. A. and Vice, R. W. (1959). An investigation of the ionospheric D-region. *J. Atmos. Terr. Phys.* **16**, 291.

Forsgren, S. K. H. (1951). Some calculations of ray paths in the ionosphere. *Chalmers tek. Högsk. Hand.* no. 104.

Försterling, K. (1942). Über die Ausbreitung elektromagnetischer Wellen in einem magnetisierten Medium bei senkrechter Incidenz. *Hochfr. Elek.* **59**, 110.

Försterling, K. and Lassen, H. (1931). Die Ionisation der Atmosphäre und die Ausbreitung der kurzen elektrischen Wellen (10–100*m.*) über die Erde. *Z. tech. Phys.* **12**, I and II, 452; III, 502.

Försterling, K. and Lassen, H. (1933). Kurzwellenausbreitung im Erdmagnetfeld. *Ann. Phys., Lpz.*, **18**, 26.

Försterling, K. and Wüster, H. O. (1951). Über die Entstehung von Oberwellen in der Ionosphäre. *J. Atmos. Terr. Phys.* **2**, 22.

Friis, H. T. *See* Schelkunoff.

Furry, W. H. (1947). Two notes on phase-integral methods. *Phys. Rev.* **71**, 360.

Furutsu, K. (1952). On the group velocity, wave path and their relations to the Poynting vector of the electromagnetic field in an absorbing medium. *J. Phys. Soc. Japan*, **7**, 458.

Gans, R. (1915). Fortpflanzung des Lichts durch ein inhomogenes Medium. *Ann. Phys., Lpz.*, **47**, 709.

Gardner, F. F. and Pawsey, J. L. (1953). Study of the ionospheric *D*-region using partial reflections. *J. Atmos. Terr. Phys.* **3**, 321.

Gerson, N. C. and Seaton, S. L. (1948). Generalised magneto-ionic theory. *J. Franklin Inst.* **246**, 483.

Gibbons, J. J. and Nertney, R. J. (1951). A method for obtaining the solutions of ionospherically reflected long waves. *J. Geophys. Res.* **56**, 355.

Gibbons, J. J. and Nertney, R. J. (1952). Wave solutions, including coupling, of ionospherically reflected long radio waves for a particular *E*-region model. *J. Geophys. Res.* **57**, 323.

Gibbons, J. J. and Rao, R. (1957). Calculation of group indices and group heights at low frequencies. *J. Atmos. Terr. Phys.* **11**, 151.

Gibson, G. A. (1929). *An Elementary Treatise on the Calculus.* London: Macmillan.

Gibson, G. A. (1931). *Advanced Calculus.* London: Macmillan.

Gill, S. (1951). A process for the step-by-step integration of differential equations in an automatic digital computing machine. *Proc. Camb. Phil. Soc.* **47**, 96.

Goubau, G. (1934). Zusammenhang zwischen scheinbarer und wahrer Hohe der Ionosphäre under Berücksichtigung der magnetischen Doppelbrechung. *Hochfr. Elek.* **44**, 17, 138.

Goubau, G. (1935*a*). Zur Dispersionstheorie der Ionosphäre. *Hochfr. Elek.* **45**, 179.

Goubau, G. (1935*b*). Dispersion in einem Elektronen-Ionen-Gemisch, das unter dem Einfluss eines äusseren Magnetfeldes steht. *Hochfr. Elek.* **46**, 37.

Goubau, G. (1942). Reziprozität der Wellenausbreitung durch magnetisch doppelbrechende Medien. *Hochfr. Elek.* **60**, 155.

Gray, H. J., Mervin, R. and Brainerd, J. G. (1948). Solutions of the Mathieu equation. *Trans. Amer. Inst. Elect. Engrs*, **67**, 429.

Guillemin, E. A. (1946). *Communication Networks.* New York: John Wiley.

Harang, L. (1936). Vertical movements of the air in the upper atmosphere. *Terr. Magn. Atmos. Elect.* **41**, 143.

Hartree, D. R. (1929). The propagation of electromagnetic waves in a stratified medium. *Proc. Camb. Phil. Soc.* **25**, 97.

Hartree, D. R. (1931*a*). Optical and equivalent paths in a stratified medium treated from a wave standpoint. *Proc. Roy. Soc.* A, **131**, 427.

Hartree, D. R. (1931*b*). Propagation of electro-magnetic waves in a refracting medium in a magnetic field. *Proc. Camb. Phil. Soc.* **27**, 143.

Hartree, D. R. (1958). *Numerical Analysis*, 2nd ed. Oxford: Clarendon Press.

Haselgrove, J. (1954). Ray theory and a new method for ray tracing. *London Physical Society, Report of Conference on the Physics of the Ionosphere*, p. 355.

Haselgrove, J. (1957). Oblique ray paths in the ionosphere. *Proc. Phys. Soc.* **70** B, 653.

Haselgrove, J. *See* Thomas.

Heading J. (1953). Theoretical ionospheric radio propagation. Ph.D. Thesis, Cambridge (unpublished).

Heading, J. (1955). The reflection of vertically incident long radio waves from the ionosphere when the earth's magnetic field is oblique. *Proc. Roy. Soc.* A, **231**, 414.

Heading, J. (1957). The Stokes phenomenon and certain *n*th-order differential equations. I and II. *Proc. Camb. Phil. Soc.* **53**, 399.

Heading, J. and Whipple, R. T. P. (1952). The oblique reflection of long wireless waves from the ionosphere at places where the earth's magnetic field is regarded as vertical. *Phil. Trans.* A, **244**, 469.

Heading, J. *See* Clemmow.

Helliwell, R. A., Crary, J. H., Pope, J. H. and Smith, R. L. (1956). The 'nose' whistler—a new high latitude phenomenon. *J. Geophys. Res.* **61**, 139.

Hilbert, D. *See* Courant.

Hines, C. O. (1951). Wave packets, the Poynting vector and energy flow. Part I: non-dissipative (anisotropic) homogeneous media. Part II: group propagation through dissipative isotropic media. Part III: packet propagation through dissipative anisotropic media. Part IV: Poynting and Macdonald velocities in dissipative anisotropic media (conclusion). *J. Geophys. Res.* **56**, 63, 197, 207, 535.

Hines, C. O. (1953). Reflection of waves from varying media. *Quart. Appl. Math.* **11**, 9.

Hines, C. O. (1957). Heavy ion effects in audio-frequency radio propagation. *J. Atmos. Terr. Phys.* **11**, 36.

Houstoun, R. A. (1938). *A Treatise on Light*, 7th ed. London: Longmans Green.

Huxley, L. G. H. (1937*a*). Motions of electrons in gases in electric and magnetic fields. *Phil. Mag.* **23**, 210.

Huxley, L. G. H. (1937*b*). Motion of electrons in magnetic fields and alternating electric fields. *Phil. Mag.* **23**, 442.

Huxley, L. G. H. (1938). The propagation of electromagnetic waves in an ionised atmosphere. *Phil. Mag.* **25**, 148.

Huxley, L. G. H. (1940). The propagation of electromagnetic waves in an atmosphere containing free electrons. *Phil. Mag.* **29**, 313.

Huxley, L. G. H. (1947). *Wave Guides (a Survey of the Principles and Practice of Wave Guides)*. Cambridge University Press.

Huxley, L. G. H. *See* Crompton.

Jackson, J. E. *See* Seddon.

Jeffreys, H. (1923). On certain approximate solutions of linear differential equations of the second order. *Proc. Lond. Math. Soc.* **23**, 428.

Jeffreys, Sir H. (1952). *The Earth*, 3rd ed. Cambridge University Press.

Jeffreys, Sir H. and Jeffreys, Lady B. (1956). *Methods of Mathematical Physics*, 3rd ed. Cambridge University Press.
Johnson, W. C. *See* Morgan.
Jones, R. E. (1951). The development of an *E*-region model consistent with long wave phase path measurements. *J. Atmos. Terr. Phys.* **6**, 1.
Keitel, G. H. (1955). Certain mode solutions of forward scattering by meteor trails. *Proc. Inst. Radio Engrs, N.Y.*, **43**, 1481.
Kelso, J. M. (1954). Group height calculations in the presence of the earth's magnetic field. *J. Atmos. Terr. Phys.* **5**, 117.
Kelso, J. M., Nearhoof, H. J., Nertney, R. J. and Waynick, A. H. (1951). The polarisation of vertically incident long radio waves. *Ann. Geophys.* **7**, 215.
Kimura, I. *See* Maeda, K.
King, G. A. M. (1957). Relation between virtual and actual heights in the ionosphere. *J. Atmos. Terr. Phys.* **11**, 209.
Landmark, B. (1955). A study of the limiting polarisation of high frequency radio waves reflected vertically from the ionosphere. *Forsv. ForsknInst. Årb.* (Oslo, Norway), Report no. 4.
Landmark, B. and Lied, R. (1957). Notes on a *QL-QT* transition level in the ionosphere. *J. Atmos. Terr. Phys.* **10**, 114.
Lange-Hesse, G. (1952). Vergleich der Doppelbrechung im Kristall und in der Ionosphäre. *Arch. Elektr. Übertr.* **6**, 149.
Langer, R. E. (1934). The asymptotic solutions of ordinary linear differential equations of the second order, with special reference to the Stokes phenomenon. *Bull. Amer. Math. Soc.* **40**, 545.
Langer, R. E. (1937). On the connection formulas and the solutions of the wave equation. *Phys. Rev.* **51**, 669.
Lassen, H. *See* Försterling.
Lepechinsky, D. (1956). On the existence of a '*QL*'-'*QT*' 'transition-level' in the ionosphere and its experimental evidence and effect. *J. Atmos. Terr. Phys.* **8**, 297.
Lewis, R. P. W. (1953). The reflection of radio waves from an ionised layer having both vertical and horizontal ionisation gradients. *Proc. Phys. Soc.* **66**, 308.
Lied, F. *See* Landmark.
Lindquist, R. (1953). An interpretation of vertical incidence equivalent height versus time recordings on 150 kc/s. *J. Atmos. Terr. Phys.* **4**, 10.
Lorentz, H. A. (1952). *Theory of Electrons* (reprinting of 2nd ed. 1915). New York: Dover publications.
Maeda, K. and Kimura, I. (1956). A theoretical investigation on the propagation path of the whistling atmospherics. *Report of Ionosphere Research in Japan*, **10**, 105.
Manning, L. A. (1947). The determination of ionospheric electron distribution. *Proc. Inst. Radio Engrs, N.Y.*, **35**, 1203.
Manning, L. A. (1949). The reliability of ionospheric height determinations. *Proc. Inst. Radio Engrs, N.Y.*, **37**, 599.
Marcou, R. J., Pfister, W. and Ulwick, J. C. (1958). Ray-tracing technique in a horizontally stratified ionosphere using vector representations. *J. Geophys. Res.* **63**, 301.

Martyn, D. F. (1935). The propagation of medium radio waves in the ionosphere. *Proc. Phys. Soc.* **47**, 323.

Meek, J. H. (1948). Triple splitting of ionospheric rays. *Nature, Lond.*, **161**, 597.

Mervin, R. *See* Gray.

Miller, J. C. P. (1946). The Airy integral. *Brit. Ass. Math. Tables*, Part Vol. B. Cambridge University Press.

Millington, G. (1938*a*). Attenuation and group retardation in the ionosphere. *Proc. Phys. Soc.* **50**, 561.

Millington, G. (1938*b*). The relation between ionospheric transmission phenomena at oblique incidence and those at vertical incidence. *Proc. Phys. Soc.* **50**, 801.

Millington, G. (1943). *Special Report D.S.I.R. Radio Research Board*, no. 17. H.M.S.O.

Millington, G. (1949). Deviation at vertical incidence in the ionosphere. *Nature, Lond.*, **163**, 213.

Millington, G. (1951). The effect of the earth's magnetic field on short-wave communication by the ionosphere. *Proc. Instn Elect. Engrs*, **98**, Part IV, 1.

Millington, G. (1954). Ray path characteristics in the ionosphere. *Proc. Instn Elect. Engrs*, **101**, Part IV, 235.

Millington, G. *See* Eckersley.

Milne, W. E. (1953). *Numerical Solution of Differential Equations*. New York: John Wiley.

Morgan, M. G. and Johnson, W. C. (1954). The observed polarisation of high frequency sky-wave signals at vertical incidence. *The Physics of the Ionosphere*, p. 74. London: The Physical Society.

Mullaly, R. F. *See* Clemmow.

Musgrave, M. J. P. (1959). The propagation of elastic waves in crystals and other anisotropic media. *Reports on Progress in Physics*, **22**, 74. The Physical Society, London.

Namba, S. *See* Yokoyama.

National Bureau of Standards (1948). *Ionospheric Radio Propagation*, Circular no. 462.

Nearhoof, H. J. *See* Kelso.

Nertney, R. J. (1953). The lower *E* and *D* region of the ionosphere as deduced from long wave radio measurements. *J. Atmos. Terr. Phys.* **3**, 92.

Nertney, R. J. *See* Gibbons, Kelso.

Newstead, G. (1948). Triple magneto-ionic splitting of rays reflected from the F_2 region. *Nature, Lond.*, **161**, 312.

Nicolet, M. (1953). The collision frequency of electrons in the ionosphere. *J. Atmos. Terr. Phys.* **3**, 200.

Ott, H. (1942). Reflexion und Brechung von Kugelwellen; Effekte 2. Ordnung. *Ann. Phys., Lpz.*, **41**, 443.

Parkinson, R. W. (1955). The night time lower ionosphere as deduced from a theoretical and experimental investigation of coupling phenomenon at 150 kc/sec. *J. Atmos. Terr. Phys.* **7**, 203.

Parkinson, R. W. *See* Davids.

Pawsey, J. L. *See* Gardner.

Pedersen, P. O. (1927). The propagation of radio waves. *Danmarks Naturvidenskabelige Samfund.* Copenhagen.

Pekeris, C. L. (1940). The vertical distribution of ionisation in the upper atmosphere. *Terr. Mag.* **45**, 205.

Perry, L. B. *See* Wait.

Pfister, W. (1949). Effect of the *D*-ionospheric layer on very low frequency radio waves. *J. Geophys. Res.* **54**, 315.

Pfister, W. (1953). Magneto-ionic multiple splitting determined with the method of phase integration. *J. Geophys. Res.* **58**, 29.

Pfister, W. *See* Marcou.

Piggott, W. R. (1954). On the interpretation of the apparent ionisation distribution in the ionosphere. *J. Atmos. Terr. Phys.* **5**, 201.

Pitteway, M. L. V. (1959). Reflexion levels and coupling regions in a horizontally stratified ionosphere. *Phil. Trans.* A, **252**, 53.

Poeverlein, H. (1948). Strahlwege von Radiowellen in der Ionosphäre. *S.B. bayer Akad. Wiss.* p. 175.

Poeverlein, H. (1949). Strahlwege von Radiowellen in der Ionosphäre. *Z. angew. Phys.* **1**, 517.

Poeverlein, H. (1950*a*). Strahlwege von Radiowellen in der Ionosphäre. *Z. angew. Phys.* **2**, 152.

Poeverlein, H. (1950*b*). Über Wellen in anisotropen Ausbreitungsverhältnissen. *Z. Naturf.* **5***a*, 492.

Poeverlein, H. (1953). Ionosphären-Grenzfrequenz bei schiefem Einfall. *Z. angew. Phys.* **5**, 15.

Poeverlein, H. (1954). Field strength near the skip distance. *Propagation Laboratory, Air Force Cambridge Research Center, Mass. Report,* AFCRC-TR-54-104.

Poeverlein, H. (1958). Low-frequency reflection in the ionosphere. *J. Atmos. Terr. Phys.* **12**, 126, 236.

Poincaré, H. (1886). Sur les intégrales irrégulières des équations linéaires. *Acta Math.* (*Stockh.*), **8**, 295.

Pope, J. H. *See* Helliwell.

Rao, R. *See* Gibbons.

Ratcliffe, J. A. (1951). A quick method for analysing ionospheric records. *J. Geophys. Res.* **56**, 463.

Ratcliffe, J. A. (1954). The physics of the ionosphere. (Kelvin lecture.) *Proc. Instn Elect. Engrs,* **101**, Part I, 339.

Ratcliffe, J. A. (1956). Some aspects of diffraction theory and their application to the ionosphere. *Reports on Progress in Physics,* **19**, 188. The Physical Society, London.

Ratcliffe, J. A. (1959). *Magneto-ionic Theory.* Cambridge University Press.

Ratcliffe, J. A. *See* Appleton, Bracewell, Farmer.

Rawer, K. (1939). Elektrische Wellen in einem geschichteten Medium. *Ann. Phys., Lpz.,* **35**, 385.

Rawer, K. (1953). *Die Ionosphäre.* Groningen-Holland (Noordhoff). English translation: Katz, L., London (Crosby Lockwood), 1956.
Robbins, A. *See* Thomas.
Rydbeck, O. E. H. (1940). The propagation of electromagnetic waves in an ionised medium and the calculation of the true heights of the ionised layers of the atmosphere. *Phil. Mag.* **30**, 282.
Rydbeck, O. E. H. (1942). A theoretical survey of the possibilities of determining the distribution of free electrons in the upper atmosphere. *Chalmers tek. Högsk. Handl.* no. 3.
Rydbeck, O. E. H. (1943). The reflection of electromagnetic waves from a parabolic ionised layer. *Phil. Mag.* **34**, 342.
Rydbeck, O. E. H. (1944). On the propagation of radio waves. *Chalmers tek. Högsk. Handl.* no. 34.
Rydbeck, O. E. H. (1948). On the propagation of waves in an inhomogeneous medium. *Chalmers tek Högsk. Handl.* no. 74.
Rydbeck, O. E. H. (1950). Magneto-ionic triple splitting of ionospheric waves. *J. Appl. Phys.* **21**, 1205.
Schelkunoff, S. A. (1938). The impedance concept and its applications to problems of reflection, refraction, shielding and power absorption. *Bell Syst. Tech. J.* **17**, 17.
Schelkunoff, S. A. (1951). Remarks concerning wave propagation in stratified media. *Commun. Pure Appl. Math.* **4**, 117.
Schelkunoff, S. A. and Friis, H. T. (1952). *Antennas: Theory and Practice.* New York: Wiley.
Scott, J. C. W. (1950). The Poynting vector in the ionosphere. *Proc. Inst. Radio Engrs,* **38**, 1057.
Seaton, S. L. *See* Booker, Gerson.
Seddon, J. C. and Jackson, J. E. (1955). Absence of bifurcation in the *E*-layer. *Phys. Rev.* **97**, 1182.
Shinn, D. H. (1953). The analysis of ionospheric records (ordinary ray), Part I. *J. Atmos. Terr. Phys.* **4**, 240.
Shinn, D. H. (1955). Tables of group refractive index for the ordinary ray in the ionosphere. *The Physics of the Ionosphere* (*Report of Conference. The Physical Society*), p. 402.
Shinn, D. H. and Whale, H. A. (1951). Group velocities and group heights from the magnetoionic theory. *J. Atmos. Terr. Phys.* **2**, 85.
Shire, E. S. (1960). *Classical Electricity and Magnetism.* Cambridge University Press.
Smith, N. (1937). Extension of normal incidence ionosphere measurements to oblique incidence radio transmission. *J. Res. Nat. Bur. Stand.* **19**, 89.
Smith, N. (1938). Application of vertical incidence ionosphere measurements to oblique incidence radio transmission. *J. Res. Nat. Bur. Stand.* **20**, 683.
Smith, N. (1941). Oblique incidence radio transmission and the Lorentz polarisation term. *J. Res. Nat. Bur. Stand.* **26**, 105.
Smith, N. *See* National Bureau of Standards (Circular 462, ch. VI).
Smith, R. L. *See* Helliwell.

Sommerfeld, A. (1909). Über die Ausbreitung der Wellen in der drahtlosen Telegraphie. *Ann. Phys.* **28**, 665.

Sommerfeld, A. (1925). Das Reziprozitäts-Theorem der drahtlosen Telegraphie. *Jb. drahtl. Telegr.* **26**, 93.

Sommerfeld, A. (1952). *Electrodynamics.* New York: Academic Press.

Spencer, M. *See* Briggs.

Stanley, J. P. (1950). The absorption of long and very long waves in the ionosphere. *J. Atmos. Terr. Phys.* **1**, 65.

Stanley, J. P. *See* Whale.

Stokes, Sir G. G. (1858). On the discontinuity of arbitrary constants which appear in divergent developments. *Trans. Camb. Phil Soc.* **10**, Part I, 106, also in *Math. and Phys. Papers*, **4**, 77, 1904.

Storey, L. R. O. (1953). An investigation of whistling atmospherics. *Phil. Trans.* A, **246**, 113.

Straker, T. W. *See* Bracewell.

Stratton, J. A. (1941). *Electromagnetic Theory.* New York: McGraw-Hill.

Suchy, K. (1952). Schrittweiser Übergang von der Wellenoptik zur Strahlenoptik in inhomogenen anisotropen absorbierenden Medien. Part I. Gleichungen für Wellennormale, Brechungsindex und Polarisation. *Ann. Phys., Lpz.*, **11**, 113.

Suchy, K. (1953). Schrittweiser Übergang von der Wellenoptik zur Strahlenoptik in inhomogenen anisotropen absorbierenden Medien. Part II. Lösung der Gleichungen für Wellennormale und Brechungsindex durch W.K.B.-Näherung. Strahlenoptische Reflexion und Alternation. *Ann. Phys., Lpz.*, **13**, 178.

Suchy, K. (1954). Schrittweiser Übergang von der Wellenoptik zur Strahlenoptik in inhomogenen anisotropen absorbierenden Medien. Part III. Gruppenfortpflanzung. *Ann. Phys., Lpz.*, **14**, 6.

Sutton, D. J. *See* Crompton.

Taylor, M. (1933). The Appleton–Hartree formula and dispersion curves for the propagation of electromagnetic waves through an ionised medium in the presence of an external magnetic field. Part I. Curves for zero absorption. *Proc. Phys. Soc.* **45**, 245.

Taylor, M. (1934). The Appleton–Hartree formula and dispersion curves for the propagation of electromagnetic waves through an ionised medium in the presence of an external magnetic field. Part II. Curves with collisional friction. *Proc. Phys. Soc.* **46**, 408.

Thomas, J. O., Haslegrove, J. and Robbins, A. (1958). The electron distribution in the ionosphere over Slough. Part I. Quiet days. *J. Atmos. Terr. Phys.* **12**, 45.

Titheridge, J. E. (1959). The determination of the electron density in the ionosphere. Ph.D. Thesis, Cambridge (unpublished).

Toshniwal, G. R. (1935). Threefold magnetoionic splitting of radio echoes reflected from the ionosphere. *Nature, Lond.*, **135**, 471.

Tuve, M. A. *See* Breit.

Ulwick, J. C. *See* Marcou.

Vice, R. W. *See* Fejer.

Wait, J. R. (1957). The mode theory of V.L.F. ionospheric propagation for finite ground conductivity. *Proc. Inst. Radio Engrs, N.Y.*, **45**, 760.

Wait, J. R. and Perry, L. B. (1957). Calculations of ionospheric reflection coefficients at very low frequencies. *J. Geophys. Res.* **62**, 43.

Walkinshaw, W. *See* Booker.

Watson, G. N. (1919*a*). The diffraction of electric waves by the earth. *Proc. Roy. Soc.* A, **95**, 83.

Watson, G. N. (1919*b*). The transmission of electric waves round the earth. *Proc. Roy. Soc.* A, **95**, 546.

Watson, G. N. (1944). *Theory of Bessel Functions*, 2nd ed. Cambridge University Press.

Watson, G. N. *See* Whittaker.

Watts, J. M. *See* Blair.

Waynick, A. H. *See* Kelso.

Weekes, K. *See* Bracewell.

Westfold, K. C. (1949). The wave equations for electromagnetic radiation in an ionised medium in a magnetic field. *Aust. J. Sci. Res.* **2**, 168.

Westfold, K. C. (1951). The interpretation of the magneto-ionic theory. *J. Atmos. Terr. Phys.* **1**, 152.

Weygandt, C. N. *See* Brainerd.

Weyl, H. (1919). Ausbreitung elektromagnetischer Wellen über einem ebenen Leiter. *Ann. Phys., Lpz.*, **60**, 481.

Whale, H. A. and Stanley, J. P. (1950). Group and phase velocities from the magneto-ionic theory. *J. Atmos. Terr. Phys.* **1**, 82.

Whale, H. A. *See* Shinn.

Whipple, R. T. P. *See* Heading.

Whittaker, E. T. and Watson, G. N. (1935). *Modern Analysis*, 4th ed. Cambridge University Press.

Wilkes, M. V. (1940). The theory of reflection of very long wireless waves from the ionosphere. *Proc. Roy. Soc.* A, **175**, 143.

Wilkes, M. V. (1947). The oblique reflection of very long wireless waves from the ionosphere. *Proc. Roy. Soc.* A, **189**, 130.

Woodward, P. M. and Woodward, A. M. (1946). Four-figure tables of the Airy function tables in the complex plane. *Phil. Mag.* **37**, 236.

Wüster, H. O. *See* Försterling.

Yabroff, I. W. (1957). Reflection at a sharply bounded ionosphere. *Proc. Inst. Radio Engrs, N.Y.*, **45**, 750.

Yokoyama, E. and Namba, S. (1932). Theory of the propagation of low frequency waves. *Rep. Radio Res. Japan*, **2**, 131.

Zenneck, E. H. J. (1907). Über die Fortpflanzung ebener elektromagnetischer Wellen längs einer ebenen Leiterfläche und ihre Beziehung zur drahtlosen Telegraphie. *Ann. Phys., Lpz.*, **23**, 846.

INDEX OF DEFINITIONS OF THE MORE IMPORTANT SYMBOLS

a	half thickness of parabolic profile	*page* 153
	(also used with other meanings)	
$\mathfrak{a}$	gradient of $-\mathfrak{n}^2$ or $-q^2$ or X in linear profile	134, 284
$\mathbf{B}$	magnetic induction in wave	15
$\boldsymbol{\mathfrak{B}}$	magnetic induction of earth's magnetic field	26
$\mathfrak{B}$	magnitude of $\boldsymbol{\mathfrak{B}}$	27
$\mathfrak{b}$	displacement of origin of height z in Epstein theory	375
C	$\cos\theta_I$, cosine of angle of incidence	85, 121
$\mathscr{C}$	cylinder function; any solution of Bessel's equation	
c	velocity of electromagnetic waves in free space	18
	(also used with other meanings)	
$\mathbf{D}$	electric displacement in wave	15
	(also used with other meanings)	
D_x, D_y, D_z	components of $\mathbf{D}$	17
D	horizontal range	180
	denominator in one form of Appleton–Hartree formula	200
	(also used with other meanings)	
$\mathbf{E}$	electric intensity in wave	13
E	magnitude of $\mathbf{E}$	21
E, E_x, E_y, E_z	components of $\mathbf{E}$	17, 146, 176
$E_\parallel$	component of $\mathbf{E}$ parallel to plane of incidence	118
E_L	longitudinal component of $\mathbf{E}$	120
e	charge on the electron	14
	(also used for the exponential)	
$F(q)$	left side of Booker quartic	122
	F is also used to denote other functions	
$\mathfrak{F}, \mathfrak{F}_o, \mathfrak{F}_x$	field variables in Försterling's equations	397
f	frequency	12
	(also used with other meanings)	
$f_H, f_H^{(e)}, f_H^{(i)}$	gyro-frequencies	27, 32
f_N	plasma frequency	25
f_p	penetration frequency	153
$\mathbf{H}$	magnetic intensity in wave	13
H_x, H_y, H_z	components of $\mathbf{H}$	16
$\mathscr{H}$	$Z_0\mathbf{H}$, alternative measure of magnetic intensity	20

$\mathscr{H}_x, \mathscr{H}_y, \mathscr{H}_z$	components of $\mathscr{H}$	*page* 20
$h, h(f)$	phase height	150
h_0	height of base of ionosphere	150
$h', h'(f)$	equivalent height of reflection	149
$\mathscr{I}$	imaginary part of	
i	$\sqrt{(-1)}$	
$\mathbf{J}$	current density in wave	14
k	$\omega/c = 2\pi/\lambda = 2\pi f/c$ propagation constant in free space	20
k_p	value of k at the penetration frequency	365
l	x-direction cosine of $\mathbf{Y}$, opposite to earth's magnetic field (also used with other meanings)	27
$\mathbf{M}$	susceptibility matrix	29
$M_{i,j}\ (i,j = x,y,z)$	elements of $\mathbf{M}$	29
$\mathscr{M}$	ray refractive index	255
m	y-direction cosine of $\mathbf{Y}$, opposite to earth's magnetic field (also used to denote an integer)	27
$\mathfrak{m}, \mathfrak{m}_e$	mass of electron	24, 32
$\mathfrak{m}_i$	mass of ion	32
N	number of electrons per unit volume	4
N_e, N_i	number of electrons and ions, respectively, per unit volume	32
n	z-direction cosine of $\mathbf{Y}$, opposite to earth's magnetic field (also used to denote an integer, and with other meanings)	27
$\mathfrak{n}$	(complex) refractive index	18, 38
$\mathfrak{n}'$	(complex) group refractive index	170
$\mathbf{P}$	electric polarisation	14
P_x, P_y, P_z	components of $\mathbf{P}$	27
P	phase path (usually for oblique incidence)	173
P'	equivalent path (usually for oblique incidence)	173
Q	effective value of q	344
q	solution of Booker quartic equation (also used with other meanings)	121, 175
$\mathscr{R}$	real part of	
R	reflection coefficient	ch. 7
R_0, R_1, R_2, R_3	values of R in specified conditions	ch. 7
$\mathbf{R}$	reflection coefficient matrix	90
${}_{\parallel}R_{\parallel}, {}_{\parallel}R_{\perp}, {}_{\perp}R_{\parallel}, {}_{\perp}R_{\perp}$	elements of $\mathbf{R}$	89
r	used to denote an integer, and with other meanings	
S	$\sin\theta_I$, sine of angle of incidence	85, 121

s	distance along ray path	*page* 279
	kz; height measured in units of $\lambda/2\pi$	433
	(also used with other meanings)	
$\mathbf{T}$	4×4 matrix	389
$T_{ij}\,(i,j = 1,2,3,4)$	elements of $\mathbf{T}$	390
$T, {}_{\parallel}T_{\parallel}, {}_{\parallel}T_{\perp}, {}_{\perp}T_{\parallel}, {}_{\perp}T_{\perp}$	transmission coefficients	87, 88
t	time	
T and t	also used with other meanings	
U	$1-iZ$	26
	group velocity	148
U_z	upward component of group velocity	148
V	wave velocity	38
	(also used with other meanings)	
V_R	ray velocity	255
X	$Ne^2/(\epsilon_0 \mathfrak{m}\omega^2)$	25
X_e, X_i	X for electrons and ions respectively	31
X_0, X_1	values of X at specified levels	
X	X_e+X_i	81
x	Cartesian coordinate	16
$\mathbf{Y}$	$e\mathbf{B}/(\mathfrak{m}\omega)$	27
Y	magnitude of $\mathbf{Y}$	27
$\mathbf{Y}_i$	$\mathbf{Y}$ for ions	32
Y_e, Y_i	Y for electrons and ions respectively	31
Y_L	nY, longitudinal component of $\mathbf{Y}$	49
Y_T	transverse component of $\mathbf{Y}$	49
y	Cartesian coordinate	16
Z	ν/ω	26
Z_e, Z_i	Z for electrons and ions respectively	80
Z_c	critical value of Z	49
Z_0	(μ_0/ϵ_0), characteristic impedance of free space	19
z	Cartesian coordinate	17
z_0	level of reflection; (complex) value of z which makes $\mathfrak{n} = 0$,	136
	or $q = 0$	142
z_P	(complex) value of height z at coupling point	417
α	coefficient of q^4 in Booker quartic	122
	gradient of f_N^2 in linear profile	150
	coefficient of z in exponent, for exponential profile	151
	angle between wave normal and ray	253
	(also used with other meanings)	

β	coefficient of q^3 in Booker quartic	*page* 122
	(also used with other meanings)	
γ	coefficient of q^2 in Booker quartic	122
	arc tan ρ_o	51
	(also used with other meanings)	
δ	coefficient of q in Booker quartic	122
ϵ	coefficient of q^0 in Booker quartic	122
	(also used for arbitrarily small quantity, and with other meanings)	
ϵ_0	electric permittivity of free space	11
$\epsilon_1, \epsilon_2, \epsilon_3$	parameters in Epstein theory	376
ζ	scaled value of height z (the method of scaling depends on the problem)	
	(also used with other meanings)	
Θ	angle between earth's magnetic field and vertical	119
	angle between earth's magnetic field and wave normal when wave normal is vertical	50
ϑ	angle between earth's magnetic field and wave normal when wave normal is oblique	245
θ	angle between wave normal and vertical, within ionosphere	121
θ_I	angle of incidence; value of θ below ionosphere	97
θ_R, θ_T	angles of reflection and transmission at sharp boundary	97
λ	wavelength in free space	20
λ_p	λ at penetration frequency	365
μ	real part of refractive index	41
μ_o, μ_x	μ for ordinary and extraordinary waves respectively	199
μ'	group refractive index	148
μ'_o, μ'_x	μ' for ordinary and extraordinary waves respectively	201
μ_0	magnetic permittivity of free space	11
ν	collision frequency for electrons	6, 25
ξ	scaled value of height z (similar to ζ)	
$\mathbf{\Pi}$	Poynting vector	22
$\bar{\mathbf{\Pi}}$	average Poynting vector	23
$\bar{\Pi}_x, \bar{\Pi}_y, \bar{\Pi}_z$	components of $\bar{\mathbf{\Pi}}$	44
ρ	wave polarisation	47
	(used with subscripts to denote polarisation of specified waves)	
σ	scaling factor in Epstein profiles	375
	wave polarisation referred to axes at 45° to magnetic meridian	434
	(also used with other meanings)	

ϕ	usually denotes (complex) phase of a wave, or polar co-ordinate angle	
χ	minus imaginary part of refractive index $\mathfrak{n}$ (also used with other meanings)	*page* 41
χ_o, χ_x	χ for ordinary and extraordinary waves respectively	199
Ψ	factor of coupling parameter	413
ψ	coupling parameter (also used with other meanings)	397
ω	$2\pi f$; angular frequency	12
ω_N	angular plasma-frequency	25
ω_H	angular gyro-frequency	27
ω_c	critical value of ν	49

NOTES ADDED IN SECOND PRINTING

Note 1. The equivalent height of reflection for a parabolic profile of electron density is given by the expressions (17.54) and (17.56). In the derivation it was assumed that the two asymptotic solutions (17.32) can be taken as the upgoing and downgoing waves. But these solutions should be multiplied by an asymptotic series which was intentionally omitted from (17.32). For the process to be valid it is necessary that the ratio of the second to the first term in the series shall be small at the top and bottom of the layer, which leads to the condition $|\frac{1}{2}n(n-1)/\zeta_a^2| \ll 1$. This is violated when $|D|$ (equation (17.43)) is large, and then the full wave solutions (17.54) (17.56) may not agree with the ray theory solutions (10.33) and (10.37). The discrepancy increases as the difference Δf (17.47), between the penetration frequency and the wave frequency is increased. The author is indebted to Dr. Kenneth Davies for pointing out this possibility.

Note 2. In §23.4 the assumption that the extraordinary wave has polarisation $1/\rho_A$ is valid only when the plane of propagation is the magnetic meridian plane, that is for propagation from magnetic north to south or south to north. The theory for other directions of propagation has been given by Budden, K. G. and Jull, G. W., 1964, *Can. Journ. Phys.*, **42**, 113.

The author is greatly indebted to numerous colleagues who, in correspondence, reviews or personal discussion, have pointed out errors in the first printing.

SUBJECT AND NAME INDEX

Page numbers in bold type refer to definitions or descriptions.

Abel, 162, 165
Abel's integral, **165**
abnormal component (of reflected wave), **489**
absorption (of energy in radio wave), 172, 195, 279, 472
absorption (of sun's radiation), 4
accumulation of energy, 480
admittance matrix, **493**, 496
adjoint matrix, **390**
aerials, 91, 177, 257, **502**, 506
Ai, *see* Airy integral function
Airy, Sir G. B., 290, 313, 512
Airy integral function, 169, 184, **283** (ch. 15), 287, **290** (fig. 15.5), 291, 303, 309, 311, 316, 320, 409, 410, 440
Al'pert, Ya. L., 278, 512
ambiguities in refractive index, 261, 264, 266
ambiguity (of terms 'ordinary' and 'extraordinary'), 69, **73**, 114, 454
Ampère's circuital theorem, 16, 96
amplitude (and imaginary part of phase), 331
amplitude (complex), **12**
amplitude-phase diagram, 309, 316
amplitude, planes of constant, **43**
analogues of progressive waves, 137
analytic continuation, 372, 373
angle of incidence (θ_I), **85**, 97, 225, 259, 283
angular frequency, **12**
angular gyro-frequency, **27**
angular plasma frequency, **25**
angular spectrum of plane waves, 92, 175
anisotropic ionosphere, 26, 29, 35 (§3.11), 47 (ch. 5), 59 (ch. 6). 88 (§§7.4–7.7), 112 (§§8.12–8.20), 143, 148, 199 (ch. 12), 225 (ch. 13), 271 (ch. 14), 385 (ch. 18), 412 (ch. 19), 450 (§§20.6–20.8), 458 (ch. 21), 482 (ch. 22)
anisotropic medium, 39, 42
antennas, *see* aerials
anti-Stokes lines, **293**, 304, 307, 311, 350, 355, 361, 441, **452**, 455, 510
antisymmetric (matrix), 399
antisymmetric (phase function), 147
aperture, diffraction by, 168, 309
apparent bearing, 250
apparent loss of energy, 479
Appleton, Sir E. V., xxiii, 8, 25, 26, 27, 94, 173, 191, 194, 209, 512
Appleton–Hartree formula, **52**, 57, **59** (ch. 6), 112, 122, 123, 144, 200, 233, 246, 256, 272, 274, 275, 397, 402, 417, 418, 438, 440, 450, 460, 472
approximations, 5, 119, *see also* asymptotic; W.K.B.
arbitrary constants (Stokes phenomenon), 292, 294
argument of complex number, 12
argument of reflection coefficient, 86, 102 (figs. 8.2–8.8)
asymptotic approximation, 295, **297**, **310**, 313, 320, 355, 360; *see also* W.K.B.
asymptotic expansions, **310**, 372
atmospherics, 29, 257, 258
attachment coefficient, 4
attenuation, 173, **250**, **280**, 506, 507; *see also* absorption
audible frequencies, 57, 257
audio frequency amplifier, 257
auxiliary sub-routine, 484
average (over small irregularities), 31
average (velocity or displacement of electron), 25
average energy flow, **23**, 44, 247
average magnetic force on electron, 28
axes, change of, 50
axis ratio, **51**, 214

backscatter, **428**, 500
Bailey, V. A., 491, 512
Banerjea, B. K., 29, 512
band width, 147, 500
Barber, N. F., 96, 127, 512
Barnes, E. W., 373, 512
Barron, D. W., xxiv, 127, 435, 513
base of ionosphere, 150, 207, 319, 322, 334, 336
basic equations, 11 (ch. 2)
bearing error, 250
Becker, W., 201, 513
Berkner, L. V., 479, 513
Bessel functions, 291, 355, 369, 461, 475
Beynon, W. J. G., 191, 194, 512, 513
Bi, *see* Airy integral function
binomial theorem, 132, 143
bivariate interpolation, 292

Blair, J. C., 220, 513
Boardman, E. M., 257, 515
Booker, H. G., xxiv, 77, 92, 121, 191, 230, 233, 238, 245, 250, 251, 388, 418, 438, 479, 491, 513
Booker quartic (equation), 117, **120**, **122**, **123**, 140, 144, 225, **226** ff., **233** ff., 246, 248 ff., 271, 274, 387, 389, 400 ff., 408, 410, 438, 439, 497, 506
boundary conditions, 69, **96**, 99, 106, 111, 115, 118, 321, 335, 422, 427, 428
boundary condition at infinity, 320, 324, 458
Bracewell, R. N., 119, 463, 513
Bradbury, N. E., 8, 513
Brainerd, J. G., 369, 513, 517
branch cut, 296, 372, 439
branch point, 94, 296, 327, 372; *see also* reflection branch point
Breit, G., 186, 514
Breit and Tuve's theorem, **186**, 188, 194, 281
Bremmer, H., 96, 228, 438, 514
Brewster angle, 94, **101**, 105
Briggs, B. H., xxiv, 10, 382, 514
broadening (of pulse), 166, 170, 204
Brown, J. N., 220, 513
Browne, J., 504, 514
Budden, K. G., 31, 96, 104, 119, 120, 127, 402, 408, 411, 435, 438, 459, 472, 485, 488, 491, 509, 513, 514
Builder, G., 209, 512
Bullen, K. E., 3, 515
Burton, E. T., 257, 515

calculus of variations, 272, 279
Caminer, D. T., 217
canonical equations for a ray, 272, **276**, 279, 281, 282
cardinal points (of compass), 250
Carson, J. R., 502, 515
Cartesian coordinates, **16**, 20, 43, **85**, 279, 282, 385
cathode ray oscillograph, 146, 166, 200, 424
Cauchy relations, 297
caustic curve, 184, 191, 313, 318
cavity definitions of **E**, **D**, 13, 15, 16
Chapman layer, **3**, **5**, 8, 134, 155, 354, 436
Chapman, S., 3, 515
characteristic equation, 387, 389
characteristic impedance of free space, **19**, 26, 492
characteristic roots, 389, 399
characteristic waves, 200, 252, 255, **388**, 402, 432; *see also* ordinary, extraordinary
charge on electron, 14, 26
Chatterjee, B., 233, 515
check of numerical methods, 465, 472
choice of dependent variable, 484
choice of solution, 320, 324, 330, 355, 360, 369, 461, 470, 474, 477, 486, 487
choice of starting solution, 483, **487**, 492, 495, 511
cigar-shaped surface, 262
circuit elements (electric), 502, 503
circuit relations, **313**, 353, **369**, 372, 375, 378
circular polarization, **48**, 60, 69, 116, 200, 214, 460, 461, 506
circularly polarised components (resolution into), 90, 116
Clemmow, P. C., xxiv, 92, 252, 396, 398, 402, 408, 411, 485, 513, 514, 515
col, 298
collision frequency, 7, **26**, 80, 85, 139, 173, 250, 280, 432
collision frequency, critical, **50**, **67**, 71, 413
collisions, 2, **25**, 40, 64, **66**, 69, 70, 113, 142, 195, 224, 248, 265, 280, 326
collisions and equivalent path, 197
collisions and equivalent height, 171, 212
collisions and group refractive index, 170 ff., 204 ff.
column matrix, 90, 389, 398, 489
compensating singularities, 291
complete reciprocity, 506, 508
complex amplitude, **12**
complex angle, 42, 43, 87, 92, 98, 100
complex angle of incidence, 87, 98
complex Brewster angle, 101
complex conjugate, 22, 30, 37, 231, 388, 393, 507
complex direction of electric field, 45
complex Fresnel integral, 308
complex group refractive index, 170, 204
complex height, 40, 41, 59, **326**, 413, 438
complex numbers and harmonic time variation, **12**, 22
complex phase, **130**, 142, 145, 331
complex point of reflection, 331
complex Poynting vector, **23**, 54, 131, 247, 388, 393
complex refractive index, 38, **41**, 43, 66, 98, 101, 133, 135, 170, 171, 199, 280, 482
complex values of X and Z, 40; *see also* complex height; complex z-plane
complex variables, 297
complex z-plane, 136, 171, 212, 285, **326**, 329, 423, 438, 473
component frequency, 147
component plane wave, 92, 175, 225, 229, 249

components of the reflection and transmission coefficient, **88** ff., 486, 489
composition of ionosphere, 1
Computation Laboratory, Cambridge, Mass., 291, 515
computing, 205, 482
condenser, 20
conduction current, 15, 21
conductivity of ionosphere, 104
cone of angles of incidence, 265
cone of radiation, 146
confluent hypergeometric functions, 464, 476
conjugate complex roots of quartic, 231, 233; *see also* complex conjugate
connection formula, **313**, 353, 356, 374, 453
constancy of energy flow, 131
constant collision frequency, 172, 327, 354, 358, 367, 460
constant electron density, 383, 464
constitutive relations, 14, **24** (ch. 3), 29, 39, 40, 48, 53, 56, 226, 396, 408, 484
continuous waves, 499
contour integral, 136, 142, 171, 174, 212, 288, 297, 302, 309, 330, 372, 423, 437
contour map, 298
convex lens, 313
conversion of ordinary to extraordinary wave, 265
convolution theorem, 168
coordinate system, 16, 96, 128, 252
Copson, E. T., 372, 374, 515
Cornu spiral, 308, 310, 313, 317
Coulomb law of force, 11
coupled equations, 385 (ch. 18), **394**, 398, 401, 412 (ch. 19), 473, 485, 500
coupled equations, matrix form, 398
coupling, 63, 144, **385** (ch. 18), 396, 401, 404, **412** (ch. 19), 418, 430, 432, 440, 450
coupling between upgoing and downgoing waves, 137
coupling branch points, 440, 442, **450**, 452, 455
coupling coefficient, 401, 409, 421
coupling echo, 405, 412, **428**, 500
coupling level, 144, 387, **412**, 425, 473
coupling parameter, ψ, **397**, 403, **412**, 417, 427, 432, 436, 473, 485, 487
coupling point, 413, **417**, 418, 452, 454
coupling region, 421, 422, 427, 428, 432, 433, 453, 454, 455, 485, 500; *see also* coupling level
coupling terms, **394**, 399, 401, 409, 410, 434
Courant, R., 36, 515
Crary, J. H., 258, 518
critical angle, 94, **101**, 105
critical case (Booker quartic), **238**, **264**, 268
critical collision frequency, **50**, 67, 71, 413
critical condition (Appleton–Hartree formula), 67, 68
critical coupling, **413**, 415, 429, 431, 455, 485
critical coupling frequency, **413**, 430
critical height, 413
critical value of Z, **49**, 450
Crombie, D. D., 96, 127, 512
Crompton, R. W., 7, 515
crystalline dielectrics, 36
crystal optics, 36, 252
cubic equation, 234, 245, 303
cubic lattice, 30
cubic law (of phase near caustic), 315
cumulative coupling, 412, **433**
cumulative phase change, 131
Cunningham, E., xxiv
curl, 17
current density, 14, 21
curvature of earth, 1, 192, 509
curvature of $h'(f)$ curve, 168
curvature of $N(z)$ profile, 5, 8, 153, 155, 157, **333**, 367
curvature of wave front, 91, 175
cusp, 233, 262
cut, 296, 372, 439
cylindrical waves, 314

Dällenbach, W., 502, 515
damping of electron motion, **25**, 27, 34, 480
Darwin, Sir C. G., 30, 515
Davids, N., 37, 485, 515, 516
De Groot, W., 163, 516
derivatives of electron density, 485
determinant, 338, 342, 387
determinant form of curl, 17
diagonal matrix, 36, 398, 497
dielectric, 14, 33
dielectric constant, 53, 104, 502, 503
differential equation for **A**, **494**, 498
differential equation for limiting polarization, 434
differential equation for **R**, 127, 491, 501
differential equation for 'vertical' polarisation, **343** ff., 349, 510
differential equation of fourth order, 67, **387**, 469
differential equation of Ricatti type, 132, 394, 435, 492
differential equations, **128**, 129, **140**, 143, 285, 327, 353, **385**, 458
differential equations for ray path, 229, 271, **276** ff.

differential equations for vertical incidence, 391
diffraction, 31, 168, 308
digital computer, 201, 205, 213, 278, 452, 482, 484, 487, 511
Dingle, R. B., 310, 516
dipole moment per unit volume, 30
dipole term (in radiation), 92
direction cosines of ray, 273
direction cosines of **Y**, 27
direction cosines of wave normal, 121, 126, 225, 245, 272
direction finding, 250
direction of energy flow, 44, 124, 247, 479
direction of ray, 253, 273
direction of wave normal, 121, 126, 225, 232, 245, 272
discontinuity of gradient, 156, **334**, 362, 368
discontinuity of the constants, 292
discrete strata, 138, 225, 228, 385
displacement current, 14, 16, 21
displacement, electric, **15**, 20, 24, 386, 397
distortion of pulses, 168, 170, 204
diurnal and seasonal variations, 8
divergent series, 301, 310
divergence theorem, 22
D-layer, 9
dominant term, **293**, 311, 320, 361, 442, 510
double linear profile, 340 ff.
double root, 144, **230** ff., 235, 237, 401; *see also* equal roots
double saddle point, 298
doubly refracting medium, 2, 98, 225; *see also* anisotropic
downgoing wave, 52, 124, 231, 393, 406, 442, 489
D-region, 9
drift of electrons, 34
dynamics, 276

earth's curvature, 1, 192, 509
earth's magnetic field, 2, **26**, 47 (ch. 5), 112, 116, 143, 199 (ch. 10), 225 (ch. 13), 458 (ch. 21), 466; *see also* anisotropic
earth's magnetic field horizontal, 115, 123, 395
earth's magnetic field vertical, 116, 406, 408, 459, 464
earth's surface, 92, 94, 101
east–west propagation, 123, 233, 236 ff., 248, 498
echos, 200
Eckart, C., 370, 516
Eckersley, T. L., 257, 331, 410, 437, 438, 516
EDSAC, xxiv, 201, 213, 452, 511
effective refractive index, 344, 481
effective value of *q*, 344, 467
Eigen values of matrix, 36, 389
Eigen vector of matrix, 390, 399
Eikonal function, 132, 272, **274** ff.
elastic waves, 3
E-layer, 8, 9, 157, 165, 363
electric displacement, **15**, 20, 24, 386, 397
electric intensity, **13**, 24
electric permittivity of free space, **11**
electric polarisation **P**, **14**, 24, 29, 30, 386
electric vector horizontal, 123
electric vector in plane of incidence, 101 ff., 343 ff., 510
electromagnetic induction, 16, 38
electromagnetic units, 12
electron (number) density, 3, 25, 85, 160 ff., 215 ff.
electron velocities, 33
electron waves, 2, 149
electrons, 209
electrons, free undamped, 24, 101
electrons prevented from moving, 466
electrons, rate of production, 4
electrostatic units, 12
ellipse in plane of propagation, 44, 215
elliptical polarisation, 19, 39, 45, **48**, 88, 129, 214, 430, 463, 486, 504
Ellis, G. R., 426, 516
energy absorbed, 35, 468, 479
energy flow, 12, **21**, 44, 46, 54, 111, 131, 328, 393, 402, 481
energy stored, 20, **33**, 135
England, ionosphere over, 120, 218
envelope of pulse, 166
envelope of refractive index surface, 266, 270
Epstein, P. S., 370, 467, 516
Epstein profile, **378**, 446
Epstein theory, **369** ff., 446, 463
equal roots of Booker quartic, **230** ff., 401, 409, 418, 439; *see also* double root
equation of motion of electron, 14, 24, **26**, 34
equation of ray, 177, **178** ff., **276**; *see also* ray path
equation of ray surface, **273**
equation of refractive index surface, **272**
equator (magnetic), 123, 127, 392
equivalent frequency at vertical incidence, 187, 194
equivalent height, 94, **149** ff., 167, 171, 172, **205** ff., 212, 215, 223, 365, 382, 424, 428, 499
equivalent height, exponential profile, 151, 207
equivalent height, linear profile, 150, 206
equivalent height, parabolic profile, 152, 209, 365

equivalent path, 173, **185** ff., 197, 279
E-region, 78
error (in Abel method), 165
error elimination, 488
error integral, 302, 308
essential singularities, 371
Euler's equations, 279
Euler's theorem, 274
evanescent wave, **41**, 43, 55, 108, 135, 321, 328, 337, 402, 454, 473, 479
exponential profile, 5, **151**, 172, 180, 207, **354** ff., 379, **460** ff., 488
extraordinary, 50, **63**, 70 ff., 81, 115, 160, 199, 202, 205 ff., 214, 215, 223, 232, 237 ff., 247, 262 ff., 387, 393, 395, 403, 406, 422 ff., 432 ff., 450 ff., 500, 506
Fabry–Perot interferometer, 109
factorial function, 358, 372, 382
failure of W.K.B. solutions, 133, 136, 141, 322 ff., 329, 410
Faraday's law, 16, 38, 96
Farmer, F. T., 173, 512, 516

Feinstein, J., 348, 516
Fejer, J. A., 9, 516
Fermat's principle, 271, **279**, 314
fictitious homogeneous medium, 53, 495
finitely conducting earth, 92, 94
first-order coupled equations, **398** ff., 402, 406, 433
fixed electron density, 383, 464
F-layer, 8, 157, 165, 220, 258, 363, 364, 424
F_1-layer, F_2-layer, 8, 160
flow of energy, 12, **21**, 44, 46, 54, 111, 131, 328, 393, 402, 481
flux, *see* flow
force on electron, 24, 26, 28
formation of layers, 3
Forsgren, S. K. H., 265, 516
Försterling, K., 346, 348, 395, 403, 408, 412, 516
Försterling's coupled equations, 394, **396** ff., 416, 418, 426 ff., 473, 485, 487
four characteristic waves, 144, 228, **387**, 394, 395, 396
Fourier analysis, 166, 347
Fourier integral, 147, 166
Fourier transforms, 168
four parts of $h'(f)$ curve, 210
four reflection coefficients, 90
four steps for finding reflection coefficient, 353
fourth-order differential equation, 2, 67, **387**, 464, 469, 483
fourth reflection condition, 472, 481
free electrons, 24, 101
free space, 19, 54, 413, 433
free space below ionosphere, 320, 413, 432, **433**, 465
F-region, 60, 78
frequency, 12, 40, 108, 147
Fresnel diffraction, 168, 309, 314
Fresnel integral, 308
Fresnel reflection coefficient matrix, 497
Fresnel reflection coefficients, **98** ff., 115, 119, 340, 380, 449, 497
frictionless slope, 162
Friis, H. T., 502, 522
full wave theory, 2, 144, 149, 155, 214, 319, 325, 334, 342, **353** (ch. 17), 364, 444, **458** (ch. 21), 474, 485, 508
functions of mathematical physics, 370, 464
Furry, W. H., 296, 517
Furutsu, K., 248, 517

Gans, R., 129, 517
Gardner, F. F., 9, 517
Gauss formula, 213, 482
Gaussian units, 12, 20
Gauss's theorem, 16
general coordinate system, 271, 272, 279
generalisation of Snell's law, 276, 277
generalisation of W.K.B. solution, 401
generalised wave admittance matrix, 491
geometrical optics 145; *see also* ray theory
Gerson, N. C., 517
Gibbons, J. J., 204, 205, 409, 426, 485, 517
Gibson, G. A., 274, 301, 517
Gill, S., 484, 511, 517
Goubau, G., 84, 206, 505, 517
Goubau's reciprocity theorem, 505, 508
gradient of collision frequency, 413
gradient of electron density, 207, 413
Gray, H. J., 369, 517
great heights, 324, 353, 355, 461, 470, 486, 487, 495
greater than critical, **429**, 456
ground wave, 1, 94
group refractive index, **148**, 153, 160, 170, 187, **200** ff., 204, 205, 212, 217, 220, 223, 254, 280, 482
group retardation, 157, 158, 212, 365
group velocity, **147**, 155, 158, 170, 187, 200, 254, 279, 480
group velocity surface, **254**, 257
guided waves, 94, 101, 257
Guillemin, E. A., 502, 517
gyrator, 503
gyro-frequency, **27**, 32, 69, 160, 206 ff., 223, 247, 256, 462

half thickness (semi-thickness), 5, **153**, 157, 191, 194, 211, 214
Hamilton's canonical equations, 276

Hankel functions, 291, 355, 461, 475
Harang, L., 424, 517
harmonics, **347**, 481
harmonic waves, **12**, 21
Hartree, D. R., xxiv, 58, 336, 340, 342, 459, 482, 517
Haselgrove, J., 218, 271, 278–81, 518, 523
Heading, J., 182, 356, 396, 398, 408, 438, 442, 445, 459, 464, 469, 471, 472, 477, 515, 518
Heading and Whipple's method, 459, **464**
Heading's rule, 442
heat, 21, 35, 351, 468, 480
heavy ions, 1, 31, 56, **78** ff., 258, 270
height, *see* equivalent height, phase height, true height, Cartesian coordinates
height as complex variable, 40, 41, 59, **326**, 413, 438
Helliwell, R. A., 258, 518
helical path, 27
Hermitian matrix, 30, 36, 390
Hermitian orthogonal, 37
Hertzian dipole, 92, 504
Hertz vector, 92
high frequencies, 60, 221
Hilbert, D., 36, 515
Hilbert space, 36, 37
Hines, C. O., 248, 258, 270, 491, 518
homogeneous functions, 274
homogeneous medium, 6, 17, 19, **38** (ch. 4), **47** (ch. 5), 108, 131, 214, 256, 495
horizontal dipole aerials, 508
horizontal direction of transmission path, 120, 250
horizontally stratified, 10, **85**, 128, 277
horizontal polarisation, 99, 105, **140**, 285, 319 (ch. 16), 336, 340, 376, 440, 493
horizontal range, 180, 181, 182, 192, 195, 281
horizontal ray, **230** ff., 260
horizontal variations, 10, 191
Houston, R. A., 111, 518
Huxley, L. G. H., 7, 26, 502, 515, 518
Huyghens's principle, 313, 318
hyperbolic secant, *see* sech
hypergeometric equation, **370**, 375, 377
hypergeometric function, 370, **372** ff.; *see also* confluent hypergeometric functions
hypergeometric series, **371**, 374

image of transmitter, 93
imaginary part of q, 124, 196, 251
imaginary part of refractive index, 40 ff., 66 ff., 173, 280
imaginary refractive index, 41, 43, 55, 102, 108
impedance, 18, 388, 491
impulse signal, 146, 166, 258
inclination of ray to vertical, 246, 258 ff.
independence of characteristic waves, 395, 398, 416, 418, 432
independent second-order equations, 129, 140, 460, 471
independent solutions, 491
India, 425
indicial equation, 346, 370, 371
induction, electric, *see* electric displacement
induction, electromagnetic, 16, 38
infinite equivalent height, 157, 159, 160, 163, 206
infinite fields, vertical polarisation, 347
infinite group refractive index, 201, 202
infinite refractive index, 59, 60, 61, 62, 64, 79, 82, 203, 206, 447, 468, 472 ff., 479
infinite root of Booker quartic, 123, 233, 285, 410
infinity and zero of refractive index, 476
inhomogeneous plane waves, **42** ff., 45, 87, 92, 98, 101, 106
initial conditions, 434; *see also* starting solutions
integral equation, 160 ff., 215 ff.
integrals of Barnes type, 373, 470
integration, 213, 300, 307, 483
intermediate inclination of earth's field, 61, 65
intrinsic impedance, 390
inverse square law, 16
inversion of integral equation, 161 ff., 215 ff., 482
inversion of matrix, 29, 490
ions, 1, 31, 56, **78** ff., 258, 270
irregularities of electron density, 10, 31, 191, 426, 428, 436
isolated infinity of refractive index, 474
isolated zero of $\mathfrak{n}$ or q, 283, 285, 334
isotropic medium, 17, **38** (ch. 4), 53, 98 ff., 252, 471 (chs. 16, 17), 491 ff., 501, 506
Jackson, J. E., 165, 522
Jeffreys, Sir H., 3, 313, 518
Jeffreys, Sir H. and Lady B. S., xxiii, 92, 129, 287, 290, 297, 310, 312, 370, 372, 519
Johnson, W. C., 436, 520
Jones, R. E., 519

Keitel, G. H., 484, 519
Kelso, J. M., 405, 426, 428, 485, 519
Kimura, I., 258, 278, 519
kinetic energy of electrons, 33, 34
King, G. A. M., 221, 519
kinks in refractive index curves, 72 ff.

Landmark, B., 431, 436, 519
Lange-Hesse, G., 37, 519
Langer, R. E., 313, 334, 519
Lassen, H., 516
lateral deviation, 94, 178, 229, 231, 233, 238, **246** ff., 261, 479
latitude variations of N, 247
ledge, 158, 212
left-handed circular polarisation, 48, 91, 460, 461
Leo Computers Ltd., 217
Lepechinsky, D., 431, 519
less than critical, **429**, 456
level line, **297**, 307, 308
level of coupling, 144, 387, **412**, 425, 473
level of reflection, 137, 143, 144, 149, 153, 176, 187, 196, 206, 215, **230** ff., 237, 260, 261, 265, 320, 333, 339, 385, 387, 409, 488
Lewis, R. P. W., 519
Lied, F., 431, 519
lightning flash, 29, 258
limiting points of spiral, 309, 317
limiting polarisation, 200, 412, 414, 416, **432** ff., 507
limiting region, **433** ff.
Lindquist, R., 405, 519
linear and homogeneous equations, 387, 487
linear electric circuit, 502
linear profile, 5, 134, **150**, 172, **179**, **188**, 206, 213, 283 (ch. 15), **319** (ch. 16), 336, 340, 440, 474
linear polarisation, **18**, 28, 39, 44, **48**, 57, 70, 77, 90, 112, 129, 137, 200, 214, 462, 466, 489
linear time base, 146, 166
lines of force (earth's magnetic), 257
lines of steepest descent, 297 ff.
logarithmic term in E_x, 347
logarithm of frequency, 221
longitudinal component of electric field, 18, 38, **53**, 67, 70, 77, 78, 117, 120, 125, 215; *see also* vertical component of electric field
longitudinal component of **Y**, **49**, 117, 126
longitudinal curves (Booker quartic), 245
longitudinal propagation, 54, **60**, **64**, **69**, 79, 80, 245, 264, 266, 460 ff.
long whistler, 258
Lorentz, H. A., 2, 30, 60, 519
Lorentz polarisation term, **30**
loss-free medium, **42**, 43, 111, 131, 390, 392
lower triangular matrix, 216
lowest ionosphere, 1, 9, 80, 354, 358, 412, 432 ff., 455, 465
low frequencies, 64, 145, 333, 353 (ch. 17), 354, 458 (ch. 21), 462; *see also* very low frequencies

Maeda, K., 258, 278, 519
magnetic energy, 21
magnetic equator, 123, 127, 233, 392, 406, 408
magnetic field of earth, 2, **26**, 47 (ch. 5), 112, 116, 143, 199 (ch. 12), 225 (ch. 13), 458 (ch. 21), 466; *see also* anisotropic
magnetic field of wave, **28**, 44, 131, 134, 320, 328
magnetic induction, **15**
magnetic induction of earth's field, **26**
magnetic intensity, 11, 12, **13**, 20
magnetic meridian, **47**, 51, 119, 215, 231, 246, 260 ff., 277, 406, 499, 508
magnetic permeability, 15, 53, 99, 127, 502, 503
magnetic permittivity, 11
magnetic pole, 247, 392
magnetic rotation, 58, 481 (Ex.)
magnetoionic medium, 19, 38, 54, 472, 481, 502; *see also* anisotropic
magnetoionic splitting, 199
magnetoionic theory, 25, 26, **47** (ch. 5), **59** (ch. 6), 121, 214, 392, 396, 432, 507
Manning, L. A., 161, 519
Marconi's Wireless Telegraph Co., Ltd., 224
Marcou, R. J., 278, 519
Martyn, D. F., 188, 197, 520
Martyn's theorem for absorption, 188, **195** ff.
Martyn's theorem for equivalent path, **186** ff., 194, 281
mass of electron, 24, 32
Mathieu's equation, 369
matrix for inverting integral equation, 216 ff.
matrix form of differential equations, **389**, 398, 494, 498
matrix form of Maxwell's equations, 226
matrix, reflection coefficient, **90** ff., 470, 489 ff., 496, 500, 505 ff.
matrix, susceptibility, 27, **29**, 32, 36, 227, 386, 391, 459
maximum in $h'(f)$ curve, 159
maximum of electron density, 5, 6, 8, 9, 153, 157, 163, 353, 358, 381, 443; *see also* penetration frequency
maximum usable frequency, **190** ff., 281
Maxwell's equations, 13, **16**, 17, 20, 22, 31, 38, 40, 42, 46, 48, 52, 94, 125, 127, 128, 133, 137, 140, 141, 143, 226, 274, 275, 320, 385, 408, 419, 484

mechanical model, 163
Meek, J. H., 424, 520
Mervin, R., 369, 517
metal sheet, 93
Miller, J. C. P., 287, 291, 310, 312, 339, 520
Millington, G., xxiv, 224, 233, 247, 249, 250, 491, 516, 520
Milne, W. E., 484, 520
m.k.s. units, 11, 491
modified Hankel functions, 291
modulus of reflection coefficient, 86, 87, 102 ff., 136, 173, 196, 197, 321, 329, 342, 351, 357, 363, 369, 380 ff., 437, 445, 461 ff., 478, 507
monotonic profile, 161, 215
Morgan, M. G., 436, 520
movements of ionosphere, 10, 436
M.U.F. *see* maximum usable frequency
Mullaly, R. F., 252, 515
multiple reflections, 109, 192
Musgrave, M. J. P., 3, 520

Namba, S., 96, 524
National Bureau of Standards, xxiv, 195, 520
Nearhoof, H. J., 405, 426, 485, 519
Nertney, R. J., 405, 409, 426, 485, 516, 519, 520
Newbern Smith, 194, 195, 281, 522
Newstead, G., 424, 520
Newton's laws of motion, 24
Nicolet, M., 7, 520
non-linear differential equation(s), 132, 142, 492, 494, 498, 501
non-singular matrix, 490, 494
normal component, **489**
normal form of differential equation, **343**, 349, 376, 427
normal incidence, 89, 108, 112, 115; *see also* vertical incidence
normal modes, xxiii, 438
normal to ray surface, 255, 273
normal to refractive index surface, 253, 254, 255, 273
northern hemisphere, 52, 67, 69, 116, 119, 247
north–south propagation, 124 ff., 238 ff., 246, 250, 260 ff., 471, 508
nose whistlers, 258
numerical methods, 127, 215, 278, 369, 482 (ch. 22)

oblique incidence, 86, 101 ff., 116, 120 ff., 138 ff., 175 (ch. 11), 225 (ch. 13), 283, 285, 319, 354, 406, 408, 464 ff.
one magnetoionic component (reciprocity), 506
one magnetoionic component, **R** for, 500
optics, 109, 111, 168, 252, 308, 313 ff.
ordinary, 50, **63**, 70 ff., 81, 115, 199, 202, 205 ff., 214, 215, 218 ff., 232, 237 ff., 247, 259 ff., 268 ff., 281, 387, 392, 395, 403, 406, 422 ff., 432 ff., 450 ff., 476, 500, 506
ordinary point of differential equation, 286, 419, 447
orthogonal, 36, 37
oscillating function, 112, 138, 288
oscillatory reflection coefficient, 500
Ott, H., 94, 520

parabolic cylinder function, $D_n(\zeta)$, 311, 359, 361, 367, 443
parabolic profile, 5, 8, 152, 157, 182, 188, 191, 194, 209 ff., 214, 281, 285, 358 ff., 381, 382, 464
parabolic ray path, 180
parallel-plate condenser, 20
parallel-sided slab, 108 ff.
Parkinson, R. W., 37, 485, 516, 520
partial fields, 394
partial penetration and reflection, 113, 138, 145, 225, 285, 342, 353, **363**, 369, 382, 500
partial standing wave, **19**, 39
particle, motion of, 162
particle velocity, 149
partitioning of matrix, 490, 494
passive electric circuit, 502
passive medium, **41**, 66
path of wave packet, 178, 196, 204, **229**, 248, 250; *see also* ray, ray path
Pawsey, J. L., 9, 517
Pedersen, P. O., 101, 521
Pedersen ray, 183, 190
Pekeris, C. L., 521
pencil of radiation, 177
penetration, 157, 160, 363, 445
penetration frequency, 8, 113, 145, 149, **153** ff., 163 ff., 172, 182, 190, 191, **208**, 214, 220, 247, 342, 364, 382, 424, 444
permeability, 15, 53, 99, 127, 502, 503
permittivity of free space, 11
Perry, L. B., 104, 119, 523
Pfister, W., 278, 425, 464, 519, 521
phase, **12**, 111, 130, 504; *see also* complex phase
phase advance on reflection, 326
phase change on reflection, 102 ff., 108, 380
phase difference between electric and magnetic fields, 42
phase height, **150** ff., 167, 171 ff., 205, 212, 365

phase integral, 136, 142, 150; *see also* phase memory
phase integral formula for coupling, 410, **423** ff., 431, **452** ff.
phase integral formula for reflection, 136, 171, 213, **329** ff., 331, 334, 336, 348, 350, 357, 368, 409, 416, 430, 442, 446
phase integral method, 136, **437** (ch. 20), 452; *see also* phase integral formula
phase memory, **130**, 144, 147, 199, 228, 275, 392
phase path, 173, 197, 507
phase, planes of constant, **43**, 46, 326
phase velocity, 255, 466
physical optics, 109, 111, 168, 308, 313 ff.
Piggott, W. R., 165, 521
Pitteway, M. L. V., xxiv, 235, 521
plane polarised, *see* linear polarisation
planes of constant amplitude, **43**, 46
planes of constant phase, **43**, 46, 326
planes of incidence, **85**, 88, 114, 231, 260
plane waves, 1, **17**, 19, 92, 225, 274
plasma frequency, **25**, 40, 113, 147, 152, 158, 160, 179, 180, 186, 205, 212, 216, 221, 223
plasma oscillations, 45
plate-like cavity, 15
Poeverlein, H., xxiv, 258, 262, 521
Poeverlein's construction, **258** ff.
Poincaré, H., 310, 311, 355, 361, 521
point of inflection (*q* curves), 244
point source, 91 ff., 175
polar coordinates, 92, 252, 278, 282
polar diagram, 281
polarisation, electric, **14**, 24 ff., 32, 36, 227, 386
polarisation ellipse, 19, 45, 49, 51 ff., 215, 463
polarisation equation, **48** ff., 57, 78, 214, 403
polarisation term, **30**
polarisation, wave, 14, 19, 45, **47** (ch. 5), 59 (ch. 6), 114 ff., 124 ff., 199, 214, 230, 248, 388 ff., 396, 424, 430, 432 ff., 456, 458, 462, 487, 500, 504 ff.; *see also* limiting polarisation
pole of refractive index, 476; *see also* infinite refractive index, infinite root
pole of ψ, 417
Pope, J. H., 258, 518
potential field (electron waves in), **2**, 149
potential function, 299
Poynting's theorem, **21**
Poynting vector, 13, **22**, 42, 44, 46, 54, 131, 135, 247, 248, 388, 393, 402
precursor, 169
prediction of M.U.F., 191, 194, 281
predominant frequency, **147**, 149, 150, 166 ff., 169
predominant plane wave, **93**
predominant values, **177**, 178, 229, 246, 253, 256
pressure of radiation, 28
principal axes, **35** ff.
principal axis components, **37**, 485
principal terms, **394**, 397, 399, 409, 410
principal value of integral, 166
probability, 25
progressive (plane) waves, **17** ff., 38, 41, 44, 45, 48, **67**, 108, **129** ff., 137, 392, 402
projected path, 249, 251
projected thickness, **338**, 342
propagation constant, *k*, **20**, 177, 438
propagation to great distances, xxiii, 104, 191 ff., 438
pulsating energy flow, 42, 135
pulses, **146**, 150, 155, 163, 166 ff., 177, 199, 200, 205, 424, 499, 506
pulse shape, 147, **166** ff.
purely imaginary refractive index, 41, 43, 55, 102, 108

q, **121** ff., 141, 144, 175 ff., **225** ff., 319 ff. (chs. 16, 17), 387, 398 ff., 438 ff.
quadratic equation for polarisation, **48** ff., 57, 78, 214, 403, 417
quadratic equation for q^2, 123, 233, 236 ff., 248, 402, 407
quadrupole terms, 92
quantum mechanics, 36
quartic equation, 117, **120**, **122**, **123**, 140, 144, 225, **226** ff., **233** ff., 246, 248 ff., 271, 274, 387, 389, 400 ff., 408, 410, 438, 439, 497, 506
quasi-Brewster angle, 103
quasi-longitudinal, 76, 117 ff., 429
quasi-transverse, 77, 429

radar, 146
radiation links (and reciprocity), 502, 503
radiation pressure, 28
radio sounding, 146, 160
random distribution of electrons, 30
random velocities of electrons, 24, 25
range, 180, 181, 182, 192, 195, 281
Rao, R., 204, 205, 517
Ratcliffe, J. A., xxiv, 10, 26, 59, 77, 163, 173, 191, 512, 513, 516, 521
rate of production of electrons, 4
rationalised units, 11
Rawer, K., 8, 375, 381, 382, 383, 521
ray, 124, 145, 177, **178**, 179, 190, 197, 253, **255**, 258 ff.
ray path, **179** ff., 190, 230 ff., 245, 247, **255**, 258 ff., 271 (ch. 14)
ray path horizontal, 230

ray path in magnetic meridian plane, 238 ff., 260 ff.
ray refractive index, **255**, 272, 273, 279
ray space, 273
ray surface, **255**, 256, 261, 263, 265, 271, 273, 274, 278
ray theory, 2, **144**, **146** (ch. 10), **175** (ch. 11), **199** (ch. 12), **225** (ch. 13), 271 (ch. 14), 364, 383, 437
ray tracing, 94, 177, 258 ff., **271** (ch. 14), 278
ray velocity, **255**, 271, 273
receiving aerial, 503
reciprocal property of surfaces, 255, 256, 274
reciprocity, 127, 230, 252, 502 (ch. 23)
reciprocity in phase, 504
reciprocity theorem, 499, **502**
reciprocity with full wave solutions, 508
recombination coefficient, 4, 5
rectangular pulse, 168
reference level for reflection and transmission coefficients, **85** ff., 93, 467, 470, 471
reflection at sharp boundary, 96 (ch. 8), 340, 380, 449, 497, 509
reflection at vertical (or normal) incidence, 85, 108, 112, 115, 146 (ch. 10), 199 (ch. 12), 459 ff., 471
reflection (branch) point, 144, 327, 330, 401, 418, **438** ff., 443, 450
reflection coefficient(s), **85** (ch. 7), 98 ff., 136, 141, 173, 196, 197, 283, 319 (ch. 16), 353 (ch. 17), 437, 444 ff., 449, 458, 461 ff., 467 ff., 478
reflection coefficient matrix, **90** ff., 470, 489 ff., 496, 500, 505 ff.
reflection level, 137, 143, 144, 149, 153, 176, 187, 196, 206, 215, **230** ff., 237, 260, 261, 265, 320, 333, 339, 385, 387, 409, 488
reflection of pulse, 149, 150, 166 ff., 205
reflection process, 129, 134, 139, 231, 283, 396, 404, 418
refractive index space, **272**
refractive index surface, **252**, 254 ff., 271 ff., 278
refractive index (wave), 3, 18, 28, **38** (ch. 4), 52, 56, **59** (ch. 6), 97 (ch. 8), 147, 148, 172, 199 ff., 391 ff., 397, 414 ff., 430, 434, 438, 450 ff., 460, 466, 472 ff., 491; *see also* group refractive index
region I, 465, 467, 471
region I(*a*), 465, 466
region II, 465, 468 ff.
regular singularities, 370
rejects, **504**, 506
relativistic effects neglected, 27
removal of electrons, 4
residuals, 218, **220**, 223
retardation, group, 157, 158, 212, 365
retarding force on electron, 2, 25, 26
reversal of sense (wave polarisation), 116
reversibility of ray path, **230**, 252, 506
reversion of series, **301**, 308
Ricatti-type equation, 132, 394, 435, 492
Riemann surface, **438** ff., 450
right-handed circular polarisation, 48, 90, 460, 461
right-handed system, 19, 50, 504
Robbins, A., 218, 523
roots of Booker quartic, **121** ff., 144, **225** ff., 387, 398 ff.
rotation in complex z-plane, 327
rotation, magnetic, 58, 481 (Ex.)
rotation of axes, 36, 50, 434
Runge-Kutta process, 483, 511
Rydbeck, O. E. H., 96, 168, 169, 363, 365, 366, 409, 424, 426, 445, 446, 522

saddle point, **298** ff.
scale height, **4**, 7, 8, 134, 154, 436
scattering, 58, 191, 426, 428
Schelkunoff, S. A., 491, 501, 502, 522
Scott, J. C. W., 248, 522
Seaton, S. L., 191, 513, 517
sech^2 profile, 154, 156, 380 ff.
second-order coupled equations, 391, 397, **408**, 411, 426, 469, 485
Seddon, J. C., 165, 522
seismology, 3
semi-thickness (half thickness), 5, **153**, 157, 191, 194, 211, 214
separation into upgoing and downgoing, 137, 486, 489 ff.
separation of differential equations, 18, 129, 140, 460, 465
series solution of differential equation, 287, 339, 346, 356, 370 ff., 419
shape of pulse, 147, **166** ff.
sharp gradient of refractive index, 134, 473
sharply bounded anisotropic medium, 68, 114 ff., 497
sharply bounded homogeneous medium, 86, 96 (ch. 5), 134, 340, 380, 449
Shinn, D. H., xxiv, 200, 201, 206, 207, 224, 522
Shire, E. S., 12, 28, 522
short whistler, 258
side-band frequencies, 169
sign convention for reflection coefficients, **88**, 470, 490, 505, 507
sign convention for square roots, 403
sign of electronic charge, 25, 27
silvered glass, 127 (Ex.)

simultaneous differential equations, 483 ff.
singular matrix, 94 (Ex.), 500
singularities of refractive index, 447, 450, 451, 476
singular point, singularity, 291, **299**, 344
singularities of ψ, 417
sink of energy, 480
sinusoidal layer, 368
size of step, 484
skip distance, **183**, 190, 281, 318
slab model of ionosphere, 108 ff.
slab of ionised medium, 45
slit, diffraction by, 168
Slough, 218
slowly varying function, 94, 176
slowly varying medium, 128 ff., **133**, 144, 228, 252, 256, 259, 271, 275, 322, 329, 358, 413, 417, 420, 479, 506
small irregularities, 31
Smith, N., 194, 195, 281, 522
Smith, R. L., 258, 518
smoothing out, 13, 31
Snell's law, **97** ff., 114, 116, 119, 121, 139, 179, 226, 259, 277, 487
Sommerfeld, A., 502, 522, 523
sound waves, 3
source, dimensions of, 2, 91, 175, 177, 504
southern hemisphere, 247
south–north propagation, 124 ff., 238 ff., 246, 250, 260 ff., 471, 508
space charge, 16
space charge waves, 24
spectrum function $F(f)$, 147, 166 ff.
specular reflection, 96 (ch. 8), 509; *see also* reflection at sharp boundary
Spencer, M., 10, 514
spherical polar coordinates, 92, 252, 278, 282
spherical waves, 1, 91 ff., 146, 175 ff., 509
spiral, 309, 316, 317
spitze, 236, **260** ff., 266 ff.
splitting, **199**, 232, 253, 424
square law increase (profile), **366** ff., 464
square root in Appleton–Hartree formula, 67, 201, 417, 450
square root, sign convention, 403
standing wave, **19**
Stanley, J. P., 463, 523, 524
starting solutions, **486**, 487, 495, 511; *see also* initial conditions
statistical mechanics, 25
stationary phase, 92, 93, 148, 149, 170, 171, 176, 178, 185, 204, 253, **307** ff.
stationary time, 271, 279, 314
steepest descents, **297** ff., **300** ff., 470
steep gradient of electron density, 134, 425, 473
step-by-step process for integration, **483**, 486
step-size, 484
Stirling's formula, 358, 365, 368, 382, 449
Stokes, Sir G. G., 293, 523
Stokes constant, **295**, 306, 313, 325, 348 ff., 356, 362, 437, 446, 510
Stokes diagram, **294**, 307, 312, 356, 361, 441, 442, 453 ff.
Stokes (differential) equation, **286** ff., 302, 312, 319 (ch. 16), 371, 409, 420, 447, 450
Stokes lines, **293** ff., 296, 305, 307, 311, 312, 350, 355, 359 ff., 440, 450 ff., 510
Stokes phenomenon, 283 (ch. 15), **292**, 302, 310, 372, 440, 452, 511
stored energy, 20, **33**, 135
Storey, L. R. O., 252, 255, 257, 523
Straker, T. W., 513
strata, 138, 225, 228, 385
Stratton, J. A., 92, 170, 204, 365, 523
strong coupling, 409; *see also* critical coupling, cumulative coupling
structure of the ionosphere, 7 ff.
subdominant term, **293**, 311, 320, 324, 339, 350, 355, 442, 454, 456, 510
successive approximations, 137, 395, 409, **421**, 426, 485
Suchy, K., 275, 523
summation convention, 389, 399
surface integral, 22
surfaces of revolution, 252 ff.
susceptibility matrix, 27, **29**, 32, 36, 227, 386, 391, 459
Sutton, D. J., 7, 515
symmetrical ionosphere, 340; *see also* parabolic profile, sech^2 profile, sinusoidal layer
symmetric (amplitude function), 147
symmetric matrix, 36, 498, 499
symmetric tensor, 502
symmetry properties of equations, 499

table for $z(f_N)$ from $h'(f)$, 218
tables of Airy-integral functions, 291
tail (of pulse), 169
Taylor, M., 523
Taylor series, 167, 298, 301, 308, 349, 417, 419
temperate latitudes, 425
thermal motions, 24
thickness of ionosphere, layer, 363, 382, 445
Thomas, J. O., xxiv, 218, 523
three equal roots, 238
tilt angle, **51**, 52
time average of energy flow, **23**, 44, 247

time of travel of pulse or wave packet, 149, 167, 170 ff., 205; *see also* equivalent height, equivalent path
time of travel of wave crest or wave front, 279; *see also* phase height, phase path
Titheridge, J. E., 224, 523
top of trajectory, 180, 230 ff.
Toshniwal, G. R., 424, 425, 523
total current density, **15**, 16
total reflection, 102, 107, 380
trains of whistlers, 258
transition through critical coupling, 429, 455
transmission coefficient, **85** (ch. 7), 98 ff., 342, 362 ff., 369, 377 ff., 381, 467, 478
transmitter, 28, 146, 167, 175 ff., 229
transmitting aerial, 91, 175 ff., 229, 253, 502 ff.
transpose of matrix, 399, 499
transverse components of **Y**, **49**, 126
transverse curves (Booker quartic), 245
transverse field components, 17, 18, 38, 39, 47, 77, 504
transverse propagation, 54, **61**, **65**, **70**, 79, 81; *see also* east–west propagation
triangulated path, **186**, 187
true bearing, 250
true height, **150**, 152, 154, 157, 205
truncation errors, 484
Tuve, M. A., 186, 514
two separate parabolic layers, 157, 210 ff.
two transmitted waves, 116 ff.

Ulwick, J. C., 278, 519
unitary matrix, 37, 91
unitary transformation, 36, 91
units, 11
University Mathematical Laboratory, Cambridge, xxiv, 201, 213, 511
upgoing wave, 122, 124, 231, 393, 405, 406, 442, 486 ff., 492, 495
upward velocity of pulse, 148, 200; *see also* group velocity

validity of W.K.B. solutions, **133**, 141, 143, **410**
variation of parameters, **426**
variations of path, 271, **279**
varying collision frequency, 383, 457 (Ex.), 464
vector diagram, 308, 309, 316
vector, refractive index as, 121, 272, **276**
vertical component of electric field, 347, 466; *see also* longitudinal component of electric field
vertical component of group velocity, 148, 200, 480
vertical dipole, 508
vertical indicence, 85, 114, 123, 128 ff., 146 (ch. 10), 299 (ch. 12), 233, 246, 365, 384, 395 ff., 402 ff., 412 (ch. 19), 438, 446, 459 ff., 473 ff., 485, 495; *see also* normal incidence
vertical magnetic field, 116, 406 ff., 411 (Ex.), 459 ff., 464 ff.
vertical motions of electrons, 347, 466
vertical polarisation, 99, 101 ff., **140**, 142, 286, **343** ff., 510
vertical tangents to q curves, 238; *see also* equal roots
very low frequencies, 9, 57, 103, 117, 119, 145, 220, 256, 353 (ch. 17), 389, 458 (ch. 21), 463, 487; *see also* low frequencies
Vice, R. W., 9, 516
Wait, J. R., 104, 119, 523
Walkinshaw, W., 438, 513
Watson, G. N., 161, 291, 346, 355, 361, 369, 370, 372, 374, 438, 475, 476, 478, 524
Watson's lemma, 310
Watts, J. M., 220, 513
wave admittance, 485, 486, 488, **491** ff.
wave admittance matrix, 493 ff.
wave crest, 150, 255, 276, 279
wave front, 2, 91, **255**, 271; *see also* ray surface
wave guide, xxiii, 87, 104, 149, 438
wave impedance, **18**, 388, 491
wave interaction, xxiii, 7, 24
wavelength in free space, 20, 484
wavelength in medium, 150, 484, 487
wave mechanics, 2
wave normal, 17, 28, 35, 41, 45, **97**, 121, 124, 231, 232, 252, 255, 272
wave packet, 146, 148, 166, 173, 175, **177** ff., 182, 185 ff., 196, 199, 204, 214, 225, 229 ff., 246 ff., 253, 259, 271, 279, 388, 479, 506
wave polarisation, 14, 19, 45, **47** (ch. 5), 59 (ch. 6), 114 ff., 124 ff., 199, 214, 230, 248, 388 ff., 396, 424, 430, 432 ff., 456, 458, 462, 487, 500, 504 ff.; *see also* limiting polarisation
wave refractive index, **148**, 200; *see also* refractive index
wave surface, 253
wave velocity, 148, **255**
Waynick, A. H., 405, 426, 485, 519
Weekes, K., xxiv, 7, 513
Weber's equation, 359, 367
Westfold, K. C., 37, 524
west–east propagation, 123, 233, 236 ff., 248, 498
Weygandt, C. N., 369, 513
Weyl, H., 92, 524

Whale, H. A., 201, 206, 522, 524
Whipple, R. T. P., 182, 459, 464, 469, 518
whistlers, 30, 32, **256** ff., 269, 278
Whittaker, Sir E. T., 161, 166, 346, 361, 369, 370, 372, 374, 476, 478, 524
Wilkes, M. V., 366, 464, 469, 524
winds, 10
W.K.B. method, 128 (ch. 9), **131** ff., 292, 323, 350
W.K.B. solution, derivation, 128 (ch. 9), **131**, 137, 405
W.K.B. solution with earth's field, 143, 199, 228, 385, 389, 392 ff., **401**, 405, 406, 412, 427, 445, 452, 465, 468, 470, 474, 485, 487, 495
W.K.B. solution, without earth's field, 128 (ch. 9), **133**, **141**, **143**, 283, 292, 303, 321 ff., 329 ff., 335, 339, 355, 362
Woodward, P. M. and Woodward, A. M., 292, 524
Wronskian, 313, 342, **426**
Wüster, H. O., 346, 348, 516

Yabroff, I. W., 127, 524
Yokoyama, E., 96, 524

zenith angle of sun, 4, 5
Zenneck, E. H. J., 101, 524
zero of *q*, 143, 231, **234**, 237, 283 ff., 320, 322 ff., 327 ff., 357
zero of refractive index, 40, **59**, 64, 78, 81, 137, 149, 438, 443, 450, 473
zero-order approximation, 421, 427
zero refractive index, 45
zeros of Airy integral function, 185
Z-trace, 209, 404, 412, 416, **424** ff., 430